SOVIET UNION
IN THE CONTEXT OF THE
NOBEL PRIZE

Facts, Documents, Thoughts and Commentaries

SOVIET UNION IN THE CONTEXT OF THE NOBEL PRIZE

Facts, Documents, Thoughts and Commentaries

Abram M Blokh
Russian Academy of Sciences, Russia

World Scientific
NEW JERSEY • LONDON • SINGAPORE • BEIJING • SHANGHAI • HONG KONG • TAIPEI • CHENNAI • TOKYO

Published by

World Scientific Publishing Co. Pte. Ltd.
5 Toh Tuck Link, Singapore 596224
USA office: 27 Warren Street, Suite 401-402, Hackensack, NJ 07601
UK office: 57 Shelton Street, Covent Garden, London WC2H 9HE

British Library Cataloguing-in-Publication Data
A catalogue record for this book is available from the British Library.

Originally published in Russian as Советский Союз в Интерьере Нобелевских Премий. Факты. Документы. Размышления. Комментарии. © 2005 by Scientific Publishing House of Biographical International Encyclopedia "Humanistica", http://humanistica.ru. English text translated by Yurii Vladimirovich Morozov, Elena Nikolaevna Korbina and Marina Aleksandrovna Blagonravova.

SOVIET UNION IN THE CONTEXT OF THE NOBEL PRIZE
Facts, Documents, Thoughts and Commentaries

ISBN 978-981-4277-97-6 (hardcover)
ISBN 978-981-4277-98-3 (ebook for institutions)
ISBN 978-981-3223-35-6 (ebook for individuals)

For any available supplementary material, please visit
https://www.worldscientific.com/worldscibooks/10.1142/7367#t=suppl

Printed in Singapore

Foreword

This book is the result of many years of meticulous work by Professor Abram Moiseevich Blokh. It will be of utmost interest to those interested in the history of science in Russia and the former Soviet Union. The book presents an overview of the relationship between the ideas expressed by the Nobel Foundation and those expressed by the autocratic and totalitarian regimes in Russia and the Soviet Union.

The system of values represented by the Nobel Prizes has strong connections with Russia. Alfred Berhnard Nobel was born in 1833 in Stockholm and in 1844 the family moved to St. Petersburg to join Alfred's father, Immanuel Nobel. During this period a number of factories were built in Russia that later became part of the Nobel family "empire". This empire was later liquidated by the Bolshevik Revolution in 1917.

Alfred Nobel and his brothers received private home education from two talented pedagogues from St. Petersburg University, believed to be Zinin and Trapp, teachers of physics and chemistry. Humanities and languages were also taught in the Nobel family house. Alfred Nobel was fluent in five languages: Swedish, Russian, English, French and German.

Nobel's parents and teachers had created an intellectual atmosphere in the house based on the philosophy of the Age of the Enlightenment and scientific optimism. Despite the oppressive regime of the emperor, Nikolai I, Alfred Nobel acquired very democratic opinions of social and humanistic questions.

The intellectual basis for the Nobel Prizes was created personally by Alfred Nobel and was based on his own philosophical views. As a result, it is not surprising that the dictators of various countries had a negative attitude to the system. The humanistic ideas were foreign to the totalitarian Bolshevik ideology. Later, the communist regime, despite its international roots, had the same attitude of revulsion towards the international cosmopolitical ideas of the Nobel Prizes. The same attitude was adopted by Nazi Germany.

During his research, Professor Blokh discovered interesting documents in Moscow, St. Petersburg and Stockholm. These documents shed new light on the difficulties that were arising during the attempts to integrate Russian and Soviet science into the world's intellectual community. I am sure that this book will make for compulsive reading for many.

Michael Sohlman
Former Executive Director, Nobel Foundation

Preface to the English Edition

It was with great satisfaction that I took the proposal to have this book *Soviet Union in the Context of Nobel Prize* published in English, at the suggestion of the laureate of the Nobel Prize for Physics, Academician Vitaly Lazarevich Ginzburg.

The contents of the book can be seen as an attempt to analyze the history of Russia within the framework of the last century, from an unusual angle, notably from the position of the attitude of Russian power structures to the unique system of the Nobel awards, formed at the turn of the 20th century. That means the attitude of both the autocratic power, which already had one foot in the grave, and that of the Soviet dominion, which replaced it to rule over the country during the greater part of the 20th century. That enabled the author to entitle the book as such.

Creating myths is a characteristic feature of any society lacking in truthful information, hence a disposition of the Soviet propaganda to constructing mythical notions as an effective means of manipulating public conscience.

In the Soviet Union, among traditional propagandist objectives for such manipulations, there was everything that could be connected with everyday activities of the Nobel Foundation. Especially after World War II, when the authorities, taking advantage of the fruits of the victory over Nazism, started gradually to substitute the ideas of internationalism with overt nationalism and to pander to xenophobic, anti-Western moods remaining in the society from pre-revolutionary times.

The Nobel Committees were accused of bias, of cultivating anti-Sovietism when choosing laureates of the most prestigious award of the century. Of course, the accusations of anti-Soviet provocations that were made during the boisterous propagandist campaigns connected with awarding the Nobel Prize to B. L. Pasternak, A. I. Solzhenitsyn, A. D. Sakharov were worked out, first of all, in the ideological departments of the CPSU Central Committee and under their patronage. However, in society, including the community of intelligentsia, these invented grudges found fertile ground. As a result, hostility towards the Nobel institutions turned into one of the hypostases of anti-Western moods insistently promoted by Party ideologists and simultaneously giving reliable feedback.

A graphic example of feedback was the Nobel Peace Prize awarded, in 1990, to the President of the USSR, M. C. Gorbachev. The reaction of the population to this award was mostly negative, although ideological structures did not take visible participation in organizing protest moods — for Gorbachev, being the president of the country, at the same time held the post of General Secretary of the CPSU Central Committee. In this case, as a result of the activities of the total propaganda, Soviet society itself was reaping the fruits of suspiciousness and hostility towards any positive step of Western countries.

The main task for this book, as regarded by the author, is to examine the reconstruction of the picture of relations between power structures of the Soviet Union and the Nobel institutions. As the reader will be able to make sure, this picture did not in the least correspond to the propagandist stereotypes used for many decades by ideologists of the Central Committee of the communist party and their assistants from the academic and writers' communities. In this examination priority will be given not to reasoning but to documentary evidence, most of which was previously inaccessible for research.

The structure of each chapter of the book is based on three parts: a descriptive part with a preliminary characteristic of the documents attached to the chapter; commenting notes, which make the contents of the first part more precise and illustrate it by additional examples, and a documentary part with its own set of required notes and comments.

The total number of the documents included in this book runs into three-digit figures. The majority of them are published for the first time. Along with the sources from national archives, mostly from the archives of the communist party, documents from the Nobel institutions are also used, which were declassified by the year of publication of the book, which is through 1956.

The Nobel movement, which received a legal basis in the last year of the 19th century (the Statute of the Nobel Foundation was signed by the Swedish King, Oskar II, on 29 June 1900) became an unprecedented phenomenon of the 20th century, one of its most significant distinctive features. The list of Nobel laureates in all its nominations will undoubtedly be considered by our successors as most valuable material for analyzing the achievements of science and literature and for perceiving the social climate of the century.

Speaking at the general meeting of the Russian Academy of Natural Sciences, in 1997, the executive director of the Nobel Foundation, Michael Sohlman, justly noted that of all Alfred Nobel's numerous inventions the most outstanding was a different, not patented one — the Nobel Prize. However, for the sake of objectivity we should elaborate this idea and recall one more achievement, closely connected with one mentioned above and also not patented, that is, the Statute of the Nobel Foundation and the regulations accompanying it. These documents in the aggregate regulate in detail the everyday activities of the Nobel institutions and reliably ensure their complete independence from any outside pressure, be it from governmental structures or from authoritative mediators from elsewhere.

The creation of a smoothly functioning mechanism of nominating, considering and electing the most worthy nominee to the most prestigious award of the century is an undeniable achievement of the Swedish academics, which has not had any equally valuable precedent in the whole history of civilization.

The Statute requirements, however, are not something unchangeable. Some of the provisions are regularly made more precise and transformed by special decrees of the Swedish government. This ensures correction of inaccuracies made earlier and giving timely response to new trends in the world, thus not falling behind newly appearing social circumstances and requirements of time that come into being.

But one thing has remained unalterable in the legal foundation that determines the activity of the Nobel institutions. It is, first of all, the legal provision of guarantees of their complete sovereignty in their decision-making, which cannot depend either on political conjuncture in their own country or, still less, on the wishes and requirements of foreign, social or governmental structures.

Like any other fruit of the human mind, the Nobel institutions are not guaranteed against erroneous decisions. But the disappointing errors that were sometimes committed were their own, purely human errors or slips, and not ones that someone coerced them into committing. Naturally, such independence of the Nobel institutions in the choice of their own stance could not arouse warm feelings towards them in foreign totalitarian regimes, repeatedly bringing these regimes to confrontation with the endowers of prizes in Stockholm and Oslo.

The opposition to the Nobel institutions was not limited to the activities of the classical totalitarian regimes of the 20th century — the ruling circles of Hitler's Germany and the communistic Soviet Union, along with its Warsaw Pact allies (for instance, Czechoslovakia, in connection with the 1984 Nobel Prize to the dissident poet Jaroslav Seifert). Along with these textbook examples there have been in the world some other cases of showing the intolerance of decisions taken by Nobel institutions. These cases received smaller responses, but nevertheless, they also provided evidence of the very essence of totalitarianism, which is intolerance of heterodoxy of their own citizens, on the one hand, and intolerance of foreign organizations giving support of their right to have their own opinion, on the other hand. Suffice it to remember the painful reaction of Burma's military authorities to awarding the 1991 Nobel Peace Prize to Aung San Suu Kyi, convicted in her motherland for her human rights activities. Or the charge of interference in internal affairs brought by the government of China against the Dalai Lama when he was awarded the 1989 Peace Prize.

Collecting archive documents was carried out in the Russian State Archive of Social and Political Studies (RSASPS), the Russian State Archive of Modern History (RSAMH), and also in the archive of foreign policy of the Russian empire and the USSR (AFP RE and AFP RF), the State Archive of the Russian Federation (SARF), the Central Archive of Social and Political History of Moscow (CASPHM), in the Moscow and St. Petersburg branches of the archive of the Russian Academy of Sciences, in the archive of the Pushkin House in St. Petersburg.

Thanks to the kind assistance of the executive director of the Nobel Foundation, Michael Sohlman, and the professor of history and editor-in-charge of the yearbook of the Nobel Foundation, Tore Frangsmyr, valuable information has been received from declassified funds of the Royal Academy of Sciences, Karolinska Institutet and the Swedish Academy in Stockholm. With the help of the head of the chair of inorganic chemistry of the Stockholm Technological Institute, Professor Ingmar Granthe, it became possible to find important documents in the personal archive of Lars Gunnar Sillen, a professor at the same institute.

The author remembers with sincere gratitude the genuine interest in the search for the required archive information on the part of the secretaries of the Nobel Committees: for Physics, Professor Anders Barany, for Physiology or Medicine, Professor Nils Ringertz and the former permanent secretary of the Swedish Academy, Sture Allen, and also the head of the Center of Scientific History of the Swedish Academy of Sciences, Urban Wrakbearg, the chief curators of the libraries of the Swedish Academy, Ake Erlandsson and of the Norwegian Nobel Committee, Anne Cecilie Kjelling. The talks with the newly appointed Permanent Secretary of the Swedish Academy, Horace Engdahl, and the eminent scholar of the works of Anna Akhmatova and Boris Pasternak, Erik Mesterton, were also very informative for this author.

Valuable information has been provided by the relatives of the Nobel laureates and people close to them: N. N. Semenov's daughter, Lyudmila Hikolaevna and his son-in-law, Academician Vitaly Iosifovich Goldansky; I. M. Frank's widow, Marina Mikhailovna, L. V. Kantorovich's son, Vsevolod Leonidovich and B. L. Pasternak's son, Evgeny Borisovich, and also by the assistant to Academician P. L. Kapitsa, Pavel Evgenyevich Rubinin; the son of the writer K. G. Paustovsky, who was a nominee of the Nobel Committee for literature in the early 1960s, Vadim Konstantinovich; Academician Vitaly Lazarevich Ginzburg; the writers Mikhail Viktorovich Ardov and Alexandr Mukhailovich Borshchagovsky; the former assistant to the President of the USSR Anatoly Sergeevich Chernyaev and many others.

In conclusion we shall note that in the Russian language of this book there are five appendices. These are not included in the English edition, since their content is of secondary importance for the subject examined here.

Abram M. Blokh

Contents

Chapter 1 • 1900–1918

The Formation of the Nobel Organizations and Sources of Presumed Russian Grievances

The death and last will of Alfred Nobel were widely covered by Russian newspapers. The interest of the press was quite understandable. The Nobel family had been invited to Russia by Great Prince Mikhail Pavlovich and had settled in St. Petersburg in 1842, creating links to industrial developments in their second homeland. These links were forcibly destroyed in the second half of 1918, after the business of the descendants of Ludvig Nobel, Alfred's elder brother, was nationalized and repressed.[1]

Until his last days, Alfred Nobel had positive feelings about Russia, where he spent his childhood, reached adolescence, and formed scientific interests that eventually brought him to the invention of dynamite, enabling him to amass an immense fortune. He continued to uphold the respected memory of Nikolai Nikolaevich Zinin, an outstanding Russian chemist who was his teacher and had helped him to evolve into a capable researcher and inventor. In 1879, Alfred became one of the four founders of the Nobel Brothers Petroleum Company and a contributor to its successful management.

The first press account of the death of the initiator of the world-famous prize appeared on the 29th November (11th December) 1896, in a central Russian newspaper, St. Petersburg-based *Novoye Vremya.* On the 8th (18th) January 1897, the Moscow newspaper *Russkie Vedomosti* reprinted from the Viennese Neue Freie Zeitung the obituary of Alfred Nobel, written by Bertha von Zuttner, a known peace activist and the former secretary and housekeeper for the deceased. The *Sankt Peterburgskie Vedomosti* published on the 15th (27th) December 1897 a detailed account of the meeting of the Swedish Peace and Arbitration League held in Stockholm on the 7th (19th) December on the occasion of the first anniversary of Nobel's death.

At that meeting, Leo Tolstoy was proclaimed the first candidate for the Peace Prize even though the statutes of the Nobel Foundation were not then legally defined, nor was the juridical force of Nobel's will recognized by the court. Indeed, the testament was fiercely contested by Nobel's relatives while the king, the government, and even the organizations cited in the testament as prize awarders did not appear to expressly agree to the terms outlined therein.[2]

The absence of an established legal base encouraged the participants of the meeting to let their imaginations run free as regards the future use of the huge prize money. The majority came to a definite agreement that the prize should be awarded to an eminent writer or thinker as "an intermediary empowered to dispose of it by his own discretion and for what he may regard as the most practical purpose."

At the meeting, the name of Leo Tolstoy was mentioned in connection with the anticipated Nobel Prize probably for the first time. But it was not news for Tolstoy himself. The writer had come to know from reports in the Swedish press that Norwegian Storting had allegedly begun to consider nominees for the Peace Prize. Being the main candidate, he immediately (on the 22 September of the same year) sent a letter to the Svenska Dagbladet stating that "the provisions specified in Nobel's will in respect of the most prominent peace activists are difficult to fulfil. People serving the cause of peace do so because they are serving God; therefore, they do not need award money and will never accept it."[3]

He further suggested that the available funds should be distributed among the poor families of Caucasian *Dukhobors* (a Christian sect of Russian origin), whose bread-winners were serving their sentences for objection to military service. "I believe," Tolstoy concluded, "that nobody did as much for the cause of peace as these people. Their families left in most unfortunate conditions [and] could by just pretend for the assets that Nobel set aside for the advocates of world peace."[4]

On the 13th (25th) September 1898, the newspaper *Birzhevye Vedomosti* reported a new initiative in connection with the as yet non-existent Peace Prize. According to the paper, a certain Peace League (based in an unidentified country) decided to "bring it to the feet of His Majesty for the generous contribution to peace building." Evidently, this suggestion was prompted by the proposal of global disarmament made by Nikolai II to the heads of European states on the 12th (24th) August 1898.[5]

For all the differences among the first reports about the posthumous will of Alfred Nobel, they testify to the interest of Russian readers in the further fate of his legacy and the possibility of involvement of their compatriots as candidates for the prestigious and weighty monetary award.

The press quickly informed the public about details of the agreement on practical implementation of Nobel's will (reached in May 1898) between the executors and three of the four prize-awarding organizations (Royal Academy of Sciences, Royal Caroline Medico-Chirurgical Institute (Karolinska Institutet), and Norwegian Storting). As early as the 11th (23rd) May, *Sankt Peterburgskie Vedomosti* and *Novosti i Birzhevaya Gazeta* published detailed reports telling their readers that the first presentation of the prizes was scheduled for the 10th (19th) December 1901, and that part of the estate left by Nobel's will would be used to establish Nobel Institutes, one for each prize discipline.

Leo Tolstoy was not the sole Russian citizen predicted by the country's press to become a Nobel Prize winner. The newspaper *Russkie Vedomosti* announced on the 9th (21st) December 1899 that the well-known Russian artist Vasily Vereshchagin (famous for the pictures depicting war "as it is") was also a candidate to receive an award. Unlike Tolstoy, Vereshchagin believed that he deserved it and had taken active steps to promote his candidacy long before the Nobel Foundation was established. In a letter to his Scandinavian friends cited by the newspaper he wrote: "I do not hesitate to state that my pictures made a tremendous impression and war will not be praised again very soon where they were exhibited."

Vereshchagin's striving to win the award is corroborated by further evidence which emerged after his death. A certain Tsabel', passing himself off as Vereshchagin's close friend, mentioned in his memoirs the author's pretensions to be awarded the Nobel Peace Prize. The recollections of Tsabel' were published by the German magazine *Deutsche Rundschau* and were cited by *Sankt Peterburgskie Vedomosti* in the issue dated 25th August (7th September) 1904 (p. 2). In a letter to Germany, Vereshchagin explained his pretensions, opposing another candidate Henry Dunant, the internationally known pacifist: "By his acts of charity, Dunant did no more than sooth sentimental souls in the grip of war. On the contrary, I acted as a wise father who did not preach moral principles to my son, but instead took him to a hospital and showed him the sufferings of the disabled..."

However, self-promotion was not the sole motive behind Vereshchagin's actions.[6] One month after the above-mentioned publication, another Russian newspaper, *Sankt Peterburgskie Vedomosti,* dated 10th (22nd) January 1900 referred to a paper based in Gothenborg, Sweden, as describing Henri Dunant, the Red Cross movement founder, to be the most likely candidate for the Nobel Peace Prize, along with the renowned peace activist Fredrik Bajer, member of the Danish Parliament, and Vereshchagin. Indeed, Dunant and Bajer were awarded the prize in 1901 and 1908 respectively. On the 16th (29th) March 1901, two days before the deadline for 1901 prize nominations,[7] the same *Sankt Peterburgskie Vedomosti* published extensive correspondence from Christiania mentioning the names of the seven most likely candidates for the Nobel Peace Prize: again, Vereshchagin was one of them. Six actually received the prize between 1901 and 1908, including the aforementioned Dunant and Bajer as well as Frederic Passy (1901), William Cremer (1903), Bertha von Zuttner (1905), and Klas Arnoldson (1908). In other words, the Russian press had in fact managed to obtain trustworthy information.

Bertha von Zuttner left an important note in her memoirs published after Vereshchagin's death in Port Arthur during the Russian–Japanese War (13th April 1904). According to von Zuttner, an influential nominator on the Nobel Committee annually recommended Vereshchagin for the prize, from the exhibition of his works in Christiania in 1900 until 1904.[8]

However, this evidence is not confirmed by archive documents from the Norwegian Nobel Institute. A thorough inquiry made at the request of the author Anne Kjelling, curator of the Institute's library, failed to discover Vereshchagin's name in the registries of nominations received for the period from 1900 to 1904. True, some findings seem to testify the interest of the Nobel Institute in the works of the Russian artist, since its library holds three catalogues of his pictures exhibited in various cities of Europe, and a monograph on Vereshchagin published in 1907 (doc. 1.1).[9]

Evidently, this discrepancy may have but one explanation. It appears that nominations in fine arts were against the letter of Nobel's will, according to which the Peace Prize should be awarded "to the person who shall have done the most or the best work for fraternity between nations, for the abolition or reduction of standing armies and for the holding and promotion of peace congresses." Naturally, a creative person seeking to achieve these goals with a brush as his main tool could hardly force his peace efforts to fit in with the procrustean bed of concrete provisions specified in the will. The Nobel Committee, bound by statutory rules, simply did not register such nominations.

Further evidence of attention from the Russian society to the rising Nobel movement is exemplified by the personality of the great physiologist Ivan Pavlov. The world press for a long period considered him to be a most likely candidate for the Nobel Prize in physiology and medicine; indeed, he was awarded the prize in 1904. But he had been on the Nobel Committee's list of preliminarily selected nominees since 1901. The probability of his being among the first winners of the Nobel Prize was so high that information to this effect appeared in the press in mid-August 1901, two months before the Nobel Assembly, consisting of professors from the Karolinska Institutet, elected the first Nobel laureate in physiology and medicine. The St. Petersburg-based *Rossia* newspaper informed its readers, with reference to its source in Stockholm, about the decision to give the prize to Pavlov for investigations into physiology of nutrition and to Niels Finsen, the Danish physiotherapist, for the development of phototherapeutic techniques. This report was later reprinted in the authoritative press organ of the Imperial Military Medical Academy.[10]

In order to account for the appearance of these premature publications one should turn back to the year 1893. At that time, three years before his death, Alfred Nobel endowed 10,000 rubles to the Institute of Experimental Medicine (IEM), St. Petersburg, laying down no conditions as to the ways the gift might be used. On the 8th (20th) August of the same year, Prince Alexander of Oldenburg, trustee of the Institute, applied to the Emperor for permission to accept the grant, and received it with a command "to convert the donation into a special fund of the Imperial Institute of Experimental Medicine."[11]

Simultaneously, Alfred Nobel sent a letter to his nephew Immanuel, who headed the Nobel Brothers Petroleum Company in St. Petersburg, in which he asked to propose that IEM experimentally test some of his ideas in the field of physiology (doc. 1.2).[12] At that time, Pavlov was the acting director of IEM. Alfred Nobel was well aware of his scientific works and thought highly of them.

The great physiologist disregarded Nobel's proposal and spent all donated money on the construction of a two-storey extension of his main laboratory building.[13] It is worth noting that Pavlov gave high praise to "several instructive experimental projects" by Nobel[14] when he received the prize in 1904, but it was nothing more than a sign of reverence for the memory of the founder of the award.[15]

The misreports concerning the alleged awarding of the Nobel Prize to Pavlov appear to have arisen from confusion common to hasty newspaper publications. The author of the publication took the prize actually awarded to the two scientists in August 1901 for the Nobel Prize that was to be officially presented more than three months later. After a week, *Moskovskie Vedomosti* established the truth by publishing a report on the 11th (24th) August saying that "...in Copenhagen, a scientific prize of 200,000 krones is awarded to physiologist I. P. Pavlov, ordinary professor of the Sankt Peterburg Medical Academy for his work on nutritional physiology; a similar prize is awarded to the Danish professor Finsen, founder of the medical institution, for the treatment of diseases by the influence of light."[16] One true fact in the report published by the *Rossia* newspaper in August was that both scientists awarded in Copenhagen were nominees of the Nobel Committee: real candidates for the 1901 Nobel Prize.

That year, Pavlov was nominated by four persons. Two of them were his Russian colleagues, pathophysiologist S. M. Luk'yanov, director of IEM, and biochemist M. V. Nentsky of the same institute. The third nomination also came from Russia and was signed by 30 professors from the Military Medical Academy, including its director V. V. Pashutin.[17] William Henry Howell, a physiologist based at Baltimore, USA, was the fourth nominator of the Russian scientist. All nominations were considered by experts on the Nobel Committee, I. Johansson and Professor Robert Tigerstedt, head of the Physiological Department at the Alexander University, Helsingfors.

I. Johansson's personality deserves special mention. In the 1880s, Nobel became interested, as an amateur, in certain aspects of human physiology. He wished experiments required to verify his conjectures to be conducted on a professional basis and decided to invite a specialist with a view to later establishing a medical research institute in Paris, where he was residing at that time. The choice fell on Johansson, a young physiologist who had moved from Stockholm to Paris in 1890.[18]

There is reason to believe that this new assistant was aware of Nobel's letter to his nephew in St. Petersburg asking to try to arouse the interest of the IEM administration in his physiological ideas. Johansson could not have been unfamiliar with the works of Pavlov, whose philosophy of experimental research lay in the sphere of Nobel's interests. Therefore, there is little doubt that Johansson, as an expert, most likely favored the candidacy of Pavlov for the prize. Another expert, Professor Tigerstedt, was in all probability equally benevolent. Suffice it to say that he nominated the Russian scientist as soon as 1902.

In 1902, Pavlov received a total of seven nominations besides that of Professor Tigerstedt. The nominators included the well-known neuropathologist Professor V. K. Rot of Moscow University and foreign scientists based at Liege, Munich, and Strasbourg (with the renowned biochemist Professor Franz Hofmeister of Strasbourg University among them).

In 1903, the majority of nominations again came from Russia, including two from the Kazan University. One of these nominators was V. I. Razumovsky, a founder of the Russian School of Neurosurgery and the future rector of Saratov University (in 1912, he was removed from this position and forced to abandon the department he headed by the Minister of Education L. A. Kasso for objections to the latter's reactionary reforms). H. G. Ushinsky, professor of pathophysiology at Warsaw University was the third nominator. The others were the former experts, Tigerstedt and Johansson.

In 1904, four nominations for Pavlov were received in Stockholm, three from foreigners and one from a compatriot (V. V. Podvysotsky, professor at the Novorossisky University in Odessa). Johansson was again summoned by the Nobel Committee to consider the nominations.

To better substantiate their opinions, Johansson and Trigerstedt came to St. Petersburg on a plausible pretext and spent three weeks there familiarizing themselves with Pavlov's research on the physiology of the digestive system. The decision of the Nobel Committee to secretly send a group of such highly competent experts to St. Petersburg shows that it had serious intentions to offer the candidacy of the Russian scientist for consideration by the Nobel Assembly the of Karolinska Institutet as a nominee for the 1904 prize. The recognized authority of one member of this "secret mission", Professor Johansson, has already been mentioned previously and the same is true of his fellow traveller, Robert Tigerstedt.

Tigerstedt had already been an influential researcher at the Karolinska Institutet when he came to head the Physiological Department in the Alexander University, Helsingfors, in 1900. He and Professor K. A. V. Moerner, rector of the University, were appointed members of the Committee set up in pursuance of Nobel's will in January 1899 to develop the statutes of the Nobel Foundation and actively participated in this work.[19] It turned out that Tigerstedt's expertise was equally highly credited among Russian diplomats in Stockholm when it came to his invitation to Helsingfors (doc. 1.3).

The experts successfully accomplished the task, and the Karolinska Institutet awarded the fourth Nobel Prize in physiology and medicine to Ivan Petrovich Pavlov on 20th October 1904.

In the period of full-blown pseudopatriotic rhetoric of the late 1940s and the early 1950s, there were numerous talks about the anti-Russian and anti-Soviet position of the Nobel organizations (see Chapter 6). Certain adherents of this view went as far as to assert that Pavlov's election "went off [...] not without some difficulty." (see Appendix 4). Such assertions are not in accordance with the facts, to say the least. Pavlov received the prize just when his time was due, and there was some logic behind the three-year delay.

The discoveries of his three predecessors awarded the Nobel Prize were at the forefront of some of the immediate problems facing mankind. Emil von Behring resolved the problem of serum therapy, especially its application against diphtheria, which saved many people destined to die prematurely.[20] Ronald Ross, the winner of the 1902 Nobel Prize, laid the foundation for successfully combating malaria, from which tens of millions of people suffered on all continents. Niels Finsen's works on the treatment of diseases with concentrated light radiation were equally important.[21]

All these products of human genius are akin to "emergency health care" for the suffering, and prizes in recognition of their importance are the gratitude for salvation. Pavlov's works awarded the Nobel Prize were quite different.

The goal of the science he represented was explicitly formulated in his Nobel lecture as "deeper and deeper insight into the functions of an organism being a highly complicated mechanism."[22] In other words, Pavlov looked ahead and sought a solution to fundamental problems that could be expected to bring benefits in the future. Pavlov's main competitor for the 1904 Nobel Prize was Robert Koch, whose investigations and discoveries concerning tuberculosis, notwithstanding their great significance for combating this terrible illness, were recognized by the Nobel Committee only the next year. Such a decision once again demonstrated the rational approach of Karolinska professors to the selection of the laureates.

The scientist's widow, S. V. Pavlova, recounted their visit to Stockholm in her memoirs published in the leading literary periodical *Novy Mir* (New World) in 1946 (No. 3, p. 134). According to her, Pavlov came to know that he had been awarded the prize from a confidential letter from Robert Tigerstedt. In the early years of the Nobel organizations, the names of the laureates were officially announced on the 10th December, the presentation day, as prescribed by the Nobel Statutes. The notice from Tigerstedt enabled the Pavlovs to leave St. Petersburg under a plausible pretext and to stay for a week in Helsingfors as the guests of their informant. After the festivities in Stockholm, including a ceremonial dinner given in honour of Pavlov by Prince Alexander of Oldenburg, Pavlov triumphantly returned home. The Prince was the trustee of the Institute of Experimental Medicine, which he had founded in 1890 at his own expense; Pavlov had been working for the Institute since its establishment, as the head of the Physiological Department. A few days later, Pavlov was met with great ovation at the meeting of the Society of Russian Physicians on the 16th (29th) December dedicated to the memory of S. P. Botkin, his teacher.

During the subsequent years up to October 1917, Russian scientists were regular participants in the nomination process. In the Soviet times, nominators from the country were equally well represented among individuals entitled to submit candidates to the Nobel Committees for both literature and the sciences.

In Russia, medicine was the leading scientific discipline developing in conformity with world standards. Soon after Pavlov, Ilya Mechnikov, who had been working at the Pasteur Institute, Paris, was awarded the Nobel Prize (1908) in recognition of his work on immunity. Thus, by the time of World War I when the Nobel Prizes were not awarded, Russia turned out to have more Nobel laureates than any other country with the exception of Germany and France.

Vladimir Bekhterev, neuropathologist and psychiatrist, had a good chance of becoming the third Russian prize-winner. In 1912, he happened to be among the eight nominees selected by the Nobel Committe from the 54 registered candidates for further consideration. However, his strained personal relations with Pavlov dated back to the period of their university studies and were further complicated by increasingly discordant scientific views,[23] creating a scandalous situation that could not be disregarded by the Nobel Committee.[24]

The life of another early Nobel winner in physiology and medicine proved to be surprisingly linked to Russia. Robert Barany, professor at the University of Vienna was called up for military service as a medical officer when the mobilization campaign was launched in Austria–Hungary in July 1914, just before the First World War. His service started in the fortress of Peremyshl close to the Russian border. The garrison of the fortress capitulated after a successful offensive of the Russian Army, carried out in March 1915. Barany was captured and placed in a prisoner-of-war camp near the ancient city of Merv (now Mary, Turkmenistan), in a remote southern district of the Empire.

Here, Barany served as a physician in the camp hospital. In addition, he organized educational activities reading lectures to his colleagues in French (he seemingly deemed it out of place to use his German mother tongue while the war continued). While in the camp, Barany received a cable from Stockholm notifying him of the decision by the Karolinska Institutet (on the 29th October 1915) to award him the 1914 Nobel Prize "for his work on the physiology and pathology of the vestibular apparatus."

The high honors given to Barany called the attention of the Swedish Red Cross Society to the fate of the imprisoned Nobel laureate. Prince Carl, head of the Society, initiated an agreement whereby Barany was transferred from Merv to the university city of Kazan, where he was afforded more comfortable conditions. In June 1916, Barany was exchanged for a captive Russian top military officer in Khaparanda, northern Sweden.[25]

Barany received his award from the king at a specially organized formal ceremony at the royal palace on the 11th September 1916. On the same occasion, he gave his lecture prescribed by the Nobel Statutes (doc. 1.4).

It appears from the available archive documents that Barany bore no malice to the country where he had been held prisoner. He maintained friendly relations with Soviet diplomats for many years after he had returned to Sweden in 1917 for the rest of his life. Also he became one of the first members of the Swedish–Soviet Society for cultural and scientific relations established in May 1935.[26]

Robert Barany transmitted his kind sentiments of the country where he had come to know about winning the Nobel Prize to his son, the future father of Andreas, who told the author the following extraordinary story in the doc. 1.4. Ernst Barany, when a university student, had contacts with the representative office of VOKS in Stockholm and was known in Moscow as a person exhibiting sentiments of goodwill towards the Soviet Union (doc. 1.5).[26a]

Table 1.1 lists Russian nominees in physics, chemistry, physiology and medicine whose names were submitted between 1901 and 1918 inclusive. Table 1.2 includes Russian nominators who submitted foreign scientists as candidates. Tables 1.3 and 1.4 present similar information concerning Nobel Prizes for literature. The tables are based on the data borrowed from the protocols of the Nobel Committees stored in the archives of the Royal Academy of Sciences, Karolinska Institutet, and the Swedish Academy, i.e. prize-awarding organizations in their respective subjects.

The tables summarize data up to 1918 because, according to the Statutes, the Nobel Committees invited nominators to submit their recommendations in September of the preceding year (e.g. nominations for the 1918 prize were invited in September 1917). It means that the majority of letters from Russia containing nominations for the 1918 prize were in all probability sent before the events of October 1917.

It follows from the tables that contacts of Russian nominators with the Nobel Committees were regular and professional. The prevalence of nominations in physiology and medicine over those in physics and chemistry once again confirms the above observation of the leading position of life sciences in Russia at that time. Worthy of note is that Russian nominators submitted for consideration not only their compatriots but also many foreign colleagues. Also, they demonstrated thorough knowledge of the situation in their fields of interest. Table 1.2 shows that 16 scientists from five research centers nominated candidates for the Nobel Prizes in physics and chemistry. They recommended 27 foreign colleagues from nine countries of which 12 eventually received the prize. O. D. Khvol'son, D. A. Goldgammer and P. I. Valden were among the most successful predictors. Three of the seven nominees submitted by Khvol'son became Nobel Prize winners. Goldgammer and Valden nominated two candidates each, and all of them received the prize.

In 1909, the physico-chemist I. A. Kablukov of Moscow was about to nominate his candidate for chemistry. He thought about Wilhelm Ostwald, whom he had known since 1889, when he worked on probation in Ostwald's department at the University of Leipzig. Ostwald entrusted the guidance of the young Russian to the even younger Svante Arrhenius, his assistant. Kablukov and Arrhenius remained friends since that time. Therefore Kablukov applied to his old friend for advice as soon as he received an invitation from the Nobel Committee to nominate his candidate.

He felt uneasy at the thought that Ostwald for some time "did not work as efficiently as before" although "his previous works [continued] to strongly influence many recent discoveries and stimulate numerous new studies." In the return letter dated 9th November 1908, Arrhenius agreed that Ostwald's accomplishments actually looked "somewhat obsolete" and that he had lately been more interested in writing textbooks than in doing research: his books contain "many original thoughts, but it would be difficult to represent them as new scientific facts."[27]

Kablukov lent an ear to his friend's comments and decided not to nominate Ostwald, nor anyone else. Paradoxically, the Royal Academy of Sciences awarded the 1909 Nobel Prize in chemistry to no one else but Ostwald "in recognition of

Table 1.1. Russian nominees for the Nobel Prizes in the sciences, 1901–1918*

Nominees; residence at the time of nomination	Year of nomination	Number of individual nominations	Number of collective nominations; figures in brackets — number of signatures	Number of nominations from the Russian Empire
Physics				
P.N. Lebedev (1866–1912), Moscow	1905	1	—	1
-”-	1912	1	—	—
Chemistry				
P.I. Valden (1863–1957), Riga	1914	1	—	—
-”-	1918	1	—	—
D.I. Mendeleev (1834–1907), St. Peterburg	1905	3	—	—
-”-	1906	4	—	—
-”-	1907	2	—	—
M.S. Tsvet (1872–1919), Nizhny Novgorod	1918	1	—	—
Physiology and medicine				
V.M. Bekhterev (1857–1927), St. Peterburg	1912	1	—	1
-”-	1914	1	—	1
S.N. Vinogradsky (1856–1953), St. Peterburg	1911	1	—	1
A.S. Dogel’ (1852–1922), St. Peterburg	1911	1	—	1
I.M. Dogel’ (1830–1916), Kazan	1907	1	—	1
A.A. Maksimov (1874–1928), Petrograd	1918	1	—	1
I.I. Mechnikov (1845–1916), Paris**	1901	3	—	—
-”-	1902	2	—	—
-”-	1903	3	—	—
-”-	1904	6	1 (13)	1
-”-	1905	8	—	1
-”-	1906	7	—	—
-”-	1907	11	—	5***

(*Continued*)

Table 1.1. (*Continued*)

Nominees; residence at the time of nomination	Year of nomination	Number of individual nominations	Number of collective nominations; figures in brackets — number of signatures	Number of nominations from the Russian Empire
-"-	1908	13	—	5***
I.P. Pavlov (1849–1936), St. Peterburg	1901	3	1 (30)	3
-"-	1902	7	—	2****
-"-	1903	6	—	4****
-"-	1904	4	—	2
B.I. Razumovsky (1857–1935), Saratov	1914	1	—	1
S.I. Chiriev (1850–1915), Kiev	1905	1	—	1

* This and the following tables of Russian nominees and nominators include only the recommendations that were submitted in perfect agreement with the requirements of the Nobel Committees and that were accepted for consideration.

** Mechnikov was on the staff of the Pasteur Institute in Paris since 1887.

*** Russian nominators included A. M. Bezredka, the Russian-born colleague of Mechnikov at the Pasteur Institute.

**** Russian nominators included Robert Tigerstedt, a former professor of Karolinska Institutet, Stockholm, who moved to Helsingfors in 1900 to head the Department of Physiology at the Imperial Alexander University.

Table 1.2. Russian nominators and their foreign nominees for the Nobel Prizes in the sciences, 1901–1918 (the names of the nominees awarded the prize are italicized; figures in brackets show the year of award)

Nominators	Year of nomination	Nominees
	Physics	
		J. Dewar, England; W. Kaufmann, Germany
N.A. Umov (1846–1915), Moscow	1904	K. Olszewski, Austro-Hungary; J.J. Thomson (1906), England
O.D. Khvol'son (1852–1934), St. Peterburg	1905	J. Dewar, England
-"-	1907	P. Duhem, France
-"-	1909	M. Planck (1918), Germany
-"-	1910	H. Poincaré, France
-"-	1911	H. Poincaré, France

(*Continued*)

Table 1.2. (*Continued*)

Nominators	Year of nomination	Nominees
-"-	1914	A. Einstein (1921), Germany
-"-	1917	N. Bohr (1922), Denmark; M. Knudsen, Denmark
D.A. Goldgammer (1860–1922), Kazan	1909	M. Planck (1918), Germany; Van der Waals (1910), The Netherlands
V.A. Ul'yanin (1863–1933), Kazan	1909	M. Planck (1918), Germany
O.A. Backlund (1846–1916), Pulkovo	1909*	H. Forman, France; G. Voisin, France; O. Wright, USA; W. Wright, USA
-"-	1911**	H. Poincare, France
B.B. Golitzin (1862–1916), Sankt Peterburg	1911	H. Poincare, France
-"-	1913	A. Righi, Italy
-"-	1914	A. Righi, Italy
N.A. Umov (1846–1915)***, Moscow	1915	R.W. Wood, USA
P.P. Lazarev (1878–1942), Moscow	1918	*J. Perrin (1926), France*
A.P. Sokolov (1854–1928), Moscow	1918	*J. Perrin (1926), France*
	Chemistry	
K.A. Bishof (1855–1908), Riga	1904	W. Ramsay (1904), England
P.I. Valden (1863–1957), Riga	1904	W. Ostwald (1909), Germany
M.F. Glasenapp (1845–1923), Riga	1904	W. Ostwald (1909), Germany
-"-	1911	A. Werner (1913), Germany
V.V. Markovnikov (1837–1904), Moscow	1902	M. Berthelot, France
N.A. Menshutkin (1842–1907), St. Peterburg	1905	H. Moissan (1906), France
K.K. Blakher (1867-?), Riga	1912	F. Haber (1918), Germany
I.S. Plotnikov (1878–1955), Moscow	1915	J.M. Eder, Austria–Hungary T. Svedberg (1926), Sweden
L.A. Chugaev (1873–1922)****, St. Peterburg	1915	W.J. Pope, England; M. Gomberg, USA
-"-	1916	W.J. Pope, England; M.Gomberg, USA

(*Continued*)

Table 1.2. (*Continued*)

Nominators	Year of nomination	Nominees
	Physiology and Medicine	
P.M. Argutinsky (1854–1911), Kazan	1907	P. Ehrlich (1908), Germany
M.B. Blauberg (1866-?), Odessa	1912	M. Rubner, Germany
N.E. Vvedensky (1952–1922), St. Peterburg	1911	W. Einthoven (1924), The Netherlands
V.S. Gukevich (1867–1933), Moscow	1908	O. Hammarsten, Sweden
A.S. Dogel' (1852–1928), St. Peterburg	1908	G. Retzius, Sweden
-"-	1911	J. Loeb, USA
V.V. Kudrevitsky (1859-?), Warsaw	1907	*T. Kocher (1909), Switzerland*
M.M. Kuznetsov (?), Warsaw	1907	T. Kocher (1909), Switzerland
E.S. London (1868–1939), St. Peterburg	1912	S. Arrhenius, Sweden*****
-"-	1914	S. Arrhenius, Sweden*****
A.A. Maksimov (1874–1928), St. Peterburg	1914	J. Loeb, USA
A.K. Medvedev (1863–1923), Odessa	1912	M. Rubner, Germany
I.I. Mechnikov (1845–1916), Paris	1906	E. Roux, France
-"-	1909	E. Roux, France
I.F. Ognev (1855–1928), Moscow	1906	P. Ehrlich (1908), Germany
N.A. Oppel' (1872–1932), St. Peterburg	1914	H. Noguchi, USA
K.M. Pavlinov (1845-?), Moscow	1906	P. Ehrlich (1908), Germany
I.P. Pavlov (1849–1936), St. Peterburg	1909	J. Loeb, USA
A.D. Pavlovsky (1857–1944), Kiev	1910	E. Roux, France
A.A. Panormov (1859–1927), Kazan	1914	B. Kronig, Germany

(*Continued*)

Table 1.2. (*Continued*)

Nominators	Year of nomination	Nominees
V.V. Podvysotsky (1857–1913), St. Peterburg	1909	J. Bordet (1919), Belgium
-"-	1910	J. Bordet (1919), Belgium
I.G. Savchenko (1862–1932), Kazan	1914	B. Kronig, Germany
I.P. Skvortsov (1847–1921), Kiev	1909	A. Kossel (1910), Germany
I.V. Troitsky (1856–1923), Khar'kov	1918	A. Jakobi, USA
N.Ya. Chistovich (1870–1942), St. Peterburg	1914	A. Wright, Great Britain
A.I. Shcherbakov (1858-?), Odessa	1907	T. Kocher (1909), Switzerland
A.I. Yarotsky (1866–1944), Yriev	1913	W. Einthoven (1924), The Netherlands

* A collective 17-page nomination was sent from Paris on 22nd January 1909 and reached Stockholm on 30th January 1909. It was signed by six nominators from France and Sweden besides Backlund, including H. Poincaré, G. Mittag-Leffler etc. The eighth nominator of the four nominees in the rising field of aeronautics was Professor I. Fredholm of Stockholm, who submitted his concise individual nomination to the Nobel Committee on 30th January 1909.
** Backlund nominated Poincaré for the 1910 prize, but his letter was sent from Pulkovo on 8th February 1910 and was received in Stockholm on 14th February 1910. Therefore its consideration was postponed until 1911.
*** Umov sent his nomination of the American physicist on 23rd December 1914; it was registered at the Nobel Committee on 4th January 1915, a week and a half before his death on the night of 15th January 1915 (see note 86).
**** L. A. Chugaev was the maternal half-brother of V. N. Ipatieff, a nominee for the Nobel Prize in chemistry over many years (see Chapter 3). Their mother abandoned her first family when Ipatieff was only five. He came to know of his blood relationship with Chugaev, professor of chemistry and a good acquaintance of his, only in 1907 when he was 40 years old.
***** This was the second nomination of Svante Arrhenius, who had already received the Nobel Prize for chemistry in 1903.

his work on catalysis and for his investigations into the fundamental principles of governing chemical equilibria and rates of reactions."

Russian nominators in physiology and medicine were as competent as their counterparts nominating candidates in physics and chemistry. There were 23 of them working in eight Russian research centers, not counting Mechnikov, who then lived in Paris. They nominated 15 candidates from eight foreign states, of whom five received the prize.

Special mention should be made of A. I. Yarotsky, a professor at the Yurievsky University.[27a] In 1913, he nominated the Dutch physiologist Willem Einthoven,

Table 1.3. Russian nominees for the Nobel Prize in literature 1901–1918

Nominees	Year of nomination	Nominators
M.Gorky (1868–1936)	1918	B.Hesselman, University of Goteborg
A.F.Koni (1844–1927)	1902	A.K.Vulfert, Military Juridical Academy, St. Peterburg
D.S.Merezhkovsky (1865–1941)	1914	N.A.Kotlyarevsky, Imperial Academy of Sciences, St. Peterburg
L.N.Tolstoy (1828–1910)	1902	E.Lichtenberger, Sorbone University, Paris
-”-	1902	M.Breal, College de France
-”-	1902	O.Levertin, University of Stockholm
-”-	1902	L.Halevy, French Academy, Paris
-”-	1903	L.Halevy, French Academy, Paris
-”-	1903	M.Berthelot, French Academy, Paris
-”-	1903	A.France, French Academy, Paris
-”-	1904	L.Halevy, French Academy, Paris
-”-	1904	A.Sorel, French Academy, Paris
-”-	1905	E.M. de Vogue, French Academy, Paris
-”-	1905	L.Halevy, French Academy, Paris
-”-	1905	M.Berthelot, French Academy, Paris
-”-	1905	A.Aubert, Academy of Sciences of Norway
-”-	1905	C.G.Estlander and 16 other members of the Finnish Academic Society
-”-	1906	L.Halevy, French Academy, Paris

the inventor of the first string galvanometer and the founder of clinical electrocardiology. Einthoven actually received the Nobel Prize in physiology and medicine 11 years later in 1924. After a considerable lapse of time, it turned out that the nominator himself was the maternal grandfather of a present key figure in the Nobel Foundation, its executive director Michael Sohlman. This showed a bizarre example of interwoven human fates that came to light after many decades and that no one could ever have foretold.

It is worth emphasizing the nomination of German physiologist P. Ehrlich by three Russians based at Moscow and Kazan. Ehrlich shared the 1908 Nobel Prize with I. I. Mechnikov. The two scientists working in the field of immunology adhered to different concepts. Mechnikov developed and advocated the phagocytic (cellular) theory of immunity that underscored the key role of specialized cells (phagocytes) that engulf and digest foreign particles. Ehrlich on the contrary gave priority

Table 1.4. Russian nominators and their foreign nominees for the Nobel Prizes in literature 1901–1918

Nominators	Year of nomination	Nominees
Great Prince Konstantin Konstantinovich (1858–1915), St. Peterburg*	1903	R.F.Glasenapp
A.A. Golenishchev-Kutuzov (1848–1913), St. Peterburg	1903	-"-
N.P. Kondakov (1844–1924), St. Peterburg	1903	-"-
A.F. Koni (1844–1927), St. Peterburg	1903	-"-
A.I. Nelidov (1835–1927), Russian Ambassador in Rome	1903	-"-
A.A. Saburov (1836–1916), St. Peterburg	1903	-"-
P. Yankovsky (?)	1903	-"-
I.F. Boduen de Kurtene (1845–1929), St. Peterburg	1905	E.Orzeszkowa
L.Yu. Shepelevich (1863–1909), Khar'kov	1905	-"-
L.K. Mazing (?)	1905	-"-
I.Yu. Mikkola (1866–1946), Helsingfors	1905	-"-

to specific antitoxic and antibacterial factors. A smaller number of nominations from Russia in support of Mechnikov[17] indicated that Ehrlich's theory enjoyed greater popularity in this country.

Half a century later, Australian immunologist M. Burnet described acquired immunological tolerance and explained why antibodies recognize foreign substances in the body and do not respond to its own tissues. The new theory integrated the rational kernels of both Mechnnikov's and Ehrlich's hypotheses. Its validity was experimentally confirmed by Sir P. B. Medawar, a British biologist. In 1960, the two scientists shared the Nobel Prize in physiology and medicine "for the discovery of acquired immunological tolerance." The modern theory of immunity proposed by Burnet and Medawar provided further evidence of the remarkable foresight and sagacity of the Nobel Assembly of the Karolinska Institutet that divided the Nobel Prize between Mechnikov and Ehrlich half a century ago.

The above data convincingly prove as false and erroneous numerous accusations of Soviet propaganda against the Nobel organizations of their being influenced by politics and having anti-Soviet orientation. They were first attacked in the propagandistic drive of 1948 that gave rise to an even more extensive campaign launched in 1949 "against cosmopolitanism and servility before the West." It included undisguised disinformation especially in connection with the claims of priority of Russian scientists in the world-famous scientific achievements.[28]

This period in the history of the Soviet Union will be highlighted in more detail in Chapter 6. Here, only a few extracts from archive documents dating from that period are cited to illustrate the attacks on the Nobel organizations during the first two decades of their work.

The first document (1.6) is represented by an extract from the draft letter of D. T. Shepilov, deputy head of the Department of Propaganda and Agitation, Central Committee of the Communist Party (CPSU(b)), addressed to I. V. Stalin. The covering note (see doc. 5.10) dates the draft at 18th March 1946. However, it had a forerunner. As early as 1945, the Department proposed to institute international prizes to be awarded in the Soviet Union (see doc. 5.1–5.3), but the prepared documents were shelved. The extract from the draft letter to I. V. Stalin (doc. 1.6) almost literally reproduces one paragraph from an earlier document (5.3).

Shepilov repeated the contents of his draft letter to Stalin in another message sent five months later to G. M. Malenkov, secretary of the Central Committee (see doc. 5.16). However, it did not include the text mentioned in document 1.6. There is reason to believe that the omission was scarcely a matter of mere chance.

The note accompanying the letter to Malenkov indicated that "the cited documents were prepared by the Department of Propaganda with the participation of comrades Vavilov, Bruevich [...]" (see doc. 5.15). It appears that S. I. Vavilov, president of the USSR Academy of Sciences, and N. G. Bruevich, secretary of one of the Divisions of the Academy, managed to persuade Shepilov of the absurdity of pretensions on the world leadership of Russian scientists and the inadequacy of the list of researchers attached to the documents. Also, it may be assumed that Shepilov accepted the arguments and asked the heads of the Academy to find out more convincing illustrations of his main theory that "Swedes ostentatiously refused to recognize achievements of Soviet scientists" (see doc. 5.12).

Otherwise, it is difficult to explain the origin of one passage (see doc. 1.7) in the draft report of S. I. Vavilov entitled "Philosophical problems of modern physics and the tasks facing Soviet physicists". Vavilov planned to deliver it at the workshop for heads of physical departments of all universities and higher education institutions of the country, to be held in the first quarter of 1949. The meeting was never convened for objective reasons, but the text of the report was published twice after Vavilov's death, in 1952 and 1968 (in a sharply curtailed form).[29]

It is worth noting that the paragraph corresponding to document 1.7 was omitted in the second publication, falling in the more liberal, post-Stalinist period. The compilers of the volume dedicated to the 75th anniversary of Vavilov's death knew better than to include this passage because the invective was beneath criticism and did not do credit to the memory of the eminent scientist. For all that, this passage can be found in the original typewritten copy of the report dated to1949. It is therefore appropriate to consider its arguments and compare them with the true state of affairs, all the more because all the scientists mentioned by the author actually deserved to become Nobel laureates.

Soviet propaganda frequently used the name of A. S. Popov, acclaimed as the inventor of radio, in its attacks on the Nobel Committees. That Popov shared priority in the development of wireless communication with an Italian, G. Marconi, was never seriously doubted. In the Russian press, this invention was invariably referred to as Popov–Marconi wireless telegraphy. Both were awarded the Large Gold Medal at the World Exhibition held in Paris in 1900. The resulting collision was most adequately characterized in 1903 by a St. Petersburg newspaper as follows: "When we speak of Marconi, we should never forget about Popov. Both did a lot in this walk of science [...]."[30]

Nevertheless, post-war Soviet propaganda continued to cultivate the far-fetched idea of Popov's exclusive priority over the invention, which was supposedly used by the "shameless Marconi".[31] Hence a natural question whose priority in this case was conceded by Popov himself "Was it really a controversial matter in the eyes of the public in the late 19th and the early 20th centuries?"

Since 1992, the author of this book has run the column "The news for 100 years" in the weekly newspaper *Poisk* (Search). In this capacity, he has for many years thoroughly looked through six or eight sets of daily newspapers of that period. There is every reason to believe that no one then turned over the above problem in one's mind even though the Russian press evidently showed great interest in all news concerning "wireless telegraph". The names of Popov and Marconi did not disappear from the newspaper front pages. Either was mentioned as an inventor independently pursuing the goal in his own way, and there was no talk of competition and attempts of one to steal ideas from the other. Each alternately enjoyed success and experienced failure as should have been expected in the exploration of the unknown. The negative attitude of some representatives of the Russian scientific community to Marconi was inspired only by the attempts of the Italian engineer and the scientific/commercial company he established in Great Britain to monopolize the rising market for the new and highly promising means of two-way information exchange.[32]

In July 1902, the two men had a chance to meet when Victor-Emmanuel, the King of Italy, visited St. Petersburg on the invitation of the Russian Emperor Nikolai II. The Italian cruiser *Carlo Alberto* dropped anchor in Kronstadt. In Italy, one of the deck-cabins had been equipped with a transmitting/receiving station to be operated by Marconi himself.[33] On the 3rd (16th) of July the two monarchs visited the scientist at his working place and listened to the story of his invention. The next day (5th July), *Novosti i Birzhevaya Gazeta* published a report according to which "the public attending the *Carlo Alberto* could talk with Marconi and see his apparatus for wireless communication. One member of the public was Professor A. S. Popov, the Russian inventor of wireless telegraph, who had a long talk with his Italian colleague." The friendly meeting of the two pioneers of radio was also reported on the same day by the *Novoye Vremya* and *Sankt Peterburgskie Vedomosti* newspapers.[34] No word of reproach was uttered against the guest, Marconi.

The Italian side was perfectly aware of the situation in country their king was to visit. Marconi, the only civilian among the crew of the warship, would hardly have been taken to Kronstadt had not the outstanding achievements of Popov been widely known throughout the world. By analogy, when Khrushchev visited London in 1956, the official Russian delegation included I. V. Kurchatov because the people involved in preparing the visit took into consideration the high level of physical sciences in Great Britain and its contribution to the peaceful use of nuclear energy.[35]

The tsarist autocratic regime habitually neglected natural endowments of its citizens; they were equally disregarded by later Soviet propaganda at the height of absurd anti-West hysteria. A striking example of such an attitude is an episode that took place in the first months of the Russian–Japanese War of 1904–1905 that bore direct relation to Popov's invention.

In March 1904, a temporary railway was laid straight over the ice of Lake Baikal that made it possible to speed up the transportation of land forces from western and central regions of the Empire to the Far East. The government decided to support these strategically important actions by means of wireless communication. The necessary equipment was ordered from Telefunken, Germany, to the detriment of Russian enterprises well-prepared to manufacture similar apparatus. It provoked a natural if outwardly restrained response in Popov, who expressed his opinion in a letter published by the newspaper *Novoye Vremya.*[36]

The sudden death of Popov on the eve of the new year of 1906 in the Julian calendar practically put an end to further work on the development and improvement of wireless telegraph in Russia, while Telefunken was granted the preferential right to supply means of communication not only to the Russian railways but also to the Navy Ministry,[37] which had formerly supported Popov's studies.

In light of these facts, all talks of the "rivalry" between Popov and Marconi look like an ill-meaning story fabricated by Soviet ideologists designed to encourage hostile attitudes toward everything coming from the West.

Marconi was awarded the Nobel Prize in 1909, almost four years after the decease of the Russian inventor on 13th January 1906. Sergei Ivanovich Vavilov more than once referred to the failure to award Popov as an example of the "systematic and intentional" prejudice of the Nobel Committees against Russian scientists (see doc. 2.2). It seems he was simply unaware that an individual might not be nominated for the prize posthumously. There is no reason to think that the thought matured in him under the influence of Soviet ideological tenets. He expressed it so many times in different audience frameworks as if he really could not reconcile himself to the fact that the Russian inventor of radio communication had not received the Nobel Prize in 1909.[38]

Ironically, Vavilov died before 1958, when a group of Soviet physicists was awarded the Nobel Prize for the discovery and interpretation of the Cherenkov effect that was routinely called the Cherenkov–Vavilov effect in the Soviet scientific literature. Had he lived that long he would have been the first candidate for the prize as Popov would have been in 1909 if he had not died a few years earlier.

Some additional comments are in order in connection with the Nobel Prize for the development of wireless telegraphy.

Marconi was first nominated for the prize as early as when the Nobel Committee in physics began to function (in 1901, 1902 and 1903; thrice during Popov's lifetime) whereas the candidacy of the Russian scientist was never submitted. It appears that Russian nominators either forgot about the breakthrough made by their compatriot or attached little importance to it despite numerous publications in the Russian press and successful demonstrations of wireless transmission experiments by the inventor in scientific institutions of St. Petersburg, Odessa and other cities.

Almost one year before the first nominations were invited by the Nobel Committee, the *Sankt Peterburgskie Vedomosti* on 10th (22nd) January 1900 had informed its readers (with reference to a Gothenborg newspaper) that "Marconi's system for wireless telegraphy" was likely to be considered for the Nobel Prize. None of the Russian physicists commented on the fact that Marconi alone was named as a candidate.

The first nomination of physicists to come to Stockholm from Russia dates to 1904 when Professor N. A. Umov of Moscow submitted four candidates; but he did not mention the name of Popov (see Table 1.2).

True, the absence of an invited nomination does not pose an insurmountable obstacle to those deserving the award. There is an "antidote" to forgetfulness. The statutes of the Nobel Foundation stipulate that the 31st January be the closing date for nominations from the outside. However, each committee member has the right to submit his own candidate on the last day of February.

A concrete example is the 1909 Nobel Prize for physics that Marconi was jointly awarded with Carl Ferdinand Braun of the University of Strasburg, whose fundamental works greatly contributed to the development of wireless communication.[39] Braun had never been nominated for the prize before. However, Professor Gustaf Granqvist of Uppsala University believed that the prize must be shared by the two scientists and as a member of the Nobel Committee in physics recommended both Braun and Marconi.

This is not the sole case when committee members efficaciously used the "last day" rule. By way of example, the English biochemist Ernst Boris Chain (having Russian-born Mikhail Khain as his father) owed his 1945 Nobel Prize to this rule. Chain had already been nominated before, jointly with Alexander Flemming and Paul Flory for the discovery and studies of penicillin, but his name was not on the list of candidates for the 1945 prize. Goran Liljestrand, secretary of the physiology and medicine committee duly corrected the omission and recommended all of the three researchers. It was not Liljestrand's first successful initiative. A year previously (in 1944), he nominated American biochemist Edward Doisy and his Danish colleague Henrik Dam for the discovery of vitamin K. The two scientists had been nominated for the first time in 1941. In the next two years, only Doisy was nominated. The announcement of the 1943 Nobel Prizes was postponed until 1944, when neither Doisy nor Dam was among the candidates. Liljestrand submitted them on 31st January. His choice was supported by both the Nobel Committee and the assembly of professors of the Karolinska Institutet that divided the 1944 prize between Doisy and Dam.

It can be concluded from the above that Popov would have had the chance to share the Nobel Prize with Marconi and Braun[40] if he had lived until 1909. However, awarding the prize for the works of these three people was out of the question between 1901 and 1905. During that period, the Nobel Committee for physics was concerned with the nominations of much stronger and serious contenders meeting all the requirements pursuant to Nobel's will, such as Roentgen, Lorentz and Zeeman, Becquerel, the Curies, Lord Rayleigh, Lenard etc.

In summary, claims concerning the absence of the name of Popov in the list of Nobel laureates are absolutely groundless.

D. I. Mendeleev died on 2nd February 1907[41] after he had been nominated for the Nobel Prize in chemistry three times: in 1905, 1906 and 1907 (Table 1.1). All his nominators were foreigners. The reasons were many, including personal ones. It will be clear from what follows that many Russian researchers, first of all those working in physics and chemistry, were not happy to recommend their colleagues for the prestigious award. It is especially true of the author of the periodic law. It is well known that Mendeleev was many times voted down at the elections for entry into the Imperial Academy of Sciences. Highly honored by colleagues in different countries and a member of many foreign academies and scientific societies, Mendeleev was ostentatiously refused advancement, corresponding to full membership of the Russian Academy.

Moreover, there was an objective reason for which Mendeleev was not awarded the Nobel Prize. In a letter of invitation to submit nominations, a Nobel Committee makes it clear what main requirements a nomination must satisfy to be in accordance with the statutes of the Nobel Foundation. During the first four decades of the work of the Nobel organizations, a principal requirement was the novelty of the candidate's work. In this context, the great scientific breakthrough made by Mendeleev could not be compelled to lie within the stringent Nobel statutory regulations.

The periodic system is believed to have been created in 1869, when the scientist published the famous paper entitled "The systematic classification of elements on the basis of their atomic weight and chemical similarity." The first three nominations of Mendeleev for the Nobel Prize reached Stockholm in 1905. The nominators were Van't Hoff, the 1901 prize winner in chemistry, Oskar Hertwig, German chemist, and Otto Pettersson, chairman of the Nobel Committee in chemistry. The nomination was prompted by the then recent sensational discovery of inert gases for which two Nobel Prizes were awarded in the previous year. One (for physics) went to John William Strutt (Lord Rayleigh) "for his investigations of the densities of the most important gases and for the discovery of argon in connection with these studies"; the other (for chemistry) to William Ramsay "in recognition of his services in the discovery of the inert gaseous elements in air, and his determination of their place in the periodic system." There was every reason for the Royal Academy of Sciences to be as generous as that. Professor Johan Cederblom, president of the academy, who presented Ramsay at the award ceremony, rightfully stated that "the discovery of a totally new group of elements, none of which was known before, is a quite unique event in the history of chemistry."[42]

The discovery of several noble gases comprising a new family of chemical elements, unknown to Mendeleev in 1869, was taken in by the world's scientific community at the borderline of two centuries as a token that the periodic system was coming to an end. However, a more detailed analysis of the principles underlying its construction enabled Rayleigh and Ramsay not only to integrate helium, neon and argon into Mendeleev's table but also to predict the existence of other elements of the same group. Thereby, they confirmed the fundamental character of these principles postulated by Mendeleev three decades before and demonstrated that they objectively reflect natural laws.[43]

The epoch-making discovery of the transient group of elements, called "zero group" by Ramsay, and the role of the periodic system as a theoretical basis of this discovery that was mentioned in its official formula by the Swedish Academy of Sciences in 1904, gave reason to the three nominators to recommend Mendeleev's candidacy for the 1905 Nobel Prize. Moreover, paragraph 2 of the statutes of the Nobel Foundation stipulated that earlier basic research could be considered for the prize along with newer discoveries if its importance was confirmed by later scientific data.

The Nobel Committee in chemistry enthusiastically accepted the first nominations of Mendeleev. His name was immediately included in the so-called "short list" of candidates selected from many others mailed to the Committee. This list also contained the names of Professor Adolf von Baeyer of Munich, known for his research in organic chemistry, and Professor Henri Moissan, an experimenter from Paris and a founder of electrometallurgy, the discoverer of the element fluorine and the creator of the electric arc furnace, in which he for the first time synthesized calcium carbide.

The Nobel Committee chose von Baeyer, whose candidacy was supported by all of its five members who emphasized in their decision of 23rd September 1905 that Mendeleev's contribution to science had been recognized worldwide and that "he had been regularly submitted for the prize by the most authoritative chemists since 1902". In the same decision, the members of the Committee formulated their opinion about two other candidates included on the short list.

The decision read as follows: "As regards the scientific services of Mendeleev and Moissan, the Committee, understanding the great importance of Moissan's experimental works carried out during the last decades and having received high acclaim around the world, cannot help noting that they do not match the achievements of Baeyer and Mendeleev in terms of their influence on the development of chemical science at large. When choosing between von Baeyer and Mendeleev, the Committee took into consideration that the periodic system of Mendeleev has lately been confirmed and supplemented by the discovery of the so-called zero group or inert gases by Ramsay and Rayleigh. This confirmation is of recent origin and awaits as wide a support both inside and outside the Committee [i.e. in the Royal Academy of Sciences awarding prizes in chemistry] as has been given to the works of von Baeyer during the few past years."[44]

This cautious formulation was designed to conceal serious discussions in the Nobel Committee between the supporters of Mendeleev and those of von Baeyer. In the end, the votes favored the German chemist, as should have been expected since he had been nominated for a fourth year in succession, whereas Mendeleev's name was included in the list of candidates for the first time. In other words, the time came for Baeyer to be awarded, while Mendeleev was, so to speak, unofficially advised to wait his turn, the two men being almost of the same age (Mendeleev was born in 1934 and Baeyer in 1935). Thus nobody doubted that Mendeleev deserved the Nobel Prize just as much as Baeyer did; the question was who should be the first to receive it.

Sven Otto Pettersson, president of the Nobel Committee in chemistry and one of the three nominators of Mendeleev in 1905, most thoroughly and convincingly substantiated his recommendation. Being a recognized expert in the chemistry of germanium, the existence of which had been predicted by Mendeleev in 1886, Pettersson presented a well-reasoned characteristic of Mendeleev's works and thereby greatly promoted the opinion of his colleagues in favor of Mendeleev and his outstanding discovery. Pettersson especially insisted that all the former successful applications of the periodic system taken together could not surpass its true value as an instrument with which a directed search for new elements triumphantly resulted in the discovery of a group of inert gases. Pettersson pointed out to further prospects opened up by Mendeleev's periodic law.

"I do not mean that all potential resources of the periodic system are fully conceived today," he wrote in his nomination. "New discoveries of radioactive substances give reason to expect even greater extension of the range of its applications in the future. But Mendeleev is unlikely to live that long. Therefore, I believe it is a duty of presently working chemists to use the opportunity, while time and excellent occasion permit, and give an honor to the author of the most profound and fruitful scientific idea."

Pettersson shared the apprehensions of the committee members about the creation of an undesirable precedent to give the prize for scientific discoveries dated very far back. Therefore, he made the following reservation: "We should not fear that conferring the laureateship on Mendeleev will entail a routine of granting awards for other old works. His discovery stands by itself in the history of science and will never give cause for concern about such kind of abuse of statutory regulations. Moreover, it is clear that even the most important discoveries take time to be appreciated at their true worth."[45]

In 1906, the three previous year's nominators of Mendeleev were joined by Professor Robert Luther of Leipzig, the closest associate of the 1909 Nobel winner in chemistry Wilhelm Ostawld. Luther and Ostwald were coauthors of the classical book *Physico-Chemical Measurements*, and the former had been a lecturer for the Saint Petersburg State Institute of Technology in the 1890s. As in the previous year, the short list was headed by Mendeleev and Moissan but included two more names, those of German physico-chemist W. Nernst and French organic chemist V. Grignard; obviously, the two latter candidates were not favorites.

The services of the two main contenders were the subject of long discussions at the meetings of the Nobel Committee. Naturally the major point of controversy was whether the discovery made by Mendeleev 30 years before met the requirements of paragraph 2 of the Nobel Statutes. In order to better acquaint the members of the committee with the situation surrounding the periodic table and the history of its development, Pettersson prepared a detailed memorandum[46] dealing with both priority matters and the basic principles of Mendeleev's law. Pettersson concluded that the nomination of Mendeleev was perfectly in agreement with paragraph 2 of the statutes of the Nobel Foundation because his system "encompassed all most essential aspects of recent important discoveries and gave evidence of the formerly unthought of possibilities for its further development."

The discussion closed on 28th September 1906 with the conclusion in which the Nobel Committee reaffirmed "its opinion expressed last year [...] that although the excellent experimental works of Moissan had found support of many scientists entitled to nominate candidates for the Nobel Prize, [these works] cannot be compared with [...] the discoveries of Mendeleev in terms of the effect they had on the development of chemical science at large." Because the analysis of the inner structure of Mendeleev's system yielded so many promising results, "the Committee thinks that this system continues to be of current importance and serves today as a guiding star for further research. Bearing this in mind, the majority of the committee members (Mrs. Hammarsten, Pettersson, Sederbaum, and Vidman) believe that paragarph 2 of the statutes of the Nobel Foundation [...] fairly well applies to the work of Mendeleev and that he should be given priority as a candidate for the prize of this year. Mr. Klason is of a different opinion and voted for Moissan."[47]

Thus the Nobel Committee recommended to the Royal Academy of Science by four votes to one to award the 1906 prize in chemistry to Dmitry Mendeleev "in recognition of his services to the advancement of science by the discovery of the periodic system of elements." At the same time, the Committee noted that were the arguments of the majority of its members called into question the alternative candidate for the prize should be Henri Moissan, taking into consideration his investigations of fluorine and other achievements obtained with the use of the electric arc furnace invented by him.

The Royal Academy of Sciences chose the second candidate and awarded the 1906 Nobel Prize to Moissan "in recognition of the great services rendered by him in his investigations and isolation of the element fluorine, and for the adoption in the service of science of the electric furnace called after him."[48]

In accordance with the statutes of the Nobel Foundation, there is no keeping the minutes of the discussions in the prize-awarding institutions and the voting ballots are destroyed after they are counted. What made the members of the Royal Academy of Sciences disregard the well-reasoned opinion of the members of the Nobel Committee in chemistry and the circumstantial conclusion of its chairman is a matter of conjecture. In all likelihood, the main reason was the fear of creating a precedent of awarding for old discoveries. It was only in the 1930s that the Academy overcame this barrier and thereby emphasized the universal character of Nobel Prizes. But the above discussions in the Academy had hardly been confined to the applicability of paragraph 2 of the Statues. The factual aspect of the nomination of the two candidates could scarcely have avoided the attention of its members.

The name of Henri Moissan first appeared in the list of the candidates as early as 1901 and 1902; in either case he was nominated by the well-known French scientist Pierre Marcelin Berthelot. In 1903 and 1904, Berthelot was joined by Ramsay and a few other foreign nominators. Mendeleev could not be nominated as early as that, i.e. before 1905 when Rayleigh and Ramsay had been awarded the Nobel Prizes for 1904.

In 1905, Mendeleev was nominated by three persons including Pettersson, chairman of the Nobel Committee in chemistry. At exactly the same time, Moissan had 22 nominations (from such personalities as the Nobel Prize winners von Baeyer, E. Fisher (1902), and Ramsay (1904) as well as the well-known Russian specialist in chemical kinetics N. A. Menshutkin).[49] The latter is the first to blame for not nominating his great countrymen, whom many foreign scientists recalled with sympathy in connection with the award of the 1904 Nobel Prizes. In the same year, 1905, the candidacy of Moissan was first supported by a new key figure in the decision-making process that ended in him being awarded the Nobel Prize for chemistry in 1906. He was professor of organic chemistry at the Technological Institute, Stockholm, Peter Klason, member of the Nobel Committee and president of the Royal Academy of Sciences. He was the only person who voted for Moissan and therefore against Mendeleev in 1906.

There were fewer nominations in favor of Moissan (only eight) for the 1906 prize, in line with a decrease in the total number of nominations delivered to the Nobel Committee in chemistry in 1905 (19 versus 41 in the preceding year). However, the nominators of Moissan were highly reputed scientists as before; specifically, they included three Nobel Prize winners, von Baeyer, Ramsay, and Lenard (awarded in 1905). Naturally, the voting members of the Academy could not disregard the opinion of such nominators.[50]

In 1907, two more nominations for Mendeleev reached the Nobel Committee from the prize winner von Baeyer and Professor A. Ladenburg of the Breslau University, chemist and science historian. The former suggested that Mendeleev should share the prize with Stanislao Cannizzaro, one of the creators of the atomic-molecular theory; the latter thought that the prize should be divided between Mendeleev and Pierre Berthelot. However, what was an "excellent occasion to show an honor to the author of the most profound and fruitful scientific idea" as Pettersson put it[45] in 1905 and 1906 was irrevocably missed in 1907; the great scientist died on 2nd February 1907. Although Mendeleev's death gave no formal cause to cease the consideration of his candidacy by the Nobel Committee because it occurred after the closing day for nominations of the current year (see note 41 to this chapter), this procedure had no sense whatsoever in the existing situation.

It is worthy of note that Moissan died as a Nobel laureate on the 20th February 1907, i.e. two and a half weeks after Mendeleev, his competitor. Being seriously ill, he could not come to Stockholm to receive the award and deliver the traditional lecture.

To conclude the story of the struggle to award the Nobel Prize to the creator of the periodic system of elements in 1906, it is worth recalling an interesting episode concerning the relations between its two principal actors. There is little doubt that Sven Otto Pettersson was greatly disappointed by the voting results from the Royal Academy of Sciences and the role played by Peter Klason. However, this controversy did not make them antagonists for the rest of their lives, partly due to the tolerance inherent in the character of the Swedes. Almost 30 years later, the elderly Pettersson nominated Klason for the Nobel Prize in 1935. True, this nomination was not supported by the members of the Nobel Committee, but this piece of information may be of some interest in the context of the present narration.

The eminent Russian scientist P. N. Lebedev was the first to experimentally observe light pressure on solids in 1900 and on gases in 1910. The latter finding aroused especially great enthusiasm in the scientific community as being of fundamental importance for astrophysics. The English physicist J. G. Pointing worked parallel to the Russian investigator and associated light pressure with the heliocentric motion of comets.

Lebedev was for the first time nominated as a Nobel Prize candidate in 1905, together with J. Dewar, by O. D. Khvol'son, a professor of physics in St. Petersburg. His one-sentence nominations read as follows: "for experimental evidence of light energy pressure" by Lebedev and "for transformation of hydrogen into a solid" by Dewar. The two nominations were concluded by the assertion that "the epoch-making importance of these works did not need any further proof."[51]

The nomination made in 1912 by the German physicist W. Wien, Nobel Prize winner for 1911, was better substantiated. It was the last one for Lebedev because he died of heart disease on 14th March 1912, a month and a half after the closing date for nominations.

In his nomination, Wien emphasized that Lebedev had managed to overcome difficulties that proved insurmountable for his predecessors: Fresner, Bartoli, and Crookes. "In thoroughly conducted and multiple modified experiments, he demonstrated how it is possible to avoid effects of interfering forces and quantitatively measure light pressure. These observations are paramount for the theory of radiation because its thermodynamic component depends in the first place on the light pressure."[52]

The nominees registered in 1912 by the Nobel Committee in physics also included the aforementioned English investigator J. H. Pointing, who had previously been nominated twice, in 1909 and 1911. The nominators were venerable scientists including Nobel Prize winners E. Rutherford and J. J. Thomson. However, the Nobel Committee disregarded the nominations of both Lebedev and Pointing.

There were several reasons accounting for the lack of attention to the Russian researcher. To begin with, the Swedish engineer N. G. Dalen was regarded by the Royal Academy of Sciences as a much stronger and serious contender among the remaining 16 nominees of 1912. He received the prize for the invention of automatic regulators to be used in conjunction with gas accumulators for illuminating lighthouses and buoys. Members of the Academy voted in his favor as opposed not only to Lebedev, but also to such distinguished scientists as H. Kamerlingh Onnes, W. Nernst, M. Planck, and A. Einstein.[53] There is little doubt that the most absurd of all conceivable reasons for not voting for Lebedev would be the assertion of "systematic and intentional prejudice against representatives of our science" included in the draft report of the president of the USSR Academy of Sciences, prepared in 1949 for presentation at an All-Union workshop (albeit never convened) (see doc. 1.7).

This assertion is refuted by an episode that took place in the final year of Lebedev's life. In 1911, Lebedev and several other professors of Moscow University retired from their positions in protest against the reactionary policy of the Minister of Education L. A. Kasso. As soon as Svante Arrhenius, then the director of the Nobel Institute, Royal Academy of Sciences, came to know about it from Lebedev's letter, he invited the Russian scientist to come to Stockholm and promised him full freedom to continue research.

It is quite probable that the nomination of Lebedev by Wien was also an act of humanity. The forced resignation of Lebedev was widely discussed in the world press, and Wien was certainly aware that the situation proved difficult for him. Being a recent (1911) Nobel Prize winner, he decided to help his Russian colleague and recommended him as a candidate for the following year.

In conclusion, it shows from Tables 1.1 and 1.2 that Professor Khvol'son was the only countryman of Lebedev who nominated him for the Nobel Prize. Other Russian nominators in physics and chemistry were also active, but they usually recommended foreign scientists. There were at least three Russian nominators that were certainly well aware of Lebedev's work. They were Professor V. A. Ul'yanin, his university fellow, Professor D. A. Goldgammer who had been trying to theoretically explain light pressure since 1901, and finally Professor B. B. Golitsin, a friend of Lebedev with whom he studied light pressure on gases and its role in the formation of comet tails. In 1911, the Nobel Committee in physics invited Golitsin to nominate his candidate, and the Russian professor recommended the French theoretician Poincaré instead of his close colleague.

Another Russian nominator was Professor O. A. Backlund, director of the astronomical observatory in Pulkovo, near St. Petersburg. Backlund was widely known for his investigations into the mechanisms underlying the motion of comets, one of which was named after him (the Enke–Backlund comet). There is no doubt that he knew perfectly well the wonderful experiments of Lebedev. Nevertheless, he (like his colleague in the Academy of Sciences Golitsin) chose to nominate Poincaré, in both 1909 and 1911.[54]

Explanations for this could be either that many of Lebedev's colleagues proved unable to appreciate the remarkable results of his studies or that they did not regard them as worthy of the Nobel Prize. This is not the sole example of "blindness" from the past history of science as will be seen more than once in the subsequent narration.

Another nomination of a Russian scientist was in 1914. The Italian nominator Giacomo Ciamician, a professor at the University of Bologna, recommended for the Nobel Prize P. I. Valden, member of the Academy of Sciences, St. Peterburg and professor at the Polytechnic Institute in Riga, known for his studies on the theory of solutions.[55]

Valden was again nominated in 1918 together with a few other candidates. Professor F. M. Jaeger of the Groningen University, The Netherlands, was invited to Stockholm to nominate his candidates and decided to favor representatives of all branches of chemistry one at a time. He nominated Valden as an organic chemist, C. Auer von Welsbach of Vienna and G. Urbain of Paris in inorganic chemistry, A. L. Dey of the Carnegie Institute of Technology, Washington, in geochemistry, H. Le Shatelier of Paris in physical chemistry, and E. Abderhalden, Germany, in biochemistry. However, this initiative was not supported by the Nobel Committee and all Jaeger's candidates were dismissed from consideration in the very beginning for the lack of acceptable formulation of their achievements.

After Latvia proclaimed independence in 1920, Valden left Riga and settled in Germany for the rest of his life. Since then he was nominated five times for the Nobel Prize for chemistry, up to 1948, being a German citizen and professor at the University of Rostok. However, these nominations never met with approval by the members of the Nobel Committee.[56]

In 1918, Russian physiologist and biochemist M. S. Tsvet became a nominee for the Nobel Prize in chemistry. In 1903, he was a young assistant professor at the Warsaw University where he developed a totally new physico-chemical method for the separation and analysis of complex organic mixtures based on differential binding of their components by specific adsorbents. Three years later, Tsvet was the first to introduce chromatography into the practice of scientific research. Having residence in Warsaw, Tsvet several times traveled abroad and visited Holland in 1911.

In 1916, during the First World War, the Polytechnical Institute was moved from Warsaw to Nizhny Novgorod. This city is indicated as the place of residence of Tsvet in his nomination for the Nobel Prize presented by Cornelius van Wisseling. This fact suggests personal contacts between Wisseling and Tsvet during the latter's visit to The Netherlands and the correspondence they kept thereafter.

Tsvet died of decompensated heart disease in 1919 at the age of 47 as professor of the University of Voronezh. The chromatographic technique developed by Tsvet did not find wide practical application during his lifetime. The breakthrough occurred in 1931 after three influential German researchers R. Kuhn (future Nobel laureate of 1938), A. Winterschtein, and E. Lederer had published their joint paper in the then leading European journal *Zeitschrift für physiologische Chemie*. Since that time, chromatography remains a universally accepted method of organic chemistry.

Some ten years later (in 1943), Professor Ch. Deret of Geneva stated that Tsvet certainly deserved to be awarded the Nobel Prize because neither Robert Kuhn nor other researchers that were given it would ever have obtained their important results if they had not used chromatographic methods developed by the Russian scientist.

Could Tsvet really receive the prize had he lived 20 years longer? There can be no doubt about it. The Nobel Prizes in chemistry were awarded as follows for the three years in succession starting from 1937: Paul Karrer of Zurich received it for his investigations on carotenoids, flavines, and vitamins A and B2; next year, Richard Kuhn of Heidelberg received the prize "for his work on carotenoids and vitamins"; finally, in 1939, Adolf Butenand of Berlin and Leopold Ruzicka of Zurich were awarded for their work on sexl hormones. Could then the Royal Academy of Sciences disregard the works of the Russian scientist had he lived at that time?

Alas, Tsvet was destined for a different fate at variance with the popular maxim that candidates for the Nobel Prize must live long.

Russian nominators for the Nobel Prizes in physics and chemistry were not as active as their colleagues entitled to recommend candidates in physiology and medicine, who sent as many as 32 nominations to the Nobel Committees in these subjects before 1918 (their total number amounted to 36 taking into account nominations from the universities of Warsaw and Helsingfors, located at that time within the bounds of the Russian Empire). This fact once again indicates that medicine was the leading scientific discipline in Russia during the first two decades of the 20th century.

It is worth specially mentioning A. A. Maksimov, a professor at the Military Medical Academy in St. Petersburg and a founder of modern haematology. His theory of stem hemopoietical cells is universally accepted by the scientific community. In 1914, Maksimov sent to the Nobel Committee in physiology and medicine his nomination of J. Loeb, USA. Four years later, he himself was nominated for the prestigious award as the author of the theory of blood histogenesis, by A. S. Dogel, his colleague at the Military Medical Academy. In the winter of 1922, Maksimov secretly crossed the Russian border assisted by a guide and left his native country for good. At that time, he was already well-known in scientific circles and soon received an invitation to occupy a vacant post of professor at the University of Chicago.

In conclusion, it is worthwhile to emphasize the following fact. Advancements of science in Russia were always highly appreciated in Sweden. It is enough to recall a report sent to St. Petersburg on 25th June (8th July) 1909 by Count F. A. Budberg, the Russian ambassador in Stockholm, in connection with the expected visit of Emperor Nikolai II. The report contained *inter alia* this quotation from the newspaper *Svenska Dagbladet*: "We know that Russian scientists are highly reputed among their European colleagues [...]"[57]

The same is true of Nobel organizations established in pursuance of their founder's will. Members of these institutions always knew and never forgot about Russian scientists that helped the initiator of the prestigious prize to evolve into a researcher and inventor. The name of Nikolai Nikolaevich Zinin, his teacher, whose memory Nobel greatly respected as long as he lived is invariably mentioned in the presentation speech of the chairman of the Board of the Nobel Foundation as he introduces the ceremony.

Indeed, certain decisions by Nobel organizations concerning awards to Russian scientists cannot be regarded other than biased either personally or politically (see, for instance, doc. 4.44). However, there are no grounds whatsoever to interpret these omissions as a malevolent scheme or collective neglect of scientific achievements in Russia.

Although the first representative of Russian literature received the Nobel Prize only in 1933, the names of Russian writers as candidates for the award had been many times before discussed at the meetings of the Swedish Academy. Leo Tolstoy was nominated to the Swedish Academy more frequently than anyone else.

Many Swedish intellectuals had little doubt that the very first prize should be awarded to the great Russian writer. True, Tolstoy himself strongly objected to any monetary prize, in accordance with his moral principles. Nevertheless, the statutes of the Nobel Foundation envisaged settling such clashes of beliefs. Despite public statements from the writer, the Swedish Academy was entitled to announce him a Nobel laureate thereby justifying him as a great figure in the literature of the late 19th century and at the same time establishing the level at which future nominees would be measured.

The decision by the Swedish Academy to award the first Nobel Prize to the little-known French lyric poet Sully Prudhomme in 1901[58] provoked numerous and vehement protestations. The immediate reaction on Swedish soil was the famous address to Tolstoy signed by 42 well-known leading authors, artists and scientists proclaiming their protest against the decision. The signatories of the address included August Strindberg, Selma Lagerlof (future Nobel laureate), Professor Oskar-Ivar Levertin, and other eminent personalities (doc. 1.10).

For all that, complaints against the decision of the Swedish Academy taken in 1901 were on the whole poorly grounded. The name of Tolstoy was absent among the 25 nominees of that year, and the Nobel organizations had no formal reason whatsoever to consider him as a candidate for the award.[59] That mistake was corrected in 1902 when Tolstoy was nominated by three Frenchmen and Levertin, who had signed the aforementioned address (Table 1.3). The French nominators included Leon Halevy, writer and dramatist, and member of the French Academy. He showed himself to be the most insistent admirer of Tolstoy, whom he had nominated for five years in succession, from 1902 till 1906, that is as long as the name of the Russian writer kept appearing in the list of candidates registered by the Nobel Committee.

French nominators of Leo Tolstoy prevailed absolutely over representatives of other nations. It is not surprising, taking into account the great popularity of the writer in France, where he was considered worthy of all possible honors and awards. Suffice it to mention the initiative of the literature and art magazine *Express*, Paris, undertaken in the late 1904. According to the *Sankt Peterburgskie Vedomosti* of 7th (20th) December (p. 4), the magazine announced the raising of funds through public subscription to erect a monument in Paris to the great Russian writer (to be sculptured by Prince P. P. Trubetskoy) and thus render homage to him in his lifetime. "The initiators of the subscription," the newspaper continued, "are organizing public festivities to involve more people in the campaign and hope to see Tolstoy himself attending the ceremony." Four days later (11th or 24th December 1904), *Novosti i Birzhevaya Gazeta* (p. 2) revealed that Trubetskoy had been officially invited by Armand Dayo, director of Arts Department, to design a monument, and outlined his "plan". The sculptor saw the "great writer of the Russian land" attired in traditional Russian shirt and top-boots sitting bare-headed on horseback.

The story of discussions concerning Tolstoy's candidacy for the Nobel Prize was highlighted in much detail in the book by Kjell Espmark,[60] a recognized expert of the Nobel movement, member of the Nobel Committee in literature since 1988 and currently its chairman. Asking himself why such an outstanding contender was rejected so many times, he answers that the fact should be ascribed in the first place to the criteria for the choice of nominees. The system of their selection and consideration was formed under the influence of the conservative views of the Committee's chairman Carl David Wirsen, permanent secretary of the Swedish Academy.

The "epoch of Wirsen", as the first decade in the history of the Nobel Committee in literature is popularly called, was characterized by the interpretation of "an ideal direction" stated in Nobel's will as a "lofty and sound idealism". This approach to the works of Tolstoy was most consistently formulated in the decision of the Nobel Committee in 1905, when the writer was nominated for the prize by 21 individuals from France, Norway, and Finland:

"In spite of the admiration for many of Tolstoy's creations, it may well be asked to what extent idealism can be sound in an author who, in his otherwise magnificent *War and Peace*, attributes to blind chance such a decisive role in the great events of world history, who in the *Kreutzer Sonata* frowns upon intimate relations between spouses, and who in countless of his works denies not only the church but the state, even the right of property (which he himself, inconsistently, enjoys) and contests the right of individuals and peoples to self-defense."[61]

The feeling of aversion to the works of Tolstoy is most straightforwardly expressed in this resolution of the members of the Swedish Academy adopted in 1905, in the "epoch of Wirsen". Such a response might just as well be provoked by an unusually large number of nominators of Tolstoy for the Nobel Prize received by the Committee that year, with a large fraction of literary critics of Scandinavian origin among them. However, a similar situation took place in 1902, when the Nobel Committee considered the very first nominations of Tolstoy and made a decision that left no hope of the Prize being awarded to him in spite of numerous compliments addressed to the writer.

The authors of the 1902 resolution do not question that Tolstoy occupies a fitting place in world literature as a writer of the "highest rank" and recognize that such "immortal works" as *War and Peace* or *Anna Karenina* would give quite enough reason "to justly yield the palm to this great Russian writer". However, their judgment becomes merciless when they pass on to the recent works of Tolstoy. *Resurrection* arises in them "a feeling of moral indignation",[62] *The Power of Darkness* "horrifies with ghastly naturalistic descriptions", and the *Kreutzer Sonata* propagates unacceptable "negative asceticism". They cannot reconcile themselves to the criticism of the state and the church by Tolstoy, nor to his interpretation of the New Testament in "half-rationalistic, half-mystic spirit."[63]

For all that, Leo Tolstoy was not the sole victim of the "epoch of Wirsen". In 1901, members of the Swedish Academy rejected Emile Zola as a potential Nobel laureate despite the priority given to him by the nominators. In subsequent years, they were equally dissatisfied with the symbolism of Henrik Ibsen. In other words, the great personality of Tolstoy was not the only one that did not fit in with their idea of "sound idealism" (see also the second paragraph in note[59]).

There is one more interesting aspect to the clash of ideas relating to the discussion of Tolstoy's works by the Nobel Committee in literature. None of his compatriots recommended him for the prize, barring the members of the Finnish Academic Society that were only nominally citizens of the Russian Empire; they all nominated Tolstoy as a candidate for the Nobel Prize of 1905. No other Russian name can be found in the registries of the Nobel Committee, not even that of A. F. Koni, despite information contained in certain Russian sources that he participated in the campaign in favor of Tolstoy's nomination for the Nobel Prize in 1906.

There are two such sources testifying to the involvement of Koni. One is a note that D. P. Makovitsky, Tolstoy's personal doctor, entered in his diary on 23rd September 1906. The note says that P. I. Biryukov, who came from St. Petersburg to spend a few days in Yasnaya Polyana, told Tolstoy that he had heard from Koni about the intention of the Swedish Academy to award the Nobel Prize to him.[64] Another source was Koni himself. It is known that he first mentioned in his memoirs of Tolstoy the discussions of the writer's candidacy for the Nobel Prize at the meetings of the Division of Literature at the Imperial Academy of Sciences. However, this episode is absent in the published version of the memoirs.[65] It appears that the author chose to exclude it when he proofed the original manuscript. Here the text is cited from a handwritten copy stored in the archives of the Institute of Russian Literature at the Russian Academy of Sciences (doc. 1.17). This document, as well as official protocols recording discussions at the meetings of the Academy (docs. 1.12–1.16), contains very significant information.

One interesting fact is the choice of Tolstoy's work proposed to be nominated for the Nobel Prize. The honorary academician K. K. Arseniev suggested that this work should be the ordinary 20-page journal article entitled *The Great Sin*, devoted to current land-use problems facing the Russian peasantry of the period.[66]

The suggestion originated from the literal observance of the requirement in the Nobel Statutes (attached to Wirsen's letter of invitation) that the work being nominated had to be published in the preceding year (in this case 1905). *The Great Sin* was actually published in 1905 and was immediately republished in the British and American press.[67] In other words, it met the formal requirements and the members of the Academy voted for its nomination.

Long discussions at three meetings of the Division of Literature lasted from November 1905 until 7th (20th) January 1906; that is, they ended only ten days before the deadline for registration of the nominees by the Nobel Committee. The translation of the nomination into French and preparation of its despatch to Stockholm also took much time.[68] As a result the relevant documents were sent to the Nobel Committee only on 1st February 1906 (see note 1 to doc. 1.16) and were hopelessly late for their consideration at the appointed time. Koni probably knew about it and decided to exclude the story of the discussion from the final text of his memoirs.

Certainly the arrival of the nomination in Stockholm in due time would hardly have changed the situation. Therefore, the most interesting aspect of this story is the circumstances around the academic discussions in St. Petersburg. Generally speaking, the atmosphere reflected the attitude prevailing in the Nobel Committee. The opinion of the members of the Swedish Academy expressed in 1902 (see above) was not significantly different from the arguments of their Russian colleagues about irrelevance and maleficence of Tolstoy's interpretation of the New Testament or their assertion that "the Academy cannot separate the political activity of the author from his literature work and, the former being unacceptable, consider him worthy of the prize." (doc. 1.17).

The inability of the members of the Swedish Academy to overcome conservatism and see in full scale the great personality of Tolstoy does not do credit to them. At the same time, one should not forget about the role of Russian intellectuals of the early 20th century in this issue; their silence objectively encouraged the negative attitude of prize-awarding voters toward the works of the Russian writer. This circumstance was not taken into account half a century later when the Bureau of the Division of Physico-Mathematical Sciences, USSR Academy of Sciences, adopted a poorly thought-out resolution (see doc. 7.19) in which the Nobel organizations were blamed for not awarding Mendeleev, Tolstoy, Gorky, and even Chekhov, who was never nominated for the Nobel Prize either by his countrymen or foreign admirers of his talent.[69]

Leo Tolstoy was the sole Russian writer to be seriously considered by the Swedish Academy as a real candidate for the Nobel Prize. The nomination of A. F. Koni in 1902 was the personal initiative of his admirer who had been occasionally invited by the Nobel Committee to recommend a candidate. It appears from the minutes of the committee meetings that this widely known Russian lawyer and biographer was nominated by A. K. Wulfert, professor of the Military Juridical Academy, for the book *Doctor Friedrich Haas. Essay on the Life of German Philathropist in Russia*, published in the German language in 1899.[70]

Merezhkovsky was first nominated in 1914 and Gorky in 1918. Gorky was nominated by Professor B. Hesselman of Heidelberg University, who simultaneously recommended the German Christian writer Pastor G. Frenssen. The Russian nominators mentioned in the preceding paragraphs will be discussed again in Chapter 2.

In conclusion, it is worth looking again at the Russian nominators and their nominees from other countries in all four categories (Tables 1.2 and 1.4).

A most remarkable fact is the high percentage of nominees who eventually received the prize. Nine of the 20 nominees in physics and three of the ten in chemistry were recommended for the award by the respective Nobel Committees. Professor O. D. Khvol'son of St. Petersburg proved to be an especially successful predictor; moreover, he was the only Russian specialist whom the Nobel Committee in physics regularly invited to recommend candidates. Eight of his three nominees from other countries were awarded the Nobel Prize at different times. The two candidates nominated by Professor D. A. Goldgammer of the University of Kazan, in 1909 also became Nobel laureates in physics. One of them, Van der Waals, was awarded as soon as the year following the nomination.

These facts indicate that Russian researchers were fairly well acquainted with the latest achievements in natural sciences throughout the world. Their fellow nominators for the Nobel Prize in literature were far less efficient.

Strictly speaking, they did not nominate any foreign candidates. K. F. Glasenapp was born and lived in Riga where he was a lecturer for the Department of German Language and Literature at the local Polytechnical Institute.[71] E. Orzeszkova, a Polish writer, lived and worked in the Grodnenskaya guberniya (province). They are distinguished here because the Nobel Committee in literature registered German and Polish writing candidates for the prize separately from Russian ones.

Glasenapp, known for his book about Richard Wagner,[72] was nominated in 1903 by seven Russians and 46 professors from Germany, Austria–Hungary, France, and Spain. Orzeszkova was nominated in 1905 by four professors from the universities of St. Petersburg, Khar'kov, Helsingfors, Yuriev, Krakow (Austria–Hungary), and three from German universities.

These facts suggest close links between philologists of different European states at that period, which facilitated the involvement of so many specialists in the nomination campaign (see note 5 to doc. 1.18). Surprisingly, none of the Russian nominators recommended candidates (including their compatriots) that had much chance of being positively accepted by the Swedish Academy, or at least of attracting its attention as outstanding literature figures.

Worthy of mention is one more fact associated with the consideration of Leo Tolstoy as a candidate for the 1906 Nobel Prize in St. Petersburg. K. D. Wirsen, chairman of the Nobel Committee in literature, sent a letter to S. F. Oldenburg, permanent secretary of the Imperial Academy of Sciences, on 8th November

1905 (doc. 1.11) in which he asked to remind members of the Academy of their right to nominate candidates for the Nobel Prize. This letter gave the Division of Russian Language and Literature of the Imperial Academy the same status as the "Swedish Academy and other academies, institutes, and societies analogous to the Academy", as stipulated by paragraph 1 of the Special Rules for awarding Nobel Prizes in literature.

There is no doubt that the governing body of the Russian Academy had been duly notified about this right of its members in autumn 1900; otherwise, Wirsen would have not asked its secretary "to remind" him that their recommendations were awaited in Stockholm. Nevertheless, Russian nominators neglected this opportunity.

Indeed, the first nominations for the Nobel Prize were sent to Stockholm by just seven Russian nominators in 1903 and by four in 1905 (see Table 1.4). These were collective nominations for Glasenapp from Riga and the Polish writer Orzeszkowa. It is difficult to say now who really initiated these nominations; Russians or their foreign counterparts. The nomination of Koni by Wulfert in 1902 did not meet the criteria of the Nobel Committee.

In other words, Russian professors and members of the Academy never used the opportunity to draw the attention of their Swedish colleagues to the writers of their country who may have deserved to be awarded the Nobel Prize. It was done by nominators from the French Academy, Finnish Academic Society, and other philological institutions of Western Europe. This situation is best illustrated by the history of the nomination of Leo Tolstoy.

A possible explanation for the passive behavior of Russian intellectuals in response to the annual invitations from Stockholm could be found by looking at the situation from 1900.

On 14th (27th) October 1900, one of the liberal newspapers in St. Petersburg published a note[73] saying that "the deceased Swedish engineer Alfred Nobel left to the special administration, subordinate to the Academy of Sciences in Stockholm[74] his capital and interest, which shall be annually divided into five prizes". It was also specified in the note where any people interested could find "detailed information about the rules regulating nomination for the prizes." The notification appeared in due time, i.e. more than three months before the closing date for the nomination of the first candidates for the prizes to be awarded in 1901. As a result the Russian public came to know about the onset of activities of the Nobel Foundation and its committees from a quite unusual source.

It was neither the Imperial Academy of Sciences, many of whose members were potential candidates for the Nobel Prize,[75] nor the Ministry of Public Education, to which all Russian universities were subordinate[76] (the Statutes of the Nobel Foundation granted their professors the right to nominate candidates for the prize). The newspaper announcement of the forthcoming award of five Nobel Prizes was signed by the Naval General Staff! Moreover, those interested were advised to apply for additional information to the Scientific Department of the Staff.

This absurd situation continued to evolve until 1902, when the full text of the Statutes of the Nobel Foundation was for the first time published in Russian in a volume issued by the St. Petersburg Institute of Communications Engineers.[77] The Statutes were placed on the last pages of the book, after a paper on multiphase electric systems under non-equilibrium loading and a report on maintaining the student hostel in 1900. Even these extraordinary events did not make the Russian Academy raise its voice as the main organization to which information from Stockholm about the new international initiative was primarily addressed.

There is more evidence that the management of the Russian Academy deliberately avoided participation in the nomination process. Members of the Division of Russian Language and Literature, Academy of Sciences, whom the Nobel Committee in literature gave the right to annually recommend their candidates (see doc. 1.11) came to know about it only in autumn 1905, as follows from the minutes of the meeting held on 8th (21th) November: "[...] the said invitation [...] only now, for the first time, reached the Imperial Academy of Sciences although the Nobel Prizes had already been awarded several times since their introduction" (doc. 1.14). In other words, the annual invitations sent from Sweden were merely shelved in the Russian Academy offices and remained unavailable for nominators.[78]

Thorough archive studies are needed to understand reasons for which the Academy practiced this unaccountable boycott of the promising international initiative. However, there are some facts that may shed some light on the immediate cause behind the blunt neglect of invitation letters. Obviously, it was underlain by political considerations and was also related to certain provisions of the Statutes governing the work of the Nobel organizations.

Incidentally, the year 1900, when the statutes of the Nobel Foundation were approved, was the period when Finland experienced its most intense wave of russification initiated by "the case for strengthening the unity of the Empire", as was stated in the tsar's manifesto: "On the introduction of Russian as the official language in certain areas of public life of the Grand Duchy of Finland", dated 7th (20th) June 1900.[79] This policy was strongly resented by both the local authorities and the population at large.[80] In response, the government issued a decree "on public meetings in the Grand Duchy of Finland" under which "public meetings organized by individuals or societies for the discussion of public, scientific, economic, and other problems [...] can be held only with permission from the governor general."[81]

This decree was signed by the Emperor on 19th June (2th July) 1900, three days after the Nobel Foundation had been established in Stockholm. Certain provisions of its Statutes could have provoked a negative reaction from St. Petersburg authorities during the restless situation of those days. Specifically, the Statutes mentioned special privileges granted to the University of Helsingfors along with the Universities of Lund and Uppsala in Sweden, the University of Copenhagen in Denmark, and the University of Christiania in Norway. Professors of these institutions were given the right to nominate annually their candidates for the Nobel Prizes in natural sciences. This may be the true cause of the negative attitude of the Imperial Academy of Sciences, headed by Great Prince Konstantin Konstantinovich, to invitations to nominate from Stockholm.[82]

That the negative attitude of the Academy management toward the Nobel movement was actually due to the situation in Finland found unexpected confirmation from a quite different source. Evidence of it can be found in the unpublished correspondence between V. K. Pleve and V. N. Lamzdorf, Russian ministers of internal and foreign affairs, in 1904 (doc. 1.19). The correspondence started through a misunderstanding that arose from the erroneous interpretation of information received by the Department of Police.[83] However, it contains an allusion by Pleve to the "Finnish question" in connection with the alleged award of the Nobel Peace Prize to the *L'Européen* newspaper, published in Paris and known for its anti-Russian position. It irrefutably supports the argument that the Russian government was concerned about the "manifestly hostile" attitudes of both the Swedish government and the Nobel organizations towards Russian policy in Finland.

However, all these diplomatic considerations based on suspicion of the St. Petersburg authorities were pushed into the background when the Nobel Prizes were awarded to Russians. An official representative of the Russian Empire, similarly to his counterparts from other countries, punctually performed his ceremonial duties and participated in all events of the presentation day of 10th December. In 1904, Baron K. Stael von Holstein, Russian envoy accredited to the Swedish government after the death of his predecessor E. K. Butzov in October that year, sent a cable to B. N. Lamzdorf with a report about the ceremony of awarding the prize to I. P. Pavlov in the presence of the king.[84]

Next year, the prize was awarded to Henryk Sienkiewicz, a Polish writer and citizen of the Russian Empire. He had been nominated by all 18 members of the Swedish Academy every year since 1901. Again, the Russian flag was hoisted and the Russian diplomat seated side by side with the laureate during the formal dinner. Stael Holstein again sent a detailed report on the 1905 festivities and statements made by Sienkiewicz for the press, emphasizing that the writer "could tactfully [...] bypass tricky points" and avoided "commenting on the relations between Russia and Poland" (doc. 1.20).

The Russian Ministry of Foreign Affairs was a ready source of supply of Russian-language literature to the libraries of the arising Nobel institutions. For example, the Second Department of the Ministry placed at the disposal of the envoy of Sweden and Norway in St. Petersburg 17 volumes of treaties and conventions concluded by the Russian government with foreign states. These materials were dispatched on 16th August 1903 in response to the envoy's request, dated 3rd (16th) June 1903 (doc. 1.21).

In 1906, Baron Stael Holstein again sent a report on Nobel festivities in Stockholm to the new minister, A. P. Izvol'sky, although there were no Russians among the Nobel laureates.[85] No other reports were received by the Ministry in subsequent years, probably because the annual prize-awarding ceremony became a routine, and regular notification about it lost meaning due to the lack of interest on the part of the government.

Notes

[1] The property of Ludvig Nobel's sons, Gustav and Emil, was nationalized, and they were arrested by Cheka but soon released on bail thanks to the efforts of the Swedish Embassy in Petrograd.

Emmanuel Nobel, the elder son of Ludvig and head of the Nobel Brothers Petroleum Company, had left for Essentuki in the summer of 1917 to undergo a course of medical treatment. Gustav remained in Petrograd as the acting director of the company while his family joined Emmanuel. The October Revolution found them in the spa town. Colonel Kazansky, mayor of Baku, occasionally met the Nobels in Essentuki and helped them to obtain forged passports and thus safely reach Finland and then Stockholm* via the Ukraine (which was under Petlura's government at that time) and Germany. Four years after the nationalization several former employees of the company were tried as "Nobel agents" by the Moscow revolutionary tribunal (24–26 July 1922). They were accused of counter-revolutionary activity, industrial espionage, and criminal collusion with Gustav Nobel in November 1918. Four of the accused were sentenced to be shot and five others to different terms of imprisonment. All of a sudden, Presidium of VCIK gave orders to delay the execution of the sentences and transfer the case to the Cassation Commission of the RSFSR Supreme Tribunal.**

* K. V. Khagelin. *Moy trudovoy put'* (My Life of Work). New York, 1945.
** *Izvestia*, 22, 23, 25, 26, 27, 28, 30 July 1922.

[2] R. Sohlman. *Zaveshchanie Al'freda Nobelya* (The Legacy of Alfred Nobel). M.: Mir, 1994. This fact is not at variance with the decision taken by the Norwegian Storting on 5 July 1897 concerning the procedure of election of the Nobel Committee as the Peace Prize-awarding organization. Sweden and Norway were involved in a "cold war" at that time that could have at any moment broken out into a full scale war. This move by "the younger brother" of the Union was designed to demonstrate the ability to take independent decisions even if they disregarded juridical aspects of the will.

[3] L. N. Tolstoy. *Polnoye sobrabie sochineniy* (Complete Works). GIHL, 1954, V. 70, p. 153.

[4] *Ibid.*

[5] In all likelihood, this Peace League was of Russian origin. The striving to show loyalty to the emperor by nominating him for the yet non-existent Nobel Peace Prize had much in common with the proposal by the mayor of Odessa to name some educational institutions of the city "in memory of the address by the Russian Emperor to other nations on 12 August 1898." To do justice to the authorities, they proved equal to the occasion, and the head of the local educational district notified the mayor that "in accordance with the opinion of the Ministry of Internal Affairs, the Minister of Education does not find it appropriate" to give such names.*

* *Birzhevye Vedomosti*, 26 September (8 October) 1899.

[6] Vereshchagin was not the sole self-advertizer. The French peace champion Frederik Passy also publicly claimed his right to the Nobel Prize in January 1901. Unlike the Russian artist, he was actually awarded it in the same year. As an illustration of his personal services to the cause of peace, Passy mentioned his own letter to Queen Victoria, in which he was calling for her abdication unless she put an end to the Boer war.*

* *Kievlyanin,* 31 January (13 February) 1901, p. 2.

[7] As an exception, the Norwegian Nobel Committee extended the deadline for 1901 nominations until 1 April. The Russian Ministry of Public Education announced this decision via its official organ *Pravitel'stvenny Vestnik* on 1st (14th) March 1901.

[8] B. von Zuttner. *Stimmen und Gestalten.* Leipzig, 1907, S. 113. This communication is indirectly confirmed by a report from Christiania published in *Sankt Peterburgskie Vedomosti* six years before the appearance of her memoirs. The reporter referred to Zuttner as saying that Vereshchagin "was recommended by a non-Russian nominator."*

* *Sankt Peterburgskie Vedomosti,* 16th (29th) March 1901, p. 1.

[9] When the artist exhibited his anti-war pictures in Christiania in May 1900, he gave an interview to a Norwegian newsman, one paragraph of which is presented below: "I was quite unexpectedly notified by Christiania that I am supposed to be a candidate for the Peace Prize and advised to exhibit my pictures of war in the city. I sent photographs but they were deemed insufficient for the purpose. Therefore, I decided to bring my pictures here."* There is certainly a kernel of rationality in those words.

* *Sankt Peterburgskie Vedomosti,* 6th (19th) May 1901, p. 1.

[10] *Izvestiya Voenno-Meditsinskoy Akademii,* V. 3, No. 1, 1901; *Rossia,* 5th (18th) August 1901, p. 2.

[11] *Polnoye sobranie zakonov Rossiyskoy Imperii.* (Complete Collection of Laws of the Russian Empire. Collection) Collection 3, V. 13, 1893, No. 9913. SPb, 1897, p. 544.

[12] This document was first published by V. S. Meshkunov in the journal *Voprosy Istorii Estestvoznaniya i Tekhniki,* No. 1, 1994, pp. 124–125.

[13] *Ibid.*

[14] I. P. Pavlov. *Polnoye sobranie sochineniy.* (Complete Works) V. 2, Book 2, 1951.

[15] Alfred Nobel was seriously interested in physiological studies. He informed Emmanuel Nobel in a letter (see doc. 1.2) that he had advanced a similar proposal to the Karolinska Institutet. It appears that the proposal failed to be accepted, as it did in St. Petersburg.

[16] *Moskovskie Vedomosti,* 11th (24th) August 1901. It should be noted that this information did not prevent a similar mistake concerning the 1901 Nobel Prize made two and a half months later by the *Russkie Vedomosti* newspaper. In a report it described the forthcoming publication of I. P. Pavlov's work in London in English as "a study [...] of digestive glands" that brought to the author an award of 100,000 rubles;* once again the reporter took the prize given in Copenhagen for the Nobel Prize.**

* *Russkie Vedomosti,* 30 October (13 November) 1901, p. 4.

** In the early 20th century, the exchange rate was one ruble to two Danish krones. In other words, the amount of prize money awarded to Pavlov in Copenhagen in August 1901 was 200,000 Danish krones.

[17] The list of Pavlov's (and Mechnikov's) nominators, including those who signed collective nominations, was published in the *Priroda* (Nature) magazine No. 7, 2001, pp. 4 and 8.

[18] Johansson built a successful career. He was a professor at the Karolinska Institutet from 1901 until 1927 and chairman of the Nobel Committee in physiology and medicine from 1918 to 1926.

[19] R. Sohlman. *Ett Testamente. Nobelstiftelsens Tillonahistoria och dess grundare.* Stockholm, Borlforlaget "Atlantis", 2001, S. 275.

[20] It would be appropriate to note that the award of the first Nobel Prize to Behring instead of Pavlov caused no irritation in the Russian press. On the contrary, when the Nobel laureate arrived in Sevastopol in a group of European and American tourists, on board a ship of the Gamburg-American line some four months later, for sightseeing at the southern coast of the Crimean Peninsula, Russian newspapers favorably informed their readers about the visit of the illustrious person.* On the way back, the yacht anchored for one day offshore at Odessa, where Behring could see the local bacteriological station and the medical clinic of the Novorossiysky University, then under construction. The daily newspaper *Odesskie Novosti* published an enthusiastic review of Behring's visit.**

* "SEVASTOPOL. A group of foreign tourists includes [...] Professor Behring, bacteriologist. After sightseeing in Sevastopol and places of historical interest, 175 tourists will move ashore to see the southern coast of Crimea." *Novoye Vremya*, 5 April 1902, p. 3.
** "The famous Professor Behring of Marburg University, whom mankind owes the development of serum therapy against diphtheria, visited Odessa and used the opportunity to visit the local bacteriological station manufacturing this serum for southern regions of Russia...", etc. *Odesskie Vedomosti*, 7 April 1902, p. 3.

[21] The method of treatment of diseases with light radiation developed by Finsen was well-known and successfully applied in Russia. The Russian press informed the readers about the "wonderful results in the management of *lupus vulgaris*" in a St. Petersburg hospital and the decision of the city health commission to "substantially extend" the possibilities for the work of the cabinet of phototherapy attached to the clinic.*

**Russkie Vedomosti*, 27 July 1903, p. 1.

[22] *Les prix Nobel en 1904.* Stockholm, 1907, p. 11.

[23] From the memoirs of L. M. Pussep, neurosurgeon: "They did not shake hands, nor did they talk with each other."*

* I. M. Guberman. *Bekhterev: Stranitsy zhizni* (Bekhterev: Pages of Life). M.: *Znanie*, p. 128.

[24] In spring 1912, Bekhterev was busy organizing an institute for the study of alcoholism. A misprint that slipped into the document prepared by his colleagues for the Ministry of Finance overstated the permissible alcohol level in the human body (g/kg) by two orders of magnitude. This made Pavlov publish a statement addressed to the Division of Physico-Mathematical Sciences of the Imperial Academy of Sciences in which he

insisted that an institute formulating such documents "has no right [...] to do research."* The statement became known to a mass audience and could not pass unnoticed by the Nobel Committee in physiology and medicine.

* I. P. Pavlov. *Polnoye sobranie sochineniy* (Complete Works). V. 6, M.: Izd-vo AN SSSR, 1952, p. 419.

[25] The exchange of Barany for a Russian officer in Sweden was unrelated to the Nobel Prize awarded to him. During World War I, the Swedish Red Cross secured the exchange of many prisoners-of-war on the condition that traveling expenses over the country's territory were borne by belligerents. After the cessation of hostilities in 1918, Soviet Russia refused to pay the debts. The conflict was settled only in 1933. It was agreed that the debts would be covered by part of the gold reserve placed by the Russian provisional government in the Bank of Sweden as a security interest in military orders.*

*A. M. Kollontay. *Diplomaticheskie dnevniki* (Diplomatic Diaries 1922–1940) V. 2, M.: Academia, 2001, pp. 170–172.

[26] A constituent assembly and banquet on the occasion of the institution of the Swedish–Soviet Society for cultural and scientific relations were held on 20 May 1935 in the Rosenbad restaurant, Stockholm. Robert Barany was among the new members admitted to the Society.*

* GARF, fond 5283, opis' 5, ed.khr. 540, l.40.

[26a] It is worthwhile noting that Barany was not the sole Nobel laureate imprisoned in Russia. Thirty years later, the Austrian zoologist Konrad Lorenz, then junior lieutenant of the Werhmacht, was captured at the Russian–German front and kept in a prisoner-of-war camp at Krasnogorsk, near Moscow, from 1944 to 1947. The administration of the camp treated 40-year-old Lorenz with dignity and gave him a reasonably good opportunity to do research. In 1973, Lorenz shared the Nobel Prize in physiology and medicine with K. von Frisch, Germany, and N. Tinbergen, Great Britain, "for their discoveries concerning the organization and elicitation of individual and social behavior patterns." The archives at the Krasnogorsk camp keep a manuscript of Lorenz's book written in the absence of necessary literature sources.*

* *Kniga istoricheskikh sensatsiy* (The Book of Historical Sensations). M.: Izd-vo Raritet, 1959, p. 139.

[27] See Yu. I. Soloviev, N. A. Figurnovsky. *Svante Arrhenius.* M.: Izd-vo AN SSSR, 1059, p. 139.

[27a] It is worthy of note that A. I. Yarotsky had been closely acquainted with the great scientist and philosopher V. I. Vernadsky, who more than once mentioned Yarotsky in his diaries. See, for instance, V. I. Vernadsky. *Dnevniki 1935–1941 v dvukh knigakh.* (Diaries. 1935–1941 in Two Books.) Book 1. 1935–1938. M.: Nauka, 2006, pp. 19, 21, 210, 211, 282, 285.

[28] See A .S. Sonin. *Fizicheskiy idealizm. Istoriya odnoy ideologicheskoy kampanii* (Physical Idealism. The History of One Ideological Campaign). M.: Izd-vo fiz-mat. nauk, 1994.

[29] S. I. Vavilov. *Filosofskie problemy sovremennoy fiziki i zadachi sovetskikh fizikov v bor'be za peredovuyu nauku* (Philosophical Problems of Modern Physics and the Tasks of Soviet Physicists in the Struggle for Progressive Science). In: *Filosofskie voprosy sovremennoy fiziki.* M.: Idz-vo

AN SSSR, 1952, pp. 5–30. S. I. Vavilov. *Ideologiya i metodologiya sovremennoy fiziki i zadachi sovetsckikh fizikov* (Ideology and Methodology of Modern Physics and the Tasks Facing Soviet Physicists). In: Istoriya i metodologiya estestvennykh nauk, vyp. 7. M.: Izd-vo Mosk. un-ta, 1968. pp. 6–22.

[30] *Sankt Peterburgskie Vedomosti*, 27 August 1903, p. 2.

[31] In the period of the pseudopatriotic rhetoric of the late 1940s and the early 1950s, the gifted Italian scientist and engineer used to be caricaturized as a vain and unscrupulous adventurer who shamlessly used the great discovery of A. S. Popov and successfully put it into practice. It is exactly in this manner that he is depicted in the film that Aleksandr Popov released in 1950* (see doc. 1.23). The USSR Academy of Sciences ostentatiously rejected the invitation from the Italian side to participate in the festivities on the 75th anniversary of Marconi's birth scheduled for 25 February 1949. At the height of the campaign against cosmopolitanism and servility before the West, a speaker at a Communist Party meeting held at the Institute of Philosophy, USSR Academy of Sciences, on 25 February 1949 artlessly said: "The Italian Academy invited representatives of our Academy to participate in the celebration in honor of Marconi, although everybody knows that radio was invented by our scientist, Popov."** Objectively, the two inventors were independently searching for the method of transmitting radio waves ("Hertz's waves") over distance. In 1895, Marconi was able to send radio signals over distances of up to 2 km.

Popov publicly demonstrated the transmission of Hertzian waves in Morse code at a meeting of the Physical Division, Russian Physico-Chemical Society, on 7 May 1895. In February 1896, Marconi moved to London to file his first patent, while Popov never did anything to claim his priority.***

* V. S. Kruzhkov, deputy head of the Department of Propaganda and Agitation, CPSU(b), sent a letter dated 22 May 1950 to G. M. Malenkov, secretary of the Central Committee, in which he recommended the support of the plans of the USSR Ministry of Cinematography to have the film Aleksandr Popov dubbed into German, Polish, Czech, Romanian, Bulgarian, Chinese, and Korean, with the quality control assured by *Soyuzeksportfilm*; RGASPI, fond 17, opis' 118, ed. khr. 887, l. 9.
** CAODM, fond 2501, opis' 1 ed. khr. 9, l. 100.
*** See, for instance, *Izvestiya*, 26 August 2005, p. 10.

[32] Professor P. S. Osadchiy of St. Petersburg University delivered a lecture on the prospects for practical application of wireless telegraph at a meeting of the Imperial Russian Technical Society late in 1902. He stated that "the company headed by Marconi seeks to monopolize the invention of wireless telegraphy and presses coastal stations and merchant ships for purchasing its apparatuses."* In other words, what he blamed was the too energetic actions taken by Marconi (much more successful than Popov's efforts, if any) to introduce his invention for practical use. No word of reproach was uttered about the incorrect behavior of Marconi in respect to his Russian colleague.

* *Novoye Vremya.* 2 December 1902, p. 4.

[33] The Italian government granted Marconi the right to use the cruiser *Carlo Alberto* for his works on wireless telegraph as long as he might need it. Besides the deck-cabin where Marconi's apparatuses were installed, the warship was equipped with a telescopic device

with which the length of a mast could be increased by 25 m.* Two months after the warship returned from Russia, it was visited by the Navy minister who discussed with Marconi prospects for the erection of a transmitting/receiving station in Italy, similar to the stations already operated in Great Britain, USA, and Canada.**

* *Moskovskie Vedomosti.* 25 October 1902, p. 4.
** *Novoye Vremya.* 6 September 1902, p. 3.

[34] *Novoye Vremya* (6 July 1902): "The inventor of wireless telegraph, Marconi, who arrived here aboard the warship *Carlo Alberto* stated that he has been commissioned to establish wireless communication between Canada and England. The other day, Professor A. S. Popov visited the warship and had a talk with Marconi, who demonstrated his apparatus." Similar information was published in *Sankt Peterburgskie Vedomosti.* It is worthy of note that the two newspapers called both Marconi and Popov inventors of wireless telegraph; that is, the Russian press treated them as scientists of equal rank.

[35] I. V. Kurchatov visited Great Britain as a member of the Soviet government delegation in summer 1956. He presented a lecture at the Atomic Energy Research Establishment at Harwell, Oxfordshire, in which he told of achievements in the peaceful use of nuclear energy in the USSR. It created an immediate sensation across the learned world.

[36] "Far from discussing the motives of the Navy Ministry for ordering from Telefunken, Berlin, the wireless communication equipment to be used at the railway passage across Baikal, and being aware of the high quality of its production, I have to state that neither the enterprises of the Navy Ministry nor myself was consulted about the placement of the orders [...] I believe that the Navy Ministry might just as well allot two stations of those manufactured for its own needs taking into consideration the military importance of the Baikal passage." Prof. A. S. Popov.*

* *Novoye Vremya.* 24 March (5 April) 1904, p. 3.

[37] In the dispatch dated 19 July (1 August) 1911, Myasoedov, the Russian *charge d'affaires* in Sweden, informed the authorities in St. Petersburg about a conflict between the Swedish Navy Ministry and the local office of British Society of Wireless Telegraphy (using Marconi's system). Citing publications in the Swedish press, Mysoedov emphasized that the Swedish Navy employed the production of Telefunken and expressed his personal opinion that "the Telefunken system adopted by the navies of many states including ours is the most sophisticated among those available at present."*

* AVP RI, fond 133, opis' 470 (1911), ed. khr. 122, l. 52–53.

[38] Vavilov was not alone in being misled by the situation around the 1909 Nobel Prize. V. A. Fok, another member of the Academy and an equally prominent figure in physics during the 20th century, almost literally repeated Vavilov's words in his nomination of P. L. Kapitsa for the 1955 Nobel Prize. In the preamble to the nomination, he expressed regret that "the Nobel Committee habitually treats Russian and Soviet scientists as unworthy candidates to the award even for the works of greatest importance." He supported his statement by reference to "the inventor of radio, Popov" among other Russian physicists disregarded by the Nobel Committee.*

* *Poisk*, No. 34–35, 2006, p. 22.

[39] At the beginning of the century, "Braun's system" was almost as much discussed in the press in connection with the problem of wireless communication as the inventions of Marconi and Popov. An example is the publication in the *Novoye Vremya*, a St. Petersburg daily newspaper, in the edition of 9th (22th) April 1903, about plans for a German expedition to the North Pole that included the establishment of a "wireless communication station and observatory equipped with Braun's wireless telegraph apparatuses between latitudes 78 and 80 degrees."

[40] Braun made the University of Strasburg one of the leading European physical centers. "Braun's tube" was the forerunner of modern oscillographs, television tubes, and computer monitors. Outstanding Russian physicists, such as P. N. Lebedev, L. I. Mandel'shtam, A. G. Gurvich, B. B. Golitsin, and N. D. Papaleksi, future Nobel nominees and nominators, worked in the department headed by Braun and under his guidance.*

* E. L. Feinberg. *Rodonachal'nik (o Leonide Isaakoviche Mandel'shtame)* (The Forefather (about Leonid Isaakovich Mandel'shtam)). Uspekhi Fizicheskikh Nauk, 172, No. 1, 2002, p. 94.

[41] Paragraph 4 of the statutes of the Nobel Foundation stipulates that nominees deceased after the closing date for current year nominations, i.e. after 31 January, are considered by the respective Nobel Committee together with all the others. There are two cases of a posthumous award, the 1931 Nobel Prize in literature to the Swedish poet E. A. Kerlfeldt, who died on 8 April of the same year, and the 1961 Peace Prize to the UN Secretary-General D. Hammarskjold, killed in a plane crash on 18 September.

[42] *Les prix Nobel 1904.* Stockholm, 1906, p. 4. The probability of receiving the Nobel Prize in 1904 and not later was so high that the world press foretold it without any doubt, much before the closing date for nominations, that is a year before the award was announced by the Swedish Academy. The daily *Novoye Vremya* published in St. Petersburg informed its readers, with reference to a source in Paris, that "the 1904 Nobel Prize will be awarded to Ramsay of Great Britain."*

* *Novoye Vremya,* 13 December 1903, p. 2.

[43] Mendeleev himself never doubted the potential resources of his system of elements. In mid-January 1903, after Ramsay had found the place for the group of inert gases in the periodic table, Mendeleev was interviewed by local journalists in Kiev and answered their questions about the recently discovered radium. In conclusion, the scientist said that this discovery once again confirmed "the validity of the periodic system of elements."*

* *Novosti i Birzhevaya Gazeta.* 25 January 1903, p. 2.

[44] Archives of the Royal Academy of Sciences, Stockholm. Fund of the Nobel Committees in physics and chemistry for 1905, p. 159.

[45] *Ibid,* pp. 213–214.

[46] *Ibid,* materials of the Committees for 1906, pp. 102–114.

[47] *Ibid,* pp. 99–100.

[48] Moissan was well-known in Russia not only among the scientific community. The Russian press emphasized that his "name [...] is associated with the most important achievements in chemistry during the last decade" when it informed the public about the award of the Gold

Medal to the scientist by the International Chemical Congress held in Berlin in summer 1903 "for outstanding discoveries in the field of experimental chemistry."*

* *Sankt Peterburgskie Vedomosti.* 22 July 1903, p. 2.

49 N. A. Menshutkin, personally acquainted with Mendeleev for many years, always paid tribute to him as an eminent scientist.* It may be supposed that, having received the invitation to nominate his candidate for the 1906 prize, he disregarded previous awards for 1904 and did not realize that they had opened the way for him to recommend his colleague from St. Petersburg.

* Menshutkin, being eight years younger than Mendeleev, suddenly died three days after his idol passed away, that is on the day Mendeleev's funeral on 23 January (5 February) 1907. Is it a mere coincidence?

50 The vote of P. Klason appears to have been decisive for securing the award for Moissan by the Royal Academy of Sciences. In 1906, Klason was president of the Academy, chaired its meetings, and in this capacity could convince other members to vote in favor of the French nominee. However, it would be wrong to see the refusal of the Nobel Prize to Mendeleev as consciously biased against Russian science and its actors.

51 Archives of the Royal Academy of Sciences, Stockholm. Funds of Nobel Committees in physics and chemistry; materials for 1905, p. 145.

52 *Ibid*, materials of the Committees for 1912, p. 277.

53 It may be supposed that the decision of the Royal Academy to award the Nobel Prize in physics to N. G. Dalen was dictated in part by humanitarian considerations. At the beginning of 1912, when Dalen tested a device for the safe use of gaseous acetylene, he was severely wounded when an explosion ruptured a gas-filled container. The 42-year-old inventor lost his sight and therefore lost the ability to work.

54 In fact, Backlund wished to have his second nomination of the French theoretician considered in 1910, after the first one submitted in 1909. However, he sent his letter from Pulkovo only on 8 February 1910 and it reached Stockholm on 14 February 1910. According to the Statutes, the nomination was registered to be considered in 1911.

55 Ciamician wanted to nominate Valden for the 1913 prize but sent his letter from Blogna only on 28 January 1913. It was received in Stockholm on 1 February 1913, i.e. the next day after the closing date. Therefore, the nomination was reserved for 1914.

56 Valden had to bear the full burden of a dramatic fate during World War II, when British planes bombarded Rostok in spring 1942. His house was demolished; the 10,000-volume library of rare editions, the art collection, and a private archive containing correspondence with eminent scientists were lost. In 1943, Valden left Rostok for Frankfurt am Main where he filled the post of extraordinary professor in history of chemistry at the local university. In 1947, he became a professor at Tubingen University. The same year, i.e. 13 years after the first nomination in 1934, his candidacy was again submitted for the Nobel Prize by Professor N. R. Dhar of Allahabad, India. Valden's memoirs, *Roads and Shelters. My life*, were published posthumously.*

* P. Walden. *Wege und Herbergen: mein Leben.* Beitrage zur Geschichte der Wissenschaft und der Technik H.13, 1974, 130 S.

[57] AVP RI, fond 133, opis' 470, ed. khr. 124 (1909), l. 20.

[58] Truly, it is not quite correct to describe Sully Prudhomme as a poorly known poet. He actually enjoyed popularity throughout the world even after his death in 1907. By way of example, his refined style was highly valued by the classical Georgian poet Titsian Tabidze. Tabidze advocated the concept of close links between oriental and Western poetics, understanding under "Western" first at all aesthetics of French symbolism as expressed in the following lines written in 1916:

"I cautiously put Hafez's* rose in the vase of Prudhomme [...]"**

Another example of the subtle understanding of Prudhomme's poetry in Russia is given by the outstanding scientist and philosopher V. I. Vernadsky. Here is an extract from his diaries dated 9 November 1919, i.e. three years after the publication of Tabidze's verse: "Yesterday, I read *La forme*, a lovely piece of poetry by Sully Prudhomme. Attempts at deep thought may be the construction of a new world outlook."*** Sully Prudhomme did not leave the award unanswered. He thanked the Swedish Academy and wrote a sonnet welcoming the new French translation of the book of verses by Oskar II, King of Sweden. Also, he visited the king when the latter stayed in Biarritz in May 1902.****

* Mohammad Shams Od-Din Hafez (1300–1389), one of the finest lyrical poets of Persia.
** T. Tabidze. *Stikhotvoreniya i poemy* (Verses and Poems). M-L.: Sovetsky pisatel', 1964, p. 7.
*** V. I. Vernadsky. *Dnevniki 1917–1921. Oktyabr' 1917–Yanvar' 1920* (Diaries 1917–1920, October 1917–January 1920). Kiev: Izd-vo Naukova Dumka, 1994, p. 179.
**** Sankt Peterburgskie Vedomosti. 24 May 1902, p. 3.

[59] The list of nominees for the 1901 prize is of interest in that it reflects the lack of experience in those invited by the Nobel Committee to nominate candidates. It appears that they did not yet realize the enormous worldwide fame and prestige the Nobel Prizes would have in the coming 20th century. Only E. Zola and H. Sienkiewicz really occupied a high place in world literature among the 25 candidates recommended for the prize, mostly by French and German nominators.* The rest were rather ordinary authors. In this situation, the choice of Sully Prudhomme as the first Nobel laureate seems to have been a matter of logical necessity, the more so that the French poet, a forerunner of the Symbolist movement, was recommended by 11 Paris-based nominators and a professor of the University of Uppsala. The names of two of the greatest writers, Leo Tolstoy and Henrik Ibsen, were paradoxically mentioned in one nomination not as candidates for the prize but in a quite different context. Christian Birch-Reichenwald of Christiania nominated Osip Lourie, a literary critic from Paris, for the articles on the works of those two literary giants of the 19th century, published in 1899 and 1900.

B. Bjornstjerne, awarded the Nobel Prize in literature in 1903, also mentioned them in his speech at the formal dinner that followed the prize-giving ceremony. He called the participants "to turn to the present and think of our two great contemporaries" without naming them, as if he wished to emphasize that the guests at the Grand Hotel in Stockholm should be well aware whom he meant. "I recollect an old friend of mine from Norway who is now ill. It was he who lit beacons along the Norwegian coast that now warn the whole world of danger. I also see another great old man whose motherland lies east of this country. He stands at the door leading to goodness and invites us to enter. The works they created for many years are permeated by a spirit that only increases with time like a flame blown by the wind."**

Six years later, when Ibsen was gone and Tolstoy lived his last year, S. Lagerlof analyzed the sources of her creative work in the Nobel lecture of 1909. That is how she spoke of her literary teachers without naming them: "Should I not be indebted to those who in my youth elevated the art of literature so that it reached unprecedented height, to the great Norwegians and the great Russians?"***

In a quarter of a century, Roger Martin du Gard, a monumentalist writer awarded the Nobel Prize in 1937, also recalled Leo Tolstoy as he echoed Lagerlof in glorifying his predecessors.

Speaking at the formal dinner in Stockholm, he emphasized the importance for a novelist "to portray his own [italicized by the author] vision of the world in his works. Here, Leo Tolstoy is a great teacher. Each of his characters is in some way or another concerned with philosophical questions [...] about the meaning of life." In conclusion, he expressed the belief that "it is the deference I paid to unattainable models and my efforts to learn the lessons from the literary genius of Tolstoy that won such high praise from the members of the Swedish Academy who crowned my writings with this award."****

* The works of Zola had nothing to do with "an ideal direction" stipulated by Nobel's will as a precondition for awarding the prize for literature. Moreover, Nobel, although a connoisseur of literature, had entirely rejected Zola's naturalism and members of the Swedish Academy were well aware of it. H. Sienkiewicz was awarded the Nobel Prize four years later.
** *Nobelevskaya primiya po literature. Laureaty 1901–2001* (Nobel Prizes in literature. Laureates of 1901–2001). SPb.: Izd-vo universiteta, 2003, pp. 6–7.
*** *Ibid*, p. 27.
**** *Ibid*, p. 80.

60 K. Espmark. The Nobel Prize in Literature. A study of the Criteria Behind the Choices. Boston, Mass., 1991.

61 K. Espmark, p. 17.

62 Sharp criticism of church observances (one reason for Tolstoy's excommunication), rejection of the state, privileges of the ruling class, and power of money in *Resurrection* discredited Tolstoy in the eyes of the conservative majority at the Swedish Academy in whom his writings inspired feelings of "moral indignation" (Espmark).* They were not alone. The same is true of certain known literary critics in France who reviewed chapters of the novel published in *Echo de Paris*. The liberal essayist Jules Lemaitre regarded the novel as a masterpiece, remarkable for the force of genius and "evangelic simplicity."** In sharp contrast, Ferdinand Brunetière, a member of the French Academy, said in an interview for the St. Petersburg daily *Novoye Vremya*: "I cannot publish it [the novel] in the Revue. We cannot treat subjects like that. A student seduces a housemaid who then finds herself in a brothel [...] Episodes in that place... Spiritual awakening in the end. Break with the infamous past and happiness with the beloved... These are routine problems in Russian literature. Take Dostoevsky's *Sonya*. Civilization of sin and sufferings. The dirtiest scenes are best written."***

* K. Espmark, p. 16.
** *Novoye Vremya*. 14th (26th) March 1899.
*** *Novoye Vremya*. 3rd (15th) March 1899.

63 K. Espmark, pp. 16–17.

[64] *U. Tolstogo. Yasnoployanskie zapiski* by D. P. Makovetski (With Tolstoy. Notes from Yasnaya Polyana by D. P. Makovetsky). Literaturnoye nasledstvo, V. 90, book. 2, 1979 (entry of 23 September 1906). Tolstoy reacted to the information brought by Biryukov in a letter to his Finnish follower A. A. Ernefelt of 25 September 1906 by asking him "to use your connections in Sweden, if any, to try to make it so that I would not be given this prize [...] being thus freed from the trouble of rejecting it."*

* L. N. Tolstoy. *Polnoye sobranie sochineniy* (Complete Works), V. 76, 1956, pp. 201–202.

[65] A. F. Koni. *Sobranie sochineniy* (Collected Works), V. 6, 1968, pp. 639–640.

[66] *Russkaya Mysl'*, kn. 7, 1905, pp. 639–640. The key thought of the author: "The main evil from which the Russian people suffer [...] is the impossibility for the majority to enjoy the natural right of each human being to use the land where he or she was born. Once the sinfulness and criminal nature of such an order is understood, it becomes clear that no political reforms will bring freedom and prosperity to the people until this [...] evil is exterminated." (p. 248).

[67] The article was reprinted from the *Russkaya Mysl'* as a separate booklet by the Posrednik publishing house and translated into English for *The Times*, in which it appeared in July 1905. The same year, *The Great Sin* was issued by several publishing houses in the United States (see *Russkie Vedomosti*). Moreover, one of the Chicago newspapers "raised one million free public subscriptions to the book. A special committee was set up for the purpose constituted by known public figures."* The article was for the first time included in Vol. 16 of the *Collected Works of Tolstoy* in 1911, which was confiscated by the censor board.**

* *Russkie Vedomosti*, 30 November (13 December) 1905, p. 3.
** From the commentaries of N. K. Gudziy; see L. N. Tolstoy. *Polnoye sobranie sochineniy*, pod red. V. G. Chertkova. GIKhL, V. 36, 1936, p. 664.

[68] Here, we leave aside the fact that collective nominations are at variance with the statutes of the Nobel Committee.

[69] More than 30 years later (1933), Anders Esterlong, literary critic and member of the Swedish Academy (and later its permanent secretary) wrote in the *Svenska Dagbladet** on the occasion of awarding the Nobel Prize to I. A. Bunin: "This year, the Nobel Prize in literature is for the first time awarded to a Russian writer. Accusations that usually follow such awards most often refer to the neglect of Russian literature. Even in 1901, when the first Nobel Prize in literature was awarded, the decision met with the indignant disapproval of world public opinion that favored the great Tolstoy as the indisputably leading candidate. Unfortunately, it did not change the situation."** Esterling also recalled Chekhov as a prime candidate for the prize with the reservation that "Chekhov was out of discussion, since he had died six years before Tolstoy."

* *Svenska Dagbladet*, 10 November 1933, s. 1.
** See doc. 1.9.

[70] The first Russian edition: *Fyodor Petrovich Haas. Biograficheskiy ocherk* (Fyodor Petrovich Haas. A Biographical Essay). SPb.: 1897.

[71] A noble ancestor of the Glasenapps came to the Baltic region from Pomerania in the first half of the 16th century.* Karl Fridrikhovich Glasenapp graduated from the Derpt (now

Tartu) University as a candidate of comparative linguistics and was promoted to the civil rank of Councillor of State. He taught at the Polytechnical Institute, Riga, from 1898.** However, his works, for which he was nominated for the Nobel Prize, were published by German publishing houses (see note 72).

> * *Russkiy biograficheskiy slovar'* (Russian Biographic Dictionary). V. 5. M.: Terra-knizhny klub, 1999, p. 189.
> ** F. V. Bukhgol'ts. *Istoricheskie i statisticheskie dannye o Politekhnicheskom Institute za vremya ot 1862–1912* (Historical and Statistical Data about the Polytechnical Institute for the Period of 1862–1912). Riga: 1912, p. 35

72 By the time Glasenapp was nominated for the Nobel Prize in 1903, he had published the encyclopedic reference book *Wagner-Lexicon*, Stuttgart, 1883, and two of the six volumes of *Das Leben Richard Wagners*, Leipzig, 1884, 1889. The complete six-volume *The Life of the Composer Richard Wagner* appeared in Leipzig only in 1905. This sole fact was enough to make a success of his 1903 nomination doubtful; moreover, mass lobbying his candidacy was widely known and never questioned (see also note 5 to doc. 1.18).

73 *Novosti i Birzhevaya Gazeta.* 14th (27th) October 1900, p. 3. A similar announcement was not found in the other seven newspapers for October 1900 including large-circulation ones, such as *Novoye Vremya* and *Moskovskie Vedomosti.* Next month, the *Russkie Vedomosti* of 1st (14th) November and *Rossia* of 8th (21th) November (also liberal newspapers) as well as the same *Novosti* of 7th (2nd) November* informed their readers that the Nobel Committees started their practical activities and distributed "to different agencies their newly-developed and rather intricate prize-awarding rules"** (see note 75). These publications did not mention the October initiative of the Naval General Staff.

> * All papers published similar texts with the exception of an additional note in the *Novosti,* which "[had] extracts from the document concerning Nobel's Prize which were at that time released to the press for "public notice" by the Ministry of Public Education. Again, conservative newspapers kept their silence.
> ** *Russkie Vedomosti,* 1st (14th) November 1900, p. 4.

74 The author's mistake. According to the Statutes approved by the King, the Nobel Foundation is an independent institution.

75 For all that, information on the beginning of the activities of the Nobel Committees and their prize-awarding procedure reached members of the Academy of Sciences in due time. It was announced at the general meeting of the Academy on 2nd (14th) September 1900, i.e. immediately after the summer holidays. After the meeting, a detailed summary of the statutes of the Nobel Foundation and its Special Rules were printed as Appendix 1 to the minutes of the meeting and were distributed to the libraries of all Divisions of the Academy.*

> * On request of the Special Division of the Department of Police, a copy of the document entitled "Foundation Nobel" was transferred to its control from the library of the II Department of the Academy of Sciences and is currently stored in GARF, fond 102, 1904, ed. khr. 14, chast' 1, lists 4 (two sides) and 5.

76 Official information about the preparation of the campaign for awarding Nobel Prizes in 1901 came from the Ministry of Public Education only after half a month.*

* See footnote* to note 73.

[77] The Collected Works of Alexander I Institute of Communication Engineers, vyp. 57, SPb.: 1902.

[78] It is worth noting that the heads of the Ministry of Foreign Affairs acted in the same way. Invitations to nominate candidates for the Nobel Prizes arriving from Christiania were sent to the archives before they were pulled out from the envelope. Such was the case with two invitations to nominate candidates for the 1903 prize addressed to the minister V. N. Lamzdorf on 4 October 1902.* There is little doubt that the same fate befell letters of invitation to recommend candidates for the Nobel Peace Prize of 1901. At the same time, an anonymous correspondent for *Sankt Peterburgskie Vedomosti* in Christiania complained that "Russia nominated no candidates [...] Was it unpardonable negligence or false modesty?"**

* AVP RI, fond 155, opis' 440, rod VI, 1900, ed. khr. 2, l, pp. 179–180.
** *Sankt Peterburgskie Vedomosti*, 16th (29th) March 1901, p. 1.

[79] *Polnoye sobranie zakonov Rossiyskoy Imperii.* (Complete Collection of Laws of the Russian Empire) Collection 3, V. 20, Part 1, No. 18772. SPb.: 1902, pp. 628–629.

[80] Here is an example of such discontent. The law prohibiting the use of Finnish postage stamps instead of Russian ones came into force on 1st (14th) August 1900. In response, Finns began to use envelopes "not only with mourning stamps but also with various mourning emblems or non-standard stamps." The Ministry of Internal Affairs prohibited the "transmission of such correspondence because the use of the said symbols is akin to a political demonstration."

[81] *Polnoye sobranie zakonov Rossiyskoy Imperii.* No. 18913, p. 842. It is worthy of note, however, that the public opinion in Finland about the role of the so-called "Russian period" in the country's history changed substantially over the course of time. One hundred years after the events described in this Chapter, Paavo Lipponen, prime minister of Finland, recognized that "our relations were not altogether negative. During the "Russian period" of Finland's autonomy within the Empire, we created a most advanced democracy in the Europe of that time. I think that Finland would have remained a marginal country had it been a province of Sweden."*

* *Izvestiya*, 27 July 2001, p. 2.

[82] Russian nominators in physiology and medicine were highly active since the very first nomination campaigns (see Table 11.1). They included university professors, researchers of the Military Medical Academy and the Institute of Experimental Medicine and were not administratively subordinate to structural departments of the Academy of Sciences.

[83] This unusual correspondence was obviously provoked by some false information obtained by the chief of secret police from his agents. The Nobel Peace Prize for 1903 was awarded to the British pacifist William Cremer, secretary of the International Arbitration League. The Russian public knew about it from the press.* The archives of the Norwegian Nobel Committee contain no evidence that the *L'Européen* newspaper was ever nominated for the Nobel Prize either in 1903 or in subsequent years. Another fact making the story of this alleged nomination even more intriguing is the absence of correspondence on this matter between Lamzdorf and the Russian envoy in Stockholm in the archives of the Russian Ministry of Foreign Affairs.

There is only one letter, No. 9, sent by Byutsov to St. Petersburg (20 March 1904) during the entire period from 16 March to 13 April (Julian calendar), i.e. between the dispatch of Pleve's message and Lamzdorf's answer to it. This letter actually touches on an annoying problem with anti-Russian publications in the Swedish press inspired by the policy of russification in the Grand Duchy of Finland. Byutsov mentions neither the newspaper *L'Européen* nor the Nobel awards (doc. 1.22-I). No other letter from him was received at the Ministry of Foreign Affairs until 6 May. The next letter, No. 10, described the ceremony of meeting King Oskar II after his long stay abroad.** During the same period, Lamzdorf also sent a single letter to the Russian mission in Stockholm (12 April 1904). It concerned a question of much greater importance than a fantastical story of the *L'Européen*'s pretension for the Nobel Prize; the Russo-Japanese War broke out in the Far East (doc. 1.22-II). No other messages from the minister were found in the archives of the Ministry of Foreign Affairs and the Russian mission in Stockholm.***

The question arises whether there was any correspondence at all between St. Petersburg and the Russian envoy in Stockholm on the mythical award in the French newspaper. It may be supposed that document 1.19-II was just a diplomatic manoeuvre of the wise and cautious Lamzdorf, who feared offending the powerful chief of secret police by recognizing the inaccuracy of the received information.**** There seems to be a grain of truth in the document, confirmed by some traits of Lamzdorf's character. He was not strong enough to persist in his opinion in the face of the "high and mighty". In the three-volume edition of sketches on the history of the Russian Ministry of Foreign Affairs issued on the occasion of the 200th anniversary of its foundation, Lamzdorf is described as a person whose "shyness and reserve sometimes made him unsociable, and prejudiced his success at court and in society."*****

* True, press reports of the 1903 prizes were not free of mistakes. For example, *Sankt Peterburgskie Vedomosty* of 30 November (13 December) 1903 told about the award of the peace prize to Bertha von Zuttner who actually received it two years later. After five days (as if apologizing for the mistake without mentioning it), the newspaper published a detailed report of its correspondent in Christiania about the ceremony of awarding the peace prize to William Cremer, a member of British Parliament, and summaries of the speeches of Berner, president of Norwegian Storting, and Levland, chairman of the Nobel Committee (*Sankt Peterburgskie Vedomosty*, 5 December 1903, p. 2). No Russian newspaper said a single word about the *L'Européen* as an alleged nominee for the Nobel Prize.

** AVP RI, fond 133, opis' 470 (1904), ed. khr. 117, l. 30.

*** Ibid, fond 193, opis' 817/1.

**** Equally unsuccessful was the search for documents relating to discussions of Nobel matters by the Swedish Minister of Foreign Affairs, Lagerheim and the Russian envoy, Byutsov in the archives of the Swedish Ministry of Foreign Affairs undertaken on the request of the executive director of the Nobel Foundation, M. Sohlman.

***** *Ocherki istorii ministrestva inostrannykh del Rossii.* T.1 (Essays on the History of the Russian Ministry of Foreign Affairs. V.1). M.: OLMA-PRESS, 2002, p. 496.

[84] AVP RI, fond of the secretariat of MID 1904, opis' 470, ed. khr. 117, l. 127.

[85] AVP RI, fond 133, opis' 470, ed. khr. 177 (1904), l.10. The typewritten text of the cable bears an inscription testifying that it was reported to the Tsar on 28 November 1906.

[86] Encyclopedic dictionaries indicate 28th (15th) January 1915 as the date of Umov's decease (see, for instance, BSE, 3rd edition, V. 27, 1977, p. 11). However, A. I. Bachinsky, a close

friend and biographer of Umov, tells in a memorial sketch published as a separate booklet in the same year (1915) that Nikolai Alekseevich died from a gastric ulcer on the night of 1–2 January 1915 (Julian calendar) after an episode of black vomit in late November 1914.* It means that Umov was bedridden while writing his nomination in December 1914.

* A. I. Bachinsky. *Nikolai Alekseevich Umov. Biograficheskiy ocherk* (Nikolai Alekseevich Umov. Biographic Essay). M.: 1915.

Document 1.1

Answers from Anne Kjelling, curator of the library of the Norwegian Nobel Institute, to the author's questions (10 August 1998 and 29 September 1998).[1]

[...] I looked through the lists of registered nominations for the period from 1899 to 1904 but unfortunately could not find the name of Vereshchagin.[2]

[...] I looked through the registries of correspondence for the period from 1901 to 1904 but did not find the name of Vereshchagin either. My impression is that he was not considered as a candidate for the award [...] Enclosed find please four references to Vereshchagin entered in our old catalogue; you will probably find them interesting.[3]

[1] Here and thereafter the spelling and punctuation of original documents are preserved.
[2] It is worthy of note that the registration of nominations for the Nobel Peace Prize began in Christiania in 1899, i.e. a year before King Oskar II signed the Statutes of the Nobel Foundation on 26 June 1900 and the nomination procedure stipulated by the Statutes came into effect.
[3] Three catalogues of the artist's exhibitions issued during his lifetime, and a monograph on his works published in 1907, i.e. three years after his death in Port Arthur (13 April 1904) during the Russo-Japanese War.

Document 1.2

Letter from Alfred Nobel to his nephew Emmanuel Nobel, from Zurich to St. Petersburg (21 June 1983).[1]

Dear Nephew,

I am writing in French this time to spare you the trouble of translation. Moreover, I wish to secure a written document exposing my views because I claim authorship of them and because they are original.

Before turning to the point of the matter, I wish to state that I offer a small gift to the Institute of the Prince of Oldenburg without making any specific terms. But if it proves desirable that I did more, I would like to see in what direction the Institute is going to develop its further research activities. I am outlining below the main themes of the desired work. I will inform the president of the Karolinska Institute in

Stockholm to the same effect. I hope that one of the two institutions will pay attention to my ideas as being serious and worthy of note.

To begin, I have been keeping a long-standing idea in my mind. I would not go so far as to predict its possible practical implications, but relevant research would certainly be interesting enough to try to carry out. Plant grafting is known to produce wonderful results; therefore, I asked myself if the same procedure could be applicable in the animal kingdom. I am aware of experiments in this field, but they are still very few.

What I propose is to investigate whether blood from a strong young animal could feed for some time an old and weak one of the same species, having blood corpuscles of equal size. Evidently, by suturing an artery of one animal to a vein of another, in the manner of two plants connected by grafting, it would be possible to transfuse blood full of vital force (unlike that from an exsanguinated animal). Skilful physiologists must learn how to perform this procedure. Meanwhile, the presence of naturally anastomosed arteries and veins in certain individuals suggests the possibility of such suturing. Moreover, arteries are not absolutely necessary for the purpose. Joining veins by mechanical means, such as pressure or cauterization, may also ensure the desired blood exchange. The question is what will be the outcome of blood transfusion from a healthy individual to a weak and old one? A very simple operation will be needed to separate the sutured vessels.

I have recently discussed my project with Professor Key,[2] the leading authority in physiology in Sweden. He found these ideas quite sound and his opinion encourages me to develop them.

There is another question of interest.

Doctors know how to evacuate gastric contents easily. Why have they not thus far tried to pour contents from a normally functioning stomach to a weakened or diseased one? It is not a complicated procedure, and studies along these lines might be interesting even if they yielded no results of practical value.

The third problem with which I am concerned was mentioned in my letter to you of 31 March. The thing is, urine contains poisonous substances, and their concentration ratio probably correlates to the development of certain diseases. I am presently studying chemical aspects of this problem and will inform you as soon as I obtain any useful or at least promising result.

It may seem that I am too presumptuous, wishing to treat such special problems. However, I am prepared to take the blame in the hope of forgiveness. I am interested in physiology, which prompts me to invent all these ideas.[3] I already blush with shame for my possible blunders; maybe this recommends me for mercy.

CGIA RF, St. Petersburg, fond 2282, opis' 11, ed. khr. 47; published by V. S. Meshkunov in the journal *Voprosy Estestvoznaniya i Tekhniki*, 1994, No. 1, pp. 124–125.

[1] Translated from the Russian version of the French letter (the original text and the Russian translation are kept as one archive file (both sides of sheets 1 and 2).
[2] Key, Alex, professor at the Caroline Medico-Surgical Institute, Stockholm, and its director from 1886 to 1897.
[3] In the will signed four and a half years later, Nobel immortalized his interest in physiology in the name of one of the scientific awards, the Nobel Prize in physiology and medicine.

Document 1.3

Correspondence in connection with the initiative of the Imperial Alexander University, Helsingfors, to invite Professor R. Tigerstedt.

I.

To: Imperial Russian mission in Stockholm
From: Director of the Second Department, Russian Ministry of Foreign Affairs
28 March (9 April) 1990, No. 2957.

In view of the initiative of the Consistory of the Imperial Alexander University in Finland to invite Professor Robert Arman-Adolf Tigerstedt of Karolinska Institute, Stockholm, to head the Department of Physiology at the said university, the Second Department has the honour to ask you, on the instruction of the Minister State Secretary of the Grand Duchy,[1] to inform it at your earliest convenience about the advisability to admit Professor Tigerstedt to the Imperial Alexander University.

AVP RI, fond 155, rod IV, ed. khr. 2/14, t.2, l. 127.

[1] Official title of the governor-general of the Grand Duchy of Finland.

II.

To: Director of the Second Department, Russian Ministry of Foreign Affairs
From: Imperial Russian mission in Stockholm
1 (24) April 1900, No. 221.

With reference to your letter of 28 March 1900, No. 2957, the Imperial mission has the honour to report that Professor Tigerstedt is equally highly respected in Stockholm as a university professor and as a man of great personal virtue. Besides his professional activities at the Karolinksa Medical Institute, he promotes education by reading public lectures for the poor classes and is a most efficient trustee of maternity homes and orphanages.

In spite of his Finnish origin, Professor Tigerstedt did not participate in pro-Finnish demonstrations in Stockholm last year.[1] Unlike many professors of the Karolinska

Institute and Swedish universities, he did not sign the address delivered to St. Petersburg last June by the international deputation for the presentation to His Imperial Majesty.
Ibid, l. 129.

[1] In connection with the manifest of February 1899 that initiated the campaign for depriving Finland of its unique status within the Russian Empire.

Document 1.4

Anders Barany, secretary of the Nobel Committee in physics, telling how his grandfather, Robert Barany, came to know about the award of the Nobel Prize while staying in a prisoner-of-war camp in Russia (31 October 2000).

Robert Barany graduated from the Medical Department, University of Vienna, in 1900 and occupied the post of assistant professor at the university ear clinic. In mid-1914, he obtained the consent of the university governing body to join as a medical officer the 120,000-strong garrison of the fortress of Peremyshl'. After the onset of hostilities, he provided care to patients with head wounds using up-to-date surgical methods.

On 28 March 1915, the garrison surrendered to the Russian Army and Barany was placed in a prisoner-of-war camp in Merv (now Mary), Turkmenistan. In the camp clinic, he treated other prisoners and servicemen of the local garrison, including its commander. Simultaneously, he read lectures in French for Russian physicians on current medical problems.

On 29 October 1915, the Nobel Assembly of the Karolinska Institutet announced the award of the 1914 Nobel Prize in physiology and medicine to Barany.[1] The cable in Russian notifying Barany about the award found him in Merv. He received it when he returned from the clinic to the camp barracks in the evening. It appears that people in Stockholm had been aware of his where-abouts in Russia before they took the decision. Barany had the cable translated to him by a fellow prisoner who spoke Russian. At approximately the same time, General Brendstrem[2] sent a cable informing Barany about his coming liberation from imprisonment; however, the local post office did not deliver it to the addressee through a misunderstanding, instead sending it back to St. Petersburg.

On 4 November 1915, Barany thanked his friend and colleague Professor Gunnar Holmgren[3] for the award and asked to convey his gratitude to the "Stockholm faculty".[4]

Through the efforts of Prince Carl, head of the Swedish Red Cross Society, Barany was transferred to Kazan University at the end of 1915 (the Prince sought to have all imprisoned scientists moved to university cities). In Kazan, Barany did surgery at the local university clinic; he was also allowed to go into private practice and had the mayor of Kazan as his first patient. University colleagues found an opportunity to introduce Barany to one of the Great Princes visiting Kazan, whom he asked for help in obtaining pre-term termination of imprisonment. During childhood Barany had suffered bone tuberculosis and was lame on one leg for the rest of his life. After the Prince left Kazan, Russian friends advised Barany to buy a thick walking stick and show more signs of lameness in order to have a stronger argument for freedom as an ex-serviceman. In the end, a conference of local doctors recognized Barany as a disabled veteran. In June 1916, Barany was exchanged for a captive Russian military officer in Khaparanda, northern Sweden.

On 16 June, Robert Barany joined his family in Vienna, and in a couple of days read his first lecture for local physicians. On 11 September, Barany received his award from the king of Sweden and delivered his lecture, as prescribed by the Nobel Statutes. In 1917, he accepted an invitation to join the University of Uppsala and moved to Sweden with his family for the remainder of his life.[5]

[1] In 1914, the Nobel Prize in physiology and medicine was not awarded; it was reserved for the next year.
[2] General Brendstrem was the special envoy in charge of representing the interests of Sweden in Russia from 1906 to 1920.
[3] G. Holmgren, otolaryngologist, was a professor at the Karolinska Institutet and its rector from 1931 to 1940.
[4] Assembly, consisting of professors of the Karolinska Institutet, that elects Nobel laureates in physiology and medicine.
[5] See also *Poisk*, N 16, 2001, p. 11.

Document 1.5

To: V. Smirnov, representative of All-Union Society for Cultural Relations with Foreign Countries, Stockholm

From: B. Byk, deputy head, 3d Western Department, All-Union Society for Cultural Relations with Foreign Countries 25 March 1935, No. 476

Because you consider advisable the trip of a group of students to Moscow, please inquire into the possibility of organizing it. You may find it convenient to have a talk with the student Barany as the group leader. Keep us in the know as regards the size of the group and the date of its arrival, if any.

GARF, fond 5283, opis' 5, ed. khr. 540, l. 195.

Document 1.6

Extract from the draft letter of D. T. Shepilov, deputy head of the Department of Agitation and Propaganda, Central Committee of the Communist Party (CPSU(b)), prepared in connection with plans to institute international Mendeleev and Gorky prizes to be awarded in the Soviet Union.[1]

According to Swedish sources, academician I. P. Pavlov is the sole Nobel laureate in Russia. They regard I. I. Mechnikov as a French laureate.

There are many facts indicating that Swedes ostentatiously refused to recognize achievements of Soviet scientists (see document 5.12). Suffice it to say that the Nobel Committee disregarded Lebedev's experiments that had led to the discovery of light pressure and astonished the world's scientific community; the works of Lyapunov, founder of the stability theory; Zhukovsky and Chaplygin, who had laid the foundations of modern aviation; and Krylov, creator of modern shipbuilding technologies and navigation. The Nobel Committee adopted a discrimination policy in respect to our science and awarded prizes to foreign specialists despite the fact that similar investigations were performed in this country.

RGASPI, fond 17, opis' 125, ed. khr. 588, l. 20–21.

[1] The full text of the letter is cited in Chapter 5 (doc. 5.17).

Document 1.7

Extract from the first version of the report by S. I. Vavilov entitled "Philosophical problems of modern physics and the tasks facing Soviet physicists", prepared for the presentation at the All-Union workshop of heads of physical departments of universities and higher education institutions in 1949 (which did not take place).

The long list of Nobel laureates, starting from the early 20th century, includes medium-rank researchers, besides any really outstanding personalities, and not a single Russian physicist. The Nobel Prize was awarded to Marconi and Braun but not to Popov. Mendeleev did not receive the Nobel Prize although it was given to persons whose contribution to science is far less significant than that of our compatriot P. N. Lebedev was universally accepted as the best experimenter in the world. The Nobel Prize for the discovery of combination light scattering was awarded to Raman of India but not to Mandel'shtam and Landsberg, who observed the same phenomenon simultaneously with Raman, as is now confirmed and recognized in the world scientific press. These examples leave little doubt that the achievements of our scientists were deliberately disregarded by the prize-awarding organizations. Such a situation is largely due to the contemptuous attitude towards our culture and science that has been for centuries cultivated in the learned world.[1] It is opportune to remember in this context that the capital of the Nobel Foundation was largely amassed by Nobel's company from the exploitation of Caucasian oil fields in pre-revolutionary Russia.[2]

GARF, fond 9396, opis' 1, ed. khr. 252, l. 29.

[1] This sentence was crossed out in the second version of the report, now in storage in the same archive file (GARF, fond 9396, opis' 1, ed. khr. 262, l. 91).
[2] This erroneous statement may be found in foreign sources too.* The initial capital of the Nobel Brothers Petroleum Company, founded in 1879, is estimated at 3 million rubles. Alfred Nobel's share was only 115,000 rubles, or less than 4% of the total. By the time of his death in 1896, this share increased to 5.23 million Swedish krones or 2.72 million rubles (at an exchange rate of 52 kopecks to 1 krone) but remained at the same level of 4%** since the total assets of the company amounted to 70 million rubles. A. Nobel had property in eight countries, and Russia ranked fifth in terms of its size following Great Britain, France, Germany, and Sweden.***

* See R. Sohlman. *Zaveshchanie Al'freda Nobelya.* M.: Mir, 1993, p. 77. (R. Sohlman. *The Legacy of Alfred Nobel.* The Bodley Head Ltd., 1983, p. 81).
** *Priroda,* 1993, No. 12, p. 116.
*** R. Sohlman, s. 110 (R. Sohlman, p. 118).

Document 1.8

Materials related to c.s. 1D. I. Mendeleev's balloting for full membership at the Imperial Academy of Sciences in 1880.

I

Imperial Academy of Sciences
Division of Physico-Mathematical Sciences
Meeting of 11 November 1880

Minutes of the election of a full member[1]

Candidate — *Dmitry Ivanovich Mendeleev, counsellor of state, professor at St. Petersburg University, corresponding member of the Academy of Sciences.*
For election to the status of *extraordinary member of the Academy in the field of technology and chemistry applied to arts and trades.*
Members of the Division residing in St. Petersburg — 16; *present at the meeting — 16.*
Number of voters — *18 (including president and permanent secretary).*
Number of votes — *19 (including 2 president's votes).*
Majority vote — *13.*
Voting results:
Candidate — *D. I. Mendeleev.*
Votes in favour of the candidate — *9.*
Votes against the candidate — *10.*
Conclusion: *blackballed.*

Bunyakovsky V., vice-president
Veselovsky K.[2]*, permanent secretary*

Arkhiv istorii nauki I nekhniki, vyp. 6, 1935, p. 324.

[1] Italicized lines and numerals are inserted by hand.
[2] Full state councillor.

II

To: Great Prince Konstantin Konstantinovich
President
Russian Academy of Sciences
From: S. Yu. Vitte
Minister of Finance
4 (16) December 1899

I have Dmitry Ivanovich Mendeleev, Russian scientist, on the staff of this ministry.[1] There is no man without sin. Maybe he is not blameless either. But he presents a striking example of a well-regarded Russian scientist. As a Frenchman, German or Englishman, he would have been long ago elected as as member of the supreme national scientific society. His name is known worldwide. I really feel pity for him and for the Academy (now that it has the honour of being under your presidency)[2] that he is not yet been elected its full member.[3]

What I have just said are my own thoughts and not a plea from Mendeleev. He is at present farther than ever from a feeling of vanity... I know too well that the promotion of any person may raise numerous objections... I am equally well aware of all the petty feelings involved in the judgement of a person by his colleagues, especially when it comes to election.[4] But I also know that sooner or later more positive thoughts and feelings prevail over negative ones to give tribute to the merits of a worthy man. Will not this time come too late for old Dmitry Ivanovich Mendeleev?"

PFA RAN, fond 6, opis'. 1, ed. khr. 16, l. 48–49.

[1] After Mendeleev had retired from St. Petersburg University because of a conflict with its administration, the Minister of Finance A. Ya. Vyshegrdasky, his university fellow, awarded him the position of inspector on the Board of Weights and Measures (10 (22) January 1892). S. Yu. Vitte, who succeeded Vyshegradsky, redesignated the Board as the Main Chamber of Weights and Measures "because it was headed by such a prominent scientist as Mendeleev, a person of great scientific and practical initiative."*

* S. Yu. Vitte. *Vospominaniya. T.2* (Memoirs. V. 2). Tallin; M.: Skif-Aleks, 1994, p. 542.

[2] Great Prince Konstantin Konstantinovich was appointed president of the Academy of Sciences by the decree of Emperor Alexander III of 3 (15) May 1889. From the very beginning, he supported Mendeleev in his election to the Academy. He wrote in his diary on 28 February (12 March) 1890: "...I presented a note on

research of Alex Onufrievicha Kovalevski at the meeting of the Division of Physico-Mathematical Sciences and managed to pass his candidacy for election. It is a victory — one more Russian is a member of the Academy. If good luck comes to me, Mendeleev will be the next [...]"* In other words, the president means that Mendeleev was blackballed by the "German party". However, this notice is out of place here. The most influential antagonist of Mendeleev was K. S. Veselovsky, the permanent secretary of the Academy, who did not belong to this party but was at the head of the most conservative adherents of old bureaucratic traditions. These people strongly opposed the election of such liberal intellectuals as Mendeleev.

* K. R. *Dnevniki i vospominaniya. Stikhi. Pis'ma* (Diaries and Memoirs. Verses. Letters). M.: Iskusstvo, 1998, pp. 152–153.

[3] Mendeleev was elected corresponding member of the Academy of Sciences on 30 November (12 December) 1876, after P. E. Lekoq de Boisbaudran discovered in 1875 the new element gallium, predicted by Mendeleev in 1870.

[4] In 1880, Mendeleev pretended to the position remaining vacant after the death of N. N. Zinin but was "blackballed" (see doc. 1.8-I). A new vacancy for a chemist appeared after the death of A. M. Butlerov in 1886. A. S. Famintsin, member of the Academy, recommended the election of Mendeleev, who was not however admitted as a candidate.

III

From the Memoirs of S. Yu. Vitte

"Mendeleev was a worldwide known Russian chemist, formerly a professor at St. Petersburg University. He had to retire from the University after qualifying for a pension because of his difficult, unaccommodating character.[1] To the shame of our Academy, he was never elected a full member; the person elected instead was quite a respectable one but had no serious scientific reputation.[2] Members of the Academy did not elect Mendeleev for his character, but that does not justify them... We began crying about the loss of the great Russian scientist only after his death.[3] Good for the nation that it gave him tribute posthumously, but Mendeleev would certainly have been much happier if he had his services recognized during the lifetime."

S. Yu. Vitte. *Vospominaniya.* T.2 (Memoirs. V. 2), p. 542.

[1] The opinion of Vitte about Mendeleev's "unaccommodating character" is not confirmed by Famintsin, who was well acquainted with Mendeleev as his university colleague. The memorandum presented by Famintsin on the occasion of Mendeleev's election to the Academy in 1886 (see note[4] to doc. 1.8-II) read as follows: "the main argument against the said candidate [Mendeleev] is his unsociable disposition. However, this argument does not stand up to impartial criticism; [...] Prof. Mendeleev has no personal enemy among the university corporation. On the

contrary, his bright intellect, devotion to science, and frank character conciliated universal love and respect."*

* *Arkhiv istorii nauki i nekhniki*, vyp. 6, 1935, pp. 327–328.

[2] The statement is incorrect. Professor F. F. Bel'shtein of the Technological Institute was elected by the Division of Physico-Mathematical Sciences on 19 November (1 December) 1882 after it refused membership to Mendeleev. However, A. M. Butlerov appealed against this decision and it was not approved by the general meeting of the Academy. In 1886, there were two new vacancies for chemists. Mendeleev did not participate in the elections and they were filled in by Bel'shtein and N. N. Beketov, professor at the University of Khar'kov.

[3] V. N. Ipatiev recollected the funeral of Mendeleev in the following words: "Mendeleev was given a state funeral by the command of the Emperor who sent a cable of condolence to his widow. Students carried the coffin as far as the grave."*

*B. N. Ipatiev. *Zhizn' odnogo khimika. Vospominaniya* (The Life of One Chemist. Memoirs). New York, 1945, pp. 323–324.

Document 1.9

To: P. N. Lebedev, Moscow, 2 May 1911.
From: S. Arrhenius
Director
Nobel Institute, Royal Academy of Sciences
(sent from Cambridge, Massachusetts)[1]

Dear Colleague,

Your letter of 29 March[2] reached me the other day. I thought over your question and arrived at the following conclusion. I could yet not see the physical journal here and therefore do not know anything about the considerations that made you retire from the Moscow University. I guess they are related to political and moral issues, and the average European can not have anything against them. Naturally, it would be a great honour for the Nobel Institute if you wished to visit and stay for work. We are prepared to provide all necessary conditions for the continuation of your research as you planned them. I believe the equipment is not too expensive. I see your situation such that if it is described to certain Swedish supporters of science they will be able to ensure your stay in Stockholm for the next two years. This time will be enough to think of the future. I can not tell now what concrete arrangements may be carried out, but I am quite sure that things will be much easier once we have overcome the difficulties of the first two years. You should have no doubt that you will be absolutely free to act in accordance with your scientific rank, which is far from that of an ordinary research scientist. I shall be back in Sweden in a month and a half and shall immediately devote myself to settling your affairs [...][3]

PFA RAN, fond 293, opis' 3, ed. khr. 5, l. 171, 172
(translated from the Russian version of the German text).

[1] Arrhenius was abroad in the first half of 1911. Lebedev's letter addressed to Stockholm was transmitted to him to the United States.
[2] Lebedev informed Arrhenius of his wish to join the Nobel Institute in a letter of 29 March 1911 written soon after he retired from the Moscow University.
[3] The Russian scientific community was not indifferent to the fate of Lebedev either. He was invited to continue research at the universities of Kiev, Warsaw, Khar'kov and in some other institutions. He was especially happy to accept the invitation from A. L. Shanyavsky at City University to organize a new physical laboratory and thereby preserve the traditions of the Moscow school of physicists created by him. Lebedev informed Arrhenius about his decision in the letter of 27 September 1911 and received an answer on 2 August 1911. Arrhenius confirmed his invitation to accommodate Lebedev at his institute "in case of necessity" and noted that "we have enough time to make the necessary arrangements and settle [...] monetary issues."* It was the last letter sent by the eminent Swedish scientist and Nobel winner of 1903 to his Russian colleague. Lebedev died at 45 at the peak of his creative powers on 14 March 1912. D. G. Pointing, his potential competitor for the Nobel Prize, did not long outlive him; he passed away at an age of 61 almost exactly two years after Lebedev (30 March 1914). The remarkable discovery of light pressure remained unawarded with the most prestigious prize of the past 20th century.

* PFA RAN, fond 293, opis' 3, ed. khr. 5, l. 174–176.

Document 1.10

Protest of a group of Swedish intellectuals sent to L. N. Tolstoy in December 1901.

In connection with the first award of the Nobel Prize in literature, we the undersigned Swedish writers, artists, and critics, wish to express our admiration for you. We see in your personality not only a highly respected patriarch of modern literature but also a powerful and sincere poet who should have been in the first place recalled on the present occasion even if you had never sought such an award for personal motives. We feel even more so addressing ourselves to you that in our firm belief the organization entitled to award literature prizes does not represent the opinions of artistic writers or that of the general public. Let people abroad know that even in our remote country we consider art true and strong that is based on the freedom of thought and creative work.[1]

[1] This address, in Swedish and French, is now in the state museum of L. N. Tolstoy, Moscow. Tolstoy sent his answer on 22 January (4 February) 1902 to Oskar-Ivar Levertin, professor at the University of Stockholm, writer and critic, and one of those who signed the address. His letter read as follows: "Dear Colleagues, I was really glad not to be awarded the Nobel Prize. First, it freed me from the trouble of deciding how to deal with the prize money that, like any money, can bring nothing but evil. Second, this incident gave me the honour and great pleasure to receive the

expression of sympathy from so many people whom I deeply respect, even if I do not know them personally. Please accept my sincere thanks and expression of good feelings. Leo Tolstoy."* The Russian translation of the above address in French and the text of Tolstoy's letter are borrowed from *The Complete Works of L. N. Tolstoy*, V. 73, 1954, pp. 204–205.

* Tolstoy's letter was first published in Russian in the newspaper *Peterburgskie Vedomosti* on 15(28) February 1902.

Document 1.11

To: S. F. Oldenburg
Permanent Secretary
Imperial Academy of Sciences in St. Petersburg
From: K. D. Wirsen
Chairman
Nobel Committee of the Swedish Academy

Because members of your organization[1] have the right to nominate candidates for the Nobel Prize in literature, I would be grateful if you could remind them about this right and familiarize them with the enclosed circulars. The Secretary of our Committee would be glad to transfer to you more copies of these circulars as soon as they are requested. I thank you in advance for what you may find appropriate to do in response to this application.

Translated into Russian from French.

PAF RAN, fond 9, opis' 5, ed. khr. 11, l. 7.

[1] Division of Russian Language and Literature, Imperial Academy of Sciences.

Document 1.12

Extract from the minutes of the meeting of the Division of Russian Language and Literature, Imperial Academy of Sciences, 5 November 1905.[1]

[...] VI. The Committee of the Swedish Academy set up for awarding Nobel Prizes[2] addressed to the permanent secretary of our Academy of Sciences[3] a written request dated 8 November[4] to remind members of our organization of their right to nominate candidates for the prize in literature and distribute the attached circulars among them.[5]

It is decided to take this request into consideration and convene a joint meeting of this Division and its Belles-Lettres Section on 8 November, Tuesday at 2 p.m.

PFA RAN, fond 1, opis' 1a-1905, ed. khr. 152, l. 303.

[1] The dates in docs. 1.11–1.14 are given in the Julian calendar. This date corresponds to 18 November 1905 in the Gregorian calendar.

[2] The Nobel Committee in literature.
[3] S. F. Oldenburg.
[4] Gregorian calendar.
[5] Statutes of the Nobel Foundation regularly attached to invitations for nomination of candidates for the Nobel Prize of a current year.

Document 1.13

To: Honorary academician K. R.[1]
From: Academician A. N. Veselovsky
Chairman at the meeting of the Division of Russian Language and Literature
Imperial Academy of Sciences
6 November 1905, No. 741.

In accordance with the decision of the Division of Russian Language and Literature, I have the honour to inform you that the joint meeting of the Division and its Belles-Lettres Section is scheduled on 8 November, Tuesday at 2 p.m. to consider nominations of candidates for the Nobel Prize in literature instituted by the Swedish Academy of Sciences.[2] You are expected to attend the meeting and participate in the discussion of the said issue.[3] Herewith please find a copy of the rules of the Nobel competition.[4]

PFA RAN, fond 9, opis' 5, ed. khr. 11, l. 12 (both sides).

[1] Great Prince Konstantin Romanov. K. R. was his pseudonym with which he signed his literary works and official documents of the Academy as its president.
[2] A mistake. What was meant was the philological Swedish Academy, counterpart of the Division of Russian Language and Literature, Imperial Academy of Sciences, in St. Petersburg.
[3] K. R., president of the Academy and honorary member of its Division of Russian Language and Literature, did not attend the meeting of 8 November. He wrote in his diary (9 November 1905): "I did not walk out today. I had to attend a meeting at the Academy but did not feel like going there because many members appear to be set against the government, if not pro-revolutionarily."*

* *Krasny arkhiv*, T.1 (44), 1931, p. 45.

[4] A copy of the circular from Stockholm containing the invitation to nominate a candidate for the 1906 Nobel Prize (see doc. 1.11).

Document 1.14

Extract from the minutes of the joint meeting of the Division of Russian Language and Literature and its Belles-Lettres Section, Imperial Academy of Sciences, 8 November 1905.

[...] IV. Academician A. N. Veselovsky, chairman of the meeting, informed members of the Belles-Lettres Section that the permanent secretary of the Academy of

Sciences received a lithographed copy (in French) of a circular[1] from the committee set up by the Swedish Academy to award monetary prizes from the interest on the capital left for the purpose by Nobel (8 November 1905, Gregorian calendar).

In this circular,[2] the chairman of the Nobel Committee C. D. Wirsen asks the permanent secretary to remind the members of the Imperial Academy of Sciences that they have the right to nominate candidates for the Nobel Prize in literature and to distribute among them the attached[3] printed copies of the rules governing the prize-awarding procedure.

Taking into account that the said invitation of the prize-awarding committee established by the Swedish Academy to use the capital allotted for the purpose by Nobel only now, for the first time, reached the Imperial Academy of Sciences and that the Nobel Prizes had already been awarded several times since their institution (1910–1905), the Belles-Lettres Section unanimously decided to include Tolstoy in the list of nominees of the Imperial Academy of Sciences and to ask the honorary member K. K. Arseniev to draw up a properly reasoned recommendation of Tolstoy and his main recent works, laying emphasis on his priority right to the next year's Nobel Prize as a contemporary author who had made the greatest contribution to world literature, for which this award was instituted.

PFA RAN, fond 1, opis' 1a-1905, ed. khr. 152, l. 328.

[1] See note 5 to doc. 1.12.
[2] A slip of the pen: it was Wirsen's autographic letter of August 1905 addressed to the permanent secretary of the Imperial Academy of Sciences (doc. 1.11).
[3] To Wirsen's autographic letter.

Document 1.15

Extract from the resolution of the meeting of the Belles-Lettres Section, Division of Russian Language and Literature, Imperial Academy of Sciences, 19 December 1905 (1 January 1906).

[...] II. The academician A. N. Veselovsky, chairman of the meeting, informed the members that the honorary academician K. K. Arseniev is unwell and can not attend the meeting; he asks to present his properly reasoned recommendation of Count Leo Tolstoy and his main recent works emphasizing his priority right to the 1906 Nobel Prize as the greatest contemporary author who has contributed enormously to world literature for which this award was instituted. The meeting decides by majority vote to ask the honorary academician N. P. Kondakov and honorary academician A. F. Koni to join honorary academician K. K. Arseniev for drawing up an extended version of Tolstoy's nomination for the Nobel Prize and to present it for discussion at the next meeting.

PFA RAN, fond 9, opis' 5, ed. khr. 11, l. 13

Document 1.16

Extract from the resolution of the meeting of the Belles-Lettres Section, Division of Russian Language and Literature, Imperial Academy of Sciences, 20 July 1906.

1. Honorary academician A. F. Koni presented the extended version of Tolstoy's nomination for the Nobel Prize drawn up by him together with honorary academician N. P. Kondakov in accordance with the decision of the Belles-Lettres Section, Division of Russian Language and Literature, and informed the members that the honorary academician K. K. Arseniev was familiarized with this final variant and approved it. The meeting decides to (1) take the extended version into consideration and approve it, (2) have it translated into the French language and ask honorary academician A. F. Koni to edit the translation, and (3) dispatch the French text of the nomination by the closing date indicated in the rules of the Nobel Prize-awarding commission (1 February, Gregorian calendar) and attach to it a printed copy of *The Great Sin*, a work of L. N. Tolstoy.[1]

PFA RAN, fond 9, opis' 5, ed. khr. 11, l. 15.

[1] The first page of the French text of the nomination bears the following inscription: "Dispatched on 19 January 1906 No. 113." (PFA RAN, fond 9, opis' 5, ed. khr. 11, 29), i.e. 1 February 1906, Julian calendar, that is after the deadline for the nominations for 1906.

Document 1.17

The unpublished fragment from A. F. Koni's manuscript entitled *Vospominaniya o L've Nikolaeviche Tolstom* (Memoirs of Leo Tolstoy) and dated to 1907; the fragment was removed by the author from the printed version[1] of the memoirs.

"Again, for the third time,[2] there were disputes about Tolstoy at the Academy in January 1906 in connection with the letter of inquiry from the Stockholm Academy[3] as to what Russian writers our Academy recognizes as worthy of the Nobel Prize. Arseniev's proposal to recommend Tolstoy met no objections and he was asked to draw up a draft letter to the Stockholm Academy.[4] He prepared rather a superficial sketch full of banalities containing a few words about *The Great Sin.* It looked more like a review than a well-reasoned recommendation.[5] The Division was dissatisfied and provoked objections from certain members to the nomination of Tolstoy. Some argued that *The Great Sin* is a weak work and were prepared to refrain from nominating at all rather than recommending it. Others maintained that the Academy can not separate political activity of the author from his literature work and, the former being unacceptable, consider him worthy of the prize. Still others did not like Tolstoy's belief that the essence of the New Testament *ist ein langst uberwunderer Punkt*[6] and regarded his biblical criticism as irrelevant and even maleficent in the "period of liberation".[7] Shikhmatov helped to calm down the opponents and thus avoid exposing the Academy to the shame of recognizing that we do not have a writer able to compete for the Nobel Prize[8] in addition to the disgrace of the loss of our army, navy[9], internal order, and external dignity. In the end, Kondakov and myself were asked to draw up the final version of the Academy's nomination[10] and its translation[11] was dispatched to Stockholm.

The original text of the nomination and its translation are attached."[12]

Archives of the Institute of Russian Literature, St. Petersburg, fond 134, opis' 1, ed. khr. 1, pp. 158–160.

[1] *Niva* (free supplement to the magazine), No. 9, 1908, columns 5–76.
[2] The nomination of Tolstoy for the 1906 Nobel Prize was discussed at three meetings of the Division of Russian Language and Literature, Imperial Academy of Sciences: 21 November (8 December) 1905, 19 December 1905 (1 January 1906), and 20 January (7 February) 1906.
[3] Swedish Academy. Secretary of the Academy and chairman of the Nobel Committee C. D. Wirsen sent his autographed letter to the permanent secretary of the Imperial Academy of Sciences on 8 November 1905 (see doc. 1.11).
[4] The Division of Russian Language and Literature, Imperial Academy of Sciences, commissioned honorary academician K. K. Arseniev to draw up the nomination of Tolstoy (8 (21) November 1905) (see doc. 1.14).
[5] PFA RAN, fond 9, opis' 5, ed. khr. 11, l. 8–9. *The Great Sin*, a publicistic article by Tolstoy, was first published in 1905.* Nominating this work for the prize, Arseniev formally observed the rule formulated in Nobel's will according to which the prizes are due to "those who, during the preceding year, have conferred the greatest benefit on mankind."

* *Russkaya Mysl'*, kn. 7, 1905, pp. 247–266.

[6] "includes something miraculous" (German).
[7] Revolutionary situation in Russia in 1905.
[8] Defeat of Russia in the Russo-Japanese War of 1904–1905.
[9] The Swedish press had suspected the probability of such collisions in Russia before the first prizes were awarded. One note to this effect was published in the *Goteborg Trade Newspaper* and reproduced in detail in *Sankt Peterburgskie Vedomosti* a month and a half before the laureates of the 1901 prize were announced.* The Swedish author dwelt on some inconsistencies in the system of selection of the worthiest candidates and expressed apprehension that "a certain candidate can be driven in his country into such a position that scientific or literary authorities will have to refrain from nominating him for the prize." He illustrated his point, which was quite understandable to Russian readers by two examples. "In particular, [...] is it possible that all Russian writers, without exception, even the most prominent and internationally known, can be nominated for the Nobel Prize by national authorities?" Or "Is there no reason to think that after the Dreyfus affair** none of the French academicians and professors will nominate Zola?" The first example surely hit the mark while the

second obviously missed it, as if to illustrate a marked difference between autocratic and democratic regimes. In fact, Zola was nominated for the Nobel Prize in 1901 by his compatriot and member of the Paris and French Academies Pierre Marcellin Berthelot.

* *Sankt Peterburgskie Vedomosti.* 8 (21) October 1901. p. 2.

** The trumped-up case of Alfred Dreyfus, officer of the French General Staff accused of espionage and sentenced to penal servitude for life in 1894, divided the French society into two uncompromising parts, Dreyfusards and anti-Dreyfusards. Zola obtained evidence of Dreyfus' innocence and joined the former. On 13 January 1898, he published his open *J'accuse...!* letter to the president of the Republic. Under pressure from high-rank anti-Dreyfusrads, Zola was prosecuted and sentenced to imprisonment, which made him flee to Great Britain. In 1899, Dreyfus was pardoned, and Zola returned to France in July 1906. Dreyfus was acquitted of all the charges.

[10] PFA RAN, fond 9, opis' 5, ed. khr. 11, l. 21–25.

[11] Ibid, l. 29–30. The French translation of the nomination is signed by F. F. Fortunatov, chairman of the Belles-Lettres Section, Division of Russian Language and Literature, Imperial Academy of Sciences (17 (30) January 1906).

[12] Neither A. F. Koni's original text nor its translation into French was attached to the essay published in 1908 (or in subsequent years).

Document 1.18

Extract from the diary of Great Prince Konstantin Konstantinovich, president of the Imperial Academy of Sciences and honorary member of the Department of Russian Language and Literature (9 (22) November 1902).

A prom was given in the Konstantinovskoye Artillery School,[1] and I was glad to go there soon after 10 p.m. dressed in the artillery uniform.[2] At last, I had an opportunity to see many of my former pupils that became their own masters [...][3] I had a talk with some of them, taking each aside. Polochanin[4] Olgerd Glasenapp was one of them [...][5]

GARF, fond 660, opis; 11, ed. khr. 50, l. 145–146.

[1] Founded in 1807 and named after Great Prince Konstantin Pavlovich to whom the author of the diary was great-nephew. Before the reform of 1894, the school was in the system of service schools as the Second Konstantinovskoye School; Konstantin Konstantinovich graduated from here.

[2] After the School was redesignated as an artillery institution, the Great Prince, who had already left it, received the title of *August konstantinovets* (member of the Konstantinovskoye School) and was allowed to wear the artillery uniform by an Emperor's edict.

[3] In March 1990, Konstantin Konstantinovich was appointed chief of military schools of the Russian Empire, in addition to his position of the president of the Academy of Sciences.
[4] Colloquial word for a graduate from the Polotsky Military School opened in 1835 and inspected several times by the Great Prince.
[5] Unidentified personality. However, a few facts give reason to think that O. Glasenapp was a relative of a nominee for the 1903 Nobel Prize (Table 1.4), maybe his son. The Great Prince had a talk with Glasenapp in late November 1902, when the nomination campaign for the 1903 prize was at its height and when he used (only once!) the right to nominate his candidate (K. Glasenapp, whom the Prince obviously favoured). The nomination of K. F. Glasenapp (author of *The Life of Composer Richard Wagner*, see note71 to this chapter) for the 1903 Nobel Prize by an unusually high number of people and the absence of nominations in subsequent years suggest an organized Europe-wide lobbying campaign on behalf of this candidate for the now prestigious award. O. Glasenapp might be just a pawn in this game. It seems very unlikely that Konstantin Konstantinovich decided to nominate K. Glasenapp by himself. His diaries, now kept in GARF (fond 660, opis' 1), do not mention either this name or any contacts of Great Prince with Nobel organizations and his participation in nomination campaigns. Nor is there evidence that K. R. highly valued literary works of the author of *The Life of Richard* Wagner, who was living in Riga. Most probably, he merely signed the nomination prepared by someone for the Nobel Committee and did not even think this episode as important enough to be entered in his diary.*

> * The Great Prince and his family liked the music of the great composer. His son Konstantin told in a letter from Pavlovsk to St. Petersburg (28 July (10 August) 1912) that he "yesterday read aloud" to his twin sister the book by E. Schure *Richard Wagner and his Musical Drama* translated from French into Russian (see *Istoricheskiy arkhiv*, No. 1, 2001, p. 125). It may be supposed that Konstantin Konstantinovich agreed to sign the nomination of Glasenapp for the 1903 Nobel Prize just because the family liked the music of the main character in his book.

Document 1.19

From documents of the Police Department, Ministry of Internal Affairs of the Russian Empire, concerning information about the Nobel Piece Prize for 1903.

I

To: B. N. Lamzdorf
Minister of Foreign Affairs
From: V. K. Pleve
Minister of Internal Affairs
16 March 1904

Confidential[1]

On 29 June 1900, Oskar II, king of Sweden, approved the Statutes of the Nobel organization (Nobel Foundation) in pursuance of the will of Alfred[2] Berngard Nobel of 27 November 1895, under which the interest from the capital specified by the testator will be annually distributed in the form of prizes to those who, during the preceding year, conferred the greatest benefit to mankind. The said interest is divided into five equal parts of which three must be awarded to the persons who made the most important discoveries within the fields of physics, chemistry, and medicine, the fourth in the field of literature for the most outstanding work in an ideal direction, and finally the fifth to the person who did the most or the best work for fraternity between nations, for the abolition or reduction of standing armies and for the holding and promotion of peace congresses. The prizes to the champions of peace are awarded by a commission of five members elected by the Norwegian Storting. The last Nobel Peace Prize was awarded to the "L'Européen" newspaper, published in Paris, which is known *to have been established exclusively for hostile propaganda and inciting the public opinion in Europe against Russia.* This paper never missed an opportunity to depict in a perverted form public events in Russia and actions of the Russian government. Of course, such a position of the "L'Européen" could not promote world peace. Nevertheless, *members of the Norwegian Storting awarded the Nobel Prize to this newspaper, certainly not for its literary merits, of which it has none. Such a decision can not be attributed to anything else but the position of the newspaper with regard to the Finnish question that finds support in Swedish society.*[3]

The award of the Nobel Prize would be of no interest for Russia if it reflected the opinion of a public group; however, it was given on behalf of the Swedish government and therefore should be considered as an inimical act against the government of the Russian empire. This makes me bring these facts to the notice of Your Excellency.

Yours respectfully, [etc.][4]

GARF, fond 102 (Department of Police, Special Division), 1904, delo (file) 14, chast' (part) 1, listy (sheets) 8–9 (both sides).

[1] This document is a typewritten draft of the letter to the Ministry of Foreign Affairs drawn up by N. A. Peshkov, counsellor for the Special Division of the Department of Police. Pleve substantially changed the draft submitted for his signing. He inserted corrections in the first paragraph with an ink pen, crossed the second and the third paragraphs, and rewrote them below. Pleve's own sentences are italicized.

[2] A mistake in the original text where Nobel is named Adolf.

[3] These phrases in the draft letter read as one: "It is worthwhile to note that this decision by the Commission might be dictated by the sympathy with the newspaper 'L'Européen' that took up a mission to defend the Finnish people as if it were oppressed by the Russian government." Pleve expressed the same thought more energetically.

[4] There is an anonymous pencil note in the bottom of the first page of the draft document (sheet 8): "true,* there was no word about the 'Nobel Prize' (underlain by the author of the note). Evidently, the note was made by one of the clerks of the Special Division after the final version of the letter was sent to Lamzdorf. That man presumably knew the primary information that gave rise to the correspondence between the two ministers but did not participate in writing Pleve's letter. It may be supposed that the source of this information received by the Department of Police and unknown to us did not tell of the award of the Nobel Prize to the 'L'Européen'; at best, it reported its nomination for the 1904 prize.** Even this supposition is unlikely because it contradicts the Statutes of the Nobel Foundation according to which the Peace Prize is awarded to persons or 'institutions and organizations' (but not to periodicals) for their direct contribution to 'fraternity between nations, abolition or reduction of standing armies or the holding and promotion of peace congresses' as indicated in Alfred Nobel's will. The cited letter might have been incited by the publication of a brief report from Stockholm in the Russian press some three months earlier: 'On the initiative of the former Finnish senator Leo Michelin and the known peace champion Frederik Passi, the Norwegian Nobel Committee allotted a grant of 10,000 krones to the daily newspaper "L'Européen', Paris.' "*** This report may account for the appearance of the pencil note on the draft of Pleve's letter made by one of the clerks.

* This four-letter word is illegible, worn away through time (only the first and the last letters can be read).

** Anne Kjelling, curator of the library of the Norwegian Nobel Institute, informed the author on 16 June 2000 that the newspaper *L'Européen* was not nominated to the Nobel Committee in 1903 and 1904.

*** *Novoye Vremya.* 11 December 1903, p. 3. On the same day similar information was published in *Russkie Vedomosti, Sankt Peterburgskie Vedomosti,* and other central newspapers. After a month and a half (29 February 1904), *Sankt Peterburgskie Vedomosti* presented a detailed analysis of this information in a large editorial article (pp. 2–3). It was indicated that the Nobel Committee had granted the subsidy to the Paris-based newspaper for the regular publications that "promoted the cause of world peace." Norwegian press organs of different political orientation unanimously blamed this initiative for a variety of reasons including the anti-Russian position of *L'Européen.*

II

To: V. K. Pleve
Minister of Internal Affairs
From: V. N. Lamzdorf
Minister of Foreign Affairs
13 April 1904
Confidential

With reference to your letter No. 3679 of 16 March 1904, I have the honour to inform Your Excellency that the Russian envoy in Stockholm drew attention of the minister of foreign affairs of Sweden and Norway to the award of the Nobel Prize to the French newspaper "L'Européen". Mr. Lagerheim expressed his sincere regret for this deplorable decision and added that he had already had a talk on this occasion with the Norwegian statsminister Ibsen.[1] The two men came to a common opinion and the statsminister promised to try and convince the chairman of the Nobel Committee, the former minister Levland, to change the decision of the committee.[2]

I consider it my duty to inform you that I brought an extract from the report of the Russian envoy in Stockholm to the notice of His Imperial Majesty.[3] Yours respectfully [etc.]

Ibid, l. 11.

[1] Sigurd Ibsen, Norwegian politician, son of the great playwright Henrik Ibsen. He was the Norwegian States Minister for Foreign Affairs until the dissolution of the union between Sweden and Norway, with permanent residence in Stockholm, the capital of the union.
[2] E. K. Byutsov, Russian envoy in Stockholm, if he actually was the informer (see note83 to this chapter), obviously misinformed his superior in St. Petersburg reporting the invented story of his conversation with the Swedish minister. Both Lagerheim and especially Ibsen knew perfectly well that the previous (1903) Nobel Peace Prize was not awarded to *L'Européen*, and Ibsen had no reason to promise Lagerheim "to try [...] and change the decision of the Nobel Committee."
[3] Thus, the emperor was aware of the information received from Stockholm.

Document 1.20

To: Count V. N. Lamzdorf
Minister of Foreign Affairs
From: Baron K. Stael von Holstein[1]
First councillor for the Russian mission in Stockholm 15 (2) December 1905

Last Sunday, two customary prize-giving ceremonies were held here in Stockholm. King Oskar and all members of the Royal family attended the ceremonies. The prizes were awarded to Professor Robert Koch, the known bacteriologist from Berlin (in medicine), Professor von Baeyer of Munich (in chemistry), Professor Lenard of Kiel (in physics), and Sienkiewicz, a Polish writer (in literature). Only Koch and Sienkiewicz came to Stockholm to receive their diplomas and medals from Royal hands.

At the formal dinner that followed the ceremonies and was chaired by Prince Gustav Adolf, H. Sienkiewicz[2] made a speech in which he gave thanks for the honour shown to him and touched on recent developments in the Polish literature and in the life of the Polish society at large whose vitality, in the opinion of the speaker, is emphasized by advances in science and literature despite the widespread belief that the country is doomed.[3] The speech produced a great impression and received applause.

During his stay in Stockholm, Sienkiewicz was frequently met with ovation. Some speeches in his honour at meetings and dinners contained hints to "the oppression of the Polish nation" and hopes for "its revival". To give him justice, he could tactfully bypass the trickier points and avoided commenting on the relations between Russia and Poland.

The so-called Nobel Peace Prize[4] awarded by the Norwegian Sorting was given to the known writer Baroness Zuttner.[5]

AVP RI, fond 133, opis' 470, ed. khr. 111 (1905), l. 258–259.

[1] At that time, Baron K. Stael von Holstein, councillor for the Russian mission in Stockholm, fulfilled the duties of *charge d'affaires* pending the arrival of the newly appointed (in May 1905) extraordinary envoy Baron F. A. Brudberg.
[2] H. Sienkiewicz was annually nominated for the Nobel Prize since 1901 by all eighteen members of the Swedish Academy.
[3] In the speech at the Nobel dinner on 10 December 1905, Sienkiewicz said: "There are many talks that say Poland is dead, exhausted, and enthralled — but here is evidence that Poland is alive and prevails."*
* *Nobelevskaya premiya po literature.* Laureaty 1901–2001 (Nobel Prize in Literature. Laureates 1901–2001). SpB, izd-vo universiteta, 2003, p. 11.
[4] It is not without reason that the Nobel Peace Prize is mentioned here in a slighting tone. This tone reflects the distrustful attitude of the Russian Ministry of Foreign Affairs to the work of the Norwegian Nobel Committee at large (see, for instance, note77 to this chapter).
[5] Today, the surname of the 1905 Nobel Peace laureate (von Suttner) tends to be spelled as Zuttner; this spelling is adopted in the present monograph.

Document 1.21

Vice-Director of personnel and management to:
The Second Department of the Ministry of Foreign Affairs, No. 66.02 of 11 August 1903.

With reference to the letter of the envoy of Sweden and Norway No. 7281[1] of 27 June 1903 containing a request to provide Russian books and documents to the library being organized by the Nobel Prize Council,[2] the Department of Personnel and Management has the honour to attach (1) a collection of treaties, conventions, and agreements concluded between Russia and foreign states in thirteen volumes and (2) a collection of treaties, conventions, and agreements concluded between

Russia and foreign states (now in force) in four volumes. I would appreciate delivering these materials to the addressee by the Second Department and notification of thereof.[3]

AVP RI, fond 155, opis' 440, rod. IV 1900, ed. khr. 2, l. 175.

[1] Ibid, l. 147.
[2] Nobel Institute of the Norwegian Nobel Peace Prize Committee.
[3] The covering letter to the Swedish-Norwegian mission No. 69326 is dated 16 August 1903; ibid, l. 176. Letter from the envoy of Sweden and Norway to the minister V. N. Lamzdorf is dated 16 (3) June 1903; ibid, l.173.

Document 1.22

Correspondence between E. K. Byutsev, Russian envoy in Stockholm, and V. N. Lamzdorf, minister of foreign affairs during the period between 16 (29) March 1904 (letter sent by V. K. Pleve, minister of internal affairs, to V. N. Lamzdorf) and 13 (26) April 1904 (Lamzdorf's answer).[1]

I

E. K. Byutsev to V. N. Lamzdorf. 20 (3) April 1904, No. 9.

Newspapers here in Stockholm have recently revealed that the Swedish minister of justice commanded that the publisher of *Keckans Nyheter* be taken to court for the violation of the press law that envisages prosecution of the author and confiscation of the edition for the use of abusive language in respect to foreign states and their heads. The said publisher printed a very rough and disrespectful article concerning His Majesty the Emperor [...][2] The leaflet was supported and edited by a certain Lille of Finnish nationality and published propagandistic materials designed to be distributed in Finland. It was printed on thin sheets of paper for easy transportation and was not on sale in Sweden.

The other day, I was told about this incident by the minister of foreign affairs, who confirmed the newspaper reports and explained that the Royal government had long ago wished to find a pretext to keep in check the Finns who misused the liberal press law of Sweden for propaganda. Also, the minister expressed hope that this time the jury selected to consider this case would refuse to show mercy to the defendant, unlike their predecessors who unfortunately would impose too mild sentences upon infringers of the press law. Their sentence must put an end to the abuse of printed word for the special purpose of exciting public opinion in Finland.

It should be emphasized that public opinion in Sweden has recently turned against Finnish agitators as evidenced by the negative response of several Swedish newspapers to the actions of Lille and his like in connection with the coming trial. Also, when Lille hinted that he may transfer his leaflet to Christiania or Copenhagen, certain Norwegian papers noted that the country would hardly be happy to be a source of articles against the form of government in foreign states or their leaders. Norwegians have positive feelings towards Finnish descendents but they also wish to maintain friendly relations with Russia and will not permit Finnish agitators to abuse their hospitality as a shelter for illegal activities, etc.

AVP RI, fond 133, opis' 470 (1904), ed. khr. 117, l. 26–28.

II

V. N. Lamzdorf to E. K. Byutzov. 12 (25) April 1904.

We have reports from our sources in Paris that Stockholm is the main centre from which the Japanese receive information about mobilization and other military preparations in this country. The Japanese embassy in Paris maintains telegraphic communication with this centre. According to these reports, a few destroyers and other warships are being built in Sweden for the Japanese and they will be hidden at different sites along the Swedish coast, ready to attack our Baltic fleet.[3] You are instructed to verify these reports thoroughly and carefully watch the actions of the Japanese embassy in Stockholm. Anyway, please alert our military attache to these facts.

Ibid., l. 1011.

[1] See doc. 1.19.
[2] Swedish authorities more than once used the press law to take to court publishers for using abuse language in respect to top officials of foreign states. According to *Moskovskie Vedomosti* on 10 (23) June 1904, the publisher of *Norrmellje Tidning* was sued in the same year "for the publication of two articles on the assassination of N. I. Bobrikov."*

* Governor-general of the Grand Duchy of Finland known for his tough russification policy was shot by a young Finn in the senate building on 3 (16) June 1904.

[3] When a squadron of Russian warships under the command of vice-admiral S. P. Rozhestvensky was about to leave Kronstadt for the theatre of war in the Pacific.

Document 1.23

Dialogue between Popov and Marconi in the film *Aleksandr Popov* (Lenfilm, 1950).[1]

I

Meeting between Popov and Marconi on board the cruiser *Carlo Alberto* (script by A. Razumovsky).

— Mr. Popov — says Marconi in an agitated tone — today, the most gifted scientists from all over the world work in my laboratory, and they all pursue one noble goal to promote the progress of mankind!

— Of your company, to be precise — Popov smiles ironically.

Popov turns round and quickly goes back to the saloon. Marconi follows him trying to keep pace with Popov and answers:

— What's wrong with that? The interests of science are as dear to me as to you. But you must agree that one can not realize a discovery unless one has adequate facilities for the purpose and a solid position on the exchange. My company guarantees this position... I need your services, Mr Popov — he adds as if forcing himself to utter it — I need your knowledge, your abilities, and experience.

— You have already used my experience in full measure — says Popov with disgust.

— Believe me, I shall not leave your services unrewarded — Marconi continues heatedly. You'll have fantastic conditions for work. It is impossible to achieve anything without money in our time, and you will have everything working with me! How much do you want for your experiments? One million? Two? You'll have as much as you need!

Popov ascends the ladder.

— I'll spare nothing for science!

Popov stops. He slowly turns his head, looks at Marconi and says very calmly:

— Don't dare to speak about science. You use it to gain profit. You have unscrupulously misappropriated another's invention and are trying to sell it. Well, maybe it is your true mission. But science, sir, is not a screen to hide dealings — he adds angrily. Then, even more sternly — Do not follow me![2]

II

From comments on the film *Aleksandr Popov.*

The history of the invention of radio by Aleksandr Popov, our countryman of genius, and of its misappropriation by the Italian adventurer scientist Marconi is highly characteristic and instructive. There have been many cases where Russian inventions were shamelessly stolen by foreign dealers who claimed them to be their own. But the story of the theft of the invention of radio is unprecedented in impudence and unscrupulousness [...] *Aleksandr Popov* is an expressive affirmation of the priority of the Russian inventor in an important field of science and technology.[3]

The film *Aleksandr Popov* was designed not only to present the biography of the scientist but also to explain to the audience the importance of Popov's invention [...]

The film exemplifies some negative features characteristic of biographical works of that time [...] Marconi was shown as a melodramatic villain [...][4] Cherkasov[5] himself had to devise some scenes to conceal the drawbacks of the script. An example is an episode where Popov for the first time sees the description of a radio set signed by Marconi in a foreign journal. Gripped by bitter feelings, he throws the journal aside, gets outside, and walks through dark and misty Kronstadt.[6]

[1] *Aleksandr Popov* was awarded the 2nd degree Stalin Prize in 1950.
[2] A. Razumovsky. *Aleksandr Popov.* M.: Goskinoizdat, 1951, pp. 77–78. The story preceding this dialogue is reduced to the fact that Marconi can not go further without Popov's assistance and arrives in Russia aboard a warship in order to convince the Russian inventor to join his company in Great Britain. This concept had nothing in common with the real situation and was in conflict with reports in St. Petersburg newspapers (see note 34 to this chapter). Their correspondents visited the Italian cruiser and could see the meeting of the two scientists on the day when the warship, decorated with lamps and pennants, was open to public attendance. It is not for nothing that neither the script nor the film itself contains a remark that the *Carlo Alberto* arrived in Kronstadt in the framework of the official visit of the Italian king to Russia, and that Marconi was on board at the invitation of the Italian officials who organized the visit.
[3] *Iskusstvo kino.* No. 3, 1950, pp. 27–28.
[4] Rumours were afloat in Moscow in 1950 that Italy entered an official protest to the USSR Ministry of Foreign Affairs concerning this episode but they are not thus far confirmed by documents.
[5] N. K. Cherkasov, the People's Artist of the USSR, played the character of Popov in the film.
[6] Yu. Gerasimov, Zh. Skverchinskaya. *Cherkasov.* M.: Molodaya gvardiya, 1976, p. 289.

Chapter 2 • 1919–1933

The Indifference of the Regime

It follows from Tables 2.1–2.3 that the situation with Russian nominations and nominators in the first 15 years of Soviet power was not significantly different from that in the preceding period described in Chapter 1. The sole difference was that substantially fewer Russian scientists still kept in touch with the Nobel organizations.

This situation was a corollary to the current developments in the country and its transition to totalitarism that was gaining strength. It is no mere chance that the first nomination from Russia that reached Stockholm after the break caused by the First World War and revolutionary upheavals dates to 1923 (to be precise, to November–December the previous year, when the majority of nominations usually arrive). The new economic policy (NEP) introduced in Russia in 1921 quickly brought its fruits and greatly contributed to the stabilization of the situation in the country. Intellectuals perceived the changes as a token of the return to normal life and the possibility of free contact with colleagues and relatives abroad as its indispensable component.

The direct relationship between nominators' activity and political events in the country is well-illustrated by the example of Professor O. D. Khvol'son, the most diligent of all Russian nominators (Tables 1.2 and 2.2). The Nobel Committee annually invited him to nominate his candidates since 1904 and he never disregarded the invitation. His contacts with the Nobel Committee were broken for six years after 1916 when he despatched his nomination for the 1917 Nobel Prize. In 1922, he resumed sending nominations; this he did virtually unceasingly up until 1930. Then, he unexpectedly ceased to respond to the invitations until his death in May 1934. In those years, Stalin had already imposed his dictatorship, and contacts abroad became increasingly dangerous. It is hardly a mere coincidence that five more Russian scientists sent their last nominations in 1930 (Table 2.2).

As before, Russian nominators in physics and chemistry tended to recommend their foreign colleagues, demonstrating perfect awareness of current events in world science and avoiding exotic nominations.[1] Six of their thirteen nominees became Nobel laureates.

It is worthwhile to once again mention Khvol'son, whose sole nomination (of the seven sent during the period under consideration) proved at variance with the decision of the Nobel Committee and the Swedish Academy of Sciences (in 1924,

Table 2.1. Distribution of nominees of Russian descent by categories of Nobel Prizes in sciences for the period 1919 to 1933

Nominees and their residence at the time of nomination	Year of nomination	Number of nominators	
		Total	Including USSR
Physics			
G.S. Landsberg (1890–1957), Moscow	1930	1	1
L.I. Mandel'shtam (1879–1944), Moscow	1930	2	2
Chemistry			
C.N. Vinogradsky (1856–1953), Switzerland*	1921	1	1
Physiology and Medicine			
V.M. Bekhterev (1857–1927), Leningrad	1925	1	1
G.A. Gurvich (1874–1954), Moscow	1929	1	—
-"-	1932	2	—
-"-	1933	3	3
V.I. Kedrovsky (1865–1937), Moscow	1933	1	1
S.P. Kostychev (1877–1931), Leningrad	1929	1	1
I.P. Pavlov (1849–1936), Leningrad	1925	2	2
-"-	1927	1	—
-"-	1929	5	5**
-"-	1930	1	1
P.G. Rozanov (1853–?), Moscow	1928	1	1***

* Vinogradsky emigrated from Russia in January 1920 and lived in Switzerland (where he bought a small house) when he was nominated for the Nobel Prize in 1921.
** It was the fifth (collective) nomination of Pavlov signed by six professors from medical institutions in Leningrad.[1]

[1] The list of nominators is published in the journal *Priroda* (*Nature*), No. 7, 2001, p. 4.

*** P. G. Rozanov was nominated by his son L. P. Rozanov, professor of the Medical Faculty, University of Byelorussia, Minsk. The nominator focused his arguments on the works of his father concerning the "epidemiology of plague and cholera" and signed the document without specifying his patronymic, whereas the name of his nominee is given in full.

he recommended to award the Nobel Prize to German scientists Friedrich Paschen and Arnold Sommerfeld (Table 2.2)). Even in this case, however, he showed a very keen insight into the state of contemporary physics and knowledge of its recent progress — a year later (in 1925), the same physicists were nominated again by the experienced Max Planck, the Nobel winner of 1918. It is noteworthy that Louis de Broglie and Chandrasekhara Raman, nominated by Khvol'son in 1929 and 1930

Table 2.2. Distribution of Soviet nominators and their foreign nominees by categories of Nobel Prizes in sciences for the period from 1919 to 1933 (nominees awarded the prize are italicized; figures in brackets indicate the year of the award)

Nominators	Year of nomination	Nominees
Physics		
Yu.A. Krutkov (1890–1952), Leningrad*	1930	E. Schrödinger (1933), Austria
N.N. Mirolyubov (?), Leningrad	1930	E. Schrödinger (1933), Austria
D.S. Rozhestvensky (1876–1940), Leningrad	1930	R.W. Wood, USA
Ya.I. Frenkel' (1894–1952), Leningrad	1930	E. Schrödinger (1933), Austria W. Heisenberg (1932), Germany
O.D. Khvol'son (1852–1934), Petrograd	1923	M. Siegban (1924), Sweden
-"-	1924	F. Paschen, A. Sommerfeld, Germany
-"-	1925	L. de Broglie (1929) M. de Broglie, France
-"-	1926	L. de Broglie (1929) M. de Broglie, France
-"-	1928	L. de Broglie, France
-"-	1929	L. de Broglie, France
-"-	1930	Ch. Raman (1930), India
Chemistry		
A.E. Favorsky (1860–1945), Leningrad	1929	S. Flexner, USA
Physiology and Medicine		
G.D. Belonovsky (1875–1950), Leningrad	1929	S. Flexner, USA
M.I. Gramenitsky (1892–?), Leningrad	1929	P. Trendelenburg, Germany
I.L. Krichevsky (1883–1943), Moscow	1933	F. d'Herelle, USA
P.S. Medovikov (1873–1941), Leningrad	1929	A. Calmette, France
D.Ya. Meerson	1933	A. Calmette, France
S.S. Khalatov (1884–1951), Leningrad	1929	G. Whipple, USA

* Yu. A. Krutkov, professor at the Leningrad University, proposed Schrödinger together with Louis de Broglie in his nomination of 5 January 1930. Thereafter, it occurred to him (23 January 1930) that de Broglie had already been awarded the Nobel Prize in 1929. He sent a letter to Stockholm asking to strike out the name of de Broglie from his nomination.

Table 2.3. Nominees of Russian descent for the Nobel Prize in literature for the period from 1919 to 1933

Nominees	Year of nomination	Nominators
K.D. Balmont (1867–1942)	1923	R. Rolland, Paris
I.A. Bunin (1870–1953)	1923	R. Rolland, Paris
-"-	1930	S. Agrel, professor of Slavonic languages, Lund
-"-	1931	S. Agrel, professor of Slavonic languages, Lund
-"-	1931	O. Broch, professor of Slavonic languages, Oslo
-"-	1931	W.A. Frantsev, professor of Slavonic philology, Prague
-"-	1931	A. Kaun, professor of Slavonic philology, Berkeley
-"-	1931	B. Pares, professor of Russian, London
-"-	1931	E. Nobel*
-"-	1931	N.K. Kulman, professor of Russian, Sorbonne
-"-	1932	O. Broch, professor of Slavonic languages, Oslo
-"-	1932	W.A. Frantsev, professor of Slavonic philology, Prague
-"-	1932	M.I. Rostovtsev, professor of classical history, University of Yale
-"-	1932	E.L. Gatto, professor of Slavonic philology, Padua
-"-	1932	S. Agrel, professor of Slavonic languages, Lund
-"-	1933	F. Braun, professor of East-European history, Leipzig
-"-	1933	M.I. Rostovtsev, professor of classical history, University of St. Petersburg
-"-	1933	W.A. Frantsev, professor of Slavonic philology, Prague

(*Continued*)

Table 2.3. (*Continued*)

Nominees	Year of nomination	Nominators
-"-	1933	O. Broch, professor of Slavonic languages, Oslo
I.A. Bunin and D.S. Merezhkovsky	1933	S. Agrel, professor of Slavonic languages, Lund
I.A. Bunin or M. Gorky	1933	S. Agrel, professor of Slavonic languages, Lund
M. Gorky (1868–1936)	1923	R. Rolland, Paris
-"-	1928	S. Agrel, professor of Slavonic languages, Lund
-"-	1928	W. von Heidenstam, Swedish Academy, Sweden
-"-	1933	S. Agrel, professor of Slavonic languages, Lund. R. Rolland, Paris
P.N. Krasnov (1869–1945)	1926	W.A. Frantsev, professor of Slavonic philology, Prague
D.S. Merezhkovsky (1865–1941)	1930	S. Agrel, professor of Slavonic languages, Lund
-"-	1931	-"-
-"-	1932	-"-
-"-	1933	-"-
I.S. Shmelev (1873–1950)	1931	Th. Mann, Berlin**
-"-	1932	N. van Wijk, professor of Slavonic languages, Leiden

* Emmanuel Nobel, nephew of the founder of the Nobel Prize, and the then head of the Nobel family.

** Th. Mann nominated Shmelev together with H. Hesse, Switzerland.

respectively, were given the Nobel Prize in the same years. In 1923, Khvol'son recommended to award the Nobel Prize to Karl Manne Siegbahn "for his brilliant discoveries and research in the field of X-ray spectroscopy" who was actually awarded by Swedish professors next year (1924).

A few words about the nomination of the eminent French physicist Louis de Broglie may be in order. He was first nominated in 1925 and 1926 together with his brother Morris. In both cases, Khvol'son was the sole nominator; due to him the two men were for the first time included in the list of nominees of the Nobel Committee in physics. Khvol'son proposed to award them for their "many fundamental works concerning Roentgen rays." In 1928 and 1929, he confined himself to nominate Louis de Broglie alone, mentioning again the results of his X-ray studies. However, the main emphasis was laid on the fundamental contribution of the nominee to wave mechanics that allowed the "resolution of controversies between supporters of the wave and corpuscular theories of light resulting in the development of a new world outlook."

Khvol'son's compatriot A. E. Favorskiy was equally successful in his nomination. In 1930, he proposed the German chemist Hans Eisen Fischer as a candidate for the Nobel award; the prize was given to Fisher in the same year.

Khvol'son also deserves special notice because he was the only Soviet specialist who remembered his countrymen (G. S. Landsberg and L. I. Mandel'shtam) when he proposed to award the Nobel Prize to the Indian Raman for the discovery of combination light scattering spectra in 1930. The story of this prize is remarkable as a striking example of the indifference to the Nobel movement in the Soviet Union of that time, not only on the part of the authorities and mass media but also of the scientific community. The decision by Swedish professors to award the prize to Raman alone did not cause any perplexity in the country and gave rise to idle talks only in a narrow circle of knowing scientists.

Combination light scattering, a landmark scientific achievement of the 20th century that found wide application in various sectors of natural sciences, had been discovered independently and almost simultaneously in two physical laboratories, in Calcutta and Moscow. In Calcutta, the effect was first observed in mid-February 1928. A short communication reporting this phenomenon was submitted to *Nature* on 16 February and published in the journal on 31 March (Vol. 121, No. 3038). The authors, C. Raman and K. Krishnan, believed that they had observed lines presently known as the Raman effect. This opinion was later questioned by many specialists in the field. By way of example, S. I. Vavilov, president of the USSR Academy of Sciences and director of its Physical Institute (FIAN) told in his unpublished address at the meeting of the Institute's scientific staff members on 25 September 1947 that the Indian scientists observed "fluorescence of sea water rather than the Raman effect, and in only a week [in fact, in two weeks or 28 February — either Vavilov's slip of the tongue or a misprint in the shorthand report] they actually obtained the Raman effect." (doc. 2.2) This incident that nevertheless gave reason for the publication in *Nature* was later many times mentioned by I. L. Fabelinsky,[2] a most authoritative expert in this field who, in his younger years, was a co-worker of G. S. Landsberg.

The Moscow physicists noticed formerly unobservable lines in a spectrogram on 23–24 February as evidenced by Landsberg's signature on the photoplate. However, they did not rush through a quick publication of their findings but chose to once again verify their conclusion in a new experiment using a more efficient light device mounted on their apparatus. Raman also had to verify the validity of his data and therefore found himself on the right track only on 28 February. But unlike the Moscow physicists, he was rather quick to give publicity to his initial interpretation of the observed phenomenon that eventually proved wrong.

The cautiousness shown by Landsberg and Mandel'shtam brought its fruits. In the very first paper in *Naturwissenschaften* submitted on 6 May 1928 and published on 13 July (Band 16, No. 28), they offered the correct interpretation of their findings. By that time, Raman had managed to publish three more articles in *Nature* but only the third (Vol. 122, No. 3062, of 7 July) contained the correct explanation of the new phenomenon.

In the meantime, the Moscow scientists familiarized themselves with Raman's earlier publications and referred to them in their own papers. Taken together, these facts (the gap between publications of the two groups and references of the Moscow authors to the works of their Indian colleagues) created the impression that Raman was the sole true discoverer of the new effect.

The first nominations for the Indian scientist reached the Nobel Committee in physics as early as 1929. One of them came from Niels Bohr, a world-renowned authority of that time. Nobody thought about the Soviet scientists, the more so that 16 (!) papers by other authors had appeared by the time of their first publication; naturally, these authors referred only to Raman's work.

In 1930, Raman was nominated for the Nobel Prize by ten scientists from different countries. The nominators included such eminent personalities as Nobel laureates N. Bohr, L. de Broglie, Ch. Wilson, E. Rutherford, J. Perrin, and J. Stark. The promotion given to the Moscow physicists was far less weak. There was a single nomination for all three pioneers and one more with the name of Mandel'shtam alone. Moreover, the nomination signed by Khvol'son looked like a humble request not to be too severe to his Soviet colleagues nor to "attach great importance to the excusable omission [...]" rather than a sound representation of their scientific achievements (doc. 2.1).

Professor N. D. Papaleksi of Leningrad was another nominator who tried to draw the attention of the Nobel Committee to the discovery of Soviet scientists. He was for the first time invited to recommend a candidate, and his nomination shows evidence of inexperience. It in the first place emphasized that the history of the discovery of combination light scattering provided an illustrious example of successful parallel research simultaneously and independently bringing new fundamental knowledge; thereafter, Papaleksi concentrated on supporting L. I. Mandel'shtam.

The essence of the nomination was reduced to the fact that the discovery made by Mandel'shtam was a result of his long-term systematic investigations into molecular light scattering, with the first publications dating back to 1907 and 1913.[3] In the opinion of Papaleksi, this gave grounds "to nominate Professor Leonid Mandel'shtam as a candidate for the Nobel Prize in physics for 1930."[4] In other words, Papaleksi just mentioned one scientist whom he considered worthy of the award instead of naming concrete contenders for the prize in connection with the discovery in question.

The nomination was registered exactly as it read in the files of the Nobel Committee together with other recommendations for the 1930 prize while other researchers racked their brains over how such a serious, competent physicist could propose the candidacy of Mandel'shtam alone, disregarding the contribution of his colleague Landsberg and the priority of Raman.

It would be certainly wrong to see the refusal of the Nobel Prize to Mandel'shtam as consciously biased against the Russian physicist. The history of science put everything into place according to honest rating. Today, nobody questions that the discovery was made independently in Calcutta and Moscow or that this fact was actually doubted in those bygone times.[5] The same journals in which articles by both groups had first appeared published detailed reports about the 6th Congress of the Association of Russian Physicists held in Moscow in the autumn of 1928. The reports were written by two reputable physicists, M. Born[6] and C. Darwin,[7] who gave an unambiguous vision of the actual state of affairs.[8] Both reports can be found in archive files of the Nobel Committee in physics, which means that they were known to the experts who considered nominations for the Nobel Prize in 1930. However, they could not have come to a different decision, guided by strict priority criteria established by the Nobel organizations.

That the Nobel Committee had no alternative is confirmed by two nominations presented by Niels Bohr in 1929 and 1930. Bohr was a most authoritative nominator of the Nobel Committee, well aware of the research carried out by Soviet physicists. Also, he knew about the discovery of the two Moscow scientists but did not include their names in his nominations. He certainly read the reports of Born and Darwin about the Moscow Congress that were published in autumn 1929, i.e. prior to the onset of the nomination campaign for the 1929 prize, but they did not help him to distinguish the priority publications of Landsberg and Mandel'shtam's from a number of secondary works by other experimenters. He did not find them worthy of the Nobel Prize, for all his benevolent attitude towards Russia and its scientists.

It is worthwhile to mention that neither Mandel'shtam nor Landsberg ever put the blame on the Stockholm experts, partly because they knew the true reason for the long delay in the publication of their sensational results that had nothing to do with the routine work of a researcher but related to the changing political situation in the USSR in the beginning of 1928. It was the time of the violent dismantling of NEP, a short period of chaotic economic developments followed by the conversion of the relatively liberal economy of those years into the totalitarian command-administrative system. Hundreds of thousands of people were groundlessly accused of economic offences and sentenced to prison terms. A close relative of Mandel'shtam fell victim to the mass arrests.

The scientist's granddaughter T. S. Mandel'shtam told the author that just at that time A. I. Gurvich's brother-in-law was arrested. Gurvich, a known biologist and nominator on the Nobel Committee in physiology or medicine (see Tables 2.1 and 3.1), was a cousin of Mandel'shtam's father. The two world-renowned experimenters had to give up their studies and do their best to help the man, who was in great trouble and had been sentenced to death by that time.[9] Their efforts and collected proofs of innocence saved the life of the accused man, who had the capital punishment commuted to administrative exile.[10] But the lost time could not be compensated, and it cost Soviet science dearly. Today, we should add to the list of guiltless victims of the totalitarian regime the lost Nobel Prize that was due to Soviet scientists in 1930 and that might have been the first such award in the Soviet period of Russian history.

Characteristically, the fact of the award of the 1930 Nobel Prize in physics to Raman alone was not used by the Soviet propaganda, even at the height of the notorious campaign "against cosmopolitanism and servility before the West" launched in the late 1940s and the early 1950s. The press of that period was full of invectives in the address of Nobel organizations and the "injustice" done to the Soviet scientists seemingly provided excellent material for the propagandistic machine.

Nor was this question raised in the course of the organization of the All-Union workshop of the heads of physical departments of universities and higher education institutions. Only S. I. Vavilov, president of the USSR Academy of Sciences, did not find it possible to pass over the matter in silence (see doc. 1.7), and I. E. Tamm, future Nobel laureate, mentioned it in his polemic speech (doc. 2.3).[11] The remark by Professor A. A. Andronov concerning the works of Landsberg and Mandel'shtam, addressed to N. S. Akulov, a conservative professor at the Moscow University, (doc. 2.4) only confirmed the concealment of the outstanding discovery of both scientists by campaigners against "cosmopolitanism."[12]

S. N. Vinogradsky, a notable microbiologist and discoverer of chemosynthesis in living nature, was initially nominated for the Nobel Prize in 1921. The nominator was the known Swedish bacteriologist Christian Barthel. In 1911, Barthel himself had been nominated for the Nobel Prize in physiology and medicine by E. S. London, who worked with Vinogradsky at the Institute of Experimental Medicine, St. Petersburg (see Table 1.1).

The discovery of the global importance of chemotrophic bacteria for the turnover of chemical elements in the Earth's crust made by Vinogradsky in the late eighties of the 19th century was recognized worldwide. The postulate of organic matter formation by chemosynthesis, in addition to photoynthesis, became the cornerstone in the system of biological sciences. It was specially emphasized by Ch. Barthel in his nomination where the contribution of Vinogradsky to biochemistry was characterized as "a universally accepted work of a genius." Although this fundamental study was published in 1890, "the main results of these investigations were confirmed by all scientists who later happened to be concerned with the problem; this makes Vinogradsky's theory of chemosynthesis absolutely steadfast" while "the discovery of free-living soil microorganisms assimilating nitrogen [...] remains one of the most remarkable achievements in biochemistry."[13]

There is little doubt that the cardinal significance of these works met strict criteria for the selection of their author as a candidate for the Nobel Prize. The trouble was the peculiarities of his mental faculties: the scientist was unable to concentrate for long periods on a subject that he regarded as a solved problem. In 1890, exactly when his fundamental discovery was recognized by world science as a strictly proved natural phenomenon, Vinogradsky lost interest in further research.

He retired in 1912 and left St. Petersburg for his patrimony in the Ukraine to be engaged in practical agriculture, although he later described his pastime, which lasted until the early twenties, as the period of a "scientist's reticent life." Thus, Vinogradsky was nominated for the Nobel Prize by Barthel in the period of his self-isolation.[14] Certainly, the Nobel Committee could not disregard the unusual behavior of the gifted naturalist.[15]

Revolution and Civil War made Vinogradsky sell his estate together with its profitable sugar refinery and emigrate. He left Odessa on board a French steamship on 12 January 1920, just in time. He recollected in his memoirs that "*chekists* [...] made a visit to my house and interrogated people [...] on the whereabouts of Professor Vinogradsky."[16] In 1922, Vinogradsky was invited to Paris by Emil Roux, director of the Pasteur Institute, and organized a laboratory of soil microbiology at the farm near Saint-Syr, owned by the Institute. He later summarized his studies in a monograph that attracted the attention of Selman Waksman of USA, a future Nobel laureate in physiology and medicine also born in southern Russia. Waksman's interest in the works of Vinogradsky was caused by the possible presence in soil of microorganisms showing antibiotic activity predicted by Louis Pasteur.[17] In the end, Waksman discovered the first antibiotic streptomycin and was awarded the Nobel Prize in 1952 (see Chapter 12). He always remained grateful to his senior colleague, with whom he frequently met (their families were on friendly terms). After Vinogradsky's death in early 1953, he wrote and published a biography of the scientist[18] that has not yet been translated into Russian.

The studies conducted by Vinogradsky during his life abroad greatly contributed to the development of microbiology. However, they could not surpass his works on chemosynthesis that in the past years had attracted attention from the world scientific community and were nominated for the Nobel Prize. His candidacy has never been again considered since 1921 with the exception of an unexpected nomination sent to the Nobel Committee in chemistry by Professor N. R. Dhar of Allahabad, India (see Chapter 6). It reached Stockholm soon before the decease of the venerable scientist on 24 February 1953, i.e. when he was only three and a half years short of his centenary.

A total of six Soviet scientists were nominated in different years for the Nobel Prize in physiology and medicine. I. P. Pavlov was nominated more frequently than others, this time for his studies on higher nervous activity.

The second nomination of Pavlov for the Nobel Prize was initiated by his colleagues in Leningrad. Professors F. Ya. Chistovich and D. A. Kamensky of the Physiological Laboratory, USSR Academy of Sciences, were the first nominators in 1925. In 1927, they were supported by the known German physiologist Professor Emil Abderhalden, professor of physiological chemistry at the University of Halle.

These nominations were most thoroughly considered in Stockholm. In 1925, the expert was Johan Johansson, chairman of the Nobel Committee, and in 1927 physiologist Hans Hertz became a member of the Committee. Both came to the conclusion that Pavlov's recent studies were not as important as those for which he had been awarded the Nobel Prize in 1904.

Nevertheless, colleagues continued to nominate the great physiologist. In 1929, nominations were especially numerous; they came to the Nobel Committee from Professors L. V. Blumenau, E. R. Gesse, M. P. Nikitin, and L. A. Orbeli (Leningrad). The latter also signed a collective nomination together with A. A. Lichachev, G. A. Nadson, S. S. Salazkin, M. V. Chernorutsky, and F. Ya. Chistovich. In 1930, Pavlov was nominated by V. A. Shtarker, head of the Department of Nervous Diseases, Omsk Medical Institute.

It is worthwhile noting that the Karolinska Institutet proved more conservative than the Royal Academy of Science, which did not hesitate to hand over the second gold medal to Marie Curie in 1911, i.e. at the very onset of its functioning as a prize-awarding institution. Up to now, no one has been given the Nobel Prize in physiology and medicine twice, while there are two such laureates in physics and chemistry (besides Curie): J. Bardeen, USA (physics) and F. Sabger, UK (chemistry).

The situation remains unaltered although the Nobel Committee almost every year receives well-grounded nominations of certain scientists for the second laureateship. Suffice it to say that four such nominations were registered by the Nobel Committee only in 1952 and 1953. In 1952, Professor H. Lobell of Munster University proposed the candidacy of his countryman G. Domagk, who had already been awarded the Nobel Prize in 1939 but was forced by Hitler's Gestapo to sign a letter declining the prize (see Chapter 9). Similarly, A. Ponsold, a professor at the same university, recommended A. Fleming of Great Britain, the 1945 Nobel laureate, for the second prize. F. M. Burnet, Australia, and J. Heyrovsky, Czechoslovakia, were for the second time nominated by their compatriots after they had previously received it in 1960 (physiology and medicine) and 1959 (chemistry) respectively.

Professor A. G. Gurvich of the Institute of Experimental Medicine, Leningrad, was nominated for the Nobel Prize in physiology and medicine thrice during the period under consideration and twice more in 1934 and 1938. His first nominators in 1929 and 1932 were physiologist Albrecht Theodore Julius Bethe, Frankfurt-am-Main (father of the future 1967 Nobel Prize winner in physics Hans Albrecht Bethe), embryologist Walter Brandt, Cologne, and histologist Siegmund Schumacher of Insbruck University.[19] In 1933, Gurvich was nominated by Professors I. L. Krichevsky, V. S. Levit, and E. I. Martsinovsky.

Professor Gurvich was nominated for the Nobel Prize as the discoverer of weak mitogenic rays of the UV spectrum in 1923. They were apparent in a living organism in the course of biochemical reactions and stimulated mitosis, i.e. cell division. In common parlance, they were called life rays. The discovery had remarkable repercussions in scientific circles and gave incentive to numerous experimental studies in laboratories of different countries.

However, attempts to reproduce the experiments of Gurvich yielded contradictory results, and the very existence of mitogenic rays was questioned by many researchers including experts of the Nobel Committee in physiology and medicine, hence their negative attitude toward the submitted nominations. Moreover, members of the Nobel Committee still had fresh in their minds a mistake made in 1926 when the prize had been awarded to the Danish microbiologist J. Fibiger "for the discovery of the Spiroptera carcinoma" that failed to be confirmed in subsequent studies.

Other nominees for the Nobel Prize in physiology and medicine were introduced by one nominator each, which in fact emphasized the casual character of recommendations. The probability of receiving the prize based on a single nomination in the 40th year of the Nobel Committees was close to zero, and experts treated such nominees accordingly.

Nominations for Gurvich, whose investigations had great repercussions among the scientific world (see above) contained comprehensive expert opinions, e.g. those signed by Professors H. Hoggkvist and E. Kulten. In contrast, nominations for V. I. Kedrovsky and S. P. Kostychev submitted by Professor A. Petterson and Professor E. Hammarstein had the form of one-sentence recommendations. The nomination for P. G. Rozanov received from the University of Byelorussia was not considered at all, one reason being that his son, Professor L. P. Rozanov, was among those who signed it.[20]

In 1925, the Nobel Committee again, after 1912 and 1914, received two nominations for V. M. Bekhterev, an outstanding neuropathologist and director of the Psychoneurological Institute he had founded in 1908 (redesignated as the State Institute of Medical Knowledge in the 1920s). One nominator was Professor E. A. Kartashevsky, pathophysiologist; the other nomination was signed by A. V. Gerver, anatomist and physiologist of the central nervous system; N. K. Rozenberg, infectionist and epidemiologist, and a few more specialists. However, Bekhterev was not already considered as a serious candidate for the Nobel Prize, and the above nominations were virtually set aside by the experts.

Nominations of foreign colleagues by Soviet medical scientists of that period were equally unsuccessful. The Nobel Committee disregarded their nominations of physiologists from the United States, France, and Germany, who failed to receive the Nobel Prize either in the same year or later. Five of the seven Soviet nominators were professors of Leningrad medical institutions. They were given the right to nominate candidates for the Nobel Prize in physiology and medicine in 1929 in accordance with the special rules for prize awards by the Karolinska Institutet (para. 5, p. 5) that stipulated the annual rotation of foreign research institutions authorized by the Nobel Committee to recommend nominees. In all probability, these rules allowed D. L. Meerson (tuberculosis expert), a professor at Odessa

University and chief physician for the Robert Koch Hospital, to act as a nominator in 1933.

An unusual fate of another Soviet nominator, physiologist B. P. Babkin, deserves to be mentioned. Babkin worked in the Physiological Laboratory headed by I. P. Pavlov and studied the secretory function of digestive glands under his guidance. Later, he wrote a fundamental biography of his teacher in the English language that was several times reprinted. During the revolution of 1917, Babkin was a professor at the Novoaleksandriysky Institute of Agriculture and Forestry in the Ukraine. In the 1920s, he was sent abroad on a scientific mission and never came back. As a professor of physiology in Halifax, Canada, Babkin nominated his British colleague Ernest Henry Starling for the 1926 Nobel Prize. In 1932, he sent from Montreal another nomination that named the outstanding American scientist Walter Bradford Cannon, honorary member of the USSR Academy of Sciences and a long-term friend of I. P. Pavlov. However, neither of his candidates attracted the attention of the Nobel Committee and the assembly of professors of the Karolinska Institutet. The same was true of Babkin's three candidates from Stockholm: Swedish biochemists Hilding Berglund, Einar Hammarstein, and Carl Gunnar Ogren. The latter nomination was sent to Stockholm in wartime (1941) when the Nobel Prizes were not awarded at all.

A former Soviet scientist who, like Babkin, had emigrated from the country, professor of chemistry at Moscow University I. S. Plotnikov, had already participated in the nomination campaign of 1915 (Table 1.2). In 1933, he was a professor at the University of Zagreb, Yugoslavia, and was known for the important monograph *General Photochemistry* published in German in Berlin. Plotnikov nominated Professor Victor Franz Hess of the Institute of Radiation Research, Insbruck, Austria. Three years later, Hess shared the Nobel Prize in physics "for his discovery of cosmic radiation" with Carl David Anderson of the California Institute of Technology, Pasadena, USA, the discoverer of the positron in this radiation.

In summary, Soviet biologists and medical researchers were generally as active in the nomination of their countrymen for the Nobel Prize during the first 15 years of the Soviet regime, as their predecessors in pre-revolutionary time. The same refers to their nomination of foreign scientists (Table 2.2). In other words, the Soviet medical profession remained an integral part of the world scientific community.

The situation with literature was different. Literature had been under ideological pressure since the mid-1920s as an important sphere of social and public influence. Naturally, this markedly hampered the contact between Soviet writers and philologists who still stayed in the country, and the Nobel organizations. None of them is listed among the nominees and nominators in Table 2.3 except M. Gorky. All the rest are emigrants.

After 1918, Gorky was nominated for the Nobel Prize three times, in 1923, 1928, and 1933 (see Table 1.3). In 1923, the nominator was R. Rolland, who named him together with I. Bunin and K. Balmont (see note 17 concerning Balmont in Chapter 12). Five years later, Gorky was nominated again by two members of the Swedish Academy, Verner von Heidenstam (Nobel laureate in literature) and Professor T. Hedberg. The information concerning Gorky that was received at the People's Commissariat of Foreign Affairs in 1931 and 1932 from A. M. Kollontai, the Soviet envoy in Stockholm, was obtained from personal talks and rumours (nobody nominated the writer in those years) (doc. 2.5 and 2.6). In 1933, Gorky was nominated by S. Agrell, professor of Slavic languages at the Lund University (southern Sweden) who had several times nominated I. Bunin. This nomination of Gorky was a sort of substitution for D. Merezhkovsky,[21] as a pair with Merezhkovsky whom Agrell had annually nominated since 1930 (Table 2.3).

The story of the consideration of Gorky's candidacy by the Nobel Committee and the Swedish Academy is analyzed in detail by K. Espmark, the present chairman of the Nobel Committee in literature. In 1918 and 1923, this outstanding Russian writer was not regarded as a serious contender for the Nobel Prize. Members of the 1918 Committee did justice to his talent but observed that Gorky's "anarchistic and often downright raw writings" were outside the framework of the Nobel Prizes.[22]

In 1928, the situation was more favorable largely due to the expert report of the eminent Slavonic authority, the professor of Slavonic philology at the University of Copenhagen Anton Karlgren. This Scandinavian specialist described Gorky's autobiography *My Childhood* as being of "remarkable historical and literary significance" and equally enthusiastically recommended two of the writer's earlier stories and the play *Na dne* (The Lower Depths). Concluding his report, Karlgren emphasized that in these works the figure of Gorky had acquired "quite a different status from before and inspires quite a different respect and sympathy, even though as a politically active person he is still quite obscure."[23]

In 1928, two members of the Nobel Committee, its chairman E. Karfeldt and A. Osterling,[24] gave their votes to Gorky while three others voted for the Norwegian writer S. Undset. The opinion of the committee's majority was supported at the meeting of the Swedish Academy.[25] The call by Karlfeldt and Osterling to award the Nobel Prize to Gorky for the above works in disregard of his remaining writings was quite reasonably opposed by one of the Academy members, P. Hallstrom, who became its permanent secretary in 1931. Hallstrom feared that the Nobel Prize, even if awarded specifically for the memoirs "would in the world at large be taken as a mark of approval of Gorky's writings in their entirety and thus provide a dubious advertisement for them, dubious also with regard to the respectability of the members of the Academy as judges of literary taste." Hallstrom characterized Gorky as a truly significant author" but concluded that a prize to him would be misunderstood by the reading public worldwide. Hallstrom specially emphasized that he had disregarded the fact that Gorky "is or was a Bolshevist" and that the Academy ought "to have enough backbone to stand against political opinion."[26]

For all that, the decision of the Academy was not altogether free from political partiality. Espmark cites extracts from a letter from F. Book, member of the Academy, to the linguist E. Tegner, his colleague in the University of Lund, dated 25 October 1925. Book informed his correspondent that he was "not willing to vote for Gorky" and that it pleased him "enormously to hear that all the members from Lund [of whom there were three] are agreed on that point." Espmark thinks that this feeling against Gorky "suggests more than literary objections."[27] A. M. Kollontai was more exacting in her characterization of the sentiments that prevailed in the Swedish society of that time and by no means contributed to the favorable consideration of Gorky's candidacy by the members of the Swedish Academy. She wrote: "The Swedes did not like tsarist Russia but Swedish nobility, landowners and feudal lords, 'respected' Russia for 'strong conservatism' and absolutism. The Soviet Union is unknown in Sweden, where they do not understand Bolshevism and are terribly afraid of it."[28]

However, is it really correct to interpret the disapproval of Gorky's candidacy for the Nobel Prize as it was described in the resolution of the Bureau of the Division of Physico-Mathematical Sciences, USSR Academy of Sciences, in 1955? (see doc. 7.19). Espmark included in his book a long list of great writers who came to the attention of the Swedish Academy arbiters between the two world wars without their greatness being duly noticed, "although their later achievements only confirmed the preconceptions of the time."[29] This list includes M. Gorky along with T. Hardy (UK), P. Valery and P. Claudel (France), S. George (Germany), H. von Hofmannsthal and S. Freud (Austria), M. de Unamuno (Spain). However, the list does not contain D. S. Merezhkovsky, Gorky's antagonist, who was nominated for the Nobel Prize ten times between 1914 and 1937! In other words, the candidacy of Merezhkovsky was considered by the members of the Nobel Committee and Swedish Academy twice as frequently as that of Gorky. Moreover, Merezhkovsky was an emigrant that should have inspired sympathy in him. Nevertheless, the outstanding figure of the "storm petrel of the revolution" caught more attention from prize-awarders despite his alien political views.

D. S. Merezhkovsky was first nominated for the Nobel Prize in 1914 by N. A. Kotlyarevsky, member of the St. Petersburg Academy of Sciences, a known scholar of the history of Russian literature of the 19th and early 20th centuries and frequent biographer of its eminent authors. It was probably not a coincidence that the nomination fell on the year when the 24-volume complete works of Merezhkovsky were published in Russia.

Thereafter, Merezhkovsky was not nominated again until 1930. Kotlyarevsky could probably not repeat his nomination during the wartime after 1914. Nor was it possible to do so after 1917 when it was simply dangerous to exhibit a positive attitude towards an emigrant and active opponent of the new regime established in Russia.[30]

However, the name of Merezhkovsky was in the list of nominees every year from 1930 to 1937. He was invariably nominated by S. Agrell, Slavonic authority from the Lund University. Until 1933, Merezhkovsky was always nominated together with I. Bunin.[31] Agrell did not forget Merezhkovsky even after Bunin had been awarded the Nobel Prize in 1933 and continued to recommend him up until his death in 1937.

The necessity to award the Nobel Prize to a Russian writer has for a long time presented itself. One of the most ardent advocates of this idea was the authoritative Scandinavian literature critic and publicist G. Brandes of Denmark, who greatly contributed to the popularization of Russian writers, from Pushkin to Gorky, in Western Europe. Brandes wrote shortly before his decease in 1927: "I was the first in Scandinavia to notice the bottomless depths of Russian literature and no one can be more disappointed that it has not thus far won recognition — the Nobel Prize."[32] I. A. Bunin was doubtlessly the first candidate for such recognition; he was always ahead of other Russian writers in terms of the number of nominations for the prestigious award.

I. A. Bunin was for the first time nominated together with M. Gorky and K. Balmont in 1923 by Romain Rolland, the Nobel laureate of 1915. The nomination was in a way induced by M. A. Aldanov, an admirer of Bunin. There is evidence that he entered into correspondence with R. Rolland in 1922 with a view to convincing him to nominate Bunin for the Nobel Prize.[33] True, Aldanov was not the first from whom Rolland had come to know the name of Bunin. A volume of Bunin's stories was published in French in the first half of 1922, and Rolland wrote a kind letter to the author after reading them (doc. 2.12-I). Nevertheless, a later message most probably addressed to Aldanov (doc. 2.12-II) leaves no doubt that the advice given to the venerable writer induced him to nominate for the Nobel Prize the author of *A Gentleman from San Francisco.*

The letter received by Bunin from the University of Lund in early 1923 suggests that the expert report concerning his candidacy was entrusted to S. Agrell (doc. 2.12-III–IV). However, it was shown by T. V. Marchenko,[34] who received access to the open materials of the Nobel Committee in literature in the mid-1990s, that the expert was actually the Slavonic scholar Anton Karlgren of Copenhagen University. Karlgren presented his comprehensive report of one and a half author's sheets and deserved the praise from the committee members for his excellent knowledge of the literary works of the nominee. The regular correspondence that was maintained between Bunin and Agrell after Bunin's first letter of 15 January 1923 made the Slavonic historian of Lund a true worshipper of the Russian writer, who nominated him for the prestigious award seven years later and thereafter repeated his nomination every year until the writer's turn came in 1933.

Olaf Brock, professor of Slavonic languages at the University of Oslo, was next to Agrell in terms of the frequency of Bunin nominations. He thrice sent his well-reasoned nominations to the Nobel Committee, from 1931 to 1933.[35] There is no evidence that the Russian writer knew about his competent supporter from the Norwegian capital. But the efforts of the even more authoritative (for the Nobel Committee) Sigurd Agrell, who greatly influenced the decision to award the Nobel Prize to Bunin, were certainly no secret to him (doc. 2.13). In 1930, the professor of

Lund was the sole nominator of Bunin. Next year, Bunin had seven nominations,[36] in 1932 five, and in the triumphant 1933 also five nominations (Table 2.3).

True, it must be said that during the two final years the nominee himself was active in searching for potential nominators, as is evidenced by his two letters to M. I. Rostovtsev, a professor at Yale University in New Haven and one of the most influential actors of the Russian humanitarian community during the first wave of emigration to the United States (doc. 2.14). These letters, now in Rostovtsev's fund in Darhem, reached their destination. The addressee became Bunin's nominator in 1932 and 1933.

The expert whom the Nobel Committee always commissioned to consider Bunin nominations was Karlgren, regarded by its members as the most competent connoisseur of the works of Russian-speaking nominees. He had studied Russian at the universities of Kazan and St. Petersburg and brought his knowledge of the language to perfection. Moreover, he had spent one winter in a village at the Dnieper River to obtain a deeper insight into the language, everyday life, and psychology of Russian peasants and described his experience in a book published in 1907. Therefore, he read Bunin's prose not only with his eyes but also with an open heart and understood it better than any other Slavonic expert whose services could be engaged by the Nobel Committee. A series of his reports finally affirmed Bunin's significance in the minds of the members of the Nobel Committee and Swedish Academy as "the last Russian prose master with the mark of great tradition."[37]

Such high praise of the creative work of Bunin and its place among the carriers of "great traditions of Russian literature" was not shared by all its actors, even those who had emigrated and had no chance to be nominated for the Nobel Prize as Bunin's rivals. Suffice it to name the outstanding poetess M. I. Tsvetaeva, who gave Bunin the last place in the triad of the Russian nominees in the early 1930s (Bunin, Gorky, and Merezhkovsky). There are no grounds to suspect her of insincerity or preconception (doc. 2.16).

The Soviet ruling elite met the announcement of the prestigious award to the exile writer with unconcealed irritation. The press was dead silent, and only *Literaturnaya gazeta* was allowed to comment on the extraordinary event. It used the permission to hold the new Nobel winner to shame as "an inveterate counterrevolutionist."[38] The ideological services of the Central Committee of AUCP(b) were perfectly aware that a clamour, if set up, would have an opposite effect and raise a new wave of natural interest in the half-forgotten exiled author. However, the extensive correspondence between the USSR embassy in Stockholm and top officials of NKVD (People's Commissariat of Internal Affairs) (doc. 2.7–2.9) gives indisputable evidence of the irritation caused in Moscow by the decision of the Swedish Academy.

The Soviet envoy in Sweden, A. M. Kollontai, reported attempts to exert diplomatic pressure in order to avert the public prize-giving ceremony or at least the reception at the French embassy.[39] She thoroughly cited explanations received from official circles about the total independence of Nobel organizations from government bodies and at the same time tried to convince the authorities in Moscow that the Swedish government lent an attentive ear to her arguments and "evidently kept the local press in check during Bunin's stay in Stockholm" (doc. 2.9).

However, all these reassuring reports addressed to Kollontai's superiors in Moscow are at variance with the notes in her personal diary (doc. 2.10) in terms of style, even though she described the same events and not infrequently used the same wording. The parallel texts are somewhat discordant, as if one pen were driven by two different persons under the influence of opposite feelings: one highly aggressive and uncompromising; the other ready to sensibly accept the situation as it is (see note 6 to doc. 2.10).

There may be one more cause for the discrepancy in the emotional content between the envoy's official reports dispatched to Moscow and intimate entries in her personal diary. Kollontai might be simply uninterested in the Nobel awarding mechanism and just performed her official duties, sending a flow of reports and explanations in response to inquiries by her irritated superiors. Otherwise, it is difficult to account for the fact that her diary, covering a long period from 1922 to 1940, most of which she spent filling top diplomatic positions in the two Nobel capitals, Christiania and Stockholm, contains but a single episode related to the Nobel Prize movement, namely the prize awarded to Bunin (doc. 2.10). Many entries in the diary are devoted to Fridtjof Nansen, for whom Kollontai had kind feelings. But even the note made two days before Nansen was handed the Nobel Peace Prize in 1922 (doc. 2.11) did not mention him as a Nobel laureate. The same refers to an old acquaintance of Kollontai, the Swedish social-democratic politician Hjalmar Branting, whom she knew since pre-revolutionary times. She mentions him many times in her diary but never as a Nobel Peace Prize winner, which he became in 1921.

For all that, Kollontai could have learned a useful lesson from the history of Bunin's laureateship. In doc. 2.18, she expressed regret that Moscow did not give proper attention to the development of cultural relations with western countries. This remark was provoked by the refusal to send the Soviet delegation to the 13th International Congress of Arts History held in Stockholm on 4–7 September 1933.[40] In 1934, Kollontai spoke outright about the necessity of broad participation of Soviet scientists and cultural workers in the Nobel movement. She concentrated efforts on neutralizing the inertia of Soviet bureaucracy and mortal fear of Communist Party ideologists, including those in academic institutions, of contacts with foreign specialists.[41] In her attempts to obtain positive decisions in favor of fostering cultural relations with the Swedish side, Kollontai was assisted by S. M. Mirny, head of the first Western Department of the USSR People's Commissariat of Foreign Affairs (doc. 2.19, 2.21, 3.3, 3.5), who had worked with her as the first secretary and *chargé d'affaires* for the USSR embassy in Norway in the late 1920s.[42] After almost 40 years (in 1971), N. A. Shcholokov, USSR Minister of Internal Affairs, submitted a memorandum to the CPSU Central Committee concerning attacks on A. I. Solzhenitsyn (doc. 11.14). He recalled the persecution of B. L. Pasternak in 1958–1959 (doc. 9.30) and referred to the unwise treatment of the Russian emigré writers including I. A. Bunin. The document contained the following closing remark: "We must learn a lesson from past experiences." (doc. 2.17)

In 1931 and 1932, I. S. Shmelev, the author of *Leto gospodne*, one of the most poetic Russian novels of the first third of the 20th century, became a nominee of the Nobel Committee in literature. Similar to other Russian emigrants seeking the prestigious award, he made much personal effort to obtain nominations although he understood perfectly well that his chances were close to zero.[43] Shmelev's attempts to secure solid support in the 1930s were not his first. He told in a letter to his regular correspondent A. I. Ylyin that in his prosperous life in Russia as far back as 1912, he had had talks with A. F. Koni, honorary academician of the Belles-Lettres Section, Division of Russian Language and Literature, Imperial Academy of Sciences, and nominator for the Nobel Committee, about his chances to be recommended for the prize.[44] The talk was brought on by the success of Shmelev's tale *Chelovek iz resotorana* (A Man from the Restaurant), published in 1911, among both readers and critics. The absence of relevant materials in the archives of the Swedish Academy indicates that the pretensions of the then young author found no response from the venerable lawyer and writer.

There is one more piece of opinion concerning the award of the Nobel Prize to Bunin that circulated in emigrant circles of Kharbin. It is akin to a vile anecdote but very characteristic of the feelings of a lost man seeking the simplest answers to the questions about the fate of the motherland lying heavy on him.

It is certain that the obviously unhealthy V. F. Ivanov, seized with the urge to search for freemasons among Russian intellectuals, published a large book soon after Bunin had received his award. The book contained the following striking passage: "Bunin is also a mason and masons helped him to receive the Nobel Prize that is, as a rule, awarded only to freemasons." To continue: "Coming back to the story of Bunin's prize given together with a diploma decorated with a five-pointed star, it needs to be pointed out that all other candidates were Russian masonic writers: Aldanov,[45] Merezhkovsky, Gorky, and Bunin; the latter received the regalia of this masonic congregation. The campaign that preceded and followed the nomination of this Nobel laureate as well as the accompanying festivities was also launched by freemasons. The clamour around this 'remarkable event' was also raised by Masonic newspapers [...]"[46]

Purely for fun, the list of Russian nominees in Table 2.3 can be supplemented by one more, self-styled, "Nobel candidate." The writer A. S. Sorokin lived in the city of Omsk, where he became notorious as a "mystifier" and troublemaker among some people but was respected as the "Don Quixote" of Siberian literature by others. Vs. V. Ivanov, another writer closely acquainted with Sorokin, described the following episode in his diary: "He [Sorokin] believed that it was impossible to force one's way into literature by ordinary means. Therefore, he resorted to an advertisement [...] Once, he ordered visiting cards reading

'Sorokin, Nobel candidate.' Why — I asked — have you ever received a Nobel Prize? He answered with a wry smile under the cropped moustache — I never told of this. The card reads: 'Candidate', and anyone can announce oneself a candidate."[47]

This manifestation of human vanity fairly well illustrates the great difference of the political atmosphere in Soviet Russia in the 1920s and some 10–20 years later, when such a jest could end in confinement in a labor camp or an even worse outcome. But the above episode from the life of the Siberian hoaxer prompts one more important thought. During the last quarter of the 20th century, the Nobel Prizes have been recognized by the Russian public as having enormous worldwide fame and prestige, in contrast to the reserved attitude toward the same institution on the part of the former monarchy and the indifference (for a time) of the Soviet regime.

Notes

[1] One such nomination was made by the physico-chemist Henry Le Shatelier, a professor at the Sorbonne, who proposed to award the Nobel Prize in physics to the Lumière brothers for the "invention of cinematography"* during the years between 1920 and 1927. Svante Arrhenius, director of the Nobel Institute of Chemistry and given the prize in 1903, made an initiative to award prizes in physics and chemistry not only to individual scientists but also to organizations, in line with the statutes of the Nobel Foundation.** In 1921, he proposed to award the prize in chemistry to the German Chemical Society; the next year, he nominated the United German Scientific Society for the prizes in physics and chemistry. However, the initiative did not meet with approval by the Nobel Committees and the Swedish Academy of Sciences.

> * Similarly, Professor V. Czerny of Heidelberg University recommended Count F. von Zeppelin (Friedrichsruh, Boddensee) and the Wright brothers (USA), constructors of the first airship and airplane, respectively, for the Nobel Prize in physics.
> ** Up to now, only the Norwegian Nobel Peace Prize Committee awards prizes to international and public organizations.

[1a] Throughout his life, S. I. Vavilov held in great reverence the memory of Mandel'shtam. His published diaries contain the following note dated 24 November 1944: "L. I. Mandel'shtam died yesterday, probably the most remarkable person among the scientists that I know in Russia. He possessed superhuman subtlety of physical mind, rare moral honesty in the most difficult situations, kindness, a good nature, and was highly cultured."*

Mandel'shtam's scientific erudition received equally high praise from V. I. Vernadsky: "Once, I went to Moscow in the same carriage as him [...] He amazed me [...] with the accuracy and lucidity of mind. His logic, sometimes formal, was quite understandable to me [...] He is a sanguine person, a thoughtful analytical experimenter [...] a philosophically educated person."**

> * S. I. Vavilov. *Mysl'o evolutsii mira — edinstvennoye absolyutnoe za chto eshche mozhno derzhat'sya soznaniem. Dnevniki* 1939–1951/(Thoughts about the world evolution — the only absolute to support the consciousness Diaries 1939–1951). Voprosy estestvoznaniya i tekhniki, No. 2, 2004, p. 23.

** V. I. Vernadsky. *Dnevniki 1935–1941 v dvukh knigakh*. Kn. 1. 1935–1938 (Diaries 1935–1941 in Two Books. Book 1. 1935–1938). M.: Nauka, 2006, p. 303.

[2] See I. L. Fabelinsky. *K istorii otkrytiya kombinatsionnogo rasseyaniya* (The History of Discovery of Combination Light Scattering). M.: Znaniye. See also *Uspekhi Fizicheskikh Nauk*, V. 168, No. 12, 1998, pp. 1341–1360 and V. 173, No. 10, 2003, pp. 1137–1144 for a more detailed description of the discovery of this phenomenon and Raman's struggle for priority by the same author. The discovery of combination light scattering in Moscow was first announced on 27 Apil 1928, i.e. a week and a half prior to the submission of the paper to *Naturwissenschaften*. G. S. Landsberg presented the discovery at the meeting of the Optical Department, Institute of Physics, USSR People's Commissariat of Health.*

* E. L. Feinberg. *Rodonachal'nik* (*o Leonide Isaakoviche Mandel'shtame*) (The Forefather (about Leonid Isaakovich Mandel'shtam)). Uspecki Fizicheskikh Nauk, 172, No. 1, 2002, p. 100.

[3] Annalen der Physik: B.23, H.4, 1907, S. 626–642 and B.41, H.3, 1913, S. 609–624.

[4] Archives of the Royal Academy of Sciences., Stockholm. Fund of the Nobel Committees in physics and chemistry. Materials of the Committees for 1930, sheets 22–23.

[5] Raman had little doubt that he would receive the Nobel Prize although he was not sure that it would be awarded to him alone. His co-worker Bhagavantam recollected half a century later that Raman "booked a steamship ticket two months before he came to know about his laureateship so as not to miss the ceremony in Stockholm". "I vividly recall his reaction when I communicated to him the first news of the award after hearing it on the telephone from one of the news agencies in Calcutta. He asked if he was the sole prize winner or did he have to share the bed with foreigners."*

* S. Bhagavantam. Proc. VII Conf. Raman Spectroscopy, V. 1, 1978, pp. 10–11.

[6] M. Born. *Naturwissenschaften*, 16, No. 39, 1928, S. 741.

[7] Ch. Darwin. *Nature*, 122, No. 3077, 1928, p. 630.

[8] Raman immediately responded to Darwin's report in *Nature* by a note dated 13 November 1928 and published in the same journal on 12 January 1929.* He actively asserted his formal priority over the works of "Russian physicists".** One of his arguments was that 16 articles by other authors from different countries were published based on his discovery during the interval between his first reports dated 31 March and 21 April 1928, and the first paper by the Moscow scientists in scientific press.

* *Nature*, 123, No. 3089, 1929, p. 50.
** In the 45-line note, Raman never calls his competitors by name as it should be in a mutually respectful scientific debate and impersonally refers to them as "Russian physicists".

[9] The story of L. I. Mandel'shtam's relative is told at length by I. L. Fabelinsky and T. S. Mandel'shtam in the journal *Priroda* (Nature) No. 6, 2002, p. 7. In connection with this, it seems appropriate to recall that the avalanche of arrests and death warrants in the USSR began to swiftly increase since the mid-1920s. B. L. Pasternak wrote to the poetess M. I. Tsvetaeva, then residing in Czechoslovakia (on 19 June 1927): "newspapers were delivered from the town [to the country house] by the weekend. There are things to which silence induced by shock and profound depression is the only natural response [...] What can I say? Tomorrow, I shall unfold

my paper and a journalist will tell me that quartering and impalement are the latest discovery of progressive society reading Marx in a library, not gnawing him in a cave; the day after tomorrow, another will show that I am an English spy because the envelope addressed to me carries English stamps; [...] the need for personal contact with people and the world is a whim in the eyes of the narrow-minded philistine that knows better than myself what I need for my own salvation."*

* *Dushi nachinayut videt'. Marina Tsvetaeva and Boris Pasternak.* Pis'ma 1922–1936 (Souls Begin to See. Marina Tsvetaeva and Boris Pasternak. Letters of 1922–1936). Moscow: Izd-vo Vagrius, 2004, pp. 352, 353.

[10] Mandel'shtam achieved abolition of the capital punishment for his relative and its substitution by the administrative exile to Vyatka through the agency of Professor A. Ya. Vyshinsky, a known lawyer and then rector of Moscow University. Paradoxically, Vyshinsky was later prosecutor-general at fabricated political trials during the period of the Great Terror.*

* E. L. Feinberg. *Uspekhi Fisicheskikh Nauk,* 172, No. 1, 2002, p. 101.

[11] S. I. Vavilov spoke of this at greater length at a Communist Party meeting in FIAN in autumn 1947 (see doc. 2.2).

[12] The cause of concealment was obvious. Anti-semitism was one of the main constituents of the campaign "against cosmopolitanism". For this reason, a group of conservative staff members of the Physics Department, Moscow University, more than once tried to defame such "anti-patriots" as Landsberg and Mandel'shtam in the course of discussion concerning further developments in modern physics. Accusations against them (including the deceased Mandel'shtam) caused his disciple Professor Andronov to make the following statement: "It was not a slip of the tongue when I called Mandel'shtam a Russian physicist. My line of reasoning is this. Anton Rubinshtein is a Russian musician, Levitan is a Russian artist, and Mandel'shtam is a Russian physicist. If a Jew tells me that Mandel'shtam is a Jewish physicist, I shall answer to this Jew that he is a Jewish nationalist. If a Russian tells me that Mandel'shtam is a Jewish physicist, I shall answer to this Russian that he is a Russian nationalist and chauvinist."*

* GARF, fond 9396, opis' 1, ed. khr. 251, l. 15–16.

[13] Archives of the Royal Academy of Sciences, Stockholm. Fund of the Nobel Committees in physics and chemistry. Materials of the Committees for 1921, sheet 55.

[14] It may be thought that the Swedish nominator Barthel knew from the press that Vinogradsky arrived in Switzerland as an exile (see note 1 to Table 2.1) and named him as a candidate for the Nobel Prize by way of rendering moral support to the man in trouble. He could not but understand that a scientist who had given up research almost ten years ago had no chance of receiving the award. Barthel became a member of the Royal Academy of Sciences but never again nominated Vinogradsky. Nevertheless, he maintained scientific contacts with him. By way of example, he sent him a kind letter (7 January 1950) after Vinogradsky published his monograph *Soil Microbiology* (see Chapter 4) in November 1949.*

* Arkhiv RAN, fond 1601, opis' 1, ed. khr. 114, l. 1.

[15] Recognizing Vinogradsky's position in the world scientific community, Nobel Committees of the Royal Academy of Sciences many times invited him to nominate his candidates for the Nobel Prize. Personal documents of Vinogradsky, now in the archives of the Russian Academy of Sciences, contain such invitations for 1948–1951.*

Ibid, ed. khr. 78, l. 1–7.

[16] *Rossiyskaya nauchnaya emigratsiya. Dvadtsat' portretov* (Russian Scientific Emigration. 20 Portraits). M.: Editorial URSS, 2001, p. 209.

[17] Waksman began to be interested in microbiology in 1914 when he was a student in an agricultural college. Vinogradsky's works became his main teaching aids. As soon as he came to know from Professor C. Barthel in 1922 that Vinogradsky had resumed his research, Waksman sent a letter to his address in Paris (12 December 1922) reading as follows: "I was really glad to hear that you feel well and eager to work again in soil microbiology [...] For my part, I did my best that you had American scientific literature delivered to you [...] also, I am sending you reprints of my own papers."* The two scientists came to personally know each other in summer 1924 during the Waksman's first visit to Europe; thereafter, they met many times in France up until 1947. Waksman was greatly impressed by the first meeting with Vinogradsky. He wrote on 29 November 1924: "You were my true teacher because I turned over and over again to you studies as classical works in our science, which badly misses pure researchers."**

* Arkhiv RAN, fond 1601, opis' 1, ed. khr. 127, l. 1.
** *Ibid*, l. 3. Under "pure researchers", the author meant scientists carrying out fundamental (as opposed to applied) research of importance in soil microbiology.

[18] S. A. Waksman *Sergei Vinogradssky: His life and work; the story of a great bacteriologist.* Rutgers Univ. Press. New Brunswick, New Jersey, 1953, p. 150. The book is prefaced with an epigraph, Pushkin's line translated into English: "I have erected for myself a monument, not moulded by human hand [...]" One volume presented by E. S. Vinogradskaya, the scientist's daughter, to the USSR Academy of Sciences together with other documents of her father is now stored in the archives of the Academy.*

* Arkhiv RAN, fond 1601, ed. khr. 192.

[19] A. G. Gurvich graduated from the Medical Faculty, University of Munich, and thereafter (from 1901 to 1905) worked in Switzerland as a laboratory assistant and then assistant professor at the Institute of Anatomy, Bern University, until he moved to St. Petersburg. He was well known in Western Europe, which accounts for his repeated nomination for the Nobel Prize in physiology and medicine by his western colleagues (up to 1938).* His directorship of the Research Institute of Experimental Biology coincided with the notorious general meeting of the Academy of Agricultural Sciences (VASKhNIL) in August 1948** with its crashing attack on genetics at which he was charged with "inability to organize research in light of progressive materialistic doctrine of biology,"*** i.e. in the light of the anti-scientific propositions of Lysenko. In the end, Gurvich had to resign (4 October 1948). In 1950 a new, absolutely absurd accusation was brought against him based on a denunciation sent to the Central Committee of AUCP(b) by a professor from Leningrad. He was described as having chaired a Jewish Masonic lodge supposedly in the northern capital in the 1930s. The Department of Propaganda and Agitation of the Central Committee took the denunciation quite seriously and reported it to Malenkov as worthy of thorough consideration.****

* See also Table 3.1.
** See Chapter 6.
*** RGASPI, fond 17, opis' 118, ed. khr. 314, l. 7.

**** G. V. Kostyrchenko. Tainaya *politika Stalina. Vlast' i antisemitism* (Secret Policy of Stalin. State Power and Anti-Semitism). M.: Mezhdunar. otnosheniya, 2001, p. 596.

[20] The thing is not reduced to tactless behavior alone (nomination of a relative). Professor L. P. Rozanov, head of the Department of Physiology, University of Byelorussia, appears to have received an invitation from the Nobel Committee as a member of the University Medical Faculty that had been selected, together with five other faculties in different countries, to give their professors an opportunity to recommend candidates for the Nobel Prize of 1928 (in accordance with special rules for prize awards by the Karolinska Institutet (para. 5, p. 5) attached to the Nobel Statutes). L. P. Rozanov may have been simply unaware of the requirements for nomination and just proposed his father, epidemiologist Professor P. G. Rozanov, who had no serious scientific works meeting the criteria of the Nobel Committee. Suffice it to say that the name index of the 36-volume Great Medical Encyclopedia containing the five-digit number of personalities does not mention P. G. Rozanov.*

* The only reference to the scientific works of P. G. Rozanov, besides his published papers, was found by the author in a copy of *Moscow Vedomosti* newspaper issued 14th (27th) January 1902 that told about his ideas concerning the spread of epidemic diseases as natural phenomena associated with the processes in soil, water, and atmosphere.

[21] Interestingly, Agrell could not help knowing that Merezhkovsky negatively appraised Gorky's works. In his sensational book *The Future Kham. Chekhov and Gorky* (SPb, 1906) Merezhkovsky characterized his fellow writers in the following words: "Generally speaking, a poetic vagabond reminds us of Smerdyakov with a guitar" (p. 45) or "Gorky's works are short of artistry." (p. 46). Gorky did not leave such invectives unanswered. He wrote in 1906: "When people like Merezhkovsky cry and whine of the necessity to defend 'cultural values', or 'legacy of centuries', you do not trust them." (*Sobranie sochineniy* (Collected works), GIKhL, 1953, p. 59). In 1928: "Dmitry Merezhkovsky, a known lover of the god of the Christian doctrine, is a petty person whose literature works are very much like typewritten copies: the type is easy to read but heartless and dull." (*ibid*, p. 337).

[22] K. Espmark. *The Nobel Prize in literature. A study of the criteria behind the choices.* Boston, Mass., 1991, p. 50.

[23] *Ibid*, p. 51. There was enough reason to characterize Gorky's views like this. By way of example, V. A. Maklakov, a known lawyer appointed by the Provisional Government as the Russian ambassador in France in 1917 (he did not return after the October Revolution) called Gorky *bol'shevisant* in a letter of 20 December 1921 to his regular correspondent B. A. Bakhmetev (ambassador of the Provisional Government to the United States and also an emigrant).*

* *Sovershenno lichno i doveritel'no! B. A. Bakhmetev — V. A. Maklakov. Perepiska 1919–1951.* T. 2 (Strictly Personal and Confidential! B. A. Bakhmetev — V. A. Maklakov. Correspondence 1919–1951, V. 2). M.: Stanford, 2002, p. 132.

[24] In the five years after Bunin received the Nobel Prize in 1933, Osterling expressed quite a different opinion. He pointed out that of all Gorky's works that had formerly attracted so much attention "only a few meet the current requirements."* In other words, Osterling, having given his vote for Gorky in 1928 in the Nobel Committee and as a member of the Swedish Academy, voted for Bunin in 1933. This change of opinion recalls the situation in 1965 when the Nobel Prize was awarded to M. A. Sholokhov. His creative activity and especially his civic

position during the pseudopatriotic campaign of the late 1940s and the early 1950s (see, for instance, footnote** to note 46 in Chapter 6), were even more at variance with the principles proclaimed by Osterling in 1933. This once again shows how the Swedish Academy periodically changes its orientation.

* Svenska Dagbladet, 10.11.1933, s. 1.

[25] A. M. Kollontai made a blunder in her published diary as regards the award of the Nobel Prize to Sigrid Undset, which suggests that she later corrected her notes or reconstructed them from memory. Being a counsellor to the Soviet minister plenipotentiary in Norway, she entered the following note into her diary dated 20 October 1922: "the writer Undset recently awarded the Nobel Prize for her historical novel [...]," etc.* In fact, Undset received the prize six years later. One more example of absent-mindedness relates to an episode that took place in the first half of 1937 when, according to Kollontai, she was present at a session of the Supreme Council during her business visit to Moscow.** The first Supreme Council was elected only on 12 December 1937, and its opening session was held a month later, in 1938.

* A. M. Kollontai. *Diplomaticheskie dnevniki 1922–1940.* T. 1 (Diplomatic Diaries 1922–1940. V. 1). M.: Academia, 2001, p. 61.
** *Ibid*, V. 2, p. 368.

[26] Espmark, p. 51.

[27] *Ibid*, p. 183.

[28] A. M. Kollontai. *Diplomaticheskie dnevniki 1922–1940.* T. 1 (Diplomatic Diaries 1922–1940. V. 1). M.: Academia, 2001, p. 465; entry of 9 May 1930.

[29] Espmark, p. 151.

[30] A popular description of Merezhkovsky in the Soviet press was "the sworn enemy" (see, for instance, the Great Soviet Encyclopedia, 1st edition, V. 39, 1938, column 9). Characteristically, Kollontai, the Soviet envoy to Sweden, called the candidacy of Merezhkovsky "politically indecent" in her comments on false rumors concerning the expected award of the Nobel Prize to this writer in 1932 (see doc. 2.6).

[31] As appears from the memoirs of contemporaries of Bunin and Merezhkovsky, the two writers were aware of their joint nomination. Bunin told Andrei Sedykh, who helped him during his stay in Stockholm that "Merezhkovsky once visited him and proposed to draw a notarial act obliging the one who may be given the Nobel Prize to pay the other 200,000 francs."* Such mercenary spirit seems to have been in the character of Merezhkovsky not only under the unsettled conditions of emigrant life but also in the time of prosperity in Russia. It is known that he basely flattered the known publisher A. S. Suvorin, seeking financial preferences. After his strange proposal to Bunin, Merezhkovsky stooped to beg the fascist dictator Mussolini (in 1936) for a grant from the Italian government to support his work on the upcoming book about Dante Alighieri.**

* Sedykh Andrei. *Dalyokie, blizkie.* (Distant and close beings) New York, 1962, p. 191.
** See *Voprosy literatury* (Literary issues), No. 2, 2001, pp. 274–278 and No. 6, 2002, pp. 185–186.

[32] See: *Russkiy New York. Antologiya "Novogo zhurnala"* (Russian New York. Anthology of "The New Journal"). M.: Russkiy put', 2002, p. 94.

[33] Minuvshee. *Istoricheskiy al'manakh,* Vyp. 8 (The Past. Historical Almanac. V. 8)."* M.: Atheneum-Feniks, 1992, p. 322; reprinted version of the volumes published under the same name by the Atheneum publishing house in Paris. It is known from this source that Aldanov proposed the same to T. Mann. However, Mann chose to nominate I. S. Shmelev (together with the Swiss writer H. Hesse) for the 1932 Nobel Prize. According to T. V. Marchenko, Mann remarked in his nomination: "I do hope that the Nobel Committee could award the prize to a Russian writer, even if just once."

Fourteen years later (in 1947), Bunin used his right as a Nobel laureate to attract the attention of the Nobel Committee to the works of Aldanov and nominated him annually until his death in 1953.

* *Izvestiya RAN,* seriya literatury i yazyka, 56, No. 6, 1997, p. 23.

[34] T. V. Marchenko. *Neizvestnye stranitsy buninskoy nobeliany po materialam shvedskoy akademii* (Unknown Pages of Bunin's Nobeliana. From the Materials of the Swedish Academy). *Izvestiya RAN,* seriya literatury i yazyka, 56, No. 6, 1997, p. 23–35.

[35] In 1928, Brock obtained a Russian visa through the agency of Kollontai, then the Soviet envoy in Oslo, and visited the USSR. He succeeded in his search for the materials of Slavonic philology in Russian libraries and returned home "a friend of the Soviet Union who helped [Kollantai] in [her] efforts to develop scientific and cultural relations with Norway" (Kollontai's memoirs).* Nevertheless, Brock for three consecutive years nominated Bunin, whose candidacy was unacceptable for Soviet Communist Party ideologists.

* A. M. Kollontai. *Diplomaticheskie dnevniki 1922–1940.* T. 1 (Diplomatic Diaries 1922–1940. V. 1). M.: Academia, 2001, p. 211.

[36] One of the nominations came from Emmanuel Nobel, nephew of Alfred Nobel. Swedish admirers of Bunin sent him to Grasse (France) a copy of Nobel's letter saying "that he [Nobel] is in favour of Bunin; has read five or six of his books and is beside himself with delight" (from *Grassky dnevnik* (Grasse Diary) by G. I. Kuznetsova; Liter. Nasledstovo, T. 84, kniga 2, 1973, p. 276). In the beginning of March 1931, Nobel had a stroke. G. P. Oleinikov, brother-in-law of E. Nobel (husband of his sister Martha) believed that it reduced Bunin's chances of winning the award in that year (ibid, p. 279).

[37] Espmark, p. 62. Karlgren's opinion is still shared by many historians of the Nobel movement. One of them, Oke Erlandsson, the main curator of the library of the Swedish Academy told the author in a conversation of 27 February 1998 that Bunin's Nobel Prize was a sort of apology to Russian literature for the disregard of Tolstoy. This remark has much in common with the opinion of A. Osterling, member of the Swedish Academy, expressed in 1933. Commenting on the award of the Nobel Prize to Bunin, Osterling mentioned Tolstoy, who "had been justly considered as the primary candidate", but "most unfortunately [...] this did not help." "The Nobel Prize of that year was a tribute of respect to the Russian classical tradition carried on in the works of Tolstoy's worthy successor."*

* *Svenska Dagbladet,* 10 November 1933, s. 8.

[38] Here is the full text: "In contrast to Gorky, whom nobody nominated and could not nominate in bourgeois society,* the white guard Olympus put forward and supported in every

possible way the candidacy of the inveterate counter-revolutionist Bunin whose works, especially recent ones featuring death, decay, and the feeling of doom in the face of the present catastrophic world crises, suit the tastes of the Swedish academic elders."** However, this paragraph from *Literaturnaya gazeta* looks a model of discretion compared with the foul language of the Swedish Communist newspaper *Ny Dag*, supported by Moscow.*** On 10 November 1933, i.e. the next day after the decision of the Swedish Academy was announced, the newspaper called Bunin an "inferior scribbler" who created "a few primitive pornographic scenes from the Russian countryside life of the tsarist time." The decision of the Academy was characterized as having been dictated by "the necessity to jump at the candidacy of any Russian white guard for the purpose of political demonstration." Paradoxically, the swearing in the *Ny Dag* had something in common with certain notes in the diary of Great Prince Konstantin Konstantinovich, poet and president of the Imperial Academy of Sciences, dated to 1910. Having read a book of verses and stories sent to him by Bunin, honorary member of the Academy, together with the Russian translation of Byron's *Manfred*, the Great Prince noted: "What a sriking disappointment. There is no poetry in the verses [...] Both poetry and prose contain immodesties bordering on pornography."****

* Gorky was nominated for the Nobel Prize four times, in 1918, 1923, 1928, and 1944; see Tables 1.3 and 2.3.
** *Literaturnaya gazeta*, 29 November 1933.
*** It was no secret. Kollontai made the following remark in her diary concerning the Swedish communist paper in June 1935: "[...] it is regarded as our mouthpiece on international policy issues" (* A. M. Kollontai. *Diplomaticheskie dnevniki 1922–1940.* T. 2 (Diplomatic Diaries 1922–1940. V. 2). M.: Academia, 2001, p. 288).
**** K. R. *Dnevniki i vospominaniya. Stikhi. Pis'ma* (Diaries and Memoirs. Verses. Letters). M.: Iskusstvo, 1998, p. 337).

[39] Bunin had a "Nansen passport" and did not wish, in principle, to be naturalized as a foreign citizen. This juridical fact allowed Kollontai to insist on the cancelation of the official reception in the French embassy and gave much additional trouble to organizers of the Nobel festivities. Specifically, it was decided not to hoist flags on the roof of the Grand Hotel, where the Nobel winners usually stay during the festivities, in the breach of tradition to adorn the hotel with the national flags of the laureates. Bunin's wife recalled this episode provoked by their stateless status: "November 9. Breakfast [...] Cable from Kalgren (in all probability, Anton Karlgren) broke our peace. He asks about Yan's citizenship. We replied: *refugie russe.** We do not know whether it is to the better or to the worse."**

* "Russian exile" (French).
** V. M. Muromtseva-Bunina. *Zhizn' Bunina. Besedy s pamyat'yu* (Bunin's Life. Talks with Memory). M.: Sovetskiy pisatel', 1989, p. 482.

[40] The Congress was held under the patronage of King Gustav I, the honorary president of the organizing committee was Prince Eugene, a known Swedish artist and decorator of the Prince's Gallery, one of the premises in the Stockholm City Hall. Prince's Gallery is the place where the king and the queen give an audience to the new Nobel laureates on the 10 December, after the formal dinner in the Blue Hall of the City Hall; the awards received by the laureates (diplomas and gold medals) are exhibited to the guests in glass cases placed along the walls.

[41] It should be said in fairness that Kollontai was not the first to try to interest the Soviet authorities in the Nobel movement. The science historian V. D. Esakov told the author that he had found among the documents of AP RF a report by N. I. Bukharin to the Political Bureau of the Central Committee about his visit to London in summer 1931 for participation in the 2nd International Congress on the History of Science and Technology. In this document addressed to the secretary-general of the AUCP(b) Central Committee, Stalin, Bukharin emphasized the "high praise" given to the Nobel Prizes in the world scientific community and the prestige they brought not only to scientists but also to the nations they represent.

[42] In a diary note made in August 1929, Kollontai characterized her assistant as follows: "Mirny is always absent-minded in everyday life but is sagacious about politics."*

> * A. M. Kollontai. *Diplomaticheskie dnevniki 1922–1940.* T. 1 (Diplomatic Diaries 1922–1940. V. 1). M.: Academia, 2001, p. 420.

[43] From the letter of 5 January 1931: "I would like to tell you in secret what I decided to try, although I foresee failure. On the one hand, I am praised by a clever man in Europe, a keen scientist. You know him as well as yourself.* On the other hand, I have something in store: 26–28 translations into a score of languages. Well, I'll take the risk,** although it all is doomed to fail. I sent Agrell four books in German and a letter, very modest [...] I also sent my kindest regards and a book to Selma Lagerlof. If Argel again nominates Bunin or Merezhkovsky, which is beyond my expectation, I shall apply to the members of the committee. Do you approve this gesture of impudence or [...] madness?*** A detailed study of numerous Shmelev's attempts to enlist the support of the members of the Nobel Committee in literature was published by Magnus Junggren, lecturer in Russian at the University of Goteborg and connoisseur of the writer's works."****

> * In all probability, Shmelev means Thomas Mann, who nominated him for the Nobel Prize in 1931 (the Russian word *to-nkiy* is spelled with a hyphen; a characteristic peculiarity of the writer's style largely encountered in his letters when he wished to draw attention to a given word by prolonging one of its sounds or, in contrast, express irony (see note 5 to doc. 2.15).
> ** The word *ryssknu* (take risk) is spelled with *y* instead of *i* to show the writer's ironic attitude towards his undertaking.
> *** *Ivan Ylyin and Ivan Bunin. Perepiska dvukh Ivanov.* T. 1 (Ivan Ylyin and Ivan Bunin. The Correspondence of Two Ivans. V. 1). M.: Russkaya kniga, 2000, p. 190.
> **** *Vsemirnoye slovo,* No. 15, 2002, p. 13–15.

[44] From the letter of 21 October 1931: "Koni wished to nominate me as early as 1912, but I was then a young writer."*

> * *Ivan Ylyin and Ivan Bunin. Perepiska dvukh Ivanov.* T. 1 (Ivan Ylyin and Ivan Bunin. The Correspondence of Two Ivans. V. 1). M.: Russkaya kniga, 2000, p. 235.

[45] M. A. Aldanov was for the first time nominated for the Nobel Prize only in 1947 (see Table 4.2), i.e. 14 years after publication of the book by V. F. Ivanov.

[46] V. F. Ivanov. *Ot Petra Pervogo do nashich dney. Russkaya intelligentsiya i massonstvo* (From Peter the First to Present Days. Russian Intellectuals and Freemasonry). Kharbin, 1934, pp. 541–544.

47 *Druzhba narodov*, No. 8, 2001, pp. 93–94. Vs. Ivanov does not specify the date of this mystification but it must have taken place when he could meet Sorokin, i.e. between the early 1920s and 1928, when Sorokin died.

Document 2.1

Nomination sent by Professor O. D. Khvol'son to the Nobel Committee in physics. Dispatched from Leningrad 6 January 1930, received in Stockholm 11 January 1930.

On 31 March 1928, the English journal *Nature* published a paper by Professor C. V. Raman written in Calcutta on 16 February in which the researcher described his discovery of a new phenomenon. It is currently known as the Raman effect.

However, there is no doubt that Professors G. S. Landsberg and L. I. Mandel'shtam of Moscow observed and explained the same phenomenon on 21 February.[1] Professor Max Born of Goettingen was well aware of things stood and even published something on the matter.[2] Unfortunately, the Russian researchers make no haste to publish their results.[3] If they had submitted them to *Nature* on 21 February, they would have appeared before Raman's paper, but we had the "Landsberg-Mandel'shtam effect" instead of "Raman effect,"[4] or it would have been named after the tree scientists. It is therefore incorrect to attach importance to the accidental omission of the Russian scientists. Therefore, I dare to propose the following: the Nobel Prize in physics for 1930 should be divided into two, one half to be awarded to Professor C. V. Raman of Calcutta and the other to Professors G. S. Landsberg and L. I. Mandel'shtam of Moscow.[5]

Translated from German into Russian

Archives of the Royal Academy of Sciences, Stockholm. Fund of the Nobel Committees in physics and chemistry. Materials of the Committees for 1930, sheet 74.

[1]1 In December 1928, Khvol'son mailed a postcard to Mandel'shtam asking for "the date when you first saw the new line. I am writing a paper and am very interested in having this information." Mandel'shtam answered on 18 December: "We for the first time noticed the new lines on 21 February 1928. The new lines were seen very clearly on the spectrogram obtained 23–24 February [...] Our unit had a weak light source, and we decided to equip it with a more powerful device. For a variety of reasons this work took more time than we had planned.* Quite good negatives were obtained with the new unit in mid-April..."**

* A hint on the troubles in connection with the arrest of Mandel'shtam's relative (see the present Chapter and note 10).

** I. L. Fabelinsky. *K istorii otkrytiya kombinatsionnogo rasseyaniya* (The History of Discovery of Combination Light Scattering). M.: Znaniye. pp. 38–39. See also *Uspekhi Fizicheskikh.*

[2] M. Born. *Naturwissenschaften*, 16, No. 39, 11928, s. 741.
[3] See note 9 to the present Chapter.
[4] The nominator seems to have meant that the distance between Moscow and London is shorter than that between Calcutta and London.
[5] In accordance with the Nobel Statutes, the prize may be equally divided between three persons if their contributions are considered to be roughly similar, or in a proportion proposed by the nominator if one person evidently contributed more than the two others. Anyway, Professor Khvol'son clearly gave preference to Raman as the primary candidate for the Nobel Prize.

Document 2.2

Extract from the address of S. I. Vavilov, president of the USSR Academy of Sciences and director of its Physical Institute (FIAN), at the meeting of the Institute's Communist Party organization (25 September 1947) convened to discuss the closed letter of AUCP(b) Central Committee concerning "the case of Klyueva and Roskin".[1]

Last year, the English journal *Discovery* published a paper describing scientific progress in the world.[2] The author chose the simplest way: he just distributed Nobel laureates by countries and nationalities.[3] The USSR turned out to be ranked lowest because we have only two laureates, Pavlov and Mechnikov. This gave the author cause to observe the poor state of science in our country even though he promises us a radiant future. Looking behind the scenes, one can see that, on the one hand, the Nobel Prizes are awarded to really worthy scientists. There are few cases where they were given to random researchers.[4] On the other hand, many equally worthy people come without being duly noticed by the Nobel Committees. Moreover, there is a system and preconception in this neglect.

For example, one of the Nobel laureates is Marconi, although it is widely known and legally confirmed, not only in this country but also in Italy (e.g. by Marconi's teacher Righi) that practical radio was discovered by Popov. Marconi could at best pretend to be Popov's co-author. Nevertheless, the Nobel Prize was awarded to Marconi instead of Popov.[5]

There is one more example. Combination light scattering was independently discovered in two cities, Calcutta and Moscow. The difference is that the conclusions of the Moscow physicists were formulated much more cautiously and exactly. Raman's success was due to his actions being in an "American manner". It was recommended today, that we should not hasten to publish our results.[6] Raman knew better. As soon as he noticed a slight spectral shift, he sent a cable to *Nature* and had his data published in two or three days. However, this first report was clearly a mistake. What he described was not what we call now the "Raman effect" but fluorescence of seawater; the Raman effect was really observed only in a week [...] Thus, scientists in two distant places of the planet quite independently discovered an important phenomenon [...] In the end, Raman was awarded the Nobel Prize, while Mandel'shtam lost it.

I can not help pointing out that, unfortunately, many in this country tend to conceal discoveries made by our scientists. I personally had to combat this tendency at such an authoritative meeting as the plenary session of the Stalin Prize Committee and submitted a relevant report.[7]

TsAODM, fond 6862, opis' 1, ed. khr. 7, pp. 95–97.

[1] See Chapter 6 for the "case of Professors Klyueva and Roskin".
[2] E. M. Friedwald. Science and political power. *Discovery*, 7, No. 10, 1946, pp. 292–298.
[3] A slip of the tongue. Friedwald listed only the laureates' countries of origin. The article had no reference to their nationality in the Soviet understanding of this word.
[4] A. P. Aleksandrov, president of the USSR Academy of Sciences spoke about the same 26 years later (in 1975) at the meeting in the Central Committee of AUCP(b) (see doc. 8.51).
[5] Vavilov several times commented on the fact that the 1909 Nobel Prize was not awarded to Popov. It appears he was simply unaware that an individual could not be nominated for the prize posthumously. Popov had died four years before Marconi received the prize (31 December 1905 or 13 January 1906).
[6] Vavilov refers to the discussion at the same meeting on the possibility to publish two papers by V. I. Veksler in *Doklady AN SSSR* in 1946. The papers concerned the discovery of the principle of autophasing in accelerators that allowed particles' energy to be increased several thousand-fold. The opinions of the participants diverged.*

* TsAODM, fond 6862, opis' 1, ed. khr. 7, pp. 41p67–48 and 55–56; see doc. 6.37.

[7] It remains unknown what discovery Vavilov meant.

Document 2.3

Extract from the speech by I. E. Tamm, corresponding member of the USSR Academy of Sciences, at the All-Union workshop of the heads of physical departments of universities and higher education institutions on 4 March 1949.

[...] An urgent recently emerged problem is the protection of our priority as regards a number of the most important scientific discoveries [...] There are a few striking examples.

The discovery of combination light scattering by Mandel'shtam and Landsberg, the priority over which was unjustly ascribed to Raman. As it is known, it caused N. Bohr to retire from his position as an expert at the commission for the Swedish Academy of Sciences in protest against the award of the Nobel Prize to Raman alone for this discovery.[1]

GARF, fond 9396, opis' 1, ed. khr. 251, pp. 94–95.

[1] It is a mistake. Bohr nominated Raman twice, in 1929 and 1930 but never recommended Landsberg and Mandel'shtam. Nevertheless, Tamm's emotional speech was not out of place. He was much obliged to Mandel'shtam, in the first place as his teacher. In the early 1920s, Tamm had been recommended to Mandel'shtam by his close relative A. G. Gurvich and moved to Odessa to be on the staff at the newly established Polytechnical Institute. Since then he became "the closest pupil, friend, and colleague of Mandel'shtam [...] lodged with his wife in the house of Mandel'shtam's family, where they occupied the vacant room of Mandel'shtam's brother-in-law, who was arrested by cheka."*

* E. L. Feinberg. *Rodonachal'nik (o Leonide Isaakoviche Mandel'shtame)* (The Forefather (about Leonid Isaakovich Mandel'shtam)). *Uspecki Fizicheskikh Nauk*, 172, No. 1, 2002, p. 98.

Document 2.4

Extract from the speech by A. A. Andronov, member of the USSR Academy of Sciences, at the All-Union workshop of the heads of physical departments of universities and higher education institutions on 2 March 1949.

[...] When D. D. Ivanenko [...] said that the first paper by Mandel'shtam and Landsberg was published in *Naturwissenschaften*, N. S. Akulov remarked that it contained "reference to Raman". When I replied that this strange observation is in conflict with the known fact (here in Soviet Russian and partly abroad) that the discovery was made before Raman, Akulov uttered a strange remark: "have a look and check up."[1]

GARF, fond 9396, opis' 1, ed. khr. 251, l. 19.

[1] Discussions on the rightfulness of the award of the 1930 Nobel Prize by the Swedish Academy of Sciences to Raman alone are still underway in Russia after three quarters of a century (see, for instance, the newspaper *Poisk* of 31 January 2003, p. 11, and *Vestnik RAN*, No. 3, 203, pp. 215–219.

Document 2.5

Extract from the diary of A. M. Kollontai, envoy of the USSR in Stockholm (7–21 December 1931).

10. XII. I attended the official ceremony of awarding the Nobel Prizes in science[1] and *belles lettres*. Had a meeting with Truger [...] and a long talk with the Spanish Envoy Pedro Garcia Konde [...] When the German envoy joined (Germans received both prizes in science), Konde said that Gorky is the primary candidate for the prize in literature as the world's most renowned writer but surprisingly had not yet received it.

AVP RF fond 0140, opis' 15, papka 114, ed. khr. 265 (1931), l. 176.

[1] C. Bosch and F. Bergius in chemistry and O. Warburg in physiology and medicine.

Document 2.6

Extract from the letter from A. M. Kollontai, envoy of the USSR in Sweden, to B. S. Stomonyakov, member of the Advisory Board for the People's Commissariat of Foreign Affairs, 15 November 1932, No. 133s.

[...] These weeks in Stockholm witnessed extensive *rapprochement* between Sweden and England. The commission carrying out negotiations on updating the agreement[1] or its substitution by a new one left for England two weeks ago [...] The Nobel Prize in literature for this year will be given to Galsworthy[2] (England). There were talks about Gorky.[3] Hostile circles[4] promoted Merezhkovsky. I had to have a few talks on this matter and pointed out through the agency of certain friendly professors that this candidacy is unacceptable for us as it is "politically indecent" and that the SD cabinet[5] should understand it.[6]

Many literary and public figures here protest in the press against the award of the Nobel Prize to Galsworthy stating that Gorky deserves it more than the English writer.[7]

However, anglophile sentiments prevail [...] There is little doubt that the Swedish bourgeois press welcomes the *rapprochement* with England and tries to impose this attitude on public opinion.[8]

AVP RF fond 0140, opis' 16, papka 115, ed. khr. 280, l. 119.

[1] Trade agreement.
[2] The Russian spelling is obsolete.
[3] Canards. The candidacy of Gorky was not considered in 1932 (Table 2.3).
[4] "Hostile circles" promoting Merezhkovsky to the laureateship were non-existent; he had been nominated since 1930 by his sole true worshipper Professor P. Agrell of the Lund University.
[5] Social-democratic.
[6] In 1932, Merezhkovsky was nominated together with Bunin. Kollontai's informants had to tell her about Merezhkovsky's mate, whose candidacy was equally "politically indecent" in the eyes of Communist Party ideologists. It appears that the envoy had incomplete information; anyway, the award of the Nobel Prize to Bunin in the next year was a great surprise for Kollontai (see doc. 2.7).
[7] Kollontai refers to the telegram sent to Gorky by a group of young writers after the announcement of the Swedish Academy's decision to award the 1932 Nobel Prize to Galsworthy. The message read as follows: "We, Swedish writers, who in vain urged the Swedish Academy* to recognize you as worthy of honour brought, as generally believed, by the Nobel Prize, are sending a message of ardent and respectful greetings to you, the great poet of Russia and the working class (*Dagens Nyheter*, 12 November 1932)." The message was signed by two future Nobel laureates in literature, Eyvind Johnson and Harry Martinson (1974), future members of the Swedish Academy Gunnar Elelef and Gunnar Lundkvist, and the most respected Swedish

prose-writers of the 20th century Vilhelm Moberg and Ivar Lo-Johansson,** etc.; a total of 21 persons.

* Their addresses were published in the press; Gorky was not nominated for the Nobel Prize in 1932.

** One of the main initiators of the collection of signatures.

[8] It is worthwhile to note that Galsworthy highly valued the works of Gorky. As organizer of the PEThe Nobel Committee in 1923, Galsworthy invited Gorky many times to attend the 1st International Congress of the new organization and regretted his refusal.*

* See: *Boris Pil'nyak: opyt segodnyshnego prochteniya* (Boris Pil'nyak: Today's View of His Writings. M.: Nasledie, 1995, pp. 137–138.

Document 2.7

To: B. S. Stomonyakov, member of the Advisory Board for the People's Commissariat of Foreign Affairs. Strictly confidential. 16 November 1933, No. 160s.

Dear Boris Spiridonovich,

[...] The awarding of the Nobel Prize to the white guard Bunin[1] is unrelated to the aforementioned feelings of alertness,[2] which I thoroughly checked. The award seems to be quite spontaneous and unexpected.[3] Anyway, the *shwedpra*[4] had no chance to prevent this move by the international committee.[5] The "elders" (the Swedish side is represented on the committee by professors and members of the Academy aged at least 80 years; the only "youth" among them is 65) first agreed on the candidacy of the Finnish writer Sillanpaa.[6] However, just a few days before the election, the relations between Swedish professors and scholars and the Finns became aggravated because of the persecution of the Swedish language in Finnish universities.[7] They nominated a certain antiquated Greek lyrical poet[8] but could not pass him by the majority of voices. Then, someone called the name of Bunin two days before the election. Thus, the problem was settled since no one "had anything against him".

The candidacy of Bunin first emerged in the press not long before the election. I had a private talk with the minister of education,[9] who seemed greatly surprised at the turn of events and told me that the committee was impossible to control because "the old grumblers" are very keen on guarding their independence against the government regardless of how inexpedient and out of place is their choice. I inquired into the matter and came to know that there were indeed cases when the prize was given despite the obvious disapproval of the government. I explained to top figures of the social-democratic party the irrelevance of awarding a white guard and could see that this decision by older members of the Academy finds no sympathy and is regarded as a manifestation of stupidity and unfriendliness towards our country on the part of a narrow circle of conservative professors. Newspapers, even *Allehadra*, are equally critical of the committee's decision and ask: "why not Gorky? [...]"[10] A. Kollontai

AVP RF, fond 0140, opis' 17, papka 116, ed. khr. 391 (1933), l. 105–106.

[1] The decision by the Swedish Academy was announced on 9 November 1933.

[2] The feeling of alertness in Swedish–Russian relations mentioned in Kollontai's letter was due to "the clamour raised around the Finnish espionage case", i.e. a new spy scandal in which the Soviet intelligence service proved to be involved that ended in the demonstrative cancelation of the previously agreed visit to Moscow by E. Klein, a journalist and the wife of the Minister of Social and Internal Affairs of Sweden. Kollontai notes: "At our crowded and boisterous reception on the 7th of November (attended by the entire diplomatic corps and many prominent figures), the Swedish Ministry of Foreign Affairs was represented by a few officials. This situation characterizes the feelings of the upper circles during the first days of 'the clamour'."*

* AVP RF, ibid, l. 105.

[3] It is not true (see, for instance, note 36 to the present chapter).

[4] Swedish government; such abbreviations were common in bureaucratic documents that were prepared for the Central Committee of AUCP(b) and the People's Commissariat of Foreign Affairs. For example, here is a phrase from the resolution of the Political Bureau of the Central Committee dated 10 November 1929, concerning the appointment of L. B. Kamenev, minister plenipotentiary of the USSR in the UK: "[...] *Sovpra* (the Soviet government) would like to know whether *velpra* (the British government) has any objection to the candidacy of Kamenev."*

* RGASPI, fond 17, opis' 162, ed. khr. 8, l. 6.

[5] It is not for the first time that Kollontai was under the delusion that the Swedish government could really influence the Nobel organizations and their decisions.

[6] This story is far from the real situation, even if regarded as rumors that might have been provoked by information leakage, always readily exaggerated by the press. However, in this case, Swedish newspapers did not mention the Finnish author at all. Only the provincial paper *Sudsvenska dagbladet* reported Sillanpaa* as a possible nominee of the Nobel Committee in its issue of 10 October, i.e. one month before the election of the laureate for 1933: "Far from trying to analyse his works in terms of their consistency with the committee's criteria, we do not hesitate to state that this Finnish prosaic author is the best writer of his country in this genre." The paper presented a critique of Sillanpaa's novel *Miehen tie* (A Man's Way) but said no word about him as a pretender for the Nobel Prize.**

* Sillanpaa was awarded the Nobel Prize in literature in 1939.

** Analysis of the publication in the Swedish press of that time was made on the author's request by Emi Vallenius, a professional translator of Swedish into Russian.

[7] One episode of the "language wars" between Sweden and Finland occurred during the festivities devoted to the 500th anniversary of the Swedish Parliament. "Finnish guests [...] distinguished themselves by delivering their salutatory addresses in

Finnish instead of Swedish so that nobody could understand them. It was regarded as a 'tactless demonstration' [...] The Swedes were angry, the Finns were laughed at for narrow-mindedness and tactlessness."*

* A. M. Kollontai. *Diplomaticheskie dnevniki 1922–1940.* T. 2 (Diplomatic Diaries 1922–1940. V. 2). M.: Academia, 2001, p. 286, entry of 1.6.1935.

[8] Kostas Palamas, Greek poet many times nominated for the Nobel Prize in literature.

[9] In the Russian text, this position is cited in an abridged form.

[10] In 1933, Agrell nominated Gorky together with Bunin. Probably, the nominator was influenced by the letter of protest published by a group of young Swedish writers discontented with the disapproval of Gorky's candidacy by the Swedish Academy (see note 7 to doc. 2.6).

Document 2.8

To: B. S. Stomonyakov, member of the Advisory Board for the People's Commissariat of Foreign Affairs. Strictly confidential. 5 December 1933, No. 175s.

[...] I already wrote you in my last letter how the Nobel Committee came to the decision to award the Nobel Prize in literature to Bunin. There is no doubt that old members of the Academy dislike us. Characteristically, however, Swedish public opinion[1] and even almost all bourgeois press disapprove of the choice of Bunin as deserving to represent Russian literature. Even *Allehandra* writes that it is strange to see the name of Bunin in the list of Nobel laureates as a representative of the country of Leo Tolstoy. Many newspapers are even more critical [...] Sandler[2] told me that he never read Bunin and does not know his works.

However reports in white-guard papers and a clamour set up by the white guards in Paris could not help finding a response in Sweden. There is a note in a newspaper that the French envoy Gossen will represent Bunin at the official prize-giving ceremony. *Poslednie novosti* (The Latest News)[3] writes that he is going to give a formal dinner in honour of Bunin and that this dinner is included in the official program of Nobel festivities. Certainly de Chassen, *press-attaché* for the French embassy of Russian origin, who left Russia before the revolution, is especially active as the chairman of the Foreign Journalist Union. He wrote foul books about our country and is an openly hostile figure. He is making a fuss of Bunin's award, using his influence in the Journalist Union[4] and taking advantage of his position as *press-attaché* for the French embassy.

That is why I met Sandler in order to talk outright to him.[5] To begin with, I pointed out how unimpressed by the awarding of the prize to Bunin and disappointed at the inability of the cabinet to prevent it. Then I asked him to influence the press so that Bunin's stay in Stockholm was not turned by hostile white guard elements into a political demonstration against the Soviet Union by representing Bunin as a "victim", etc. I also asked Sandler to inquire "privately" if the French embassy is really involved in the formal prize-giving ceremony. Sandler answered that the Swedish press can not be influenced by the Ministry of Foreign Affairs[6] but promised "to do whatever possible" and "privately" inform me about the festivities being prepared. He added that the award of the Nobel Prize to Bunin highly displeased the Swedish government. In a few days, Baron Hamilton, Sandler's deputy, asked me to call the ministry and said that no flags of the nations represented by the laureates will be hoisted during Nobel festivities, at variance with tradition. He added that "Bunin, as a sick man" may be unable to participate in the official ceremonies and, if so, the prize will be given to him in the hotel.

I came to know that the Ministry of Foreign Affiars tried to avert Bunin's visit but it proved impossible. Hamilton told me that the French embassy is not officially involved in the festivities but added that the French are going to celebrate the award of Bunin in a "private manner". It is certain that de Chassen organized an evening party for foreign journalists in honour of Bunin. Of course, this party will not be attended by our TASS (Telegraph Agency of the Soviet Union) correspondent comrade Zeifert and I shall not take part in the prize-giving ceremony.[7] Hamilton asked me: "Would you come if Bunin were ill this day?"

After my talk with Sandler, the Ministry evidently tried to smooth things over with respect to the tactless decision of the Academy.

The Nobel Committee in physics awarded the prize to a German who emigrated to England from Hitler's regime. He has the sympathies of the cabinet as having been persecuted under fascist rule [...][8]
A. Kollontai

AVP RF, fond 0140, opis' 17, papka 116, ed. khr. 301 (1933), l. 120–122.

[1] Swedish public opinion.
[2] Swedish Minister of Foreign Affairs.
[3] Daily Russian paper published in Paris and edited by the historian P. N. Milyukov (former leader of the constitutional-democratic party in pre-revolutionary Russia) until German troops entered Paris on 14 June 1940.
[4] Members of the Foreign Journalist Union.
[5] A year before, Kollontai made her first visit to the Minister of Foreign Affairs in the new social-democratic government and gave him the following uncomplimentary characteristic in her diary (15 October 1932): "Sandler is dry and cold. He looks like a dried moth." At the same time, he is "well-educated and clever," translated Marx's *Das Kapital* into Swedish.*

* A. M. Kollontai. *Diplomaticheskie dnevniki 1922–1940.* T. 2 (Diplomatic Diaries 1922–1940. V. 2). M.: Academia, 2001, p. 131.

[6] Interestingly, Oskar II, King of Sweden, had to explain the same to the newly accredited Russian ministry plenipotentiary Baron F. A. Budberg, who presented his credentials on 12 June 1905. Budberg wrote to St. Petersburg: "[...] The king added

that I can rely on the support and correctness of his government after I enter upon my duties. The only sphere that is unfortunately beyond his control is the press. He deeply regrets its attacks on our country and asks to see in them no more than they actually are, i.e. the manifestation of extreme views of radical parties which he has to take into consideration."*

* AVP RF, fond 133, opis' 470 (1905), ed. khr. 111, l. 97.

[7] Andrei Sedykh,* Bunin's secretary in Stockholm, explained the absence of Kollontai at the formal prize-giving ceremony with unwarranted self-confidence although it was just an expression of discontent frequently resorted to in diplomatic practice. He wrote: "Journalists greeted Bunin at a border station. They were interested in who would introduce Bunin to the king. Traditionally, it should be the ambassador of the laureate's homeland. But Kollontai was the Soviet ambassador. I uttered something very ungracious during her address. The morning Stockholm newspapers turned this episode into a sensation, and Kollontai stated that she would not attend the prize-giving ceremony."*

* Pseudonym of the writer Ya. M. Tsvibak who appears to have coined it from the name of the symbolist poet of the "Silver Age," Andrei Bely (B. N. Bugaev) by transforming its second half (*Bely* — Russian for "white" to *Sedykh* — Russian "grey-haired").

Sedykh Andrei. *Dalyokie, blizkie.* New York, 1962, pp. 195–196.

[8] E. Schrödinger, one of the creators of quantum mechanics.

Document 2.9

To: B. S. Stomonyakov, member of the Advisory Board for the People's Commissariat of Foreign Affairs. Strictly confidential. 28 December 1933, No. 178s.

[...] As regards Bunin, I would like to remind you (see my last message) that practically the next day after the award of the Nobel Prize to this writer was announced I visited Sandler and told him outright that this decision of the Nobel Committee will be taken in Moscow in an unfavourable way. Sandler looked at a loss and disappointed by the choice of Bunin but confirmed that the government could not influence the committee. In other words, I have already fulfilled your instructions. I can add that the Swedish government satisfied my request related to the Nobel festivities and in a way checked up the local press during Bunin's stay in Stockholm.[1] At any rate, there were no political escapades in our address and publications were reduced to literary panegyrics to Bunin. The semi-official newspaper *Socialdemokraten* reported the festivities in a somewhat ironical tone.

The protocol of the Nobel ceremonies included the usual dinners, tea-parties, reception of the laureates by the king, etc. Certainly, we did not participate. *Svenska Dagebladet* noticed the absence of a Soviet representative at the prize-awarding ceremony.

However, I can not help informing you that two embassies, the Czechoslovak (it is their right)[2] and the French, gave tea-parties to honour Bunin. This is unacceptably incorrect, the more so that I inquired at the Ministry of Foreign Affairs whether the French were going to organize a reception in honour of Bunin (as reported in my previous letter). Certainly, the French embassy is aware of my talks.

The most active white guard figures residing here in Sweden showed themselves up during this Nobel campaign. We took note of them. Usually, they are noticeable once a year when they organize a party in support of the Russian church assisted by their Swedish "patrons" having connections at the court. The French embassy still participates in this activity. Evidently the same de Chassen (Kheisin), French *press-attaché*, plays a role here; he promoted Bunin and introduced himself to our TASS correspondent as a "personal friend" of Bunin.

During this "Nobel week," I became even more convinced than ever that we made a mistake not sending our delegates to the Congress of Arts History[3] last autumn. Such meetings and our participation in them are highly useful for the strengthening of connections with the scientific world and enhancing our prestige in this area.

AVP RF, fond 1040, opis' 18, papka 117, ed. khr. 322, l. 3–5 (1934).

[1] It is no more than personal interpretation by Kollontai of the mood that prevailed among the Swedish press corps during the Nobel festivities of 1933. It is noteworthy that the Soviet diplomat describes the state of affairs in virtually the same words as King Oskar II 30 years ago when he spoke about the attacks of the Swedish press on the Russian policy in his conversation with the Russian envoy E. K. Byutsov in 1902. However, Kollontai depicts quite an opposite situation. The king said that the authors of anti-Russian publications made use of "the freedom of speech granted by *law which the government has no power to control*" [italicized by the author]*; see also note 6 to doc. 2.8.

* AVP RI, fond Kantselyarii MID for 1902, opis' 470, ed. khr. 94, l. 116.

[2] In 1933, Czechoslovakia had no diplomatic relations with the Soviet Union; hence there were no legal grounds to object to the participation of its official representatives accredited in Stockholm in the celebration in honor of Bunin. The benevolent attitude of Czechoslovakia toward the triumph of the Russian emigrant was in line with the support rendered by the country's president and government to Russian literary and scientific refugees in the 1920–1930s. The council of Russian professors was given an official status and had the right to invite scientists settled in other European states ensuring their material security. Students of Russian educational institutions in Czechoslovakia received scholarship, free text-books, medical care, clothes, etc. The government covered large related expenses from that part of the Russian gold reserves that had been captured after the seizure of Kazan by the Czechoslovakian corps in 1919 during the Civil War and transferred to Europe via Vladivostok.*

* V. P. Maslov. *Nazovite eto chudom. Pervaya volna russkoy nauchnoy emigratsii v Chekhoslovakii* (Call It a Miracle. The First Wave of Russian Scientific Emigration in Czechoslovakia); newspaper *Poisk*, No. 18, 2000, p. 12.

[3] Correspondence between the USSR mission in Stockholm and Moscow concerning the International Congress of Arts History in Stockholm (see doc. 2.11–2.14).

Document 2.10

Extract from the diary of Kollontai, USSR minister plenipotentiary in Sweden. 13 December 1933.

Diplomatic circles watch with great interest the *rapprochement* between Beck[1] (Poland) and Hitler. Heads of foreign missions pay reciprocal visits, make inquiries, and look for sources of information. The French ambassador and Khurban (Czechoslovakia) look especially agitated. The former met with a cold reception from me. But Khurban represents a friendly state — why did he not curb his personal feelings of sympathy with emigrants and gaie the reception and evening party in Bunin's honour? The Nobel Committee of *belles-lettres*[2] awarded the prize to Bunin as the best Russian writer. Why not to Gorky, why not to Sholokhov?[3] He is known and loved in Scandinavia.

Sandler behaved as if justifying himself for the prize to Bunin, who was living in Paris. "Who is Bunin?" he asked me. "I've never heard this name in literature circles." "Unfortunately," he added, "the cabinet has no power to control the choice of candidate. They would simply not listen to us. I hope you'll find words to explain it to your government."

Those in Moscow will certainly be disappointed that the Swedes chose to award an emigrant, but the fact is that the Nobel Committee[2] is constituted by arch-reactionary old grumblers, each over 80, and their choice many times caused annoyance in Sweden. I wonder who prompted the name of Bunin[4] to these archaic elders from the Nobel Committee.[2] His books are only recently on sale here. They say that Chossen (in fact, Kheisin), *press-attaché* for the French embassy and Russian emigrant, took much pains to persuade the Committee members.[5] Anyway, the Swedes might show more tact when such serious negotiations are underway [...][6]

A. M. Kollontai. *Diplomaticheskie dnevniki 1922–1940.* T. 2 (Diplomatic Diaries 1922–1940. V. 2). M.: Academia, 2001, pp. 198–199.

[1] Polish Minister of Foreign Affairs.

[2] Swedish Academy, the prize-awarding institution in the field of literature.

[3] Kollontai again speaks about Gorky and Sholokhov as candidates for the Nobel Prize in December 1934 (doc. 3.9).

[4] An answer to this puzzled question can be found in Table 2.3.

[5] Totally absurd conjecture displaying Kollontai's incompetence as regards the mechanism of consideration of the Nobel nominees and selection among them the final candidate to be awarded the prize for a given year.
[6] Negotiations between the Soviet Union and Sweden were initiated in November 1933 in a very complicated international situation caused by severe economic crisis that affected all developed countries. The Soviet Union sought a large credit worth 100 million krones for purchasing Swedish industrial equipment. The lengthy note in the diary dated 10 December (prize-giving day) concerns only those negotiations carried out in Stockholm between the official delegations of the two countries.* It contains not a single word about Bunin... Even the afore-cited lines make up only half of the entry dated 13 December. It may look strange bearing in mind the extensive correspondence between the envoy and her seniors in Moscow concerning the 1933 Nobel Prize that lasted as long as until 16 January 1934 (doc. 2.7–2.9, 3.1). The thing is that Kollontai understood perfectly well that all *demarches* from the Soviet side would be in vain, and all her official messages were no more than attempts to soothe the displeased ideologists in the Central Committee...
* Kollontai, pp. 195–197.

Document 2.11

Extract from the diary of Kollontai. 8 December 1922.

[...] Another interesting piece of news of these days is the decision by the Soviet government to give Nansen the authority to coordinate international aid to German and Austrian internees kept in detention in Siberia since the last war.[1] The Norwegians are very pleased and flattered by this decision. We thus show that Soviet people remember Nansen and appreciate his friendly services during the misfortune (drought) that befell the Republic in 1921. He helped us to combat famine.[2] We must use this fact as soon as possible to strengthen confidence in our country among the wider Norwegian public [...] Nansen's name is a banner for them. The fact that the Soviet republic entrusted him with this delicate mission (to protect the interests of citizens of the nations that quite recently were at war with us) can be in full measure used here in our favour [...][3]

A. M. Kollontai. *Diplomaticheskie dnevniki 1922–1940.* T. 1 (Diplomatic Diaries 1922–1940. V. 1). M.: Academia, 2001, pp. 73–74.

[1] Kollontai speaks about the repatriation of 500,000 German and Austrian soldiers from Russia. Nansen, acting on behalf of the League of Nations, concluded an agreement with the Soviet government under which the prisoners of war were brought to seaports, from which they were transported to their respective countries aboard trophy German ships, lent to Nansen by Great Britain.

[2] In summer 1921, Nansen was appointed Supreme Commissar on refugee affairs for the League of Nations. In this capacity, he managed to raise financial support from many governments and charities in different countries with which he saved millions of starving people in the Volga region, where a severe drought had completely destroyed crops. In addition, he granted part of his 1922 Nobel Peace Prize to the fund set up to ensure adequate relief actions.
[3] The name of Nansen many times occurs in the diary of Kollontai, who invariably writes of him with great enthusiasm. For example, entries dated 8 January 1923 and 12 April 1923 (pp. 84–85 and 109–111 respectively) describe her visits to the great humanist. Although he "devised the so-called Nansen passports used by runaway Russian white guards",* he "was and remains our friend", etc. Still, there is no word about Nansen's Nobel Prize.

> * When Nansen unexpectedly died seven years later (17 May 1930), Kollontai wrote that "our relations worsened by virtue of the so-called Nansen passports issued for the former Russian white guards. We kept aside from him and were about to refuse a visa to him not long ago. But he sought our friendship and understood us in his own way better [...] than many friends." (p. 469) See also note 39 to this chapter.

Document 2.12

Materials related to the nomination of Bunin for the Nobel Prize in 1923.

I

Romain Rolland to I. A. Bunin, 10 July 1922.

Let me tell you about the artistic pleasure that I experienced reading your collected short stories in a volume entitled *A Gentleman from San Francisco.*[1] Probably, we are separated by many different views or, to be precise, must be separated as a matter of course.[2] But I don't care. I can see only one thing: the inspiring beauty of your stories, the renovation of Russian art that is already so rich and made even richer, both in from and content, by your efforts [...] I am astonished at the vitality of human nature: it is trampled down, broken, soiled but always comes to life again and always in new forms [...]

Translated from French into Russian.

Novosti literatury, Berlin, 1922, book 1, p. 48.

[1] A volume of collected short stories entitled *A Gentleman from San Francisco* in the French language was issued by the Bossard publishing house in 1922.
[2] Rolland professed left political views.

II

Romain Rolland to M. A. Aldanov (?),[1] no later than the end of 1922.

Certainly, I admire Ivan Bunin. In my opinion, he is the greatest literary artist of our time. I am ready to support the candidacy of Mr. Bunin for the Nobel Prize but not his nomination jointly with Merezhkovsky.[2]

I must emphasize that I shall support Mr. Bunin only if I am sure that Gorky himself does not want to be nominated.[3] If Gorky were proposed, I should in the first place vote for him. I would be especially glad if two candidates, Bunin and Gorky, were simultaneously nominated: it would give indisputable evidence that politics are not involved.[4] I am aware, though, that Gorky has a profound respect for Bunin; he recently wrote to me about it — he venerates him as the most gifted of all modern Russian writers.[5]

Translated from French into Russian.

Voprosy literatury, No. 6, 1965, p. 255; published by A. Baboreko.

[1] A. Baboreko writes that Bunin familiarized one of his correspondents with Rolland's letter. However, it was certainly not addressed to him. The most probable addressee was M. A. Aldanov, who maintained correspondence with Rolland, trying to induce him to nominate Bunin for the Nobel Prize (see note 33 to this chapter).

[2] It appears that Rolland's correspondent proposed that Bunin be nominated together with Merezhkovsky.

[3] The question of Gorky's nomination for the Nobel Prize was never mentioned in the correspondence between Rolland and Gorky, comprising a total of 218 letters.* It means that Rolland strictly observed confidentiality required by the Nobel Statutes. It may be supposed that he was thinking of some other way to be in the know as regards Gorky's attitude towards nomination.

* M. Gorky and R. Rolland. *Perepiska (1916–1936). Arkhiv Gorkogo*, T. 15 (M. Gorky and R. Rolland. Correspondence. Gorky's Archive. V. 15). M.: Nasledie, 1995.

[4] That is exactly what Rolland did when he sent his nomination in 1923 (but added the emigrant poet Balmont to the names of Bunin and Gorky) (see Table 2.3). Later, however, the relations between Bunin and Balmont and their nominator became strained. On 12 January 1928, they published a letter in the Paris-based *Avenir* newspaper containing severe criticism of Rolland's message of greetings on the occasion of the 10th anniversary of the October Revolution. Rolland responded to his opponents in "Europe" on 15 February. His letter was reprinted in the Soviet Union under the title *Romain Rolland and Soviet Russia. Answer to Konstantin Balmont and Ivan Bunin.**

* *Ibid*, pp. 145 and 405–406 (footnote 5).

[5] Gorky did justice to Bunin's talent even when he criticized him for political statements. Telling that at one of the emigrant disputes "Ivan Bunin swore in the most obscene Russian language that I ever heard," he nevertheless calls him a "wonderful literary artist."*

*From the letter of 21 March 1925; ibid, p. 124.

III

I. A. Bunin to S. Agrell, 15 January 1923.

Dear Sigurd Vasilievich,[1]

M. F. Khandamirov[2] told me about your request to send you all my literary works and the critiques on them to enable you to introduce me to the Swedish Academy. I am deeply touched by the interest of the Academy in Russian literature and am cordially grateful to you for addressing me. I immediately fulfilled your wish... The packets are as follows: the registered letter to M. F. Khandamirov, three packages with books (for yourself) and one with critical articles (also for you)... I could not yet find the remaining books here in Paris and do not think I shall find them even if I do my utmost; but I am sure you have quite enough to be able to judge my writings.

The situation with the critical articles, both in Russian and foreign languages, leaves much to be desired. I had a full chest of them (Russian) but it was left in Russia. Here, I managed to borrow only a few and random ones from friends and acquaintances; most of them date back to the period when my works had been somewhat different from what they became after 1908–1909, i.e. less vivid and diverse; this made some of my critics describe me as a melancholic lyrical writer poetizing the life of landed gentry, an idyllic author...

I am yet quite a new author in Europe and abroad. Only two of my books have thus far been published in the French language, the third will be out in March. One was published in German (by Fischer)[3] and one in English. Two books will be published in America only in summer 1923; also in Japan [...]

Mosty, Munich, vyp. 15, 1970, pp. 390–391.

[1] Russian correspondents addressed S. Agrell using the russified patronymic.
[2] M. F. Khandamirov taught Russian language at the University of Lund.
[3] Fischer publishing house.

IV

I. A. Bunin to A. V. Tyrkova-Vilyams;[1] not later than 20 February 1923.

Dear Ariadna Vladimirovna, you are a friend, an old and true one, who will excuse me this somewhat indelicate letter (let it remain between you and me). I came to know that Romain Rolland had taken advantage of his position as the Nobel Prize winner and officially nominated my candidacy for the same award (in 1923) to the Swedish Academy of Sciences;[2] the Academy applied to Professor Agrell of Lund (Sweden) for comments on my works (Agrel[3] is a Slavonic philologist). However, this nomination needs the support of the European press because I am not yet known to the world; what is more important, my position in Russian literature is unknown.

May I ask you for an article in *The Times*, even if a very short one (under the pretext that my works are coming to be widely translated)? Two volumes will be soon out in America, and talks to the same effect are underway in Japan. The first volume of my collected works in the German language (printed by the Fischer publishing house) was sold out within three weeks. You are aware of the French. The Swedes also translate... Excuse me, please, and try to understand: I do not seek to win fame, my aim is to avoid hunger. The prize means salvation. I am paid a pittance for translations [...]

Minuvshee. Istoricheskiy al'manakh, No. 15 (The Past. Historical Almanac. No. 15). M.: Atheneum-Feniks, 1994, pp. 182–183.

[1] A. V. Tyrkova-Vilyams was a prominent representative of Russian liberal thought in British society. She was Bunin's regular correspondent since the early 1920s until the breakdown of their relations in summer 1925 (for an unknown reason).
[2] Most likely a slip of the pen. Bunin correctly names the prize-awarding institution in literature, the Swedish Academy, in the afore-cited letter to Agrel.***
[3] The name of Agrell is given in the Russian spelling.

V

I. A. Bunin to A. V. Tyrkova-Vilyams; 7 March 1923.

Dear friend, thank you very much indeed. I am afraid of mentioning R. Rolland and Agrell... it would make fellow authors boil with anger.[1] Also, it will become known to the English novelist Galsworthy who appears (from some secret talks) to be my rival for the 1923 prize (in fact, he is burdened with pounds and might just as well wait till 1924, unlike myself, burdened only with centimes).[2] A list of my works in foreign languages? Alas, it yet is too short [...][3]

ibid., 183.

[1] Bunin had no illusions about the attitude of his "fellow authors;" see doc. 2.15.
[2] Bunin had incorrect information about the nomination of Galsworthy for the 1923 prize. The name of the English writer was not mentioned in any of the 20 proposals submitted to the Nobel Committee and officially registered in its files. True, Galsworthy received the Nobel Prize before Bunin, but only in 1932.
[3] The attached list enumerated Bunin's works published in foreign languages.

Document 2.13

I. Bunin to S. Agrell, 20 November 1933.

Dear Sigurd Vasilievich, I apologize for writing to you only now; the latest developments knocked me off my feet and carried me away like a man who fell down the Imatra waterfall![1] I cordially thank you for all you have done for me during the last years.[2] I am leaving for Sweden around the 5th (of December) and will be truly happy to meet with you either in Stockholm or in Lund.

Yours sincerely Iv. Bunin

Mosty, Munich, vyp. 15, 1970, p. 395.

[1] The award of the Nobel Prize to Bunin was announced on 9 November; that is, the letter was written a week and a half after.
[2] Bunin really had more than one reason to thank his persistent nominator. The prize brought both honor and money to settle his difficult financial situation. He recollected the time after the award in a letter to his old friend in Moscow, the writer N. D. Teleshov, seven and a half years later (8 May 1941): "I was 'rich' but am as poor now as Job, so fate willed it."*

* *Literaturnoye nasledie*, T. 84, kn. 1. M.: *Nauka*, 1099, p. 623

Document 2.14

Extracts from Bunin's letters to M. I. Rostovtsev, a professor at Yale University.

I

March 1931.

[...] I resort to your help. At last, the Nobel Committee came to think about Russian literature [...] It is known that Professor Agrell nominated Merezhkovsky[1] and myself for the 1930 prize and that his opinion was strongly supported by Professor Anton Karlgren of Copenhagen, who read my *Life of Arseniev* in Russian and stated that it is "the most remarkable *belles-lettres* work written in the last decades." However, it all ended in a complete fiasco because members of the committee could not make the choice between the two of us... Now, I am again nominated for the 1931 prize... However, my well-wishers (i.e. nominators) need support. Almost all Russian writers rushed to propose themselves... I think that the members of the Nobel Committee must be at a loss now because they do not know modern Russian literature well enough, never read much of it, and have no idea about our relative statuses... In a word, it is necessary that some widely-known Russians [...] expressed (absolutely honestly and without the slightest anti-Soviet slant)[2] their thoughts and opinions. Dear Mikhail Ivanovich, please support me if you can and if you consider me most worthy of that!

II

November 1932.

My dear friends,* [...] It makes me feel awkward to apply to you, but I can't do otherwise. Please, help me once more — propose my candidacy again, write to Sweden.

From G. M. Bongrad-Levin: *Skifsky roman, ili zhizn' Mikhaila Ivanovicha Rostovtseva* (Scythian Novel, or Life of Mikhail Ivanovich Rostovtsev). In: *Rossiyskaya nauchnaya emigratsiya. Dvadtsat' portretov* (Russian Scientific Emigration. Twenty Portraits). M.: Editorial URSS, 2001, p. 312.

[1] This request is dictated by Bunin's apprehension that speculations based on his critical attitude towards the Soviet regime might, for political reasons, bias the decision of the members of the Nobel Committee as regards the candidacy of the Russian emigrant.

[2] Such caution, increasingly noticeable in subsequent years, found favourable response in the upper circles of the post-Stalinistic Soviet Union. Suffice it to say that members of the USSR embassy in France handed out 50,000 francs to Bunin's widow as an extraordinary grant from the Soviet government. At approximately that time (late 1955), the Department of Culture, Central Committee of CPSU, submitted to the CPSU Secretariat a recommendation to buy Bunin's archive "in order to disable the use of the writer's unpublished documents by reactionary circles for hostile purposes." The Secretariat was asked "to instruct the USSR Ministry of Foreign Affairs to carry out talks with V. N. Bunina, through the agency of its embassy in Paris, about the purchase of the entire I. A. Bunin's archive."*

* RGANI, fond 5, opis' 17, ed. khr. 535, l. 124.

Document 2.15

Extract from the letter of M. A. Aldanov to A. V. Amfiteatrov. 27 November 1933.

[...] Like yourself, I am very glad that Bunin received the Nobel Prize; he certainly deserves it. As for the next one, I thank you sincerely for the flattering prophecy.[1] But I assure you that none of the Russian emigrants will be given another such prize. I am really sorry for Merezhkovsky. He is old, and it is a sad blow to him although he endures it with dignity. To recall, when Sinkler Luis[2] was awarded the same Prize, his major contender Theodore Dreiser committed bodily assault on him.[3] In contrast, Merezhkovsly sent a cable, and Gippius published an article.[4] True, Dmitry Sergeevich did not attend yesterday's festivities at the Théatre des Champs Elysées, but Remezov and Balmont were absent too.[5]

Minuvshee..., 1997, No. 22, p. 569.

[1] A hint at Aldanov's right to pretend for the Nobel Prize. Bunin started to annually nominate Aldanov since 1947 and did it until he himself died in November 1953.
[2] Nobel laureate of 1930.
[3] In 1930, T. Dreiser was nominated by A. Osterling. In 1931 and further, Dreiser was no longer nominated, perhaps after his shabby behavior.
[4] The poetess Z. N. Gippius was D. S. Merezhkovsky's wife. She published an article in the Paris-based emigrant newspaper *Poslednie novosti* (The Latest News) on 16 November 1933 in which Bunin was characterized as "an unparalleled literary artist" and "the personification of the remaining most valuable part of Russia that can not be taken away from us." The article emphasized "his miraculous picturesqueness" and "great power of speech."*

* *Minuvshee...* p. 570.

[5] Shmelev also pretended to the Nobel prize (see Table 2.3), but he came and heartily welcomed the new laureate.* However, he expressed his irritation at the decision of the Swedish Academy in a letter to the same Amfiteatrov thee weeks later (19 December 1933). Recollecting the public ceremony of seeing off Bunin for Stockholm, he spoke biliously: "He is marching as if we are revering him [...]** He put on a short European frock-coat and keeps banqueting, etc. [...]*** I. A. Ilyin, a close friend and regular correspondent of Bunin, knew about Shmelev's jealousy and wrote in a letter of 14 November 1933, adapting himself to Shmelev's low spirits: "All these days, we celebrated the new Nobel award. I don't think that our views of and attitudes towards this event are much at variance."

Meanwhile, the "views and attitudes" of Shmelev should have been quite opposite if he had been governed by different moral principles. Shmelev experienced hard times in Germany where he first settled. Bunin invited him to move to France where living conditions for Russian emigrants were much more tolerable. On his arrival in the country in 1923, Shmelev lived at Bunin's villa in Grasse for two summer months. Bunin organized literary parties for Shmelev, introduced him to publishers, etc. However, it was all forgotten with time, effaced from memory. Even eight years later, Shmelev was convinced that he had wrongly "lost" the Nobel Prize and did not dissemble his irritation. He again wrote about his disappointment in a letter to his congenial friend, O. A. Bredius-Subbotina, dated 20 October 1941: "If I were not myself and showed more flexibility [...]believe me Olya, I would have been long ago a 'laureate' [...] Bunin had been promoted over 12 years by Nobel himself [Alfred Nobel's nephew Emmanuel], a Swedish Archbishop, certain members of the Nobel Committee, protégés of Jewish Masonry [...]"**** At the end of his life, Shmelev let himself sink as low as the obscure V. F. Ivanov of Kharbin in search of the simplest answers to the questions with which he was concerned (see this chapter and notes 45 and 46).

*I. S. Shmelev and O. A. Bredius-Subbotina. *Roman v pis'makh* (I. S. Shmelev and O. A. Bredius-Subbotina. A Novel in Letters). M.: Izd-vo "ROSSPEN," 2003, p. 694.

** These words in Russian are spelled with a hyphen. In a letter to O. A. Bredius-Subbotina, Shmelev explained his love for hyphens in the following way: "I for one like to space some words, that is to make intervals between letters. I think it important when I wish to draw the attention of the reader to a given word or emphasize its peculiar sense." Ibid, p. 542.
*** *Minuvshee,* No. 22, 1997, p. 574.
**** Ivan Ylyin and Ivan Bunin. *Perepiska dvukh Ivanov.* T. 2 (Ivan Ylyin and Ivan Bunin. Correspondence of Two Ivans. 2. 1). Isd-vo "Russkaya kniga," 2000, p. 411.
***** I. S. Shmelev and O. A. Bredius-Subbotin, p. 179.

Document 2.16

Extract from the letter of M. I. Tsvetaeva to A. A. Teskova.[1] 24 November 1933.

Nobel Prize. On the 26th,[2] I shall be seated on a stage and will celebrate Bunin's award. Evasion would be taken as a protest. I do not protest, I simply disagree, because Gorky is incomparably greater, more humane and original. Gorky is an epoch, and Bunin is its end. But such is politics, the king of Sweden can not decorate Communist Gorky with orders [...] By the way, the third candidate was Merezhkovsky, who is much worthier of the Nobel award than Bunin. If Gorky is an epoch and Bunin is its end, then Merezhkovsky is the epoch of the epoch's end and his influence both in Russia and abroad is immeasurably greater[3] than that of Bunin, who had never had any influence either here or there. It's a shame that *Poslednie novosti*[4] compared his style with the style of Tolstoy and put the former higher than the latter (as if the matter concerned a "stylus", i.e. a pen with which we write). Of course, one has to keep silence [...] I have not yet seen Bunin. I don't like him: he is a cold, cruel, presumptuous barin [...][5]

M. I. Tsvetaeva. *Sobranie sochineniy v 7 tomakh.* T. 6 (Collected works in 7 volumes. V. 6. M.: Ellis Lak, 1995, p. 407.

[1] Czech writer and translator of Tolstoy, Dostoevsky, Merezhkovsky, and other Russian authors; a regular correspondent of M. I. Tsvetaeva in the 1930s.
[2] Celebration in honor of Bunin was organized by Russian emigrants in Paris on 26 November 1933.
[3] An obvious overstatement in conflict with real popularity of Bunin and Merezhkovsky among the readers.
[4] Paris-based emigrant newspaper published in the Russian language.
[5] The personal attitudes of Tsvetaeva toward Bunin and Merezhkovsky were virtually identical. She wrote in the same letter: "Merezhkovsky and Gippius [his wife] are enraged [...] Everyone is afraid of them because both are... fierce. As fierce as ghosts." However, arbitrarily equating the two outstanding representatives of the Russian emigrant literature is at least incorrect. Here is an example of Bunin's

unreserved admiration for the talent of his young colleague absolutely alien to the character of Merazhkovsky. L. D. Lyubimov, a Russian "Parisien," recollected that, after having read the novel of V. V. Nabokov *The Defence of Luzhin* in the journal *Sovremennye zapiski* (October–December 1929), Bunin exclaimed: "This boy snatched out a pistol and killed all his elders on the spot with a single shot. Including myself [...]"*

* L. D. Lyubimov. *Na chuzhbine* (Abroad). *Novy mir*, No. 3, 1957, p. 167.

Document 2.17

Extract from the memorandum of N. A. Shcholokov, USSR minister of internal affairs, submitted to the CPSU Central Committee; dateless, but no later than early October 1971.

[...] The inflexible attitude towards big Russian writers in the past made them emigrate. These were Bunin, Andreev, Kuprin, etc. History showed that the way we treated these authors was wrong. We wanted from them what they could not give us on account of their class affiliation and class-consciousness. Today, our position with respect to these people is much better defined. But we must learn a lesson from past experiences. This lesson teaches that we should be more flexible, tolerant, and far-looking as far as our creative forces are concerned [...]

Kremlevsky samosud (sbornik dokumentov). Sostaviteli A. V. Korotkov i dr.(Kremlin's Self-Condemnation. Compiled by A. V. Korotkov etc.). M.: ROSSPEN, 1994. p. 171. The original document is stored in AP RF, fond 80.

Document 2.18

A. M. Kollontai to N. Ya. Raivid, head of the First Western Department, USSR People's Commissariat of Foreign Affairs. 13 May 1933.

This is at least my third application to the People's Commissariat of Foreign Affairs for information about a delegation to the International Congress of Arts History to be held this summer in Stockhom [...] To recall, I dispatched in due time detailed records of talks with the Crown Prince,[1] who wished to know about our participation and a coming delegation.[2]

I feel awkward when asked shortly before the congress who is coming, if anyone. Please convey the necessary data [...]

The letter carries a note in pencil: "To comrade M.[3] Make sure who is coming. Send a cable. N. Raivid. 21/V."

AVP RF, fond 0140, opis' 17, papka 116, ed. khr. 314, l. 3.

[1] Seemingly, a slip of the pen. Kollontai means Prince Eugene, the honorary president of the Organizing Committee of the 13th International Congress of Arts History (see also note 40 to this chapter).
[2] Unpublished.
[3] In all likelihood, S. M. Mirny, Raivid's assistance.

Document 2.19

From: S. M. Mirny, assistant to the head of the First Western Department, USSR People's Commissariat of Foreign Affairs.

To: Kollontai A. M., USSR envoy in Sweden
No. 15062, 27 August 1933.

Narkompros[1] informs us that the proposal of our participation in the conference on arts history has been submitted to a higher authority.[2] The issue has not been discussed yet. We shall inform you by telegraph as soon as the relevant decision is taken.

AVP RF, fond 0140, opis' 17, papka 116, ed. khr. 314, l. 7.

[1] People's Commissariat of Education.
[2] Top officials in the Central Committee of AUCP(b). Such allegorical formulation had been invented long before the Soviet bureaucracy came to power. It was used by court and government officials in pre-revolutionary Russia. The difference lies in terminology. The tsar and higher officialdom were collectively referred to as "spheres" in bureaucratic slang.

Document 2.20

From: A. M. Kollontai
To: A. S. Bubnov, people's commissar of education of RSFSR
August 1933 (after a copy sent to the People's Commissariat of Foreign Affairs; received 18 August 1933).

With reference to the International Congress of Arts History to be held in Stockholm on the 2nd of September. I earnestly ask you, first, to propose our delegates and immediately inform me by telegraph. Second, speed up the dispatch of displays for the exhibition to be run during the congress.

AVP RF, fond 0140, opis' 17, papka 116, ed. khr. 314, l. 9.

Document 2.21

From: Berezov, head of the First Western Department, USSR People's Commissariat of Foreign Affairs, and S. M. Mirny, assistant to the head.

To: Kollontai A. M., USSR envoy in Sweden
No. 15332, 7 September 1933

You are right that the issue of our participation in the Congress of Arts History created inconveniences. We phoned daily to *Narkompros* and pointed out to the necessity to take a decision promptly about our participation in higher authorities. *Narkompros* assured us that a positive decision is coming and even reported that such a decision was made on the 20th August, indicating that certain names of professors in the initial list were substituted by others. However, we did not send you a cable until we were sure the final list of delegates had been approved [...] On the 26th of August, we were officially informed of the negative decision of the authorities.[1] We have already sent you an official report about our refusal to participate in the congress.

AVP RF, fond 0140, opis' 17, papka 116, ed. khr. 314, l. 13.

[1] Reasons for such decisions were never given.

Chapter 3 • 1934–1944

Awakening Authorities

As mentioned in Chapter 2, the story of the Nobel award to Bunin was an instructive lesson taught by Aleksandra Kollontai, Soviet representative in Sweden. Mingling in the midst of cultural life of Stockholm, she came to understand what enormous worldwide fame and prestige the Nobel Prizes had acquired for over three decades since their institution. She could see that the award of the Nobel Prize objectively contributed to the glory of the nation under whose flag a laureate received honours.

Since 1934, Kollontai more insistently than ever tried to convince Moscow authorities that in the first third of the 20th century intellectuals had to live and act in a new reality. A new global center of attraction came into being, thus far disregarded by Soviet political leaders, to which they would sooner or later have to give most serious attention for the benefit of the country.

True, she once again turned back to the events around the award of the Nobel Prize to Bunin in her message to the People's Commissariat of Foreign Affairs dated 16 January 1934. She still believed that the Swedish Academy was about to award the prize to the Fnnish writer Frans Sillanpaa and only in the last moment decided to give it to the Russian emigrant (doc. 3.1). But it all was none other than an echo of the past storm. Now, Kollontai was concerned with other things. The main one was to arouse interest from Moscow in the Nobel movement and create favorable conditions for the full-scale participation of Soviet scientists and writers in the struggle for the Nobel Prizes with their foreign colleagues.

Convinced that the Soviet leadership pursued erroneous and short-sighted policy with respect to the Nobel movement, Kollontai launched in 1934 a wide-scale campaign designed to revive and strengthen cultural relations between the two countries and involve Soviet scientists and writers in the contest for Nobel Prizes. A good opportunity was provided by the organization of the VII Mendeleev Congress in Leningrad in September 1934 to commemorate the centenary of the great scientist.

In particular, she was concerned about the invitation of the Swedish chemist V. Palmer, member of the Nobel Committee in chemistry, favorably disposed toward the Soviet Union (doc. 3.2–3.4). She did not conceal "another aspect" of the invitation, that is to use Palmer's stay in Leningrad "to instil in him the thought that one prize (e.g. in chemistry, in connection with the Mendeleev Congress) should be awarded to a Soviet scientist" (doc. 3,4).[1]

This idea was certainly prompted by good intentions. At the same time, it demonstrates the total incompetence of the high-ranking Soviet diplomat. The wish to obtain the prize "in connection with the Mendeleev Congress" could be attributed to the deep-rooted Soviet tradition to time awarding orders, commissioning new facilities, overfulfilling plans, etc. to one or another red-letter day.[2] But the ambassador representing her country in Sweden had to know that the closing date for nominations for the 1934 Nobel Prize (if they really had to have anything to do with the Mendeleev Congress) was 31 January, i.e. it expired half a year before the letter was sent to Moscow (2 August).

This is not the sole example. In some letters (doc. 3.6, 3.9, 3.14, 3.19), Kollontai does not distinguish between the Academy of Sciences and the Swedish Academy, takes the four Nobel Committees in Stockholm for one awarding prizes in science and literature, calls the polar researcher O. Yu. Shmidt worthy of the Nobel Prize (for a few years up to 1937) as if being unaware that this award is not given for geographic discoveries, etc.

Even worse, the published archive documents show that Kollontai did not know basic provisions of the Nobel Statutes requiring strict confidentiality of nominations and forbidding to collectively discuss them or involve anybody from the outside, to say nothing of government bodies. One can imagine the feelings of the committee members with whom the official representative of the foreign country happened to discuss these issues...

Diplomats in the Soviet mission in Stockholm appear to have inherited the belief of their predecessors in the possibility to impose opinions of outside circles on the Nobel organizations. E. P. Rymko, *press-attaché* and thereafter councillor for the Soviet embassy in Stockholm in the 1950–1960s, wrote in connection with the award of the Nobel Prize to Mikhail Sholokhov in 1965: "This Nobel Prize has a long and complicated history. The issue dates back to the thirties of this century and has *not* until recently *yielded to solution*[3] [italicized by the author]. That is, he had as little doubt about the possibility to influence the Nobel organizations as Kollontai 65 years before (see also doc. 10.9).

For all that, undertakings of the Soviet representative in Stockholm that became especially active in 1934, after the award of the Nobel Prize to the Russian emigrant, were not altogether in vain. They objectively contributed to the involvement of the Soviet Union in the Nobel movement and the development of relations between Soviet scientists and the Nobel organizations. Signals from Stockholm initiated messages from the People's Commissariat of Foreign Affairs to the USSR Academy of Sciences and other government bodies concerned with scientific developments and sooner or later reached researchers.[4]

It is not the fault of Kollontai that her efforts brought no fruits and the number of nominations from the Soviet Union progressively decreased. The assassination of S. M. Kirov in 1934 was followed by the period of Great Terror. Unauthorized correspondence with foreign organizations became mortally dangerous (see note 1 to doc. 3.15).

True, some Soviet scientists still maintained contacts with the Nobel organizations, even if markedly reduced, in the first years of the Great Terror period. At least they could nominate their countrymen for the Nobel Prize in physiology and medicine. A total of five nominees were proposed to the Nobel Committee in the framework of 1934–1937 nomination campaigns (i.e. up to the closing date of 31 January 1937) (Table 3.1).

A total of 19 nominations of Soviet scientists were received by the Nobel committee in physiology and medicine including 16 from the Soviet Union and three from abroad. The nominators recommended the worldwide-known investigations of A. G. Gurvich into mitogenic rays (see Chapter 2) and N. N. Anichkov's research on experimental pathology. All three foreign nominators were German scientists. The candidacy of Anichkov was proposed by Professor F. Hoff (University of Wurzburg) and the candidacy of Gurvich by Professor T. K. Studnicka (Brunn) and Professor A. T. U. Bethe (Frankfurt-am-Main, for the second time, see Chapter 2). The fourth nomination of Gurvich reached Stockholm in 1935; he was nominated

Table 3.1. Distribution of nominees of Russian descent by categories of Nobel Prizes in sciences for the period from 1934 to 1944

Nominees and their residence at the time of nomination	Year of nomination	Number of nominators	
		Total	Including USSR
Physics			
A.G.Gurvich (1874–1954), Moscow	1935	1	—
G.A.Gamow (1904–1968), Washington	1943	1	—
Chemistry			
V.N.Ipatiev (1867–1952), Chicago	1938	1	—
-"-	1941	1	—
-"-	1942	1	—
Physiology and medicine			
N.N.Anichkov (1885–1964), Leningrad	1937	1	—
A.G.Gurvich (1874–1954), Moscow	1934	3	2
-"-	1938	1	—
P.P.Lazarev (1878–1942), Moscow	1934	1	1
L.A.Orbeli (1882–1958), Leningrad	1934	3	3
A.D.Speransky (1888–1961), Moscow	1934	2	2
-"-	1936	6	6
E.S.London (1869–1939), Leningrad	1939	2	2

for the Nobel Prize in physics by Professor L. A. Zender of the Basel University, Switzerland. Zender was a known expert in invisible radiation and for a long time maintained correspondence with Roentgen on this issue; the second edition of his fundamental monograph was published in Basel in 1937.

These were the last nominations for A. G. Gurvich. V. I. Vernadsky, who personally knew and highly regarded him as "a big scientist and man of moral honesty" wrote in his diary on 7 January 1938 that controlled experiments carried out in the USA in 1937 "finally compromised his life-long studies on mitogenesis and mitogenic rays" and described this pessimistic conclusion as the tragedy of the scientist...[4a].

The name of L. A. Orbeli is first found in the list of nominees for the Nobel Prize in physiology and medicine in 1934. His candidacy was proposed by three physiologists from Leningrad: the Nobel laureate I. P. Pavlov and professors of medical institutes A. A. Likhachev and F. E. Tur.[5] Next year, Orbeli was nominated by physiologist G. V. Fol'bort, professor of the Medical Institute in Khar'kov.

I. P. Pavlov recommended to the Nobel Committee Orbeli's studies on the adaptive and trophic functions of the sympathetic nervous system. Two years before (22 December 1931), Pavlov informed the Presidium of the USSR Academy of Sciences about his younger co-worker's successful investigations: "A basic problem is about to be solved that has for half a century escaped interpretation by physiologists."[6]

The nomination of Orbeli signed by V. P. Volgin, permanent secretary of the USSR Academy of Sciences, and sent to the Nobel Committee on 27 January 1935 (doc. 3.12) was declined as should have been expected. It did not meet the requirements of the Nobel Statutes, being signed by an Academy official instead of the nominator specially invited by the committee; moreover, this official was a historian and had no right to nominate candidates for the prize in physiology and medicine.[7]

Pavlov's nomination of 1934 included, besides Orbeli, his other disciple, A. D. Speransky, an outstanding pathologist. In the same year, Speransky was nominated by S. P. Fedorov, a famous surgeon and professor of the Military Medical Academy in Leningrad. In 1936, Speransky had six nominations submitted by professors of Kazan University V. S. Gruzdev, A. D. Gusev, Z. I. Malkin, P. V. Manenkov, V. V. Miloslavsky, and A. G. Teregulov. However, these nominations, unlike the two presented in 1934 and duly considered by the Nobel Committee, were not included in the so-called short list. The expertise of the 1934 nomination did not bring a favorable result. Three more nominations of Speransky (all from Kazan) were received in Stockholm in 1936 and one (from Moscow) in 1942. However, all these nominations were entered in the registry of the Nobel Committee with a note "disproved for nomination" without specification of concrete requirements disregarded by the authors.

There were no Soviet scientists among the nominees for the Nobel Prizes in physics and chemistry. V. N. Ipatiev, nominated for the prize in chemistry in 1938, 1941, and 1942, did not return to the Soviet Union from a business trip to the USA in 1930 and was later denaturalized by the Soviet authorities.[8] In 1997, Yu. I. Solovyev, working in the Archives of the Russian Academy of Sciences, published an article under the title "Why did not academician V. N. Ipatiev become a Nobel laureate?"[9] This publication calls for a more detailed consideration of Ipatiev's nomination.

V. N. Ipatiev was actually an outstanding scientist known for sensational success in the application of high pressure in organic chemistry. In 1903, he devised an apparatus, a prototype of modern autoclaves, for the study of catalytic reactions under high pressure and temperature. In 1908, he summarized his results and became widely known in scientific circles. In 1915, P. I. Valden, B. B. Golitsin, and N. S. Kurnakov, members of the Imperial Academy of Sciences, proposed him for election to the Academy. They compared his works with the studies of the French chemist P. Sabatier and concluded that they are "more diverse than Sabatier's research for which he was awarded the Nobel Prize in 1912."[10]

Specifically, P. I. Valden wrote: if Sabatier received the prize for "a single catalytic reaction, [...] the works of Ipatiev are sure to deserve the same award because he more widely used catalysts in various reactions [...] and introduced an absolutely new high pressure technique that allowed him to hydrogenate such compounds for which the method of Sabatier was inapplicable."[11]

Solovyev refers to the award of the 1931 Nobel Prize in chemistry to German researchers C. Bosch and F. Bergius "in recognition of their contributions to the invention and development of chemical high pressure methods" and arrives at the conclusion that "the Nobel Committee did not wish to give the prize to the Russian scientist [...] because he actively participated in the development of the Russian economy in 1918–1927. The decision was influenced by political preconception..."[12] However, materials presented below do not justify such categorical statement. To begin with, here are some facts about the nomination of Ipatiev for election to the Imperial Academy of Sciences in 1915.

Two of the three scientists who proposed the candidacy of Ipatiev to the Academy were nominators of the Nobel Committee in chemistry (Valden and Golitsin). Valden nominated his candidates twice (in 1904 and 1911) and Golitsin thrice (in 1911, 1913, and 1914), but neither mentioned Ipatiev.

At the same time, P. Sabatier and V. Grignard, who shared the Nobel Prize in 1912, were many times nominated by their colleagues, largely their compatriots. As shown in Table 3.2, Sabatier and Grignard were nominated four and three times respectively. Every year, each candidate had two nominations at the most; that is, not many.

In other words, Ipatiev would have had a good chance to compete with them had he been nominated by Valden and Golitsin who, however, proposed other candidates in 1911. Why did they not think of the most important results obtained by their colleague, of which they were perfectly aware? Certainly, the Nobel Committee had no reason to consider Ipatiev in the absence of written nominations.

A similar situation took place in 1931 when the German scientists C. Bosch and F. Bergius were awarded the Nobel Prize. Bosch was nominated six times since 1916, and the number of his nominators totalled 20 including A. Einstein. Bergius was nominated twice, each year by one nominator. No one nominated Ipatiev...

Table 3.2. Nominations of the Nobel laureates in chemistry P. Sabatier and F. Grignard

P. Sabatier		F. Grignard	
Year of nomination	**Number of nomination**	**Year of nomination**	**Number of nomination**
1907	2	1909	1
1909	2	1910	1
1911	1	1912	2
1912	2		

In addition, Ipatiev himself made a mistake. In his memoirs, he mentions a report that he delivered at the International Congress of Industrial Chemistry held in Strasburg in July 1928, i.e. three years before the Nobel Prize was awarded to Bergius. In that report, the scientist presented a detailed analysis of his earlier high pressure hydrogenation studies. In conclusion, he said that "all Bergius' patents (1911) are based on the results of my works done in 1903–1904, and my method developed for various organic compounds was applied to hydrogenation of resins and hydrocarbons." The Society of Industrial Chemistry, organizer of the Congress, recognized Ipatiev's scientific achievements, and the representative of the French government in the minister rank solemnly handed over to him the Berthelot medal, the highest award of the Society.[13]

What then was the reason for which the Nobel Committee refused the award to the Russian scientist in 1931? The trouble was he did not patent his results. Ipatiev repeated the mistake made by many Russian inventors including A. S. Popov (see Chapter 1, note 31), who did not care to patent their discoveries in due time. The Nobel Committee would hardly have disregarded his contributions if he had a patent. Even in the absence of nominations from the outside, someone from the committee could use his right to recommend him shortly before the deadline for nominations, as was the case with Carl Braun in 1909 (see Chapter 1) and some other researchers.

This inference is confirmed by the data borrowed from nomination lists of the Nobel Committee in chemistry. In the period from 1916 to 1929, Bosch and Bergius were nominated only by their compatriots. No nomination came from Germany in 1930, and V. Palmer, chairman of the Nobel Committee, was their sole nominator. In 1931, the candidacy of Bosch was supported by nine Swedish professors in the absence of nominations from Germany. One of the nine, E. Hegglund (member of the committee) simultaneously nominated Bergius. It may be thought that he would have not disregarded Ipatiev had the Russian chemist had a legally valid patent...

The name of Ipatiev first appeared in the minutes of the Nobel Committee in 1938 as professor of the Northwestern University, Chicago. He was nominated by the Dutch chemist and specialist in chemical catalysis H. I. Waterman of the Delft University. In 1941, Ipatiev was nominated by G. Dupont of the Sorbonne and in

1942 by Professor E. K. Rideal of Cambridge University.[14] But his time had passed. He could really have become the first Russian Nobel laureate in chemistry but failed for the mere neglect of fellow colleagues in his own country.

The above materials unambiguously demonstrate the inadequacy of the conclusion about "political bias" of the Nobel Committee in chemistry made by the author of the aforecited article. It is not only a far-fetched conclusion, it is an unconscientious one. The attitude of Ipatiev toward the regime established in Russia is as well-known as his sound pragmatism in the issues related to economic developments in the country. Suffice it to recall his words at the meeting of the Physico-Mathematical Division of the Academy of Sciences held for the first time after the October revolution. He said, in response to the call of some members of the Academy, including I. P. Pavlov to protest against usurpation of power by Bolsheviks: "We intellectuals make up a thin stratum in the midst of masses; having no foothold, we should not at present come out against anything [...] The attitude of each of us to the Bolshevist government is the 'holy of holies' [...] Many of those present were not in sympathy with the monarchic rule, but it did not prevent us from honestly discharging our duties to the country and productively perform our functions."[15]

If members of the Nobel Committee had been guided by their political views (as supposed by Solovyev), they would have made their decision in favor of the potential opponent of the dictatorial regime. However, they acted in accordance with quite different criteria having nothing in common with the incompetent assertion of Solovyev.[16]

G. A. Gamow was another emigrant from the Soviet Union nominated for the Nobel Prize in the period under consideration. Gamow graduated from the University of Leningrad where he stood out against the mass of his fellow students. His close friends included a man of genius L. D. Landau, and D. D. Ivanenko, who was the first to postulate that atomic nuclei consist of protons and neutrons. This trio was known in the university as the "three musketeers." Having received his university diploma, Gamow was sent abroad for advanced training. He worked in Goettingen with Max Born, in Copenhagen with Niels Bohr, in Cambridge with Rutherford, and in Leiden with Paul Erenfest. A result of Gamow's work with these eminent physicists was his quantum hypothesis of alpha decay. This idea made the young physicist widely known in scientific circles. In 1931, when he was only 27, Gamow was elected a corresponding member of the USSR Academy of Sciences by the recommendation of V. I. Vernadsky and V. G. Khlopin.

In September 1933, Gamow participated as a member of the Soviet delegation in the Solvay Congress held in Brussels from which he did not return to his country.[17] In a year, he moved to the United States and very soon became a highly reputed theoretical physicist and organizer of popular annual physical conferences in Washington (the first was held in 1935).

In 1943, Gamow was nominated for the Nobel Prize by Professor Hannes Alfven of the Royal Technological Institute, Stockholm, the future winner of the same prize for basic research in plasma physics (1970). This nomination contained, besides

Table 3.3. Distribution of Soviet nominators and their foreign nominees by categories of Nobel Prizes in sciences for the period from 1934 to 1944 (nominees awarded the prize are italicized; figures in brackets indicate the year of the award)

Nominators	Year of nomination	Nominees
Physics		
A.K.Valter (1905–1965), Khar'kov	1936	*J.Chadwick (1935)*, Great Britain
-"-	1936	*E.Fermi (1938)*, Italy
B.V.Ilyin (1888–1964), Moscow	1941	*O.Hahn (1944)*, Germany
Chemistry		
A.R.Kizel (1982–1942), Moscow	1935	*P.Karrer (1937)*, Switzerland
S.Z.Roginsky (1900–1942), Leningrad	1937	F.London, France
A.E.Favorsky (1860–1945), Leningrad	1937	*L.Ruzicka (1939)*, Switzerland
Physiology and medicine		
I.P.Pavlov (1849–1936), Leningrad	1935	G.Kato, Japan
-"-	1935	K. Kure, Japan

Gamow, the name of Donald Kerst, professor of nuclear physics at the University of Urbana, Illinois, and the inventor of betatron, a prototype of accelerators of charged heavy particles of the 1940–1950s.

In the same period, Soviet scientists nominated their foreign colleagues for the Nobel Prizes. There were six such nominators (Table 3.3) including five that submitted their proposals to Stockholm prior to the onset of Stalin's Great Terror, i.e. between 1934 and January 1937. The sixth was B. V. Ilyin, professor of the Moscow University, whose strange nomination was sent to Stockholm half a year before the Nazi troops invaded the Soviet Union.

The best known among the five nominators was certainly I. P. Pavlov, Nobel winner in physiology and medicine of 1904, whose proposals look somewhat surprising. To begin with, his first nomination of 1909 (see Table 1.3) and the next one (1934), when he proposed the candidacy of his disciples L. A. Orbeli and A. D. Speransky, are separated by an unbelievably long period of a quarter of a century! Highly unusual for a Nobel laureate...

Pavlov's last nomination submitted in 1935, i.e. one year before his death, is even stranger. He recommended two professors from Tokyo, Geniti Kato and Ken Kure. These Japanese scientists were simultaneously nominated by Pavlov's colleagues from research medical institutions of Tokyo. The number of nominations was really astonishing; 23 for Kato and 20 for Ken Kure. Each of the two long lists of nominators in the registry of the Nobel Committee is closed by the name of the Soviet scientist.[18] The fact of the nomination of Japanese colleagues by the elderly Nobel laureate gives food for analysis to connoisseurs of Pavlov's heritage; Ivan Petrovich has never during his lifetime been disposed to participate in collective actions.[18a]

It may be supposed that such unusual symbiosis with Japanese nominators originated at the International Physiological Congress held in Rome in 1932; the two nominees were among its participants. The works of Geniti Kato on microphysiology of nerves and muscles lay within the scope of Pavlov's interests and might attract his attention. Maybe some of the members of the Japanese delegation asked the patriarch to support researchers from Tokyo...

Even such mass support did not help them to win the prize. Three other Soviet nominators in physics and chemistry (A. K. Valter of Khar'kov, A. R. Kizel of Moscow University, and A. E. Favorsky of Leningrad) showed a keener insight; all their nominees were awarded Nobel Prizes.

Professor A. K. Valter of the Ukrainian Physico-Technical Institute left a marked trace in the history of Soviet physics as one of the first scientists in the country (jointly with his co-workers A. I. Leipunsky, K. D. Sidel'nikov, and G. D. Latyshev) who conducted nuclear fission by artificially accelerated protons. Well aware of current research in nuclear physics of that time, he recommended to the Nobel Committee in physics the English scientist James Chadwick of Liverpool and the famous Italian Enrico Fermi. True, there was some confusion about Chadwick's nomination because Valter hastened to send it to Stockholm. His letter was sent on 5 November 1935 and received on 23 November 1935, to be registered in the committee's nomination list for 1936. Unable to duly receive adequate information about the activities of the Nobel organizations via the Soviet press, Valter did not know that Chadwick had already been awarded the Nobel Prize "for the discovery of the neutron" and was to receive it from the Swedish king in a month after the nomination was dispatched from Khar'kov. Fermi received the prize two years later, in 1938, "for his demonstration of the existence of new radioactive elements produced by neutron irradiation, and for his related discovery of nuclear reactions brought about by slow neutrons."

Professor A. R. Kizel of Moscow University organized the Department of Plant Biochemistry, the first in this country, and jointly with A. N. Byelozersky (discoverer of DNA in plant tissues) nominated for the 1935 Nobel Prize in chemistry, Paul Karrer of Switzerland for his important investigations into vegetable pigments and elucidation of the nature of vitamins A and B2. Karrer was born in 1889 in Moscow where his father was a dentist and was moved to his parents' motherland at the age of three years. He was awarded the Nobel Prize in 1937.

Kizel himself did not have such a good luck. Professor N. D. Zelinsky, the leading chemist in the Soviet Union, recommended him for the membership in the USSR Academy of Sciences a year after Karrer received his prize in 1938. However, he was not elected.[18b]

Leopold Ruzicka was born in Vukovar on the Danube River (Austro-Hungary, then Yugoslavia) in 1887. He was known worldwide for the synthesis of male sex hormones and deciphering the structure of their main representative, testosterone. A. E. Favorsky, member of the USSR Academy of Sciences, nominated him for the Nobel Prize in chemistry in 1937. Favorsky was one of the founders of the Soviet school of organic chemistry and formulated theoretical principles for the commercial production of synthetic rubber in the USSR. He was able to guess who of the contemporary researchers were most likely to receive the Nobel Prize for important contributions to organic chemistry.

Ruzicka was awarded in 1939. The award ceremony was cancelled since the Second World War had already broken out. The Gold Medal and the diploma were handed over to Ruzicka in Zurich by the Swedish ambassador to Switzerland in 1940. His Nobel lecture was delivered in Stockholm only in 1945.

During the war, Ruzicka did not forget his motherland and regularly supported the resistance movement in Yugoslavia both via Swiss charities and by establishing the Swiss–Yugoslavian Society for assistance to war victims; the society continued to function in the post-war period.

In 1937, the future corresponding member of the USSR Academy of Sciences S. Z. Roginsky, Leningrad-based specialist in the theory of catalysis, sent to Stockholm his nomination of physico-chemist Fritz London, who had emigrated from Nazi Germany to France. London was one of the creators of the theory of valent bonds developed by L. Pauling in the thirties of the 20th century. In 1934, he suggested a theory explaining the behavior of superconductors in weak magnetic fields. In 1950, these postulates were generalized and extended to strong magnetic fields by Soviet physicists V. L. Ginzburg and L. D. Landau; the former was awarded the Nobel Prize for these studies in 2003 (see Chapter 12). In 1950, London published the first volume of his fundamental monograph *Superfluidity*. The second volume was out only after four years, in 1954 when the gifted scientist died before he was 55.

London did not receive the Nobel Prize. After Roginsky's proposal, he was nominated only twice, viz. at the height of the World War II (1942), when the prizes were not awarded at all, and in the fatal year of 1954. His nominators were authoritative American physico-chemist Kazsmir Fajans of Ann-Arbor and Professor Robert Marshak of Rochester University, USA, respectively. The latter nominated London together with L. D. Landau for his pioneering theory of quantum nature of superfluid helium developed in 1941.

It may be supposed that members of the Nobel Committee in Stockholm waited for the second volume of London's book to appear but in vain. Nor did the author live long enough to see it; he died in January that year...

It is worthwhile to note the far-looking nomination of the Soviet physicist S. Z. Roginsky, the first nominator of Fritz London, who was able as early as 1937 to see a serious pretender to the prestigious award in the then young scientist.

January 1937 put an end to the contact between Soviet nominators and the Nobel Committees that were broken for the long 15 years until Stalin's death. The merciless totalitarian regime prevailed over the vast country terrorizing its population by permanent threat of arrests and physical annihilation at the whim of the despot and his satraps.

In this situation, B. V. Ilyin, professor of Moscow University, proposed the German nuclear physicist Otto Hahn as a candidate for the 1941 Nobel Prize (see Table 3.3) in recognition of his "work on the fission of uranium and thorium nuclei and the formation of barium radioactive isotopes during their neutron irradiation".[19] The nomination was sent to the Nobel Committee in physics half a year before Hitler's Germany attacked the Soviet Union and was registered in Stockholm on 28 December 1940.

It was not the first nomination of Hahn for the Nobel Prize. The earliest one dated back to 1914. By 1941, he had been nominated 20 times by 13 scientists from different countries. Until 1939, the nominators focused on Hahn's studies on weak radioactivity of such chemical elements as potassium and rubidium, the discovery in 1918 of the long-living protactinium isotope 231Pa having a half-life of 32,400 years, and the description of an absolutely new physical phenomenon of nuclear isomerism in natural radioactive elements, etc.

The discovery of uranium and thorium nuclear fission was announced in 1938 and began to be reported to the Nobel Committee from 1939. The first nominator was the known specialist in nuclear chemistry Theodor Svedberg, Nobel Prize winner in chemistry in 1926, who subsequently headed the Gustav Werner Institute of Nuclear Chemistry at the University of Uppsala. In 1940, Hahn was nominated again by the Nobel laureates in physics James Franck (1926) and Arthur Compton (1927), supported by Professor Dirk Coster (the Netherlands), the discoverer of hafnium. James Franck repeated his nomination of Hahn in 1941; other nominators were Arne Westgren, secretary of the Nobel Committee in physics and chemistry for many years, Professor A. Goudsmit of Ann Arbor, Michigan, and Professor B. V. Ilyin of Moscow.

Hahn had a total of nine nominations from six persons during the period from 1942 to 1945.[20] For foreign scientists invited by the Nobel Committee, nomination for Nobel Prizes was a routine. On the contrary, contacts of the Soviet professor with the Nobel organizations were far from that, the more so that military actions were already in full swing in Europe. There was one more point worth mentioning: all the above nominators of Hahn were well-known figures in the field of atomic physics and chemistry. Ilyin was outside this circle.

It appears from his CV that the main sphere of his scientific interests was intermolecular (van der Waals) interactions and the nature of adsorption forces.[21] Radioactivity lying outside this sphere could probably have interested Ilyin as a layman. But it was hardly enough to induce him to nominate a candidate from quite an alien field of science for the most prestigious award. Moreover, the eight-page text of the nomination was very formal and revealed the author's incompetence. It contained nothing like true scientific substantiation of the proposal.[19] References to the published works of the nominee included two marginal papers in *Naturwissenschaften* for 1939 (Bd. 27, s. 11 und 89) but not his article of 1938 (in co-authorship with Fritz Strassman) describing the very first observations of uranium nuclear fission with the release of energy under the effect of neutron bombardment.

There may be only one more or less acceptable explanation for the appearance of such a strange nomination. It appears that Professor Ilyin was chosen by a certain organization[22] as an instrument of nomination for reasons of having political rather than scientific implications. The real incentive was to once again demonstrate faithfulness to the Nazi ally, and the friendship and non-aggression pact concluded in 1939. A continuous sequence of trainloads of corn and ore bound for Germany served the same purpose. They went there until the fatal night of 22 June 1941; the last one crossed the border station of Brest past 2 a.m. 22 June, i.e. 45 minutes before the Nazi troops invaded the Soviet Union. The same is true of the shameful modification of standard schoolbooks of history initiated by the Central Committee of AUCP(b) to please the newly acquired friends in Berlin (doc. 3.35).

The nomination sent by Ilyin to Stockholm as well as all those loadtrains of strategic raw materials would have been impossible if not approved by the highest authorities. However, anonymous initiators of Otto Hahn's nomination for the Nobel Prize made a serious mistake. They appear to have forgotten the story of awarding the 1935 Nobel Peace Prize to the pacifist and concentration camp inmate Carl von Ossietzky, in retaliation for which Hitler banned German citizens from even mentioning the Nobel movement (see also note 1 to doc. 3.16).[23]

It would be highly naive to think that the Moscow professor acted at his own discretion in the tense atmosphere of that time. The free will of Soviet intellectuals was so oppressed by the ruthless regime that they could see unauthorized contacts with the Nobel organizations only in nightmarish dreams. Here is a documented example of the impossibility of independent decision-making even by a high-ranking official that illustrates the level of oppression under the totalitarian rule of the pre-war period.

In April 1941, i.e. a few months after Professor Ilyin sent his nomination to the Nobel Committee, the President of the USSR Academy of Sciences V. L. Komarov informed G. M. Malenkov, secretary of the AUCP(b) Central Committee, about improper actions of the head of the Academy Staff Department and asked "for instructions to dismiss from office this worthless functionary," nominally his subordinate (doc. 3.22). It means that even the president of the Academy could not get rid of a corrupted worker without the permission of the highest party organ. How then could a university professor virtually unknown abroad think of displaying his own initiative?

To conclude the review of nominations of foreign scientists by their Russian colleagues, it is appropriate to mention S. N. Vinogradsky, who found himself in emigration in France in the early 1920s (see Chapter 4 and footnote* to note 4 in doc. 4.40). In 1935, Vinogradsky nominated to the Nobel Committee in chemistry the known biochemist Gabriel Bertrand, foreign member of USSR Academy of Sciences.

There were only two Russian nominees for the Nobel Prize in literature, both emigrants (Table 3.4). None of the Soviet writers was nominated for the Nobel Prize either by foreigners or countrymen.

S. Agrell, the Slavonic philologist from Lund (who died in 1937) continued to nominate Merezhkovsky as long as he lived. Another specialist in historical and philological sciences from the University of Lund, professor of theoretical philosophy

Table 3.4. Nominees of Russian descent for the Nobel Prize in literature during the period from 1934 to 1944

Nominees	Year of nomination	Nominators
N.A.Berdyaev (1874–1948)	1942	A.Nyman, professor of philosophy, Lund
-"-	1943	-"-
-"-	1944	-"-
D.S.Merezhkovsky	1934	S.Agrel, professor of Slavonic languages, Lund
-"-	1935	-"-
-"-	1936	-"-
-"-	1937	-"-

A. Nyman, started to regularly nominate philosopher and thinker N. A. Berdyaev.[24] His first nomination dated to 1942. Thereafter, he was as persistent as his deceased colleague Agrell in nomination of Merezhkovsky and continued to annually propose the candidacy of Berdyaev until the Russian philosopher died in 1948.

However, it was virtually impossible to attract attention from the Nobel Committee and Swedish Academy to the works of the Christian sage, a singer of philosophy of history[25] and champion of anthropocentric absolute as opposed to Christian theodicy canons. The existentialism of the 1930–1940s was alien to the aesthetics of prize-awarders in literature. The candidacy of Berdyaev was disapproved together with that of M. de Unamuno,[26] while prizes to A. Camus and J.-P. Sartre were yet to come.[27]

The awakening interest of the Soviet regime to contacts with the Nobel organizations was evidenced by the piety shown to foreign cultural and scientific figures awarded the Nobel Prizes.

Selma Lagerlof, Nobel laureate of 1909, celebrated her 75th birthday in 1933. Her literary works, highly popular in pre-revolutionary Russia, were practically not reprinted in the subsequent period.[28]

The reason was purely ideological; the writer was blamed for "Christian mysticism"[29] etc. Nevertheless, jubilee events were touched upon in diplomatic correspondence (doc. 3.23). The author of the document recommended to pay homage to the patriarch of Swedish literature but could not help making ritual propagandistic remarks in her address.

In 1944–1945, top Soviet authorities gave much attention to Albert Szent-Gyorgyi, Hungarian researcher awarded the Nobel Prize in physiology and medicine in 1937. Members of the USSR Academy of Sciences P. L. Kapitsa and Ya. O. Parnas informed the Central Committee of AUCP(b) about the scientist's desperate situation after Nazi troops invaded Hungary in 1944 (doc. 3.24). Prompt actions of the Soviet government probably saved the life of Szent-Gyorgyi (doc. 3.25–3.26), but the fate of both intercessors was dramatic.

Kapitsa fell into disfavor in summer 1946. He was discharged from office as the head of Glavkislorod and director of the Institute of Physical Problems, which he had founded. The government decree announcing the dismissal was signed by Stalin on 17 August. (Kapitza again returned to his duties as the director of the Institute only on 28 January 1955.) His name was no longer mentioned in public and scientific press. Even his living apartments in Moscow were taken away and he had for ten years to reside in a timber house on the outskirts of Moscow (Nikolina gora).

A response to this sort of exile of the researcher and public figure widely known in scientific circles and among the lay public was his nomination for the Nobel Prize by the Polish physicist Ch. Bialobrzeski in 1952, after breaks in 1949, 1951, and 1952 (Table 6.1). Kapitsa was nominated together with such eminent scientists as Max Born, Louis Alvarez, and Pierre Auger. Nominations of these three were submitted in due form, whereas Kapitsa was described as a citizen of the Socialist camp without reference to his affiliation and exact whereabouts. The nominator simply had no relevant information because the Polish mass media did not comment on the fate of Kapitsa after the example of "big brother"...

Parnas also fell victim to Soviet ideologists in 1947 (doc. 3.34). However, it was only the beginning. After the Communist Party newspaper *Pravda* published an article on 29 January 1949 that started a campaign "against cosmopolitanism" (see Chapter 6), Parnas was arrested and died on the same day from cardiac failure in the prison of Moscow KGB headquarters...[29a]

Special attention was given to Swedish scientists, including Nobel laureates and members of the Nobel organizations at the end of the Second World War in Europe (doc. 3.28–3.30). This campaign appears to have been launched in connection with the choice of foreign guests to be invited to the jubilee session of the USSR Academy of Sciences. The choice was explicitly biased by political considerations. The main criterion for invitation was political views of the guest and his attitude toward the Soviet Union rather than his scientific reputation. The same criterion was used by staff departments of the Soviet research institutions when they selected between candidates for filling vacant positions. Evidently, it was planned to build up relations between Soviet and Swedish scientists on the same basis.

In 1945, a representative of the All-Union Society of Cultural Relations with Foreign Countries still tried to find "extenuating circumstances" when a scientist could not be introduced straightforwardly as "a friend of the Soviet Union". Two to three years later, the selection criteria became much more rigorous. Certain characteristics of foreign scientists are really humorous. Professor A. Tiselius, chairman of the Nobel Committee in chemistry and future Novel laureate, did not show direct evidence of great sympathy with the Soviet Union and had to be described in a biographic note as "never being a Nazi" (doc. 3.30).

Of interest is a document issued a week and a half before German troops attacked the Soviet Union in 1941. The document concerns E. Hemingway, a future Nobel Prize winner. A year before, he published the presently well-known novel *For Whom the Bell Tolls* displaying signs of his anti-Communist and anti-Soviet sentiments. The AUCP(b) Central Committee was immediately informed about it by one of the top officials of its Department of Propaganda and Agitation in the memorandum of 13 June 1941 addressed to A. A. Zhdanov, secretary of the Central Committee. He insists on the necessity to ban the printing of the novel in the Russian language. "Despite sympathy with the struggle against fascism,"[30] the aversion of Communist ideology by the author makes it unfeasible to publish the novel in the Soviet Union.

This story continued over 14 years during the Khrushchev Thaw, when the writer was awarded the 1954 Nobel Prize in literature. The decision of the Swedish Academy gave reason to the "Foreign Literature" publishing house to include the long ago made translation of the novel in its publication plan for 1956. The newspaper *Literaturnaya gazeta* announced the coming sensation on the book market. But things turned out otherwise...

Domestic ill-wishers of Hemingway received unexpected support from France. It came from the journalist Simona Teri, member of the French Communist Party and wife of a Spanish emigrant who took part in the civil war and left the country after the defeat of the republican army in March 1939 and the establishment of General Franco's regime. As soon as Teri came to know about the coming publication of Hemingway's book in the Soviet Union, she wrote a verbose letter[31] to Boris Polevoy, then a member of the World Peace Council. Teri insisted that *For Whom the Bell Tolls* is a slanderous novel "caricaturizing the war of the Spanish people." The letter[32] was transferred to the AUCP(b) Central Committee on 15 June 1955 together with the cover note by Polevoy, who shared the opinion of its author.

The response of ideological services of the AUCP(b) Central Committee was quite predictable. On 27 July 1955, the Department of Science and Culture, AUCP(b) Central Committee, informed its Secretariat about "a serious mistake by *Inoizdat*" and recommended that *Literaturnaya gazeta* "denounced the approach of *Inoizdat* to the choice of [...] foreign literature works" for publication in the Soviet Union (doc. 3.32).

This time, however, secretaries of the Central Committee did not support an ideological initiative. The year of 1955 was a "honeymoon" in the developing relations with the West and dissociation from Cold War principles. The success of the Geneva Summit, closed on 23 July with the participation of USSR, USA, Great Britain, and France, gave rise to what the world mass media defined as "the spirit of Geneva." In this context, conflicting proposals formulated in doc. 3.32 and submitted for approval four days after the Summit could not be supported.

No article was published in the *Literaturnaya gazeta*. But the novel by the Nobel laureate in Russian translation did not appear either. Even the wise K. I. Chukovsky yielded to the euphoria of the relaxation of international tension reached in Geneva and fell victim to his own illusions of the coming "rehabilitation" of the great writer.[33] The hidden enmity of Party ideologists and their accomplices among intellectuals towards the landmark literary figure of the second third of the 20th century did not disappear.[34]

An example is information transferred by the Department of Science and Culture, AUCP(b) Central Committee, to the Secretariat AUCP(b) Central Committee on 25 January 1958: "Translators and their associates persistently recommended to publishing houses Hemingway's novel *For Whom the Bell Tolls*, describing the Civil War in Spain in 1936–1938 from positions to progressive forces."[35] Four years later (26 February 1962), the known publicist Yuri Zhukov wrote to the AUCP(b) Central Committee about an interview of the young Soviet author Vasili Aksyonov given to the Italian newspaper *Il Giorno*. Zhukov emphasized that Aksyonov described Hemingway as the "greatest writer of this century" and made other statements "unacceptable from our standpoint."[36]

Forty years later, A. A. Belyaev, instructor at the Department of Culture in the 1960s, exposed one more obstacle that hampered publication of the novel in the Soviet Union. In 1962, the head of the Department D. A. Polikarpov was visited by K. M. Simonov, a known author, influential secretary of the Union of Soviet Writers and recently member of the Central Revision Commission. Simonov asked for permission to publish the illustrious work of the Nobel laureate the main idea of which was the feeling of personal responsibility for the fate of an alien country where the future of his own motherland was at stake.

Polikarpov made a helpless gesture saying that publication of Hemingway's book was hampered by a certain "stubborn reader" who "vigorously opposes it in her strongly-worded letters to the First Secretary of the AUCP(b) Central Committee…" This reader was Dolores Ibarruri, secretary-general of the AUCP(b) Central Committee of the Spanish Communist Party, who then lived in Moscow as an emigrant.

It may be supposed that Ibarruri had a hand in preventing another attempt to publish the long-suffering manuscript in 1964. *Literaturnaya gazeta*, then a mouthpiece of liberal ideas, not only briefly announced the coming publication of *For Whom the Bell Tolls* (as nine years ago) but even reprinted the closing chapter of the novel in its issue of 24 August 1964. But things did not go further. Soviet readers could read the book only after four years.

According to A. A. Belyaev, officials of the International Department of the AUCP(b) Central Committee in the end managed to prevail over Ibarruri; they phoned and said that she remained of the same opinion but no longer objected to the publication of *For Whom the Bell Tolls* in the collected works of Hemingway. "Draw up a note and our head will sign it."[37] The novel was published in 1968,[38] that is seven years after its author had died...[39]

The complicated fate of the Russian version of *For Whom the Bell Tolls* to the readers had repercussions for the publication of later works of Hemingway including his great *The Old Man and the Sea*, which brought the author the Nobel Prize in 1954. A year and a half later, Ilya Erenburg was strongly criticized by ideologists of the AUCP(b) Central Committee. The list of absurd accusations included his "boundless enthusiasm" in the naturalistic and wingless story *The Old Man and the Sea* by Hemingway (doc. 3.33).

To conclude, here is one more example of the interference of traditionalists from abroad in book-publishing projects in the Soviet Union. In 1960, A. T. Tvardovsky, editor-in-chief of the journal *Novy mir,* decided to publish the famous novel *The Plague* by Albert Camus, Nobel laureate of 1957, after his tragic death in a car accident. Foreseeing censorial problems, he wanted to ensure support from the then famous Louis Aragon, member of the AUCP(b) Central Committee of the French Communist Party and laureate of the Lenin Peace Prize. His initiative proved a complete fiasco. With a cable in hand asking him to write a short preface to the novel, Aragon visited the Soviet embassy and expressed surprise over the intention to promote the book by the writer known for his "anti-Communist views."[40] In the end, the Department of Culture, AUCP(b) Central Committee, prohibited the printing of *The Plague* in the journal, and it was out in the Russian translation only in 1969.

It seems appropriate to turn back to the above episode of the nomination of the German physicist Otto Hahn for the Nobel Prize of 1941 by B. V. Ilyin, professor of the Chemical Department, Moscow University. To recall, this chapter covers the period encompassing three important events: Hitler's accession to power, the Great Terror in the Soviet Union as a result of totalitarian Stalin's rule, and World War II. The nomination of the poorly known Moscow professor of chemistry sent to Stockholm in 1941 turns out to be linked by invisible threads to all these events.[41] Let us compare a few historical dates.

On 13 November 1940, the People's Commissar of Foreign Affairs V. M. Molotov arrived in Berlin at the invitation of the German government. Negotiations carried out in the capital of the Third Reich by no means slowed down the thorough preparation of the offensive against the Soviet Union known as "Barbarossa plan" (after Friedrich the First, emperor of the Holy Roman Empire). Elaboration of the plan had begun three and a half months prior to Molotov's arrival by the General Headquarters of the German land forces by Hitler's order dated 21 July. In other words, Germany was ready to start the war, and negotiations with the Moscow minister were needed to distract the attention of top Soviet authorities from war preparations.

Nobel Committees distributed invitations for nominations in September and October. In 1940, military actions were already in full swing in Europe. Delivery of correspondence was delayed. We do not know the date when the letters from the Nobel Committees reached Moscow but it was sure before Molotov left for Berlin.

Ilyin sent his nomination to Stockholm on 15 December 1940 and it was registered by the Nobel Committee on 28 December. Just in this time, the development of operation "Barbarossa" was completed and Hitler signed on 18 December.[42] In other words, the cunning scheme of Ilyin's patrons to please their Nazi ally was a shattering failure.

This episode provides one of the examples of political short-sightedness in line of many other fatal miscalculations by Stalin in the face of the coming aggression. These errors cost the Soviet people an unprecedented number of lives, the exact number of which still remains unknown and is estimated with variations from plus to minus millions.

...Could Professor Ilyin with his poor knowledge of heavy nuclear fission refuse "the honorable commission of the Party and the Government?" It was out of the question. The regime did not forgive such refusals, and there were very few brave enough to resist the threat.

Notes

[1] In two decades, another generation of Soviet diplomatists continued to cherish the vain hope to work for the benefit of scientists and give them a chance to receive a Nobel Prize. To this effect it was planned to take advantage of the official visit of the Swedish Prime Minister T. Erlander to Moscow in 1956 (see Chapter 7 and note 30 therein).

[2] A curious example related to the ephemeral hopes of Kollontai to use the VII Mendeleev Congress for assuring a Nobel Prize to a Soviet researcher is given by an earlier episode. The decision of the Swedish Academy to award the 1902 Nobel Prize in literature to the German historian Theodor Mommsen for his capital work *The History of Rome* did not receive unambiguous approval. A correspondent for the French paper *Figaro* wrote from Stockholm (his article was reprinted in *Sankt Peterburgskie Vedomosti* on 7 (29) December 1902) that "rumors had ascribed the award to Bjornson Bjornstjerne on the occasion of his 70th birthday* that the Norwegians celebrated with great enthusiasm.** These hopes fell to the ground, and the award of the prize to Mommsen is met with great disappointment [...]"***

> * Bjornstjerne was 70 on 8 December 1902, two days before the prize-giving ceremony.
> ** The pathos of this phrase is quite in order. After Bjornstjerne died on 26 April 1910, the Russian envoy in Finland A. N. Krupensky wrote that "all Norwegians mourned the death of the national writer whom they worshipped [...] Bjornstejrne was certainly a first-class writer and poet who greatly promoted national self-consciousness of the Norwegian people." The Russian envoy concluded with a touch of irony that "honors given to Bjornstejrne by Norway are somewhat exaggerated because it exalted itself in the deceased poet in a sincere but somewhat naive belief that his death was an irreplaceable loss for the whole world"; AVP RI, fond 133, opis' 470, ed. khr. 31 (1910), l. 106–107.
> *** Bjornstejrne received the Nobel Prize a year later (in 1903), i.e. regardless of his jubilee.

[3] *Obshchaya gazeta*, No. 22, 200, p. 10.

[4] I. P. Pavlov was aware of Kollontai's initiatives (see doc. 3.13).

[4a] V. I. Vernadsky. *Dnevniki 1935–1941 v dvukh knigakh*. Kn. 1. 1935–1938 (Diaries 1935–1941 in Two Books. Book 1. 1935–1938). M.: Nauka, 2006, pp. 179, 183.

[5] It follows from the minutes of the Nobel Committee in physiology and medicine that nomination of Orbeli in 1935 was considered by Professor G. Hetz.

[6] In: *Neopublikovannye i maloizvestnnye materialy I. P. Pavlova* (Unpublished and Poorly Known Materials of I. P. Pavlov). Leningrad: Izd-vo "Nauka", 1975, p. 129.

[7] Para. 7 of the Nobel Statutes reads: "No person may be considered eligible for an award unless nominated in writing by a person competent to make such a nomination."

[8] V. N. Ipatiev and A. E. Chichibabin were sent to Berlin (29 June 1930) to take part in the International Energy Congress and permitted to stay abroad for a year for medical treatment. Ipatiev had more than one reason not to come back. He was a member of the Presidium of All-Union National Economic Council (VSNKh SSSR), the central planning organ of the country's economic developments in the 1920s, and the head of its Scientific and Technical Department. In this capacity, he was closely associated with many top economic and political figures that one by one fell into disfavor and were repressed as Stalin's regime gained momentum. In 1929, he felt that he was unsafe too.* There is evidence, that Ipatiev regarded his stay abroad as temporary. He never thought of purchasing a permanent lodge and lived with his wife in a Chicago hotel for 22 years of his stay in the USA, from 1930 to 1952 when he died.** While in Chicago, he wrote and published a fundamental monograph on chemical catalysis under high pressure and temperature. In 1936, it was printed in the Soviet Union in accordance with the plan of scientific publications of the USSR Academy of Sciences.***

Ipatiev received an unexpected order to return to the Soviet Union, where mass repressions started. A. A. Gromyko, the Soviet ambassador to the United States recollected that in 1944 Ipatiev applied to him for the permission to come back to the USSR. He received an official refusal,**** which certainly saved him from arrest and maybe from being killed. This request by Ipatiev looks like a gesture of despair. In the very beginning of the Great Patriotic War (July 1941), his son V. V. Ipatiev and former co-worker M. S. Nemtsov were arrested in Leningrad and charged with espionage for France.*****

It should be mentioned that even in Khruschev's liberal time, the name of the outstanding Russian scientist caused idiosyncrasy in Soviet ideologists. The letter from the Department of Propaganda and Agitation of the CPSU Central Committee sent to its Secretariat on 23 April 1959 informed the Party leadership about ideological mistakes by the editorial board of the Great Soviet Encyclopaedia during the work over the additional volume 51 of its second edition. The authors of the letter pointed out that "certain biographical notes look apolitical" including the article about Ipatiev, an organic chemist who did not return to the USSR, worked in the USA, and came out against the Soviet Union.******

* *Rossiyskaya nauchnaya emigratsiya. Dvadtsat' portretov* (Russian Scientific Emigration. Twenty Portraits). M.: Editorial URSS, 2001, p. 196.

** *Ibid*, p. 2203.

*** V. N. Ipatiev. *Kataliticheskie reaktsii pri vysokikh temperaturakh i davleniyakh. 1900–1933 (s predissloviem prof. Willshtattera)* (Catalytic Reactions Under High Temperatures and Pressures). M.-L.: Izd-vo AN SSSR, 1936, 774 pp. 1700 copies.

**** A. A. Gromyko *Pamyatnoye.* T. 1 (Memorable. V. 1). M.: Politizdat, 1986, pp. 151–154.

***** E. I. Grakina. *Uchyonye Rossii v gody Velikoy Otechestvennoy Voiny 1941–1945* (Scientists of Russia During the Great Patriotic War 1941–1945). M.: Izd-vo Instituta rossiyskoy istorii RAN, 2000, p. 18.

****** RGANI, fond 11, opis' 1, ed. khr. 367, l. 47–48. See also *Ideologicheskie komissii 1958–1964.* Dokumenty. M.: POSSPEN, 1998, p. 155.

[9] Vestnik RAN.

[10] *Ibid*, p. 627.

[11] V. I. Ipatiev. *Zhisn' odnogo khimika. Vospomonaniya* (Life of a Chemist. Memoirs). New York, 1945, V. 1, p. 486. See also PFA RAN, fond 788, opis' 4, ed. khr. 34, l. 33–34.

[12] Vestnik RAN, p. 630.

[13] Ipatiev, V. 2, pp. 509, 510. Richard Willstatter, awarded the Nobel Prize in chemistry in 19, gave the following complimentary evaluation of the work of our compatriot was: "There was no a greater person than Ipatiev in the history of chemistry."* This statement certainly allows for some exaggeration; nevertheless, the outstanding figure of Vladimira Nikolaevicha, a scientist and engineer, receives an adequate appraisal...

* See: E. L. Nitoburg. *Russians in the USA. Histories and Fates.* Izd-vo "Nauka", 2005, pp. 365–366.

[14] During the World War II, in 1941 and 1942, the Nobel Prizes were not awarded.

[15] Ipatiev. T. 2, p. 48. It is worth citing the letter of V. A. Maklakov sent from Paris on 7 December 1920 to B. A. Bakhmetev in Washington:* "[...] General Ipatiev, who was formerly a member of our State Defence Conference where he had been appointed the gunpowder dictator and perfectly well accomplished his functions [...] this Ipatiev was invited by Bolsheviks to fulfil the same role; he consented on the condition that, first, his plant would not have any Communism smell at all and, second, he would be given back his estate. Bolsheviks promised to comply with both demands, and they now produce explosives on a permanent basis."**

* See note 3 to Chapter 2 for this correspondence and its participants.
** Sovershenno lichno i doveritel'no! B. A. Bakhmetev – V. A. Maklakov. Perepiska 1919–1951. T. 1 (Strictly Personal and Confidential! B. A. Bakhmetev – V. A. Maklakov. Correspondence 1919–1951, V. 1). M.: Stanford, 2002, pp. 309–310.

[16] Solovyev might miss the published memoirs of Ipatiev, but the same statement of the scientist dated to the end of 1917 could be found in the biographical essay* mentioned under number 6 in the list of references to his article.

* V. I. Kuznetsov, A. M. Maksimenko. Vladimir Nikolaevich Ipatiev. M.: "Nauka", 1992, p. 51.

[17] The thought of not coming back to Russia first occurred to Gamow in 1932 when he was refused, without an explicit clarification, the trip to the 1st International Congress of Nuclear Physics in Rome.* His report on "quantum mechanics of nuclear structures" was presented by his friend from Copenhagen Maxim Delbruck, future Nobel Prize winner in physiology and medicine, awarded to him in 1969. Two others of the trio stayed in the Soviet Union and were subjected to repressions. Ivanenko was arrested in 1935, Landau in 1938.** Gamow's refusal to return to the USSR had a boomerang effect since it affected many Soviet scientists. In what follows, the situation is commented on by P. L. Kapitsa, who spent his 1934 summer holidays in the Soviet Union (for the fifth time) and was not allowed to leave for Cambridge where he had for over ten years worked with Rutherford while remaining Soviet citizen. In the letter of 4 December 1934 he wrote to his wife, who remained in Britain: "The behavior of Gamow and co. [...] did more harm than anything to healthy links of our science with the West [...] his egoism and complete lack of ethical principals taken together with his scientific gift bring great misfortune. It will be very difficult to restore confidence in scientists at large and in myself in particular. Everything seems to be against us."***

* *Rossiyskaya nauchnaya emigratsiya. Dvadtsat' portretov* (Russian Scientific Emigration. Twenty Portraits). M.: Editorial URSS, 2001, p. 78.
** *Ibid*, p. 80. See doc. 8.34 and notes related to the arrest of Landau and his liberation through the agency of P. L. Kapitza.

*** V. D. Esakov, P. E. Rubinin. Kapitsa. Kreml' i nauka (Kapitsa. The Kremlin and Science). M.: Izd-vo *Nauka*, 2003, V. 1, p. 92.

[18] Such mass nomination of scientists by their countrymen was not infrequently recorded by the Nobel Committee in physiology and medicine even though it was rare in physics and chemistry. It may be accounted for by the fact that the number of nominees and, accordingly, nominators in physiology and medicine was always significantly greater than registered by the Nobel Committees of the Royal Academy of Sciences. By way of example, the lists of nominees submitted to the Karolinska Institutet in 1951 and 1952 comprised 89 and 102 names respectively compared with 50 and 44 nominees in physics and 47 and 55 in chemistry (see Table 6.4). The most recent example of such mass nomination is provided by the candidacy of the French surgeon and physiologist Rene Leriche in 1952 and 1953. Leriche headed the Department of Experimental Medicine, College de France and died at 76 in 1955. In 1951, he was nominated by two professors of pathology and histology from Toulouse and Besanson respectively. Next year, the number of his nominators sharply increased to 24! Seventeen persons sent their proposals from a small town of Montpelier, southern France, and five from Lyons, where the pretender had taken courses in the local university. The last nominator was the Nobel laureate Paul Muller of Basel (Switzerland). In 1953, the number of nominators dropped to 11 but remained rather high. It is worthy of note that the candidate obtained no sensational results in physiology either soon before 1952 or during the five or six previous years.

Evidently, such situations, as in the case with Japanese professors, arise from the efforts of a few active researchers to attract attention of a Nobel Committee to their colleague by lobbying for him and involving prominent figures from the outside, such as I. P. Pavlov in 1935 or P. Muller in 1952.*

* To recall, similarly, and even more massive, lobbying was organized in 1903 to nominate the music critic K. F. Glasenapp for the Nobel Prize in literature. Then, the number of proposals amounted to 53, which arrived in Stockholm from five European countries (see Chapter 1).

[18a] Here is a concrete example of inaccessibility for Pavlov, in principle, of such collective actions. It is borrowed from the same Congress held in Rome in 1932. The opening of the Congress had to be delayed because the Italian dictator Mussolini was somewhat late. When he at last appeared in the Presidium, Italian Blackshirts jumped up all at a time and raised their hands in the fascist salute. At this moment, the audience heard the voice of Pavlov: "*Eine bedingte Reflex* (conditional reflex)."*

* *Novy mir*, No. 8, 1982, p. 185.

[18b] V. I. Vernadsky. Dnevniki 1935–1941 v dvukh knigakh . Kn. 1. 1935–1938 (Diaries 1935–1941 in Two Books. Book 1. 1935–1938). M.: Nauka, 2006, p. 355.

[19] Archives of the Royal Academy of Sciences, Stockholm. Fund of the Nobel Committees in physics and chemistry. Materials of the Committees for 1941, sheets 20.

[20] Hahn received his Nobel Prize for 1944 in 1945 (see also note 7 to doc. 5.6). To conclude the issue of Hahn's award so oddly reflected in the mirror of Soviet Union actions during pre-war months of 1941, let us briefly consider the history of the discovery that brought about

the prestigious award. In 1938, Hahn jointly with F. Schtrassman showed that bombardment of uranium by neutrons released nuclei of alkaline earth metals, specifically, barium nuclei. In 1939, L. Meitner of Austria (who had previously worked with Hahn) and her colleague O. Frisch proposed a theoretical explanation of the phenomenon observed in Berlin by showing that the nucleus of 235U isotope could be split into two smaller fragments. They introduced quite a new notion of nuclear division in analogy with cell division in biology. For this reason, the very first nomination of T. Svedberg for the 1939 Nobel Prize in physics contained the names of Hahn and Meitner. Later, they were again proposed together by other nominators. After the Anschluss of 1938, Meitner moved to Stockholm and was continued to be nominated for the Nobel Prizes in physics and chemistry up to 1949, either alone or together with Frisch, by such authoritative scientists as the Nobel laureates M. von Laue, J. Franck, A. Compton, M. Planck, L. de Broglie, and Hahn himself (in 1946 and 1948) who received his prize for 1944 in 1945. It seems that members of the Royal Academy of Sciences realized their mistake. Anyway, the prize in physics in 1958 was awarded not only to P. A. Cherenkov, the discoverer of Cherenkov's effect but also to I. E. Tamm and I. M. Frank who provided theoretical explanation of the new phenomenon (Chapter 8). When Cherenkov alone had been nominated in 1952 and 1955 (Table 6.3), his candidacy was disproved by the Nobel Committee in physics.

[21] L. V. Levshin. Dekany fizicheskogo fakul'teta Moskovskogo universiteta (Deans of the Physical Department of the Moscow State University). M.: Izd-vo fizfaka MGU, 2002, pp. 219–223. Ilyin left a trace in the history of Soviet physics as a founder (together with P. P. Lazarev) of the journal *Uspekhi Fizicheskikh Nauk* in 1918 and was the scientific secretary of the journal.* Moreover, he was on friendly terms with future eminent representatives of Soviet science, such as brothers N. I. and S. I. Vavilov.**

* In 1918, only three issues of the journal were published; thereafter, it was closed due to the impossibility of having scientific publications from abroad under conditions of "War Communism".

** In: Sergei Ivanovich Vavilov. Ocherki i vospominaniya (Sergei Ivanovich Vavilov. Essays and Memoirs) (pod red. I. M. Franka). M.: "Nauka", 1991, pp. 162, 175.

[22] There is no doubt that this nomination was initiated by the apparatus of AUCP(b) Central Committee, in the first place its Department of Propaganda and Agitation. Ilyin was most likely listed in the AUCP(b) Central Committee nomenclature. This inference is confirmed by the following circumstance. Eight months after the nomination was dispatched to Stockholm (28 December 1940), the Physical Department of Moscow University was evacuated to Ashkhabad and only a small section was left in front-line Moscow. Ilyin was appointed the dean of this minor part.* Such an appointment would have been impossible without permission of the Department of Propaganda and Agitation.

* Voprosy istorii estestvoznaniya i tekhniki, No. 2, p. 239.

[23] It should be noted that Hitler had felt that the Nobel Prizes were peculiar long before the award was given to Ossietzky. There were fears in Germany that Nobel gold medals may be confiscated by the new regime, and certain Nobel laureates took them abroad and handed them over to N. Bohr for storage. D. S. Danin, who spent a few weeks in Copenhagen to work on his book about Bohr mentioned the names of W. Heizenberg and J. Franck with

reference to this episode.* In a quarter of a century, this story was told in *Science* mentioning M. von Laue instead of Heizenberg.** The further fate of these medals is similarly described by the two sources. After the Nazis occupied Denmark in 1940, George de Hevesey, Bohr's co-worker, chose to dissolve the medals in *aqua regia*. The flask with the gold solution stood openly on a laboratory shelf together with other chemical reagents and did not attract attention even from German officers disposed at the territory of the Institute in 1943 in order to look for evidence of the development of atomic weapons. After the end of World War II, the Nobel Foundation used the preserved gold to stamp replicas of the medals.

* D. S. Danin. Niels Bohr. M.: Molodaya gvardiya, 1978, p. 403.
** R. Stone at 100, Alfred Nobel's Legacy Retains its Lister. *Science*, 294, No. 5541, 2002, p. 291.

[24] It would be incorrect to call Berdyaev an emigrant. He did not leave Russia at his own discretion but was forced to go after his arrest and imprisonment in the Moscow KGB headquarters. There, he was informed of the Kremlin's decision to permanently exile him abroad, together with other Russian intellectuals. They were instructed to submit written applications for "permission to go abroad" of their own accord.* The steamship *Oberburgermeister Hacken*,** notoriously known as the "philosophers' steamship" with the exiles on board left Russia on 28 September 1922. Berdyaev first settled in Berlin but thereafter moved to Paris (in summer 1924) where he remained for the rest of his life. In the Soviet Union use of the name Berdyaev was banned until the second half of the 1980s, when Gorbachev began to promote the policy of *glasnost*. There is a letter by the head of the Leningrad department of *Glavlit* (the top censorship board) prescribing to confiscate all Berdayev's books from second-hand bookshops, regardless of the publication date.***

There was a positive aspect to the exile of the most eminent Russian representatives of the humanities by the Bolshevist regime. For many of them, including Berdyaev, the exile saved lives. The high probability of their death in Russia was cynically referred to by one of the Communist leaders L. D. Trotsky in a talk with the American journalist A. L. Strong, one month before the exile: "Those elements that we exile or will exile [...] are potential instruments in the hands of our enemies. In case of new military action, we'll have to shoot them under the law of war. That is why we prefer to exile them in advance while the situation is still calm. And I hope that you do give us credit for this humane precautionary measure..."****

* From the application of Berdyaev: "To the Board of GPU. In conformity with provision 4 of the SO Department of GPU concerning my exile, I ask the Board of GPU to permit my going abroad together with my family and of my own accord [...]"; see *Novy mir*, No. 5, 1998, p. 189.
** Detailed description of the "philosophers' steamship" case (based on archive materials of security agencies of that time) can be found in the article of V. S. Khristoforov, historian and head of the Department of Registration and Archive Funds of RF FSB (Federal Security Service): "*Filosofskiy parokhod*". *Vysylka uchenykh i deyatelei kul'tury iz Rossii v 1922* ("Philosophers' steamship". Exile of scientific and cultural figures from Russia in 1922). *Novaya i noveishaya istoriya*, No. 5, 2002, pp. 121–141.

*** GARF, fond P-8425, l. 24; see also: Ideologicheskie komissii TsK CPSS 1958–1964. Dokumenty (Ideological Commissions of CPSU Central Committee 1958–1964. Documents). M.: ROSSPEN, 1998, p. 97.

**** *Izvestiya.* 30.8.1922. p. 1.

[25] "[...] history and historical are not a mere phenomenon... They make the most radical prerequisite for the philosophy of history [...] Philosophy of history is certain spirituality and transformation of the historical process."*

* N. A. Berdyaev. *Smysl istorii. Opyt filisofii chelovecheskoi sud'by* (The Meaning of History. Essay on Philosophy of Human Fate). Berlin, 1923, pp. 24, 27.

[26] Espmark considers the neglect of Unamuno as one of the most noticeable mistakes made by the Swedish Academy in the interwar period (Espmark, p. 151).

[27] Camus was awarded the Nobel Prize in 1957 and Sartre in 1964.

[28] The catalogue of the Russian State Library, Moscow, lists three editions of the collected works of S. Lagerlof in 12 volumes each dated to 1909–1911 and 99 of her books published by private publishing houses till the early 1920s. State-owned publishing houses issued only five of her books up to 1941 (four of them were a popular fairytale *The Wonderful Journeys of Niels with the Geese*).* Critics and scientists from time to time published notes about Lagerlof's works, and her death in 1940 was only briefly commented on in the mass media. It may be accounted for by the writer's civic deed: in December 1939, after the beginning of the "winter war" between the Soviet Union and Finland, she donated her Nobel gold medal to the Swedish Assistance Fund for Finland. This demonstrative step might be dictated by her long-standing negative attitude toward Russia, not only to its ruling elite but to the people in general. It formed as long ago as 1912 when the writer visited St. Petersburg and Moscow. She mentioned it 14 years later at the meeting of the Swedish Academy on 29 February 1926: "[...] I know that Russia is a terrible and dangerous country. I would not like to stay there. It frightens me."**

* An analysis of book publishing policy in the Soviet Union with respect to the works of S. Lagerlof is also presented in the journal *Vsemirnoye slovo.* No. 15, 2002, pp. 46, 48.
** See: O. V. Chernysheva. Shvedy i russskie. Obraz soseda (Swedes and Russians. Image of a Neighbor). Moscow: Iizd-vo "Nauka", 2004, p. 154.

[29] See for instance *Great Soviet Encyclopaedia,* 1st ed., V. 1, 1937, column 652: "Lagerlof dreams of the restoration of former patriarchal Sweden [and] sinks into Christian mysticism." Thomas Ventslov, a Lithuanian dissident poet told an unusual story. One of his acquaintances was arrested and sentenced to eight-year imprisonment in a labor camp because he had works of Lagerlof and other "subversive literature" at his home. It happened in the 1950s, after Stalin's death.*

* *Literaturnoye obozrenie,* No. 3, 1999, p. 56.

[29a] Certainly, no information about the tragic death of the world-renowned scientist appeared in the mass media during Stalin's time and later, until the onset of *perestroika.* This fact was long unknown abroad including in Sweden. It accounts for the presence of the name of Ya. O. Parnas in the list of Soviet members of the Division of Chemical Sciences, USSR

Academy of Sciences, prepared by the Nobel Institute of Chemistry in 1955 for inviting them to recommend nominees for the 1956 Nobel Prize.*

* Archives of the Royal Academy of Sciences, Stockholm. Fund of the Nobel Committees in physics and chemistry. Materials of the Committees for 1955, sheet 10.

[30] It is an interesting illustration of opposite attitudes toward Nazi Germany in the office correspondence of Party functionaries from ideological services of the AUCP(b) Central Committee on the one hand and in the mass media on the other, where any critical hint in the address of the Hitler regime was severely punished after the two Molotov–Ribbentrop pacts had been concluded in 1939. It is not the sole example of such dualism, especially shortly before the war broke out. Here is another fact of duality. On 23 May 1941, AUCP(b) Central Committee severely punished journalists of the newspaper *Komsomol'skaya pravda* (two of them were sacked) for the publication of the article (21 May) "Teaching Lenin and Stalin about War" that reflected the thoroughly veiled anxiety about the situation of those days and the coming war.* In a week (1 June 1941), the Secretariat adopted a resolution (initiated by the Department of Propaganda and Agitation of the Central Committee) on "non-critical overestimation of sport achievements in Germany."**

* RGASPI, fond 17, opis' 116. ed. khr. 90, p. 13. The Editorial Board was accused of "infringement of the established regulations for publishing articles concerning foreign policy and [...] distribution of politically harmful concoctions akin to carelessness and political thoughtlessness."
** *Ibid*, ed. khr. 92, p. 13d. The cover letter of A. Puzin, deputy of the Department's head (*Ibid*, opis' 117, ed. khr. 282, l. 133–135) draws attention to the fact that volumes of the collected articles *Sports Abroad* are "largely filled with German propaganda," and their editors are therefore guilty of "a serious political mistake."

[31] RGANI, fond 5, opis' 17. ed. khr. 539, l. 105–110.

[32] *Ibid*, l. 104.

[33] K. I. Chukovsky wrote to his son Nikolai on 14 August 1955: "Our great diplomatic victory (relaxation of international tension reached in Geneva) makes itself felt even in *Goslit*: Coldwell and Hemingway are being published..."*

* *Znamya*, No. 1, 2004, p. 162.

[34] At the beginning of 1960, the Department of Culture, CPSU Central Committee, analyzed the state of literary criticism for the past year and emphasized that "certain central and provincial publishing houses have recently printed and reprinted too many works by bourgeois authors" and that "all this activity did not receive an adequate assessment of literary criticism." Hemingway* headed the list of "bourgeois authors" enumerated in the document. It was followed by another signal to top officials. The Literature Museum "has recently [...] organized two literary parties devoted to the works of Hemingway", that proved "unreasoned and poorly prepared."**

* RGANI, fond 5, opis' 36. ed. khr. 119, l. 14.
** *Ibid*, l. 22.

[35] *Kul'tura i vlast' ot Stalina do Gorbacheva. Ideologicheskie komissii TsK CPSS 1958–1964. Dokumenty* (Culture and Authorities from Stalin till Gorbachev. Ideological Commissions of CPSU Central Committee 1958–1964. Documents). M.: ROSSPEN, 1998, p. 36.

[36] RGANI, fond 5, opis' 36. ed. khr. 140, l. 52.

[37] A. A. Belyaev. Na Staroi ploshchadi (At the Old Square). *Voprosy literatury*, No. 3, 2002, p. 245.

[38] E. Hemingway. Sobranie sochineniya v. 4 tomakh. T. 3 (Collected Works in four volumes. V. 3). M.: Khudozhestvennaya literatura.

[39] Dualism with respect to the works of Hemingway in Soviet criticism had been apparent long before the novel *For Whom the Bell Tolls* was published. An agonizing search by the "lost generation" for genuine values in the period between the two World Wars proved that they were beyond the narrow frameworks of social realism and induced allergy in its adherents. This attitude to the writer's reflections was especially brightly expressed in the pejorative words of M. Gorky about "all sorts of Hemingways"* in his article entitled "About Formalism" published in *Pravda* on 9 April 1936. A few years later, this epithet of Gorky was reinterpreted in the lengthy letter by M. M. Rozental', editor-in-chief of the journal *Literatrny kritik*, addressed to the secretaries of AUCP(b) Central Committee (29 March 1940). Asking for support in his conflict with *Literaturnaya gazeta*, Rozental' drew their attention to the fact that this newspaper published articles containing positive comments about Hemingway's works "that some are trying to imitate."**

> * "Unnecessary ornamentation and detalization in the literature inevitably leads to the obscuring of the sense of facts and images. Those interested should read Ecclesiastes, Shakespeare, Pushkin, Tolstoy, and Flaubert in parallel with Marcel Proust, Joyce, Dos-Passos, and all sorts of Hemingways."
> ** TsAODM, fond 3, opis' 51, ed. khr. 115, l. 62.

[40] A. T. Tvardovsky. Rabochie tetrady 60-ykh godov (Work Note-Books of the 1960s). Znamya, N 9, 2000, p. 117; the compiler's note to the entry of 24 March 1963. For a long time, Tvardovsky could not pardon Aragon for the offence. He comes back to this episode in the entry of 15 January 1965. Speaking about Aragon on a certain occasion, he adds: "he played a swinish trick on me — his behavior in response to my request for a preface to Camus' novel."* In a quarter of a century, V. Ya. Lakshin, an old friend and co-worker of Tvardovsky on the Editorial Board of *Novy mir*, tried to analyze in more detail the quick response of Communist Party ideologists to the denunciation of Aragon. He published the essay "Albert Camus. Purgatory" first in a curtailed form in *Izvestia* (22 November 1991) then in full (three years later) in the journal *Rossiyskaya provintsia*.** Lakshin pointed out that "the ideas of the resistance movement [against the German occupation of France during World War II] could be misinterpreted if applied to the situation in the Soviet Union. Such authors were regarded as 'fumblers' and their writings as confusing our citizens. Ideologists saw Camus as a typical 'fumbler' and his misty existentialism as casting shadow on the self-evident truths of domestic Marxism."** Therefore, the signal from Paris was like an "egg at Easter" for the Department of Propaganda and Agitation…

> * *Znamya*, No. 12, 2001, p. 159.
> ** "Possiyskaya provintsiya", No. 5, 1994, p. 171.

[41] Moscow University appears to have been chosen by the Nobel Committees of the Royal Academy of Sciences (in compliance with the annual rotation principle required by the Nobel Statutes), one of the six foreign institutions whose professors received the right to nominate their candidates for the 1941 Nobel Prizes in physics and chemistry.

[42] L. A. Bezymensky. Razgadannye zagadki Tret'ego Reikha. 1940–1945 (Resolved Puzzles of the Third Reich). Smolensk.: Izd-vo "Rusich," 2001, p. 61.

Document 3.1

Extract from the letter of A. M. Kollontai, plenipotentiary of the USSR in Sweden, to B. S. Stomonyakov, member of the Advisory Board, People's Commissariat of Foreign Affairs, 16 January 1934, No. 7c.

[...] Past traditionally close relations between Sweden and Finland have evidently changed during the last two years (since the onset of the Lapua movement);[1] anti-Swedish tactics of Finnish reactionaries is highly annoying in Swedish public opinion.[2]

Here is a characteristic example: the Swedish Academy had to disapprove the candidacy of the nominated Finnish writer one week before the election of the Nobel laureate [...]"[3]

AVP RF, fond 0140, opis' 18, papka 117, ed. khr. 322, l. 12.

[1] Kollontai seems to have overestimated the influence of the nationalist movement of the late 1920s (named after the Lapua area) on Swedish–Finnish relations. A correspondent for the newspapaer *Dagens Nuheter* told in the article dated 28 August 1933 "[t]he Lapua movement is infantile but safe"; "members of this movement are naive young people, and they do not do anything unlawful."
[2] The Finnish government tried to limit the use of Swedish as a teaching language in educational institutions of the country.
[3] F. E. Sillanpaa; the writer was annually nominated by Finnish Academy members and professors of the Helsinki University since 1930; he received the Nobel Prize in literature in 1939.

Document 3.2

From: Bel'ko, first secretary, plenipotentiary representation of USSR in Stockholm
To: Berezov, head of the First Western Department, USSR People's Commissariat of Foreign Affairs. April 1934 (dateless). No 4c.

We would not recommend the inclusion of Professor K. Benediks[1] in the list of Swedish specialists, whom the USSR Academy of Sciences plans to invite for participation in the Mendeleev Congress.[2] He is certainly above medium rank but far from being an "outstanding physicist and metallographer". According to the available information, he is not very popular in local scientific circles.

Instead, Professor Palmer Wilhelm should be one of the more desirable guests. He has a big and popular name. The professor of inorganic and theoretical chemistry now works for the Nobel Committee;[3] formerly, he was secretary and then chairman of the Nobel Academy[4] Professor Palmer deeply sympathizes with our country.

The remaining candidates proposed by the Academy are OK. They all are widely known scientists and Nobel Prize winners[5] of moderate political views. They display a positive attitude towards our country.

AVP RF, fond 0140, opis' 18, papka 117, ed. khr. 336. l. 5.

[1] Informants of the Soviet representation in Stockholm appear to have used unreliable sources. Carl Benediks was director of the metallographic Institute, Stockholm, and a leading member of the International Association for Testing Materials. That was quite enough to be popular in scientific circles. Organizers of the Mendeleev Congress disregarded the recommendations from the Soviet representation in Stockholm and Scandinavian Department, People's Commissariat of Foreign Affairs dated (see doc. 3.3). Professor Benediks actively participated in the work of the Congress. In subsequent years, he greatly contributed to the involvement of Soviet scientists in international cooperation (doc. 3.15).
[2] VII Mendeleev Congress was held in Leningrad on 10–13 September 1934.
[3] The Nobel Committee in chemistry, where W. Palmer was a member from 1926 until his death in 1942; he was a professor of the Royal Technological Institute in Stockholm.
[4] The Nobel Committee in chemistry; in 1934, Palmer was its secretary.
[5] T. Swedberg was the only Nobel laureate among the three members of the Swedish delegation at the Mendeleev Congress. Another Nobel laureate, participating in the Congress was Professor H. O. Wieland, from Germany.

Document 3.3

From: Berezov, head of the First Western Department, USSR People's Commissariat of Foreign Affairs, and S. M. Mirny, assistant to the head.
To: Academician I. V. Grebenshchikov, secretary-general of the Organizing Committee of the VII Mendeleev Congress.
No. 16334c, 23 July 1934.

The Consular Department of the USSR People's Commissariat of Foreign Affairs informed academician Volgin in its letter No. 27104 of 7 May 1934 about the request made by the plenipotentiary representation of the USSR in Stockholm to include Professor Palmer Wilhelm in the list of participants in the Mendeleev Congress [...] Simultaneously, the representation objects to the participation of Professor C. A. Benediks, as he is poorly known in Swedish scientific circles.[1]

The People's Commissariat of Foreign Affairs did not receive a reply to its request to include Professor Palmer. The USSR minister plenipotentiary once again emphasizes the importance of the invitation to Professor Palmer. You are asked to make a decision at your earliest convenience taking into consideration the request of the plenipotentiary representation of USSR in Sweden.

AVP RF, fond 0140, opis' 18, papka 117, ed. khr. 336. l. 31.

[1] See note 2 to doc. 3.2.

Document 3.4

From: A. M. Kollontai, plenipotentiary of the USSR in Sweden
To: B. S. Stomonyakov, deputy people's commissar of foreign affairs.
No. 7c, 16 January 1934.

1. I many times wrote and told the People's Commissariat of Foreign Affairs about the forthcoming Mendeleev Congress in the Soviet Union. I pointed out that the existing list of known Swedish chemists must be supplemented by the name of Professor Palmer, member of the local Academy of Sciences, a big scientific and public personality of radical views, and our common friend. I see an additional aspect to this matter (don't you think that one of the scientific prizes, e.g. in chemistry in connection with the Mendeleev Congress, should be awarded to a Soviet scientist?). Palmer is just the man who could be very helpful in this respect. That is why I ask you to insist that the Organizing Committee of the Congress invite Palmer [...]

2. In April or May, I was addressed by the bureau responsible for publishing of a book about Sven Hedin[1] on the occasion of his 70th birthday with a request for an article of a Soviet scientist to be included in the book. I asked the People's Commissariat of Foreign Affairs whether they consider our participation desirable and whether they are able to organize the receipt of such an article here. I believe it would be useful if our scientists participated in this international project along with world-renowned authors, the more so that Sven Hedin is on friendly terms with us.[2]

I'll greatly appreciate advice at your earliest convenience because they start drawing up the volume in September and need relevant information.[3]

AVP RF, fond 0140, opis' 18, papka 117, ed. khr. 336. l. 35–36.

[1] Hedin — an outstanding traveler to Central Asia and Far East in the late 19th and early 20th centuries.
[2] Hedin highly appreciated the attention given to him by the Soviet government on the occasion of his 70th birthday and became a member of the Society for Cultural and Scientific Relations between Sweden and the Soviet Union, organized in May 1935.* After World War II, the Soviet government bodies sharply changed their attitude towards Hedin (see doc. 6.19 and note 3 therein).

* GARF, fond 5283, opis' 5, ed. khr. 540. l. 40.

[3] This wish of Swedish publishers is quite natural. Hedin had many contacts with Russian authorities during his travels and enjoyed their support. He knew Russian, frequently visited St. Petersburg to report his scientific observations, and was elected an honorary member of the Imperial Russian Geographic Society. In 1885, Hedin was a tutor for the family of Robert Nobel (brother of Alfred Nobel and one of the founders of the Nobel Brothers Petroleum Company) in Baku, from whence he started his first journey to Persia in the same year.

Document 3.5

From: S. M. Mirny, assistant to the head of the First Western Department, USSR People's Commissariat of Foreign Affairs.
To: A. M. Kollontai, plenipotentiary of the USSR in Sweden.
No. 16469c, 5 September 1934.

1. We were informed from Leningrad that the invitation to Professor Palmer had already been sent by the Organizing Committee of the Mendeleev Congress. Palmer accepted the invitation and will arrive in Leningrad to participate in the Congress.

2. The Academy of Sciences did not yet tell us the name of a Soviet specialist who might write an article for the jubilee volume on the occasion of Hedin's 70th birthday. We shall inform you by telegraph as soon as we know it.[1]

AVP RF, fond 0140, opis' 18, papka 117, ed. khr. 336. l. 39.

[1] The author of the article in the jubilee volume* was the known polar explorer Professor G. L. Samoilovich. It was published under the title "Exploration of the Polar Part of the USSR in 1934 and Sedov's expedition."
* Hyllnisskrift tillagnad pa hans 40-arsdag den 19 Febr. 1935. Stockholm, 1935.

Document 3.6

From: A. M. Kollontai, plenipotentiary of the USSR in Sweden.
To: B. S. Stomonyakov, deputy people's commissar of foreign affairs.
No. 118c, 24 October 1934.

Dear Boris Spiridonovich,

I would like to draw your attention to a special question concerning a Nobel Prize for a Soviet scientist.

It is too late to raise this question this year. Nominations for the Nobel Prizes must be received by the Nobel Committee by February. Hence, if we wish one of our scientists to receive the prize in 1935 (in December 1935), then the proposal should be made in January, and not later than the end of January. We can hardly expect that the superconservative Nobel Committee constituted by eighteen elders[1] with the only 65 year-old "youth" among them will be ready to award the Nobel Prize to one of our prosaic writers. They will never do it in the belief that all our literature is nothing else but agitation and propaganda of Communist ideas; they are frightened even by the name of Gorky.[2]

However, there are many chemists, physicists, clinicians, and other scientists[3] who recognize the success and achievements of Soviet science. Therefore, I think that we must set an objective to have one 1935 Nobel Prize for some of our outstanding physicists, chemists or physicians. It should be borne in mind that such a man should be well known abroad and his discoveries and scientific works made not too long ago.

If Moscow considers it desirable that the Nobel Prize be awarded to a Soviet scientist, please get in touch with the Academy of Sciences and send a list of several big researchers as soon as possible.

I've already told that many Swedish specialists recognize the achievements of our science and have discussed this issue with me.[4] In my turn, I shall send you a list of Soviet scientists who their radical Swedish colleagues regard as deserving the Nobel Prize. Our supporters here include such widely known figures as professors of chemistry Palmer[5] and Swedberg[6], polar researcher Professor Ahlmann,[7] and professor of medicine Israel Holmgren.[8] I believe that a physician or polar researcher has the best chance of being awarded. The candidacy of Professor Otto Shmidt is not good because hostile members of the prize-giving committee would object;[9] it must be a man who made a discovery.

I attach great importance to this issue as a tool of political propaganda.[10] It is for you to decide whether we must work along these lines.

Another question with which I am concerned and which makes me apply directly to you is the visit of a group of known Soviet physicians to Sweden. I've already written about this to Moscow.

Professor Israel Holmgren participated in the Congress of Rheumatism in the Soviet Union. He visited me to remind me about his agreement with comrade Kaminsky[11] that a few our professors must go to Sweden and deliver lectures for university students and practitioners. Such a trip would take two weeks at most and require from us only travel expenses because the Swedish side is prepared to cover all accommodation expenses.

It is now almost two months since I wrote about it to Moscow; still, I have no answer.[12] I consider it a great mistake that we do not even respond when we receive such proposals. Politically, we are the losers.

AVP RF, fond 0140, opis' 18, 117, ed. khr. 336. l. 41–42.

[1] Eighteen members of the Swedish Academy, the prize-awarding institution in literature.

[2] Frequent references to the name of Gorky in the correspondence concerning Nobel Prizes are attributable to Kollontai's feeling of gratitude to the writer for the help in a difficult situation. When the Provisional Government arrested Kollontai in 1917, Gorky and L. B. Krasin, future member of the Soviet government and diplomat, paid their money for her release on bail.*

* A. M. Kollontai. Diplomaticheskie dnevniki 1922–1940. T. 1 (Diplomatic Diaries 1922–1940. V. 1). M.: Academia, 2001, p. 274.

[3] Members of prize-awarding institutions in sciences.

[4] Kollontai probably means Soviet nominators to whom the Nobel Committees sent invitations to recommend candidates for the Nobel Prizes, as well as the small number of their countrymen that these nominators proposed (especially in physics and chemistry).

[5] See doc. 3.2–3.4.

[6] Nobel Prize winner in chemistry in 1926.

[7] Professor at the University of Stockholm, scientific secretary of the Swedish Geographic Society, president of International Geographic Society in 1956–1960. In 1945, Ahlmann was present at the jubilee session of the USSR Academy of Sciences on the occasion of its 220th anniversary. In 1969, he was awarded the gold medal of the USSR Geographic Society.

[8] In the list of initiators of the Society for Cultural and Scientific Relations between Sweden and the Soviet Union, Holmgren is characterized as a "luminary of Swedish medicine." A month and a half before Kollontai* sent her letter to the People's Commissariat of Foreign Affairs dated 10 September 1934, she wrote in her diary: "Today — visit of Professor Holmgren [to the Soviet representation]; he officially invited Soviet physicians and professors to read lectures in the Medical Society of Stockholm."

* A. M. Kollontai. Diplomaticheskie dnevniki 1922–1940. T. 1 (Diplomatic Diaries 1922–1940. V. 1). M.: Academia, 2001, p. 256.

[9] An incompetent statement. "Hostile elements" had little to do with it. Nobel Prize categories do not simply include geographic sciences.

[10] The addressee underlined in pencil the word "propaganda" and put question and exclamation marks in the margins.

[11] RSFSR People's Commissar of Health and USSR chief sanitary inspector in 1934.

[12] The addressee underlined in pencil the words "have no answer." The first page of the document has the inscription: "To comrade Berezov for execution and report. Stomonyakov, 15.11."*

*In Berezov's hand: "1. Find the letter. 2. Flarify the matter. Either positive or negative answer must be given." (dateless). In L. Gaikis' hand: "Letters are sent to 1) Volgin, 2) Bronner. 22/XI."

Document 3.7

From: Berezov, head of the First Western Department, USSR People's Commissariat of Foreign Affairs, and L. Gaikis, assistant to the head.
To: V. P. Volgin, permanent secretary of the USSR Academy of Sciences
No. 16677c, 22 November 1934; urgent.
cc: Scientific Committee of the USSR Central Executive Committee.[1]

The closing date for nominations for the Nobel Prizes is 1 February. The People's Commissariat of Foreign Affairs considers it opportune to discuss the feasibility of recommendation of a Soviet scientist for this prize.

Because the award of the prize to a Soviet writer looks unlikely taking into account the conservatism of the members of the Nobel Committee,[2] it appears more expedient to think of the candidacy of one of our specialists representing exact sciences and medicine or a polar researcher.[3]

It goes without saying that the candidate must be well known abroad and his discoveries and scientific works made not very long ago.

The issue needs an urgent solution and the list of possible candidates must be prepared as soon as possible, since the deadline for nominations is very close and our representatives have much to do in the way of promoting Soviet nominees among the circles close to the Nobel Committee.

Please keep the People's Commissariat of Foreign Affairs informed about your further actions.

AVP RF, fond 0140, opis' 18, 117, ed. khr. 336. l. 44.

[1] Supreme Soviet of the USSR since January 1938.
[2] It follows from this text that the authors of the letter think of a single Nobel Committee constituted by "conservative members" that awards Nobel Prizes of all categories, from exact sciences and medicine to literature. Relevant documents of the post-war period reveal a similar misunderstanding.
[3] See note 9 to doc. 3.6.

Document 3.8

From: Berezov, head of the 1st Western Department, USSR People's Commissariat of Foreign Affairs, and L. Gaikis, assistant to the head.
To: Bronner, director of the All-Union Bureau of Foreign Sanitary Information, RSFSR People's Commissariat of Health.
No. 16676c, 22 November 1934.

This is to inform you, with reference to your letter No. 922c of 1 August 1934, that the opinion of comrade Kollontai, our minister plenipotentiary in Sweden, concerning organization of lectures of our physicians in Stockholm, is stated in her letter to comrade Kaminsky (No. 122 of 24 October 1934). We confirm once again the advisability of such a trip by our specialists to Sweden and ask you to discuss concrete candidates with Volgin, permanent secretary of the USSR Academy of Sciences. Also, comrade Volgin is expected to recommend Soviet scientist candidates for nomination for the Nobel Prize.

If such a person is actually nominated (and maybe of medical profession), it would be reasonable to include him in the delegation of our specialists going to Sweden.

Please keep the People's Commissariat of Foreign Affairs informed about your further actions.

AVP RF, fond 0140, opis' 18, papka 117, ed. khr. 336. l. 43.

Document 3.9

From: A. M. Kollontai, plenipotentiary of the USSR in Sweden.
To: L. Gaikis, assistant to the head of the 1st Western Department, USSR People's Commissariat of Foreign Affairs.
No. 135c, 3 December 1934.

Dear comrade Gaikis,

With reference to your letter dated 27 November, I once again ask to be advised, at your earliest convenience, as regards our possible candidate for the Nobel Prize. According to the Statutes of the Nobel Prize Institute,[1] the Nobel Prize has the following categories: medicine,[2] physics, chemistry, literature, and peace. Inventions and new works in these fields must be published in one of the widely used European languages (German, English or French).[3] Moreover, discoveries and scientific research must be made recently.

It appears from my talks with Swedish scientists that the candidacy of a physician has the best chance to be supported, e.g. someone from Pavlov's Institute. It is important that we know the candidate as early as December because the closing date for submitting nominations is 1 February 1935. Naturally, we must have enough time to inform those interested to know about our candidate and find local specialists ready to nominate and support him.[4]

As regards the Nobel Prize in literature, I have a feeling that Sholokhov might have some chance. His books are read here and readers give high praise to his talent. The candidacy of Gorky would be opposed by the conservative elders constituting the Nobel Committee.

I ask you once again to remind Kaminsky that I am waiting for the name of at least one physician who could come to Sweden and read a few lectures here. It is urgent to have either a positive or a negative answer.

I share your opinion that a physician who may happen to be nominated for the Nobel Prize should be included in the delegation to be sent to Sweden.

I myself reminded Kaminsky about it some two months ago but certainly received no answer. Try to settle this issue with him by phone and convince him to answer at least some of our questions.

AVP RF, fond 0140, opis' 18, papka 117, ed. khr. 336. l. 46.

[1] The Nobel Foundation.
[2] The exact name of the award is the Nobel Prize in physiology and medicine.
[3] It is not true. The Nobel Statutes contain no such requirement.
[4] The author of the letter naively thinks that, knowing the name of the candidate, she will be able to find a nominator for him in Stockholm.

Document 3.10

From: Berezov, head of the First Western Department, USSR People's Commissariat of Foreign Affairs, and L. Gaikis, assistant to the head.
To: V. P. Volgin, permanent secretary of the USSR Academy of Sciences
No. 16775c, 20 December 1934; very urgent.

With reference to our letter No. 166677 of 22 November 1934, we additionally inform you that in accordance with the Statutes of the Nobel Prize Institute,[1] the Nobel Prize has the following categories: medicine, physics, chemistry, literature, and peace. Inventions and new works in these fields must be published in one of the widely used European languages (German, English or French).

It appears from the talks of comrade Kollontai, our plenipotentiary in Sweden, with Swedish scientists that the candidacy of a physician has the best chance to be supported, e.g. someone from Pavlov's Institute.[2]

As regards the Nobel Prize in literature, the feeling is that Sholokhov might have some chance. His books are read in Sweden and the readers show high praise of his talent.[3]

It is important that we know about the candidate as early as possible to have enough time to inform those interested about our choice and find local specialists ready to nominate and support him.

AVP RF, fond 0140, opis' 18, papka 117, ed. khr. 336. l. 49.

[1] See note 1 to doc. 3.9.
[2] Institute of Physiology, the USSR Academy of Sciences.
[3] Kollontai wrote in her diary under 22 October 1934: "the two most popular literary names in Sweden today are Sholokhov and Vera Inber." *

* A. M. Kollontai. Diplomaticheskie dnevniki 1922–1940. T. 1 (Diplomatic Diaries 1922–1940. V. 1). M.: Academia, 2001, p. 266.

Document 3.11

From: A. M. Kollontai, plenipotentiary of the USSR in Sweden.
To: B. S. Stomonyakov, deputy people's commissar of foreign affairs
No. 149c, 24 December 1934.

[...] Comrade Sholokhov came here and made a good impresssion. All newspapers including the right-wing ones published favourable reviews. Sholokhov has a recognized name here in Sweden [...] I am asking once again whether we should try to nominate him for the Nobel Prize in literature in 1935 although it would be easier for me to struggle for a scientific prize, e.g. in chemistry.[1]

The editorial board postponed printing the volume dedicated to Sven Hedin pending articles from the Soviet Union. It is good that one is coming.

AVP RF, fond 0140, opis' 18, papka 117, ed. khr. 322, l. 123.

[1] Either a slip of the pen (because earlier correspondence concerned medicine, see doc. 3.6 and 3.9) or the author recalled her good relations with the secretary of the Nobel Committee in chemistry Professor W.Palmer.

Document 3.12

From: V. P. Volgin, permanent secretary of the USSR Academy of Sciences
To: 1st Western Department, USSR People's Commissariat of Foreign Affairs
No. 20c, 15 January 1935.[1]

With reference to your letter No. 16677/c[2] of 22 November 1934 and relevant telephone talks, enclosed find please brief information about the corresponding member of the Academy, Professor Leon Abgarovich Orbel.

L. A. Orbeli was born in 1882 and educated at the Military Medical Academy under the guidance of I. P. Pavlov. At present, L. A. Orbely heads the Laboratory of Animal Physiology, USSR Academy of Sciences, and is the deputy director of the P. F. Lesgaft State Institute of Natural Sciences. In 1926 and 1929, Orbeli participated in the work of international physiological congresses in Stockholm and Boston as a member of Soviet delegations. Today, L. A. Orbeli is the vice-chairman of the Organizing Committee of the International Congress of Physiologists to be held in the Soviet Union in 1935.

In 1932, L. A. Orbeli was elected full member of the German Academy of Sciences in recognition of his extraordinary contribution to physiology. In 1934, the Soviet government awarded him the title of Honoured Scientist.

The candidacy of L. A. Orbeli is proposed for the election to full membership in the USSR Academy of Sciences and will be discussed at the forthcoming session of the Academy scheduled for February this year.

Find attached the list of scientific publications of L.A.Orbeli on fourteen pages.[3]

AVP RF, fond 0140, opis' 19, papka 119, ed. khr. 358, l. 16.

[1] Note by hand: "Transferred to Sweden on 27/1" (signature is illegible).
[2] See doc. 3.7.
[3] Not published here.

Document 3.13

Extract from the letter of Berezov, head of the 1st Western Department, USSR People's Commissariat of Foreign Affairs, and L. Gaikis, assistant to the head, to A. M. Kollontai, plenipotentiary of the USSR in Sweden. No. 12088c, 27 January 1935.

1. This is to inform you that some questions mentioned in your previous letters (viz. visits of Soviet physicians and Professor Vise[1]) have not yet received a positive response even though their solution would certainly contribute to the strengthening of our cultural relations with Sweden. We are trying to accelerate their solution but without apparent success.

2. I. P. Pavlov recommends for the Nobel Prize his disciple Professor Orbeli Leon Abgarovich, corresponding member of the USSR Academy of Sciences [...][2]

Because the candidacy of Orbeli is not yet approved here, we are sending some data about him for your notice and preliminary inquiries as to its acceptability (if approved) for the Swedish circles concerned [...][3]

AVP RF, fond 0140, opis' 19, papka 119, ed. khr. 359, l. 2.

[1] A known polar researcher. Organization of his visit to Stockholm related to the aforementioned hopes to obtain a Nobel Prize for a Soviet scientist who distinguished himself in Arctic explorations (see note 9 to doc. 3.6).
[2] In 1934, Pavlov proposed Orbeli for the Nobel Prize on his own initiative unrelated to the efforts of Kollontai (see the present chapter).
[3] Enclosed is information about Orbeli borrowed from doc. 3.12.

Document 3.14

Extract from the letter of A. M. Kollontai, plenipotentiary of the USSR in Sweden, to Berezov, head of the 1st Western Department, USSR People's Commissariat of Foreign Affairs. No. 26c, 21 February 1935.

Professor Pletnev's visit was most welcome and successful. All newspaper reports concerning his stay in Sweden were in favour of our country.[1] Pletnev read several scientific lectures in Swedish universities, addressed a crowded meeting of the Society of Physicians, was invited to a number of dinners, etc. I also had dinner with the minister of education, certain Swedish public figures, and leading representatives of the medical profession. Holmgren, who had visited the Soviet Union before and received a most favourable impression, asked me to organize another trip of Soviet physicians to Sweden. In a word, the visit of Pletnev should be regarded as successful and to our credit.[2]

In a few weeks, the initiative group organizing the Society for Cultural and Scientific Relations Between Sweden and the Soviet Union convenes its first meeting to elect the executive body in accordance with our recommendations.[3] The Society will be chaired by our good friend Professor Palmer, secretary of the Nobel Committee,[4] highly reputed chemist and public figure. His daughter[5] is also our "friend"[6] and an active worker. Attached find please the list of members of the initiative group of the Society [...][7]

The known geographer and orientalist Sven Hedin was 70 on 19 February. He is still in China.

AVP RF, fond 0140, opis' 19, papka 119, ed. khr. 358, l. 19.

[1] D. D. Pletnev (1872–1941) was a prominent cardiologist. Two years after his triumphant visit to Sweden, he fell victim to the Great Terror policy. In 1937, the newspaper *Pravda* published a slanderous editorial entitled "Professor — ravisher, sadist" in which Pletnev was charged with sexual abuse of patient B.* In the entry to his diary dated 7 July 1937, i.e. one month after the publication, V. I. Vernadsky calls out the name of the "heroine": "The woman blackmailing Pletnev is Braude** of the *Pravda* staff." The editorial concluded: "Soviet physicians and public opinion [...] will pass severe sentence on the offender who abused the patient's confidence, showing himself an animal, sadist, and ravisher where he had to be human." Pletnev was prosecuted as a criminal and condemned. In 1938, he was again brought to court in connection with the faked political trial of the "anti-Soviet right-trotskyist block, accused of poisoning the writer M.Gorky, and was sentenced to a 25-year imprisonment. In his last word on 12 March 1938, Pletnev told that he wrote a 10–12 sheet monograph during his confinement prior to the commencement of the trial (not found).*** He had at his disposal 20 volumes in four languages delivered to him for the purpose from his private library. Pletnev was imprisoned in the city of Orel. Shortly before the Nazi troops entered Orel in September 1941, all political prisoners were shot. In 1985, Pletnev was rehabilitated posthumously.

* *Pravda*, 8 June 1937, p. 3.

** "Istoricheskiy arkhiv", No. 5, 2002, p. 62. This note is prefaced by the following remark: "The entire editorial board of *Pravda* appears to be *sans foi at loi* (dead from shame and with no conscience)."

*** Afterword to the book "D. D. Pletnev. Izbrannoye (D. D. Pletnev. Selected works)," M.: Meditzina, 1989, p. 342.

[2] V. Smirnov, representative of the All-Union Society of Cultural Relations with Foreign Countries in Sweden, sent to the Society's central office (25 March 1935) a

few press-cuttings that evidenced the great success of Pletnev's visit, who "was paid homage as a luminary of Soviet science."* The Soviet press reported that the Swedish Minister of Education thanked Kollontai for the help in familiarizing "Swedish scientific circles with the progress in Soviet science in the person of one of its outstanding representatives."**

* GARF, fond 5283, opis' 5, ed. khr. 540, l. 109.

** "Klinicheskaya meditsina," No. 4, 1935, p. 614.

[3] "Yesterday [20 May 1935] the new Society for Cultural and Economic Relations Between Sweden and the Soviet Union solemnly held its first meeting. It was created for the promotion of its title activities. The organization took much energy. Preparations started at the end of the last year [...] On 4 May, the initiative group was set up, the Society developed its structure and adopted the charter. A huge amount of work was done even if it was invisible to outsiders! [...] I was elected an honorary member and I presented a lecture about Soviet culture (literature, theater, painting). It was a lively and friendly gathering, more crowded than expected [...] Thus, it's in the bag, and the Society is satisfied."*

* A. M. Kollontai. Diplomaticheskie dnevniki 1922–1940. T. 1 (Diplomatic Diaries 1922–1940. V. 1). M.: Academia, 2001, p. 285; entry dated 21 May 1935.

[4] Nobel Committee in chemistry,

[5] Eva Palmer, member of the Swedish Communist Party since 1929, chaired the Swedish–Soviet Society after the death of her father in 1942. Her husband, Gustav Johansson, edited the Communist newspaper *Ny Dag* and was a member of the Swedish *Riksdag*. Eva Palmer, educated in chemistry, was an expert in metal corrosion.

[6] sic.

[7] Unpublished here.

Document 3.15

Letter of V. Smirnov, representative of the All-Union Society of Cultural Relations with Foreign Countries in Sweden, to the Society's central office.
No. 80, 4 February 1935.

I was addressed by Professor Carl Benediks, known here as the director of the Institute of Metallography, Stockholm. He visited the Soviet Union at the invitation of the USSR Academy of Sciences to participate in the festivities on the occasion of the Mendeleev jubilee and favourably described his experience in the press. In Moscow, he made the acquaintance of Professor Rebinder and wrote to him but received no answer. Professor Benediks is concerned with the involvement of one of our institutions in the International Association for Testing Materials. I am asking you for assistance in this issue; ensure please that Professor Rebinder answered Benediks.[1] Otherwise, we find ourselves here in an awkward position.

GARF, fond 5283, opis' 5, ed. khr. 540, l. 116.

[1] It was quite clear why P. A. Rebinder, corresponding member of the USSR Academy of Sciences, kept silence. Since early December 1934 (after the assassination of C. M. Kirov), i.e. two months before this letter was sent from Stockholm, a wave of arrests covered the country affecting intellectuals among others, and the fear of being accused of "having relations abroad" prevailed over everyone.

Document 3.16

Extract from the letter of A. M. Kollontai, plenipotentiary of the USSR in Sweden to N. N. Krestinsky, deputy people's commissar of foreign affairs
No. 16c, 10 February 1937.

[...] Self-respecting Swedes were especially annoyed by the story of Ossietzky's Nobel Prize[1] and the insolent tone in which Hitler spoke to Norway. The Swedish press, even right-wing newspapers, stood up in defense of Sweden against the impudence of fascist Germany [...]

AVP RF, fond 0140, opis' 21, papka 120, ed. khr. 389, l. 21.

[1] The antifascist journalist was awarded the Nobel Prize in 1936 when he served his sentence in the concentration camp at Zonnenburg. This decision by the Norwegian Nobel Committee provoked a violent reaction in the Nazi leadership in Germany. As a result, Hitler prohibited German citizens from receiving foreign awards without permission of the government or even mentioning the Nobel movement and Novel laureates in the press. In the Soviet Union, the decision by the Nobel Committee received high praise; *Pravda* wrote in the issue of 23 November 1936 that "in this decision the Nobel Committee expressed the opinion of progressive mankind." The ceremony in Oslo on 10 December was attended by the entire diplomatic corps including the *charge d'affaires* of the USSR in Norway.*

* Unfortunately, the same methods of Nazi propaganda were acquired by the Soviet press some 10–15 years later, at the height of the Cold War. For example, the journal *Novoye Vremya* published a favourable review of the book by the English physicist P. Blackett criticizing the short-sighted policy of the USA in the field of practical use of nuclear energy (January 1949) but did not even mention that the author received the prestigious Nobel Prize in physics a month and a half prior to the publication (see Chapter 6 and notes 42 and 43).

Document 3.17

Extract from the letter of A. M. Kollontai, plenipotentiary of the USSR in Sweden to V. P. Potemkin, deputy people's commissar of foreign affairs
No. 49c, 9 August 1937.

[...] An important aspect of our current work relates to the visit of Otto Yulievich Shmidt, coming here from Paris, i.e. in October.[1] He was invited by Aeroclub and Geographic Society (polar explorers) to read lectures. Of course, we shall use this opportunity for the purposes of the Society for Cultural and Scientific Relations Between Sweden and the Soviet Union.[2] But I ask you to inform us at your earliest convenience whether Shmidt's visit to Stockholm in October is approved. It is important because we are going to raise the question about awarding him the Nobel Prize and pave the way for it.[3] We must know that Moscow approves in principle this intention of ours [...]

AVP RF, fond 0140, opis' 21, papka 120, ed. khr. 389, l. 52–53.

[1] The decisions concerning participation of O. Yu. Shmidt in the International Physical Congress, Paris, and his visit to Stockholm for lecturing were taken by the *Politbureau* on 4 and 27 July 1937 respectively.*

* *Akademiya nauk v resheniyakh politburo* TsK RKP(b)-VKP(b)-KPSS. 1922–1952 (Academy of Sciences in the Decisions of Politbureau of the Central Committee of RKP(b)-AUCP(b)-CPSU. 1922–1952). M.: POSSPEN, 2000, p. 260.

[2] See doc. 3.14.

[3] See note 9 to doc. 3.6.

Document 3.18

From: Bezhanov, acting head of the 1st Western Department, USSR People's Commissariat of Foreign Affairs, and Iliynsky, assistant to the head.
To: A. M. Kollontai, plenipotentiary of the USSR in Sweden.
No. 12942c, 1 September 1937.

1. It is expected that comrade O. Yu. Shmidt will be able to visit Stockholm in the second half of October because during its first half he will be staying in Paris (please take this as preliminary information).

2. Please clarify the question raised in your letter of 9 August, No. 49/c, concerning the possibility of nominating comrade Shmidt for a Nobel Prize in sciences. The fact is we had a talk with comrade Shmidt and he was highly sceptical about this matter because, to his knowledge, the Nobel Prizes are currently awarded only for the most important discoveries in chemistry.[1]

We could discuss this issue here again after obtaining comprehensive information from you [...]

AVP RF, fond 0140, opis' 21, papka 121, ed. khr. 400, l. 7.

[1] Another evidence of poor knowledge of the structure and functions of Nobel organizations in the Soviet Union (including its diplomatic services) of the 1930s.

Document 3.19

From: A. M. Kollontai, plenipotentiary of the USSR in Sweden
To: Bezhanov, acting head of the 1st Western Department, USSR People's Commissariat of Foreign Affairs.
No. 61c, 8 September 1937

I have already informed the Aeroclub and Geographic Society about the forthcoming visit of Shmidt this October and will little by little prepare the future audience of his lectures.

As regards nomination of comrade Shmidt for the Nobel Prize, the secretary of the Nobel Committee[1] and friendly personalities here find this question difficult. The prizes are awarded for discoveries in chemistry, physics, and medicine but do not cover Arctic explorations in general, and meteorology in particular. Because we are thinking of the prize for 1938, I will return to this issue later to discuss it in more detail.[2]

AVP RF, fond 0140, opis' 21, papka 121, ed. khr. 400, l. 8.

[1] W. Palmer, secretary of the Nobel Committee in chemistry.
[2] The letter concerning the 1938 Nobel Prize has not been found in foreign policy archives. It may be thought that Kollontai at last understood the hopelessness of securing Nobel Prizes for Soviet scientists via diplomatic channels.

Document 3.20

From: A. M. Kollontai, plenipotentiary of the USSR in Sweden.
To: Bezhanov, acting head of the 1st Western Department, USSR People's Commissariat of Foreign Affairs.
No. 72c, 9 October 1937

[...] It is really a pity that O. Yu. Shmidt will not be able to come.[1] We can not influence wide public circles[2] here friendly disposed to our country even without regular visits of its representatives.

The Germans worked indefatigably, sent here their most well-known specialists for lecturing in the local German–Swedish Society. This season will be opened with the participation of General von Epp, the Bavarian governor-general, who will present to the said society a report on "the importance of territories and raw materials for the state."[3]

Many people here waited for Shmidt to come. I am saying that his visit is postponed.

AVP RF, fond 0140, opis' 21, papka 120, ed. khr. 389, l. 68.

[1] The foreign trip of O. Yu. Shmidt, who then headed *Glavsevmorput'* (Main Northern Seaway Administration), was canceled by *Politbureau* (15 September 1937) at his

request. He wrote in a letter addressed to Stalin and Molotov: "Now is the most critical moment in the search for comrade Levanevsky* for which I am responsible. My absence, even if for ten days, would hardly be beneficial for the entire expedition."**

* The plane H-209, with a crew of six under command of S. A. Levanevsky, was lost on 12 August 1937 during the non-stop flight from Moscow to the United States via the North Pole.

** *Akademiya nauk v resheniyakh politburo* [...], pp. 260–261.

[2] Swedish public opinion.

[3] Seemingly unintentional but very characteristic parallelism between the propagandistic efforts of the two totalitarian regimes.

Document 3.21

From: A. M. Kollontai, plenipotentiary of the USSR in Sweden.
To: V. G. Dekanozov, deputy people's commissar of foreign affairs.
No. 72/272c, 13 November 1944.

After a long break during wartime, the Swedes announced the award of Nobel Prizes to Danes, Americans,[1] and Canadians for 1944. The Danes received the prizes in medicine and literature. Henrik Dam was awarded for the discovery of vitamin K (catalyst of blood coagulation, the deficiency of which leads to coagulation disorders). He shared the prize with Professor E. Doisy of Canada, who elucidated the chemical nature of vitamin K. The 1944 Nobel Prize in literature was awarded to the Danish writer Johanness V. Jensen. Jensen is a medium-rank bourgeois writer. It was expected in the progressive circles of Sweden and Denmark that the Nobel Prize for 1944 would be awarded to Andersen-Nexe. This was not the case, however, the Swedish Academy of Sciences[2] being constituted by the most conservative scientists. Nevertheless, the Danes are satisfied to have laureates in two nominations. Christmas Moller[3] regards the award of the Prize to Jensen as a big event. This opinion is shared by other members of the Danish government in England.[4] The main work of Jensen is the historic and ethnograhic novel *Long Journey*.

AVP RF, fond 0140, opis' 28, papka 35, ed. khr. 7, l. 68.

[1] O. Stern and I. Rabi, Nobel laureates in physics.

[2] The author confused the Royal Academy of Sciences and the Swedish Academy, the prize-giving organization in literature.

[3] Danish politician, prime minister of the government-in-exile during the occupation of Denmark by Germany.

[4] When the 1944 Nobel Prizes were awarded, Denmark remained under occupation, and its government, together with governments of other occupied European countries, was based in Great Britain.

Document 3.22

Materials on the dismissal of a corrupt staff member from office in the Presidium of the USSR Academy of Sciences.

I

V. L. Komarov, president of the USSR Academy of Sciences to G. M. Malenkov, secretary of the AUCP(b) Central Committee; 12 April 1941.[1]

Dear Georgiy Maksimilianovich,

Knowing how extraordinarily busy you are, I have long forborne from bothering you on a matter that makes me anxious. It is about comrade Kurochkin, head of the Academy Staff Department, who proves utterly unsuitable to fill such an important position.[2] Now, I have undeniable evidence that he uses his official standing as an unscrupulous money manager (as comrade Kuzmin, director of our publishing house, informed me and your assistant, comrade Butova). I feel that further silence will put the blame on me and it will be entirely my fault that a key position in the Academy is occupied by a corrupt civil servant.

I therefore apply to you with an urgent request to help the Academy and give instructions to dismiss from office this worthless worker. The Academy would greatly benefit from such a decision [...]

Malenkov's instructions: "To comrade Shatalov. Kurochkin must be sacked. Prepare draft resolution of s-t[3] to-day. Propose a new candidate. Malenkov. 14/4."

RGASPI, fond 17, opis' 117, ed. khr. 264, l. 2.

II

Resolution of the Secretariat of the AUCP(b) Central Committee. 19 April 1941 (protocol 80, p. 211).

Decided: To release comrade Kurochkin G. D. from his duties as head of the Staff Department, USSR Academy of Sciences.[4]

[1] Komarov was over 70 at that time. I. M. Frank recollected that the president was frequently ill and "as a result, started to lose power, which passed almost entirely to the bureaucratic machinery of the Academy. A popular saying of that time stated that the Academy was governed by Komarilla (Camarilla) rather than by Komarov."* The above letter, found among archive documents of the AUCP(b) Central Committee, illustrates Komarov's attempts to overcome, with the help of top authorities, the unbearable situation that oppressed him as a scientist and administrator.

* In: Sergei Ivanovich Vavilov. Ocherki i vospominaniya (Sergei Ivanovich Vavilov. Essays and Memoirs) (pod red. I. M. Franka). M.: "Nauka", 1991, p. 18.

[2] The position of the Staff Department head in the USSR Academy of Sciences and other large government bodies was under the authority of the AUCP(b) Central Committee. Usually, it was filled by officers of USSR NKVD-NKGB or functionaries of the AUCP(b) Central Committee. They were appointed by a resolution of the Central Committee Secretariat and were practically insubordinate to the heads of the respective bodies.

The invasion of academic and other research institutions by secret officers of NKVD reached its climax during the Great Terror period of the late 1930s. These people occupied a greater variety of positions than those created to control the institutions' staff policy or foreign relations. To illustrate, here are CV data of A. I. Aseev, who graduated from the Moscow University (Economic Division of Historical Faculty) and was recruited by NKVD in April 1932. In 1938, he was appointed head of the program department in the Presidium of the USSR Academy of Sciences and in 1940 rector of the then famous Institute of Philosophy and Literature (resolution of the All-Union Committee on Higher Education). Yu. P. Sharapov, the former student of the Institute, noted half a century later that "the best arts school of that time was for almost two years headed by the NKVD officer! Such were grimaces or, to be precise, realities of the epoch."*

*Yu. P. Sharapov. Litsey v Sokol'nikakh. Ocherk istorii IFLI (Lyceum in Sokol'niki. Essay on the History of IFL). M.: AIRO-XX, 1995, p. 23.

[3] Secretariat of the Central Committee.

[4] A rather cynical formulation of the resolution of Secretariat of the AUCP(b) Central Committee. It involuntarily exposes the situation created in the country by the totalitarian regime. A formally unrelated body dismissed (not sanctioned or recommended dismissal, but dismissed!) an employee of the Presidium of the USSR Academy of Sciences.

Document 3.23

Extract from the letter of M. A. Apletin of the All-Union Society of Cultural Relations with Foreign Countries in Sweden to Smirnov, representative of the Society in Sweden, 7 May 1934.

[...] The question of Selma Lagerlof stands by itself. We congratulated her by telegraph on the occasion of her 75th birthday and she briefly thanked us.

We would like to know your opinion about the advisability of more regular contact with her. Her political face is quite clear and certainly deeply alien to us.[1]

However, her participation in the recent action of Swedish intellectauls against fascist terror [may be] a good starting point for making her interested in cultural developments in the Soviet Union and organizing personal correspondence with one of the Soviet authoresses...

AVP RF, fond 0140, opis' 18, papka 117, ed. khr. 336, l. 6.

[1] That is how *Pravda* characterized Lagerlof in more liberal times (21 November 1958, p. 6): "S. Lagerlof had erroneous ideas of the possibility of peaceful 'coexistence' between the rich and the poor, oppressors and oppressed, prosperous and wretched. She dreamt of a society based on 'Christian' principles where the supreme power is given to a kind heart."

Document 3.24

Academician Ya. O. Parnas to academician P. L. Kapitsa, 24 September 1944.

Dear Petr Leonidovich,

I am concerned with the following problem. To my mind, the Red Army together with Rumanian troops will enter Szeged in southern Hungary within the next few days. The University of Szeged has Albert Szent-Gyordyi[1] among its professors, one of the most prominent biochemists and outstanding scientists of our time; moreover, he is a true democrat, oppressed by the fascist government of his country.

You must know him because he worked at Cambridge at the same time as yourself. A few words are enough to characterize his scientific services. It was he who discovered ascorbic acid (vitamin C) in 1933 and was awarded the Nobel Prize for this discovery [...]

We have some information about his situation in wartime. In winter 1941–1942, he made a democratic speech at a meeting of the Hungarian Supreme Chamber (where he had been a member since the Nobel award was given to him) and was expelled from it. In winter 1943–1944, Professor Florey of Oxford staying here[2] told that English scientists are concerned about his fate, being informed that fascists put Szent-Gyordyi into an asylum. I had a talk with Professor Broish of Istanbul (S.-Gyordyi's pupil), who is equally concerned because he was told that S.-Gyordyi no longer works at the university and remains in disguise. The other day, Broish informed me that S.-Gyordyi is still at the university but not allowed to keep foreign correspondence.[3]

I am writing to you in full confidence that you will advise the respective executives to instruct our military units entering Szeged to give this noble man as much consideration as he really deserves from the Red Army and the Soviet Government [...]

RGASPI, fond 17, opis' 125, ed. khr. 362, l. 7–8.

[1] Modern spelling: Szent-Gyorgyi.
[2] In the Soviet Union.
[3] In 1944, Kapitsa headed *Glavkislorod* and was permanently in touch with top officials including M. Malenkov, secretary of the AUCP(b) Central Committee, to whom he transferred this letter.

Document 3.25

Memorandum on the political views of Professor Albert Szent-Gyorgyi (based on telegraph reports of the Political Department, 2nd Ukrainian Front) signed by Major General M. Burtsev, head of the VII Department of GlavPURKKA,[1] 26 January 1945.

The most well-known Hungarian scientist and Nobel Prize winner Professor Albert Szent-Gyorgyi[2] (University of Szeged) was persecuted by the Gestapo for his antifascist activity and had to live undeground in Budapest till Szeged was liberated by our troops.

According to him, he first joined political actions in February 1943 when he decided to facilitate the withdrawal of Hungary from the war, being deeply impressed by the outcome of the Stalingrad operation. To this effect, he visited Turkey (with the consent of Kallay)[3] to have talks with English and American diplomats. These talks had no practical results because they became known to Hitler, who demanded the extradition of Szent-Gyorgyi from Horti.[4]

Horti was aware of the above negotiations and managed to safeguard Szent-Gyorgyi by convincing Hitler that he had been given false information.

In the beginning of 1944, Szent-Gyorgyi tried to set up a new "radical-democratic" political party. The German occupation ruined his plans, certain organizers were arrested, and Szent-Gyorgyi himself was under police or Gestapo surveillance at one time and had to live in disguise at another.

On 24 August 1944, Szent-Gyorgyi was called by Horti to be informed about his decision to withdraw Hungary from the war after the capitulation of Romania in order "to avoid great swinishness."

Szent-Gyorgyi suggested that the regent should empower him to go to Moscow and conclude a truce with the Soviet Union. Horti approved the initiative and sanctioned his flight to Moscow on 25 August. However, he asked Szent-Gyorgyi on the morning of the next day to consider their last talk of no consequence. Horti is supposed to have changed his mind after negotiations with the German ambassador, who convinced him that German troops would be able to defend Hungary from the Red Army [...]

Szent-Gyorgyi described his personal view of the political situation in Hungary as follows: "I am really happy that we have signed a truce with the Soviet Union, England, and the United States. It means the resuscitation of Hungary [...] My personal activity may develop along the following lines:

(1) I could resume my research and teaching and thus influence only a narrow circle of my pupils and disciples.
(2) I could participate in the political life of this country, but [...] I am no politician and simply unfit for everyday political work due to some intrinsic psychological

deficiencies. However, I could be a banner for a certain wide political movement, elaborate and implement some political ideas.
(3) Finally, I can consider such a form of my collaboration with Russian and Anglo-American representatives as being simply a friend of theirs. Imagine that I visit them once a week or so to give advice, inform them about prevailing sentiments in public opinion, suggest my proposals, wishes, etc. [...]"

RGASPI, fond 17, opis' 125, ed. khr. 362, l. 4–6.

1 Main Political Department of the Workers and Peasants' Red Army.
2 Original spelling of the document is preserved.
3 Prime Minister of Hungary.
4 Hungarian dictator during World War II.

Document 3.26

From: G. F. Aleksandrov, head of the Department of Propaganda and Agitation, AUCP(b) Central Committee.
To: G. M. Malenkov, secretary of the AUCP(b) Central Committee.
15 February 1945

The world-known biochemist and Nobel Prize winner Professor Albert Szent-Gyorgyi worked at the University of Szeged, Hungary. Professor Szent-Gyorgyi is one of the most prominent biochemists in the world. In 1933, he discovered ascorbic acid (vitamin C) and purified it from plant tissue; this work brought him the Nobel Prize [...] After the Red Army entered Szeged, Professors Kapitsa and Parnas applied with a request to find Professor Szent-Gyorgyi and render to him as much assistance in his research as he deserves. Professor Szent-Gyorgyi is a democrat and was persecuted by German authorities and the Gestapo. In 1943–1944, he visited Turkey (with the consent of Kallay and Horti) and undertook unsuccessful attempts to withdraw Hungary from the war. He had to visit Moscow on 25 August last year. Szent-Gyorgyi had connections with the leaders of the "Hungarian Resistance Front" but never took active participation in the activity of this organization.

Comrade Susaikov, member of the Military Council of the 2nd Ukrainian Front, provided the following information (by telegraph): "The professor was found in a basement room in Budapest and is at present with us. During a talk, Szent-Gyorgyi stated his wish to go to Moscow to establish personal contacts with Soviet scientists. Advise, please, whether he should be brought to Moscow and, if so, specify his concrete destination. Please answer at your earliest convenience because Szent-Gyorgyi can not stay with us too long. I am waiting for instructions."

A later typewritten text reads as follows: "To comrade Molotov V. M. Please, give attention to this information. G. Malenkov. 16. II." and "To comrade Malenkov. 1. Szent-Gyorgyi should be given care and his political activity should be thoroughly clarified (responsibility of comrade Fitin).[1] 2. Szent-Gyorgyi may be brought to Moscow (with his consent) and settled accordingly;[2] if he wishes to stay in Hungary, however, let him choose by himself, but he must be given all necessary care and his political actions must be known (the responsibility of comrades Voroshilov, Pushkin[3] and Fitin). 3. People's Commissariat of Foreign Affairs will seek the advice of Voroshilov on the best way to treat Szent-Gyorgyi (comrade Dekanozov). V. Molotov. 17 February 1945."

This text is crossed by the note in Malenkov's hand: "Comrade Aleksandrov. I need a talk with you on this issue. 17. II." A separate sheet bears a note by Aleksandrov: "To be stored in the archives. Comrade Malenkov is in the know. 22 February 1945."[4]

RGASPI, fond 17, opis' 125, ed. khr. 362, l. 1–2.

[1] P. M. Fitin — head of the First (Intellegence) Department, People's Commissariat of State Security.
[2] In June 1945, Szent-Gyorgyi took part in the jubilee session of the USSR Academy of Sciences on the occasion of its 220th anniversary.
[3] After Szent-Gyorgyi returned from Moscow, he became close enough to G. M. Pushkin, Soviet minister plenipotentiary in Hungary, and could ask him for the liberation of his nephew, Imre Szent-Gyorgyi, from a prisoner of war camp in the Soviet Union (see doc. 3.27).
[4] In 1947, when candidates for the election of foreign corresponding members of the USSR Academy of Sciences were discussed in the AUCP(b) Central Committee, Szent-Gyorgyi was described as an "antifascist persecuted by German aggressors."*
*RGASPI, fond 17, opis' 117, ed. khr. 834, l. 204.

Document 3.27

From: S. N. Kruglov, people's commissar of internal affairs.
To: V. M. Molotov, vice-chairman of the council of people's commissars.
No. 927/k, 14 March 1946.

In the letter dated 11 January 1946, No. 38/bs, comrade Dekanozov informed the People's Commissariat of Internal Affairs that the known Hungarian scientist and Nobel laureate Albert Szent-Gyorgyi applied to G. M. Pushkin, Soviet minister plenipotentiary in Hungary, with a request to release his nephew, Imre Szent-Gyorgyi (born in 1912 in Budapest, a post-clerk, recruited in September 1944 to No. 14 artillery detachment and imprisoned in May 1945 in the city of Brunn).

Szent-Gyorgyi is kept in No. 338 camp of NKVD (Saratov region). NKVD has no materials discrediting Szent-Gyorgyi and does not object to his release. I am waiting for further instructions.

RGASPI, fond 9401, opis' 2, ed. khr. 142, l. 50.

Document 3.28

Extract from the letter of L. D. Kislova of the All-Union Society of Cultural Relations with Foreign Countries to V. Razin, representative of the Society in Sweden, 21 October 1944.

[...] Advise please with the soonest dispatch on political characteristics of Professors Carleman (presumably, a mathematician), Borgkvist, Bedberg, Stolfeld, and Swedberg with special reference to their attitude towards the USSR during the Great Patriotic War.[1] This information is urgently needed to decide who may be proper contact persons for our scientists.[2]

AVP RF, fond 0140, opis' 29, papka 128, ed. khr. 28, l. 37.

[1] A letter in reply from Razin has not been found.
[2] This request was most probably related to the organization of the jubilee session of the USSR Academy of Sciences on the occasion of its 220th anniversary that was held in summer 1945. Its foreign participants included Swedish scientists T. Swedberg and A. F. Westgren.* The latter name is mentioned (along with Swedberg) in Razin's letter dated 10 April 1945 (doc. 3.30).
* Yubileinaya sessia Akademii Nauk SSSR. 15 iyunya-3 iyulya 1946. T. 2 (Jubilee Session of the USSR Academy of Sciences. 15 June–3 July 1946. V. 2), M.-L.: Izd-vo USSR Academy of Sciences, 1947, p. 142.

Document 3.29

From: A. Volkov, head of the Central-European Department, All-Union Society of Cultural Relations with Foreign Countries.
To: S. G. Suvorov, head of the Department of Science, AUCP(b) Central Committee.
No. 63c, 24 April 1945.

As agreed in telephone conversations, please find enclosed data on Swedish and Finnish scientists provided by the USSR mission in Stockholm and UCC[1] in Helsinki.

RGASPI, fond 17, opis' 125, ed. khr. 362, l. 24.

[1] Union Control Commission in Finland.

Document 3.30

Extract from the letter of W. Razin, representative of All-Union Society of Cultural Relations with Foreign Countries in Sweden, of 10 April 1945 sent by the head of the Central-European Department of the Society, to S. G. Suvorov, head of the Department of Science, AUCP(b) Central Committee, 24 April 1945.

1. Professor Westgren was for some time member of the Swedish–Soviet Society but left it with the onset of the Finnish–Soviet War of 1939–1940 because "he did

not understand the character of this war" (as specified in primary information). At present, he changed his opinion (like all Swedes) and holds pro-British views. Westgren is secretary of the Swedish Academy of Sciences.
2. Professor Swedberg was a member of the Swedish-Soviet Society but left it with the onset of the Finnish–Soviet War. He changed his views, similar to Westgren. He remains an antifascist but does not sympathize with the Soviet Union.[1]
3. Esmark is unknown here. A misspelling?
4. No information concerning Professor Bottforse is available.
5. Professor Tiselius works together with Professor Swedberg. He is not a Nazi.
6. Professor Lise Meitner, working together with Professor Siegbahn,[2] is half or one-quarter German Jew who had emigrated from Germany. Emerging from the *Deutsche-nazionale* circles, she is a nationalist in the German spirit and an anglophile;[3] her attitude to the USSR is one of scepticism and distrust. Meitner emigrated to Sweden shortly before the war, in 1939. She became antifascist by necessity (non-Aryan).[4] Meitner is described as a gifted researcher.[5]

RGASPI, fond 17, opis' 125, ed. khr. 362, l. 29.

[1] A typical disillusion of Soviet functionaries, who could not distinguish between the attitude of foreigners towards a nation and the concrete actions of its rulers. Three months after this document, V. Razin sent to the Society of Cultural Relations the Russian translation of Swedberg's article published in the Communist newspaper *Ny Dag* of 7 July 1945 about his stay in the Soviet Union during the jubilee session of the USSR Academy of Sciences on the occasion of its 220th anniversary. The article was headed "They First of All Thought of Science in the Soviet Union" and was full of enthusiastic comments on the system of state support of science in the country.*

* GARF, fond 5283, opis' 20, ed. khr. 253, l. 66–73.

[2] Manne Siegbahn, Nobel laureate in physics for 1924.

[3] A strange characteristic. It is impossible to be a German nationalist and an anglophile at a time.

[4] A far-fetched statement at variance with German biographic sources concerning the outstanding atomic physicist; see, for instance, Beitrage zur Physik und Chemie des XX Jh. Brauschweig, 1959; D. Crawford. Lise Meitner. *Atomic Pioneer.* N.–Y., 1969.

[5] Too modest praise for the scientist 17 times nominated for the Nobel Prize in physics and chemistry between 1924 and 1949. Meitner had a total of 33 nominators including Nobel laureates de Broglie, N. Bohr, Heisenberg, Compton, von Laue, Planck, Swedberg, J. Franck, and O. Hahn.

Document 3.31

Correspondence in connection with the publication of E. Hemingway's novel *For Whom the Bell Tolls* in the USA.

I

T. Rokotov, acting editor-in-chief of the journal *Inostrannaya literatura*, informs G. F. Aleksandrov, head of the Department of Propaganda and Agitation, AUCP(b) Central Committee, (7 February 1941) about the publication of E. Hemingway's novel *For Whom the Bell Tolls* in the USA.[1] The novel contains a chapter "slanderously depicting Andre Marti[2] under his own true name as the chief commissar of international brigades."[3]

Rokotov asks for permission to publish a critique of the novel.

II

From: A. Puzin, head of the Department of Propaganda and Agitation, AUCP(b) Central Committee.
To: A. A. Zhdanov, Secretary of the AUCP(b) Central Committee
13 June 1941.

[...] One of the characters in the novel is Andre Marti acting under his own name and depicted as a cruel, narrow-minded person whose actions impede the struggle of the Spanish people. Journalist Karkov (in whom Hemingway embodied M. Kol'tsov),[4] cynical and immoral, is shown as a mouthpiece of Communist ideology [...] The ideological content of the novel reduces to the demonstration of moral superiority of bourgeois-democratic principles over Communist ideology. Therefore, it is unfeasible to publish the novel in the Soviet Union despite the writer's sympathy with the struggle against fascism. It is equally out of place to criticize it in *Inostrannaya literatura* because [...] any article by a Soviet author would be regarded as our official dispute with Hemingway.[5]

RGASPI, fond 17, opis' 125, ed. khr. 62, l. 1–4.

[1] The novel was first published in the United States in October 1940 and was rapidly translated into Russian in the Soviet Union. Anyway, it appears from the letter of the editor-in-chief of the *Inostrannaya literatura* that the translation was already available by the beginning of February 1941, i.e. less than four months after the publication in the United States. Ilya Erenburg told about the circulation of handwritten copies of the translation in his memoirs covering 1941.* Once more, information about a handwritten copy was mentioned in the notes of the writer Daniil Danin, published posthumously.**

* "[...] air-raid warning was given in Moscow almost every night in the end of July 1941; all had to go down to bomb shelters. We (B. M. Lapin and I) needed sleep and decided to spend a night in the vacant country-house of Vishnevsky in Peredelkino. I had a handwritten copy of Hemingway's novel translated into Russian. We did not sleep at all but read throughout the night, passing one page

after another between us. Anti-aircraft guns thundered but we kept reading [...]" Ilya Erenburg. Sobranie sochineniy. T. 9. "Lyudi, gody, zhizn' (Collected Works. V. 9. People, Years, Zhizn')". M.: Khudozhestvennaya literatura, 1967, p. 154.

** Danin happened to come to near-front Moscow on 15 October 1941, the day before it was panic-stricken by the real threat of Nazi troops breaking the defenses and entering the city. On 16 October, he dropped into the empty building of *Goslitizdat* and found there a single woman who "wandered along a corridor [...] with a thick paper-case in her hands [...] and told that it was a manuscript of the translation of *For Whom the Bell Tolls*. "Voprosy literatury," No 6, 2005, pp. 254–255.

[2] French Communist, member of *Politbureau* of the French Communist Party. At the plenary session of the Central Committee held 3–4 September 1952, Marti was for the first time publicly charged with nationalistic views, opportunism, refusal to implement resolutions of the XII Congress of the French Communist Party, and other faults. After a time, he was expelled from the party. R.Orlova, literary critic, wrote in 1964: "[...] history confirmed that Hemingway better and with deeper insight understood Marti than many of his brothers-in-arms [...] and generalized in his character certain traits of the personality cult."* Erenburg was of a similar opinion: "The novel was blamed by different people and from different standpoints; [...] some disliked [...] pages devoted to Andre Marti (once a writer tells anything 50 years or even a day before it becomes generally known, he is attacked by everybody...).**

* Zhurnal "Zvezda," No 1, 1964, p. 213.

** "Lyudi, Gody, Zhizn'", p. 158.

[3] During the Civil War in Spain, 1936–1939.

[4] Writer and journalist, participated in the Civil War, wrote the book *Ispanskiy dnevnik* (The Spanish Journal) published in 1938. Negative characteristics are given to him by the author of the document, which is attributable to the fact that Kol'tsov was repressed in 1940 (rehabilitated posthumously).

[5] There were no debates about Hemingway for political reasons, but his works were not published in the USSR from 1940 to 1955 (barring a small story *Voyage to Victory*). To compare, 45 *belles-lettres* and publicistic works of Hemingway were published from 1934 to 1939 inclusive.

Document 3.32

From: A. M. Rumyantsev, head of the Department of Science and Culture, AUCP(b) Central Committee; P. A. Tarasov, deputy-head, and V. I. Ivanov, sector head.
To: Secretariat of the AUCP(b) Central Committee.
27 July 1955

[...] Inclusion of Hemingway's novel *For Whom the Bell Tolls* in the publication plan of the "Foreign Literature" publishing house for 1956 and the announcement about it in the newspaper *Literaturnaya gazeta* is a serious mistake of the direction of Inoizdat and its *belles-lettres* sector [...] as well as of the editorial board of *Literaturnaya gazeta*. We consider it necessary to point out to this mistake [...] and recommend that *Literaturnaya gazeta* denounce the approach of Inoizdat to the choice of foreign literature works for publication.[1]

RGANI, fond 5, opis' 17, ed. khr. 539, l. 113–115.

[1] As a result, the award of the Nobel Prize to Hemingway in 1954 was indirectly marked only by three editions of *The Old Man and the Sea* (1956);* these editions put an end to a long (since 1949) break in the publication of Hemingway's works in the Russian language.

* Two books were out in Moscow (*Biblioteka ogon'ka*, No. 24 and Isdatel'stvo *Detskaya literatura*) and one in Kiev (Goslitizdat of the Ukraine). One year before, the story was published in the journal *Inostrannaya literatura*, No. 3, 1955.

Document 3.33

Extract from the letter of D. A. Polikarpov, head of the Department of Culture, CPSU Central Committee, and V. I. Ivanov, sector head, to the CPSU Central Committee, 4 January 1956.

In early 1955, Erenburg, then member of the editorial board of the journal *Inostrannaya literatura*, tried to impose his views and ensure publication of the authors of his choice.[1] At the meetings of the editorial board, he spoke with boundless enthusiasm about the naturalistic and wingless story *The Old Man and the Sea* (in which, in his opinion, "even the weak places surpass the average things published in our journals").[2] Erenburg recommended as "true" writers Faulkner, a formalistic and obscure author, and Mauriac, a reactionary catholic writer from France [...][3]

Kul'tura i vlast' ot Stalina do Gorbacheva. Apparat TsK CPSS i kul'tura. 1953–1957. Dokumenty (Culture and Authorities from Stalin till Gorbachev. Apparatus of CPSU Central Committee 1953–1957. Documents). M.: ROSSPEN, 2001, p. 467

[1] This remark was provoked in the first place by publication, on the recommendation of Erenburg, of the first Russian translation of Hemingway's story awarded the Nobel Prize for 1954 (March 1955).

[2] Two years later (25 January 1958), heads of the Department of Culture turned back to the Nobel-winning story in a memorandum prepared for the Secretariat of the CPSU Central Committee. Its previously negative appraisal was extended as follows: "Certain authors praise to the skies Hemingway's *The Old Man and the Sea*. In fact, it is an apolitical work in an individualistic spirit."*

* Kul'tura i vlast' ot Stalina do Gorbacheva. Ideologicheskie komissii TsK CPSS 1958–1964. Dokumenty (Culture and Authorities from Stalin till Gorbachev.

Ideological Commissions of CPSU Central Committee 1958–1964. Documents). M.: ROSSPEN, 1998, p. 37.

[3] In other words, the leadership of the Department of Culture, CPSU Central Committee, had an aversion to W. Faulkner and F. Moriac, besides Hemingway, who were awarded the Nobel Prize in 1949 and 1952 respectively.

Document 3.34

Extract from the letter of B. D. Petrov, sector head of the Staff Department, AUCP(b) Central Committee, to A. A. Kuznetsov, secretary of the Central Committee, 2 October 1947.

Medgiz[1] has just published D. L. Rubinshtein's book *General Physiology* (over 40 author's sheets, circulation 10,000) presenting a bright example of servility to foreign science, neglect of the achievements of our science, and the irresponsible attitude of the publishing house to the choice of literature [...]

It is no coincidence that Professor Rubinshtein works at the Institute of Biological and Medical Chemistry, Academy of Medical Sciences, headed by academician Parnas. This Institute already published the book entitled *Progress in Biology* (1947)[2] with an article by Professor Parnas, where foreign science was praised as being opposed to Soviet science; the author wrote about the "insignificant role of the Soviet Union in the general development of biochemistry making enormous progress during the period under consideration."[3]

RGASPI, fond 17, opis' 127. ed. khr. 1670, l. 162, 165.

[1] State Medical Publishing House.

[2] Correct heading: *Recent Progress in Biology.*

[3] There is no such phrase in the book. Nor is there an article signed by this author among the 13 included in the volume. Parnas wrote only a short preface to the book as its editor-in-chief.* Here is an extract from the preface: "During the Great Patriotic War and the evacuation of research institutions, many researchers had no access to scientific literature and published few works; conferences and reports about them were rare, foreign publications were unavailable for the majority of institutions. For these reasons, many fell behind in scientific development and were unaware of recent achievements. In order to enhance the level of knowledge, I organized conferences at which topical problems and progress in the most advanced scientific disciplines were reviewed."**

* In the library copy of the volume from which this citation is borrowed, the name of Parnas as the author of the preface and the editor-in-chief (at the title page) is thoroughly stained with black Indian ink. It suggests that Glavlit (the top censorship board in the Soviet Union) distributed relevant instructions to the libraries after the arrest and the death of the scientist.

** Uspekhi sovremennoy biologii. Sbornik trudov ob'edinyonnykh nauchnykh konferentsiy Instituta biologichesloy i meditsinskoy khimii AMN SSSR and Laboratorii fiziologicheskoy khimii AN SSSR. T. 1. (Recent Progress in Biology. Transactions of Joint Scientific Conferences of the Institute of Biological and Medical Chemistry, USSR Academy of Medical Sciences, and Laboratory of Physiological Chemistry, USSR Academy of Sciences, V. 1). M.: Medgiz, 1947, p. 5. In other words, the author of the above letter deliberately committed forgery and showed negligence and slovenliness in citing the title of the book.

Document 3.35

From: G. F. Aleksandrov, head of the Department of Propaganda and Agitation, AUCP(b) Central Committee.
To: A. A. Andreev and A. S. Shcherbakov, secretaries of the Central Committee. 7 October 1941.

In pursuance of the decisions of AUCP(b) Central Committee, Narkompross RSFSR[1] had to prepare several textbooks in history [...] The textbook for new history for the ninth term of the secondary school edited by Professors L. I. Zubok and M. V. Khvostov was written before the war with Hitler's Germany against the Soviet Union. Naturally, it contains many statements that can not be published today. These politically wrong statements include sharp criticism of the foreign policy of England and America and pass over in silence the imperialist policy of Germany. What was *right* [italicized by the author] before 22 June 1941 is not right now.[2] There are about 30 such statements in the book. All of them are marked in the advance volume and need correction. I therefore ask you is (1) to permit postponing the circulation of the printed textbook for new history for the ninth term of the secondary school and (2) to introduce necessary changes in its text taking into account the current situation.[3]

RGASPI, fond 17, opis' 117. ed. khr. 298, l. 196–197.

[1] People's Commissariat of Education.
[2] Most cynical reasoning.
[3] Secretariat of the AUCP(b) Central Committee took the following decision with respect to Aleksandrov's application: "Postpone circulation of the printed textbook for new history for the ninth term of the secondary school. Charge the authors with introducing necessary changes in its text to take into account the current situation."*

* *Ibid,* opis' 116, ed. khr. 100, p. 88.

Chapter 4 • 1945–1947

From Awakening to an Ideological Clinch

After the end of World War II, the Nobel organizations resumed annual awards in all categories of the Nobel Prize. However, connections with Soviet scientists seemed to have been totally lost although invitations were regularly sent to them from Stockholm. The Nobel Committees were seriously concerned about the situation and undertook unusual steps to recover contacts with Soviet colleagues through diplomatic channels that they traditionally avoided using.[1]

The Committees in physics and chemistry, acting through the agency of the Swedish embassy in Moscow, passed on to L. D. Kislova, member of the Executive Board of the All-Union Society for Cultural Relations with Foreign Countries (VOKS), 30 letters inviting Soviet nominators to propose their candidates for the Nobel Prizes of 1946. The letters were readdressed to the Presidium of the USSR Academy of Sciences (doc. 4.1) as recorded in the "List of Swedish correspondents of VOKS for 1945" (doc. 4.2).

The letters eventually reached the addresses, as follows from the fact that VOKS archives contain the original nominations presented by A. A. Lebedev (doc. 4.6) and A. N. Terenin (doc. 4.4), two of the 30 invited persons listed in the appendix to doc. 4.1. L. A. Orbeli received an invitation directly from the Nobel Committee in physiology and medicine sent to Leningrad by post.

Simultaneously, Swedish scientists tried to personally re-establish contacts with their Soviet colleagues interrupted by the war. Theodor Svedberg, Nobel Prize winner in chemistry, was especially active. Even before the end of the war with Germany, Professor A. N. Frumkin transmitted to Svedberg (via VOKS) reprints of his works for the last years (doc. 4.7). In a few months, Ya. I. Frenkel, corresponding member of the USSR Academy of Sciences, also sent to Svedberg reprints of his published papers (doc. 4.8).

The Physico-Chemical Institute of the University of Uppsala, where Svedberg and another Nobel laureate, Manne Siegbahn, were based, addressed VOKS for assistance in establishing contacts with N. N. Semenov. K. F. Vinogradov, representative of VOKS in Stockholm, sent the application to Moscow in May 1945 but received an answer only after four months. The answer was positive (doc. 4.10), but researchers in Uppsala were no longer interested in the matter. The jubilee session of the USSR Academy of Sciences on the occasion of its 220th anniversary was held in the Soviet capital from 15 June to 3 July 1945. T. Svedberg was among eminent foreign scientists from different countries invited to participate and had an opportunity to make acquaintance with his Soviet colleagues, including Semenov.[2]

The Nobel Statutes envisage one more category of foreign nominators, besides those receiving individual invitations from the respective Committees to nominate candidates for a given year (listed in the appendix to doc. 4.1). These are professors chosen by the rectors of foreign universities that a given Nobel Committee selected as nominating organizations.[3]

The University of Leningrad along with other five universities was given such a right in 1946. The initial list of ordinary professors, physicists and chemists, prepared by its rector Professor A. A. Voznesensky at first confused VOKS officers (doc. 4.9). In the end, the list was sent to Stockholm. But the acting chairman of the Executive Board A. V. Karaganov tried to avoid responsibility, asking the VOKS representative in Sweden to decide himself whether the list should be delivered to the respective Nobel Committees (doc. 4.10).

There is an anecdotic detail in the story of this list. The head of the Scandinavian Department of VOKS, E. P. Meleshko, was clearly puzzled, as appears from a marginal note in doc. 4.10 ("[...] ask [...] for what purpose the Nobel Committee needs these data"). The bewilderment was shared by A. V. Karaganov, who wrote in doc. 4.10 "[...] we do not know what purpose the committee pursued by asking for these data [...]".[4] Probably, the Swedish side was aware of these doubts. Anyway, Staffan Sederblum, the Swedish envoy in Moscow, sent five copies of the Nobel Statutes to VOKS in March 1946 and a request to convey them to those institutions that might be interested in procedures preceding prize awards (doc. 4.14).

The Goddess of History Cleo often acts with an unkind smile curling her lips. The cover letter of Sederblum was dated 14 March. A week and a half before (5 March 1946), W. Churchill made his famous speech in the American town of Fulton that is generally believed to have initiated the "cold war."[5] Therefore, the attempt of the Swedish side to enlighten Soviet bureaucrats and eliminate suspicion about certain hidden motives behind the interest of the Nobel organizations in the names of university professors was doomed to failure from the very beginning.

However, the cold wind that swept away all hopes to re-establish contacts between the Nobel organizations and Soviet scientists had been felt in the USSR for some time before Churchill's speech. The demarcation line lies between 1945 and 1946. The fears of VOKS functionaries in summer 1945 were trifling compared with what followed.

Here are a few facts to depict the situation.

As mentioned above, the notorious list of professors of Leningrad University was sent to Stockholm in September 1945. In late October, the Soviet nominators received invitations to propose their candidates for the Nobel Prizes in physics and chemistry. However, the nominations (at least those of Professors A. A. Lebedev and A. N. Terenin) never went outside the VOKS office (late December 1945 to early January 1946) (see docs. 4.3 and 4.5).

Even more trouble was given at the request of Professor G. Liljestrand, secretary of the Nobel Committee in physiology and medicine, to present a list of professors of the Medical Faculty, Moscow State University. The committee decided to grant them the right to nominate candidates for the Nobel Prize of 1947 in accordance with para. 5, p. 5 of the Special Rules for the Karolinska Institutet, the award-giving organization in these disciplines (doc. 4.20).

A characteristic detail is worth mentioning. Before despatching the above request, Liljestrand, in the capacity of professor at the Department of Pharmacology, Karolinska Institutet, sent a personal letter and a reprint of his scientific paper to N. N. Anichkov, a reputed Soviet pathologist. The packet safely reached the addressee, who thanked his correspondent, as follows from the registry of sendings to the VOKS office in Sweden dated 6 April 1946 (doc. 1.18). However, the spirit of time was already noticeable: the special department of VOKS demanded that the address of the Second Leningrad Medical Institute be crossed off the letter (doc. 4.19).[6]

In other words, Liljestrand had some experience of postal communications with Soviet scientists and obviously waited for a positive and timely response to his request. However, his were vain expectations. Political considerations certainly prevailed in Moscow. Certainly, habitual negligence also played a role.

Members of the Nobel Committee were unaware that Medical Faculty of the Moscow University was redesignated as a separate First Medical Institute in 1930. They had to be informed about this as soon as the VOKS office in Stockholm received the letter of Liljestrand. However, the letter was not read with due attention, and the work over the list in Moscow was from the very beginning heading in the wrong direction (see note 1 to doc. 4.22). The first list drawn up by the Medical Section of VOKS (doc. 4.22) included 15 candidates of which only three (A. I. Abrikosov, V. N. Vinogradov, and N. A. Semashko) were professors of the 1st Medical Institute, the assignee of the former university faculty; seven others lived in Leningrad and Kiev. In other words, had the approved list ever reached Stockholm, it would have been rejected all the same by the Nobel Committee as failing to meet the Statutes requirements.

In fact, no one was going to send the list to Sweden in the situation of 1946. The letter of the secretary of the Nobel Committee became an object of bureaucratic manipulations.[7] Sham activity proceeded along two lines, via VOKS[8] and along diplomatic channels.

The letter from Liljestrand was received by the USSR mission in Sweden on 20 March 1946 and sent not only to VOKS (doc. 4.21) but also to the 5th European Department of the USSR Ministry of Foreign Affairs (MID USSR)[9] (see note 3 to doc. 4.20). In two months (8 June 1946), the head of the Department A. N. Abramov asked the mission "about the Nobel Committee in Sweden and its address." The professional diplomat responsible for the maintenance of relations with Scandinavian countries could not understand why Stockholm needed the names of Soviet professors. It did not take much time to have an answer from Stockholm.

Acting *charge d'affairés* I. S. Chernyshev[10] wrote to Abramov on 4 July 1946 about the absolute necessity to positively respond to the initiative of the Nobel Committee in physiology and medicine, although prefaced his convincing argumentation with ritual rhetoric about "systematic disregard [...] of representatives of Russian science and culture" (doc. 4.24).

A copy of Chernyshev's letter was sent to VOKS but induced no apparent activity in this organization. The Nobel Committee was concerned about Moscow's silence because invitations to nominators should have been distributed during September. It appears from doc. 4.25 that the Committee asked to send the list of Moscow professors by telegraph before 17–18 September.

This date attracted attention from the heads of VOKS. A. V. Karaganov, vice-chairman of the Executive Board signed the prepared document on 17 September but sent it to the G. F. Aleksandrov, head of the Department of Propaganda and Agitation, AUCP(b) Central Committee, instead of the Nobel Committee (doc. 4.25).

The delay was hardly a result of usual bureaucratic negligence. The protraction would have been immediately punished because they did not forgive such carelessness in the Central Committee and strictly controlled the movement of documents. In fact, it was nothing more than bureaucratic games under the scheduled scenario. The underlying motive of all these delays was the decision by ideological services to secretly block the initiative of the Nobel Committee.

... By mere coincidence, the dispatch of the list of potential nominators from VOKS to the AUCP(b) Central Committee fell on the period when ideological authorities of the country started to increasingly demonstrate intention to curtail scientific and cultural relations with foreign states. The first episode occurred in 13 September 1946 when members of the Secretariat all of a sudden cancelled the visit of Professor A. N. Terenin to Paris for the participation in the International Optical Congress to be held on the occasion of the 25th anniversary of the French Institute of Theoretical and Applied Optics; the visit was previously approved by all authorities including L. P. Beria, then deputy prime minister, and M. A. Suslov, head of the Department of Foreign Policy, AUCP(b) Central Committee (doc. 4.26).[11] On 1 October, the Politbureau described as unsuitable the participation of Professor N. D. Papaleksi in the general meeting of the International Scientific Union of Radio (doc. 4.27).[12] It is worth mentioning that both scientists were regarded by the Nobel Committees in physics and chemistry as nominators for the 1946 prize (doc. 4.1)...

One more month had elapsed after VOKS forwarded the copy of Chernyshev's message and the cover letter of A. V. Kaganov to the Central Committee of AUCP(b) before G. F. Aleksandrov, the head of Department of Propaganda and Agitation, applied to A. A. Zhdanov, secretary of the Central Committee, with the proposal to "use the opportunity to strengthen international cultural relations" (doc. 4.29). On receiving the outdated document, Zhdanov did not punish the careless bureaucrats but wrote for additional information to the Ministry of Health and the Military Medical Department.

The new addressees, unaware of the ongoing game, responded quickly and suggested concrete proposals (doc. 4.29), unlike Aleksandrov who wrote to his superior again only after three months (in late January 1947), i.e. close to the deadline for nominations in the current year (doc. 4.30).

The comedy would have ended in accordance with the designed scenario, had not the Nobel Committee continued to insist. On 12 February 1947 (i.e. a month earlier than in the preceding year), Professor Liljestrand informed the Soviet side about the wish of the Nobel Committee to "invite our colleagues from Russia" to join the Nobel movement involving scientists all over the world. The cover letter of VOKS representative in Sweden also reminded that "almost a year has passed [...] but we still do not have an answer." (doc. 4.31)

However, Soviet ideologists continued to realize the same scenario. Aleksandrov presented the documents from Stockholm to Zhdanov only at the end of June and noted, as if surprised, that this "issue remains topical even today." (doc. 4.31) The final stroke in this shameful epistolary intercourse was made in a note by S. G. Suvorov, subordinate to Aleksandrov, addressed to Zhdanov's Secretariat. The author cynically stated that the proposal of the Nobel Committee "needs not to be further considered because September" is the last month for nominations to be submitted.

The sad outcome of two of Liljestrand's applications looks quite logical in the situation that took place in the Soviet Union in 1946–1947. A benevolent response to the proposal of the Nobel Committees would be unnatural in the political atmosphere of that time, characterized by a series of unprecedented anti-intellectual campaigns launched in the country.

In June 1947, when functionaries of the Department of Propaganda and Agitation were preparing the document concerned with the second application of the Nobel Committee in physiology and medicine, another project was being elaborated in relation to the then forthcoming election of foreign members of the USSR Academy of Sciences. By the decision of Politbureau dated 10 June, the names of outstanding scientists from the USA, Great Britain and Mexico were crossed off the list of candidates presented by the Presidium of the Academy; four of them were Nobel laureates. The argument was that Soviet scientists "were not elected to scientific organizations analogous to our Academy in the last years." (doc. 4.37) Instead, Politbureau proposed to elect to the Academy the Indian scientist Raman; Frederic and Irene Joliot Curies, known for their pro-Communist political views; and Szent-Gyorgyi of Hungary.[13]

The day before, Presidium of the USSR Supreme Council adopted a decree toughening responsibility for "the divulgence of state secrets". The main purpose of this decree was to maximally extend the range of information to be regarded as "secret". Four months later, Professor P. P. Shirshov, director of the Institute of Oceanography, addressed the following statement to the Communist Party meeting: "There are facts suggesting a misunderstanding of what is secret and what is not [...] that all earlier published data are not secret. It is a delusion because any data properly collected and treated may become secret."[14]

It was not the first experience of classifying as secret information that was actually not so. The resolution of AUCP(b) Central Committee dated 22 May 1942 obliged censors to strike off paragraphs containing criticism of the country's leadership in foreign scientific journals received in the Soviet Union via subscription. The exception was made for members of the USSR Academy of Sciences to whom such journals were available as "secret". The subscribers could read them in special premises following "the established procedure for the use of secret materials." Naturally, censorial pressure roused indignation among the subscribers that ended in a scandal in late 1945. P. L. Kapitsa was the first to raise the protesting voice (doc. 4.42).

A logical continuation of the campaign designed to heighten tension in intellectual circles was the decision by the Politbureau of 14 July 1947 to cease publication of three natural science journals in the English language (doc. 4.38).[15] The decision was initiated by the letter of G. F. Aleksandrov and his deputies S. G. Suvorov and P. N. Fedoseev addressed to A. A. Zhdanov, A. A. Kuznetsov, and M. A. Suslov. The letter described the publication of English-language journals as "servility of certain organizations to the West. The fact that some researchers are eager to have their papers published in these journals reflects reverence of our intellectuals for the foreign practice to judge the value of scientific results depending on whether they are recognized abroad or not."[16]

Such a measure as the shutdown of English-language journals of the USSR Academy of Sciences had no rational explanation and therefore remained a matter of concern to Communist Party ideologists. For this reason, the Party organization of the Presidium of the USSR Academy of Sciences was instructed in April 1947 (i.e. long before the above letter was written) to hold a meeting to discuss the forthcoming measure. Some participants doubted the usefulness of such an initiative, but leaders of the Academy managed to ensure its "unanimous support" (doc. 4.39). After the strange (to put it mildly) decision of the Politbureau was taken on 14 July, only words of approval were heard at Party meetings held in academic and other research institutions...[17]

One of the speakers in the Mathematical Institute, the USSR Academy of Sciences, even suggested an explanation of "servility to the West" in mathematical sciences (15 September 1947). He recalled that the mathematical series of the oldest journal *Izvestiya akademii nauk* was founded in 1832 with the title in French (a parallel Russian title appeared in 1895). The speaker expressed satisfaction that "recently, the French heading was removed under the decision of the government and Party Central Committee." Simultaneously, publication of articles in foreign languages was stopped in the popular periodical *Matematichesky sbornik* whose portfolio was "full of materials," according to the same speaker.[18]

The "retroactive force" principle being alien to Soviet law, the formerly permissible publications in English-language academic journals were later regarded as punishable. For example, Professor A. A. Zhukovitsky, director of research for the L. Ya. Karpov Physico-Chemical Institute, was dismissed from his position in October 1948 (i.e. half a year after the shutdown of foreign-language journals). One of the reasons was his work as an editor of *Acta physico-chemica.* In April 1948, he addressed a letter to A. A. Kuznetsov, secretary of AUCP(b) Central Committee, in which he recognized the fact of "sending reprints of some of his papers to foreign scientists,"[19] that is practicing generally accepted exchanges of information with colleagues from abroad. However, it did not help him to retain his position.

The shutdown of English-language journals was logically followed by the cessation (since September 1947) of publication of English abstracts in all scientific and technical journals that existed in the Soviet Union.[20] This absurd situation lasted almost eight years, until June 1955. Even then, however, English abstracts were allowed only in 20 academic journals and a special decision from the CPSU Central Committee was needed to expand this circle further.[21]

In 1948, president of the USSR Academy of Sciences S. I. Vavilov tried to secure cancellation (even if partial) of this savage decision. He applied to the Council of Ministers with a request to make an exception for *Astronomichesky ezhegodnik SSSR* founded in 1922. This journal published tables with only small amounts of parallel explanatory texts in Russian, English, and French and provided an efficacious instrument for the collaboration of Soviet astronomers with their foreign colleagues. Bearing in mind that all the above restrictions were largely directed against the UK and USA, Vavilov was ready to reconcile with saving the French version at the expense of the English one (doc. 4.48).

However, this roundabout maneuver did not help and the president's initiative failed.

Scientists in the West could not see any reason for such restrictions. Some of them happened to come face to face with the totalitarian regime. Max Born, a future Nobel winner, who had emigrated from Nazi Germany to Great Britain in 1933 tried delicately to draw attention to the absurdness of the situation. In November 1948, he wrote a letter to the Soviet embassy in London mentioning the lack of works of Soviet authors in foreign languages. He wished to know whether it was possible to "renew such publications" (doc. 4.50). However, the letter remained unanswered, exactly as Vavilov's application…

The shutdown of foreign-language journals was followed by accusation of the Foreign Literature publishing house, the work of which was described as unsatisfactory in the resolution of the AUCP(b) Central Committee dated 25 July 1947. The director of the publishing house B. L. Suchkov, a known literary critic, was charged with malevolent actions, such as "ordering books [...] intended to invade the Soviet Union with counter-revolutionary literature and demonstrate that science and technology are developing only in the USA and England."[22] One more accusation concerned the inclusion of works by Russian emigrant authors in the publishing plan. One such work appeared in the first half of 1947. It was the monograph *Antagonism of Microbes and Antibiotic Compounds* by Z. A. Vaksman,[23] the 1952 Nobel laureate in physiology and medicine, who had left Russia in 1910.[24]

The same resolution mentioned two more books translated into Russian and included in the publishing plan for 1947. One was G. A. Gamow's *Biography of the*

Earth containing the first scientific substantiation of the "hot universe" concept and the other *Genetics and the Origin of Species* by F. G. Dobrzhansky.[25] In the late 1920s, both authors, based at Leningrad, were invited to continue studies at foreign universities as recipients of the Rockefeller Brothers Fund scholarship. The fate of the two books in Russia was not as happy as of Vaksman's monograph; they never appeared in print and the typesetting was stopped.

As a result of these actions, works by foreign scientists practically ceased to be translated into Russia by 1949, and Soviet scientific publications were no longer translated into foreign languages. Apart from other things, this created a situation that could not avoid attention from functionaries of the AUCP(b) Central Committee. The Department of Propaganda and Agitation informed G. N. Malenkov, secretary of the Central Committee, on 15 January 1949 that "services of translators in the USSR Academy of Sciences can not be properly used because they have nothing to translate."[26]

One of the few exceptions in the last period of Stalin's rule was the publication of the monograph *Soil Microbiology* by S. N. Vinogradsky, a nominator on the Nobel Committees, who had emigrated from Russia and lived in France. This event, akin to a small miracle, became possible owing to the insistence and resourcefulness of S. I. Vavilov, president of the USSR Academy of Sciences and (after his death) of A. A. Imshenetsky, corresponding member of the Academy and director of the Institute of Microbiology.

Vavilov made an agreement with the author about the translation of his yet unpublished book on the advice of Professor V. L. Isachenko, a stalwart of Vinogradsky,[27] in the beginning of 1946. Vavilov even managed to help the old scientist, who had financial problems after German occupation, by remitting him 100,000 francs against the author's fee.[28]

The monograph in French was out in November 1949 and aroused great interest among the world scientific community. N. R. Dhar of Allahabad nominated the author for the Nobel Prize in chemistry for 1952. However, the organization of the Russian translation encountered difficulties. Correspondence between Imshenetsky and N. N. Sushkina, microbiologist and one of the selected translators, was forcibly interrupted (doc. 4.40-III). Preparation of the Russian version of the book was resumed only nine months after the book had been published in French. It may be conjectured that Vavilov and Imshenetsky reached a compromise with ideologists by consenting not to mention the émigré status of Vinogradsky. Anyway, neither the long article about the scientist published as the preface to the monograph (see doc. 4.40 and notes 8 and 10 therein) nor his biographical note in volume 8 of the *Great Soviet Encyclopaedia* mentioned that Vinogradsky was a professor of the Pasteur Institute and lived in France for almost 30 years.[29]

The monograph was published in 1952 at the peak of the ideological clinch during the late Stalinist period, when it seemingly could not be allowed by definition. It was really a civic feat of S. I. Vavilov and A. A. Imshenetsky...

Absurd restrictions and prohibitions also affected scientific publications in the Russian language; in the first place newsletters designed for exchange of information with foreign research institutions. Demands from ideologists from the Central Committee concerning international data exchange were sometimes made even more absurd by local Communist Party organizations. By way of example, one of them (at the Institute of Geophysics, the USSR Academy of Sciences) decided to control regular exchange of information about stratospheric changes, physics of clouds, and the like, of primary importance for long-term weather forecasting. The logic of such decisions was sometimes ridiculous (doc. 4.5).

The wall between the Soviet Union and Western democracies grew very quickly. Systematic anti-West propaganda directed in the first place against the United States could not but influence public opinion and psychology at large. Support for the "policy of the Party and the government" came not only from common people but also from intellectuals who were not always motivated by the fear of authorities. The personal fund of I. V. Stalin contains a letter from a Navy captain reported to him by the secretary. The officer asks Stalin whether his "definition of Lenin's work style that implies an American businesslike manner given in 1924 still holds."[30] The officer answers himself: "We, practical workers, are now strong enough to do without resorting to American practical school [...] A contemporary Leninist worker far surpasses an American businessman in terms of efficiency."[31] Any further comment is superfluous.

The blockade of contacts with the Nobel organizations looked like a logical element of the country's lifestyle formed by hysterical propaganda. It could not be otherwise at the early stage of the post-war ideological clinch, which encompassed all aspects of contemporary Soviet society. Half a century later, A. N. Yakovlev, an architect of Gorbachev's *glasnost* in the capacity of the head of the Department of Propaganda and Agitation and secretary of the CPSU Central Committee, recognized that "ideology was a steel hoop of the system; everything was subordinate to ideological dogmas."[58] Therefore, the break between the Soviet authorities and the Nobel organizations was inevitable, and the potential was fully realized.

The Nobel movement is in its essence incompatible with any totalitarian regime. Ten years before the period under consideration, Hitler displayed perplexity toward the Nobel Prizes after one was awarded to the anti-fascist journalist Carl von Ossietzky while he served his sentence in a concentration camp (see note 1 to doc. 3.16). Since 1946, Stalin's regime also started to demonstrate aversion toward freedom-lovers from Stockholm and Oslo who did not in principle tolerate any pressure from the outside. It was latent in the beginning, but very soon, ideologists from the Central Committee came to recognize the incompatibility of the Nobel movement with Soviet reality and their discontent splashed out (see Chapter 6).

There was one more reason for the negative attitude of Soviet ideologists toward the Nobel movement clearly apparent in the first post-war years. It was Sweden's neutrality in World War II. Moscow instructed its representatives in Sweden to collect data about Swedish scientists for testing of their political reliability (see Chapters 3 and 6). The talk between V. S. Kemenov, VOKS chairmen, and Rolf Sulman, the newly appointed Swedish envoy in Moscow, in 1947 testifies to the same (doc. 4.41).

Another painful point in Swedish–Soviet relations influencing Moscow's policy (even if indirectly) was the fate of Raoul Wallenberg, a Swedish diplomat who had managed to save tens of thousands of Jews from death in Auschwitz concentration camp. Traces of the man were lost in Hungary in January 1945. At approximately the same time (late 1945 and early 1946), when the letters of nomination written by Professors A. A. Lebedev and N. A. Terenin were detained by VOKS and the Swedish envoy tried in vain to persuade Soviet officials to resume interrupted contacts with the Nobel organization (docs. 4.3–4.6, 4.14–4.17), the Swedish government took the first steps in the hope to ascertain the fate of Wallenberg. Soviet authorities were not prepared to tell the truth, as appears from the Stalin's talk with Sederblum.[32]. It was an additional motive not to restore relations with the Nobel organization.[33]

Ideological services were especially concerned about the nomination of Wallenberg for the Nobel Peace Prize of 1948 by three members of the Swedish Rickstag. A parallel immediately suggested itself with the award of the 1936 Nobel Peace Prize to the anti-fascist journalist Carl von Ossietzky, a prisoner at Hitler's concentration camp. It made Soviet ideologists allow Wallenberg's name to be for the first time mentioned in press. The unsigned article (i.e. editorial) under the title "Against disinformation and slander" was published in the weekly *Novoye vremya* in January 1948. The two-page article said no word about Wallenberg's mission in Budapest but alluded to "bootlickers of Swedish and foreign reactionary circles who initiated the so-called Wallenberg's case." (doc. 4.43-I.)

Paradoxically, Moscow's animosity was paralleled by a substantial rise in nominations of Soviet scientists for the Nobel Prizes that started to arrive in Stockholm after the end of World War II. The reason was obvious. The civilized world was grateful to the Soviet people for their invaluable contribution to the victory over Nazism, and nominators from different countries conferred these feelings onto their colleagues in the Soviet Union.[34]

Another contributing factor was the jubilee session of the USSR Academy of Sciences held on the occasion of its 220th anniversary soon after the end of the Great Patriotic War. Many foreigners were invited to participate including widely known Swedish scientists and members of the Nobel Committees A. F. Westgren and T. Svedberg (see note 2 to this chapter), English physico-chemist C. N. Hinshelwood, etc. These people became personally acquainted with their Soviet colleagues and visited research institutions in Moscow where they could see for themselves the high level of science in the USSR.

As shown in Table 4.1, N. N. Semenov and P. L. Kapitsa were the first nominees in chemistry and physics from the Soviet Union in 1946 (they received the Nobel Prizes in 1956 and 1978 respectively). Their nominators were British scientists C. N. Hinshelwood[35] and P. Dirac, A. N. Frumkin was another nominee in chemistry; he was nominated by three professors from the University of New Haven, Connecticut. L. A. Orbeli nominated V. A. Engelgardt and M. N. Lyubimova for the Nobel Prize as the authors of the outstanding discovery of oxidative phosphorilation in the late 1930 that laid a basis of modern bioenergetics.[36] In a talk with the author

Table 4.1. Scientists of Russian descent nominated for the Nobel Prizes in science from 1945 to 1947

Nominees and their whereabouts	Year of nomination	Number of nominations	
		Total	In the USSR
Physics			
G.A.Gamow (1904–1968), Washington	1946	1	—
P.L.Kapitsa (1894–1984), Moscow	1946	1	—
-”-	1947	4	—
V.I.Veksler (1907–1966), Moscow	1947	1	—
D.V.Skobeltsin (1892–1990), Moscow	1947	1	—
Chemistry			
N.N.Semenov (1896–1986), Moscow	1946	1	—
-”-	1947	1*	—
A.N.Frumkin (1895–1976), Moscow			
Physiology and medicine			
V.A.Engelgardt	1946	1	1
M.N.Lyubimova	1946	1	—

* Collective nomination of three scientists from the University of New Haven, USA.

of this book, the long-standing chairman of the Nobel Committee in chemistry B. G. Malmstrom expressed regret that Engelgardt had not received the Nobel Prize and described this omission as an “obvious mistake” by the Committee (doc. 4.44).

In 1947, Kapitsa had four nominations from N. Bohr, W. Pauli, T. Svedberg, and W. Watson, University of New Haven, Ct., USA. Svedberg also nominated Semenov and Hinshelwood for the Nobel Prize in chemistry.[37]

In the same year, V. I. Veksler and D. V. Skobeltsin were nominated for the Nobel Prize in physics. The former was proposed by Professor L. B. Loeb, University of California, Berkeley, USA (together with the future Nobel laureate in chemistry for 1951, E. M. McMillan of the same university) and the latter by Cz. Bialobrzheski, professor of theoretical physics and astrophysics from Warsaw.[37a]

All these nominations (excepting that of Skobeltsin) were rapidly included on the so-called “short list” of candidates.

Accordingly, they were considered by experts of the Nobel Committees in physics and chemistry who expounded their well-grounded opinion about each nominee.

Special attention of the Nobel Committees was given to P. L. Kapitza, who was nominated for the discovery of helium superfluidity by P. Dirac, the Nobel laureate of 1933. The experts emphasized that abnormally low viscosity of helium II was independently described by J. Allen and D. Misener in Cambridge (see Chapter 8 for details). The Committee concluded that “the definitive opinion about the contribution of Kapitza to the study of helium II properties is difficult to ascertain because of the yet poor understanding of this phenomenon"; therefore, "consideration of this nomination for the 1946 Nobel Prize in physics appears unfeasible."[38]

In 1947, Kapitza was nominated by four persons including three Nobel Prize winners, which made the Nobel Committee more thoroughly consider his candidacy. Moreover, the nominators motivated their proposals by referring to Kapitza's other scientific results, besides the discovery of helium superfluidity to which Dirac confined his nomination. Bohr, Pauli, and Watson emphasized the importance of the nominee's fundamental contribution to the development of experimental techniques for superlow temperature studies. Professor Svedberg suggested dividing the 1947 prize between Kapitza and Willem Keesom, a professor at the University of Leiden, who described phase transition in liquid helium at a temperature of 2.19 K and a jump of heat conduction below it. N. Bohr recommended taking into consideration Kapitza's investigations into effects of very strong magnetic fields on the state of matter.

Members of the Nobel Committee underscored again fundamental importance of the works of Kapitza but referred to the contribution of other scientists to the studies of helium II. Their final verdict contained a promise to "thoroughly consider new important studies, if any, that could help to better assess the personal contribution of Kapitza." Meanwhile, the Committee "chose to wait [...], the more so that other nominations appear more topical."[39] The 1947 Nobel Prize was awarded to Edward Appleton "for his investigations into the physics of the upper atmosphere, especially for the discovery of the so-called Appleton layer."

Nominations for Semenov for the prize in chemistry in 1946 and 1947 gave hope for a promising outcome. Professor A. Fredga, member of the Nobel Committee and expert for the very first nomination of Semenov emphasized that the works of the Soviet scientist "produce an impression of integrity and are noticeable for a convincing inner logic, having an object to substantially modify or critically revise the current concept of reactions in gaseous media, especially in the process of combustion, as indicated by the nominator." (The nominator was S. Hinshelwood, with whom Semenov shared the Nobel Prize in 1956.) Fredga was the first to suggest that "the prize should be divided between Semenov and [...] Hinshelwood,"[40] although nobody nominated the British scientist in 1946.

T. Svedberg, Nobel laureate in chemistry for 1926, who nominated both scientists, proposed the same in the next year. He also thoroughly considered the works of his nominees at the request of the Nobel Committee and arrived at the conclusion that studies of the two scientists on chemical chain reactions were of paramount importance, i.e. fully met the requirements of the Nobel Committee. Svedberg emphasized that our knowledge in this area would be poor without the works of the nominees. The studies of both of them "fairly well supplement each other and aim at the solution of related problems. It would be only fair if they shared the Nobel Prize."[41]

The Nobel Committee agreed with Svedberg's opinion but decided to postpone the award to Hinshelwood and Semenov. The prize for 1947 was given to the recent guest in Moscow Professor Robert Robinson of Oxford University "for his investigations on plant products of biological importance, especially the alkaloids."

The nomination of A. N. Frumkin sent to the Nobel Committee in 1946 was met far less favorably than that of Semenov. Arne Tiselius, an expert for the Nobel Committee, gave credit to the nominee's fundamental contribution to electrochemistry with special reference to his studies on surface potential and surface layer chemistry at large. At the same time, the future winner of the Nobel Prize (1948) for the investigations on electrophoresis and adsorption noted that the works of Frumkin "are not as significant and original as required to justify their proposal for the Nobel Prize."[42]

Professor Ya. P. Stradyn' (Latvia) suggested that the candidacy of Frumkin did not attract attention from the Nobel Committee because the Swedish experts underestimated the importance of electrochemistry in the system of chemical sciences in the second half of the twentieth century.[43]

The works of V. I. Veksler and E. M. McMillan were described by the Nobel Committee in 1947 as "a recent sensational breakthrough in acceleration of high-energy charged particles." However, evaluation of practical applications of the autophasing principle discovered by the two scientists was possible only from the data obtained on American accelerators because all such data were classified as secret in the Soviet Union since 1945 (doc. 6.36). It was the absence of documented materials from the USSR that might have influenced the final decision of the Committee "to wait for new experiments, especially on the use of the principle [autophasing] in the so-called synchrophasotron, before awarding the prize for this important contribution to science."[44]

In fact, there was no need to "wait for new experiments" even in 1947 because the first 30 MeV synchrophasotron had already been built up and commissioned under the guidance of Veksler at the Physical Institute, the USSR Academy of Sciences. The second machine (250 MeV) was ready to operate in two years. In 1949, S. I. Vavilov and V. I. Veksler initiated the construction of a principally new accelerator of elementary particles with an incredibly high (at that time) output energy of 10 GeV! The accelerator was commissioned in 1957...

Thus, in the late 1940s, accelerators in Moscow were the first on which totally new results were obtained that confirmed the conclusion of the "important contribution to science" of the Veksler and McMillan's discovery" made by the Nobel Committee in 1947. Suffice it to say that studies on the second Soviet synchrophasotron demonstrated for the first time the photobirth of mesons and laid the foundation for the physics of hadrons,[45] elementary particles involved in strong electromagnetic interactions.

In other words, there was every reason to award the Nobel Prize in physics to the Soviet scientist V. I. Veksler and his American colleague McMillan in 1949–1950. But many data obtained on Soviet synchrophasotrons were declassified only after

Stalin's death when it was too late to discuss the candidacy of Veksler again. In 1951, the Nobel Prize in chemistry was awarded to E. McMillan and Glenn Seaborg "for their discoveries in the chemistry of the transuranium elements."

The chairman of the Nobel Committee Arne Westgren stated at the official ceremony in Stockholm that the works of McMillan aimed at the development of particle accelerators contributed to further progress in this field of chemistry. Thus, the Nobel Prize crowned the epoch-making discovery that brought about a wealth of essentially new scientific information. The Soviet researcher did not receive it only because his sensational data obtained in 1951 remained unknown to the scientific community as it was classified by security services, even though they contained neither state nor military secrets.

Later nominations of the Soviet scientist either alone or together with McMillan in the 1950s and 1960s did not attract attention from the Nobel Committee. The importance of the autophasing principle for in-depth investigations of matter was once recognized, and the Nobel Prizes are not given twice for one and the same discovery.

The lack of interest in the nomination of D. V. Skobeltsin among the members of the Nobel Committee was predetermined by objective reasons. The scientist made an important contribution to modern physics, especially high-energy physics. He was virtually the first to use Wilson's chamber and observe traces left in the atmosphere by charged particles with an energy much higher than that emitted from radioactive sources (in 1927). He interpreted the phenomenon as experimental evidence of cosmic radiation. In other words, Skobeltsin created an experimental basis for the verification of the hypothesis of extraterrestrial fluxes of particles suggested by the Austrian physicist V. F. Hess in 1912. Also, Skobeltsin made the pioneer observation of cosmic particles transported through space in coherent groups (showers). Today, this observation makes up a key principle in high-energy physics.

Finally, Skobeltsin was the first to report the birth of electron-positron pairs in cosmic rays even though he failed to explain the phenomenon in the context of P. Dirac's prediction of particles' interconversion (i.e. the existence of a new elementary particle with the mass of the electron and the positive charge). In 1932, American physicist C. D. Anderson, who shared the 1936 Nobel Prize with Hess, reported the presence of positron in cosmic rays. This miss of Skobeltsin was noted by V. I. Vernadsky in his diaries (4 December 1938).

After a time, Skobeltsin explained his caution by the difficulty to announce the existence of "a new elementary particle differing from all known ones." (doc. 4.53) He dared not give the new phenomenon the name that offered itself and therefore lost the fame due to him as the discoverer of positron. Naturally, the Nobel Committee had no reason to consider his candidacy in 1947, i.e. 11 years after the awarding of the Nobel Prize to Anderson and Hess, as a contributor to the development of high-energy physics...

G. A. Gamow, graduate of the Leningrad University (see Chapter 3), was also nominated in 1946 by Professor J. Thibaut of Lyon, but his candidacy did not attract the attention of Swedish professors. Nevertheless, many authoritative scientists regarded Gamow as a most worthy candidate for the prestigious award (see, for instance, the afterword to this book by Ilya Prigogine, p. 841).

As before, all nominators of Soviet physicists and chemists were foreigners. It was not because Soviet researchers did not wish to use their rights as nominators. The true reason was the same that prevented the proposals of A. A. Lebedev and A. N. Terenin from reaching Stockholm. Of these two nominations detained by security services in 1946 (docs. 4.4 and 4.6), that of Lebedev aroused the greatest suspicion since it concerned two German scientists.[46] Lebedev's choice demonstrated his adherence to high moral principles. The war that had just ended took millions of lives, including some dear to the nominator. Nevertheless, scientific ethics prevailed, and Lebedev recommended those whom he considered most worthy. It is a remarkable example of an unprejudiced opinion and allegiance to professional duty...[47]

The list of nominees for the 1947 Nobel Prize in literature included the poet B. L. Pasternak, the first Soviet writer nominated after a long break since 1933, when M. Gorky had been for the last time proposed as a candidate for the award.[48]

Pasternak was nominated again in 1947. In the same year, Professor H. Ollson, member of the Swedish Academy, for the first time mentioned the name of M. A. Sholokhov, the author of *Tikhii Don* (And Quiet Flows the Don). The two writers were widely known in the West and their inclusion on the list of contenders looked quite natural.

Western intellectuals were familiar with the writings of the two authors as early as the 1930s (see, for instance, doc. 3.9). It was much easier for a prose writer like Gorky to gain popularity among foreign readers than for a Russian-speaking poet. Nevertheless, Pasternak was known in Europe, as evidenced by a remark made by Romain Rolland when he visited the Soviet Union in summer 1935.[49]

Post-war lists of candidates for the Nobel Prize in literature also contained names of Russian emigré writers. Professor A. Nyman of Lund continued to nominate N. A. Berdyaev. The candidacy of M. A. Aldanov[50] was submitted by I. A. Bunin in 1947 (Table 4.2). The Nobel winner continued to recommend Aldanov till his death in 1953 (see Table 6.2). The behavior of Bunin should be accounted for not only by the merits of Aldanov as the author of many historical epics[51] but also by the gratitude the Nobel laureate always showed to the man that had made so much for the recognition of his talent by the Swedish Academy.

Moscow did not respond to nominations of the former countrymen, either for the lack of reliable information or rather because the brilliant representatives of Russian literature were nearer the end of their lives and were no longer regarded as serious rivals by the post-war ideologists. Moreover, the probability of their being awarded a Nobel Prize after the peak of their creative activity had passed decreased every successive year.

Those who stayed in the country but remained unacceptable for ideological services as carriers of alien principles were considered more dangerous. This in the first

Table 4.2. Writers of Russian descent nominated for Nobel Prizes in literature from 1945 to 1947

Nominees	Year of nomination	Nominators
N.A.Berdyaev (1974–1948)	1945	A.Nyman, professor of philosophy, Lund
-"-	1946	-"-
-"-	1947	-"-
M.A.Aldanov (1886–1957)	1947	I.A.Bunin, Nobel laureate
B.L.Pasternak (1890–1960)	1946	C.M.Bowra, professor of ancient Greek literature, Oxford
-"-	1947	-"-
M.A.Sholokhov	1947	Professor H. Olsson, literature critic, University of Stockholm

place applied to the great poet Boris Pasternak who happened to be a nominee of the Nobel Committee (see Chapter 9).

There was one more painful point, i.e. the fate of the writers who died in KGB prisons during the Great Terror. One piece of evidence of this concern was a message sent to the Soviet mission in Stockholm at the beginning of 1947 (doc. 4.45). It referred to the anticipated consideration of the candidacy of B. A. Pil'nyak, supposedly killed in 1937, by the Nobel Committee in literature.[52] The message was provoked by an application from the Swedish side to the Soviet mission containing the naive request to provide a collection of published works by the author. The request was most likely dictated by the desire to have these books in the library of the Swedish Academy because the writer was rather popular in Scandinavian countries.[53] Anyway, the name of Pil'nyak is absent on the list of nominees for the 1947 Nobel Prize.[54]

Document 4.45 is interesting for its undercurrent aspect. This short message is a quintessence of dramatic circumstances during Stalin's totalitarian rule, when the authorities were absolutely sure about their omnipotence ("take all measures [...]" etc.). It was total state terror under which the term "people's enemy" was coined and entered into everyday usage, and nobody was safe from being labeled as such. The situation is best exemplified by the fate of A. V. Karaganov, the author of the message, who was sacked nine months later on denunciation.[55]

The activities of the Norwegian Nobel Committee were also in the focus of Moscow authorities for a few post-war years, first and foremost because the candidacy of A. M. Kollontai was considered in the framework of the campaign for the selection of a contender for the 1946 Nobel Peace Prize. A detailed report concerning this issue reached Moscow in September 1946, i.e. a month before the new laureate was announced (doc. 4.46). The document contains little known data about the position of the Nobel Committee and its financial means during the occupation of Norway by German troops. Also, it gives political characteristics to the members of the Nobel Committee.[56]

Simultaneously, the Norwegian Nobel Committee sent to the USSR People's Commissariat of Foreign Affairs a few copies of the document announcing the campaign for nominations for the 1946 Nobel Peace Prize (doc. 4.47). It follows from the now available archive materials that none of the Soviet nominators used the invitation to make proposals; Kollontai was recommended only by foreigners.[57]

Notes

[1] A basic principle of the work of Nobel organizations consists of maintaining direct contacts with nominators in different countries and universities whose ordinary professors are invited to submit candidates for the Nobel Prizes in a current year (see note 3 for details). In other words, contacts with government bodies of foreign states are not encouraged as far as nomination issues are concerned.

[2] It was the second visit of Svedberg to the USSR, the first being in 1934, when he participated in the work of the VII Mendeleev Congress as a foreign guest. In his lecture, delivered at the joint meeting of the Division of Chemical Sciences, the USSR Academy of Sciences, and the All-Union D. I. Mendeleev Chemical Society, Svedberg stated that he "came in the hope of learning many new facts, but what I could see and hear surpassed my expectations."* There is little doubt that "what he could see and hear" made him the following year (1946) nominate Soviet scientists P. L. Kapitsa and N. N. Semenov for the 1947 Nobel Prizes in physics and chemistry respectively.

> * Yubileinaya sessia Akademii Nauk SSSR. 15 iyunya-3 iyulya 1945. T. 2 (Jubilee Session of the USSR Academy of Sciences. 15 June–3 July 1945. V. 2), M.-L.: Izd-vo AN SSSR, 1947, p. 151.

[3] Special rules for awarding Nobel Prizes in sciences (para. 1 for the prizes in physics and chemistry and para. 5 for the prizes in physiology and medicine) establish six categories of specialists invited annually as nominators. Of interest for our purpose are two categories: 5 — professors of at least six universities and institutes, besides those stipulated by p. 4 of the relevant paragraphs, chosen by a prize-awarding organization for a current year so as to achieve their even distribution in terms of geographic location; 6 — other scientists at the discretion of an awarding organization. The list of Soviet physicists and chemists (doc. 4.1) had to include scientists from category 6 and the list of professors for the Nobel Committee in physiology and medicine was drawn up so as to contain candidates from category 5.

[4] An example of the low professional level characteristic of many functionaries in the sphere of cultural relations with foreign states.

[5] "Cold war" was openly proclaimed in Churchill's speech of 5 March 1946 [...] (*Great Soviet Encyclopaedia*, 3rd edition, V. 28, 1978, column 1026).

[6] The demand to strike off the address was totally absurd and shows more evidence of the low educational level of the VOKS censor. The address of the 2nd Medical Institute (the former Psychoneurological Institute founded by V. M. Bekhterev in 1907) was known all over the world, and the Karolinska Institutet, to which N. N. Anichkov sent his letter, regularly invited its specialists (including Anichkov himself) to nominate candidates for the Nobel Prizes.*

* Personal letters sent via VOKS to its representative in Sweden were for the last time mentioned in the registry on 6 August 1947 with the handwritten notes "hand over if appropriate" or "without personal address". Thereafter (up to the first half of 1953), only printed matter, gramophone disks, etc. were registered (GARF, fond 5283, opis' 20, ed. khr. 270, l. 35). At approximately the same time (26 August 1947), the two last personal letters were sent via VOKS to its representative in Norway (*ibid,* ed. khr. 100, l. 158).

[7] An example of a deliberate delay of an answer to the invitation is provided by an episode having no direct relation to the issue under consideration, but characteristic in the context of Soviet–Swedish cultural relations. The Swedish mission in the USSR addressed VOKS to inquire whether Lieutenant-General A. A. Ignatiev, author of the then bestseller *Pyat'desyat let v stroyu* (50 Years in Service), could visit Stockholm and present a lecture (9 April 1946). G. Shumilov, senior officer for the Central European Division of VOKS, was instructed to get in touch with the USSR Ministry of Foreign Affairs to clarify the situation. It appears that the Ministry was not eager to give consent to the visit because Shumilov drew up a draft letter informing the mission that Ignatiev received the invitation (which was a lie) but "will not be able to come in spring, as was planned, which should be brought to the notice of the Society." The draft text is crossed by the following phrase in the handwriting of Karaganov: "Delay the answer and inform them verbally about the refusal, as the situation permits." (18 April 1946)* The practice of delays was a very common instrument of the Soviet bureaucratic machinery even in the post-Stalinist period. Suffice it to mention the story of arrest of the Swedish diplomat Raoul Wallenberg by Soviet secret agents after Hungary was liberated from German troops in 1945. During Stalin's life, the fact of the arrest was simply denied (see note 32 to this chapter and doc. 4.43-1). When it became impossible to conceal the fact, the post-Stalinist leaders resorted to the strategy of delays including fabrication of the texts for verbal statements that the Soviet ambassador in Sweden N. M. Lun'kov had to make during his meetings with Unden, Swedish Minister of Foreign Affairs, in May and July 1956 (doc. 4.43-II).

* GARF, fond 5283, opis' 20, ed. khr. 265, l. 93–94.

[8] There is no doubt that VOKS had regular mandatory contacts with the country's top officials. An interesting example was recorded by V. S. Kemenov, chairman of the VOKS Executive Board, after his meeting with the newly appointed Swedish ambassador Rolf Sulman. The ambassador wanted to know Kemenov's opinion about a reception that the Swedish side planned to hold at its Moscow mission to celebrate the election of the composer S. S. Prokofiev as an honorary member of the Swedish Musical Academy. Sulman told that he had had a talk with M. S. Vetrov, head of the Scandinavian Department, the USSR Ministry of Foreign Affairs, who advised him to address VOKS. Kemenov wrote: "Sulman asked whether the Soviet side had any objections. I knew from Vetrov that the Ministry resolved the issue in a positive manner and said that I find the idea acceptable."*

* AVP RF, fond 1040, papka 146, ed. khr. 50, l. 9.

[9] The law adopted by the USSR Supreme Council on 15 March 1946 redesignated the Council of People's Commissars as the Council of Ministers, and People's Commissariats as Ministries.

[10] Plenipotentiary since 1947. The attaché of the Soviet mission in Sweden A. M. Aleksandrov-Agentov characterized him as a man of "great intellect, delicacy, genuine modesty, and the ability to get on with people, both his countrymen and foreigners [...]"*

* A. M. Aleksandrov-Agentov. "*Ot Kollontai do Gorbacheva* (From Kollontai to Gorbachev)". M.: Mezhdunarodnye otnosheniya, 1994, p. 40.

[11] In a month and a half, Terenin was not included in the small Soviet delegation leaving to participate in the work of the Association of Scientists in India (RGASPI, fond 17, opis' 117, ed. khr. 675, l. 3–5).

[12] Unlike the case of Terenin, the resolution concerning Papaleksi was preceded by the letter from G. F. Aleksandrov dated 14 September 1946 that described the visit as undesirable. Zhdanov wrote: "Agree." It may be supposed that the order to stop all visits abroad came on 12 September from the top, maybe from Stalin himself. The apparatus of the Central Committee was unaware of it on the morning of 13 September; therefore, Suslov gave his consent to Terenin's trip. When the resolution concerning Papaleksi's visit was to be taken on 14 September, the order was already known to the decision-makers.

[13] The decision of Politbureau was taken on the initiative of Stalin, as evidenced by his signature in blue pencil on Zhdanov's facsimile below, who signed all minutes of Politbureau and AUCP(b) Secretariat sittings since 1947. The choice of the candidates instead of the turned-down ones was pretty self-explanatory. Szent-Gyorgyi represented countries of people's democracy* and Raman, from India, which after two months (15 August 1947) became an independent state friendly to the Soviet Union, while Frederic Joliot Curie and his spouse always supported Communist ideas. Appointed Supreme Commissar of Atomic Energy by the French government in 1945, F. J. Curie ensured the commissioning of the first experimental nuclear reactor that the American journal *Time* called a "Communist reactor".**

* The generally accepted name of Eastern European countries liberated by the Red Army in the last two years of World War II. They were subsequently governed by Communist Parties or their analogs, and were collectively referred to as the Socialist camp.
** *Voprosy istorii*, No. 1, 2004, p. 6.

[14] CAODM, fond 2971, opis' 1, ed. khr. 1, l. 26. See also note 22 to Chapter 6.

[15] It was not the first case of the shutdown of natural science periodicals published in foreign languages in the Soviet Union. A similar situation took place in the late 1930s; the publication of *Physikalische Zeitschrift der Sowjetunion* in German was all of a sudden interrupted after its regular issue was out in February 1938. The journal was published by the Physico-Technical Institute, Kharkov, since 1932 and enjoyed popularity in the West. Its circulation of 1300–1700 was rather big for a scientific edition distributed via subscription. The publication was stopped quite unexpectedly. The last issue (B. 13, No. 2, 1938) contained no announcement of the coming shutdown. On the contrary, the two inner covers listed subscription addresses in the USSR and 20 foreign states, and the prices in the currency of the respective countries. Such a surprising event could not be spontaneous since thriving periodicals do not cease to exist in the midst of a subscription year unless a *force majeur* event. The decision to close the journal might be related to a series of arrests of some members of the editorial board and researchers that took place in Kharkov in 1937.*

* Some chronological coincidence draws attention. The last issue was approved for printing on 4 April 1938. Therefore, the editorial board had the whole of April to prepare the next one. During the night of 27–28 April, L. D. Landau was arrested in Moscow; he had worked in the Physico-Technical Institute in Kharkov from 1932 to 1937. Naturally, he knew fairly well the members of the editorial board, including those repressed in the preceding year.

[16] RGASPI, fond 17, opis' 117, ed. khr. 868, l. 7–9. It is quite probable that G. F. Aleksandrov used the publication in *Pravda* (1936) as a source of information about the meeting at the Institute of Organic Chemistry, the USSR Academy of Sciences, held in connection with "Luzin's case". The speaker was I. M. Gubkin, vice-president of the Academy. He drew attention to the "traditions of servility to Western Europe"* in scientific circles and called on "Soviet researchers to have self-respect. The Russian language does not yet possess full rights in the international arena. We strive to have our papers translated into foreign languages and published in Germany, France, and England earlier than in our own country...* We must and can make the Russian language and articles written in Russian to be as wanted as those in English or German."**

> * An outmoded way of thinking. The main object of xenophobia in pre-revolutionary Russia was Western Europe because relations with the United States were highly irregular and regarded as "exotic". In the period between the two World Wars, the situation started to change. After the Second World War, the American leadership in the Western world became indisputably established; since then, the USA have turned into a source of evil for Soviet authorities and the main objects of their political attacks.
> ** *Pravda*, 15 July 1938, p. 4.

[17] Here is an example of the approval. L. Ya. Karpov, a speaker at the Communist Party meeting at the Physico-Chemical Institute, Moscow (29 July 1947), emphasized "the correctness of the decision of AUCP(b) to close scientific journals published in the USSR in foreign languages."*

> * CAODM, fond 3013, opis' 1, ed. khr. 5, l. 22.

[18] CAODM, fond 7334, opis' 1, ed. khr. 2, l. 16.

[19] "[...] Unfortunately, I made a mistake... I used to send reprints of some of my articles published in the Soviet press to foreign scientists. But I had neither personal relations nor correspondence with these people. My unpardonable mistake was the dispatch of reprints to two specialists in the UK and USA after I had been familiarized with the closed letter of AUCP(b)* and the shutdown of *Acta Physicochemica URSS*. This mistake of mine was condemned by the Institute's Party organization."**

> * Closed letter of AUCP(b) concerning "Klyueva and Roskin's case", 16 June 1947 (see Chapter 6 and note 12 therein).
> ** RGASPI, fond 17, opis' 134, ed. khr. 15, l. 14. See also note 3 to doc. 4.38.

[20] Esakov V. D., Levina E. S. Delo "KR" (Iz istorii goneniy na sovetskuyu intelligentsiyu ("KR" Case. From the History of Persecution of Soviet Intellectuals)). Kentavr, No. 3, 1994, p. 116. The campaign launched by Central Committee ideologists against the use of foreign languages in the Soviet press was not restricted to the prohibition of English abstracts of scientific papers. Somewhat later (29 November 1947), the AUCP(b) made a decision that finally led to the shutdown of the *Moskovskie novosti* (Moscow News) newspaper published in English. The newspaper was initially designed as a tool for the propaganda of the "Soviet lifestyle" in the West but proved to be inefficient for the purpose.* It was closed at the decision of Politbureau of AUCP(b) on 20 January 1949, a few days before the editorial in *Pravda* launched the campaign against "cosmopolitanism and servility to the West."**

> * RGASPI, fond 17, opis' 118, ed. khr. 95, l. 43–46.
> ** *Ibid*, ed. khr. 290, l. 22.

[21] On 7 May 1955, the Central Committee of CPSU allowed the publication of a new journal *Geokhimiya* (Geochemistry). It was not included in the list of journals where abstracts in English were permitted. The presidium of the USSR Academy of Sciences had to specially apply to the Central Committee for the inclusion of *Geokhimiya* in the list.*

* RGANI, fond 50, opis' 35, ed. khr. 16, l. 69.

[22] RGASPI, fond 17, opis' 117, ed. khr. 917, l. 36–41. The document entitled "On the anti-Party behavior of E. I. Koval'chik"* noted that she "was on friendly terms with Suchkov, a disclosed American spy." Suchkov was rehabilitated after 1953, appointed director of the Institute of World Literature, the USSR Academy of Sciences, and elected corresponding member of the Academy; he died in 1974.

* Koval'chik was the secretary of the editorial board of the newspaper *Kul'tura i zhizn'* (Culture and Life) published by the Department of Propaganda and Agitation, AUCP(b). In March 1948, she was sacked on denunciation (RGASPI, fond 17, opis' 118, ed. khr. 6, l. 89).

[23] A. A. Vaksman. Microbial Antagonisms and Antibiotic Substances (with the preface by G. F. Gauze, D. Biol. Sci.). M.: IL, 1947. Translation of the English version (New York, 1945).

[24] The episode with Vaksman, the discoverer of streptomycin, illustrates a sharp change in the attitude of Soviet authorities towards Western scientists, especially those of Russian descent. Vaksman was born in Priluki in 1988 and had to leave Russia, where he had no chance to receive higher education (up to 1917 a percentage quota limited the right of Jews to education). He passed his secondary school exams without attending lessons (in Odessa in 1910) and emigrated to America the same year.* He came to his motherland again in 1946 on the invitation of the USSR Academy of Sciences to read a series of lectures on "Antibiotics, their nature, synthesis, and applications" (from 8 to 12 August). In 1947, his name was already on the list of "non-returnees and deserters." Such a sharp change was related to the so-called "Klyueva and Roskin's case" fabricated by Stalin and Zhdanov in 1947 (see Chapter 6). Vaksman was described as one involved in the "case," his name was mentioned in the "Closed letter of AUCP(b) concerning "Klyueva and Roskin's case," 16.6.1947 (see Chapter 6 and note 12 therein). As a result of this campaign, volumes of *Les Prix Nobel* (annually published by the Nobel Foundation), with the biographies of the Nobel laureates for 1952 (including Vaksman), became available to readers only in specialized departments of the libraries under a special permission.**

* See also Chapter 12 and note 4 therein.
** See Chapter 12. The *Les Prix Nobel* yearbooks are kept in closed archives.

[25] A note addressed by M. Shcherbakov, head of the Staff Department, to A. A. Zhdanov, secretary of AUCP(b) Central Committee (28 August 1947), commented on the inclusion of the monographs by these three emigrant scientists in the publishing plan in the following way: "The Department believes that the works of such authors are unsuitable for publication. Whenever they are worth publishing, they must have strictly limited circulation for specialists,* that is labeled as 'classified' or 'for administrative use'."

* RGASPI, fond 17, opis' 117, ed. khr. 917, l. 36.

[26] RGASPI, fond 17, opis' 118, ed. khr. 295, l. 14.

[27] After a time, Isachenko decided to publish the book by Vinogradsky in Moscow not only in Russian but also in French because the post-war economic situation in Western Europe

(including France) made difficult the publication of scientific literature. Isachenko obtained support for his initiative from the presidium of the USSR Academy of Sciences. Z. A. Vaksman wrote to Vinogradsky from America (26 June 1947): "I received a few letters from Isachenko. He complains [sic!] that the USSR Academy of Sciences is prepared to immediately start publication of your book in both French and Russian. You should send them the manuscript at your earliest convenience (these are his words) [...]"*

* Arkhiv RAN, fond 1601, opis' 1, ed. khr. 127, l. 57.

[28] G. A. Zavarzin. Sergei Nikolaevich Vinogradsky v emigratsii (Sergei Nikolaevich Vinogradsky in Emigration). In: "Rossiyskaya nauchnaya emigratsiya. Dvadtsat' portretov (Russian Scientific Emigration. Twenty Portraits)." M.: Editorial URSS, 2001, p. 220. Financial support from Moscow was mentioned by Vaksman in the letter of 15 March 1946: "It is a pity you are going to remit to Manon* those scanty funds granted to you by Moscow; I hope it will not happen. There is no doubt that the book will sell well enough to cover all the expenses..."**

* A slip of the pen. Vaksman meant the "Masson et Cie" publishing house in Paris with which the publication of Vinogradsky's book was negotiated.

** Arkhiv RAN, fond 1601, opis' 1, ed. khr. 127, l. 54.

[29] Vinogradsky was well aware of the difficulties facing the publishers because of his emigré status. In the letter of 8 May 1950, he advised Vavilov: "I do not object to your intention to write a 'preface' with characteristics of the author's scientific works. Moreover, I believe your article might substitute the entire 'Introduction' in which biographic data were needed in the context of recent studies. These data may be useless to cite in Russia, where I am fairly well known."*

* Arkhiv RAN, fond 1601, opis' 1, ed. khr. 93, l. 9.

[30] RGASPI, fond 558, opis' 11, ed. khr. 876, l. 1; from the collection of letters reported to Stalin from January to July 1949. The cited letter contains the following words from Stalin's lecture read in early April 1924: "What are its [Lenin's style] main features? These are two: a) Russian revolutionary sweep and b) American efficiency. The essence of Leninism is the combination of these two approaches to Party and state work."*

* I. V. Stalin. Voprosy Leninisma (Problems of Leninism), 9th edn. M.: Partiynoye izdatel'stvo, 1932.

[31] RGASPI, fond 558, opis' 11, ed. khr. 876, l. 5.

[32] See J. Birman. Pravednik. Istoriya of Raule Vallenberge, propavshem geroye Kholokosta (The Righteous Man. The Story of Raoul Wallenberg, Missing Hero of the Holocaust), M.: Rudomino, 2001, p. 136. The Swedish side for the first time officially addressed the USSR Ministry of Foreign Affairs in connection with the disappearance of Wallenberg before the capitulation of Germany. S. Sederblum visited A. N. Abramov, head of the 3rd European Department, on 30 April 1945. Abramov's diary (now in AVP RF) contains the following emotional phrase from Sederblum: "I am plied with questions and telegrams [...] Wallenberg is from a family of world-known bankers, and I am sometimes deep in despair and unable to answer numerous questions."*

* See B. N. Sopel'nik. Zagadka Wallenberga. Pravda i lozh vokrug dela shvedskogo diplomata (The Puzzle of Wallenberg. The Truth and the Lie About the Swedish Diplomat). Rodina, No. 1, 2002, p. 78.

[33] Soviet–Swedish relations on the borderline between 1945 and 1946 were aggravated, besides the "Wallenberg's case", by the demand by the Soviet side to repatriate some Soviet citizens who had found refuge in Sweden. The first step was taken by I. S. Chernyshev, the Soviet envoy in Stockholm, on 17 December 1945.* In autumn 1946, he unambiguously stated at the meeting with the Prime Minister, T. Erlander, that the refusal to fulfill this demand would "seriously aggravate Swedish–Soviet relations."**

* Birman, p. 275.
** *Ibid.*, p. 295.

[34] The English intellectual Isaya Berlin, on the staff of the British embassy in Moscow in 1947, described the euphoria of those days in the following words: "Unprecedented heroism shown by Soviet people and the terrible misfortunes they had to endure during the war with Hitler raised such a great wave of sympathy for Russia in mid-1945 that made silent many critics of the Soviet system and its methods."*

* I. Berlin. From: "Vstrechi s russkimi pisatelyami" (Meetings with Russian Writers). In: A. A. Akhmatova. Requiem. M.: MPI, 1989, p. 198.

[35] During his stay in Moscow, Hinshelwood visited the Institute of Physical Chemistry, the USSR Academy of Sciences, headed by N. N. Semenov (21 June 1945). He spoke at a meeting of the Scientific Council and told about the works in Oxford and his own chemical kinetic studies.* When back at home, Hinshelwood gave credit to the high professional level of his Soviet colleagues.** He twice nominated N. N. Semenov for the Nobel Prize, in 1946 and 1950. Hinshelwood also visited the Institute of Organic Chemistry together with R. Robinson, vice-president of the Royal Society, and J. Trefuel, director of the Pasteur Institute, Paris. The guests had talks with Soviet researchers including A. A. Balandin, corresponding member of the USSR Academy of Sciences and well-known organic chemist. Next year (1946), Balandin was elected a full member of the Academy and received the Stalin Prize for scientific achievement, but was arrested by officers of the USSR Ministry of State Security three years later under a far-fetched charge of "anti-Soviet activity", and was sentenced to a ten-year imprisonment. Here is an excerpt from the minutes of the interrogation dated 9 March 1950:

— When did you establish contacts with Robinson, Hinshelwood, and Trefuel?
— At the jubilee session of the USSR Academy of Sciences held in Moscow in 1945. Robinson, Hinshelwood, and Trefuel visited Institute of Organic Chemistry as guests of the Academy... We discussed scientific problems in my room at the Institute.
— [...] Tell all about your criminal relations with foreigners!
— I had no criminal relations with foreigners.***

* Yubileinaya sessia Akademii Nauk SSSR. 15 iyunya-3 iyulya 1945. T. 2 (Jubilee Session of the USSR Academy of Sciences. 15 June–3 July 1945. V. 2), M.-L.: Izd-vo AN SSSR, 1947, p. 151.
** Two years later, S. M. Kuyumdzhi, deputy-director of the Institute, recollected that "Hinshelwood wrote an article in which he praised our results, and it was met with satisfaction." TsAODM, fond 8099, opis' 1, ed. khr. 8, l. 36.
*** Tsentr. Arkhiv FSB RF. Sledstvennoye delo P-24199, l. 144.

[36] "Biokhimiya", 4, vyp. 6, 1939, pp. 716–735; *Nature* 144, No. 3650, 1939, pp. 668–669. The discovery is described at greater length in "Biokhimiya", 7, vyp. 5–6, 1942, pp. 205–229.

[37] In 1956, Semenov and Hinshelwood shared the Nobel Prize in chemistry "for their research into the mechanisms of chemical reactions."

[37a] In 1910 (i.e. before Poland was granted independence, Bialobrzeski taught physics in Saint Vladimir University, Kiev, and his name was cited in documents with his patronymic, Czeslaw Feofilievich. V. I. Vernadsky twice mentioned his reading the fundamental work of Bialobrzeski "Atom" published in *Kievskie universitetskie izvestiya*, No. 4, 1915, pp. 73–134.*

* V. I. Vernadsky. Dnevniki 1917–1921, Yanvar' 1920-mart 1921 (Diaries 1917–1921. January 1920–March 1921) Kiev.: Naukova Dumka, 1997, pp. 80, 92.

[38] Archives of the Royal Academy of Sciences, Stockholm. Fund of the Nobel Committees in physics and chemistry. Materials of the Committees for 1946, sheet 159.

[39] *Ibid*, Materials of the Committees for 1947, sheets 75–79.

[40] *Ibid*, Materials of the Committees for 1946, sheets 294–295.

[41] *Ibid*, Materials of the Committees for 1947, sheets 179–180.

[42] *Ibid*, Materials of the Committees for 1946, sheet 293.

[43] In: "Aleksandr Nikolaevich Frumkin. Ocherki. Vospominaniya Materialy (pod redaktsiey Ya. P. Stradynya) (Aleksandr Nikolaevich Frumkin. Sketches. Memoirs. Materials). M.: Nauka, 1989, p. 251.

[44] Archives of the Royal Academy of Sciences, Stockholm. Fund of the Nobel Committees in physics and chemistry. Materials of the Committees for 1947, sheets 81–82.

[45] Baldin. "K stoletiyu akademika D. V. Skobeltsina (D. V. Skobeltsin. On his 100th Anniversary)." Dubna, 1992, p. 4.

[45a] "Skobeltsin is an excellent experimenter, but he missed, did not understand some phenomena he discovered."*

* V. I. Vernadsky. Dnevniki 1935–1941 v dvukh knigakh. Kn. 1. 1935–1938 (Diaries 1935–1941 in Two Books. Book 1. 1935–1938). M.: Nauka, 2006, p. 364.

[46] One of the two nominees recommended by Lebedev was Ernst Ruska, the inventor of the electron microscope, who was awarded the Nobel Prize in physics in 1986. Characteristically, this nomination of the Soviet scientist was the first for the future Nobel laureate even though it never reached the Nobel Committee.

[47] There is one more example of Lebedev's high moral position. He did not hesitate to state regret over disassembling of equipment at C. Zeiss works, Jena, in the Soviet occupation zone, whither he was sent after capitulation of Germany as a member of the repatriation commission.*

* RGASPI, fond 17, opis' 118, ed. khr. 29, l. 161.

[48] The poet's son recalled that the leak of information from the Nobel Committee about the nomination of Pasternak for the 1946 Nobel Prize in literature caused "great concern" in bureaucratic circles of the Soviet Writers Union.* The secretary-general of the union A. A. Fadeev stated at the meeting of Moscow writers on 17 September 1946 (soon after A. A. Akhmatova and M. M. Zoshchenko were expelled from membership of the Union) that "Akhmatova is not the sole representative of ideologically uncommitted poetry. Take

the works of Pasternak. They are equally uncommitted and politically indifferent [...] Nevertheless, we know some critics and writers who admire Pasternak's poetry as if it were an ideal for us, heirs of great Russian literature."** It is easy to see what consequences the award of the most prestigious prize in literature would have had for the poet who had fallen victim to a new defamation campaign...***

* E. B. Pasternak Boris Pasternak. Biografiya (Boris Pasternak. Biography). M.: Tsitadel, 1997, p. 613.

** "Literaturnaya gazeta", 21.9.1946, p. 1. One year before (26 March 1945), Fadeev made a similar statement at the meeting of Soviet and British authors in London. Answering a question about Pasternak who was awaited in the UK, Fadeev said: "Pasternak was never popular in the USSR among the wide reading public because his poetry is highly individualistic and difficult to understand for its complex versification. He wrote two things of social importance (*The Year of 1905* and *Lieutenant Shmidt*). Unfortunately, he chose to proceed along a different path." A. A. Fadeev. Sobranie sochineniy v 5 tomakh. T. 5 (Collected Works in 5 volumes. V. 5). M.: Goslitizdat, 1961, pp. 494–495.

*** One indispensable feature of a totalitarian society is the duplicity of its servants. The above speech by Fadeev at the meeting of Moscow writers in 1956 was equally insincere. Ilya Erenburg wrote in his memoirs: "I remember a meeting with Fadeev after his speech in which he denounced certain writers including Pasternak. Once, we met on the street [...] Aleksandr Aleksandrovich invited me to a café [...], ordered cognac, and said: Ilya Grigorievich, would you like to listen to a piece of genuine poetry? [...] He started to read Pasternak's verses by heart and could stop only to ask me: Is it good? He was not a hypocrite; it was the tragedy of a man who consecrated his life to what he considered to be the just cause" (Lyudi, gody, zhizn'. N. 3 (Lyudi, gody, zhizn'. V. 3)) M.: Sov. pisatel', 1990, pp. 128–129. Fadeev's ambivalence was not a secret for Erenburg. The literary critic A. K. Gladkov recollected Erenburg saying to him: "Personally, Fadeev treats me fairly well, but if he is ordered to quarter me, he will do it diligently and report the deed to his superiors; thereafter, being drunk again, he will speak that I was a good man and that he pities me." In: "Vospominaniya o Borise Pasternake (Memoirs of Boris Pasternak)." M.: Slovo, 1993, p. 336.

[49] The writer refers in his "Moscow Diary" (25 June 1935) to his frustrated meeting with Pasternak at the then border railway station of Negoreloye,* passed by passenger trains bound from Western Europe for the Soviet Union via Poland: "Doctor Levin** told me that he went by the same train as Pasternak, sent by Moscow for participation in the work of the International Congress of Writers in Paris.*** Pasternak was really disappointed that he could not stay at Negoreloye for a few hours to meet with me. The Congress drew to a close, and he could not spare the time to stop!**** Our trains just passed each other, and that was how our paths intersected."***** According to E. B. Pasternak, his father owed the correspondence acquaintance with Rolland to his future wife, M. P. Kudasheva. Pasternak had known her in Moscow since the 1920s. In the letters to Rolland, Kudasheva described Pasternak as 'an inspired poet, a living personification of time,' whose verses were translated into French by Elena Izvolskaya as early as 1925 and published in *Commerce.* Also, Kudasheva arranged to have Pasternak receive a book from Rolland that he sent to him with a dedicatory inscription: 'To Boris Pasternak, a younger brother on the side where the sun rises from his elder brother on the side were it sets down. Good day, I wish you a good year and a good passage from one side to the other. Vilnev, the year that goes away.'******

* Until autumn 1939, when the state border was shifted westward after the formerly Polish territories of Western Byelorussia were annexed by the USSR.
** Rolland visited the USSR as an honorary guest of the Soviet government. Professor Levin, a highly reputed cardiologist, was to accompany the elderly writer from Negoreloye to Moscow as his personal doctor. In 1937, Levin was accused as a member of the "anti-Soviet right-trotskyist block" in the framework of the political process fabricated by Stalin, sentenced to capital punishment, and shot dead.
*** The anti-fascist Congress of Writers was held for five days in the Mutualite Hall, Paris, in June 1935. Pasternak was reluctant to go but eventually had to consent. Sholokhov, also listed as a member of the delegation, proved more obstinate and managed to "get off," as evidenced by the known Musicologist Solomon Volkov (*Znamya*, No. 3, 2006, p. 134).
**** "Babel and Pasternak arrived in the midst of the Congress. He [Pasternak] came to know that he had to go to Paris when he was in a rest home"; Ilya Erenburg. Sobranie sochineniy V. 9 tomakh T. 9. (Collected Works in 9 volumes. V. 9). M.: Khudozhestvennaya literatura, 1967, p. 67.
***** "Moskovsky dnevnik Romena Rollana (Moscow Diary of Romain Rolland)". "Voprosy literatury" No. 3, 1989, p. 210.
****** E. B. Pasternak. Boris Pasrenak. Na tom beregu. Perepiska (Boris Pasternak. On the Other Side. Corespondence), *Znamya*, No. 11, 2002, pp. 162–168.

[50] By the time Aldanov was first nominated for the Nobel Prize, his name had been for nine years in the dossier of A. F. Kerensky started by the Main Department of State Security, USSR People's Commissariat of Internal Affairs, in 1938. One document in the dossier evidenced that relations maintained by Kerensky "beyond the Party circle" included Aldanov, a friend of his since their life in Russia.*

* V. V. Gurzhiy. "Razvedka NKVD protiv A. F. Kerenskogo (NKVD Intelligence Service Against A. F. Kerensky)." "Novaya i noveishaya istoriya", No. 6, 2002, p. 73.

[51] Bunin wrote to Aldanov in October 1945: "I have recently read *Mogila voina* (The Tomb of the Warrior), for the third or fourth time. How good it is! It is worthwhile to note that Aldanov was a keen observer, capable of foretelling future cataclisms. Suffice it to say that he predicted Hitler's rule as early as 1935: '[...]* Hitler is now supported by millions of people and he may be very soon able to set the world on fire.'**

* The novel *The Tomb of the Warrior* (1939) tells about Byron and his mysterious death.
** Russkoye zarubezhie, Zolotaya kniga emiigratsii. Pervaya tret' XX veka (Russian Exiles. The Golden Book of Emigration. The first third of the twentieth century). "Entsiklopedichesky biografichesky slovar". M.: ROSSPEN, 1997, pp. 20–21.

[52] Pil'nyak was arrested in October 1937. According to his son, B. B. Abdrinikashvili-Pil'nyak,* who saw evidence, the writer was shot dead on 21 April 1938.

* The afterword to the three-volume collected works of B. Pil'nyak. Moscow: 1994, p. 569.

[53] The *Swedish Encyclopedic Dictionary* published a 16-line biographic note of the writer in 1931.* A fundamental monograph of his creative activity appeared in Denmark in 1979, i.e. over 40 years after the writer's dramatic death.**

* Nordisk Familljebok. Enciklopedi och Konversationslexikon 15, s. 995–996. Stockholm, 1931.

** P. A. Jensen. *Nature as Code. The achievement of Boris Pilnjak, 1915–1924.* Copenhagen, 1979.

[54] The panicked reaction to any mention of the repressed writer was not surprising during the last 15 years of Stalin's regime. Requisition of publications containing portraits or positive characteristics of those who were persecuted as "enemies of the people" was a permanent concern for hundreds of thousands of editors, censors, and other workers of the "ideological front." This work could not be done totally free of faults, for the number of victims continued to increase; these, in turn, provoked new denunciations.* Even the *Kratky kurs istorii VKP(b)* (Short Course on the History of AUCP(b)), the "New Testament" of Stalin's regime, was thoroughly reviewed when translated into various languages for the elimination of the names of the dictator's former companions shot down in KGB prisons. The situation is illustrated by an application of a top official responsible for propaganda and agitation to the chief and Stalin's resolution (doc. 4.52). Both are examples of the criminal power in the absurd pursuit of the shadows of former associates.

* Here is an example. A worker of the Foreign Languages publishing house feeling hurt by her superior reported to the AUCP(b) Central Committee that the marketing department published in April–August 1940 and sold to bookshops the book by E. G. Erenburg *Zhizn' avtomobilya* (The Life of a Car) with the portrait of Kh. G. Rakovsky, a well-known politician brought to trial in the political process of 1938. Also mentioned in the report were *Volume of Songs of International Brigades* and *Children and International Brigades* with photographs of the Spanish general Miaja, who surrendered Madrid to Franco in 1939. RGASPI, fond 17, opis' 125, ed. khr. 10, l. 61.

[55] A. V. Karaganov was dismissed from his position of the deputy chairman of the VOKS Executive Board, and the proposal to expel him from AUCP(b) was presented by the Central Committee Secretariat to the consideration of the Party Control Committee on 14 October 1947.* The dismissal followed a report by the USSR Ministry of Foreign Affairs saying that Karaganov had concealed the social origin of his father and made anti-Soviet notes in the diary during his studentship years.** In a month and a half, he was released from the position of editor of the journal *Soviet Literature*, published in English*** and expelled from the AUCP(b) membership "as deserving no credit from the Party."**** As a consequence, many other people suffered for keeping relations with Karaganov. One of them was M. V. Urnov, a connoisseur of English literature, whom Karaganov, "a politically dishonest person expelled from the Communist Party" had accepted for employment at VOKS in the capacity of the head of the Anglo-American Department. Urnov was also expelled from AUCP(b).*****

* RGASPI, fond 17, opis' 116, ed. khr. 322, l. 445.

** *Ibid*, opis' 117, ed. khr. 938, l. 41–45.

*** *Ibid*, ed. khr. 997, l. 35.

**** *Ibid*, opis' 11, ed. khr. 1, l. 25.

***** *Ibid*, ed. khr. 285, l. 193–194.

[56] This document for the first time informs Soviet authorities about the work of the Norway Nobel Committee. Archive materials of VOKS representatives in Norway do not mention the existence of a Nobel Committee in Oslo.

[57] See note 2 to doc. 4.24.

[58] A. N. Yakovlev. Sumerki (The Twilight). M.: Materik, 2003, p. 503.

Document 4.1

From: L. D. Kislova, member of VOKS Executive Board.
To: N. G. Bruevich, academician-secretary of the USSR Academy of Sciences.
1 November 1945

Dear Nikolai Grigorievich,
The second secretary of the Swedish mission in Moscow handed over to me at a reception in VOKS a pile of letters from the Nobel Committee of the Swedish Academy[1] and asked me to pass them to the addressees. In these letters, the Nobel Committee invites Soviet scientists to report (before 1 February 1946) the most important discoveries in different fields of science worthy of the Nobel Prize, awarded as you know for the works that conferred the greatest benefit on mankind.[2] Please acknowledge receipt of these letters and inform VOKS about your decision and the measures you are going to take. Enclosed find please 30 letters addressed to Soviet scientists.

GARF, fond 5283, opis' 20, ed. khr. 266, l. 87.

[1] The Nobel Committees in physics and chemistry at the Royal Academy of Sciences.
[2] The author of the letter did not understand the request. The Swedish diplomat asked Soviet scientists to nominate candidates for the Nobel Prizes in physics and chemistry for 1946 rather than their own discoveries.

Appendix to doc. 4.1:

List of the members of the USSR Academy of Sciences to whom the Nobel Committees addressed invitations:

Vavilov S. I., Grebenshchikov I. V., 3. Ioffe A. F., 4. Kapitsa P. L., 5. Krylov A. N., 6. Krylov N. M., 7. Kurchatov I. V., 8. Lebedev A. A., 9. Papaleksi N. D., 10, Fok V. A., 11. Khristianovich S. A., 12. Tamm I. E.,[1] 13. Professor Landau L. D.,[2] 14. Arbuzov A. E., 15. Bakh A. N., 16. Dubinin M. M., 17. Chernyaev I. I., 18. Khlopov V. G., 19. Zelinski N. D., 20. Kistyakovsky V. A., 21. Nametkin S. S., 22. Nesmeyanov A. N., 23. Porai-Koshits A. E., 24. Rodionov V. M., 25 Semenov N. N., 26. Terenin A. N., 27. Favorsky A. V., 28. Frumkin A. N., 29. Alikhanov A. I., 30. Bernstein S. N.

G. Shumilov, senior officer of the Central European Department

GARF, *ibid.*, p. 88.

[1] Corresponding member; elected full member of the Academy in 1953.
[2] Elected full member in 1946.

Document 4.2

Excerpt from "the list of Swedish correspondents of VOKS for 1945."[1]

Sheet 7. The Nobel Committee of the Swedish Academy of Sciences forwarded via the second secretary of the Swedish mission, Mr. Jerring, a large number of letters inviting Soviet scientists to report (before 1 February 1946) the most important discoveries in different fields of science worthy of the Nobel Prize.[2]

Dispatched to the academician-secretary of the USSR Academy of Sciences.

GARF, fond 5283, opis' 20, ed. khr. 247, l. 7.

[1] The handwritten "list" in an account-book is arranged into two columns containing the first and the second paragraph respectively.
[2] The clerk filling in the columns repeats the mistake made by the author of doc. 4.1.

Document 4.3

From: K. Karkhan, scientific secretary, section of natural sciences and mathematics, VOKS.
To: Department of Culture, VOKS.
19 December 1945

The section of natural sciences and mathematics forwards to you the letter from Professor A. N. Terenin to be dispatched to Stockholm at your earliest convenience.

In the top left corner (hand-inked): "Please dispatch. 20 December. Vorobieva."
At the bottom (handwritten in pencil): "Do not dispatch." (Unsigned.)

GARF, fond 5283, opis' 20, ed. khr. 263, l. 253.

Document 4.4

From: A. N. Terenin, member of the USSR Academy of Sciences.
To: Nobel Committee in chemistry.
8 December 1945

With reference to your invitation to recommend a candidate for the 1946 Nobel Prize in chemistry, I am glad to propose the outstanding American chemist Professor Gilbert Newton Lewis of the University of California, Berkeley.

His works are so well known to physicists and chemists that I may confine myself to the enumeration in brief of the most important ones:

1. G. N. Lewis expounded fundamentals of the theory of covalent bonding in electron pairs in his book *Valence and the Structure of Atoms and Molecules* (1931).[1] His findings found confirmation in quantum mechanics.

2. G. N. Lewis was the first to obtain heavy water that finds application in nuclear energetics. Fragments of this work were published in the *Journal of the American Society* 55 (1933), 1297.
3. G. N. Lewis suggested a new concept of acids and bases (Franklin Institute, 226 (1938) 293) and successfully applied it to the study of intermolecular communications.
4. In the last years (1939–1945), G. N. Lewis investigated absorption and luminescence spectra of dyes and set up a new line of research in this field; he discovered elementary photooxidation via electron separation, geometric isometry, and biradical state of dyes (see J. Am. Chem. Soc., May 1939–1945).

G. N. Lewis was the first to propose a universal rule for the identification of spectral positions of all dyes; this proposal is as important as the formulation of the principle of compatibility of atomic spectra (C. N. Lewis. *Valence and the Structure of Atoms and Molecules.* N. Y., 1931).

Translation from Russian into English.

In one corner (handwritten in pencil, dateless): "Cannot be dispatched." Signed: M.[2]

[1] G. N. Lewis. *Valence and the Structure of Atoms and Molecules.* New York, 1931.
[2] Signed in all probability by a person controlling documents outgoing from VOKS. In Soviet offices, such people usually headed the so-called First Department subordinate to the state security service.

Document 4.5

From: A. Khokhulina (signature illegible), acting head of the Department of Scandinavian Countries, VOKS.
To: T. M. Kovalenskaya, head of the Department of Culture.
2 January 1946

Attached find please a letter from A. A. Lebedev, member of the USSR Academy of Sciences, to the Nobel Committee. The Department of Scandinavian Countries asks for comments of the concerned Section prior to the departure of the letter to the addressee.

This text is followed by a note in ink: "Tatyana Mikhailovna, I don't think such letters (written by full members of the USSR Academy of Sciences) need to be commented on by any Section. 11 January 1946. Karkhan."

In the left top corner (in ink): "Please submit to the approval of VOKS Executive Board. T. Kovalenskaya. 15 January 1946."

The text is crossed by the following phrase, handwritten with pencil: "Do not dispatch. M."[1]

GARF, fond 5283, opis' 20, ed. khr. 263, l. 255.

[1] See note 2 to doc. 4.4.

Document 4.6

From: A. A. Lebedev, member of the USSR Academy of Sciences.
To: Nobel Committee in physics.
26 December 1945

Thank you for your kind invitation to nominate a candidate for the 1946 Nobel Prize in physics.

I believe that the invention of the electron microscope is important enough to be awarded the Nobel Prize even though it was made rather long ago. A thorough discussion is strongly needed to evaluate the significance of this invention and its great contribution to the developments in many scientific disciplines. Over 500 works have been published since the first report of the electron microscope appeared dealing with the design of the instrument and its applications; a society for its improvement was set up in the USA.

Thus, I propose to award the Nobel Prize in physics for 1946 to E. Ruska of the Siemens Laboratory, Berlin, and E. Bruche, AEG Research Institute, Berlin, for their fundamental contribution to the creation and further development of the electron microscope."[1]

Back translation from Russian into English.

GARF, fond 5283, opis' 20, ed. khr. 264, l. 256.[2]

[1] A. A. Lebedev, the father of electron microscopy in Russia was awarded the 2nd degree Stalin Prize in 1947.* The idea to recommend the invention of the electron microscope for the 1946 Nobel Prize was probably prompted to him not only by his own studies but also by the agreement concluded with the United States in 1945 under which the up-to-date version of this instrument was to be delivered to the Soviet Union. The first three electron microscopes for the USSR Academy of Sciences were brought to Arkhangelsk in June 1946.**

* *Pravda*, 7 June 1946, p. 2.

** TsAODM, fond 992, opis' 2, ed. khr. 14, l. 104.

[2] It turns out that this letter that never reached the Nobel Committee in physics was the first nomination of E. Ruska* for the Nobel Prize. He received the award in 1986, i.e. 40 years after his nomination by the Soviet scientist A. A. Lebedev.

* To be precise, E. Ruska was first nominated by Professor L. Villa as early as 1942, but that nomination was in physiology and medicine. This nomination was the only one among those registered by the Nobel Committee at the Karolinska Institutet up to 1955 inclusive.

Document 4.7

Extract from the letter of L. D. Kislova, member of VOKS Executive Board, to the VOKS representative in Stockholm V. F. Razin, No. 18.
9 March 1945

[...] 2. The letter from Professor Frumkin to Professor Svedberg and twelve reprints of Frumkin's publications. Whether these materials should be handed to the addressee [...] is left at your discretion.[1]

GARF, fond 5283, opis' 20, ed. khr. 250, l. 112.

[1] Professor Rolf Svedberg received the letter of Frumkin.*

* "One week after the end of the war, I received a letter from my friend Frumkin, who asked me to send to him a copy of the description of my laboratory because his own had been lost during the war." *Ny Dag* 7 July 1945 (GARF, fond 5283, opis' 20, ed. khr. 253, l. 72.

Document 4.8

From: L. D. Kislova, member of VOKS Executive Board.
To: K. F. Vinogradov, representative of VOKS in Stockholm.
No. 68. 13 September 1945

Enclosed find please the following materials:

[...] 4. A letter and reprints of published works of Ya. I. Frumkin, corresponding member of the USSR Academy of Sciences, to Professor Svedberg (Uppsala).

GARF, fond 5283, opis' 20, ed. khr. 250, l. 74.

Document 4.9

From A. A. Voznesensky, rector of Leningrad University to VOKS.
3 May 1945.

Please transfer this letter to the Nobel Committee in physics, Stockholm 50.[1]

Resolution in the left margin (handwritten with pencil): "To comrade Voinov. It is a very serious question. It needs to be agreed with the USSR Academy of Sciences and comrade Suvorov.[2] 2 June 1954. Meleshko."

In the right margin: "To comrade Sheinina. Write to the rector of the university and ask him about the purpose for which the Nobel Committee needs these data. Meleshko."[3]

GARF, fond 5283, opis' 20, ed. khr. 250, l. 70.

[1] See doc. 4.10.

[2] Head, Division of Science, Department of Propaganda and Agitation.

[3] See note 2 to doc. 4.10. Paradoxically, the question is asked by the head of the VOKS Department of Scandinavian Countries.

Document 4.10

Extract from the letter of A. V. Karaganov, acting chairman of VOKS Executive Board, to K. F. Vinogradov, representative of VOKS in Stockholm, No. 73.
28 September 1945

[...] On 15 March, the Nobel Committee in physics (Stockholm 50) applied to the Order of Lenin, Leningrad State University for the names of ordinary professors in physics and chemistry. The university presented the enclosed list of its professors.[1]

Being unaware of the purpose for which the Nobel Committee needs these data,[2] we leave it at your discretion whether to transfer it to the addressee or not.

On 24 May 1945, you sent us a letter from the Physico-Chemical Institute, Uppsala, in which the Swedish side expresses the desire to resume scientific correspondence with Professor Semenov. Please inform Mr. Margit Herdner,[3] librarian at the Institute and author of the letter, that Professor Semenov does not object to having contacts with the Physico-Chemical Institute, Uppsala. Professor Semenov currently keeps the position of director of the Institute of Physical Chemistry, the USSR Academy of Sciences (Moscow, Vorobievskoye shosse, 2).

GARF, fond 5283, opis' 20, ed. khr. 250, l. 69.

[1] There is no list of professors in the archive file.
[2] Shocking incompetence by a top VOKS official. Even in the 1940s, there was hardly a cultured person unaware of the purposes for which the Nobel Committees had been established and why they annually distributed invitations to selected nominators all over the world to recommend their candidates for the Nobel Prizes.
[3] Judging by the name, the position of the librarian was occupied by a lady.

Document 4.11

From: F. Hanshen, chairman of the Nobel Committee in physiology and medicine.
To: L. A. Orbeli, member of the USSR Academy of Sciences.
31 January 1946

Dear Sir,

I have the honour to inform you that the Nobel Committee in physiology and medicine received your letter of January 1946 concerning the Nobel Prize. The Committee is obliged for your interest in this matter [...]

Back translation from Russian into English.

GARF, fond 5283, opis' 20, ed. khr. 266, l. 73.

Document 4.12

From: K. F. Vinogradov, representative of VOKS in Stockholm.
To: V. S. Kemenov, chairman, VOKS Executive Board.
No. 8-/94. 7 February 1946

Enclosed find please the letter of Folke Hanshen, chairman of the Nobel Committee in physiology and medicine,[1] addressed to academician Orbeli, Hero of Socialist Labor, to confirm that the letter of Orbeli was received in the Nobel Committee. Please pass the enclosed letter to the addressee.

Under the text (handwritten in pencil): "To comrade Shumilov. Please dispatch. M. 18/II.46."[2]

In an upper corner (in ink): "Dispatch via the Department of Culture. 20/II.46."

GARF, fond 5283, opis' 20, ed. khr. 266, l. 72.

[1] Nobel Committee in physiology and medicine.
[2] See note 1 to doc. 4.4.

Document 4.13

From: A. Khokhulina, acting head of VOKS Department of Scandinavian Countries (signature is illegible).
To: Kalinin, head of the VOKS Department of Culture (dateless).

Attached find please a letter from Folke Hanshen, chairman of the Nobel Committee in physiology and medicine, addressed to academician Orbeli, Hero of Socialist Labor. Please pass the attached letter to the addressee.

In an upper corner (in ink): "on science, 22 February 1946." The signature is illegible but matches that under the text of the letter.

GARF, fond 5283, opis' 20, ed. khr. 266, l. 71.

Document 4.14

From: S. Sederblum, Swedish envoy in the USSR.
To: V. S. Kemenov, chairman, VOKS Executive Board.
14 March 1946

Dear Mr. Kemenov,

Enclosed find please five copies of the Nobel Statutes to be used by the All-Union Society of Cultural Relations with Foreign Countries and research institutions that may be interested in this document.

I use this opportunity to send you some publications about my country [...]

In the left top corner: "resolution of E. P. Meleshko, acting head of the Department of Scandinavian Countries entrusted with drawing up the draft reply letter."

GARF, fond 5283, opis' 20, ed. khr. 265, l. 108.

Document 4.15

Draft reply letter from V. S. Kemenov, chairman, VOKS Executive Board, to S. Sederblum, Swedish envoy in the USSR (unsigned by Kemenov).

Dear Mr. Sederblum,

Thank you very much indeed for the books about Sweden that you kindly dispatched to me under your cover letter of 14 March. We passed the copies on the Nobel Statutes to the USSR Academy of Sciences and the remaining publications to other research institutions for which they may be of great interest.

This text is crossed out and edited with pencil by Kemenov. In the top left corner: "It needs to be drier." The typewritten text is followed by new writing in Meleshko's hand in ink, taking into account Kemenov's proofs (see doc. 4.16).

GARF, fond 5283, opis' 20, ed. khr. 265, l. 105.

Document 4.16

Reply letter of V. S. Kemenov, chairman of the VOKS Executive Board to S. Sederblum, Swedish envoy in the USSR.
No. 22. 30 March 1946

Dear Mr. Sederblum,

Thank you for the books about Sweden that you dispatched to me under your cover letter of 14 March. Copies on the Nobel Statutes were transferred to the USSR Academy of Sciences and the remaining publications to other research institutions for which they may be of interest.

This text is endorsed by Meleshko.

GARF, fond 5283, opis' 20, ed. khr. 265, l. 104.

Document 4.17

From: E. P. Meleshko, acting head of VOKS Department of Scandinavian Countries.
To: N. G. Bruevich, academician-secretary of the USSR Academy of Sciences.
No. 99. 5 April 1946

Enclosed find please find copies of the Nobel Statutes received by comrade Kemenov V. S. from Mr. Sederblum, Swedish envoy in the USSR.

GARF, fond 5283, opis' 20, ed. khr. 263, l. 141.

Document 4.18

Extract from the letter of E. P. Meleshko, acting head of the Department of Scandinavian Countries to K. F. Vinogradov, representative of VOKS in Stockholm; the letter contains the list of printed and handwritten materials sent to Sweden, No. 34, 6 April 1946.

[...]* Letter from Professor Anichkov addressed to Professor Liljestrand.

GARF, fond 5283, opis' 20, ed. khr. 260, l. 208.

Document 4.19

Letter of N. N. Anichkov, member of the USSR Academy of Sciences, to Professor Liljestrand (dateless).[1]

To Professor Liljestrand, Karolinska Institutet, Stockholm.

Thank you for your kind letter. I am now with the 2nd Leningrad Medical Institute, ul. Kurakina, 1/3, Leningrad, USSR. Yours sincerely.

Back translation from Russian into English.

Below this text: "without home address." The signature is illegible, the first letter is "M."[2]

[1] The letter is dateless because Anichkov could not know when it would be dispatched.
[2] See note 2 to doc. 4.4; "home address" is mentioned by mistake of the translator from English.

Document 4.20

From: G. Liljestrand, secretary of the Nobel Committee in physiology and medicine.
To: USSR mission in Stockholm.
11 February 1946.
(Translated from English by E. Kondakova; the original text in English has not been found.)

The Nobel Statutes stipulate that members of certain medical faculties annually selected by the staff of the Karolinska Institutet are invited to propose their candidates for the Nobel Prize. This year members of the medical faculties of Moscow University are likely to be invited.[1]

However, the Nobel Committee could not obtain a list of these faculties. Textbooks usually used for the purpose[2] are outdated (the last edition dates to 1938) and obviously unsuitable. We would therefore greatly appreciate the mission providing us with the necessary data.

GARF, fond 5283, opis' 20, ed. khr. 266, l. 59.[3]

[1] Medical Faculty of the Moscow University was redesignated as the First Moscow Medical Institute in 1930.
[2] To select professors to be invited for nomination.
[3] Copy of doc. 4.20 sent to the USSR Ministry of Foreign Affairs is now in AVP RF, fond 0140, opis' 030, papka 40, ed. khr. 18, l. 43.

Document 4.21

From: K. F. Vinogradov, representative of VOKS in Stockholm.
To: V. S. Kemenov, chairman, VOKS Executive Board.
No. 83/110. 2 March 1946

Enclosed find please a letter from Mr. Liljestrand, secretary of the Nobel Committee in physiology and medicine, in which he asks the mission to provide a list of members of medical faculties of Moscow University, to enable the Committee to invite them in accordance with the Nobel Statutes to personally participate in nominating candidates for the Nobel Prize in physiology and medicine.[1]

Below the text (handwritten in pencil): "Attn: comrade Meleshko, via the USSR Academy of Sciences. 8.4. M." and "To be executed by comrade Kislova. Meleshko. 10.4."

GARF, fond 5283, opis' 20, ed. khr. 266, l. 58.

[1] See note 2 to doc. 4.4.

Document 4.22

From: G. Shumilov, Head, VOKS Department of Scandinavian Countries (signed by Pisarevskaya).
To: Kalinina, head, VOKS Department of Culture.
18 April 1946

Mr. Liljestrand, secretary of the Nobel Committee in physiology and medicine, applied to the USSR mission in Sweden for a list of members of medical institutions in Moscow[1] to enable the Committee to invite them in accordance with the Nobel Statutes to personally participate in nominating candidates for the Nobel Prize in physiology and medicine. The Department of Scandinavian countries asks you to inform the Medical Section of VOKS about this request and provide the list of the candidates.

GARF, fond 5283, opis' 20, ed. khr. 266, l. 60.

[1] A mistake stemming from a superficial reading of Liljestrand's letter. The author asked for the list of professors from the concrete higher education institutions.

Document 4.23

From: Professor V. V. Parin, vice-president of the VOKS Medical Section and academician-secretary of the USSR Academy of Medical Sciences, and N. G. Egorov, Secretary of the Medical Section.
To: G. Shumilov, Head, VOKS Department of Scandinavian Countries.
25 April 1946

With reference to the request of Mr. Liljestrand, Secretary of the Nobel Committee in physiology and medicine, Bureau of the VOKS Medical Section asks to report to him the following list of full members of the USSR Academy of Medical Sciences who could participate in the nomination of candidates for the Nobel Prizes in physiology and medicine:

A. I. Abrikosov (vice-president of the Academy)
N. N. Anichkov
A. A. Bogomolets
V. N. Vinogradov
Yu. Yu. Dzhanelidze
P. A. Kupriyanov (vice-president of the Academy)
L. A. Orbeli
E. N. Pavlovsky
A. V. Palladin
V. V. Parin (academician-secretary of the Academy)
O. Ya. Parnas[1]
N. A. Semashko
A. D. Speransky
V. K. Khoroshko
S. S. Yudin[2]

GARF, fond 5283, opis' 20, ed. khr. 266, l. 61.

[1] The name of Parnas is crossed out from the typewritten list and substituted by N. F. Gamaleya. The substitution was most likely introduced after the reception of discreditable materials from the USSR Ministry of Internal Affairs. Parnas was born in Galicia (Austria-Hungary), graduated from the Higher Pedagogical School, Berlin, and occupied the position of director of the Institute of Medical Chemistry, University of Lvov, since 1920. He became a Soviet citizen after the West Ukraine territories formerly belonging to Poland were annexed by the USSR in 1939. Parnas and a group of other scientists were evacuated to the city of Ufa in June 1941 shortly before the invasion of Lvov by Nazi troops. In 1942, Parnas was elected member of the USSR Academy of Sciences and moved to Moscow in May 1943. In 1942, he was in contact with Kot, the former Polish ambassador in the USSR, and was under surveillance of security agents since that time. In December 1945, he was not allowed to go to Paris.* Parnas was arrested by the USSR Ministry of Internal Affairs on 29 January 1949 and died of heart failure at KGB headquarters in Moscow.**

* L. P. Beria, head of KGB, wrote to V. M. Molotov, deputy prime minister, that he "considered the visit of O. Ya. Parnas to Paris undesirable." (GARF, fond 9401, opis' 2, ed. khr. 105, l. 107–108).

** By mere coincidence, Parnas was arrested the next day after *Pravda* published an editorial that launched the campaign "against cosmopolitanism" (see Chapter 6).

[2] The list of the members of the USSR Academy of Sciences drawn up by the Medical Section of VOKS was a direct result of the mistake of the author of doc. 4.22 (see note 1 to doc. 4.22).

Document 4.24

From: I. S. Chernyshev, acting *charge d'affairés* of the USSR in Sweden.
To: A. N. Abramov, head, 5th European Department of the USSR Ministry of Foreign Affairs.
No. 284. 4 July 1946.

With reference to your letter No. 90 of 8 June 1946 concerning the Nobel Committee in Sweden and its addresses to certain Soviet scientists, I would like to remind you that our memorandum "Awards and Handing of Nobel Prizes in 1945" dated 8 January, contains the general characteristic of the institute of Nobel Prizes and Swedish organizations awarding them in different fields of culture. In addition, enclosed find please a note with information about the Nobel Committee drawn up by the second secretary of the mission comrade Aleksandrov.

As regards the nomination of Soviet scientists and writers for the Nobel Prizes, the following considerations may be in order.

It would be unwise to lose an opportunity to nominate them despite the fact that the Nobel Committees in Sweden has thus far systematically disregarded representatives of Soviet science and culture.[1] It should be borne in mind that the Nobel Prizes are regarded as the most prestigious awards in science and literature all over the world; hence, they automatically bring fame to both the winners and the countries they represent.

Also, the Swedes themselves are well aware of the obvious pre-eminence of anglo-saxon candidates among the nominees during the last years and would like to somewhat change this trend by maintaining closer contacts with Soviet scientists and writers, as evidenced by the invitation to many of them to propose their compatriots for the Nobel Prizes, and the campaign now underway in Norway, Sweden, and Finland in favour of A. M. Kollontai as a candidate for the 1946 Nobel Peace Prize.[2]

The political reputation of the Nobel Prizes is that of "militant humanism"; they acquired it after Hitler prohibited German citizens from receiving Nobel Prizes following the award of one to the anti-fascist journalist Ossietzky in 1936.[3]

RGASPI, fond 17, opis' 125, ed. khr. 452, l. 90

[1] This inference is attributable to the fact that in 1946 the author could not know archive documents of the Nobel Committees (available today) that provide evidence of the very low activity of Soviet nominators in terms of submitting their countrymen for the Nobel Prizes (see Chapters 1–3).
[2] A. M. Kollontai was nominated for her contribution to the conclusion of the truce agreement between the USSR and Finland in 1944 under which Finland withdrew from the war as an ally of Germany and the Soviet Union bound itself not to occupy its territory. Kollontai was nominated by 23 members of the Norwegian Storting, 21 Swedish members of Parliament, and 28 Finnish statesmen including President Paasikivi (since March 1946), Prime Minister Enkel, and a leading figure of the Finnish social democratic party, Kerstin Hesselgren.*

* Redegjoresle for Mobels Fredspis, Oslo, 1946, s. 8–9.

[3] See also doc. 3.16.

Document 4.25

From: A. V. Karaganov, deputy chairman, VOKS Executive Board.
To: G. F. Aleksandrov, head, Department of Propaganda and Agitation, AUCP(b) Central Committee.
No. 2213c. 17 September 1946

Enclosed find please a copy of the letter received from comrade I. S. Chernyshev, acting *charge d'affairés* of the USSR in Sweden, concerning participation of Soviet scientists in the nomination of candidates for the Nobel Prizes. The Nobel Committee asks for the names of Soviet scientists who could be invited to propose candidates for the Nobel Prize in physiology and medicine. The answer is expected by telegraph before 17–18 September. Comrade Dekanozov V. G. shares the opinion of Chernyshev and asked me to apply to you for advice.

I think the following Soviet scientists might be recommended to the Nobel Committee:

1. A. I. Abrikosov
2. N. N. Anichkov
3. V. N. Vinogradov
4. Yu. Yu. Dzhanelidze
5. P. A. Kupriyanov
6. L. A. Orbeli
7. E. N. Pavlovsky
8. A. V. Palladin
9. V. V. Parin
10. N. F. Gamaleya
11. A. D. Speransky

12. V. K. Khoroshko
13. S. S. Yudin[1]

RGASPI, fond 17, opis' 125, ed. khr. 452, l. 82.

[1] This list differs from that proposed by the Medical Section of VOKS (doc. 4.23); it does not include A. A. Bogomolets (deceased 19 July 1946) and N. A. Semashko (for an unknown reason, but probably his age of 72 years).

Document 4.26

Materials on the participation of A. N. Terenin, member of the USSR Academy of Sciences, and other scientists in the International Optical Congress, Paris.

I

From: D. F. Ustinov, USSR minister of defence.
To: L. P. Beria, deputy prime minister.
No. M-9/7041. 28 June 1946

State Optical Institute, USSR Ministry of Defence, received an invitation from the Institute of Theoretical and Applied Optics, Paris, to send a delegation to the International Optical Congress to be held in Paris from 4 to 19 October 1946. The Congress is convened to celebrate the 25th anniversary of the establishment of the Institute of Optics in Paris. It is highly desirable to have contacts with this institute. I therefore ask for permission to send to the Congress the following prominent Soviet scientists:

1. A. N. Terenin, full member of the USSR Academy of Sciences, Stalin Prize[1] laureate;
2. G. G. Slusarev, D. Phys.-Math. Sci, Stalin Prize laureate;
3. T. P. Kravets, corresponding member of the USSR Academy of Sciences;
4. V. P. Linnik, full member of the USSR Academy of Sciences, Stalin Prize laureate.

In the upper left corner: "AUCP(b) Central Committee. Attn: comrade Zhdanov A.A. 8/VII. L. Beria."
In the upper right corner: "Attn: comrade Efimov. M. Suslov, 10/VII."

RGASPI, fond 17, opis' 117, ed. khr. 648, l. 8.

[1] Stalin Prizes instituted in 1939 were awarded until 1954. Since 1957, Lenin Prizes were awarded again after being cancelled in 1935. Stalin Prize winners received new diploma and badges of State Prize laureates.

II

From: M. A. Suslov, head, Department of Foreign Policy, AUCP(b) Central Committee.
To: A. A. Zhdanov, secretary of Central Committee.
No. 25-C-3057. 13 September 1946

The State Optical Institute, USSR Ministry of Defence, received an invitation from the Institute of Theoretical and Applied Optics, Paris, to send a delegation to the International Optical Congress to be held in Paris from 4 to 19 October 1946. The Congress is convened to celebrate the 25th anniversary of the Institute of Optics in Paris. The Ministry of Defence considered the invitation and proposed to send a delegation of four Soviet scientists for participation in the Congress. The USSR Ministry of Foreign Affairs (comrade Dekanozov) does not object to the visit of the Soviet delegation to Paris. It is known that over 1500 physicists from 31 states accepted the invitation to come to the Congress. The program envisages reading and discussing theoretical lectures, exhibition of physical instruments, visits to laboratories and commercial companies. The Congress would give Soviet scientists an opportunity to familiarize themselves with recent results and achievements of foreign opticians; also, our participation would promote strengthening cultural relations between France and the USSR.

Therefore, the Department of Foreign Policy of AUCP(b) Central Committee supports the proposal of the Ministry of Defence but recommends to send to Paris a delegation consisting of A. N. Terenin, full member of the USSR Academy of Sciences, and A. T. Ashcheulov, Ph.D.[1] Attached find please a draft resolution of the AUCP(b) Central Committee.

RGASPI, fond 17, opis' 117, ed. khr. 648, l. 7.

[1] The substitution of an unknown candidate of physico-mathematical sciences for the three eminent investigators was most likely dictated by the secret services.

III

Draft resolution of the AUCP(b) Central Committee drawn up by its Secretariat and endorsed by M. A. Suslov's signature.

1. AUCP(b) Central Committee permits USSR Ministry of Defense to send a delegation consisting of A. N. Terenin, full member of the USSR Academy of Sciences, (head) and A. T. Ashcheulov, Cand. Phys.-Math. Sci., for the participation in the International Optical Congress to be held in Paris.
2. This decision is subject to the approval of Politbureau.

Voting results: four signatures of the members of the Secretariat of the Central Committee; Zhdanov, Kuznetsov, and two others (signatures are illegible) voted against the proposal.

RGASPI, fond 17, opis' 117, ed. khr. 648, l. 6.

IV

Extract from the minutes of the meeting of the AUCP(b) Secretariat, No. 278, item 460, 27 September 1946.

AUCP(b) Secretariat considers impractical the proposal of the Department of Foreign Policy, AUCP(b) Central Committee, to send a delegation of opticians (comrades A. N. Terenin and A. T. Ashcheulov of the State Optical Institute) to the Optical Congress in Paris.

RGASPI, fond 17, opis' 117, ed. khr. 648, l. 5.

Document 4.27

Materials on the scheduled visit of N. D. Papaleksi, member of the USSR Academy of Sciences, to the meeting of the International Scientific Union of Radio, Paris.

I

Extract from the minutes of the meeting of the AUCP(b) Secretariat, No. 278, item 461, 27 September 1946.

On the visit of N. D. Papaleksi, member of the USSR Academy of Sciences, to Paris.

The AUCP(b) Secretariat considers impractical the visit of N. D. Papaleksi, member of the USSR Academy of Sciences, to the general meeting of the International Scientific Union of Radio in Paris.

This decision is subject to the approval of Politbureau.

At the bottom (a handwritten note by Ilyina, a clerk, dated 27 September 1946): "Passed to PB"[1]:

1. The letter to comrade A. A. Zhdanov signed by comrade Aleksandrov and dated 14 September 1946, with resolution: "Approved. Zhdanov."
2. A copy of the letter from the USSR Academy of Sciences No. 1989c of 4/9 September 1946 addressed to comrade Beria, 1 page, No. 67187.

RGASPI, fond 17, opis' 117, ed. khr. 648, l. 18.

[1] Politbureau.

II

Extract from the minutes of the meeting of AUCP(b) Politbureau, No. 55, item 89, 1 October 1946.

On the visit of N. D. Papaleksi, member of the USSR Academy of Sciences, to Paris.

The AUCP(b) Politbureau considers impractical the visit of N. D. Papaleksi, member of the USSR Academy of Sciences, for the participation in the general meeting of the International Scientific Union of Radio in Paris.

RGASPI, fond 17, opis' 3, ed. khr. 1062, l. 25.

Document 4.28

From: G. F. Aleksandrov, head, Department of Propaganda and Agitation, AUCP(b) Central Committee.
To: A. A. Zhdanov, secretary of Central Committee.
14 October 1946

The Swedish Nobel Committee asks the Executive Board of the All-Union Society for Cultural Relations with Foreign Countries to give them the names of Soviet scientists who could be invited to propose candidates for the Nobel Prize in physiology and medicine. The Department of Propaganda and Agitation, AUCP(b) Central Committee, considers advisable the participation of Soviet scientists in the nomination of candidates for the Nobel Prize in physiology and medicine.[1]

The Nobel Prizes are regarded in worldwide public opinion as the most prestigious awards in recognition of the high level of the laureate's works; they raise the reputation of the prize-winner's country among the world scientific community [...] There is every reason to use this opportunity to strengthen our international cultural relations and give Soviet scientists the possibility to nominate their candidates for the Nobel Prizes; it will not require their visits abroad.

It appears appropriate to instruct the Executive Board of the All-Union Society for Cultural Relations with Foreign Countries (comrade Karaganov) to send to the Nobel Committee the following list of Soviet scientists for participation in nominating candidates for the Nobel Prize in physiology and medicine: A. I. Abrikosov, N. N. Anichkov, V. N. Vinogradov, Yu. Yu. Dzhanelidze, Kh. S. Koshtoyants, P. A. Kupriyanov, L. A. Orbeli, E. N. Pavlovsky, A. V. Palladin, V. V. Parin, I. P. Razenkov, V. K. Khoroshko, S. S. Yudin.[2]

In the upper left corner (handwritten in pencil by Zhdanov): "Ask the opinion of comrades Mitirev and Smirnov."

RGASPI, fond 17, opis' 125, ed. khr. 452, l. 84.

[1] An openly hypocritical phrase that imposed no obligation on the author. The document was drawn up and dispatched one month after the deadline set by the Swedish side (see doc. 4.25). It is nothing more than imitation of activity according to the prescheduled scenario.
[2] The list of potential nominators (doc. 4.25) is supplemented by Koshtoyantz and Razenkov, whereas the name of Speransky is stricken off for an unknown reason. It may be supposed, judging by the previous changes in the list (exclusion of Parnas (see note 1 to doc. 4.23) and Yudin (note 1 to doc. 4.34)), that Speransky was crossed off for political or ideological reasons too.

Document 4.29

From: G. A. Mitirev, minister of public health, and E. I. Smirnov, head, Central Medical Department of USSR Military Forces.
To: A. A. Zhdanov, secretary of AUCP(b) Central Committee.
28 October 1946

Participation of Soviet scientists in the nomination of candidates for the Nobel Prize in physiology and medicine is highly desirable as a means of promoting prestige of our country. The USSR must be represented by a small group of researchers widely known in the world for their works in biological and clinical medicine.

We suggest that the following list of them should be forwarded to the Nobel Committee via VOKS:

1. N. N. Anichkov, member of the USSR Academy of Sciences.
2. L. A. Orbeli, -"-
3. E. N. Pavlovsky, -"-
4. A. V. Palladin
5. S. S. Yudin, member of the USSR Academy of Medical Sciences
6. V. V. Parin, -"-
7. N. D. Strazhesko, -"-

In the upper left corner (handwritten in pencil by Zhdanov): "Attn: Aleksandrov. Take into account the opinion of Mitirev and Smirnov."

RGASPI, fond 17, opis' 125, ed. khr. 452, l. 89.

Document 4.30

From: G. F. Aleksandrov, head, Department of Propaganda and Agitation, AUCP(b) Central Committee.
To: A. A. Zhdanov, secretary of Central Committee.
30 January 1947

In pursuance of your instructions, comrades Mitirev and Smirnov were asked to name scientists who might be recommended as nominators for the Nobel Prize in physiology and medicine.

Comrades Mitirev and Smirnov did not object to the candidates enumerated in the list submitted by comrade Karaganov, deputy-chairmen of the VOKS Executive Board, but suggested that the list be shortened; in addition, they proposed to include academician Strazhenko [...] The list presented by comrades Mitirev and Smirnov needs to be supplemented by the names of A. I. Abrikosov and Yu. Yu. Dzhanelidze, members of the USSR Academy of Medical Sciences. Both are authoritative researchers whose works are well known in the Soviet Union and abroad.

There is no need to substantially shorten the above list, the more so that the Nobel Committee will most likely approve only a few of those enumerated whom it will find most acceptable. These candidates are agreed upon with the Ministry of Public Health (comrade Kuznetsov) and E. I. Smirnov, head, Central Medical Department of USSR Military Forces.

RGASPI, fond 17, opis' 125, ed. khr. 452, l. 88.

Document 4.31

From: F. N. Mal'gin, first secretary of the VOKS mission in Stockholm.
To: V. S. Kemenov, chairman, VOKS Executive Board.
No. 830/105. 14 March 1947

The letter from Mr. Liljestrand, secretary of the Nobel Committee, was sent to you on 20 March 1946. Mr. Liljestrand asked for a list of members of the medical faculties at Moscow State University, whom the Nobel Committee could invite, in conformity with the Nobel Statutes, to personally nominate candidates for the Nobel Prize in physiology and medicine. Almost a year has elapsed since then, but we still have no answer.

Enclosed find please a new letter of the secretary from the Nobel Committee on the same matter dated 12 February 1947. I am looking forward to hearing from you at your earliest convenience.

AVP RF, fond 200, opis' 32, papka 35, ed. khr. 23, l. 72.

Document 4.32

From: Liljestrand, secretary of the Nobel Committee in physiology and medicine
To: USSR mission in Stockholm
12.2.1947 (translated from English by E. Kondakova)

The Nobel Statutes stipulate that members of medical faculties annually selected by the Karolinska Institutet are invited to nominate personally candidates for the Nobel Prize in physiology and medicine. This year, members (ordinary professors) of the medical faculties of the Moscow State University were supposed to be invited. However, the Nobel Committee thus far failed to obtain the list of these members. Textbooks usually used for the purpose are outdated (the last edition dates to 1938) and obviously unsuitable. Having no answer to our request, we unfortunately could not invite our Russian colleagues. We would therefore greatly appreciate the mission providing us with the necessary data about members of the medical faculties in Moscow.

RGASPI, fond 17, opis' 125, ed. khr. 452, l. 91.

Document 4.33

From: A. V. Karaganov, deputy-chairman, VOKS Executive Board.
To: G. F. Aleksandrov, head, Department of Propaganda and Agitation, AUCP(b) Central Committee.
No. 816c. 8 April 1947

With reference to your letter No. 2213c of 17 September 1946 enclosed find please a copy of a translated letter from Mr. Liljestrand, secretary of the Nobel Committee, in which he asks for a list of members[1] of Soviet scientists working in physiology and medicine who could be invited to nominate candidates for the Nobel Prizes in these disciplines. The USSR mission in Stockholm needs to have an answer as soon as possible because it is the second application by Liljestrand concerning the same matter.

RGASPI, fond 17, opis' 125, ed. khr. 452, l. 83.

[1] Sic!

Document 4.34

From: G. F. Aleksandrov, head, Department of Propaganda and Agitation, AUCP(b) Central Committee.
To: A. A. Zhdanov, secretary of Central Committee.
30 June 1947

The Department of Propaganda and Agitation, AUCP(b) Central Committee, once again considered the list of physiologists and medical scientists who could be invited to nominate candidates for the Nobel Prizes in these disciplines. This issue appears topical even today because VOKS has recently received another letter concerning the same matter.

We propose that the Executive Board of All-Union Society for Cultural Relations with Foreign Countries should report to the Nobel Committee the names of the following Soviet scientists who could participate in the nomination of candidates for the Nobel Prizes: N. N. Anichkov, L. A. Orbeli, E. N. Pavlovsky, A. V. Palladin, Yu. Yu. Dzhanelidze, N. D. Strazhesko [...] Professor S. S. Yudin, included in the earlier list, should now be crossed out as a person who proved unable to behave with dignity due to being a Soviet scientist in contact with foreign specialists.[1]

The upper part of this text bears a note handwritten by Zhdanov in pencil: "S-t," meaning that the document was to be discussed by the AUCP(b) Secretariat; the discussion never took place, however (see doc. 4.36).

RGASPI, fond 17, opis' 125, ed. khr. 452, l. 92.

[1] A direct result of the campaign "against antipatriotism" (see Chapter 6). A year and a half later (the night of 22–23 December 1948), Yudin was arrested and spent

three years in solitary confinement in the Lefortovo prison in Moscow. Thereafter, he was exiled to Berdsk, near the city of Novosibirsk (in spring 1952) but rehabilitated in 1953.

V. V. Parin, academician-secretary of the USSR Academy of Medical Sciences was arrested on the night of 17–18 February 1947 as an "American spy" ten days after returning from a five-month stay in the USA; his name was also in the list of scientists prepared for the Nobel Committee (docs. 4.3, 4.25, 4.29). He was arrested immediately after the evening meeting of the Politbureau on 17 February at which the so-called "Klyueva and Roskin's case" was discussed (see Chapter 6). Parin was charged with the unauthorized transfer of these scientists' manuscript "Biotherapy of Malignant Tumours" and their anti-tumour "KR" preparation to American specialists. Closing the meeting, Stalin said: "I do not trust Parin." V. V. Parin was sentenced to a 25-year imprisonment and served his term in a prison at Vladimir until rehabilitation in 1953. He shared a cell with L. L. Rakov, historian and arts critic, and D. L. Andreev, philosopher. While in prison, the three men wrote an amusing book *The Newest Plutarchus.**

Rehabilitation of Parin proceeded in two stages. First, the initial accusation of high treason (article 58-1a of the RSFSR Criminal Code) was reclassified as divulgation of a state secret (an article from the decree of the Presidium of the USSR Supreme Council dated 15 November 1943) and the prison term was reduced from 25 to five years; he was immediately released on 29 October 1953. In a year and a half, complete rehabilitation came as proposed by the Office of the Prosecutor-General to the USSR Council of Ministers (doc. 4.35).

* Published by the "Moskovsky rabochiy" publishing house in 1991. The story of his arrest is told by Parin's widow N. D. Parina in the afterword to the book (pp. 293–303).

Document 4.35

From: P. V. Baranov, acting Prosecutor-General of the USSR.
To: N. A. Bulganin, chairman of the USSR Council of Ministers.
19 March 1955

In August 1947, V. V. Parin, then the academician-secretary of the USSR Academy of Medical Sciences, gave permission to the members of the Soviet–American Society Lesley and Madd (invited by VOKS) to visit the laboratory of Professor Klyueva, who was developing the anti-tumour "KR" drug. In October 1946, Parin visited USA with a group of medical scientists, where he handed over the manuscript of Kluyeva's book and ampoules of "KR" to American specialists (November 1946). (According to Parin, he followed instructions received from the former USSR minister of public health Mitirev). The chapter, containing the description of "KR" manufacture technology was deleted from the manuscript prior to its transfer [...]

After his return from the USA, Parin was arrested by Abakumov[1] (17 February 1947), accused of high treason, and imprisoned for 25 years by the decision of the Special Conference,[2] Ministry of State Security, dated 8 April 1948.

On 13 October 1953, comrade Kruglov, minister of internal affairs, and comrade Rudenko, prosecutor-general, sent a joint letter No. 12/14k to the CPSU Central Committee proposing to revise V. V. Parin's case. The Presidium of the CPSU Central Committee approved the proposal to reclassify accusation of high treason (article 58-1a of the RSFSR Criminal Code) as divulgation of a state secret (an article from the decree of the Presidium of the USSR Supreme Council dated 15 November 1943), to reduce the prison term from 25 to five years, and to release Parin as he had already served this sentence.

It appears from the letter by comrade A. P. Bakulev, president of the USSR Academy of Medical Sciences, that Klyueva and Roskin's discovery was not confirmed. With this in mind, I believe that Parin's case should be revised with a view to his rehabilitation.

In: Reabilitatsiya: kak eto bylo. Dokumenty prezidiuma TsK KPSS i drugie materialy. Mart 1953-Fevral' 1956 (Rehabilitation: How it Proceeded. Documents of the Presidium of CPSU Central Committee and Other Materials. March 1953–February 1956). M.: 2000, pp. 205–206.

[1] USSR Minister of State Security.

[2] Special Conference of NKVD SSSR (USSR Ministry of State Security since 15 March 1946) was established by the USSR Central Executive Committee and the USSR Council of Ministers (11 November 1934) as a quasi-judicial body to maximally expedite consideration of charges and sentencing without the right of appeal. Its sittings were held in the absence of the accused and their advocates; the sentences were announced after the chairman and the two members signed them. The abolition of the Special Conference was discussed by the Presidium of the CPSU Central Committee on 12 August 1953 and it ceased to exist on 1 September 1953. A total of 442,531 people were sentenced by the Special Conference to capital punishment or exile to remote regions of the country; 10,101 persons were shot.*

* In: Reabilitatsiya [...], pp. 72–73.

Document 4.36

From: S. G. Suvorov, head, Division of Science, Department of Propaganda and Agitation, AUCP(b) Central Committee.
To: A. A. Zhdanov's Secretariat.
11 October 1947

The All-Union Society for Cultural Relations with Foreign Countries (comrade Karaganov) proposed to report to the Nobel Committee the names of Soviet scientists who could participate in nominating candidates for the forthcoming Nobel Prize in physiology and medicine.

At present, there is no need to submit a draft resolution on this issue for the consideration of the AUCP(b) Central Committee because the Nobel Committee ceased to register candidates for the said Nobel Prize in physiology and medicine in September.[1]

The material is to be shelved.[2]

RGASPI, fond 17, opis' 125, ed. khr. 452, l. 94.

[1] A mistake. In September, the Nobel Committees cease to register nominators invited to propose candidates for the Nobel Prizes to be awarded the following year.
[2] The delay was deliberate and virtually undisguised. G. F. Aleksandrov's letter to Zhdanov (doc. 4.34) was dated 30 June. Zhdanov's dateless resolution forwarding the document to the consideration of the Secretariat was certainly made in the same days. Under the procedure established in the AUCP(b) Central Committee, the Secretariat considered documents endorsed by a secretary within a few days or two weeks at most. It means that doc. 4.34 was sent back to the Department of Propaganda and Agitation, AUCP(b) Central Committee, by mid-July and shelved for three months, probably under a verbal agreement between the functionaries.

Document 4.37

Extract from minutes No. 59 of the meeting of AUCP(b) Politbureau, 10 June 1947.[1]

1) Turn down the proposal by the Presidium of the USSR Academy of Sciences to elect American chemist Langmuir, English scientists Robinson (chemist), Blackett (physicist), Keilin (biochemist), Appleton (radiophysicist), and Mexican physicist Vallarta[2] corresponding members of the Academy.
2) Recommend academicians S. I. Vavilov and N. G. Bruevich to submit to the general meeting of the Academy the following candidates for the election corresponding members of the USSR Academy of Sciences: Raman S[3].V., Frederic and Irene Joliot-Curie (France), Albert Szent-Gyorgyi (Hungary), Zdenek Needly (Czechoslovakia), Todor Pavlov (Bulgaria), Constantin Parhon (Romania), and Kazimierz Nitsch (Poland).

RGASPI, fond 163, opis' 4, ed. khr. 963, l. 101.

[1] This decision by Politbureau is signed by Stalin (with blue pencil) under Zdanov's facsimile.
[2] The rejected candidates included Langmuir (Nobel Prize winner in chemistry for 1932), two future laureates in physics, Blackett (1948) and Appleton (1947), and future laureate in chemistry Robinson (1947).*

* President of the London Royal Society R. Robinson participated in the jubilee session of the USSR Academy of Sciences celebrating its 220th anniversary. He delivered a welcoming speech on behalf of British scientists at a plenary session.

[3] A mistake. Raman's name was Chandrasekhara.

Document 4.38

Extract from minutes No. 59 of the meeting of AUCP(b) Politbureau, 10 June 1947.[1]

The Central Committee of AUCP(b) finds that publication of Soviet scientific journals in foreign languages does harm to the Soviet state and provides ready-to-use materials for foreign intelligence services. Publication of scientific journals in foreign languages by the USSR Academy of Sciences while no other country publishes them in Russian demeans the Soviet Union and does not encourage patriotism in Soviet scientists.[2]

Taking the aforesaid into consideration, the AUCP(b) Central Committee decides to stop publishing foreign language versions of the following journals of the USSR Academy of Sciences *Doklady akademii nauk*, *Fiziko-khimichesky zhurnal*, and *Zhurnal po fizike*.[3]

I. Stalin, secretary of Central Committee.

RGASPI, fond 17, opis' 117, ed. khr. 868, l. 5.

[1] The corresponding decision of the USSR Council of Ministers No. 2544-776c is dated the same day.

[2] Party functionaries were concerned not only about natural scientists. At the height of the war with Nazi Germany in which the Soviet Union had the UK and USA as allies, the Russian language journal *Internatzional'naya literatura* (International Literature)"* was shut down by the decision of AUCP(b) Central Committee** for "Anglo–American propaganda" (the last issue dates to January 1943). Its former editor-in-chief B. L. Suchkov, an eminent literary critic, was arrested by the USSR Ministry of State Security in 1947 and found guilty of such "propaganda."

* RGASPI, fond 17, opis' 117, ed. khr. 933, l. 33.

** Resolution of the AUCP(b) Secretariat dated 13.2.1943 (RGASPI, fond 17, opis' 116, ed. khr. 116, l. 274).

[3] Until 1947, *Doklady akademii nauk* was published in both Russian and English. The other two journals were published once every two months only in English under the titles of *Acta physicochimica URSS* and *Journal of Physics*. They were founded in 1934 and 1939 respectively. In 1947, only two issues of each appeared. Ready-to-print issues for May and June were never published.

Document 4.39

Extract from the minutes of the closed Party meeting at the Presidium of USSR Academy of Sciences, 10 April 1947.

I

Discussion of the report by the academician-secretary N. G. Bruevich.

E. A. Asratyan, corresponding member of the Academy: "[...] I see nothing bad in publishing journals simultaneously in English or any other language in order to familiarize the world public opinion with our scientific achievements."[1]

B. I. Ressin (Institute of Physiology, the USSR Academy of Sciences) remarked that in pre-war time *Byulleten eksperimental'noy biologii i meditsiny* was published in foreign languages.[2] "Now, it proved impossible to resume such publication despite all the efforts of academician L. S. Shtern."

TsAODM, fond 992, opis' 2, ed. khr. 26, l. 110.

II

Extract from the reply speech of N. G. Bruevich, academician-secretary of the USSR Academy of Sciences.

There are many sorts of evidence that publications in Russian are fairly well known abroad when their use can make money, and the ignorance of Russian is no obstacle in such cases [...] Our literature is closely followed up and studied. Therefore, publication in Russian does not render our works unavailable to foreigners even if it makes the understanding somewhat difficult. The Americans ask "send your papers and we shall publish them." There is every reason for them to do this; it gives a possibility to be in the know of new ideas and paths along which to go further; it is a very important problem.[3] Thus, the main thing is not to strive to have your works published abroad (it is not forbidden) but to strengthen the reputation of Soviet journals and to interest foreigners in publishing their papers in them. Then, we shall be aware of scientific developments abroad at the stage of publication [...] I therefore do not think that comrade Asratyan is totally right.

Ibid, l. 136–138.

[1] This thought would certainly be supported by N. S. Shatsky, corresponding member of the USSR Academy of Sciences, if he were member of AUCP(b) and present at the meeting.* He publicly spoke of the inevitable globalization of science a year before the English language journals published in the USSR were closed. In 1949, he was dismissed from his position of deputy-director at the Institute of Geological Sciences under the pretext of this and other manifestations of his liberal thoughts unacceptable in the "dark kingdom."

* TsAODM, fond 8156, opis' 1, ed. khr. 9, l. 125.

[2] To save paper after the outbreak of the war, the AUCP(b) Central Committee temporarily closed 174 central journals with a total circulation of 4,762,000 copies (3 July 1941).* The list included the English version of *Byulleten of eksperimental'noy*

biologii i meditsiny. It started to be published again only in 1956. The articles were translated into English and published in New York.

* RGASPI, fond 17, opis' 116, ed. khr. 99, l. 136 (from minutes No. 98 of the meeting of AUCP(b) Secretariat).

[3] An example of logic forgery. The speaker ignores a main goal of a modern scientific publication to ensure priority of an important discovery not only for its author but also for the country he represents.

Document 4.40

Correspondence related to the publication of the Russian version of the monograph *Soil Microbiology* by S. N. Vinogradsky in the USSR.[1]

I

From: S. I. Vavilov, president of the USSR Academy of Sciences.
To: S. N. Vinogradsky, professor of the Pasteur Institute, Paris.

I have just received your letter dated 28 October in which you inform me that your book *Soil Microbiology* to be translated into Russian is about to be out. Please send a volume of it to the USSR Academy of Sciences via our embassy. We shall do our best to translate and publish it as soon as possible.

Arkhiv RAN, fond 1601, opis' 1, ed. khr. 125, l. 3.

II

From: A. A. Imshenetsky, corresponding member of the USSR Academy of Sciences.
To: S. N. Vinogradsky (dateless).[2]

I have received your kind letter.[3] The other day, Professor S. I. Vavilov received your book *Soil Microbiology*, and we initiated its translation.[4] It will be printed by the USSR Academy of Sciences publishing house in the near future. Of course, your request will be fulfilled and you will have an opportunity to look through the Russian translation and proof-read it [...] I would like to preface the book with an article containing your main biographic data. I hope you will have no objections to this scheme.

Ibid., ed. khr. 142, l. 3.

III

From: E. S. Vinogradskaya, the scientist's daughter and secretary.
To: the Consular Department, USSR Embassy, Paris.
6 November 1950 (handwritten copy)

1) I take liberty to address you on behalf of my father Professor S. N. Vinogradsky, honorary member of the USSR Academy of Sciences.
2) Would you be so kind as to clear up the following questions:

a) The USSR Academy of Sciences proposed that *Collected Works*[5] of Professor S. N. being published at that time in French here in Paris be translated and published in Moscow; the proposal was readily accepted by the author.
b) The French version was out in November 1949, and a copy was immediately sent to the USSR Academy of Sciences via the Soviet embassy as advised by the president of the USSR Academy of Sciences.
c) Professor S. N. exchanged opinions on this matter with Sergei Ivanovich Vavilov and other members of the academy as well as its research scientists, Imshenetsky and Nadezhda Sushkina, who is also a professor of the Moscow University.[6]
d) A reply letter was sent to Professor Sushkina by air mail two months ago but was turned back [...] Two letters of S. N. addressed to Imshenetsky also remained unanswered.

Thus, correspondence on the matter of interest for the USSR Academy of Sciences is interrupted, which may adversely affect the publication being planned.

This makes the author ask the Soviet mission whether it is possible to further keep correspondence through the agency of the USSR Embassy to ensure delivery of letters to the addressees.

Ibid., ed. khr. 87, l. 2.

IV

A. A. Imshenetsky to S. N. Vinogradsky, 13 December 1951.

I have received your letter and would like to inform you that we managed to overcome the difficulties related to making drawings, etc. The manuscript of *Soil microbiology* has been sent to the printers [...] The biographic article with the analysis of your work is also ready to print.[7] I would like to reassure you that the translators (microbiologist professors) were especially particular about the choice of terminology.

Ibid., ed. khr. 142, l. 1.

V

A. A. Imshenetsky to S. N. Vinogradsky, 24 April 1952.

The publication of *Soil Microbilogy* was slightly delayed, but it is now set up and the proofs of the whole book are available (50 author's sheets). I looked through half of the book. The drawings are engraved perfectly well, and I am convinced that the design will be superior to that of the French edition. I think that the book will be out by 15 June,[8] and the author's copies will be dispatched to you. There are few people as hard-working as you are at your age.[9] I would greatly appreciate your sending me reprints of your works with autographs. The *Great Soviet Encyclopaedia* published an article about you.[10] You can probably read it because this edition must be available in Paris libraries.[11]

Ibid., l. 2.

[1] The correspondence is borrowed from the personal archive of Vinogradsky, now in Brie-Comte-Robert near Paris. His daughter E. S. Vinogradskaya gifted the archive to the USSR Academy of Sciences. It keeps original letters from Moscow and author's drafts from Brie-Comte-Robert.

[2] This letter was sent to France via the Department of Foreign Relations of the USSR Academy of Sciences. Such letters were usually dateless to conceal the large discrepancy between the time of writing and the date of dispatch on the postmark.* The letter being the answer to Vinogradsky's letter of 25 February 1950, it was written not later than March or April.

* The letter of N. N. Sushkina, professor of Moscow University, addressed to Vinogradsky and sent via the same department, bears the name of the month instead of the date ("July 1950"); *Ibid*, ed. khr. 170, l. 1.

[3] The letter of Vinogradsky was sent to Moscow on 25 February 1950 (see the draft).*

* *Ibid*, ed. khr. 98, l. 1.

[4] There is evidence that translation of Vinogradsky's book was started only on 9 August 1950.*

* G. A. Zavarzin. Sergei Nikolaevich Vinogradsky v emigratsii (Sergei Nikolaevich Vinogradsky in Emigration). In: *Rossiyskaya nauchnaya emigratsiya. Dvadtsat' portretov* (Russian Scientific Emigration. Twenty Portraits). M.: Editorial URSS, 2001, p. 220.

[5] *Soil Microbiology.*

[6] Arkhiv RAN, fond 1601, opis' 1, ed. khr. 170, l. 1.

[7] A. A. Imshenetsky. *S. N. Vinogradsky i ego tvorchestvo* (S. N. Vinogradsky and His Works). See note 8.

[8] S. N. Vinogradsky. *Soil Microbiology. Problems and Methods. Fifty Years of Research.* Preface by A. A. Imshenetsky (15 pages). M.: Izd-vo AN SSSR, 1952. p. 795, 50 illustrations, circulation 4000.

[9] S. N. Vinogradsky was 95 at that time. He died soon after his monograph was published in the Russian language (24 February 1963).

[10] *Great Soviet Encyclopaedia*, 2nd ed., V. 8, pp. 123–124. The 50-line biographical note (in two columns) was supplemented by S. N. Vinogradsky's portrait measuring 98 × 70 mm (instead of the usual 40 × 32 mm). It was certainly an initiative of S. I. Vavilov, editor-in-chief of the second edition. The note did not mention the emigré status of the scientist and contained no data on his work in France as professor of the Pasteur Institute during the last 30 years; it would not be passed by censorship if it did.

[11] The volume of the *Great Soviet Encyclopedia* was approved for printing on 16 November 1951.

Document 4.41

Extract from the diary of V. S. Kemenov, chairman of the VOKS Executive Board, concerning his talk with R. Sulman, Swedish envoy in Russia, 2 June 1947.

Sulman [...] said that he attaches great importance to the strengthening of cultural relations between Sweden and the USSR and also to his contacts with VOKS. I told him that we are also interested in the development of cultural relations between Sweden and the USSR, the more so that they had been virtually non-existent during many years, which had facilitated the distribution of anti-Soviet slander and disinformation in Sweden. Sulman answered that Sweden maintained diplomatic relations with the Soviet Union during the war when many other countries broke them. I noted that he apparently alluded to Sweden's neutrality during the war against fascist Germany. The Swedes appear to be proud of their neutrality, but this position does not seem as convincing to the Soviet people. I don't think that neutrality should be a matter of pride. Sulman said that they could not behave otherwise when their country was surrounded by Germans on all sides and threatened even from Norway. I answered that this argument does not make me change my position [...]

AVP RF, fond 0140, opis' 38, papka 146, ed. khr. 50, l. 8.

Document 4.42

From: P. L. Kapitsa, member of the USSR Academy of Sciences.
To: G. M. Malenkov, secretary of AUCP(b) Central Committee, and the reply letter.

I

The letter from P. L. Kapitsa dated 13 November 1945.

Enclosed find please an issue of *Nature* as it is after it fell into the hands of our censors. As known:

1. The Charter of the USSR Academy of Sciences stipulates that the books and journals sent to its members are not subject to censorship.
2. Our censors appear to hold a very low opinion about the moral principles of Soviet scientists and believe that some pieces of information can lead them astray.
3. The censors are too dull-witted to understand that our country is so strong that any statement against it can rouse laughter rather than fear. I therefore think that the censor must be punished and made to stand in the corner like a guilty schoolboy until he learns by heart three chapters by Karl Marx.[1] Also, I ask that the journal be delivered to me as it was out from the printing-house.[2]

II

From: G. F. Aleksandrov, head, Department of Propaganda and Agitation, AUCP(b) Central Committee.
To: G. M. Malenkov, secretary of Central Committee.
20 November 1945

Department of Propaganda and Agitation, AUCP(b) Central Committee made an inquiry in relation to the letter of P. L. Kapitsa and arrived at the following conclusions:

1. Members of the USSR Academy of Sciences receive literature from abroad under the third category. In accordance with the directions of AUCP(b) Central Committee of 2 May 1942, Glavlit[3] can make cuts in such editions. Regulations for members of the USSR Academy of Sciences are somewhat different, by way of exception. If a book or a journal coming to them contains criticism of the soviet leadership, Glavlit classifies such an edition as secret and transmits it to the library of the Academy, where scientists can read it as stipulated by the procedure established for the use of classified publications.[4]
2. P. L. Kapitsa is entitled to receive all foreign publications without their preliminary censoring by Glavlit. But the subscriber to *Nature* is Glavkislorod,[5] and Kapitsa is personally not on the list of its recipients. Glavkislorod, like other organizations directly subordinate to the USSR Council of People's Commissars, receives foreign literature under the third category, i.e. preliminary censored by Glavlit.

Comrade Dobrosel'skaya, head of the Glavlit Foreign Department, instructed a censor to strike off certain paragraphs from an article in *Nature*, No. 3947, 1945[6] (see the attached translation of these paragraphs and the intact copy of the journal). It appears that the censor was somewhat overzealous and cut a few passages from the article "University reconstruction" without any good reason.[7] Certainly, a bourgeois journal can not be expected to treat ideological questions in the spirit of Marxism.
A full copy of *Nature* was sent to P. L. Kapitsa.[8]

Ibid., l. 241–242.

III

Full translation of the text cut by censors from *Nature*, No. 3947.

1. "By rejecting any real attempt to discover and then to teach a unified concept of life, the liberal university ceases to be a university; in the second part of the book, Mr. Nash maintains that students at the European continent are so readily open to Nazist and Communist ideas because totalitarian philosophies offer the sole (well-defined) life choice as opposed to the confusion and chaos of the liberal view of the world." (p. 740)
2. "In the second part of the book, he offers an even more challenging explanation of the ways by which European students came to accept totalitarian philosophies.

Many readers will hardly like his explanation of Nazism and Marxism. He holds that both Nazists and Communists are right rejecting the thesis that science has no *a priori* ideas, that knowledge must serve any purpose other than knowledge for the sake of knowledge and disputing the assertion that any human knowledge is determined by irrational factors." (p. 741)[9]

Ibid., l. 242–243.

[1] A very risky (under Stalin's rule) ironical note mentioning the name of a founder of Marxism. It may be supposed that it was recalled by those who dismissed Kapitsa from his position as the head of Glavkislorod and director of the Institute of Physical Problems he founded at the USSR Academy of Sciences.

It is opportune to add that Kapitsa was famous for the unceremonious treatment of his opponents regardless of their position and personal attitude to them. E. Rutherford, his teacher whom he held in great reverence, characterized him as "a highly gifted experimental physicist very naive in other matters and so easily carried away that he needed to be from time to time hit on the head to be brought to his senses." This remark was made by Rutherford during the talks he had with a representative from Moscow on the selling to the Soviet government of scientific equipment of the Mond Laboratory which Kapitsa had used during his stay in Cambridge (until he was prohibited from coming back to England by the decision of top authorities taken in summer 1934).*

* V. D. Esakov, P. E. Rubinin. Kapitsa, Kreml' i nauka Tom. 1. Sozdanie instituta fizicheskikh problem. 1934–1938 (Kapitsa, Kremlin, and Science. Foundation of the Institute of Physical Problems. 1934–1938). M.: Nauka, 2003, pp. 313–314.

[2] It may be supposed that the harsh tone of the letter was dictated by recollections of events dated from a decade ago. The note in *Nature* informing that Kapitsa was barred from a new visit to the UK was cut from the copies of the journal allowed for circulation among Soviet subscribers. Kapitsa asked his wife staying in England to bring this fact to the attention of P. A. Gregory, the then editor-in-chief (to whom a letter was sent on 21 December 1935).*

* *Ibid*, p. 597.

[3] Chief Agency for the Protection of State Secrets in Press at the USSR Council of People's Commissars.

[4] A subscriber could read his own copy classified as secret only in special premises of the library and enter the needed extracts into a notebook kept in the special department; each time he wished to re-read them, he had to sign for the use.

[5] In 1945–1946, Kapitsa was the head of the Main Department of Oxygen Industry, Council of People's Commissars (Glavkislorod).

[6] R. Brightman. University reconstruction. *Nature*, 155, No. 3947, 1945, pp. 740–741.

[7] A rather liberal conclusion for the head of Department of Propaganda and Agitation, AUCP(b) Central Committee [...] It would be absolutely out of the question after 1946 when the all-round opposition to the West and its values became the official policy; such a statement by a staff member of the AUCP(b) Central

Committee even in the departmental correspondence might have very serious consequences for him.

[8] Kapitsa was deprived of all these privileges after he was dismissed from his position.

[9] Department of Propaganda and Agitation, AUCP(b) Central Committee, did not pioneer such methods of censorship in scientific and technical publications. A similar approach was practiced by secret services in pre-revolutionary Russia, at least during the Russo-Japanese War in 1904–1905. However, at those times some information about their wrongful actions was one way or other published in the liberal press whereas such publications were absolutely impossible under the Soviet regime. For example, Russian newspapers informed their readers that the Imperial Technical Society did not tolerate the interference of censorship and addressed to top officials (as Kapitsa did 40 years later) the demand to "cancel censorial restrictions imposed on the circulation of foreign technical journals."*

* "Novosti i birzhevaya gazeta," 14 June 1905, p. 2.

Document 4.43

"Wallenberg's case" — from mocking negation to partial recognition.

I

Extract from the editorial in the weekly *Novoye vremya* under the heading "The legend of Wallenberg."

[...] Servitors of Swedish and international reaction pulled out of an anti-Soviet dust-heap and put into circulation the so-called "Wallenberg's case". The case is reduced to an episode during the fight for the capital of Hungary in early 1945 when Rauol Wallenverg, member of the Swedish mission in Budapest, was missing [...] A thorough inquiry into Wallenberg's fate carried out by Soviet authorities at the request of the Swedish government gave no result. Initial information that he was found by Soviet military representatives proved incorrect [...]

Two years after Wallenberg had disappeared (in the beginning of 1947), certain right-wing Swedish newspapers made a sensation (to be precise, provocation) of this deplorable (but not infrequent during hostilities) case by telling a myth of "Soviet secret police" that continues to detain Wallenberg [...][1]

Those who fabricated "Wallenberg's case" act without scruples [...] In early January 1948,[2] three deputies of Rickstag (Frisen, Ture Nerman, and W. Lundstedt) addressed a proposal to the Nobel Committee of Norwegian Storting to award the 1948 Nobel Peace Prize to Wallenberg. The newspaper *Arbetaren* exposed the secret aim of this initiative: "The nomination of Wallenberg for the Nobel Peace Prize would draw the attention of world public opinion to its activities and disappearance behind the Russian iron curtain." In this way, Swedish reaction tries to involve neighbouring Norway in its anti-Soviet gambles. Soviet peoples express indignation at this provocation and hope that Swedish democratic circles will correctly assess and repulse the mean actions of Swedish "step-brothers" of American war-mongers.

Novoye vremya, No. 4, 1948, pp. 24–25.

[1] It was recognized in the memorandum of A. A. Gromyko, deputy-minister of foreign affairs, forwarded to the Swedish Ministry of Foreign Affairs on 6 February 1967 that Wallenberg had been kept in the prison of KGB headquarters in Moscow and died from heart failure on 17 July 1947. The memorandum read: "The Soviet government expresses deep regret over this fact and presents its condolences to the Swedish government and relatives of Raoul Wallenberg."*

*J. Birman. Pravednik. Istoriya of Raule Vallenberge, propavshem geroye Kholokosta (The Righteous Man. The Story of Raoul Wallenberg, the Missing Hero of the Holocaust), M.: Rudomino, 2001, pp. 159–160.

[2] According to Swedish sources, the nomination for the Nobel Peace Prize dates to December 1957.*

*J. Birman, p. 141.

II

From: V. M. Molotov, USSR minister of foreign affairs, and I. A. Serov, head of KGB, USSR Council of Ministers.
To: CPSU Central Committee.
24 April 1956

During a visit by the Swedish government delegation to Moscow, the Swedish minister of internal affairs, Hedlund, presented to the Soviet side testimonial evidence of certain German war criminals repatriated from the USSR to FRG in late 1955 concerning the Swedish diplomat Raoul Wallenberg. The evidence gives reason to the Swedish government to believe that Soviet authorities in Budapest arrested Wallenberg and kept him for a long time in a Moscow prison [...] Evidence provided by these and other people presented by the Swedish side coincide in many details with actual circumstances of Wallenberg's arrest and imprisonment in the USSR.

Taking into consideration the importance of settling this problem with the Swedish side and the fact that they will continue to inquire into the case, it seems appropriate to little by little inform the Swedish government about the fate of Wallenberg. It appears desirable to drag out the inquiry and give the final answer two or three months after parliamentary elections in Sweden scheduled for September this year. With this in mind, we should apply, via our embassy in Stockholm, for additional materials at the disposal of the Swedish side, e.g. more detailed evidence, Wallenberg's photos, etc. Such applications would be regarded by the Swedish government as an indication that our authorities actually intend to thoroughly study the case of Wallenberg based on the materials available from Sweden (Appendix 1). After a time (e.g. in July), our embassy in Stockholm could be instructed to inform the Swedish government that we will verify the presented materials, cross-examine potential witnesses, etc. and will bring the results of the inquires to their notice (Appendix 2).[1] The final statement about the fate of Wallenberg could be made two or three months after the parliamentary elections in Sweden are over.

The draft resolution of CPUS Central Committee is attached for consideration.[2]

In: Reabilitatsiya: kak eto bylo. Fevral' 1956 — nachalo 1980 godov. Dokumenty (Rehabilitation: How It Proceeded. February 1956 — the early 1980s. Documents). M.: 2000, pp. 81–82.

[1] The two Appendices containing verbal statements to be made by the Soviet ambassador to the Swedish Minister of Foreign Affairs are not included in the present publication.
[2] The draft resolution of CPSU Central Committee is not included in the present publication.

Document 4.44

Extract from the author's interview with Professor G. B. Malmstrom, Geteborg University, 24 February 1998.

[...] the fact that outstanding Russian chemists and especially biochemists were not awarded the Nobel Prize is certainly a mistake [...], it equally refers to Engelgardt, the discoverer of ATP.[1] One reason is the Nobel Committee in chemistry did not consider biochemical studies until the 1970s. True, the Committee had two member-biochemists by the time I came to chair it, but they were rather far from the main lines of research in this field [...] I think that such omissions at the early stages of the work of the Nobel Committee should be attributed to the insufficient competence of its members.

Published in the newspaper *Poisk*, No. 10, 1998, p. 14.

[1] Aerobic resynthesis of adenosine triphosphoric acid (ATP) coupled to cellular respiration.

Document 4.45

From: A. V. Karaganov, deputy-chairman, VOKS Executive Board.
To: A. N. Abramov, head, 5th European Department of the USSR Ministry of Foreign Affairs.
29 January 1947

Please inform comrade Ilyin, press-attaché of the Soviet mission in Stockholm, of the following:

"The literature and music agency of VOKS received your telegram with the request of the Swedish Nobel Committee to send them the works of Pil'nyak. It should be brought to your notice that Pil'nyak is the people's enemy. Take all measures to prevent consideration of his candidacy for the Nobel Prize."[1]

AVP RF, fond 0140, opis' 31, papka 138, ed. khr. 35, l. 20.

[1] It appears that the author of this letter did not know that Pil'nyak was shot in 1938 or that the Nobel Prizes are not awarded to the deceased.

Document 4.46

Extract from the note "The Nobel Committee" drawn up by M. Cherkasov, 1st secretary of the Soviet embassy in Norway, in September 1946.

The Committee actually dissolved when Norway was occupied by the German troops because some members emigrated and others quitted it for a variety of reasons [...] This made Quisling[1] state in 1942 that he debarred emigrants from participation in the Committee for "high treason" and appointed new members. Thereafter, he demanded that the personnel of the Committee be implicitly obedient to the government and turn over the available funds to the minister of finance.

However, this plan was not realized because the Swedish mission in Oslo immediately interfered. The Swedish consul took buildings of the Committee and the Institute under his protection. In 1943–1944, Quisling tried once again to establish control over the Nobel Committee but failed as before [...]

Members of the Nobel Committee appointed by the Norwegian Storting are ardent supporters of pro-British and pro-American orientation in the foreign policy of Norway (Hambro, Gunnar, Tranmel,[2] etc.). One of the candidates for the Nobel Peace Prize of 1946 is Kollontai. In a talk with me, Trummel said that Kollontai has a greater chance to be awarded than other pretenders.[3]

[...] Quisling tried to expropriate the funds and assets of the Nobel Institute during German occupation of Norway but failed as he did in a similar situation with the Nobel Committee. The Swedes helped to take the Institute's funds from a Norwegian bank and transfer them to Sweden and those from French banks to England and Switzerland.

AVP RF, fond 116, opis' 29, papka 23, ed. khr. 20, l. 16–19.

[1] Prime Minister of Norway during the German occupation of the country.
[2] The author of the note took the name of Gunnar Jan, chairman of the Norwegian Nobel Committee, for his surname. Carl Hambro was deputy chairman and Martin Tranmel member of the Committee.
[3] In 1946, the Nobel Peace Prize was awarded to E. G. Balch, British peace champion, and J. Mott, American missionary and public figure.

Document 4.47

From: R. Moe secretary of the Norwegian Nobel Committee.
To: V. M. Molotov, USSR people's commissar of foreign affairs.
30 October 1945[1]

You will receive a few copies of the letter concerning nominations for the Nobel Peace Prize for 1946.[2] I dare to ask you to distribute these documents among those in the People' Commissariat of Foreign Affairs who may be interested in them. (translated into Russian from French).

AVP RF, fond 116, opis' 28, papka 21, ed. khr. 15, l. 68

[1] The letter was delivered to the People' Commissariat of Foreign Affairs through the agency of the Norwegian embassy in the USSR.
[2] Special rules of the Norwegian Nobel Committee (para. 3, p. 3) granted the right to nominate candidates for the 1946 Nobel Peace Prize to members of the USSR diplomatic service. In 1947, the right to nominate candidates for the 1948 prize was given to professors of politology, law, history, and philosophy of the State University of Riga, Latvia (see doc. 6.17).

Document 4.48

From: S. I. Vavilov, president of the USSR Academy of Sciences.
To: V. M. Molotov, deputy chairman of the USSR Council of Ministers.
No. SV-2966c. 12 June 1948

The USSR Academy of Sciences publishes *Astronomicheskie ezhegodniki*, of great practical importance, compiled jointly with astronomic institutions of other countries. *Astronomicheskie ezhegodniki SSSR* containing only tables with small amounts of relevant explanatory text were annually published in Russian, English, and subsequently in French. Taking into account the leading position and high reputation of *Astronomicheskie ezhegodniki SSSR*, I ask for permission to continue its publication in the Russian and French languages, as before.

Document 4.49

From: L. F. Ilyichev, deputy head, Department of Propaganda and Agitation, AUCP(b) Central Committee.
To: G. M. Malenkov, secretary of Central Committee.
30 August 1948

Publication of the works of the USSR Academy of Sciences in foreign languages was prohibited by the decision of Politbureau of 16 May 1947. With reference to this decision I think that the request of Vavilov should be turned down [...] I ask for permission to bring this to the notice of Vavilov.

RGASPI, fond 17, opis' 118, ed. khr. 151, l. 18.

Document 4.50

From: M. Born, professor of the University of Edinburgh.
To: V. Karavaev, USSR embassy in London.
16 November 1948 (translated from English).[1]

Professor Max von Laue, the famous physicist who discovered X-ray diffraction, wrote to me in connection with the following. He was elected an honorary member of the USSR Academy of Sciences in 1928 and received printed matter published by the Academy until 1933, when the German censorship office put an end to this practice. He asked me to try and clarify the situation since I am also an honorary member of the Academy. He appears to be as proud as myself of being a member of the USSR Academy of Sciences and wishes to maintain relations with this organization. I have not received printed matter from the Academy either in the recent years. I was told that it ceased to publish physical papers in English and German.[2] Nevertheless, I would greatly appreciate your asking the Academy whether it is possible to resume the exchange of publications.[3]

RGASPI, fond 17, opis' 117, ed. khr. 369, l. 141–142.

[1] The original English text was not found.
[2] Speaking about German language journals, the author meant the monthly *Physikalische Zeitschrift der Sowjetunion* published in 1932–1938, first in Kharkov and thereafter in Moscow (since mid-1936) (see note 15 to this chapter).
[3] M. Born's interest in the situation in the USSR was not without reason. In 1933, he had to leave his motherland and settled in Cambridge, where he made friends with P. L. Kapitsa. When Kapitsa was busy organizing the Institute of Physical Problems in Moscow, he tried to obtain permission to invite Born as a leading theoretical physicist to fill a vacant position. Born and his wife received visas and bought ship tickets. However, a vacancy for a professor became available in Edinburgh just before their departure, and the scientist decided to stay in Great Britain.*
* V. D. Esakov, P. E. Rubinin. Kapitsa, Kreml' i nauka Tom. 1. (Kapitsa, Kremlin, and Science. V. 1.). M.: Nauka, 2003, pp. 407–411.

Document 4.51

Extract from the report of the Communist Party organization of the Geophysical Institute for the period from 20 May 1948 to 23 February 1950.

The Party Bureau totally neglected the problem of data exchange with foreign institutions. The bulletin published by the institute was sent to 68 stations of different countries compared with 59 received from them. The bulletin was sent to

Cambridge, known as a centre of active anti-Soviet propaganda carried out by hostile forces. Also, it was forwarded to ten stations in the USA and only seven of them dispatch their materials to our institute. A bulletin that came from fascist Spain,[1] boycotted by all democratic states, should have been returned back unpacked. All these facts indicate that some of our scientists still strive to gain cheap popularity abroad and cringe before foreigners.

CAODM, fond 2969, opis' 1, ed. khr. 6, l. 55.

[1] Spain under the rule of Franco.

Document 4.52

From: G. F. Aleksandrov, head, Department of Propaganda and Agitation, AUCP(b) Central Committee.
To: I. V. Stalin, secretary-general of Central Committee.
7 October 1940

Kratky kurs istorii VKP(b) (Short Course on the History of AUCP(b)) is now published in the languages of newly formed Soviet socialist republics (Estonia, Latvia, and Lithuania).[1] The department of Propaganda and Agitation, AUCP(b) Central Committee, asks for permission to omit in these editions of the *Short Course on the History of AUCP(b)* (p. 234) the phrase "Comrade Ezhov was preparing the soldier to mass revolt at the western front, in Byelorussia" and (p. 313) "from the report of comrade Ezhov, secretary of Central Committee."[2]

In the upper left corner (in Stalin's handwriting in green pencil): "Just so. I. Stalin."[3]

RGASPI, fond 17, opis' 125, ed. khr. 10, l. 111.

[1] Lithuania, Latvia, and Estonia lost independence in summer 1940. The official propaganda depicted the annexation as the Baltic states voluntarily joining the Soviet Union in the capacity of Soviet Republics in accordance with the free will of their peoples. The strange wording of the head of Department of Propaganda and Agitation, AUCP(b) Central Committee, ("newly formed Soviet Socialist Republics") was at variance with the terminology adopted in the Soviet mass media and betrayed a Party functionary who fairly well understood (at the level of Freudist subconsciousness) that the results of the recently held "nationwide polls" in the Baltic states were counterfeited. Stalin himself was not especially scrupulous in his statements concerning this question. Stalin's personal fund keeps a volume of *Iosif Vissarionovich Stalin* (short biography) published in 1939 and proof-read by Stalin himself for the forthcoming edition of 1940. Page 83 has the typewritten editor's supplement containing the following sentence: "[...] Later, Soviet Baltic Republics were admitted to the USSR." Stalin struck off the word "admitted" and put down the more emphatic "brought back."*

* RGASPI, fond 558, opis' 11, ed. khr. 1281, l. 65.

[2] N. I. Ezhov, USSR people's commissar of internal affairs, one of the chief organizers and actors of Great Terror in the period of 1937–1938 known as *ezhovshchina.*

He was dismissed from his post in autumn 1938 to be succeeded by Beria. Soon after the dismissal, he was arrested and shot in January 1940 under the sentence of the Military Board of the USSR Supreme Court.

[3] Another example of the chief's cynicism.

Document 4.53

How Skobeltsyn did not dare to discover the positron:

[...] Now that observers have found an ensemble of elementary particles in cosmic rays [...] the introduction of a new particle in the "inventory" of physical notions does not seem extraordinary. However, it was not that easy to take the first step to "recognize" a new elementary particle as distinct from the known ones. The existence of the positron was deduced from the consistent development of Dirac's theory formulated in 1928–1930. But there was a fair amount of scepticism concerning this idea at that time, and the author of the theory first tried to interpret it so as to avoid the necessity to postulate the existence of a new hypothetical particle. The positron achieved general recognition only after K. Anderson, P. Blackett, and G. Occhialini found this particle in cosmic rays [...][1]

D. V. Skobeltsin. Otkrytie Iskusstvennoy Radioaktivnosti i eyo rol' v razvitii fiziki za dvadtsat' let (Discovery of Artificial Radioactivity and Its Role in the Development of Physics During Twenty Years). "Voprosy filosofii," No. 4, 1954, pp. 97–98.

[1] Three decades later, I. S. Shapiro, who started his scientific carrier under the guidance of Skobeltsin and was later elected corresponding member of the USSR Academy of Sciences, showed as much caution as his teacher. In 1956, he took interest in the problem of parity non-conservation and beta-fission during weak interactions. However, he did not dare to publish his conjecture in the scientific literature. One reason was that L. D. Landau did not approve his ideas. In a few months after Shapiro spoke with Landau, similar ideas were put forward by young Americans of Chinese descent, Lee and Yang. They were soon confirmed in the famous experiment by Wu. In 1957, the two scientists were awarded the Nobel Prize in physics.

Czech physicist Fr. Yanouch, a graduate of Moscow State University, stated: "through Landau's fault, Soviet physics lost one Nobel Prize." L. P. Pitaevsky, a disciple and close associate of Landau, rejoined: "I don't know why Shapiro did not try to publish his ideas. That such a world [with odd relations] that seemed disgusting to Landau should not have discouraged Shapiro. I think Shapiro actually did not believe in parity non-conservation, while Lee and Yang did."* Similarly, Skobeltsin did not believe in the reality of positron, but Anderson did and received the Nobel Prize [...]

* G. E. Gorelik. Mify, legendy i dokumenty iz zhizni Lva Landau (Myths, Legends, and Documents of the Life of Lev Landau). In: "Nauchnoye soobshchestvo fizikov SSSR. 1950–1960 i drugie gody." Vyp. 2. Eds. V. P. Vizigin, A. V. Kessenikh (in press).

Chapter 5 • 1945–1949

An Attempt to Institute Alternative Prizes in the USSR

The idea to institute international prizes for outstanding scientific achievements first emerged in the USSR in 1945 (doc. 2.5), when the country was gripped by the euphoria of victory over Nazism[1] and the philosophy of confrontation with the West started to prevail in all aspects of political and public life. The realization of the idea was entrusted to Suvorov, deputy head of the Department of Propaganda and Agitation, AUCP(b) Central Committee, as the key figure through whose agency the department controlled scientific developments in the country.

It appears that Suvorov was the first to propose that the new prizes be regarded as an alternative to the Nobel Prizes. A list of Nobel laureates for the period from 1901 to 1944 was attached to the draft resolution of the AUCP(b) Central Committee concerning this issue (doc. 5.1). This inference is confirmed by doc. 5.3 (both 5.1 and 5.3 are dateless).[2] Document 5.3 is actually the cover letter to the draft resolution of the Central Committee.[3] It contains a program of confrontation with the Nobel organizations in Stockholm. However, the situation in the Department of Propaganda and Agitation substantially changed by mid-1947.

The head of the department G. F. Aleksandrov, his deputies A. M. Egolin and V. S. Kruzkov, and some of their subordinates were caught attending an underground brothel and sacked.[4] Before that (autumn 1946), G. M. Malenkov, head of the AUCP(b) Central Committee Secretariat, had fallen into disfavor in connection with the arrests of the leaders of aviation industry.[5] He was substituted by A. A. Zhdanov, the main actor in the recent (and first after the war) ideological campaign against journals *Zvezda* and *Leningrad*.[6] In his new position, Zdanov focused the activity of the Secretariat on the propaganda of the struggle against "antipatriotism" and "servility to Western science and technology."[7]

In late 1947, S. G. Suvorov had to retire from his post too and left the Central Committee apparatus in January 1948.[8] He drew up the note of 9 December 1947 (doc. 5.2) after he was already relieved of his official duties as the deputy head of the department. Further work related to the institution of alternative international prizes was continued by D. T. Shepilov, the newly appointed deputy head of the Department of Propaganda and Agitation.[9]

Shepilov broadly depicted his activity as a Party functionary and a statesman in his memoirs published posthumously in the journal *Voprosy istorii* (Nos. 3–12).

When working on his memoirs, he did not hesitate to uphold views and actions that might look ambiguous to readers. For all that, he says no word about the institution of international prizes as an instrument of the policy of hard-edged confrontation with the Nobel movement. This fact may be regarded as silent condemnation of the adventurous idea and recognition of an *a priori* vain attempt to undermine the indisputable authority of the Nobel Prizes that they by that time had won all over the world.

Alas, the understanding came 20 years after the young and educated Shepilov had started to implement the task set up by the Party leadership.[10]

The wish of A. A. Kuznetsov to have a "slightly modified" project instead of the one prepared by Suvorov in 1945 (see doc. 5,2) is not confirmed by archive documents; the desired changes were probably formulated verbally. The document signed by Shepilov contained the following propositions at variance with what was suggested by Suvorov:

a) the prizes were named after Lenin;
b) they were to be awarded on Lenin's birthday (April 22), instead of Victory Day;
c) the number of prize categories was increased from five to eight (the additional categories included social sciences, literature, and arts);
d) the prize money increased from 300,000 to 500,000 rubles.

In all other aspects (barring minor details), the new proposals did not differ from the 1945 ones. Even the time of the first award remained the same (1949), logical in 1945 but somewhat strange in 1948. This, however, did not confuse Shepilov...

Shepilov participated in closing sittings of the Stalin Prize Committee at which Stalin was as a rule present and took an active part in the discussion; therefore, he was well aware of voluntarism in the choice of the winners. He recollected in his memoirs some episodes where the final word belonged to Stalin. One of them was the award of the 1949 prize to G. S. Milyutin for the music of operetta *Trembita*.[11] The same was true of the prizes awarded for the design of the first skyscrapers built up in Moscow; no professional architects were invited to participate in the discussion.[12]

The latter voluntaristic decision reflected the interest that Stalin always showed to building activities in Moscow. Stalin took all important decisions related to the architectural aspect of the capital[13] whether they concerned living apartments and public buildings on the central streets, subway stations, tree felling on boulevards etc.

The design of the main facade of the hotel Moskva looking out on the Manezhnaya square gives an illustrative example of the then established practice of approval by Stalin of the most important building projects. The two symmetric corner towers of the building, occupying a huge area in the center of Moscow, have different architectural decoration. The reason became apparent during the "Khrushchev thaw," when A. V. Shchusev, an outstanding Moscow architect, was already dead. It turned out that he designed two variants of architectural decoration of the towers and presented them to the consideration of the chief in a picture showing one on the right and the other on the left side of the common facade, for comparison. Stalin did not understand that he had to choose a single variant and approved the project as it was for implementation. Shchusev did not dare to apply for an explanation.

Here is another story illustrating the approach to awarding Stalin Prizes. Playwrights M. D. Volpin and N. R. Erdman, victims of political repression in the 1930s, became co-authors of the script of the film *Brave people* in the post-war period. The film was a great success, and Stalin saw it. The main character in the film was a dashing horseman. Nobody dared to nominate the film for the prize bearing in mind the past accusations of the playwrights. According to M. V. Ardov, his father V. E. Ardov told him that Stalin looked through the list of the films recommended for the award and said: "Bold horses also deserve it."[14] Members of the committee supported the proposal...[15]

Before doc. 5.4 was presented for consideration of the Central Committee Secretariat, the USSR Academy of Sciences was committed to design medals and diplomas of International Lenin Prize laureates and the USSR Ministry of Foreign Affairs to summarize information about organizations awarding the Nobel Prizes by 1 January 1948. Both tasks were rapidly fulfilled. The Academy submitted for approval the list of artists and sculptures to be involved in the work and draft letters to them within a week (13 December 1947) after Shepilov familiarized himself with the documents prepared by Suvorov (doc. 5.5). The Department of Scandinavian States, the USSR Ministry of Foreign Affairs, submitted the requested data (doc. 5.6) on 2 January 1948. The contents of this document are worth expounding at some length.

Both its style and mode of presentation of facts revived Cold War rhetoric. Besides manifest inconsistencies and data garbling (see notes to doc. 5.6), it contained downright misinformation about interference of Swedish authorities in the activity of prize-giving organizations and its "subjection to the purposes of Swedish policy." By way of forestalling, it should be noted that the 5th European Department of the USSR Ministry of Foreign Affairs responsible for relations with Scandinavian countries, regardless of who was at the head of it, always (up to 1953) showed a distinct aversion to the Nobel movement.[16]

On 2 January, the secretary of AUCP(b) Central Committee A. A. Zhdanov sent the typewritten text of the Nobel Statutes to Stalin and four other members of the Politbureau. It was translated from the official publication in English on the instruction of S. G. Suvorov (doc. 5.7) in 1945. In addition, an order was given to translate the description of the prize-awarding ceremony obtained from the V. I. Lenin Library[17] after 5 January 1948 (doc. 5.9).

As follows from doc. 5.5, the afore-cited doc. 5.4 was drawn up in December 1947[18] in connection with the projected institution of the Lenin Prize. Its conversion to Mendeleev and Gorky Prizes first mentioned in doc. 5.10 dates to January or February 1948. Both documents are dateless which suggests that they were never submitted to AUCP(b) Central Committee by the Department of Propaganda and Agitation that had prepared them. Secretaries of Central Committee were certainly aware of their existence but preliminary discussions probably took place during unofficial business meetings.

It may be thought that one or another secretary (most likely Zhdanov or Suslov) verbally advised Shepilov to give up the idea of international prizes at one of such meetings.

The reason was obvious. Lenin Prizes were instituted in the Soviet Union in 1925 but ceased to be awarded in 1935.[19] Stalin Prizes, as the highest monetary award, were instituted on the occasion of the dictator's 60th birthday in 1939. The more experienced courtiers fairly well understood the danger of Shepilov's incautious initiative. Restoration of Lenin Prizes, even if as an international award, would objectively diminish the value of Stalin Prizes and was fraught with unpredictable consequences bearing in mind Stalin's pathological jealousy. Shepilov himself recalled a characteristic episode illustrating how highly Stalin appreciated the value of the prizes after his name:

"At a meeting [of the Stalin Prize Committee], one nomination was found poorly substantiated. Stalin asked someone (perhaps minister Kaftanov): 'Which do you think more important: Nobel Prize or Stalin Prize?'.[20] The answer was obvious. 'Then,' said Stalin 'nominations must be properly grounded. We are not here to distribute alms.[21] We appreciate achievements at their true value.'"[22]

This ridiculous dialogue brings back to memory the old parable telling of a passerby who stopped to ask the crucified how he feels. "All right," the answer was, "only it hurts to smile [...]"

Further discussion of proposals attached to the cover letter (doc. 5.10) published below brought about more comments and advice that Shepilov took into consideration. A selection of relevant documents was submitted to M. A. Suslov, secretary of the Central Committee and head of its Department of Propaganda and Agitation (18 March 1948). Next day, Suslov passed them to the head of the Secretariat, A. A. Zhdanov.

The selection contained, besides the cover letter (doc. 5.11), draft resolutions of the AUCP(b) Central Committee and the USSR Supreme Council on the institution of Mendeleev and Gorky Prizes (doc. 5.12) to be approved by the Central Committee, the draft letter to Stalin (5.13), and proposals for the ceremony of handing over the international prizes scheduled for 8 November (doc. 5.14).

Preambles of the two last documents contain a detailed description of the current situation in which "the arbiter of all achievements in science and literature in all countries is [...] Sweden" which awards the Nobel Prizes to the "enemies of people's democracy and malicious anti-Soviet slanderers" etc. These statements are followed by enumeration of the "sins" of the Nobel organizations disregarding representatives of Soviet science.[23] The preambles are closed by the conclusion that "institution of international prizes [...] will consolidate pro-democratic scientists, writers, and art workers and increase the influence of the Soviet Union on intellectuals in the friendly countries of Eastern Europe."

It may be thought that the documents were favorably accepted at the Secretariat of the AUCP(b) Central Committee; otherwise, they would hardly be set going and reach the office of P. N. Pospelov, editor-in-chief of the newspaper *Pravda*.[24]

Further events occurred at the background of cataclysms that shattered ideological structures of AUCP(b) Central Committee. Their immediate cause was organization of the session of the All-Union Academy of Agricultural Sciences (VASHNIL) that was held from 31 July until 7 August 1948 under the patronage of Stalin, and ended in the destruction of genetic science in this country, one of the most dramatic episodes in the history of world science in the mid-twentieth century.

Direct participation of Stalin in the preparation of Lysenko's notorious speech at this session is confirmed by the documents available now from the archives of the AUCP(b) Central Committee. Suffice it to mention the original text of the report "On the situation in Soviet biological science" with remarks and insertions of the chief. This document was discovered after the death of Lysenko, when KGB seized his personal archive. The finding was reported by KGB chairman Yu. V. Andropov to the CPSU Central Committee on 8 December 1976.[25]

It should be noted for the sake of objectivity that genetics had weak points that paved the way for manipulation of an army of ignoramuses and time-servers. At that time, there was no comprehensive definition of a gene and its action on the protein substance and thereby on heredity at the molecular level. The definitive solution of the problem was given in the letter of James Watson, Harvard University, and Francis Crick, Cambridge, to *Nature* (171, No. 4356, pp. 737–738; 25 April 1953) where they postulated the DNA structure in the form of a "double helix." The idea was enthusiastically accepted by the scientific community. Half a century later, Professor Martin Kemp of Oxford will poetically call the double helix "the Mona Lisa of modern science" in the same journal (421, No. 6921, 2003, pp. 416–420) and equate this symbol of biological knowledge with other values common to all mankind.

In 1962, this achievement brought the Nobel Prize in physiology and medicine to the authors of the two-page paper and their colleague Maurice Wilkins for "their discoveries concerning the molecular structure of nuclear acids and its significance for information transfer in living material." The world-recognized discovery of the "double helix" sounded as a knell for Lysenko and his associates. Next year, the rule of the "people's academician" in Soviet genetics collapsed and his anti-scientific activities came to an end...

Lysenko's ignorant views of heredity in living nature were opposed not only by scientists and philosophers but also by some officials in the Department of Propaganda and Agitation including Shepilov and Yu. A. Zhdanov, son of the secretary of the AUCP(b) Central Committee A. A. Zhdanov,[26] who was appointed to head the Sector of Science on 8 December 1947, after the dismissal of S. G. Suvorov.

Before filling this position, Yuri Zhdanov was a postgraduate student at the Institute of Philosophy, the USSR Academy of Sciences, and a lecturer at the chemical faculty, Moscow State University. He was well aware of the situation in Soviet genetics and pseudoscientific theories of Lysenko. He expounded his views in the report "Debatable problems in modern biology," delivered at the All-Union seminar of the secretaries of regional Communist Party committees held at the Polytechnical Museum on 10 April 1948.

This action was sharply rebuked by Stalin at a sitting of Politbureau and a commission was set up to examine it.[27] The commission was to be chaired by G. M. Malenkov, but he soon had to substitute A. A. Zhdanov as the head of the AUCP(b) Central Committee Secretariat;[28] Zhdanov died of myocardial infarction two months after retirement from this position. The death came in a Valdai residence (Novgorod region) on 31 August 1948 under circumstances that put some puzzling questions, in the first place related to the bedside conflict between L. F. Timashuk, head of the ECG cabinet in the Kremlin Health Department, and her superiors. Timashuk diagnosed myocardial infarction after the very first heart attack on 28 August. Head of the department Professor P. I. Egorov, Stalin's personal doctor Professor V. N. Vinogradov, and the attending physician G. I. Maiorov did not agree on the diagnosis of infarction and insisted on "functional disturbance of cardiac activity."[29] (See also doc. 5.24.)

According to Shepilov, his relations with Malenkov left much to be desired. As a result, Malenkov did not approve certain proposals from Shepilov. The selection of documents concerning Mendeleev and Gorky prizes was forwarded to Central Committee Secretariat on 19.8.1948 (docs. 5.15–5.19). The next day, the Secretariat instructed the Department of Propaganda and Agitation[30] "to prepare documents to be reported to the Politbureau." (doc. 5.20)

Such formulation implied a disguised disapproval of the submitted documents. Usually, the Secretariat either approved or disapproved a document. If it needed to be additionally authorized by Politbureau, the closing paragraph of the resolution contained the standard phrase: "The decision is subject to the approval of Politbureau", to which the document and the resolution of the Secretariat were submitted. Today, it is impossible to say what caused the fiasco of international prizes, either the relations between Malenkov and Shepilov that eventually led to the latter's dismissal from the apparatus of the AUCP(b) Central Committee or understanding the lack of prospects for using them as a tool to "increase the influence of the Soviet Union on intellectuals"; there are no relevant documents besides those published in this chapter, barring a few notes confirming that they were shelved.[31]

Meanwhile, the USSR Academy of Sciences continued the work to design gold metals and diploma of laureates of the Mendeleev and Gorky although the Central Committee gave up discussing this matter. Nine months after the Secretariat disapproved the respective documents, the USSR Academy of Sciences received new instructions from the Central Committee concerning the project[32] (doc. 5.21). Evidently, Shepilov still hoped to realize his idea.

The USSR Academy of Sciences accomplished the task and forwarded to the Secretariat of the Central Committee the now unneeded materials concerning the awards (doc. 5.22) along with the individual opinion of its president, S. I. Vavilov (doc. 5.23). Paradoxically, the status of the Soviet international prize was approved on the same day by the Presidium of the USSR Supreme Council under the name of the Stalin Prize "For strengthening peace among peoples", to celebrate Stalin's forthcoming 70th birthday. Thus, the ambitious project to compete with the Nobel movement was reduced to the contest with the Nobel Peace Prize.

To conclude the story of Soviet international prizes, it is appropriate to turn back to the motives behind Malenkov's blockade of the plans of Shepilov. It is very likely that his actions were dictated not only by confrontation with the head of the Department of Propaganda and Agitation but also by common sense.

After Malenkov once again occupied the key position in the Secretariat, he could more rationally estimate the lack of prospects for the competition with the long-standing system of Nobel awards and their world-recognized authority. At the same time, he was well aware of the prize-awarding practice in the USSR and understood the negative consequences of its submission to the judgment of world public opinion. One more illustration of the said practice is provided by a formerly unpublished document of the Secretariat of the AUCP(b) Central Committee, testifying to the corruption of members of the Stalin Prize Committee.

According to the inspector-general of the USSR Ministry of State Control, inspection of the cheese-making company in the city of Uglich in 1943 showed that D. A. Grannikov, chief engineer of the Department of Butter-making and Cheese-making Industry organized an exhibition of the company's produce in the premises occupied by the Stalin Prize Committee. Over half a ton (660.5 kg) of cheese varieties were delivered to Moscow for the purpose in December 1942, when the army and the population suffered a tremendous deficit of victuals. According to the document dated March 1943, the cheese was consumed by participants of the plenary session of the Stalin Prize Committee [convened to discuss and elect candidates for the award] and continued to be consumed by top officers of the All-Union Committee of Higher Education in whose building the exhibition was held.[33]

The hospitable Grannikov was awarded the 2nd degree Stalin Prize for "the development of technology and production of high quality varieties of cheese" on 22 March 1943.[34] It should be thought that this decision by the Committee was based on the gustatory sense of its members ("knowing no shame," to cite Anna Akhmatova) rather than objective evaluation of product quality...

However, corruption (which occurs in many established democratic societies too) is not the main feature in this shameful episode. It demonstrates the vulnerability of the Soviet prize-awarding system to the interference of concerned persons and organizations predetermined by the lack of confidentiality, collective discussions in competent institutions and mass media coupled to boundless opportunities for lobbying.[35] In this context, the Soviet and Swedish models are diametrically opposite and have no common ground.

Organic defects intrinsic in the Soviet system were apparent not only in such areas as literature, arts, social sciences etc., comprehensible to the lay public but also in exact sciences, where unverified results of research could not by definition be submitted for a prize without preliminary discussion in the narrow circle of specialists.[36] For all that, such results had real chances to be awarded. One example was presented by A. V. Topchiev, chief secretary of the USSR Academy of Sciences, at a meeting of its Presidium, in 1950 (doc. 5.26).

The failure of Communist Party ideologists to institute international scientific prizes in the Soviet Union proved favorable for the image of the country because the society was spared another profanation in the politicized competition with the well-established practice of the Nobel organizations.

Notes

[1] The jubilee session of the USSR Academy of Sciences celebrating its 220th anniversary gave an incentive to the idea to institute international scientific prizes in the Soviet Union. The session held in Moscow in June 1945 attracted many foreign scientists. It means that docs. 5.1 and 5.3 most likely appeared in mid-1945.

[2] The lack of date suggests that the two documents (5.1 and 5.3) were not passed to A. A. Zhdanov, secretary of AUCP(b) Central Committee, and remained in the Department of Propaganda and Agitation.

[3] A marked difference in the numeration of the two documents (No. 93879 for doc. 5.1 and No. 97319 for doc. 5.3) does not contradict the close connection between them. These are incoming numbers under which the documents were registered in the technical division of the Orgbureau among huge amounts of other documents that came from Central Committee departments and from the outside. It is clear, however, that doc. 5.1 was received (and registered) earlier than doc. 5.3. Hence, the phrase in doc. 5.3 that doc. 5.1 was reported to A. A. Kuznetsov, secretary of AUCP(b) Central Committee, in early December 1947 without the cover letter, i.e. unofficially.

[4] This dirty story had no serious consequences for the former head of Agitprop.* He was dismissed from his position but received the post of director of the Institute of Philosophy, the USSR Academy of Sciences, and was appointed Minister of Culture in 1954. In this capacity, he fell back into his old ways, together with his old friend Egolin. This time, their orgies were held in the sanatorium "Uzkoye" in the south-west outskirts of Moscow.** In 1955, he was exiled to Minsk to work in the local Institute of Philosophy. He died there of hepatic cirrhosis caused by chronic alcoholism in 1961.***

> * A widely used abbreviation for the Department of Propaganda and Agitation.
> ** The closed letter of CPSU Central Committee "On the misbehavior of comrades Aleksandrov, Egolin, and others" was read at Communist Party meetings in May and April 1955.
> *** D. T. Shepilov. Vospominaniya (Memoirs), "Voprosy istorii," No. 5, 1998, pp. 6–7.

[5] Malenkov was entrusted by the AUCP(b) Central Committee to control military industry during the war. He promoted many specialists that were later arrested on the order of Abakumov,* Minister of State Security. The resolution of the AUCP(b) Central Committee dated from 4 May 1946 charged Malenkov with the inability to fulfill his duties as "the person responsible for the development of the aviation industry and aircraft production for the air forces. [...] It was considered necessary to expel comrade Malenkov from the Secretariat of the AUCP(b) Central Committee."** In 1951, Malenkov returned to power and succeeded in persuading Stalin to arrest Abakumov.***

* A. G. Malenkov. O moyom otse Georgii Malenkove (About My Father Georgiy Malenkov). M.: Izd-vo NTTs "Tekhnoekos", 1992, pp. 52–53.
** In: "Politbyuro CK VKP(b) i Soviet Ministrov SSSR. 1945–1953." M.: ROSSPEN, 2002, p. 207.
*** A. G. Malenkov, p. 55.

[6] Resolution of the AUCP(b) Central Committee: "On the journals *Zvezda* and *Leningrad* of 14 August 1946; see also Chapter 6 and notes 86–88 therein."

[7] See Chapter 4; the term "antipatriotism" was substituted by "cosmopolitanism" in the beginning of 1949.

[8] See note 1 to doc. 5.2.

[9] Document 5.1 bears the clerk's note testifying that it was reported to Shepilov on 8 December 1947, i.e. three days before Suvorov presented his memorandum to the Secretariat of the Central Committee (doc. 5.2). Shepilov occupied his post in the Department of Propaganda and Agitation* on 18 June 1947.

* D. T. Shepilov. *Voprosy istorii*, No. 5, 1988, p. 12.

[10] Shepilov basically completed his memoirs in 1966.

[11] From an interview with journalists (23 February 1989). In: "I primknivshiy k nim Shepilov. Pravda o cheloveke, uchonom, voine i politike (Shepilov Who Had Joined Them. The Truth About a Person, Scientist, Warrior, and Politician)." M.: Zvonnitsa-MG, 1998, p. 127.

[12] D. T. Shepilov. "Voprosy istorii", No. 5, 1988, p. 24.

[13] "When the first ambitious building projects began to be implemented in Moscow, he used to advise architects and sometimes took seemingly fantastical but actually competent decisions. According to Albert Schpeer, a gifted fascist architect* responsible for German military industry, "Hitler, like a youth, took offence at Stalin, who he believed to be stealing his architectural ideas [...]** Not a single project of underground stations could be realized without being personally approved by Stalin [etc.]"***

* Schpeer was the author of the cyclopic building of Rechskanzlei erected in 1939, on the eve of World War II. According to him, Hitler told at the stage of projecting about "the power that a small fry can have over other people if he is able to appear against such an imposing background." An added: "Such halls with their great historical past raise even a worthless successor to the level of a historical personality." *Izvestiya*, 30 March 2002, p. 8.
** A. Schpeer. *Vospominaniya* (Memoirs). Smolensk-Moskva, 1997, p. 226.
*** B. S. Ilizarov. Stalin. Shtrikhi k portretu na fone ego biblioteki i arkhiva (Stalin. A Few Strokes to His Portrait in His Library and Archive)". "Novaya i noveishaya istoriya", No. 4, 2000, p. 160.

[14] Stalin many times demonstrated an inborn disposition toward improvisation and paradoxical formulations. There are many sorts of evidence of such inclination, especially in political matters. The known Georgian journalist M. G. Sturua, son of a top Party functionary (chairman of the Presidium of the Supreme Council of Georgia in the 1940s), told that after the war his father "wrote memoirs about when he actually rehabilitated all Caucasian Bolsheviks repressed in 1937–1938." They became known to the leaders of the Georgian Communist

Party. The memoirs were submitted to the consideration of Politbureau. It was planned to punish Sturua and relieve him of his duties. Stalin all of a sudden remarked: "The old man appears to have lost his mind; let us dismiss him since Georgian comrades can not properly work with him." Sturua was discharged for "nostalgia for Trotskyism."*

* *Izvestiya*, 11 March 2002, p. 3.

[15] *Novy mir*, No. 4, 1994, p. 32. Stalin believed in cinematography as an efficient tool whereby to influence the wider public and not infrequently interfered in the routine work of a film director or a script-writer. B. Z. Shumyatsky, a head of Soviet cinematography, left archive notes dated to the mid-1930s when one of his duties was to regularly select documentary and feature films for private night-time viewing by Stalin and his close associates in the Kremlin. Once (24 December 1934), after a news film and a documentary film featuring Kirov's funerals* were shown, those present wished to see a feature film. Shumyatsky** said that he had none completed and that only silent films were currently being produced at GUKF*** studios. Koba****: "Never mind. It is even better that they are not yet ready. We could meddle and give you advice" (!? — A.B.).*****

* The leader of the Leningrad Bolsheviks, S. M. Kirov, was assassinated in Leningrad on 1 December 1934 under circumstances that thus far remain unclear. He was buried at Red Square, Moscow, on 6 December 1934.
** B. Z. Shumyatsky speaks about himself in the third person.
*** Main Department of Movie and Photo Industry, USSR Council of People's Commissars, headed by Shumyatsky.
**** Stalin's nickname in the period of his underground revolutionary activity.
***** RGASPI, fond 558, opis' 2, ed. khr. 898, l. 85.

[16] A. N. Abramov, who succeeded M. S. Vetrov (the author of doc. 5.6) as head of the department in March 1948, displayed an equally negative attitude toward the prize-giving organizations of the Nobel Foundation (see doc. 7.2).

[17] Presently Russian State Library.

[18] Documents 5.4 and 5.5 are kept in the same archive file No. 497 together with other documents dated from 1947. More documents concerning the subject matter of Chapter 5 and dated from 1948 are kept in RGASPI in file No. 588. Therefore, doc. 5.4 could not have been drawn up in January 1948.

[19] *Great Soviet Encyclopaedia*, 3rd edition, V. 14, 1973, column 972.

[20] Iosif Vissarionovich was not the first to so disrespectfully speak about Nobel Prizes and put them below the prize he named after himself. Forty years before, I. I. Mechnikov also held them in very low esteem even though his reasons were quite different and he meant a concrete award (in physiology and medicine) rather than the entire system of Nobel Prizes. Mechnikov visited L. N. Tolstoy in Yasnaya Polyana on 30 May (12 June) 1909 and noted in the talk with the writer that the Karolinska Institutet giving Nobel Prizes in physiology and medicine "has no outstanding scientists" and that "he prefers the Moscow Prize* awarded to him by the committee of the international congress of physicians to the Nobel Prize" that he received soon after it.**

* The Moscow Prize was instituted by the city *duma* (government) to commemorate the opening of the 12th session of the International Medical Congress in the Bolshoy Theater in August 1897. The prize was to be awarded every four years to outstanding

representatives of world medicine at the regular sessions of the congress. The committee of the Twelfth Congress awarded the first Moscow Prize to Henry Dunant, the founder of the International Red Cross and the first winner of the Nobel Peace Prize.
** U Tolstogo. Yasnoplyanskie zapiski D. P. Makovitskogo (With Tolstoy. Notes from Yasnaya Polyana by D. P. Makovitsky). "Literat. nasledstvo", T. 90. Kn. 3. M.: Nauka, 1979, p. 423.

[21] The attribution to Stalin of the comparison of the Nobel Prizes with "alms" looks plausible. The anti-Nobel campaign launched in the Soviet Union in the late 1940s to early 1950s and splashed over the front pages of newspapers was surely initiated by Stalin himself, like other ideological campaigns.

[22] This episode (*Voprosy istorii*, No. 5, 1998, p. 26) could take place at one of the sittings convened to approve the lists of Stalin Prize laureates for 1948 and 1949. Shepilov was relieved of his duties as the head of the Department of Propaganda and Agitation and dismissed from the AUCP(b) Central Committee* in summer 1949, after Politbureau adopted the resolution "On the journal *Bolshevik*" (13 July 1949); therefore, he could no longer participate in such meetings.

* *Voprosy istorii*, No. 6, 1998, pp. 19–20.

[23] The 1948 list of persons "offended" by the Nobel Committees did not contain the names of P. L. Kapitsa, G. S. Landsberg, L. I. Mandel'shtam, and A. P. Orekhov (present in doc. 5.3 drawn up in 1945). The name of Orekhov might have been omitted for objective reasons (his works did not meet the criteria of the Nobel Committees*), but three other scientists were obviously excluded from the list for political considerations. Stalin dismissed Kapitsa from all his administrative posts in 1946,** while Landsberg and Mandel'shtam were ranked (posthumously) as "antipatriots" in the framework of the just another ideological campaign launched in 1948.***

* Moreover, Orekhov was never listed among the nominees for the Nobel Prize (see note 10 to doc. 5.3).
** See "Petr Leonidovich Kapitsa. Vospominaniya. Pis'ma. Dokumenty. Sostaviteli E. L. Kapitsa i P. E. Rubinin". (Petr Leonidovich Kapitsa. Recollections. Letters. Documents). M.: Nauka, 1994.
*** See note 12 to Chapter 2.

[24] The notebook of P. N. Pospelov contains the following record dated 13 April 1948: "Mendeleev c[ommitt]ee (intern[ational] prizes)
Gorky c[ommitt]ee (intern[ational] prizes)*
The record leaves no doubt that in mid-April the editor-in-chief knew about the project to institute international prizes in the Soviet Union elaborated by the Department of Propaganda and Agitation.

* RGASPI, fond 629, opis' 1, ed. khr. 97, l. 98.

[25] "Arkhivy Kremlya i Staroy ploshchadi", No. 5, 1993, p. 109.
The complete shorthand record of the VASHNIL session including criticism of Lysenko's report and shouts of his opponents from the audience was rapidly published for mass circulation in the same 1948. The decision of Orgbureau of the AUCP(b) Central Committee (13 December 1948) instructed the "Inostrannaya literatura" publishing house to publish as soon as possible the text of the shorthand report in German (by 20 February 1949), English (by

25 February 1949), and French (by 27 February 1949) in 20,000; 10,000; and 10,000 copies respectively.*

* RGASPI, fond 17, opis' 116, ed. khr. 402; minutes of the meeting of Orgbureau of AUCP(b) Central Committee, 13.12.1948, p. 12.

[26] In the second half of the 1950s, the son of another high-ranking person (this time, the head of state N. S. Khrushchev) had domestic debates with his father concerning the anti-scientific views of Lysenko. S. N. Khrushchev told to a correspondent of *Izvestiya* some 50 years later: "I tried to prove that genetics exists whereas he was convinced that his advisers knew better that it was non-existent.* He all but turned me away from home."** The collapse of Lysenko's political career fell on 26 June 1964. That day, the general session of the USSR Academy of Sciences blackballed his protégé N. I. Nuzhdin at the elections of its new members, despite Lysenko's demand to postpone the secret vote and the threat to immediately apply to the CPSU Central Committee.

* It should be noted that even V. I. Vernadsky, a great scientist and philosopher, fell under the influence of Lysenko in the beginning of 1936 (as follows from his diary). Three years later, however, he saw things more clearly: "He passes himself off as a successor of Darwinism — Michurinian [sic] scientific methodology, sharply attacks genetics that he does not know about." (V. I. Vernadsky. Dnevniki 1935–1941 v dvukh knigakh. Kn. 1. 1935–1938 (Diaries 1935–1941 in Two Books. Book 1. 1935–1938). M.: Nauka, 2006, pp. 74, 335).
** *Izvestiya*, 22 August 2002, p. 2.

[27] A month and a half later, Yu. A. Zhdanov spoke of a book by one of Lysenko's associates at a meeting of Politbureau considering scientific works submitted for the Stalin Prize (see entries dated 31 May–1 June 1948 in the diary of V. A. Malyshev, deputy chairman of the USSR Council of Ministers). He emphasized that he expressed his personal opinion. Stalin responded: "There can be no personal opinions and personal views in the Party, only the Party's views exist."*

* *Istochnik*, No. 5, 1997, p. 135.

[28] Malenkov was ousted from his position of secretary of the AUCP(b) Central Committee and head of its Secretariat and Orgbureau as a result of new Stalin's provocation, which ended in the arrest of leaders of the Aviation Industry and Air Command.* Malenkov was entrusted by the Central Committee to accept military equipment for the Air Forces. Stalin accused him of heedlessness and inefficiency with respect to control of aircraft supplies for the Army.**

* At the legal trial held on 10–11 May 1946, the panel of military judges of the USSR Supreme Court sentenced A. I. Shakhurin, aviation minister, A. A. Novikov, chief air marshal and Air Force commander-in-chief, and a few other people to long-term imprisonment. One of them, air marshal S. A. Khudyakov, was charged with russifying his true name and surname Armenak Artemievich Khanferyants. Such charges became more common somewhat later, after Stalin initiated a campaign "against cosmopolitanism" in 1949 (see Chapter 6).
** R. G. Pikhoya. Sovetsky Soyuz: istoriya vlasti, 1941–1991 (Soviet Union: The History of Power, 1941–1991). Novosibirsk: Izd-vo "Sibirsky khronograf", 2000, pp. 35, 36, 45.

[29] The refusal of leading specialists of the Kremlin Health Department to validate the results of ECG (confirmed at autopsy) made Timashuk apply to supervising organs. Her first letter to N. S. Vlasik, head of the Central Guards Department, USSR Ministry of State Security, is

dated 29 August 1948 (doc. 5.24). Two more letters addressed to A. A. Kuznetsov, secretary of the AUCP(b) Central Committee in charge of staff matters, were sent after Zhdanov's death. Timashuk had no response to her applications although her first letter was forwarded to Stalin via V. S. Abakumov, Minister of State Security, on 30 August. G. V. Kostyrchenko, an authoritative investigator into the last period of Stalin's life, wrote: "suspicious and ready-to-punish Stalin this time did not order an inquiry [...] but with his own hand passed Timashuk's letter to archives."* Egorov and Maiorov permitted Zhdanov to get up, walk, and see films two days before his death (29 August).** The letters of Timashuk were shelved for four years, up to the second half of 1952, when Stalin gave order to initiate his last large-scale campaign known as "the doctors' case" (see Chapter 7 and note 2 therein). Timashuk was announced the heroine who "exposed" the killer doctors and received the Order of Lenin (20 January 1953) for "the assistance rendered to the government in the discovery of the terrorist group of physicians." However, Timashuk was not guilty of the fate of the majority of victims of this openly anti-Semitic action.***

There is no doubt that Timashuk correctly interpreted the events of 28–30 August 1948.**** This gives reason to think that top figures in the Soviet leadership were interested in hastening Zhdanov's death probably because the "Leningrad case" was already being designed to end in mass arrests of former and present-day Party functionaries in 1949 and execution of capital punishment for them in 1950. Stalin might not plan to prosecute the leader of the Leningrad Party organization***** if for no other reason than that Yu. A. Zhdanov married his daughter Svetlana soon after his father died. Stalin and his closest associates had many means at their disposal to make leading specialists of the Kremlin Hospital renounce professional honor and overlook obvious facts. Hence, the strange "blindness" of Kremlin medical luminaries...******

* G. V. Kostyrchenko. Tainaya politika Stalina. Vlast' i anti-Semitism (Secret Policy of Stalin. Power and Anti-Semitism). M.: Mezhdunar. otnosheniya, 2001, p. 640.

** Extract from Zhdanov's medical card (in Egorov's handwriting, 28 August 1948): "Recommended [...] to intensify motion, car drives since 1 September; discuss the trip to Moscow on 9 September". Ibid.

*** Information about the "arrest of a terrorist group of physicians" was published under the heading "Chronicle" in Soviet newspapers on 13 January 1953. Nine leading figures of the medical profession were mentioned and only three of them were ethnic Russians (Vinogradov, Egorov, and Maiorov), the remaining were Jews. After the action was withdrawn and the doctors were released, the award of Timashuk was declared invalid in the report of 4 April 1953 running: "The Presidium of the USSR Supreme Council decided to annul the decree of 20 January 1953 as incorrect in light of the now known circumstances."

**** Verification of evidence undertaken by new leaders of the USSR Ministry of Internal Affairs after Stalin's death revealed inconsistency of accusations under the "doctors' case". Vinogradov wrote to the minister L. P. Beria from prison (27 March 1953): "It should be recognized that A. A. Zhdanov had myocardial infarction and the denial of this diagnosis by Professors Vasilenko and Egorov, doctors Maiorov, Karpai and myself was a mistake. However, we had no malicious intent." Kostyrchenko, p. 642.

***** Stalin appointed Zhdanov to head the Leningrad Party organization after the assassination of its leader S. M. Kirov in 1934. Zhdanov occupied this position until 1944. All main figures of the "Leningrad case" (N. A. Voznesensky, A. A. Kuznetsov, P. S. Popkov, M. I. Rodionov etc.) were in close business relations with Zhdanov; many were his protégés. Evidently, initiators of the "case" could not have managed without the evidence of this key figure. This inference is confirmed by the fact that his associates started to be discharged from office only half a year after Zhdanov's death. They were arrested in August–October 1949 and shot on 1 October 1950. Transcripts of interrogation and applications for rehabilitation of those sentenced to long-term imprisonment after Stalin's death suggest the same conclusion. One of them, I. M. Turko, former secretary of the Yaroslavl regional Communist Party committee (previously 2nd secretary of the Leningrad regional committee) told to the procurator-general R. A. Rudenko: "Putintsev [government investigator for the USSR Ministry of State Security] called me and required that I signed the prepared transcript of interrogation. I said that the transcript contains a lie and casts aspersions on Zhdanov. Putintsev replied that they hold an investigation without regard for ranks. I refused to sign the document; Putintsev beat me and put me in the punishment cell. In the end, [...] I could no longer resist and signed all that the investigator offered me." See: "Politbyuro TsK VKP(b) i Soviet Ministrov SSSR. 1945–1953 (Politbureau of AUCP(b) Central Committee and the USSR Council of Ministres. 1945–1953)". M.: ROSSPEN, 2002, p. 309.

****** The letter of Timashuk came to be in demand only after four years. Arrests of physicians who treated Zhdanov in summer 1948 started in September 1952. Also, Stalin ordered to involve Vlasik (arrested 16 December 1952) and Kuznetsov (posthumously; he was imprisoned on 13 August 1949 and executed on 1 October 1950). Egorov was subjected to torture and gave all required evidence (7 February 1953). The available documents dated from 1952 do not mention that Abakumov reported to Stalin Timashuk's first letter addressed to Vlasik, nor that Stalin himself ordered to shelve it (a note in his handwriting: "To be shelved"). See. J. Brent, V. Naumov. Poslednee delo Stalina (The Last Action of Stalin). M.: Izd-vo "Prospekt", 2004, p. 238.

[30] The Central Board of Propaganda and Agitation was reduced to a lower rank of "Department" on 10 July 1948 as a result of the scandal caused by the criticism of Lysenko by Yu. Zhdanov and the related discontent of Stalin.

[31] The date of shelving Shepilov's materials on the order of A. A. Kuznetsov, secretary of the AUCP(b) Central Committee, is informative in itself. It happened on 21 July 1948 (see docs. 5.8 and 5.11), i.e. when Malenkov returned to the AUCP(b) Central Committee in July 1948 and a month before Shepilov passed a selection of documents on international prizes for their official consideration by Central Committee Secretariat (19 August 1948). In other words, the AUCP(b) Central Committee had decided to give up further attempts to further try and compete with the Nobel organizations long before the Secretariat adopted the resolution of 20 August 1948 (doc. 5.20). Therefore, Kuznetzov passed Shepilov's documents to archives as soon as new people came to lead the Central Committee.

[32] In December 1947, the USSR Academy of Sciences was focused on designing medals and diplomas for the winners of the international prize that was supposed to be named after Lenin (see doc. 5.5).

[33] RGASPI, fond 17, opis' 118, ed. khr. 405, l. 222, 223.

[34] Resolution of the USSR Council of People's Commissars (22 March 1943): "On the award of the Stalin Prizes in 1942 for a) outstanding discoveries and b) for radical improvement of production methods."*

* *Pravda*, 24 March 1943, p. 2.

[35] Even pretenders to the prize took part in lobbying campaigns. An illustrative example is the nomination of the Georgian writer K. S. Gamsakhurdiya. In 1944, he came to know that his candidacy was not approved by the Stalin Prize Committee and asked L. P. Beria, deputy chairman of the USSR Council of People's Commissars with whom he was personally acquainted, to influence members of the committee and have the decision revised (doc. 5.25). He failed only because in 1944 the government did not give prizes for 1943. They were awarded for two years (1943 and 1944) at the end of January 1946.

[36] This absurd situation was created by the Stalin Prize Statute approved by the USSR Council of People's Commissars on 23 May 1940. One of the paragraphs of the Statutes reads as follows: "Works in the fields of science and military knowledge and descriptions of inventions are submitted in any language in three copies, either printed or typewritten." S. V. Kaftanov, deputy chairman of the Stalin Prize Committee in sciences and inventions drew the attention of V. M. Molotov, chairman of the USSR Council of People's Commissars, to this intolerable circumstance in a letter dated 1 June 1946 and asked to envisage evaluation of "scientific works only after their publication" in the new Statute being prepared.* Kaftanov had more than enough arguments in favor of his proposal. By way of example, Professor N. I. Kobozev of Moscow University sent a letter to the AUCP(b) Central Committee (2 April 1945) saying that the Stalin Prize Committee refused to consider his work "The principle of ensemble and aggregation in catalysis". The letter was readdressed to the Department of Propaganda and Agitation. The reply letter to G. M. Malenkov, secretary of the Central Committee was signed by the head of the Department and Kaftanov. The authors of the letter reported that the section of chemical sciences of the committee considered the work of Kobozev and unanimously voted "to postpone the decision [...] until the work is published in press and better known to the scientific community." The letter closed with the following phrase: "There is no reason to reconsider the verdict of the section and the Stalin Prize Committee."**

* RGASPI, fond 117, opis' 125, ed. khr. 400, l. 49–50.
** *Ibid.*, ed. khr. 363, l. *Ibid.* 57–59.

Document 5.1

Draft resolution of the AUCP(b) Central Committee "On the institution of International Prizes of the Soviet Union for the highest scientific achievements"; the document was received by the technical secretariat of the AUCP(b) Central Committee Orgbureau on 6 December 1947; No. 93879.

1. The Central Committee of the AUCP(b) approves the following text of the decree of the Presidium of the USSR Supreme Council "On the institution of International Prizes of the Soviet Union for the highest scientific achievements":

"With a view to promote progress of mankind in its struggle for mastery over natural forces and to motivate outstanding cultural workers throughout the world, the Presidium of the USSR Supreme Council decrees:

1. To institute annual International Prizes of the Soviet Union for the highest scientific achievements in the fields of a) physics, b) chemistry, c) biology, d) medicine, e) geology, f) mathematics.

2. To establish that each prize of the Soviet Union comprises:
    a) the award money equivalent to 300,000 rubles in Soviet or foreign currency;
    b) a diploma;
    c) gold breastplate;
    d) table gold medal featuring the portrait of M. V. Lomonosov for prizes in physics, D. I. Mendeleev for prizes in chemistry, I. P. Pavlov for prizes in biology, N. I. Pirogov for prizes in medicine, A. P. Karpinsky for prizes in geology, N. I. Lobachevsky for prizes in mathematics.
3. To determine that International Prizes of the Soviet Union for the highest scientific achievements are awarded by the special Committee for International Prizes of the Soviet Union in natural sciences whose decisions are subject to the approval of the Presidium of the USSR Supreme Council.[1]
4. To hand over the prizes annually on Victory Day, 9 May, the first prizes to be awarded in 1947.
5. To approve the Statute of International Prizes of the Soviet Union in natural sciences."[2]

Pencil note in the bottom left corner of page 1: "Attn: comrade Shepilov. Vedernikova. 6. December 1947."

RGASPI, fond 17, opis' 125, ed. khr. 497, l. 1–2.

[1] In doc. 5.3, the date of the first prize award is shifted to 1948.

[2] Attached to the draft resolution of the AUCP(b) Central Committee are the Prize Statute (ibid, pp. 3–5) and the table showing distribution of Nobel Prize winners by countries for 1901–1944 (ibid, p. 6) (not published here).

Document 5.2

Memorandum No. 93879 (doc. 5.1) forwarded by S. G. Suvorov, the former head of the Sector of Science,[1] to the technical secretariat of the AUCP(b) Central Committee Orgbureau; 9 December 1947.

This document was drawn up by myself, together with the unattached note,[2] in 1945 and presented to the Department of Propaganda and Agitation; however, it was not reported to the AUCP(b) Central Committee. The document was shown to comrade A. A. Kuznetsov[3] on his request; he instructed me to prepare a new document on the same issue, with slight modifications.

I think that the present document should be either shelved or declassified and destroyed.

RGASPI, fond 17, opis' 125, ed. khr. 497, l. 7.

[1] Suvorov, educated as a physicist, was appointed to head the newly organized Sector of Science in the Department of Propaganda and Agitation in 1942 (shifted from the

position of director of Gostekhizdat (the "Technical literature" publishing house). He was relieved of his duties by the decision of Central Committee Orgbureau (4 November 1947) for a misdeed, severely reprimanded and dismissed from the apparatus of the Central Committee on 15 January 1948.* The day before he prepared the aforementioned document (8 December 1947), Yu. A. Zhdanov succeeded him as head of the Sector of Science by the decision of Central Committee Secretariat.**

* RGASPI, fond 17, opis' 117, ed. khr. 954, l. 140; the discharge was officially confirmed by the resolution of Central Committee Secretariat of 2.4.1948 (ibid, opis' 116, ed. khr. 344, p. 251).

** RGASPI, fond 17, opis' 117, ed. khr. 954, l. 132.

[2] It appears from the context that the author means doc. 5.3 under the absent note.

[3] Secretary of the AUCP(b) Central Committee.

Document 5.3

From: G. F. Aleksandrov, head, Department of Propaganda and Agitation, and S. G. Suvorov,[1] head, Sector of Science.
To: A. A. Zhdanov, secretary of Central Committee.
Dateless, No. 97319.

Department of Propaganda and Agitation asks the AUCP(b) Central Committee to approve the decree of the Presidium of the USSR Supreme Council "On the institution of International Prizes of the Soviet Union for the highest scientific achievements".

Our science had made numerous discoveries over 200 years,[2] laid the foundation of many theories, and gave to the world outstanding scientists of whom our people are proud by right [...]

Despite the important role of our science and its great influence on scientific developments in the world, especially in the last decades, this influence is virtually unsupported by organizational measures. Many countries, the United States, England, France etc., instituted international prizes for scientific work in the form of diplomas and gold medals named after outstanding national scientists. Sweden having a relatively modest scientific potential acts as an arbiter in the assessment of the value of scientific research throughout the world via awarding "Nobel Prizes" to scientists of different states working in the fields of physics, chemistry, physiology and medicine, and also for the works of art and "promotion of the cause of peace".

The Nobel Prizes are highly reputed among the world scientific community; the award of a Nobel Prize testifies to the highest scientific level of the winner's work. Indeed, Nobel laureates are the most prominent scientists.

However, international recognition of the Nobel Prizes can not conceal the fact that Swedish Nobel Committees have pursued pro-German and anti-Russian policy for over 40 years[3] even when the matter concerns natural sciences, to say nothing about literature and the promotion of peace.

Germans predominate among Nobel laureates. There are 38 of them among the total of 132 or almost 32%! England ranks next in the number of Nobel laureates (22 or some 17%). Sweden is fifth on the list (of five laureates) whereas Russia shares 13th–17th positions with such countries as Tunisia, Spain, Hungary, Yugoslavia[4] and India. Germany received almost half of all prizes in chemistry and about a third of them in physics.

According to the Swedish statistics, I. P. Pavlov is the sole Russian prize winner. They believe that I. I. Mechnikov represents France since he received the 1908 Nobel Prize as the director of the Pasteur Institute[5] after he had lived in that country for more than 25 years. Moreover, Mechnikov shared the prize with Ehrlich, Germany.

They could not help giving the Nobel Prize to I. P. Pavlov since his works had been widely recognized all over the world before the Swedish Nobel Committees awarded it.[6] Some of our scientists argue that I. P. Pavlov received the Nobel Prize after Swedish democratic student groups organized demonstrations in his support.[7] The works of many others of our scientists were never recognized by the Swedes. Suffice it to mention the exquisite experiments of P. N. Lebedev on light pressure[8] that astonished the world scientific community, works of K. A. Timiryazev (plant photosynthesis), A. M. Lyapunov (founder of the theory of stability), N. K. Zhukovsky and S. A. Chaplygin (development of the theoretical basis of modern aviation), A. N. Krylov (founder of the theory of modern shipbuilding and navigation). All these works contributed to the treasury of world science but failed to be recognized by the Nobel Committees.[9] Moreover, the Nobel Committees awarded prizes for similar works to foreign scientists but did not notice the more important achievements of Soviet researchers. For example, the 1913 Nobel Prize was awarded to the Dutchman Kamerling-Onnes for low-temperature research whereas the more valuable studies of Kapitsa performed a few years later at even lower temperatures and ending in the discovery of the essentially new phenomenon of helium superconductivity were disregarded.[10] In 1930, the Nobel Prize was awarded to Professor Raman, India, for the discovery of combination light scattering despite the universally known fact that this phenomenon had been theoretically predicted and experimentally observed by Soviet scientists, members of the USSR Academy of Sciences Mandel'shtam and Landsberg. True, their works were published one or two months after Raman's[11] but in other similar cases the Nobel Committees found it possible to share the prize between the researchers. Also disregarded were the wonderful studies by Professor Orekhov,[12] who discovered tens of new alkaloids and described their structure. Karrer (Switzerland) and Haworth (Germany) shared the Nobel Prize for similar research.[13]

In other words, our science is depicted to the world in the distorting mirror of the Nobel Committees[14] whose opinion is taken into consideration by foreign researchers.

It appears advisable that the Soviet Union take on itself the role of the arbiter in the assessments of achievements in world science by instituting international prizes for the highest scientific achievements to promote the progress of mankind in its struggle for mastery over natural forces and to motivate outstanding cultural workers throughout the world. The institution of such prizes would give Soviet researchers to establish relations with their colleagues abroad and provide an organizational basis for strengthening these relations. Also, it will give intellectuals in the West a clear idea of the high level of Soviet science.

The overwhelming majority of the scientists in democratic countries will meet the institution of International Prizes of the Soviet Union as new evidence of its progressive role and readily apply for them.

The materials submitted propose the following principal provisions of the Statute of International Prizes:[15]

> [...]
> 4) the decisions of the Committee are subject to the approval of the Presidium of the USSR Supreme Council; the prizes are handed over on Victory Day, 9 May, on an annual basis, the first prizes to be awarded in 1947;
> 5) candidates for the award can be proposed by large Soviet organizations listed in the Statute, viz. the USSR Academy of Sciences and the Academies of Sciences of union republics, the largest universities and research institutes; the right to nominate candidates is also granted to laureates of International Prizes of the Soviet Union;
> 6) each candidate must receive the prize personally (whenever possible) and has to read a public lecture on the developments and prospects in the field of interest with special reference to his or her own studies awarded by the prize. The lecture is delivered in the Soviet Union within half a year after the announcement of the award [...]

The Nobel Prize comprises the award money equivalent to some 160,000 rubles by the current exchange rate, a diploma, and a gold medal featuring the portrait of Nobel. Nomination and awarding procedures are deliberately chosen to be identical to those established for the Nobel Prizes.[16]

Attached documents: Draft resolution of the AUCP(b) Central Committee approving the decree of the Presidium of the USSR Supreme Council "On the institution of International Prizes of the Soviet Union for the highest scientific achievements";[17] list of Nobel Prize laureates in sciences for 1901–1944; table showing the distribution of Nobel Prize winners in natural sciences by countries; extracts from the Nobel Statutes.[18]

RGASPI, fond 17, opis' 125, ed. khr. 497, l. 8–12.

[1] The document is signed by Suvorov, but the signature of G. F. Aleksandrov, head of the Department of Propaganda and Agitation is absent. Therefore, the document could not be forwarded to a secretary of the AUCP(b) Central Committee. Suvorov could hardly draw up the document without G. F. Aleksandrov's knowledge. In all probability, leaders of the Central Committee lost interest in the idea of international prizes by the time the document was ready for consideration in 1945. G. F. Aleksandrov did not endorse it with his signature and handed it over to records.
[2] In 1945, the USSR Academy of Sciences celebrated its 220th anniversary.
[3] Once again, the document was most likely drawn up in mid-1945 (see note 1 to this chapter), i.e. soon after the victorious outcome of the war in Europe. Hence the allusion to the "pro-German" policy. In a year or two, at the onset of the Cold War, this propagandistic stock phrase gave way to invectives against the recent allies, Great Britain and the USA, and the Nobel organizations were regarded as servants of Anglo-American imperialists.
[4] An obvious mistake; there are no citizens of Yugoslavia among Nobel laureates.
[5] See notes 2 to doc. 5.6 and 1 to doc. 5.12.
[6] Only one Nobel Committee (in physiology and medicine) awarded the prize to I. P. Pavlov.
[7] A far-fetched note. There were no demonstrations in support of Pavlov in Sweden.
[8] See Chapter 1.
[9] None of these scientists, excepting P. N. Lebedev, was ever nominated for the Nobel Prize.
[10] This excursion into history of science reveals the low competence of the author of the document. Kamerling-Onnes observed superconductivity in 1911 when Kapitsa was 17 and had just left the school. Liquid helium was obtained even earlier, in 1908. Kapitsa described "the essentially new phenomenon" (helium superfluidity erroneously called superconductivity in the document) in 1938, i.e. 12 years after the death of Kamerling-Onnes. Kapitsa was awarded the Nobel Prize for fundamental inventions and discoveries in low-temperature physics in 1978.
[11] The first paper of Landsberg and Mandel'shtam reporting their discovery of combination light scattering was published three months after Raman's first report (see Chapter 2).
[12] A. P. Orekhov, member of the USSR Academy of Sciences, was not nominated for the Nobel Prize.
[13] The Nobel Prize in chemistry for 1937 was shared between Karrer and Professor Haworth of Birmingham, UK.
[14] Such prejudice against the Nobel organizations is still alive in this country in the 21st century. G. S. Osipov, member of the USSR Academy of Sciences (Institute of Socio-Political Studies), asked himself: "Why do we have so few Nobel laureates?" and answered: "Many of our scientists did not receive them for political considerations." He alluded to the names of Korolev, Keldysh, Aleksandrov, Vavilov etc.* It is worthwhile to note that A. P. Aleksandrov had quite a different opinion of the

activities of the Nobel organizations. He wrote in 1975: "[...] maybe our scientists are awarded less frequently than Americans. Still, the awards are generally given according to deserts (see doc. 8.51). S. I. Vavilov could not receive the 1958 Nobel Prize because he died in 1951 (see note 2 to Chapter 8). The outstanding achievements of S. P. Korolev, M. V. Keldysh, and A. P. Aleksandrov went beyond the fields covered by the Nobel Statutes. The same is true of the legendary American physicist Robert Oppenheimer, who could not receive the Nobel Prize despite his great contributions to science.

* *Izvestiya — Nauka*, 27 December 2002, p. 11.

[15] The first three items, identical to the corresponding paragraphs in doc. 5.1, are not reproduced.

[16] An erroneous assertion. The Nobel Statutes stipulate that only individual nominations of candidates for the prizes are admitted; candidates submitted by organizations are not considered, at variance with p. 5 of the present document.

[17] See doc. 5.1.

[18] Not published.

Document 5.4

From: D. T. Shepilov, deputy head, Department of Propaganda and Agitation of AUCP(b) Central Committee.
To: A. A. Zhdanov, secretary of Central Committee (dateless).[1]

The Department of Propaganda and Agitation submits for approval by the AUCP(b) Central Committee the draft resolution "On the institution of the International Lenin Prize of the Soviet Union for the highest scientific achievements in science, technology, art, and literature [...]"[2]

The materials submitted propose the following principal provisions of the Statute of the International Lenin Prize:

1. To institute the annual International Lenin Prize of the Soviet Union for the highest scientific achievements in the fields of a) social sciences, b) medical and biological sciences, c) physico-mathematical sciences, d) chemical sciences, e) geologo-geographic sciences, f) technology, g) literature, h) fine arts and music;
2. The prizes are awarded on 22 April, birthday of Vladimir Ilyich Lenin;
3. Each prize comprises the award money equivalent to 500,000 rubles, a diploma, and a chest gold medal featuring the portrait of Lenin; persons awarded the prize are given the rank of Lenin Prize laureate of the Soviet Union [...][3]
4. Proposals to be considered annually may include works done in the last five years. The first prizes will be awarded in 1949;
5. Lenin Prize laureates are invited to read in Moscow a public lecture on the developments and prospects in the field of their scientific interest.

The draft resolution of the AUCP(b) Central Committee envisaged the establishment of the Lenin Prize Committee comprising 30 members.

The Presidium of the USSR Academy of Sciences was committed to design gold medals and diplomas of Lenin Prize laureates.

The drafts of the resolution of the AUCP(b) Central Committee and the decree of Presidium of the USSR Supreme Council "On the institution of the international Lenin Prize of the Soviet Union" and the draft Statute of the international Lenin Prize of the Soviet Union are attached.[4]

RGASPI, fond 17, opis' 125, ed. khr. 497, l. 32–34.

[1] The document was received in the technical secretariat of the AUCP(b) Central Committee Orgbureau on 27 January 1948.
[2] The text that follows is identical to that in doc. 5.3, after the first paragraph to the one starting with the words "The materials submitted [...]" (pp. 289–291).
[3] Paragraphs 2–7 of this document are not published here.
[4] Not published.

Document 5.5

From: N. G. Bruevich, academician-secretary of the USSR Academy of Sciences.
To: Yu. A. Zhdanov, head, Sector of Science, Department of Propaganda and Agitation of the AUCP(b) Central Committee.
No. 3123c. 13 December 1947.

Enclosed find please a draft letter to be sent to sculptors and artists invited to design medals and diplomas of the International Lenin Prize.[1] Please bring this letter to the notice of comrade Shepilov and inform me about your decision.

Names of artists and sculptors on trial are given in the attached drafts of the letters[2]

RGASPI, fond 17, opis' 125, ed. khr. 497, l. 29.

[1] Synopsis of draft letters (see also l. 30–31): deadline for presenting designs of medals and diplomas — 1 January 1948; the design of each medal is presented as actual-size sketch and a sculptural image 150 mm in diameter; the design of the four-page diploma must include a flyleaf and binding; job rates: medal design — 10,000 rubles (15,000 if approved), diploma — 6,000 rubles (10,000 if approved).
[2] Not published.

Document 5.6

Extract from "Information about the Nobel Prize Foundation" drawn up by M. S. Vetrov, head, 5th European Department of the USSR Ministry of Foreign Affairs, for the AUCP(b) Central Committee. 2 January 1948.

[...] I. Swedish Nobel Committee.[1]

The Committee has been established for the technico-administrative management of assets of the Nobel Foundation. The Committee is constituted by eight members (Swedes) including chairmen appointed by the king and his deputy. Other members represent Swedish prize-giving organizations. One member of the Committee is elected to be the director who manages virtually all proceedings of the Nobel Foundation.

The Swedish Nobel Committee has been chaired since 1929 by the known Swedish lawyer Hjalmar Hammarskjold. His deputy is Birger Ekkeberg, member of Swedish Academy, president of the Court of Appeal of Central Sweden and chairman of the Swedish International Help Committee. Sven Hammarskjold, member of the Court of Appeal of Central Sweden and a distant relative of the chairman of the Committee has been its director since May 1946. Other members of the Committee are Professor Arne Westgren, secretary of the Swedish Academy of Sciences, Professor Carl Iogann Bergstrand of the Karolinska Institutet, Doctor Axel Gavelin, professor of literature Martin Lamm, member of the Swedish Academy, etc.

Since 1901 until now, the Nobel prizes in science and literature have been awarded to the following Russian scientists and writers:

I. P. Pavlov (physiologist) in 1904, I. I. Mechnikov (in medicine) as a representative of French science[2] (his Russian descent was mentioned only in brackets) in 1908, and emigrant writer Ivan Bunin (literature) having French citizenship.[3]

The Swedish Nobel Committee invariably ignores representatives of Soviet science and literature after the October revolution. It demonstratively awarded the Nobel Prize in literature to the Finnish writer Sillanpaa[4] in 1939 during the Soviet–Finnish War.

[...] the reserved 1945 prize in chemistry was awarded to Professor Otto Hahn,[5] German investigator of the atomic nucleus, director of Keizer Wilhelm Institute, who was engaged in the development of the atomic bomb during the war, on the instruction of Hitler. After the capitulation of Germany, Hahn was taken by the allies to the United States and continues to work for the American Army. The Americans continuing to conceal Hahn's whereabouts,[6] he had not yet received the prize.[7]

The Finnish Nobel winner Wirtanen, who arrived in Stockholm to receive the prize in 1945, showed himself as the bitterest enemy of the USSR. He stated at a press conference: "I am highly critical of the USSR. I believe that we can not cooperate with the Soviet Union as long as the conquest of the neighbouring territory (Karelia) is considered possible but the consent to conquer is not given to Germany [...] I don't wish to have anything to do with scientists from the USSR [...]"

The aforesaid fairly well characterizes the position of the entire institute of "Nobel Foundation".

Reactionary traditions in the selection of candidates for Nobel Prizes are not surprising bearing in mind that the "18 immortals" of the so-called Swedish Academy comprise inveterate fascists, such as Sven Hedin, personal friend of Hitler, literary critic Professor Beek, and Professor Niels Anlund, participant of the anti-Soviet campaign launched in connection with the extradition of interned Hitler's soldiers to the Soviet Union[8] etc.

Professors of the Karolinska Institutet include Jesta Hegquist, participant of "international expertise" organized by Gebbels in the Khatyn forest.[9] Many members of the Swedish Academy of Sciences are reactionaries.

It should be borne in mind that prize-awarding and giving are official acts controlled by the Swedish authorities. The king himself hands over the awards and government bodies choose candidates in accordance with political considerations.[10]

The institute of Nobel Prizes is used by the Swedish ruling class to increase popularity of the country abroad and win the sympathy of scientific and cultural circles in different states depending on circumstances.

II. Norwegian Nobel Committee.[11]
[...] Since the Nobel Peace Prize was instituted, it has never been awarded to a representative of pre-revolutionary Russia or the Soviet Union. In 1946, a few groups of statesmen and public figures in Finland, Norway, and Sweden for the first time nominated A. M. Kollontai for the Nobel Peace Prize. However, the prize was awarded to two Americans who nobody knows.[12]

In 1947, the candidacy of Kollontai was again submitted for consideration along with Eleanor Roosevelt. On 10 December 1947, "the peace prize" was given to two Quaker organizations (English and American) as "pioneers" (since 1810) of the peace movement.

AVP RF, fond 0140, opis' 39, papka 154, ed. khr. 51, l. 44–48.

[1] The author means Nobel Foundation.
[2] This statement is either a lie or evidence of the author's incompetence. Mechnikov was actually deputy-director of the Pasteur Institute, Paris, where he had worked since 1887 on the invitation of Lois Pasteur. But he never lost the Russian citizenship and received the prize as representative of Russia. He could not come to Stockholm to participate in the prize-giving ceremony on 10 December 1908* and was represented (in accordance with the Nobel status) by Baron F. A. Brudberg, the Russian envoy.** Also, at the official dinner, Brudberg delivered the gratifying speech on behalf of the laureate addressed to the Karolinska Institutet, the prize-awarding organization.***
Mechnikov was always regarded as a Russian scientist, notwithstanding his work in Paris during many years. The newspaper *Russkie vedomosti* informed the readers

about the forthcoming World Exhibition of 1904 in St. Louis, USA, and invitation of "*Russian scientists*" [italicized by the author] Professors I. I. Mechnikov, I. P. Pavlov, F. F. Martens, and S. N. Vinogradsky as its participants.****

* Mechnikov visited Stockholm and read the Nobel lecture envisaged by the Nobel Statutes in spring 1906.

** *Les prix Nobel en 1908*, Stockholm, 1909, s. 7, 46.

*** Ibid, s. 47.

**** *Russkie vedomosti*, 17 December 1903, p. 3.

[3] Bunin did not have French citizenship (see Chapter 2 and note 39 therein); France is indicated as his place of residence in the minutes of the Swedish Academy; see also note 1 to doc. 5.12.

[4] The author had good reason for such a statement. Indeed, a political bias was observable in awarding the Nobel Prize to Sillanpaa on 10 November 1939 at the height of the diplomatic conflict between Finland and the Soviet Union, preceding the onset of hostilities on 30 November. However, this award had another aspect too. The Finnish writer was first nominated in 1930 and continued to be annually nominated since that time. In 1935, C. Heidenstam, the Nobel laureate in literature of 1916, proposed to share the prize between Sillanpaa and the Finnish–Swedish author J. Hammer (Espmark, p. 104).

[5] A mistake. In 1945, O. Hahn received the Nobel Prize in chemistry for 1944.

[6] German physicist C. Funk was one more nominee for the 1946 Nobel Prize in chemistry whose whereabouts were not specified.

[7] A mistake. O. Hahn arrived in Stockholm to receive the Nobel Prize accompanied by his wife and English major Ronald Freiser, appointed his bodyguard by Western occupation authorities in Germany. This fact (with reference to the newspaper *Dagens Nyheter* of 5 December 1946) is fixed by E. Afanasieva, member of the Soviet mission in Sweden, in her "Chronicle of events for December 1946 extracted from the Swedish press". She also reported the arrival of German physicist O. Schtern, Nobel Prize winner of 1943, for the same purpose.*

* GARF, fond 5283, opis' 20, ed. khr. 257, l. 5, 6.

[8] In all probability, citizens of Baltic States that fought at Germany's side.

[9] The expertise was organized by Nazi authorities in the occupied territory of Smolenskaya *oblast'* (province) in spring 1943, where they found burial places of Polish servicemen imprisoned by the Red Army in 1939 and shot by NKVD in 1940 in the Khatyn forest.

[10] Rubbish! The Nobel Foundation and prize-awarding organizations are totally independent of government bodies; the king performs a purely representative function during the prize-giving ceremony.

[11] By this subtitle, the author legally equates (erroneously) the Nobel Committee in Oslo with the executive board of the Nobel Foundation in Stockholm. In fact, the committee in Oslo and four other committees in Stockholm are constituent units of the Nobel Foundation.

[12] Emily Balch (England) and John Mott (USA) were well-known peace champions. See note 2 to doc. 4.24 for details concerning Kollontai.

Document 5.7

The Nobel Statutes, 29 June 1900.

The typewritten text of the Statutes follows (not published here). The last page has a handwritten note under the text: "Retyped from the English version published by the Nobel Committee. Ser(gei) Suvorov" (dateless).
On the first page (above the title): "Send to comrades Stalin I. V., Molotov B. M., Beria L. P., Mikoyan A. I., Malenkov G. M., Voznesensky N. A. 2. I-48. Zhdanov (facsimile)."

The note in pencil below is the shelving date: 11 May 1948.

RGASPI, fond 17, opis' 125, ed. khr. 588, l. 66–74.

Document 5.8

To: comrade Stalin. I am sending the Nobel Statutes. 2 January 1948 (unsigned).[1]
The note in pencil below: "To shelve. 21.7.48.A.K."[2]

RGASPI, fond 17, opis' 125, ed. khr. 588, l. 75.

[1] Probably, the draft of the cover latter to the text of the Nobel Statutes prepared for Stalin.
[2] A. A. Kuznetsov, secretary of the AUCP(b) Central Committee.

Document 5.9

Nobel Prize-giving ceremony.

The text is not published here.

The typewritten text below reads: "From the book *Junk V. Die Nobelpreistrager. Dreissing Jahre Nobelstifumg. Wien-Leipzig,* M. Winkler, 1930, s. 21–22. Translated from German and verified by comrade Klivinskaya, V. I. Lenin State Library. 5 January 1948."

RGASPI, fond 17, opis' 1257, ed. khr. 588, l. 76–77.

Document 5.10

From: D. T. Shepilov, deputy head, Department of Propaganda and Agitation.
To: A. A. Kuznetsov, secretary of Central Committee.
(dateless)[1]

Attached find please the drafts of the letter to comrade Stalin;[2] decree of the Presidium of the USSR Supreme Council "On the institution of the International Mendeleev Prize of the Soviet Union for the highest scientific achievements" and the Statute of International Mendeleev Prize Committee in science; decree of the Presidium of the USSR Supreme Council "On the institution of the International Gorky Prize of the Soviet Union for the highest achievements in literature" and the Statute of International Gorky Prize Committee in literature.

The note in pencil below is the shelving date: 30 March 1948.

RGASPI, fond 17, opis' 125, ed. khr. 588, l. 1.

[1] The letter was certainly written before 18 March 1948 (see doc. 5.11).
[2] The first variant of the draft letter to Stalin (an extract see in doc. 1.3). The drafts of this letter and the attached documents (RGASPI, ibid, l. 2–15) are not published here.

Document 5.11

From: D. T. Shepilov, deputy head, Department of Propaganda and Agitation.
To: M. A. Suslov,[1] head of the Department, secretary of Central Committee.
18 March 1948.

Attached find please the draft resolutions of the AUCP(b) Central Committee "On the institution of International Mendeleev Prize of the Soviet Union for the highest scientific achievements" and the Statute of the International Mendeleev Prize Committee in science; resolution of the AUCP(b) Central Committee "On the institution of International Gorky Prize of the Soviet Union for the highest achievements in literature" and the Statute of International Gorky Prize Committee in literature.[2]

Handwritten note in the upper left corner: "Attn: Zhdanov A. A. M. A. Suslov, 19/III."
Handwritten in pencil in the lower left corner: "To shelve. 21.7.48. A. K."[3]

[1] M. A. Suslov succeeded G. F. Aleksandrov in the position of the head of Department of Propaganda and Agitation and was elected secretary of the AUCP(b) Central Committee (17 September 1947).
[2] The drafts of the attached documents are not published.
[3] A. A. Kuznetsov, secretary of the AUCP(b) Central Committee.

Document 5.12

From: D. T. Shepilov, deputy head, Department of Propaganda and Agitation.
To: A. A. Zhdanov, secretary of Central Committee.
(dateless)

The Department of Propaganda and Agitation submits for approval by the AUCP(b) Central Committee the drafts of the resolutions "On the institution of International Mendeleev Prize of the Soviet Union for the highest scientific achievements" and "On the institution of International Gorky Prize of the Soviet Union for the highest achievements in literature".

As known, Sweden having a relatively modest scientific and literary potential acts as an arbiter in the assessment of the value of scientific research throughout the world via awarding "Nobel Prizes" to scientists of different states working in the fields of physics, chemistry, physiology and medicine, and also for the works of art and "promotion of the cause of peace".

Throughout 46 years of its existence the Nobel Committee has pursued hostile anti-Russian and anti-Soviet policy even when the matter concerned natural sciences, to say nothing about literature and the promotion of peace. Germans predominate among Nobel laureates followed by representatives of the Anglo–American block. There are 38 Germans among the total of 132 laureates given the Nobel Prizes for the period from 1901 to 1944, or almost 30%. England has 22 laureates or some 17%, USA and France 18 laureates (14%) each. Russia shares 13th–17th positions with such countries as Tunisia, Spain, Hungary, and India, behind Switzerland, Austria, and other minor countries.

According to Swedish statistics, I. P. Pavlov is the sole Russian prize winner. They believe that I. I. Mechnikov represents France.[1]

There are many facts that the Swedes demonstratively did not recognize works of our scientists. Suffice it to mention the exquisite experiments of P. N. Lebedev on light pressure[2] that astonished the world scientific community, works of K. A. Timiryazev (plant photosynthesis), A. M. Lyapunov (founder of the theory of stability), N. K. Zhukovsky and S. A. Chaplygin (development of theoretical basis for modern aviation), A. N. Krylov (founder of the theory of modern shipbuilding and navigation). The Nobel Committee pursues the policy of discrimination with respect to our scientist giving prizes for similar works to foreign researchers.[3]

The Nobel Prizes are demonstratively awarded to enemies of people's democracy and anti-Soviet slanderers like Andre Gide.[4]

It appears advisable that the Soviet Union, as the most progressive country in the world, take on itself the role of the arbiter in the assessments of the achievements of science and culture in different countries by instituting the International Mendeleev Prize of the Soviet Union for the highest scientific achievements and International Gorky Prize of the Soviet Union for the highest achievements in literature and fine arts to be awarded every other year on an alternative basis.

The institution of international prizes in science and culture would further increase the authority of the Soviet Union in the eyes of progressive people, speed up the process of differentiation of intellectuals in capitalist states, unite workers of science, art, and literature having democratic views, and enhance the influence of the Soviet Union on intellectuals in friendly countries of Eastern Europe.

The materials submitted propose the following principal provisions of the Statute of International Mendeleev and Gorky Prizes:

1. To institute the International Mendeleev Prize of the Soviet Union for the highest scientific achievements in each of the following fields: a) social sciences, b) medical and biological sciences, c) physico-mathematical sciences, d) chemical sciences, e) geologo-geographic sciences;
2. To institute the International Gorky Prize of the Soviet Union for the highest scientific achievements in each of the following fields: a) literature, b) cinema, c) fine arts, d) music;
3. The International Mendeleev and Gorky Prizes are awarded every other year on an alternative basis;
4. The prizes are awarded on 7 November, the day of the Great October Revolution;
5. Each prize comprises the award money equivalent to 300,000 rubles, a diploma, and a gold medal featuring the portrait of Mendeleev or Gorky; persons awarded the prize are given the rank of Mendeleev and Gorky Prize laureate of the Soviet Union, respectively;
6. The prizes are awarded for research contributing to the progress of science and technology, improvement of people's well-being, for works of art and literature promoting the development of humanistic ideas and strengthening of democratic peace between nations;
7. Only published scientific and literature works may be submitted to the competition; art works are submitted to the Committee for a preview;
8. The prizes are awarded by special Committees, approved by the Presidium of the USSR Supreme Council, and handed over to the laureates by the chairman of the Presidium of the USSR Supreme Council in the presence of members of the government, representatives of the diplomatic corps, scientific, literary, art, and public organizations of the Soviet Union;
9. A candidate for a prize may be any person regardless of nationality, race, gender, and citizenship;
10. Proposals for each next competition may include works done in the last 5 years. The first Gorky Prize will be awarded in 1948, Mendeleev Prize in 1949;
11. Mendeleev and Gorky Prize laureates are invited to read in Moscow a public lecture on the developments and prospects in the field of science or culture where the laureate works.

The draft resolutions of the AUCP(b) Central Committee envisage the establishment of International Mendeleev Prize Committee comprising 21 members and International Gorky Prize Committee comprising nineteen members.

The draft decrees of the Presidium of the USSR Supreme Council "On the institution of International Prizes of the Soviet Union" and the draft Statutes of International Prizes of the Soviet Union are drawn by the Department of Propaganda and Agitation with the participation of S. I. Vavilov, A. N. Nesmeyanov, N. G. Bruevich, and N. N. Anichkov, members of the USSR Academy of Sciences, S. V. Kaftanov and A. A. Fadeev.

RGASPI, fond 17, opis' 125, ed. khr. 588, l. 20–23.

[1] The number of Nobel laureates in physiology and medicine from Russia is deliberately underestimated. In fact, there were two of them. Mechnikov worked in Paris but preserved his Russian citizenship. He was not the sole laureate who worked abroad but received the Nobel Prize as a representative of his country. Suffice it to mention the Dutch chemist Debye (Keizer Wilhelm Institute, Berlin), Austrian medical scientist Landschteiner (Rockefeller Institute, New York) etc. The statement that the Swedes "believe that I. I. Mechnikov represents France" is simply deceitful. None of the publications of the Nobel Foundation, from *Les Prix Nobel* to the biannual *Nobel Foundation Directory*, mentions the name of the scientist without the word "Russia" immediately behind it. At the prize-giving ceremony of 1908, the Russian envoy F. A. Brudberg represented the country on equal footing with his counterparts from Great Britain, Germany, and France whose citizens also received Nobel Prizes in that year. The Russian flag was hoisted side by side with the flags of these states. The Russian citizenship of Mechnikov was always taken into account in all statistical reports, e.g. in an article in the English magazine *Discovery** published in 1946, two years before the appearance of the document under consideration.

* *Discovery*, 7, No. 10, 1946, p. 296.

[2] See Chapter 2.

[3] Symptomatically, the author does not mention Soviet scientists G. S. Landsberg and L. I. Mandel'shtam although he means in the first place the Nobel Prize in physics awarded in 1930 for the discovery of combination scattering spectra (see Chapter 2). There is reason to believe that their names are deliberately omitted in connection with the anti-Semitic campaign that was already in full swing in 1948 (see also note 12 to Chapter 2).

[4] This statement is attributable to the publication of Gide's book *Return from the USSR* at the end of 1936. The writer described his observations during the trip across the country from 17 June to 24 August when he had an audience with Stalin and spoke from Lenin's mausoleum on Red Square at the funeral of M. Gorky. Many illusions he had about the Soviet society were dispelled after his contacts with Soviet leaders and common people. He could no longer believe in the USSR as a country where "Utopia becomes reality"* and came to think that "[...] soon nothing will remain of this great heroic people so worthy of love, besides profiteers, butchers and their victims."** The leaders of the Soviet Union met so radical a change in the views of the writer with painful disappointment*** which predetermined a similar reaction

12 years later.**** Gide's book was first published in the USSR in the Russian language in 1989.*****

* A. Gide. Return from the USSR. M.: Izd-vo politich. literatury, 1990, p. 64.
** Ibid, p. 121.
*** An extract from the editorial in the newspaper *Pravda* of 3 December 1936 entitled "Tears and laughter of Andre Gide": The known French writer Andre Gide cried much and laughed much when he visited our country last summer. He laughed out of happiness and cried of affection [...] He thinks that by mixing smiles and tears of delight with malicious slander he will retain respect of working people and the honorary title of a friend of the Soviet Union. But he is already welcomed in the trotskyist-fascist camp."
**** Even less favorable characteristics dated from 1951 are cited in Appendix 4.
***** Journal *Zvezda*, No. 8, 1989.

Document 5.13

To Comrade Stalin I. V.[1]

Submitted for your approval are the draft resolutions of the AUCP(b) Central Committee "On the institution of International Mendeleev Prize of the Soviet Union for the highest scientific achievements" and "On the institution of International Gorky Prize of the Soviet Union for the highest achievements in literature and fine arts..."[2]

We are asking for your consent.

RGASPI, fond 17, opis' 125, ed. khr. 588, l. 42–26.

[1] The document has neither date nor signature. In all probability, it was a variant of the draft letter that was not passed to the addressee.
[2] The text analogous to doc. 5.12 follows.

Document 5.14

Prize-giving ceremony for the International Mendeleev and Gorky Prizes of the Soviet Union (draft).

1. In accordance with the decree of Presidium of the USSR Supreme Council of ....1948, International Mendeleev and Gorky Prizes of the Soviet Union are handed over to the Laureates[1] by the chairman of the Presidium of the USSR Supreme Council.
2. The prizes are awarded at the State Academic Bolshoy Theater on the afternoon of 8 November.
3. The prizes are handed over in the presence of 1) members of the Presidium of the USSR Supreme Council, USSR ministers of foreign affairs, higher education, public health, and cinematography; chairmen of USSR State Committees

of architecture and fine arts; 2) diplomatic corps; 3) members of Presidiums of the USSR Academy of Sciences, Union of Soviet Architects, Union of Soviet Writers, Union of Soviet Artists, Union of Soviet Composers, All-Union Society for spreading political and scientific knowledge, All-Union Society of Cultural Relations with Foreign Countries, USSR Academy of Arts, USSR Academy of Architecture, V. I. Lenin All-Union Academy of Agricultural Sciences, USSR Academy of Medical Sciences, Rectorate of the Moscow State University, representatives of other scientific and public organizations and higher education institutions; 4) Laureates of Mendeleev and Gorky Prizes of the past years.

4. The prize-giving ceremony is opened with the USSR anthem. Thereafter, the chairman of the Committee of the International Mendeleev or Gorky Prize of the Soviet Union presents a brief report characterizing the Laureates. The decree of the Presidium of the USSR Supreme Council is read out and the prizes are handed over to the Laureates. The chairman of the Presidium of the USSR Supreme Council addresses a short welcoming speech to the Laureates. Laureates may present a short replay speech. The ceremony is closed with the state anthem.
5. On 5, 9, 10 November, the Mendeleev or Gorky Prize Committee organizes public lectures by the new Laureates in the House of Scientists or Chaikovsky Concert Hall. Presentation of the lectures is closed by welcoming speeches of representatives of the USSR Academy of Sciences and Moscow Lomonosov State University or representatives of the Union of Soviet Writers, Union of Soviet Composers or Union of Soviet Artists, who congratulate the Laureates and wish further successful work.
6. Chairman of Presidium of the USSR Supreme Council holds a reception in honour of the Laureates of International Prizes of the Soviet Union in the Georgievsky Hall of the Great Kremlin Palace.

RGASPI, fond 17, opis' 125, ed. khr. 588, l. 63–64.

[1] Here and hereinafter, the word Laureate is written with a capital latter (as in the original typewritten text).

Document 5.15

From: D. T. Shepilov, deputy head, Department of Propaganda and Agitation.[1]
To: G. M. Malenkov, secretary of Central Committee.
19 August 1948.

Attached find please the draft resolution of the AUCP(b) Central Committee and draft decrees of the Presidium of the USSR Supreme Council "On the institution of the International Mendeleev Prize of the Soviet Union for the highest scientific achievements" and "On the institution of the International Gorky Prize of the Soviet Union for the highest achievements in literature". These documents are drawn by the Department of Propaganda and Agitation[2] with the participation of comrades

S. I. Vavilov, N. G. Bruevich, Fadeev, and N. N. Anichkov, president of the USSR Academy of Medical Sciences.

RGASPI, fond 17, opis' 118, ed. khr. 132, l. 157.

[1] The Division of Propaganda and Agitation was redesignated as the Department and its former departments as sectors (resolution of the AUCP(b) Central Committee, minutes 64, 10 July 1948, p. 117).
[2] The Division of Propaganda and Agitation of the Central Committee of All-Russia Communist Party (Bolsheviks).

Document 5.16

From: D. T. Shepilov, deputy head, Department of Propaganda and Agitation.*
To: G. M. Malenkov, secretary of Central Committee.
19 August 1948.

The Department of Propaganda and Agitation submits for approval by the AUCP(b) Central Committee the drafts of the decree of the Presidium of the USSR Supreme Council "On the institution of the International Mendeleev Prize of the Soviet Union for the highest scientific achievements" and the Statute of the International Mendeleev Prize Committee in science; decree of the Presidium of the USSR Supreme Council "On the institution of the International Gorky Prize of the Soviet Union for the highest achievements in literature" and the Statute of the International Gorky Prize Committee in literature [...][1]

The materials submitted propose the following principal provisions of the Statute of the International Mendeleev and Gorky Prizes:

1. To institute the International Mendeleev Prize of the Soviet Union. Six prizes are awarded biannually for the highest scientific achievements, one in each of the following fields: a) physico-mathematical sciences, b) chemical sciences, c) medical and biological sciences, d) technical sciences, e) geologo-geographic sciences, and f) socio-humanitarian sciences.
2. To institute the International Gorky Prize of the Soviet Union. Two prizes are awarded biannually for the outstanding works of literature.
3. The prizes are awarded on 13 November, the foundation day of the USSR Academy of Sciences.[2]

International Mendeleev Prizes of the Soviet Union are awarded every odd year, International Gorky Prizes every even year.

4. Each International Mendeleev Prize of the Soviet Union and each International Gorky Prize of the Soviet Union comprises the award money equivalent to 300,000 rubles, a diploma, and a gold breast medal featuring the portrait of Mendeleev or Gorky; persons awarded the International Mendeleev Prize of the Soviet Union are given the rank of Laureate of the International Mendeleev

Prize of the Soviet Union; persons awarded the International Gorky Prize of the Soviet Union are given the rank of Laureate of the International Gorky Prize of the Soviet Union.

5. The International Mendeleev Prize of the Soviet Union is awarded for research and inventions promoting mastery over natural forces, improvement of people's health and welfare, progress of mankind. The International Gorky Prize of the Soviet Union is awarded for literature works in the spirit of humanism promoting progress of civilization, strengthening cooperation and democratic peace between nations.
6. A candidate for the International Mendeleev Prize and the International Gorky Prize may be any person regardless of nationality, race, gender, or citizenship.
7. The International Mendeleev Prize and International Gorky Prize are awarded by the Presidium of the USSR Supreme Council to which they are submitted by the International Mendeleev Prize Committee in the former case and by the International Gorky Prize Committee in the latter case. The prizes are handed over to the Laureates by the chairman of the Presidium of the USSR Supreme Council in the presence of members of the Soviet government, representatives of the diplomatic corps, scientific, literary, art, and public organizations of the Soviet Union and other countries.

The draft decrees envisage the establishment of International Mendeleev Prize Committee comprising 22 members and International Gorky Prize Committee comprising thirteen members.

The USSR Academy of Sciences is committed to design diplomas of Laureates of the International Mendeleev Prize of the Soviet Union and of the International Gorky Prize of the Soviet Union and gold breast medals featuring the portrait of Mendeleev for Laureates of the International Mendeleev Prize and the portrait of Gorky for Laureates of the International Gorky Prize.

The drafts of the decree of Presidium of the USSR Supreme Council "On the institution of International Mendeleev Prize of the Soviet Union for the highest scientific achievements" and the Statute of the International Mendeleev Prize Committee in science; decree of the Presidium of the USSR Supreme Council "On the institution of the International Gorky Prize of the Soviet Union for the highest achievements in literature" and the Statute of the International Gorky Prize Committee in literature[3] are attached.[4]

RGASPI, fond 17, opis' 118, ed. khr. 132, l. 156–161.

[1] The text analogous to doc. 5.12 follows.
[2] By order of Peter I, "*leib-medic*" L. L. Blumentrost submitted the project of the establishment of "the Academy of Sciences and Curious Arts". The emperor approved the project on 22 January (2 February) 1724. The first meeting of the Academy was held under Catherine I on 12 (23) November 1725.* Therefore, the date of 13

November is incorrect. If the USSR Academy of Sciences proper is meant, it was redesignated from the "Russian Academy of Sciences" in July 1925.**

**Entsiklopedicheskiy slovar' Brokgauza I Efrona,* T. 1, SPb, 1890, p. 263.

[3] Not published.

[4] See docs. 5.17 and 5.18.

Document 5.17

Statute of the Committee for the International Mendeleev Prize of the Soviet Union in sciences (draft).

1. The International Mendeleev Prize Committee is set up by decree of the Presidium of the USSR Supreme Council of ......1948 "On the institution of the International Mendeleev Prize of the Soviet Union for the highest scientific achievements". The Committee is constituted by the most prominent scientists of the Soviet Union.
2. The International Mendeleev Prize Committee considers scientific works proposed for the award of the International Mendeleev Prize of the Soviet Union and submits the best of them for the approval by the Presidium of the USSR Supreme Council.
3. The International Mendeleev Prizes of the Soviet Union are awarded on a competitive basis for the highest scientific achievements, one in each of the following fields: a) physico-mathematical sciences, b) chemical sciences, c) medical and biological sciences, d) technical sciences, e) geologo-geographic sciences, and f) socio-humanitarian sciences. Either published scientific works or manuscripts[1] in any language may be submitted for the competition.
4. Proposals for the award should include works done in the last two years and be submitted one year before the next competition.
5. In order to completely and comprehensively assess the works submitted for the International Mendeleev Prize of the Soviet Union, the Committee has the right

   a) to organize sections constituted by specialists in different research areas and invite authoritative scientists from the outside to participate in expert advisory panels and discussions at the meetings of the Committee;
   b) to ask for the opinion of academies, research institutions, higher education schools, scientific societies, and individual scientists in the Soviet Union and abroad;
   c) to publish information about works submitted for the prize in newspapers and journals.[2]

6. Meetings of the Committee are convened by its chairman as appropriate and the discussions shall be valid if at least three-quarters of its members are present. Decisions of the Committee are taken by simple majority votes (secret).
7. Decisions of the Committee either on the submitted works or related issues are entered in the minutes signed by its chairman, his deputies, and all members of the Committee.

8. The prize awarded for a work done by two or more persons is shared between them. The Committee is entitled to determine the number of the authors and the relative contribution of each of them to the joint work.
9. The award can be given to the Laureate in foreign currency (if that be his or her will) based on the current exchange rate.
10. The Committee covers travelling and accommodation expenses of the Laureates related to their visit to Moscow to receive the prize.
11. The Committee has the right to publish the reference book *Laureates of International Mendeleev Prize of the Soviet Union* containing their concise biographies and characteristics of the awarded works; lectures of the Laureates of the International Mendeleev Prize; booklets popularizing works of the Laureates of the International Mendeleev Prize of the Soviet Union.
12. The Committee has a secretariat.
13. The budget of the Committee and the staff of its secretariat are approved by the USSR Council of Ministers.
14. The International Mendeleev Prize Committee has an official stamp.

RGASPI, fond 17, opis' 118, ed. khr. 132, l. 166–167.

[1] Unpublished works, i.e. unknown to the scientific community. This proposition by the developers of the International Mendeleev Prize was in sharp contrast to the universally accepted practice but consistent with the Soviet tradition of awarding Stalin Prizes (see note 36 to this chapter).
[2] In neglect of the fundamental principle of confidentiality adopted by the Nobel organizations.

Document 5.18

Statute of the Committee for the International Mendeleev Prize of the Soviet Union in sciences (draft).

1. The International Gorky Prize Committee is set up by the decree of the Presidium of the USSR Supreme Council of ....1948 "On the institution of the International Gorky Prize of the Soviet Union for the highest achievements literature". The Committee is constituted by the most prominent literary workers of the Soviet Union.
2. The International Gorky Prize Committee considers writings proposed for the award of the International Gorky Prize of the Soviet Union and submits the best of them for approval by the Presidium of the USSR Supreme Council.
3. The International Gorky Prizes of the Soviet Union are awarded on a competitive basis for the highest achievements in literature. Either published writings or manuscripts[1] in any language may be submitted for the competition.
4. Proposals for the award should include works done in the last two years and be submitted one year before the next competition. Literary works for the 1948 competition are submitted to the Committee not later than 15 October 1948.[2]

5. In order to completely and comprehensively assess the works submitted for the International Gorky Prize of the Soviet Union, the Committee has the right
   a. to organize sections constituted by specialists in different genres of literature and invite authoritative scientists from the outside to participate in expert advisory panels and discussions at the meetings of the Committee;
   b. to ask for the opinion of local literature associations, academies of sciences, research institutions, universities, and individuals in the Soviet Union and abroad;
   c. to publish information about works submitted for the prize in newspapers and journals [...][3]

RGASPI, fond 17, opis' 118, ed. khr. 132, l. 168–169.

[1] Thus, the right to submit unpublished works concerns not only scientific research (see note 1 to doc. 5.17) the results of which may be at least partially known to the authors' colleagues from reports at scientific councils, symposia etc. but also the works of literature that the reader can appraise only after they are published in full.
[2] One more strange proposition implying that the "complete and comprehensive assessment" of the submitted works is possible within three months (according to doc. 5.16 endorsed by Shepilov on 19 August 1948, the prizes are to be awarded on 13 November 1948; the same refers to the draft resolution of the Central Committee Secretariat (doc. 5.19)).
[3] Pages 6–14 are not published; they are identical to pp. 6–14 in doc. 5.17.

Document 5.19

Draft resolution of Secretariat of AUCP(b) Central Committee on the institution of the International Mendeleev Prize of the Soviet Union for the highest scientific achievements and the International Gorky Prize of the Soviet Union for the highest achievements in literature endorsed by D. T. Shepilov, head, Department of Propaganda and Agitation, 19 August 1948.

1. The Secretariat of the AUCP(b) Central Committee hereby approves the drafts of the decree of the Presidium of the USSR Supreme Council "On the institution of the International Mendeleev Prize of the Soviet Union for the highest scientific achievements" and the Statute of the International Mendeleev Prize Committee in science; decree of the Presidium of the USSR Supreme Council "On the institution of International Gorky Prize of the Soviet Union for the highest achievements in literature" and the Statute of the International Gorky Prize Committee in literature.
2. This decision is subject to the approval of Politbureau.

RGASPI, fond 17, opis' 118, ed. khr. 132, l. 156.

Document 5.20

Resolution of the Secretariat of the AUCP(b) Central Committee; 20 August 1948 (minutes 369, item 33-c).

On the institution of the International Mendeleev and Gorky Prizes of the Soviet Union for the highest achievements in science and literature.

The Department of Propaganda and Agitation is committed to prepare materials to be submitted to Politbureau.

RGASPI, fond 17, opis' 118, ed. khr. 132, l. 155.

Document 5.21

Extract from the shorthand record of the meeting chaired by A. V. Topchiev, member of the USSR Academy of Sciences, chief scientific secretary of the Academy. 26 July 1949.

A. V. Topchiev:

I would like to remind you that two months ago we were committed to design medals of Gorky and Mendeleev that, according to the preliminary plans of higher organs, will be awarded for the outstanding works in literature and science respectively. We asked comrades Manizer, Merkurov, Konenkov, Mukhina, and other sculptors to take part in the competition. Some of these works are now presented for our consideration by Manizer, Mukhina, Merkurov, Konenkov, Mograchev. We must assess them and present our recommendations (together with the models of the medals) to higher organs for further consideration [...][1]

RGASPI, fond 17, opis' 132, ed. khr. 270, l. 61–62.

[1] See Appendix 2 for the full text of the shorthand record.

Document 5.22

From: S. I. Vavilov, president, and A. V. Topchiev, chief scientific secretary, the USSR Academy of Sciences.
To: G. M. Malenkov, secretary, AUCP(b) Central Committee.
No. 1–398, 20 December 1949.

In pursuance of instructions of the AUCP(b) Central Committee, the Presidium of the USSR Academy of Sciences presents models of Mendeleev and Gorky gold medals designed at the request of the Presidium by sculptors S. D. Merkurov, V. I. Mukhina, M. G. Manizer, S. T. Konenkov and artist I. F. Rerberg.

Also attached are the models of diplomas of laureates and hard covers for them made after the sketches of A. E. Lopukhin.

RGASPI, fond 17, opis' 132, ed. khr. 270, l. 59.

Document 5.23

Separate opinion of S. I. Vavilov, president of the USSR Academy of Sciences.

To my mind, the second variant of the medal featuring Gorky's portrait by S. D. Merkurov is the best one; the reverse side is much worse and needs to be remodeled.

The design of the Mendeleev medal by S. D. Merkurov can not be accepted. Mendeleev looks quite different from his known portraits and more closely resembles an ecclesiastic than a scientist. This medal must be remade.

All other models of Gorky and Mendeleev medals seem unacceptable.

RGASPI, fond 17, opis' 132, ed. khr. 270, l. 60.

Document 5.24

From: L. F. Timashuk, head of the ECG cabinet in the Kremlin Health Department.
To: N. S. Vlasik, head of the Central Guards Department, USSR Ministry of State Security.
29 August 1948[1]

Head of the Health Department Professor Egorov instructed me to perform an ECG in comrade A. A. Zhdanov on 28 August 1948.[2] The same day, I went to the patient by plane together with Professor Egorov, academician Vinogradov, and Professor Vasilenko. The ECG made in the afternoon was indicative of "myocardial infarction involving the left ventricle and interventricular septum". I immediately reported this diagnosis to the counseling professors. Professor Egorov and doctor Maiorov regarded it as erroneous, did not agree with me, and stated that the patient had "functional disturbance related to sclerosis and hypertensive disease" rather than infarction. They wanted me to rewrite my conclusion without mentioning "myocardial infarction" and use more cautious terms, as doctor Karpai did after the previous ECG.

A. A.[3] had another heart attack on 29 August (after he left the bed), and I was again brought to him from Moscow. However, the academician Vinogradov and Professor Egorov delayed the ECG until 30 August and again demanded that I make a diagnosis without mentioning myocardial infarction [...] I am sure that the consultants and the attending physician Maiorov underestimate the severity of the patient's condition, allowing him to stand up, walk in the park, and see films.[4] Taken together, these led to the relapse of the heart attack and may have a fatal outcome. Although I had to repeat the ECG without diagnosing "myocardial infarction" by order of my superiors, I still keep to my opinion and insist on the necessity for the patient to be confined to bed [...]"[5]

From: V. B. Malkin. "Sem' pisem[6] Lidii Timashuk (Seven Letters of Lidiya Timashuk)". "Novoye vremya", No. 28, 1993, p. 39.

[1] Timashuk handed the above letter to A. L. Belov, head of Zhdanov' guards; it thereafter passed to N. S. Vlasik and further on to V. S. Abakumov, Minister of Internal Affairs. Abakumov reported information from Timashuk to Stalin who wrote on her letter: "To be shelved."* Stalin changed his view in four years when the "doctors' case" began to be framed by the USSR Ministry of State Security after the arrest of Abakumov and his close associates on the initiative of the deputy minister M. D. Ryumin. He approved the accusation of "Abakumov and Vlasik, who sacrificed Timashuk to [...] foreign spies and terrorists Egorov, Vinogradov, Vasilenko, and Maiorov" and added, when editing the document: "Zhdanov did not simply die, he was killed by Abakumov." In the course of investigation, Egorov, the former head of the Kremlin Health Department, gave the following evidence: "There is no doubt that had Abakumov and Vlasik verified the statement of Timashuk immediately after she made it, we, responsible for the death of Zhdanov, would have been disclosed as early as 1948."**

* R. G. Pikhoya. Sovetsky Soyuz: istoriya vlasti, 1941–1991. (Soviet Union: The History of Power, 1941–1991). Novosibirsk: Izd-vo "Sibirsky khronograf", 2000, p. 50; see also note 29 to this chapter.

** *Ibid*, p. 90.

[2] Electrocardiogram.

[3] A. A. Zhdanov.

[4] See also footnote** to note 29 of this chapter.

[5] Arguments from Timashuk's letters testifying to the incorrect treatment of Zhdanov were almost literally repeated four years later in the *Pravda* editorial "Base spies and killers under the guise of medical professors".* It told that "the criminals [...] intentionally kept secret the fact that comrade Zhdanov had myocardial infarction and prescribed the regime contraindicated in this condition that resulted in the death of comrade Zhdanov."

* *Pravda*, 13 January 1953, p. 1.

[6] Four other letters in which Timashuk tried in vain to prove her non-involvement in the "Zionist" aspect of the "doctor's case" were written after Stalin's death.

Document 5.25

How the writer K. S. Gamsakhurdiya lobbied the award of the Stalin Prize to himself.

I

Extract from the letter of K. S. Gamsakhurdiya to L. P. Beria, deputy chairman of the USSR Council of People's Commissars. 14 May 1944.

[...] My literary enemies in Tbilisi transmitted false information to Alexey Tolstoy, who I deeply respect [...] I was told that he spoke at a meeting of the Committee and said: "In his novel, Gamsakhurdiya combats Christianity and advocates paganism

[...] Had the disapproval of the Stalin Prize Committee been the matter of hurting my personal pride or the matter of money, I would have never troubled you; but it casts a shadow on the greatest hero of Georgian history [...] Moreover, everybody in Georgia knows that my books were submitted for the prize on the recommendation of the Central Committee of the Georgian Communist Party. It is also a delicate topic. Finally, such unjust belittling of my two works[1] leads to moral depression that hinders the hard work over another two books [...]

You must know everything about us, good and bad. Therefore, I would like to tell you the following: I was about to go to Moscow in January to settle things related to the publication of my books in the Russian language. But my works were submitted for the prize just at that time, and I did not go to avoid talks of my enemies that I specially visited Moscow to obtain support of Lavrentiy.[2] You must be aware that matters of importance in our literature are often arranged by "standing a treat." I never resorted to such tricks and must pay for it.[3]

RGASPI, fond 17, opis' 125, ed. khr. 234, l. 118–120.

II

From: S. I. Kavtaradze, deputy minister of foreign affairs.
To: L. P. Beria.
15 May 1944

[...] Gamsakhurdiya asked me to pass his letter addressed to you because the matter in which he is interested must be resolved within a few days,[4] and he is afraid that you will not receive the letter in time in any other way.

I ask you to read the letter. On my part, I think that Gamsakhurdiya is the sole modern Georgian writer [...] whose historical novels can be called works of high art.

Handwritten in the top left corner: "Attn: Comrade Malenkov G. M. Please, be interested. L. Beria. 16/V 1944."

Ibid., l. 117.

III

From: G. F. Aleksandrov, head, Department of Propaganda and Agitation, AUCP(b) Central Committee.
To: G. M. Malenkov, secretary of Central Committee.
19 May 1944

In pursuance of your instruction, comrade Egolin (head of the Literature Section) and I read the novel by K. Gamsakhurdiya [...] Novels by Gamsakhurdiya did not receive a majority of the votes (ten of 29). It appears that none of the members of the Stalin Prize Committee read the novels by Gamsakhurdiya, besides A. N.

Tolstoy, who proposed to award the prize to him [...][5] I regard the decision of the Stalin Prize Committee in literature to turn down the novels of Gamsakhurdiya as erroneous. It seems appropriate to raise again the question of awarding the prize to historical novels by Gamsakhurdiya and discuss it on the Stalin Prize Committee.

Ibid., l. 111–112.

IV

Handwritten note by G. F. Aleksandrov dated 5 October 1944.

Attn: comrade Ergolin. Please, give attention to the discussion of Stalin Prizes.[6]

The note of the addressee in ink: "Had a talk with comrade Khrapchenko, 26 March1945."

Ibid., l. 121.

V

Typewritten note by A. M. Ergolin dated 18 August 1945.
To be shelved. It is brought to the notice of the Stalin Prize Committee (comrade Khrapchenko) that the works of K. Gamsakhurdiya need to be considered once again.[7]

Ibid., l. 122.

[1] Novels *David Stroitel'* (David the Builder) and *Desnitsa velikogo mastera* (The Hand of the Great Master) were nominated for the Stalin Prize in 1944.
[2] L. P. Beria.
[3] The letter was written in Moscow.
[4] The prizes for 1943 were not awarded in May 1944 as expected, probably because Stalin was very busy as the supreme commander-in-chief preparing offensive actions in Byelorussia that started on June 23. As known, Stalin always took part in the final discussions of candidates for the prizes in science and literature.
[5] A slip of the pen. It follows from doc. 5.25-I that A. N. Tolstoy objected to awarding the prize to Gamsakhurdiya.
[6] The style of the author.
[7] The name of K. S. Gamsakhurdiya was not mentioned in the decisions on awarding Stalin Prizes for 1943 and 1944. Obviously, Stalin had the last word. Khrushchev said at the meeting of the Presidium of the CPSU Central Committee (25 April 1963): "Stalin did not like Gamsakhurdiya."*

*Arkhivy Kremlya. Presidium TsK KPSS. 1954–1964. T. 1 (Kremlin Archives. Presidium of CPSU Central Committee. 1954–1964. V. 1). M.: POSSPEN, 2003, p. 711.

Document 5.26

Extract from the address of A. V. Topchiev, chief scientific secretary of the USSR Academy of Sciences, at the meeting of Communist Party organization of its Presidium, 26 January 1950.

Last year, a professor and doctor of sciences received the Stalin Prize. His work was not discussed in the institute. Its director, a respectful member of our Academy, promoted consideration of the work on the Stalin Prize Committee, where it met with approval based on publications and a few reviews. The work was awarded with the prize but subsequent examination did not confirm its value, and the researcher himself recognized that the effect he had expected was not obtained. Who is to blame? Some would take the easiest way, that is to sack the man and send him to Kem.[1] We are not bold enough to admit to being guilty of not verifying the results and discussing them.

TsAODM, fond 992, opis' 2, ed. khr. 23, l. 10.

[1] Kem is a town in Karelia, one of the Gulag centers in the 1930s–1950s. The speaker hints at suggestions to sentence the researcher to exile for an experimental mistake. The hidden sense put by the speaker in the name of the town may be interpreted in a different way. A legend holds that the name originates from a round oath uttered by Peter the Great after a night he spent at the place, tortured by mosquitoes during his journey to the White Sea. By alluding to the coarse word, the speaker implied that some people wanted to see the unlucky laureate severely punished. In fact, punishment was due to the designers of the system of awarding prestigious prizes open for meddling from the outside. It is in sharp contrast with the situation in the Nobel organizations well protected from such interference by the Nobel Statutes, an unparalleled creation of Swedish academics [...]

Chapter 6 • 1948–1952

Apotheosis of the Ideological Clinch

In order to better understand sources of ideological cataclysms that shook the country during the last five years of Stalin's rule, it is necessary to turn back to the notorious "Klyueva–Roskin case". This case started a series of noisy propagandistic campaigns against "servility to Western science and technology", "antipatriots" etc. that involved an undeclared boycott of Nobel organizations.

In the late 1930s, the Moscow microbiologists N. G. Klyueva and G. I. Roskin synthesized a compound later referred to as "KR" preparation that suppressed the growth of cancer cells in experimental animals. The lack of funds, absence of necessary imported equipment, and inattention of scientific bureaucracy slowed down further studies.

The work remained inefficient until one of the friends and colleagues lent a helping hand. V. N. Viktorov had been at the head of provincial health department in Nizniy Novgorod where A. A. Zhdanov led the local Communist Party organization. Viktorov addressed a letter to Zhdanov, now an influential member of Politbureau and head of the Central Committee Secretariat, and told about difficulties facing the scientists.

Attached to the letter was a copy of Klyueva's application to A. I. Mikoyan, deputy chairman of the USSR Council of Ministers, dated from 22 December 1945, which remained unanswered. Klyueva wrote about "evidence that our method is used in the USA that outstrips us [...] in prejudice of interests of the USSR; neither we nor the country can be indifferent to priority matters; we have to abstain from replying to the US National Institute of Cancer and publishing our data".

Klyueva told how Americans came to know about their research. "One of us (G. I. Roskin) published a short note in our scientific journal distributed abroad in 1940. In September 1945, the National Institute of Cancer informed us via the USA embassy that they found Roskin's work very interesting and had initiated similar studies at a larger scale; they therefore wanted us to provide more details about our research and consult their further work on the development of the agent of interest."[1]

This information was met with due attention in Central Committee headquarters. On 3 April 1946, the Secretariat instructed the minister of public health "to consider the letters of comrades Victorov, Klyueva, and Roskin, take necessary measures, and report the results to the Central Committee".[2] The extract from the resolution was sent to G. A. Mitirev, the minister, and A. N. Poskrebyshev, Stalin's secretary.[3] In means that Stalin was from the very beginning aware of the discovery and could follow the progress of research.

On the 22nd of April G.A. Mitirev reported to the Central Committee that in the Institute of Epidemiology and Microbiology a special laboratory is in the process of creation. In the Academy of Science a new special department is supposed to be created, the head of the new department is Professor Klyueva. The question about the purchase of the new equipment is considered at the same time.[4]

In the summer, the US ambassador and several American scientists including Z. A. Vaksman, the future Nobel laureate, visited the laboratory of Klyueva and Roskin. Soviet authorities allowed these visits (unusual at those times). Relevant documents were thoroughly analysed by V. D. Esakov and E. S. Levina.[5]

In the same year (1946), the USSR Academy of Medical Sciences published the book by Klyueva and Roskin *Biochemistry of Malignant Tumours* with a circulation of 10,000 unprecedented for scientific monographs. Permission for the publication was given by the Central Committee Secretariat on 18 February 1947. Simultaneously, the Department of Propaganda and Agitation was instructed to "organize reviews of the book in scientific and public press."[6]

A month before, a popular scientific magazine in France published the article of both scientists about "KR" preparation[7] (January 1946). There is no doubt that the manuscript was sent abroad legally, either by authority of the Department of Science of the AUCP(b) Central Committee or on its initiative. This inference is indirectly confirmed by the letter from Professor S. Johansson of Heteborg addressed to Klyueva and dispatched via the Swedish embassy in Moscow and VOKS. Had the publication of the article been unauthorized by Soviet authorities, VOKS would have acted quite differently, and the letter would have never reached Klyueva...

The period from late 1946 to early 1947 was like another "honeymoon" for the married couple. Nina Georgievna Klyueva was in the good graces of ideological services that were preparing her for the role of "a true daughter of the Soviet people". On 9 February 1947, she was elected deputy of the RSFSR Supreme Council from the Frunzensky constituency, Moscow, with its numerous clinics and other medical facilities. The official election campaign started on 2 January, which means that her candidacy was approved by Communist Party organs not later than the second half of December 1946. During registration of Klyueva as a candidate in the beginning of January 1947, members of the local election committee characterized her as a daring innovator.[8]

All of a sudden, the good-natured "flirt" of Central Committee ideologists with the scientists gave way to "sacral sacrifice". Immediately after the praising article in *Pravda*, the leading newspaper of the AUCP(b), (24 March 1947) the two scientists were accused of antipatriotism, neglect of interests of the Soviet people, and disclosure of their discovery to Americans.[9] However, such an abrupt reversal seemed surprising only to those unfamiliar with the ways of the "Kremlin court" and its insidious sovereign. The change could already be anticipated in winter 1947 as illustrated by the attempt of Sholokhov to obtain the permission to visit "America or England".

The writer applied for permission to Zhdanov who "did not advise going to these countries". In half a year (29 July 1947), Zhdanov wrote to Stalin that "subsequent events" confirmed the correctness of his [Zhdanov's] advice (see doc. 10.5)...

On 25 March, i.e. the next day after the panegyric publication in *Pravda*, Zhdanov wrote to Stalin reporting fulfillment of his instructions to organize "courts of honour". In another three days (28 March), the USSR Council of Ministers and the AUCP(b) Central Committee adopted the resolution "On courts of honour in USSR ministries and central departments". One paragraph of the resolution stipulated that "courts of honour must be first organized within two weeks" in three ministries including the Ministry of Health.[10]

The court of honour in this ministry was the first in the country to consider the case of "antipatriotism" and N. G. Klyueva and G. I. Roskin turned out to be the first accused. Court sittings lasted over half a month and were strictly watched over by the AUCP(b) Central Committee as appears from the "Statement" to the court written by Zhdanov and approved by Stalin (see Appendix 1). The court found the scientists guilty of "antis-tate" and "anti-patriotic" actions and "publicly rebuked them."[11]

Zhdanov was the author of the resolution on the courts of honour and the closed letter of the AUCP(b) Central Committee concerning the "Klyueva–Roskin" case (16 June 1947),[12] but Stalin was the true inspirator of the campaign against the country's scientific community.[13] Since early 1946, Stalin pursued the course toward suppressing scientific thought and confining it within the walls of secret research institutions. This is what he stated a year later at the meeting of Politbureau held on 11 June 1948: "The 'Klyueva–Roskin case' showed that some of our scientists lack the feeling of national honour and patriotism. There are talks about 'internationalization of science' [...] The idea of 'internationalization of science' is an idea of spies. We must beat the Klyuevas and Roskins."[14]

It is not for nothing that Stalin named the scientists in plural. Researchers in all fields of knowledge were "beaten" on a mass scale. They were accused of "antipatriotism" in hundreds of cases that were not infrequently used to pay back for personal offence, to ruin an opponent etc.[15] Frightened authors crossed out references to foreign publications in manuscripts, excluded co-authors with non-Russian names, stopped to exchange reprints with their colleagues abroad.[16] In other words, Stalin[17] initiated the nationwide "witch-hunt."[18]

Speakers at numerous meetings held in connection with the "Klyueva–Roskin case" and organization of "courts of honour" proposed to classify all research and developments as secret disregarding the inevitable loss of priority of discoveries. M. A. Sadovsky, future member of the USSR Academy of Sciences and deputy director of the Institute of Chemical Physics in 1947, said: "Foreign science is 100% hostile to us, and we must work to help our country without thinking of priority abroad. Today, we must regard any work as secret. Striving for priority may prove logically dangerous if underlain by the subconscious thought that 'my priority' is at the same time the priority of Soviet science. It should be borne in mind that positive results of recognition of our priority abroad are reduced to nought by the use which capitalist hosts of foreign countries make of the discoveries of Soviet scientists."[19]

It is appropriate to recall in this context that one of the victims of ideological campaigns was young V. L. Ginzburg, now member of the USSR Academy of Sciences and Nobel Prize winner (2003). The *Literaturnaya gazeta* published an article entitled "Against servility" (4 October 1947) mentioning Ginzburg along with A. R. Zhebrak, a long-standing opponent of Lysenko who had to appear before a court of honour in accordance with the decision of the AUCP(b) Orgbureau. Ginzburg was charged with "concealing discoveries of Soviet science" in his two papers published in academic journals. In other words, the newspaper incited authorities to launch one more shameful campaign against scientists (see note 3 in doc. 6.38 and footnotes**** and***** therein).

By 1948, courts of honour became an efficacious tool for intimidation and oppression. Designed originally to pursue scientists and technicians, they gradually spread influence farther over the sphere of arts and culture.[20]

The early signs of the forthcoming ideological campaign "against antipatriotism" can be traced to 1936. A signal to persecute one or another person was usually given by a publication in *Pravda*, the leading newspaper of the AUCP(b) Central Committee. Such was the case of N. N. Luzin, a world-known mathematician and member of the USSR Academy of Sciences, accused of "servility to foreigners".[21] At that time, however, the anti-intellectual component of repressions was not yet as distinct as in the late 1940s, disguised by mass political arrests and executions.

In 1949, the campaign against "antipatriots" (now called "cosmopolitans" and "vagrants without passport") was paralleled by another one having the purpose to classify all previous scientific publications (even those available in foreign libraries).[22] Exchange of scientific literature was substantially restricted;[23] in particular, it was impossible to find publications of the Nobel Foundation in public libraries as appears from the correspondence between VOKS and its representatives in Stockholm (docs. 6.5–6.6).

Foreign researchers could not receive Soviet scientific journals or even popular science ones designed for lay readers (docs. 6.7–6.8). The fear of imputation of "antipatriotism" after the "Klyueva–Roskin case" prevailed over everyone from leading scientists of the USSR Academy of Sciences to VOKS officers, whose official duty was to develop and maintain scientific and cultural relations with those circles abroad that showed interest to achievements in the Soviet Union.

The absurd situation in which scientific materials that had passed censorship and appeared in the open press were classified as secret was further aggravated by the impossibility to refer to them in new publications; such references could give cause for accusation of espionage,[24] a most serious criminal offence of those times. Such was the everyday reality of the late 1940s in which many eminent Soviet scientists had to work.

Paleontologist A. G. Vologdin, corresponding member of the USSR Academy of Sciences, was arrested in the framework of the fabricated "case of professor geologists" on 31 March 1949 (see note 39 in this chapter). The arrest of Vologdin and some other scientists was authorized by the Politbureau at the meeting of 30 March chaired by Stalin. Vologdin was interrogated at the USSR Ministry of State Security on 25 April. The interrogator: "It appears from your correspondence including fifteen letters from Americans that your scientific works were published abroad. How did you manage *to give them a going* [italicized by the author] over to the foreign press?"[25] It was also maintained that the Ch. D. Walcott bronze medal awarded to the scientist for outstanding discoveries in paleontology by the US National Academy of Sciences in 1947 and "the dollars were a reward for the work done for American intelligence services"...[26]

Another investigator interrogated the known paleontologist B. K. Likharev arrested together with Vologdin. He brought an even more absurd accusation of "subversive activities in the 1930s–1940s and transmitting to Americans during twelve years scientific data on paleontology and stratigraphy of different regions of the USSR, including those that are *presently* [italicized by the author] regarded as secret."[27] It should be emphasized that all these data were published in the Soviet scientific press after they passed censorship and were available to everybody via subscription or retail trade.

The lack of Soviet scientific literature abroad puzzled foreigners. John Bernal, member of the Communist Party of Great Britain and vice-chairmen of the British–Soviet Friendship Society wrote to S. I. Vavilov, president of the USSR Academy of Sciences, in 1948: "Virtually all scientific journals that we had received [...] by way of exchange ceased to come since the beginning of 1947, for unexplained reasons. Letters from British institutions to Soviet research institutes [...], same as addresses to individual scientists, remain unanswered."[28] Vavilov could not say anything intelligible in reply. "As far as I understood from your letter the main difficulty is the delay in delivery of publications in the course of exchange", which Vavilov had to ascribe to "an unassignable cause."[29]

Bernal should be given credit for understanding the situation of the distinguished Soviet scientist under the weight of the cruel regime. Two years later, he expressed condolences on the premature decease of Vavilov in the obituary published in *Nature*.[30]

Published scientific works including books by foreign authors were subject to strict ideological control. The attention of ideologists was focused on biological literature in connection of persecution of geneticists initiated in 1948. Not infrequently, charges were underlain by vulgar political considerations. For example, genetic studies at the Institute of Roentgenology and Radiology, USSR Ministry of Health, were found "harmful and apolitical."[31] The classical book of E. Schrödinger, Nobel Prize winner and foreign member of the USSR Academy of Sciences, translated into Russian in 1947[32] was described as "idealistic in its essence" for the proposed molecular approach to the study of genes in the living cell. Many key figures in the "Inostrannaya literatura" state publishing house were dismissed from their positions.[33]

Alternating sequences of ideological cataclysms in the sphere of science initiated by the AUCP(b) Central Committee or researchers themselves were especially common in biology (e.g. the so-called "*lysenkovshchina*", after the name of Lysenko, the most notorious figure of that time). They created favourable grounds for a variety of retrograde speculations, such as the pseudoscientific theories of G. M. Bosh'yan and O. B. Lepeshinskaya popularized in numerous publications (see notes 1 and 6 in doc. 6.39). In this case, the exposure of ignorance was greatly facilitated by the indifference of authorities to these studies (in contrast to the unreserved support they provided to the "people's academic" Lysenko).

Professor V. N. Orekhovich, director of the Institute of Biological and Medical Chemistry, USSR Academy of Medical Sciences, was one of the most ardent opponents of the pseudoscientific arguments of Bosh'yan. In summer 1950, Orekhovich addressed a letter to the AUCP(b) Central Committee (doc. 6.39.I). Somewhat later, he was supported by Professor S. R. Mardashev. The two scientists acted in defiance of the opinion of E. I. Smirnov, minister for health, and the Presidium of the Academy of Medical Sciences (doc. 6.39-III). The collapse of the pseudoscientific theory of Bosh'yan put an end to the carrier of Lepeshinskaya, whose experiments designed to prove the cell-free structure of living matter turned out to be a bluff.

It should be recalled that all the above events date to the dark, late period of Stalin's rule that was especially difficult for basic science and its actors. Their miraculous victory over obscurants was actually "a ray of light in the land of darkness".

Ideological attacks on biological sciences were followed by similar fault-finding in physics. The idea to convene the All-Union workshop of heads of physical departments of universities and higher education institutions with the participation of the Division of Physico-Mathematical Sciences, the USSR Academy of Sciences, emerged in January 1949. The idea was based on the assertion that "physics is taught in many higher education institutes without due regard to dialectic materialism" and that "certain books of bourgeois physicists are translated without critical remarks." Under "bourgeois physicists" were meant the pride of contemporary physics: Bohr, Heisenberg, Schröedinger etc. (doc. 6.9). The eminent American physico-chemist Linus Pauling, Nobel Prize winner in chemistry (1954), was described as an adherent of Mach's philosophy and subjective idealist.[34] He was criticized for the resonance or hybridization theory of chemical bonds in aromatic compounds efficiently employed for the prediction of properties of organic substances.[35]

Fundamental principles of modern physics were "disproved" at the primitive level in the form of critical attacks on prominent Soviet scientists, such as members of the USSR Academy of Sciences A. F. Ioffe, A. F. Kapustinsky, T. N. Kravets, S. Z. Roginsky, and Ya. K. Syrkin. At the Communist Party meeting in the Presidium of the USSR Academy of Sciences (4 July 1952), the secretary of the Party organization accused them of "idealism" for the arguments in favour of "identity of substance and energy, conversion of substance to energy, energy as substance, etc."[36] The speaker openly criticized not only Soviet scientists but also drove at Albert Einstein and his theory of relativity with the famous formula $E = mc^2$, one of the symbols of the 20th century.[36a] In a few days, critical articles against Einstein and his teaching appeared in newspapers (see below and note 38 in this chapter).

Criticism of physics and physicists soon extended to philosophy and natural sciences at large. The authoritative academic journal *Voprosy filisofii* was among the first to be affected. Its editor, Nobel Committee chief V. M. Kedrov, a widely known historian of science, was substituted by a party functionary, D. I. Chesnokov, who suggested the dismissal of his deputy, M. E. Omel'yanovsky, specialist in philosophical aspects of quantum physics. Chesnokov wrote to the secretary of the AUCP(b) Central Committee, G. M. Malenkov, that "the views of Omel'yanovsky leave no hope that he will be able to protect materialistic principles in natural sciences."[37] Finally, certainly A. A. Maksimov was appointed member of the editorial board instead of Omel'yanovsky; very soon he became notoriously known as the author of the ignorant article "Against the Einsteinian doctrine in physics" that aroused indignation and protests of Soviet physicists and led to a scandal very unusual under Stalin's regime.[38]

Attacks on intellectuals during the last years of Stalin's life revived seemingly forgotten accusations in "trotskyism" and "non-information of the trotskiyst past" of friends and relatives widely practiced in the 1920s–1930s. Members of the Communist Party organization at the Institute of Organic Chemistry, the USSR Academy of Sciences, took delight in verbally tormenting their colleague R. Kh. Freidlina, a leading specialist in chemistry of metallo-organic compounds, after she informed the Party bureau about the arrest of her husband by the USSR Ministry of State Security for his speech at a meeting of the Young Communist League a quarter of a century ago (doc. 6.10; see also note 34 in this chapter and footnote**).

Contacts with foreign missions were severely persecuted even if a scientist had nothing to do with state secrets (doc. 6.11). Actions formerly approved by authorities were subject to severe punishment (see, for instance, doc. 6.38). Not infrequently, a person had to bear responsibility for the actions of his predecessor. Stalin encouraged such accusations and openly told about it, taking no trouble to veil his cynicism.[39]

No wonder, the mass media kept silent about the "Klyueva–Roskin case" in such a situation. Hundreds of thousands of people turned out to be involved in the discussion of this case in the second half of 1947 and the beginning of 1948 but any mention of it in the press was prohibited under the threat of confiscation of the edition (doc. 6.12).

Simultaneously, authorities decided to cut off one more information channel: Russian language broadcasts from abroad. The USSR Council of Ministers instructed the Ministry of Communication to implement an expensive project "to jam anti-Soviet radio broadcasts from the UK and USA" (19 April 1949). In February 1950, as many as 190 big radio stations (total power up to 1,800 kW) and 210 smaller ones were involved in the fulfilment of the "governmental program."[40]

People visiting foreign countries to perform official duties, including newspaper reporters, lived under constant fear of being prosecuted; some of them failed to endure the stress and committed suicide. P. N. Pospelov, editor of *Pravda*, mentioned in his notebook (30 March 1948) the suicide of the known journalist M. R.

Galaktionov who developed persecution-mania stemming from the thought that "people changed their attitude toward him after he returned from America."[41]

Ideological services practically reduced to nothing contacts with the Nobel organizations. In 1945, invitations from the Nobel Committees were still received by certain addressees (see Chapter 4), and their nominations reached Stockholm (see doc. 4.11). The situation changed dramatically in the post-war period. In 1949, the Nobel Committees in physics and chemistry addressed four Soviet nominators (including two future Nobel laureates P. L. Kapitsa and N. N. Semenov). The letters were forwarded via the Swedish embassy in Moscow but were shelved in the USSR Ministry of Foreign Affairs (docs. 6.13–6.15). The ministry informed the Presidium of the USSR Academy of Sciences about the receipt but the leadership of the Academy refused to accept the letters under the pretext that "the Academy was not going to nominate candidates for the Nobel Prizes" (doc. 6.15).

The letter of invitation sent by the Norwegian Peace Prize Committee to the University of Latvia on 15 December 1947 suffered a similar fate. It was dispatched directly to Riga, from where it was immediately forwarded to the USSR Ministry of Higher Education and thereafter to the USSR Ministry of Foreign Affairs. It was filed in the archive of the Ministry on the day of reception.

The anti-Nobel movement in the Soviet Union gained momentum in the post-war time. Even East European countries where ideological atmosphere was formed under the influence of Soviet Party leaders tried to maintain relations with the Nobel Committees. Suffice it to say that in 1950 the Nobel Committees received 13 nominations from Poland, Hungary, and Rumania and none from the Soviet Union (as in the previous three years). Authorities in these countries generally showed more liberal attitudes to scientists than in the USSR (see note 64 to this chapter).

The Soviet press of that time avoided mentioning the Nobel Prize among the awards of a person, even when it positively characterized him. An article published in the weekly *Novoye vremya* in January 1949 under the title "The book about the policy of atomic imperialists" was actually a review of the recently published monograph *Military and political consequences of the use of atomic energy* by Patrick Blackett, later elected honorary member of the USSR Academy of Sciences.[42] The review ended as follows: "Blackett is neither Marxist nor communist. He is a big natural scientist who tried to look honestly into imperialist policy, which brings new disasters to mankind, science and culture."[43] The author does not say a word about the fact that Blackett was awarded the Nobel Prize in physics only a month and a half before the review was published.

One of the elements of the "witch-hunt" in the Soviet Union was the overall check of political reliability of all foreign members of the USSR Academy of Sciences and their attitude to the Soviet Union during World War II.[44] In 1947, such testing encompassed biochemist Hans von Euler-Chelpin, Nobel Prize winner in chemistry for 1929, and geneticist Niels Hermann Nilsson-Ehle. These two Swedish scientists were elected foreign members of the USSR Academy of Sciences in 1927 and 1932 respectively. Extensive bureaucratic correspondence between the Academy, the USSR Ministry of Foreign Affairs, and the AUCP(b) Central Committee (docs. 6.18–6.24) ended in the proposal to deprive "Swedish biochemist Hans Euler of his membership of the Academy" (doc. 6.24).

It should be noted that the two scientists were actually known as Hitler sympathizers, and their membership of the USSR Academy of Sciences did not give it much credit. But a similar check-up involved a few other Swedish researchers that were not members of the USSR Academy of Sciences. One of them was paleontologist Erik Stensi, member of the Royal Academy of Sciences. The USSR Ministry of Foreign Affairs instructed its representatives in Sweden to collect relevant information, and the man was characterized as an "apolitical bourgeois scientist" who "never made public political statements" and "spoke well of Soviet scientists". (doc. 6.25)

There may be only one sound explanation for the motive behind this large and seemingly purposeless work. Stensi as a member of the Royal Academy of Sciences participated in its annual meetings convened to elect Nobel laureates in physics and chemistry. It may be supposed that ideologists of the AUCP(b) Central Committee collected data on all people that had a vote in order to use them later one way or another for propaganda in the USSR and abroad. Also, this work could be related to the institution of alternative international prizes in the Soviet Union (see Chapter 5). Documents dating from the late 1940s, mentioned other Swedish professors, besides Stensi. One of them was Sven Hedin, member of Swedish Academy that awarded Nobel Prizes in literature. He was also a foreign member of the USSR Academy of Sciences and attracted attention from Soviet ideologists (see, for instance, doc. 6.19 and note 3 to it, Appendix 3, etc.).

That was the time of growing confrontation between recent allies in the war against Nazism underlain by imperial ambitions of the Soviet leadership. An example is an archive document concerning a meeting of A. A. Zhdanov, secretary of the AUCP(b) Central Committee, with the officials responsible for the organization of the 1st International Congress of Slavists that had to be held in Moscow in 1948 (doc. 6.26).

Lofty disdain of everything one way or another related to Western culture cultivated by Soviet authorities naturally extended to the Nobel movement. In the first post-war years, anti-Nobel propaganda was largely a matter of secret office correspondence. In 1949, it became a theme of articles in the public press as a component of unprecedented campaign designed "to fight cosmopolitanism and servility to the West".

The campaign was launched by the editorial in *Pravda* of 28 January 1949 entitled "On one anti-patriotic group of theatre critics."[45] The article called to "purify the atmosphere [...] from antipatriotic philistines" and gave rise to propaganda of xenophobia and narcissism[46] that lasted several months and encompassed not only mass media but also public meetings inspired by authorities. Practically all epoch-making discoveries were ascribed to Russian scientists and inventors while the names of true pioneers were barefacedly defamed. It was then that the protective reaction of ordinary people gave birth to the biting phrase: "Elephants originate from Russia"...

The attitude toward many representatives of Western culture was revised. Names of foreign writers and playwrights well-known to Soviet readers and spectators disappeared from publishing plans and playbills. They were all of a sudden labelled as "advocates of imperialist banditry" and their works as decadent. Such were Nobel laureates Maurice Maeterlinck and Bernard Shaw, among others.[47]

Another aspect of nationalistic psychosis was renaming hundreds of settlements at the former autonomous territories in 1941–1944 (Volga-German Republic, Kalmykia, Crimea, republics of Northern Caucasus). This process affected even natural geographic objects. The Presidium of the USSR Academy of Sciences received a letter signed by K. E. Voroshilov, deputy chairman of the USSR Council of Ministers, (31 May 1950) asking for a list of objects to be renamed.[48]

Propaganda of hatred of all elements of Western culture began to influence the everyday life of all citizens. Tangos, foxtrots, and even waltzes popular among young people were no longer played at dancing rings substituted by long-forgotten polkas, polonaises, and other ball-dances.[48a] Jazz bands were called orchestras. Hence the popular saying: "He who loves jazz today will be a traitor tomorrow"... French rolls in bakeries turned into "town rolls", a habitual "match" into "competition", etc.

The same tendency manifested itself in the sphere of education. The Joule–Lentz physical law, formulated in 1841 by the former and next year by the latter author, was taught at schools as the Lentz–Joule law because J. P. Joule was an Englishman and E. H. Lentz a German born in the Russian Empire and elected to the St. Petersburg Academy of Sciences in 1830. The teachers did not say that Lentz spent his last years abroad and died in Rome. The *Great Soviet Encyclopaedia* was equally silent (see second edition, V. 24) although the volume was approved for printing after Stalin's death (8 December 1953).

The idiotic situation of those days was reflected in an anecdotic discussion that took place at a meeting of the Communist Party organization of the public and political journal *Novoye vremya*. It was initiated by the remark by one of the participants that the journal had missed the opportunity to keep pace with time and use the term "cosmopolitanism" in its earlier publications (doc. 6.27). This word had not been in common use and remained practically unknown to the wide public until it appeared in an article published in *Pravda*.

The campaign "against cosmopolitanism and servility to the West" in the mass media was preceded by the closed sitting of Orgbureau of the AUCP(b) Central Committee on 24 January 1949. One of the main issues discussed by its participants was the preparation of the second edition of the *Great Soviet Encyclopaedia* (GSE). The documents submitted for the discussion contained the draft resolution of the AUCP(b) Central Committee endorsed by D. T. Shepilov, deputy head of the Department of Propaganda and Agitation, and S. I. Vavilov, president of the USSR Academy of Sciences, on 5 January. Certain phrases of this document were later almost literally reproduced in mass media.

A characteristic paragraph is cited below. The authors of the document noted that the first edition of GSE "contained serious theoretical and political mistakes" and continued: "Volumes of the encyclopaedia very poorly illustrate the role of Russian and especially Soviet science in the development of world science, ignore the legacy of outstanding Russian scientists and cultural workers. Many articles of the encyclopaedia are permeated with the spirit of cringing and servility to values recognized as such abroad [...]"[49]

One of the victims of the ideological campaign "against cringing and servility to values recognized as such abroad" was eminent physico-chemist A. N. Frumkin who had to retire from the position of director of the Institute of Physical Chemistry, the USSR Academy of Sciences. The ridiculous charges brought against him were in sharp contrast to the fact that both compatriots and foreign colleagues had many times nominated him for the Nobel Prize in chemistry since 1946 (see Chapters 4 and 8). D. T. Shepilov and Yu. A. Zhdanov, heads of the Department of Propaganda and Agitation and the Sector of Science respectively, wrote in a letter to G. M. Malenkov, secretary of the AUCP(b) Central Committee, that works by Frumkin "only supplemented data, views, and theories already published abroad."[50] and that he "made antipatriotic mistakes" in the article "Physical chemistry" for GSE.[51]

Some of the charges brought against Frumkin in 1949 were made open to the public in a memorial volume of articles published by his colleagues 40 years later. Certainly, accusations were not formulated in the AUCP(b) Central Committee. They were proposed by Frumkin's foes in academic circles, who tried to use the opportunity to pay him back for personal offence, either true or imaginary. One "sin" of Frumkin and his co-author was that they "did not mention A. M. Butlerov as the author of the theory of chemical structure and underestimated the role of M. V. Lomonosov as the founder of physical chemistry" in their article "Physical chemistry", written for GSE.[52]

The future Nobel laureate in chemistry N. N. Semenov, member of the USSR Academy of Sciences, was also ranked as "cosmopolitan". His "servility to the West" consisted in that he dedicated his monograph *Chain Reactions* published in 1934 to the memory of Svante Arrhenius and Jacobus Van't Hoff...[53] The ideological mistakes of Semenov were discussed at a meeting at the Institute of Chemical Physics, the USSR Academy of Sciences, which he had founded and headed. The secretary of the Communist Party organization said at the meeting that "the commission set up at the institute applied to the Moscow State University (MGU) and receives relevant materials."[54]

Two years later (in Spring 1952), the same group of conservative professors from MGU assisted by Communist Party organs managed to organize another discussion of the theory of chain reactions. The discussion was advertized as a public dispute between the author of the theory and the university professor N. S. Akulov. One of the participants, Professor G. I. Barenblatt, recollected that one of Akulov's supporters tried to strengthen his political argumentation by alluding to the recently published paper of V. I. Gol'dansky, future member of the USSR Academy of

Sciences, where "the closest and favourite disciple of N. N. Semenov" many times cited S. Hinshelwood.[55] Such citation was ranked as criminal "servility to foreigners" unworthy of a Soviet scientist. It is worthwhile to note that Hinshelwood twice nominated Semenov for the Nobel Prize and shared it with him in 1956...

When no apparent candidate for public punishment could be found in a scientific organization, Party organs demanded to look him (or her) up and expose him (or her) as a cosmopolitan. The statements of the then party leader at the Mathematical Institute (1951), the USSR Academy of Sciences, look very much like a malicious caricature to the modern reader. But there were many such ones at those times, as appears from the minutes of a meeting of the Institute's party organization preserved in the archives (doc. 6.28).

Attacks on individual members of the Academy in the framework of the campaign "against cosmopolitanism" were soon extended to the whole of it. The Academy was about to lose one of the basic provisions of its Charter, i.e. regular elections of new members on an alternative basis.

After elections in November 1946, the next were scheduled for autumn 1949. Preparation took its normal course, and in due time (16 July) the USSR Council of Ministers issued decree No. 3152 by which 20 new vacancies for full members and 30 for corresponding members were open. Based on this decision, the Presidium of the USSR Academy of Sciences set up an expert commission in each division and started to accept applications from the contenders.

The first alarm signal was the delay in the approval of the membership of election commissions by the Sector of Science, AUCP(b) Central Committee, from which it was transmitted to the Secretariat only in December 1949, i.e. after the scheduled time of elections. The Sector of Science introduced many changes and proposed to postpone the elections until the end of March 1950. In particular, it was proposed to exclude N. N. Semenov from the expert commission in chemistry and A. F. Ioffe from the commission in physics along with several other scientists whose candidacies proved unacceptable for ideological considerations.

New delays followed before the Secretariat considered the list of pretenders for election and gave permission to publish it (28 April 1950). In a week (5 May), eight more vacancies were open. On the same day, almost all members of the Politbureau signed this new document (doc. 6.29).

It had to be approved by Stalin, but he never gave his consent and the next elections took place only on 23 September 1953, i.e. half a year after the death of the dictator and twice later than stipulated by the Charter of the Academy.

The elections of 1949–1950 were frustrated for two reasons.

First, Stalin was discontented with the roll of candidates seeking to fill the available vacancies because some of them seemed unacceptable in the context of the campaign "against cosmopolitanism" that he had recently initiated. However, this was not the main cause because a new revision of the list was after all just a matter of time.

The problem was rooted deeper than that. Elections on an alternative basis were incompatible with the totalitarian regime. Despite the strict control of Party organs over this rudiment of democratic freedom, it opened up the possibility of expressing free will unacceptable for the Soviet leaders. A ballot was considered valid only if the voter crossed out one or more alternative candidates that might happen to be those approved by the authorities.

The dictator had tolerated this tiny element of freedom for a quarter of the century but did not wish to have it any longer and decided to put an end to it once and for all...[56]

The series of repressions against the most prominent intellectuals of the country in the post-war years was a permanent concern for the publishers of the second edition of the *Great Soviet Encyclopaedia.* The first volume was out in 1950, and the next were naturally designed much before they were published. There was always a possibility that someone whose name was included in the word-list would all of a sudden be described as "an enemy of the people".

One such case dates to the beginning of 1953, when several eminent representatives of medical science were arrested and interrogated in Moscow (see Chapter 7). V. F. Zelenin, a reputed clinician and the founder of the Soviet school of clinical electrocardiology (widely known among the lay public owing to "Zelenin's drops", an efficacious medicine for the treatment of heart neurosis) was arrested on 25 January 1953. His biography was to appear in volume 16 of GSE, approved for printing on 18 October 1952. In the last moment, page 616 was reprinted and the note about Zelenin was substituted by a new one, the 12-line description of *Zelyonaya laygushka* (Green frog).

In the heat of ultrapatriotic passion, the Nobel organizations proved the best objects for propagandistic attacks. They were accused of prejudice, russophobia, serving American capital, and other mortal sins. Past "offences" were recalled and current activities of prize-giving institutions criticized without proof. Almost each new Nobel winner was labelled as "a servant of new World War mongers", "a puppet of American imperialism", etc. No distinction was made between politically active scientists and those staying far from politics. Simultaneously, the founder of the Nobel Foundation and his assistants were defamed. Specifically, Soviet propaganda described Ragnar Sulman as an agent of Dupont's military-industrial concern (see note 5 to Appendix 4). Sulman was a close associate of Alfred Nobel during his last years and executor of his will, and greatly contributed to its fulfilment (together with Alfred's nephew Emmanuel, head of the Nobel Brothers Petroleum Company in Russia).

Some Nobel laureates regarded by the AUCP(b) Central Committee as "loyal" and beyond criticism did not avoid persecution of Soviet ideologists. One of them was nuclear physicist Cecil Powell, awarded the prize in 1950 (see also Table 6.2 and the accompanying text in the present chapter). Powell held left views and successfully combined scientific and public activity as the vice-president of the national Peace Committee, and a member of the Bureau of the World Peace Council. For mere negligence, his name was mentioned in an article published in the *Literaturnaya gazeta,* among other Nobel laureates of 1950 who "received the prizes from the same murderous hand that drops bombs on defenceless women and children in Korea" (Appendix 3). The AUCP(b) Central Committee immediately adopted a resolution in the strength of which workers for the newspaper were either rebuked or sacked (docs. 6.30–6.33).[57]

However, these strict measures were a response to a marginal mistake and by no means led to cessation of the anti-intellectual campaign and attacks on the Nobel organizations. Another anti-Nobel libel appeared in the weekly *Novoye vremya* (Appendix 4) within three months after the aforementioned article.[58]

It was placed under the heading "Critique and bibliography" as a commentary on the book dedicated to the 50th anniversary of the Nobel Foundation published in Stockholm in 1950.[59] The author disparaged Nobel winners awarded over 50 years, nominees who did not receive the award, and people associated with prize-giving institutions.

The choice of the laureates is of interest. 24 of a total 247 Nobel winners were mentioned, i.e. every tenth one, of whom 21 were covered in mud. Most of them were Nobel laureates in literature (11, or each fourth of the 44 writers). Here are some epithets used by the author: "petty reactionary writer" Sully Prudhomme, "literary pigmy" Eugene O'Neill, "dirty decadent" Andre Gide, "cannibal nuclear warmonger" Bertrand Russell[60], etc.

The founder of the Prize is characterized as a multimillionaire, "satiated with wealth and power, flattery of parasites and ministers, sceptic and cynic despising people and human ideals".

Who, then, was the author of the article signed by a certain S. Maksimov? As mentioned above, the journal *Novoye vremya* was regarded in the Soviet Union and abroad as a semi-official edition. It frequently published articles that the AUCP(b) Central Committee deemed necessary to bring to the attention of the wider public, not via its official newspaper but through "an independent press organ" under which the journal was camouflaged. The authors of such articles were top Party and government officials that signed them by pseudonym (usually consisting of elements of the name or surname of the author), a tradition dated back to pre-revolutionary times.[61]

There was no "S. Maksimov" among the known international publicists of that time[62]; therefore, it was most likely the pseudonym of a high-ranking worker of the USSR Ministry of Foreign Affairs. This guess of mine was confirmed by N. M. Lun'kov, the attorney for the Soviet embassy in Stockholm between 1954 and 1957 (see doc. 7.37). Moreover, the papers with S. Maksimov's article cut out of the journal were attached to a note by A. N. Abramov, head of the 5th European Department of the USSR Ministry of Foreign Affairs concerning the work of the Nobel organizations (see doc. 7.2).

Visits by Soviet scientists abroad forbidden since 1946 were still restricted to rare trips to East European countries and participation in various propagandistic forums. Curiously, a group of specialists including A. M. Prokhorov, the future Nobel laureate in physics (1964), did not receive permission to take part in the international conference on radioscopy held in Amsterdam in 1950 (doc. 6.34).

The reason for the prohibition was typical of that time. The refusal was formulated in meaningless words and substantiated by the failure of the members of the delegation proposed by the USSR Academy of Sciences "to meet the existing requirements" and "the absence of other worthy candidates".

The same year 1950 was marked by the scandalous boycott of the International Mathematical Congress in Cambridge, Ma., held 14 years after the previous one in Oslo (1936), where it had been decided to hold the next congress in the United States in 1940. World War II frustrated this plan. The boycott was initiated by ideologists of the AUCP(b) Central Committee.[63] In order to sugar the pill for scientists from East European countries, a parallel congress was quickly (for three months) organized in Budapest at the same time as in Cambridge (27 August–3 September) (doc. 6.35) as if to emphasize the division of the civilized world into two opposite scientific as well as political camps (see also note 30 to Chapter 12).

The leading scientists of the Cambridge Congress accepted the rules of the game and sent a welcoming address to their East European colleagues, which was reciprocated. Several Soviet participants signed the reply letter. This unauthorized action became subject matter of the discussion at a meeting of the Communist Party organization at the Mathematical Institute, the USSR Academy of Sciences (23 March 1951). The secretary of the organization emphasized that certain (unnamed) scientists "seek popularity abroad [...] by signing the welcoming address either to the Mathematical Congress in the USA or to one of the American mathematicians." Another speaker at the meeting mentioned "facts of servility to foreigners. One of them is the welcoming address sent to the Mathematical Congress in America."[64]

Alternating noisy campaigns of the "fight for peace" and "against warmongers" sometimes had results quite opposite to those they were designed for.

Here is a "strictly secret" episode illustrating the cynical duplicity of the organizers of these campaigns. N. A. Mikhailov, secretary of the Young Communist League, informed G. M. Malenkov, secretary of the AUCP(b) Central Committee, on 16 January 1952 about "a serious mistake of the editorial board of the newspaper *Komsomol'skaya pravda* that published the music and words of 'The song of peace champions', by A. I. Khachaturyan and S. G. Ostrovoy, 'popularizing bourgeois-pacifist views alien to our youth.'"[65]

However, all these manifestations of omnipotence of the state power could not totally suppress intellectual life in the country; its prominent actor continued to work despite isolation from contacts with other representatives of world science. A new discipline of experimental physics and technology (quantum electronics) arose in the Soviet Union at that time of new repression and ideological pressure. The epoch-making discovery that led to the creation of a self-sustaining molecular generator of radiation was reported at the All-Union Conference of Radiospectroscopy in May 1952. The report was delivered by A. M. Prokhorov (who had not been allowed by the AUCP(b) Central Committee to participate in the international meeting on radiospectroscopy two years ago) and his younger co-worker N. G. Basov.[66]

Table 6.1. Nominees from the Soviet Union by categories of Nobel Prizes in sciences for 1948–1952

Nominees and their residence at thetime of nomination	Year of nomination	Number of nominators	
		Total	Including USSR
Physics			
Veksler V. I. (1907–1966), Moscow	1948	2	—
- " -	1951	1	—
Kapitsa P. L. (1894–1994), Moscow	1948	2	
- " -	1950	1	
- " -	1953*	1	
Cherenkov P. A (1904–1990), Moscow		1	
Chemistry			
Semenov N. N. (1896–1986), Moscow	1948	1	—
-"-	1950	1	—
Braunshtein A. E. (1902–1986), Moscow	1952	1	—
Physiology or Medicine			
Timofeev-Resovsky N. V. (1900–1982)**	1950	1	—
Filatov V. P. (1875–1956), Odessa (1874–1954)	1950	1	2

* Kapitsa was nominated in 1952 but the nomination reached Stockholm after 31 January, and consideration of his candidacy was postponed until 1953, in accordance with the Nobel Statutes.
** In 1952, the nominee served term in a Gulag camp in the Urals (see note 79 of this chapter).

An equally important discovery was made on 26 July 1951. Excavations near Novgorod Veliky made by the archaeological expedition led by Professor A. V. Artsikhovsky brought about the first birch bark writing and opened for the world the source of previously unknown written documents dated from the 11th to the 15th centuries.[67] The discovery would certainly win the Nobel Prize had Alfred Nobel mentioned in his will history and archaeology along with physics, chemistry, and medicine.

Next year (1952), the young linguist Yu. V. Knorozov, a World War II veteran, became the first man to decipher the hieroglyphic writings of Maya. He was able to read manuscripts of the 12th to the 15th century saved from the Inquisition by a miracle when it was most active after the Spanish conquest of Mexico in the 16th century. Due to this work, totally new data became available to scientists about the ancient civilization unknown in the Old World until Columbus opened America...[68]

The innate talent of the people found a road to realizing the natural need for cognition of the unknown even in the most difficult situations. The world knew and appreciated the high level of science in the Soviet Union, despite the Iron Curtain. As shown in Table 6.1, Soviet scientists continued to be nominated for the Nobel Prizes despite their boycott by official authorities. The nominators were foreign specialists as in the two post-war years. No nominations of countrymen or foreign scientists came from the USSR.

It can be seen from Table 6.1 that almost all nominees were the same as in 1946 and 1947, viz. V. I. Veksler and P. L. Kapitsa in physics and N. N. Semenov in chemistry.

Veksler was nominated in 1948 by Professor I. Platt of the Rochester University, New York, specialist in nuclear and meson physics, and M. Siegbahn, Swedish Nobel winner of 1924. The latter nominated Veksler together with McMillan and additionally proposed Donald Kerst[69] in anticipation that the candidacy of Veksler would not find support from the Nobel Committee. Professor Kerst of the University of Illinois constructed the first circular accelerator of electrons (betatron) and suggested the colliding beam principle in 1940. In 1951, Veksler and McMillan were nominated by Professor J. Rotblat of London while Siegbahn proposed McMillan alone.

Kapitsa and Willem Keesom, a specialist in low temperatures from the Netherlands, were nominated by T. Svedberg, Nobel laureate of 1926 (as in the previous year). One more nomination for Kapitsa in 1948 came from another Nobel Prize winner Niels Bohr. Svedberg nominated Semenov and Hinshelwood for the Nobel Prize in chemistry. In 1950, the two Soviet scientists were again recommended by the same nominators who proposed their candidacies in 1946.

Paul Dirac nominated Kapitsa. He reminded about his detailed nomination of 1946 but drew attention to the altered situation in the nominee's country whereupon "I have presently no access to the recent data because contacts with the Soviet Union became difficult." Certainly, Dirac meant not only the tendency to make secret practically all results of research and restricted supply of Soviet scientific literature to foreign institutions in the late years of Stalin's rule (see this chapter and note 28 to it), but also disfavour of the nominee dismissed from all his official positions. Dirac concluded his 1950 nomination by emphasizing the fundamental importance of Kapitsa's brilliant experiments for the understanding of superfluidity and superconductivity and characterized them as a "first-rank achievement".

Dirac again mentioned Kapitsa's helium liquifier (as he did in 1946) that yielded first production in April 1934 and thereby created prerequisites for further progress in ultralow temperature research. Today, it may be noted that Swedish professors awarding Kapitsa in 1978 not only used the arguments of Dirac in support of outstanding scientific achievements of the Soviet scientist but also emphasized the value of his "fundamental inventions".

The nomination of Semenov by Cyril Hinshelwood in 1950 contained no reference to the political situation in the USSR and actually repeated his arguments presented four years ago. Hinshelwood himself was not nominated in 1950. Next year the situation changed to the opposite. Professor R. P. Linstead of London proposed the candidacy of Hinshelwood; it was again submitted in 1952 by Professor M. G. Evans of the University of Manchester.

In 1953, Hinshelwood was recommended by three nominators, viz. Linstead, R. Robinson of Oxford University (Nobel Prize winner in 1947), and S. Bottcher of Leiden University, Netherlands. The name of Semenov was absent on the list of nominees for three successive years. Meanwhile, there was every reason in 1953 that the two equally merited pretenders be included in one nomination to enhance the likelihood of their selection for the award.

Such nomination was undertaken by Professor L. G. Sillen of Technological Institute, Stockholm, who knew Russian well enough to understand scientific texts (see, for instance, doc. 7.3 and note 3 to it). In order to make his nomination more convincing, Sillen needed additional facts about the Soviet scientist and reprints of his recent publications absent in Stockholm libraries.

He applied to the USSR embassy in Stockholm for these materials and assistance in 1952. His application was sent to the USSR Ministry of Foreign Affairs, but Sillen did not receive any reply. A. N. Abramov, head of the Department of Scandinavian States, cynically blocked a positive response to the initiative of the Swedish professor. This shameful episode, dated to the last weeks of Stalin's life, is considered at greater length in Chapter 7; doc 7.2 gives an example of the psychic marasmus of the late Stalinist epoch.

It is appropriate to mention here the prerequisites for the award of the Nobel Prize in chemistry to N. N. Semenov and S. N. Hinshelwood as early as 1953. The relevant documents of the Nobel Committees available today are summarized in the table for convenience.

Table 6.2 presents a summary of statistical data borrowed from the nomination lists of the Nobel Committees in physics and chemistry concerning the nominees that received the award in 1949–1953, i.e. during the last five years for which documentation is now open for researchers after having been kept for 50 years in the archives. We shall start from physics.

It can be seen that all these nominees were supported by some authoritative Swedish nominators in the year of award whereas such support was absent in previous years when they were nominated but did not receive the prize. The sole exclusion that however confirms the tendency is the candidacy of Cecil Powell, UK.

In 1947, the unknown physicist from Bristol found pi-mesons in cosmic rays with the aid of his sensitive emulsion. In 1949, he was nominated for the Nobel Prize by very many researchers including the theoretical physician Torsten Gustafsson of the University of Lund; it was amid a burst of enthusiasm about the discovery of the long-awaited elementary particle predicted by Hideki Yukawa in 1935.

It may be supposed that the Nobel Committees privately maintained relations with Swedish nominators. Their proposals are like signals to the members of the Nobel Committees indicating the nominees of previous years most worthy to be considered at autumn sessions of the Royal Academy of Sciences. Recommendations for these nominators usually have the form of very short letters presented to the Committees shortly before the deadline.

Table 6.2. Statistical data on nomination of Nobel laureates in physics and chemistry in 1949–1953

Year of award	Laureates	Total number of nominations for the previous four years	Including Swedish nominators	Number of nominations in the year of award	Including Swedish nominators	Swedish nominators in the year of award
Physics						
1949	H. Yukawa, Japan	3	none	5	1	T. Svedberg
1950	C. Powell, UK	8	1	12	1	T. Svedberg
1951	J. Cockroft, UK	3	none	4	1	T. Svedberg
	E. Walton, Ireland	1	none	1	none	
1952	F. Bloch, USA	10	none	6	1	K. Siegbahn
	E. Purcell, USA	5	none	5	1	K. Siegbahn
1953	F. Zernike, Germany	6	none	3	1	K. Siegbahn
Chemistry						
1949	W. Giauque, USA	4	1	3	3	A. Tiselius, K. Siegbahn, G. Hagg, A. Olander (joint letter)
1950	O. Diels, Germany	2	1	2	none	
	K. Alder, Germany	1	1	2	none	
1951	E. McMillan, USA	1*	none	3	2	A. Tiselius,
	G. Seaborg, USA	14	none	8	2	T. Svedberg
1952	A. Martin, UK	1	1	2	1	J. Jorpes
	R. Synge, UK	1	1	1	1	J. Jorpes
1953	H. Staudinger, Germany	14	1	8	2	O. Samuelson, M. Gralin, E. Alder (joint letter), K. Myrback

* McMillan was twice nominated for the Nobel Prize in physics before 1951 (in 1947 and 1948 together with Soviet scientist V. I. Veksler). In 1951, he was simultaneously nominated for the Nobel Prize in chemistry by Svedberg.

This conjecture may be illustrated by the example of Kai Siegbahn, son of the Nobel laureate Manne Siegbahn (1924), who received the same award in 1981. Kai worked as an assistant researcher at the Nobel Institute of Physics for almost ten years (until 1951). Thereafter, he filled the position of professor at the Technological Institute, Stockholm, and became a regular nominator for the Nobel Committee in physics from 1952.

His first nomination, dated 18 January 1952 and received in the Committee a week later (25 January 1952), recommended American physicists Felix Bloch and Edward Purcell who had developed the nuclear magnetic resonance (NMR) technique and the efficacious method for in-depth investigations into the inner structure of bedrock and ore samples without their destruction. Details of the development of resonance methods for diagnostics of substances and contribution to this work made by different inventors (including Soviet scientist E. K. Zavoisky) are given in Chapter 8. Here, we shall focus on the analysis of nominations presented to the Nobel Committee by the young professor.

The first one is a three-page typewritten text that exposes deep knowledge of the subject by the author. The nominator is well aware of both the NMR technique and the relevant publications. It could not be otherwise because his personal interests lay in the sphere of electron spectroscopy and the adjoining field of resonance methods including NMR.

The second nomination for the Nobel Prize of 1953 is different. It recommends the candidacy of Dutch physicist Frits Zernike, the author of the phase contrast method, and is no match to the first one in terms of size, contents, and date of submission. The nomination of 1952 appears to have cost the scientist much time and labour for which reason he wrote it beforehand and sent it to the Nobel Committee in mid-January. The second nomination dated 27 January reached the Nobel Committee only on 29 January, i.e. two days before the deadline.

It contains only three sentences instead of three pages: "for the invention and development of phase contrast methods". No details or references to relevant publications are given, no word is said that the invention permitted to markedly improve the resolving power of light microscopes and enabled researchers to observe native cells and bacteria without their preliminary staining. All necessary data could be found in other nominations received by the Nobel Committee from abroad. The task of the Swedish nominator was just to confirm the value of the submitted work...

A similar signalling system, even if less explicit, was operated in the prize-awarding practice in chemistry. On the one hand, it involved such prominent figures of natural science in Sweden as Theodor Svedberg and Arne Tiselius; on the other hand lesser known professors supported the nominees in the year when they received the Nobel Prize. In 1949, William Giauque, USA, was nominated for his discoveries in the sphere of ultralow temperatures by Professors Gunnar Heg and Arne Erlander, besides Svedberg and Tiselius. In 1952, Archer Martin and Richard Sing, UK, were recommended to the Nobel Committee for the development of analytical chromatographic methods by biochemist Erik Jorpes of the Karolinska Medico-Surgical Institute (he also nominated them two years before, in 1950).

There is one more difference from the nominations in physics. Nominees in chemistry were supported by regular Swedish nominators during four years preceding the award, even in cases not as unusual as that of Powell. The discrepancy between the procedures adopted by the two related Nobel Committees may be attributed to the following circumstance.

The long-standing secretary of the Nobel Committee in physics Anders Barani told the author in 2000 that the committee in chemistry traditionally distributes twice as many invitations for nominators as the Physical Committee. It may be thought that many of them are intended for Swedish professors. Hence the unpredictable numerical predominance of their nominations compared to those of physicists. This inference is illustrated by Table 6.3.

History does not accept subjunctive mood. Nevertheless, there is every reason to think that in 1953 members of the Nobel Committee in chemistry would have met L. G. Sillen's nomination of Semenov and Hinshelwood as favourably as they did three years later...

Despite all obstacles, Soviet scientists were nominated for the Nobel Prizes by their foreign colleagues in 1952 and 1953, unlike 1949. The name of P. A. Cherenkov of the Physical Institute, the USSR Academy of Sciences, first appeared among nominees for the Nobel Prize in physics in 1952. His candidacy was submitted by Leon Rosenfeld, professor for the University of Manchester, a disciple and co-worker of Niels Bohr, keeper of his Copenhagen archive and historiographer.

The nomination contained a brief characteristic of Cherenkov's discovery of the new physical effect (see Chapter 8 for details). It was concluded by a remark

Table 6.3. The total number of nominators in the categories of physics and chemistry during 1949–1953 and the number of nominations from Swedish professors

Year of nomination	Prize category	Total number of nominators	Number of Swedish nominators
1949	Physics	46	3
	Chemistry	40	6
1950	Physics	40	1
	Chemistry	62	4
1951	Physics	50	2
	Chemistry	47	4
1952	Physics	44	2
	Chemistry	55	5
1953	Physics	51	2
	Chemistry	58	6

reminiscent of that made by Dirac and related to what was the matter of concern for Sillen: "I am afraid that I can not present copies of Cherenkov's works where this effect is described. The only thing I can do is to attach the list of references to them."

Professor E. Obel of Paris nominated A. N. Braunshtein, professor at the Institute of Biological and Medical Chemistry, USSR Academy of Medical Sciences, for the Nobel Prize in chemistry in 1952. Braunshtein was known among the world scientific community as the discoverer of biochemical conversion of amino acids and its role in natural metabolism of nitrogen, one of the most important contribution to the theory of cellular metabolism.

The name of Kapitsa was again included in the nomination list in 1953. His candidacy was proposed by Professor Cz. Bialobrzeski who had nominated D. V. Skobeltsin in 1947 (see Table 4.1). He recommended his first candidates as long ago as 1933; they were W. Heisenberg, P. Dirac, E. Schrödinger (all received the Nobel Prize in the same year) and Paul Langevin, the teacher of de Broglie. His 1953 nomination contained the names of L. Alvarez (Berkeley, California), M. Born (Edinburgh), and P. Auger (Paris); the latter scientist discovered the effect of atomic ionization and atmospheric showers of cosmic radiation ("Auger showers"). Twenty years after his first nomination, the Polish professor showed a striking predictive ability as in 1933: three of his four nominees became Nobel winners in different years!

For the sake of precision, it should be noted that Ch. Bialobrzeski sent his 1952 nomination from Warsaw by mail (23.1.1952). It reached the Nobel Committee only on 5.2.1952; therefore, its consideration was postponed until 1953.[69a] This rule was strictly followed by the Nobel Committees even if this sometimes led to unusual situations. By way of example, the Nobel Committee in physics had two nominations for Albert Einstein by Max Planck (Nobel laureate of 1918), one dated 1921 the other intended for consideration in the previous year (it was dispatched from Berlin on January 25 but registered in Stockholm on 2 February 1920)...

Candidacies of Semenov, Kapitsa, and Cherenkov who were eventually awarded Nobel Prizes will be discussed in Chapters 7 and 8 below. Here, it is appropriate to trace events that did not allow Veksler to win the prize.

As mentioned in Chapter 4, the principle of autophasing that allowed energies attainable in accelerators of elementary particles to be increased tens and tens of thousands of times was independently proposed by V. Veksler in February 1944 (doc. 6.36) and E. McMillan one year later. The priority of the Soviet scientist was never doubted. As soon as McMillan published a brief note in *Physical Review* in 1945, many colleagues sent to him photocopies of Veksler's papers published in *Doklady Akademii nauk SSSR* unknown to the American researcher (the journal was then printed in both Russian and English). The Nobel laureate in physics Ernest Lawrence, creator of the first cyclotron confirmed the indisputable priority of Veksler.[70]

There was no doubt that the discovery deserved the Nobel Prize. The two researchers were several times nominated together for the Nobel Prize after 1947. The Royal Academy of Sciences set the value of such discoveries by awarding the

Nobel Prize to Lawrence in 1939.[71] The next step in the construction of particle accelerators based on an essentially different principle opened up inexhaustible possibilities for in-depth studies of nuclear processes and the search for the family of transuranium elements.

It is very likely that the Nobel Committee did not give due attention to the nominations of Veksler and McMillan because it had no opportunity to compare the contribution of either author: these investigations were classified as secret in the Soviet Union in contrast to the situation in the United States. The publication of Veksler's paper in the open press was akin to a small miracle (see doc. 6.36).

An accelerator to be operated based on the autophasing principle was started to be built under Veksler's guidance in late 1944 to early 1945. In three years, the first Soviet synchrotron was commissioned, and the first scientific results were obtained.[72] However, the lack of available materials from the Soviet Union precluded consideration of Veksler as a candidate for the Nobel Prize while giving the prize to McMillan alone or jointly with Kerst would be unfair.[73]

The nomination of Kerst by Siegbahn[74] reflected the feeling of uncertainty that pulsed inside the Nobel Committee with respect to the prize for the works of Veksler and McMillan. Manne Siegbahn, member of the Royal Academy of Sciences and a prize-giving person in physics and chemistry, was certainly well aware of the discussions of nominees for the 1947 prize; therefore, he proposed Kerst as a reserve candidate in the event of rejection of Veksler by the members of the Committee. On the one hand, he was convinced that the Royal Academy had no reason to ignore the epoch-making discovery in physics of the mid-20th century. On the other hand, he could not help sharing the doubts of those who had no information on whether the idea of the Soviet scientist remained purely theoretical or was put into practice. Therefore, he suggested an alternative variant that all the same did not satisfy the Nobel Committee.[75]

In 1951, the names of Veksler and McMillan appeared in the nomination list again. This time they were recommended by Joseph Rotblat, professor of nuclear physics from London[76] and the author of the fundamental monograph on atomic energy published in 1954. It was however too late to award the sensational discovery of the Soviet scientist seven years after it had been made, and the proposal had no chance to be met favouarbly by the members of the Nobel Committee. As mentioned in Chapter 4, the Royal Academy of Sciences awarded the 1951 Nobel Prize to McMillan and Glenn Seaborg, his colleague at the University of California, "for their discoveries in the chemistry of transuranium elements".

The two prize winners were nominated by Ernest Orlando Lawrence (USA) and Swedish scientists T. Svedberg and Arne Tiselius (all Nobel Prize winners). Their proposal was unanimously supported by the members of the Nobel Committee at the very first meeting. True, Seaborg alone was nominated several times since 1948 after the synthesis of americium and curium but only by his compatriots from different research centres of the United States[77] In 1951, his nominators included leading members of the Royal Academy of Sciences. It appears that by 1951 they came to understand that further waiting for news from Moscow was hopeless. It was decided to award the Nobel Prize to a creator of synchrotron as a contributor to the fundamental discovery of a family of superheavy metals because his machine was used in the synthesis of berkelium, californium, and other transuranium elements.

This decision by the Royal Academy of Sciences demonstrated once again that the Soviet scientists had good chances to receive the Nobel Prize if it were not for the mania to classify as secret all scientific results in the post-war Soviet Union...

The year of 1948 gave a start to a new series of nominations of B. N. Ipatiev, who refused to return to the Soviet Union from the United States, from where he had been sent to do research in 1936 (see Chapter 3). That year his candidacy for the Nobel Prize in chemistry was submitted by four colleagues from the University of Evanstone near Chicago. In 1949, he was nominated by Professor G. Dupont (for the second time after 1941). Dupont repeated his nomination in 1950 and was joined by L. Hackspill, another chemist from Paris.

In 1952, one more nomination of S. N. Vinogradsky by Professor Dhar of Allahabad, India, was registered by the Nobel Committee in chemistry, the first one after a long break since 1921. In this way, Dhar responded to the publication of the fundamental monograph of Vinogradsky on soil microbiology in Paris in November 1949. The Russian language version of the book was out in 1952 (see Chapter 4). However, this nomination was none other than a tribute of respect to the researcher who had made the outstanding contribution to soil science. His time as a pretender for the Nobel Prize had passed long ago (see Chapter 2)...

In 1950, Soviet scientists were nominated for the Nobel Prize in physiology and medicine for the first time after 1946. The famous ophthalmologist V. P. Filatov, member of the USSR Academy of Medical Sciences and professor of the University of Odessa, was nominated by Professor G. Nastase of the University of Yassy, Romania.[78] The author specially emphasized the works of Filatov related to transplantation of corneal and other tissues in patients with eye diseases and some skin pathologies.

Biophysicist N. V. Timofeev-Resovsky, the legendary "*Zubr*" (Auroch)[79], who had worked in Germany since the 1920s until the end of World War II was nominated by radiologist Boris Raevsky, director of the Max Planck Institute and also born in Russia (in the city of Chigirin). The nominator introduced his candidate as "the former head of Department at the Kaiser Wilhelm Institute in Berlin-Buch"; he did not mention that Timofeev-Resovsky had been deported to the Soviet Union and did not know his whereabouts at the time of nomination. This explains why the nominee was described as having residence in Berlin-Buch (north-eastern outskirts of Berlin) in the list of pretenders for the Nobel Prize in physiology or medicine for 1950. The Kaiser Wilhelm Institute for Brain Research (renamed the Max Planck Institute for Brain Research after the war) was built up near the village of Buch in 1930. One of its laboratories was redesignated in 1937 as the Institute of Genetics and Biophysics where Timofeev-Resovsky worked till September 1945. Raevsky nominated his colleague for experimental studies of in the field of radiation genetics.

One more nominee of the Nobel Committee in physiology and medicine in 1950 was B. P. Babkin, a physiologist from the University of Montreal, Canada, mentioned in Chapter 2. His candidacy was submitted by Professor Erzji-Kaulbertz in Medical Academy of Krakow. He repeated his nomination in three years (1953).

The list of Russian nominees in literature in 1948–1953 did not change substantially compared with the preceding two years (see Table 6.4). Nikolai Berdyaev was nominated for the seventh and last time in 1948, again by Professor Alf Nyman of the University of Lund. The new nominator of Boris Pasternak was Martin Lamm,[80] the literary critic from Stockholm who had already submitted him for the prize in 1948 and 1950. Professor Cecil Bowra of Oxford again nominated Pasternak in 1949. It was his last nomination of Pasternak for the Nobel Prize; probably, Bowra decided that his candidate had no chance to be accepted by the Nobel Committee. He did not nominate anybody in the next two years but proposed the candidacy of Juan Ramon Jimenez in 1952; the Spanish poet received the Nobel Prize in 1956, i.e. two years before Pasternak.

Table 6.4

Nominees	Year of nomination	Nominators
M. A. Aldanov (1886–1957)	1948	I. A. Bunin, Nobel laureate
- " -	1949	- " -
- " -	1950	- " -
- " -	1951	- " -
- " -	1952	- " -
- " -	1953	- " -
N. A. Berdyaev (1874–1948)	1948	A. Nyman, professor of philosophy, Lund
L. M. Leonov (1899–1994)	1949	V. Kiparsly, professor of Russian language and literature, Helsinki
- " -	1950	- " -
B. L. Pasternak (1890–1960)	1948	M. Lamm. literary critic, professor of Stockholm University
- " -	1949	C. Bowra, professor of ancient literature, Oxford
- " -	1950	M. Lamm. literary critic, professor of Stockholm University
M. A. Sholokhov (1905–1984)	1948	Nobel Committee for literature
- " -	1949	V. Kiparsky, professor of Russian language and literature, Helsinki
- " -	1950	- " -

When Martin Lamm prepared his nomination of Pasternak for the Nobel Committee in January 1948, the "Sovetsky pisatel" publishing house in the USSR sent to its printing works the manuscript of Pasternak's verses to be published in a volume of "selected poems". The volume was part of a jubilee series under the title *The library of selected works of Soviet literature (1947–1947)* dedicated to the 30th anniversary of the October revolution. However, the volume of Pasternak's poems never reached readers. It was printed with a circulation of 25,000 in the first quarter of 1948 but soon destroyed by the decision of the Executive Board of the Union of Soviet Writers. This barbarian act reminiscent of the times of Holy Inquisition was initiated by ideologists of the AUCP(b) Central Committee as appears from the letter from the Department of Propaganda and Agitation addressed to the secretary of Central Committee G. M. Malenkov, 14 December 1948, under the heading "About a serious mistake of the 'Sovetsky pisatel' publishing house".

The letter was in the first place directed against the most popular satirical novels by Ilya Ilf and Evgeny Petrov, *The Twelve Chairs* and *The Little Golden Calf.* According to the authors of this document, Soviet people were represented in these novels as "narrow-minded, dull, and ridiculous philistines creating a favourable medium for swindlers and rogues to act in". A few more books were mentioned as ideologically inadmissible; the volume of Pasternak's poems was among them.[81]

As a result, the last volume of Pasternak's verses published in the lifetime of the great poet of the twentieth century was a small book dated to 1945...

Mikhail Sholokhov was nominated again in 1948 to 1950. In 1947, his candidacy had been submitted by Henry Olsson, future member of the Swedish Academy. Next year his collective nominator was the Nobel Committee in literature that probably undertook this unusual action because no one else proposed the author of *Tikhy Don*, who was regarded by the committee as the most worthy candidate for the Nobel Prize. In 1949 and 1950, Sholokhov was nominated by Professor Vladimir Kiparsky of Helsinki.

No Soviet writers were listed among pretenders for the Nobel Prize in literature in 1951–1953. The absence of Pasternak had a more or less plausible explanation (see note 2 to Chapter 9) whereas it was much more difficult to account for the lack of new nominations for Sholokhov. In all probability it was due to political reasons since Sholokhov sometimes actively participated in shameful ideological campaigns of the late Stalinist period...

In the same years Kiparsky recommended one more Russian candidate, Leonid Leonov, for whom it was the first nomination for the Nobel Prize. Leonov somehow came to learn about this nomination but was very sceptical about his chances (personal communication of Daniil Granin to whom Leonov told this story in 1951).

It follows from the first nomination by Kiparsky forwarded to the Nobel Committee on 25 January 1949 that he met Leonov for the first time in Helsinki when the Soviet writer visited the Finnish capital in February 1945. The nomination contained no serious literary analysis; in fact, it was a list of the writer's works for the last two decades from the really important novels *Barsuki* (1925) and *Vor* (Thief) (1927) to precocious works created during the Great Patriotic War, such as the play *Nashestvie* (Invasion) and the novel *Vzyatie Velikoshumska* (The Taking of Velikoshumsk) (1944). Naturally, this superficial nomination could not arrest attention from the members of the Nobel Committee despite flattering comparison of Leonov's style with that of Fyodor Dostoevsky, Amadeus Hoffman, and Andre Gide.[82]

One cannot but mention the closing part of Kiparsky's nomination. Having no experience (his name never appeared among nominators until 1949), he listed a few Russian-speaking writers who, in his opinion, could not pretend for the Nobel Prize (reverse nomination, so to say)... Soviet writers were represented in this list by Ilya Erenburg[83] and Aleksandr Fadeev (for their politicized works). Others included the emigrant authors Mark Aldanov, Mikhail Osorgin, and Vladimir Nabokov (without explanation of the negative attitude to their works).

At the same time, Ivan Bunin continued to submit the candidacy of Aldanov up to his death in November 1953. In his last nomination, Bunin recommended to the Nobel Committee Aldanov's last novel *Live as You Like* published in New York in Russian and English in 1952. He also mentioned Aldanov's novel *The Beginnings* and referred to encyclopaedias (including *Encyclopaedia Britannica*) containing notes about Aldanov. The response of the Nobel Committee was as terse as usual: "The proposal is turned down as several times before"...

Such a heavy sentence of the Nobel Committee is quite understandable. Bunin himself could hardly object to the honest rating given to his protegé if he knew about this conclusion. Bryan Boyd, the most competent biographer of V. V. Nabokov, unaware of the nomination of Aldanov for the Nobel Prize and the attitude of the Nobel Committee to this author very clearly described the situation.

Talking about the sincere friendship between Nabokov and Aldanov "despite their absolutely different temperaments", he noted: "Aldanov regarded with reverence the talent of Sirin (one of Nabokov's pseudonyms) but was afraid of him as something biting and unpredictable. Sirin, in his turn, respected the sceptical intellectuality and the logical composition of Aldanov's novels but knew perfectly well that they were untouched by the magic of art."[83a]

The Nobel Committee was no longer interested in the candidacies of Russian writers who had emigrated over 30 years ago. A new generation of gifted poets arose in the Soviet Union during the time elapsed since 1917 that could successfully compete with Western writers for the Nobel Prizes. They started to attract attention from the Nobel Committee since the mid-1940s as worthy heirs of great Russian literature. Their countrymen were expected to take initiative and nominate them in accordance with the requirements of the Nobel Statutes.

However, there was nobody to take the lead under the Stalinist regime. The very thought about contacts with the Nobel organizations was akin to a bad dream after which one wakes up in a cold sweat. Moreover, invitations for nomination did not reach the addressees after the end of 1946.

Nevertheless, members of the Swedish Academy and its Nobel Committee in literature knew fairly well (even without proposals from the USSR) who of the

Russian writers was worthy of the award. One of them was Boris Pasternak, first nominated in 1946 by C. M. Bowra, an English intellectual and connoisseur of the Russian literature of the 20th century (see Chapters 4 and 9).

Moscow also showed an interest in Pasternak, a very specific one. Documents of the CPSU Central Committee dated from 1948 and now available for researchers reveal an example of this interest. One of the documents was prepared in the Department of Propaganda and Agitation in connection with the appointment of the literary critic A. K. Tarasenkov, editor-in-chief of the "Sovetskiy pisatel" publishing house.

The candidacy of Tarasenkov was approved even though he was subjected to criticism for "mistakes he made in his literary articles, especially for his uncritical attitude toward bourgeois individualistic poetry of Akhmatova and Pasternak."[84]

Obviously, this characteristic was borrowed from earlier documents. It was in the spirit of the age to remind of all penalties imposed on a person by Communist Party organs of all levels, even if they were later cancelled as erroneous. In other words, a person once punished had to wear the brand for the rest of his (or her) life.[85] There is little doubt that the above formulation dates from 1946.

The resolution of the AUCP(b) Central Committee concerning the situation in Soviet literature was issued in August 1946. The main blow had to strike Anna Akhmatova[86] and Mikhail Zoshchenko.[87] According to an unwritten law, such attacks had to be unanimously supported by the working masses.[88] Speakers at numerous meetings of members of the Union of Soviet Writers did not confine themselves to criticism of Akhmatova and Zoshchenko but "denounced" other colleagues too who were not named in the resolution of the Central Committee. Pasternak was one of the main targets of demagogic criticism because his poetry had little to do with the model designed and supported by Party ideologists.

Anatoliy Tarasenkov was "passionately fond"[89] of the poet and therefore could not avoid being condemned in the heat of discussions provoked by the above document of the Central Committee. However, he had never written about Akhmatova, as evidenced by his close friend Daniil Danin, the known connoisseur of Pasternak's poetry. He could not be publicly charged with an "uncritical attitude" to the poetess because such a statement would betray the ignorance of the speaker.[90]

It appears that the entry of Akhmatova's name together with Pasternak's was backdated in the above characteristic of Tarasenkov in order to more closely associate him with the disgraced writers mentioned in the 1946 resolution.[91]

To conclude, it is appropriate to mention a document related to the Nobel laureate Romain Rolland, a left-wing writer and long-standing friend of the Soviet Union. A. A. Fadeev, secretary-general of the Union of Soviet Writers, informed the AUCP(b) Central Committee (6 September 1949) about the wish of the writer's widow to have photocopies of Rolland's letters to Gorky for his museum being then organized in Paris. Fadeev noted that "almost 120 such letters" were written during twenty years of continuous correspondence between the two writers.[92]

According to Fadeev, "these letters generally demonstrate the friendly feelings of the writer toward the Soviet people, Gorky, and his creative work. However, some of them [...] bear evidence of his political mistakes and hesitation. They contain incorrect estimates of the Russian people and other wrong opinions."[93] This overview of Rolland's epistolary legacy was supplemented by an allusion to the positive characteristics of "certain persons who later turned out to be enemies of the people." This gave Fadeev reason to recommend sending to Paris photocopies of only 56 letters of the 122 available. This recommendation was approved by the AUCP(b) Secretariat...

Notes

[1] RGASPI, fond 17, opis' 117, ed. khr. 598, l. 116–122. It was shown in the mid-1950s that preparation "KR" was inefficient; see doc. 4.35.
[2] *Ibid.*, opis' 116 , ed. khr. 256, p. 44.
[3] Esakov V. D., Levina E. S. Delo "KR" (Iz istorii goneniy na sovetskuyu intelligentsiyu ("KR" Case. From the History of Persecution of Soviet Intellectuals)). Kentavr, No. 3, 194, p. 57.
[4] RGASPI, fond 17, opis' 117, ed. khr. 598, l. 123.
[5] *Kentavr*, No. 2, 1994, pp. 58–60.
[6] RGASPI, fond 17, opis' 117, ed. khr. 699, l. 86. The sheet with the resolution of the Central Committee (18 February 1947) bears the following additional text: *Kultura i zhizn* published the review of A. Strukov "The outstanding discovery of Soviet science", No. 7 (26), 11 March 1947. *Pravda* published the review of Professor N. Petrov, full member of the USSR Academy of Medical Sciences, No. 72 (10463), 24 March 1947."*

* The article in *Pravda* was entitled "The outstanding success of Soviet science". The identical headings of the two articles betray the same conductor's hand, although they were written by authoritative representatives of Soviet medicine, A. I. Strukov and N. N. Petrov.

[7] Medecine de France, 7, No. 2, 1947.
[8] *Pravda*, 9 January 1947, p. 3.
[9] It was a far-fetched accusation for which apologies were presented only in twelve years. On 15 October 1959, the scientists were invited to the CPSU Central Committee to be familiarized with the decision of its Presidium, rehabilitating them from political accusations.* Their incorrectness was recognized in the memoirs of Shepilov, deputy head of the Department of Propaganda and Agitation of the AUCP(b) in 1947.**

* *Kentavr*, No. 3, 1994, p. 118.

** "Soviet scientists Klyueva and Roskin were charged with the alleged [underlined by the author] disclosure of their anti-tumour preparation to the Americans"; *Voprosy istorii*, No. 5, 1998, p. 25.

[10] *Izvestiya TsK KPSS*, No. 11, 1990, p. 135.
[11] *Kentavr*, No. 3, 1994, p. 114.
[12] Complete text of the "Closed letter" is published in *Kentavr*, No. 2, 1994, pp. 65–69. Some half a year later, Orgbureau of the AUCP(b) Central Committee considered the results of discussions of this letter at institutions of the Ministry of Aviation and Electrical Industry and adopted a resolution (25 February 1947). It was emphasized that the instructions of the Central Committee were fulfilled unsatisfactorily. Specifically, the importance and necessity of "radical improvement of political work among intellectuals, elimination of all manifestations of servility to foreigners, and education of our intellectuals in the spirit of Soviet patriotism" were underestimated.*

*RGASPI, fond 17, opis' 116, ed. khr. 331, l. 8–9.

[13] Some speakers at the meetings convened to discuss the "closed letter" of the Central Committee suggested that Klyueva be recalled from the RSFSR Supreme Council. For example, one such person at the Institute of Philosophy, the USSR Academy of Sciences, stated that "voters of the respective constituency had to be proposed to recall the deputy Klyueva from the Supreme Council as faithless".* This idea was realised as usual, i.e. "at the request of working masses". Paradoxically, all actions ascribed to the two scientists as well as the situation around them, dated to mid-1946. In other words, it had been known to the authorities long before the AUCP(b) Secretariat approved the candidacy of Klyueva for the election of deputy to the Supreme Council in December that year. The thing is, Stalin decided to launch a new anti-intellectual campaign after the parliamentary elections were over...

* TsAODM, fond 2501, opis' 1, ed. khr. 3, l. 272.

[14] D. T. Shepilov. Vospominania (Memoirs). "Voprosy istorii, No. 5, 1998, p. 25. Stalin frequently used the word "beat" and similar sayings in his everyday speech, both when angry and good-humoured. Here are a few examples.

In 1935, certain films that Stalin liked were criticised in the press. Stalin remarked after a night-time viewing of new films in the Kremlin: "[...] if one has something to say about our developing cinematography, he is welcome. But we always beat and always will beat those who can do nothing but critisize."*

S. A. Novikova, daughter of the Air Chief Marshal A. A. Novikov, related an episode at the reception in Kremlin on the eve of the new 1945 year (Novikov organized a highly efficient air blockade of Paulus' army, surrounded at Stalingrad in winter 1942–1943). "Stalin said in his peculiar tone: 'Novikov is a very good marshal. He built up impressive air forces. But as soon as he works badly we shall hang him.'"**

V. A. Malyshev, a leading figure in Soviet industry, often met Stalin over fourteen years. He made the following note in his diary (28 December 1946): "Comrade Stalin warned us [heads of industrial ministries] that we must fulfil plans and added: 'Otherwise I will beat you myself.'"***

Stalin's minions tried to imitate his style. Vyshinsky (bearer of the same name as that of the ill-fated prosecutor-general) spoke as a member of the Communist Party bureau at a meeting at the Institute of Philosophy (24 February 1949) at the height of the campaign "against cosmopolitanism". He expressed his hatred in the following words: "We do not dispute with cosmopolitans, we must beat, crush, and denounce them!"****

Stalin acted in this manner regardless of the impression it made on his listeners, even if they were leaders of foreign states. At a banquet during the Teheran summit (4.2.1943), Stalin proposed a toast "to the shooting of 50,000 of German-Prussian officers".***** He probably recalled his own experience with the organization of the senseless execution of thousands of Polish officers in 1940.

* RGASPI, fond 558, opis' 2, ed. khr. 829, l. 13; from the notes of B. Z. Shumyatsky (see note 15 to Chapter 5) dated 13.3.1935.

** "Izvestiya", 3.2.2003, p. 3.

*** "Istochnik", No. 5, 1997, p. 131.

**** TsAODM, fond 2501, opis' 1, ed. khr. 8, l. 79.

***** L. A. Bezymensky. Razgadannye zagadki Tret'ego Reikha. 1940–1945 (Resolved Puzzles of the Third Reich). Smolensk.: Izd-vo "Rusich", 2001, p. 277.

[15] At exactly the same time, T. D. Lysenko was preparing to crush Soviet genetics under the patronage of Stalin. He immediately used the "Klyueva–Roskin case" to dispose of his main opponent, A. R. Zhebrak, an outstanding selectionist, professor of Timiryazev Academy in Moscow and president of the Byelorussian Academy of Sciences. In 1944, the Antifascist Committee of Soviet Scientists* asked him, on behalf of the Department of Propaganda and Agitation, to write an article about the situation in Soviet biological science for the American journal *Science*; the article was designed as a response to the critical paper of the American biologist Karl Sax on the same theme.** In November 1947, Lysenko initiated the court of honour at the USSR Ministry of Higher Education that accused Zhebrak of "servility to bourgeois science, an attempt to belittle Soviet science and defame his compatriot T. D. Lysenko." His article was described as "faulty in essence and the behaviour antipatriotic."***

* The activity of USSR Antifascist Committees (of Soviet scientists, women, youth, etc.) after World War II was focused on propaganda of the government policy and the Soviet way of life among the foreign audience.

** "Science", 99, 2572, 1944, p. 289 (the paper by Sax) and 102, 2649, 1945, p. 357 (Zhebrak's paper); see also doc. 6.38.

***RGASPI, fond 17, opis' 117 , ed. khr. 955, l. 193–195.

[16] I. E. Glushchenko, D. Biol. Sci., speaking at the extended meeting of the Presidium of the USSR Academy of Sciences on 24.8.1948 devoted to the situation in biological science, told that "the list of references in a paper published by the *Zhurnal obshchey biologii* contained 165 publications of which 28 were by Russian and the remaining ones by foreign authors". The speaker described this fact as "bowing and scraping for reactionary foreign scientists".* Another anecdotic example is the anxiety caused by the non-Russian name of the eminent Soviet scientists Ivan Avgustovich Oding, recommended by the Presidium of the USSR Academy of Sciences for the position of deputy director of the Institute of Mechanical Engineering. Bureaucrats from the Staff Department of the AUCP(b) Central Committee who submitted his candidacy for the approval of the Secretariat emphasized that "all his works were published in the Russian language".**

* "Vestnik AN SSSR", No. 9, 1948, p. 65.

** RGASPI, fond 17, opis' 117 , ed. khr. 1018, l. 20; CV of I. A. Oding attached to the submitted document describes him as Russian by birth.

[17] D. T. Shepilov asserted that no ideological campaign in the Soviet Union could be launched without Stalin's personal initiative; he emphasized that "these were the years of his personal absolute dictatorship".*

* In: "I primknivshiy k nim Shepilov. Pravda o cheloveke, uchonom, voine i politike (Shepilov Who Had Joined Them. The Truth About Person, Scientist, Warrior, and Politician)". M.: "Zvonnitsa", 1998, p. 135.

[18] Here are a few extracts from addresses in the courts of honour created in 1947 by the instruction of the AUCP(b) Central Committee in all ministries and other government bodies for the denouncement of local "antipatriots".

Committee of Architecture, USSR Council of Ministers. Architect Khiger wrote a book "praising the advantages of American housing construction." The book was not approved by Glavlit* and could not be published. "Aleksandrov and Burdin of the Moscow Architectural Institute returned from a business trip to America and "tried to popularize among students 'principles' of constructivism in architecture long ago denounced in this country."**

USSR Ministry of Heavy Industry Building. Minister P. A. Yudin mentioned engineer Dobrynin who "admired houses of steel and aluminium in England and America and recommended to use these materials in our country disregarding the capitalist nature of these structures."***

USSR Ministry of Geology. Participants in the discussion of the closed letter of the AUCP(b) Central Committee concerning the "Klyueva–Roskin case" drew attention to "the use of foreign terms in geology while Soviet scientists are far ahead of their colleagues abroad in many other fields [...] Special literature cultivates the pseudoscientific 'theory' of foreign origin that the basis of the mining industry in this country was laid by a Finnish tribe [...] the case of Professor Chupilin had been submitted to the court of honour but the hearing was postponed."****

Central Administration of the Northern Sea Route, USSR Council of Ministers. Report of the Department of Propaganda and Agitation, the AUCP(b) Central Committee, emphasized that "Party members revealed evidence of the thoughtless attitude of certain workers toward keeping state secrets. For example, employees of the Administration let out data concerning preparation of the ice-breaker *Litke* to a special expedition. These data became known in England where the ice-breaker was repaired and the works were deliberately dragged out. As a result the fulfilment of the important task set by the government is at risk of aggravation."*****

Finally, one more example of the feeble-mindedness shown by courts of honour. A favourable weather forecast was given for the red-letter day of 1 May 1947. However, it was a cloudburst after the military parade at Red Square and the mass demonstration had to be cancelled. As a result, the head of the Hydrometeorological Service, E. K. Fedorov, was brought to the court of honour and relieved of his duties.******

* USSR top censorship board.

** RGASPI, fond 17, opis' 117 , ed. khr. 913, l. 64–66.

*** *Ibid.*, l 75–77.

**** *Ibid.*, ed. Khr. 917, l. 58–59.

***** *Ibid.*, l. 66.

****** "Izvestiya" 29.4.2003, p. 10.

[19] TsAODM, fond 8099, opis' 1, ed. khr. 8, l. 37 (speech at the Communist Party meeting, Institute of Chemical Physics, the USSR Academy of Sciences, 30.7.1947) and l. 60 (*ibid.*, 14.10.1947). Such coarse wording from Professor M. A. Sadovsky was probably accounted for by his direct involvement in a super-secret atomic bomb project where even minor details were classified. He was responsible for the development of measuring devices to evaluate the consequences of an explosion. His instruments were successfully employed at the testing ground near Semipalatinsk, Kazakhstan, during the very first nuclear explosion on 29.8.1949.*

Sometimes, such tough statements at courts of honour attracted the attention of the secret services. Admiral of the Fleet N. G. Kuznetsov, the Navy commander-in-chief during the Great Patriotic War, and three other distinguished naval commanders were taken to the court of honour (15 January 1948) under an accusation of transferring to allies a secret torpedo that had actually been declassified by that time. Stalin ordered to pass materials of the trial to the Military Board of the USSR Supreme Council that demoted Kuznetsov from the highest rank of Admiral of the Fleet to rear admiral and sentenced three other officers to 4–10 years of imprisonment.**

* "Novaya i noveishaya istoriya", No. 5, 2001, p. 134.

** N. G. Kuznetsov. Krutye povoroty. Iz zapisok admirala (Sudden Turns. From Recollections of an Admiral). M.: Molodaya gvardiya, 1995, pp. 23–29, 230–231.

[20] P. I. Lebedev, chairman of the Arts Committee suggested in a letter of 12.3.1948 to A. A. Kuznetsov, secretary of the AUCP(b) Central Committee, to take to the court of honour a group of musical critics comprising D. Zitomirsky, I. Martynov. I. Belza, G. Shneerson, A. Mazel, and S. Shlifshtein.* It was recommended "to take to court three of them in the first place and interrogate the others as witnesses". Martynov was said "to display servility to the perverted music of Western European composers Berg and Mahler, positively assess the influence of renegade Stravinsky and German formalist Hindemith on Shostakovich [...], imply close links of Shostakovich's music to the traditions of Chaikovsky and thereby distort creative image of the great Russian classic."** (RGASPI, fond 17, opis' 118 , ed. khr. 65, l. 51–52).

* This proposal followed the resolution of the AUCP(b) Central Committee "About the opera *Great friendship* by V. I. Muradeli" adopted on 10.2.1948 and published the next day in all newspapers of the country. The resolution "emphasized the intolerable situation in Soviet musical critique where leading positions belong to the opponents of Russian realistic music and adherents of decadent formalistic music".

** The same resolution mentioned that "anti-popular music of Shostakovich was sharply criticized in the editorial 'Muddle instead of music', published in 1936 in *Pravda*, the newspaper of the AUCP(b) Central Committee. The article appeared in connection with his opera *Lady Macbeth of Mtsensk* to expose the harm and the danger of such compositions for the development of Soviet music...

[21] In July 1936 *Pravda* published several editorials concerning Luzin under the headings "Enemies under the mask of a Soviet citizen" (3 July), "The unmasked enemy. In the Academy of Sciences commission on *Mister* [italicized by the author] Luzin's case" (14 July), "Academician Gubkin about the so-called academician Luzin" (15 July). On 6 August, the newspaper published the resolution of the Presidium of the USSR Academy of Sciences, adopted 5.8.1936, "About N. N. Luzin, member of the USSR Academy of Sciences" that forestalled the decision of the special commission. Simultaneously, the editorial on the same issue was published under the heading "The dignity of Soviet science". The article accused Luzin of "servility to great bourgeois scientists, publication of his main works abroad and only the less important ones in Soviet journals."* In addition, it was recalled that in 1930 Luzin had refused to sign the address of a group of mathematicians and physicists of the Moscow University to French scientists calling them to protest "against interventionist plans of certain states with respect to the Soviet Union"** and that "he gave consent to be elected a member of a presidium of a reactionary congress abroad."*** P. L. Kapitsa responded sharply to the very first publication in the letter to V. M. Molotov, chairman of the USSR Council of People's Commissars. This letter had a positive effect despite an even more coarse reply from the addressee (doc. 6.4). A Soviet citizen called "Mister" in the main newspaper of the country where mass repressions were developing was doomed to be arrested. He was probably saved due to the courageous interference of Kapitsa.

A month before *Pravda* started the campaign against Luzin, another newspaper, *Leningradskaya pravda*, accused scientists of the Pulkovo Astronomical Observatory of "servility to publications in foreign journals" (4.6.1936).

* Luzin lived in Paris from 1928 until 1930, where he published his fundamental work "Leçons sur les ensembles analytiques et leurs applications" (Paris, 1930). The book was

translated into Russian and published under the title "Lectures on Analytical Manifolds and Their Applications" in 1953, after the death of the author; the Russian translation contains no reference to the French edition.
** At that time (from 25 November to 7 December), the falsified trial of "Prompartia" was held in Moscow that induced a negative reaction in the West.
*** In all probability, under the "reactionary congress" is meant the International Mathematical Congress held in Bologna, Italy, in August 1928 where Luzin represented the Soviet Union and was elected vice-president. Alternatively, it was the conference of Polish mathematicians in Lvov where Luzin was a key figure as a world-renowned scientist.

[22] Here are two personal recollections of such practice. The large-scale map of the Donets coal basin published in the first half of the 1930s and sent to large world libraries was classified as secret in 1949. Those attending a public library to have the book *Metallogenic Epochs and Regions of Siberia* published simultaneously in the Russian and German languages (in Moscow and Germany respectively) in 1926 had to present a certificate from the place of employment indicating that he or she really needed to read it. Two more striking examples of absurd secrecy refer to agriculture. In September 1949, an assistant to the deputy minister of agriculture was sacked and brought to trial for the loss of "top secret" documents, such as data on land under cultivation of *kok-sagyz* (Russian dandelion)* and its crop capacity in the Ukraine, the situation at *kok-sagyz* plantations in the Kiev region, and the draft resolution of the AUCP(b) Central Committee of the Ukraine on *kok-sagyz*.** The Head of the Central Department of Field-Protective Forestry had to explain in the Agricultural Department, AUCP(b) Central Committee, his behaviour and recognize his mistakes after he "had told members the Bulgarian delegation about damage inflicted by rodents to young oak seedlings in state field-protecting forest plantations."***

One more resolution of the AUCP(b) Central Committee dated 24.6.1949 reads: "With reference to the publication of secret data on the production of footwear and hosiery in the journal *Lyogkaya promyshlennost'* (Light Industry), comrades Shepilov, Pegov, and Chesnokov are due to present within three days suggestions on the strengthening of the editorial board"****, i.e. the discharge of the editor-in-chief as a minimum...
* A rubber-bearing plant cultivated in the Soviet Union since 1933 but forgotten in the post-war period due to the production of synthetic rubber.
** RGASPI, fond 17, opis' 118 , ed. khr. 561, l. 81–82.
*** *Ibid.*, ed. khr. 924, l. 69.
**** *Ibid.*, fond 17, opis' 116 , ed. khr. 441, l. 46.

[23] Inspection of book exchanges between astronomical and geophysical institutions of the USSR Academy of Sciences and their partners abroad undertaken by the AUCP(b) Central Committee has shown that "[...] Soviet organizations put at the disposal of their counterparts not only experimental but also theoretical works superior to foreign ones". Based on this conclusion, the Central Committee Secretariat set up a commission (26.3.1948) "to regulate conditions for the exchange of non-subscription editions published by Soviet scientific institutions."* This decision practically put an end to book exchanges at that period.
* RGASPI, fond 17, opis' 118, ed. khr. 18, l. 105–107.

[24] The example that follows illustrates the practice of backdated accusations. It is borrowed from evidence considered by the USSR Ministry of State Security in 1949. The accused was

Professor S. A. Yakovlev, a geologist, member of the International Association for the study of the Quaternary period in Europe. He was charged with disclosure of secret data about Quaternary deposits of the USSR in his reports at the meetings of the Association. "This area of Quaternary geology being later classified as secret, the reports of Yakovlev should be considered as communication of secret materials to foreigners" (verdict of senior investigator lieutenant colonel A. P. Leonov).*

* Tsentral'ny arkhiv FSB RSFSR, sledstvennoye delo P-24208, t. 1, l. 336–337.

[25] *Ibid.*, 1.110.

[26] The Ch. D. Walcott great bronze medal instituted by the Walcott Fund in April 1928 has been awarded by the US National Academy of Sciences (NAS) "for outstanding discoveries in paleontology, Cambrian and Precambrian history" since 1934. Vologdin was awarded the medal in April 1947 and had to receive it at the autumn session of NAS to be held in Washington on 17–19 November 1947 (or 26–28 April 1948 if more convenient for the laureate). When interrogated on 25.4.1949, Vologdin said that he "asked S. I. Vavilov, president of the USSR Academy of Sciences, for permission to go to the USA. I was given permission but decided not to go to avoid possible consequences."* In the end, the medal and the diploma were passed to the Presidium of the USSR Academy of Sciences via the Soviet embassy in Washington and the USSR Ministry of Foreign Affairs. S. I. Vavilov handed them over to Vologdin in his private office in the second half of December 1948. This unusual event was not reported in the press. The formal act was preceded by the decision of Politbureau dated 9.12.1948 "to approve the proposal of the Presidium of the USSR Academy of Sciences (comrades Vavilov and Bruevich) to present the Charles Doolittle Walcott medal to A. G. Vologdin, corresponding member of the Academy, awarded to him by the US National Academy of Sciences."** The medal was confiscated when Vologdin was arrested on 31.3.1949, expropriated by the state in November 1950 and thereafter disappeared.

In half a century (in 2003), Bruce Alberts, president of NAS, came to know about the sad fate of the award from N. P. Laverov, vice-president of the Russian Academy of Sciences, and undertook to have a replica of the medal made for the family of the deceased scientist. N. P. Laverov handed it over to the laureate's daughter, E. A. Vologdina, together with copies of the diploma and related documents (21.4.2004).***

* Tsentral'ny arkhiv FSB RF, sledstvennoye delo P-24208, t. 1, l. 124–124. Vologdin probably took this decision under pressure from the outside. It could be related to the resolution of the Communist Party organization of the Institute of Paleontology adopted at its meeting (12.11.1947) following the discussion of the "Closed letter of the AUCP(b) Central Committee concerning the Klyueva–Roskin case" of 16.6.1947 (see note 12 to this chapter). The resolution emphasized that "some of our scientists keep faulty opinion" making them "strive to have their works published abroad, keep extensive correspondence with foreigners, and *be elected in foreign scientific societies*" [italicized by the author]. (TsAODM, fond 7939, opis' 1, ed. khr. 1, l. 28).
** RGASPI, fond 17, opis' 3, ed. khr. 1073, p. 165.
*** See also the author's article "Arrested award" in the newspaper *Poisk*, No. 19, 2004, p. 13.

[27] Tsentral'ny arkhiv FSB RF, sledstvennoye delo P-24208, t. 1, l. 85–86. Naturally, geological data reported in the works sent by Likharev to his foreign colleagues were not secret at that

time; otherwise, they could not be published. After the trumped-up "geologists' case" and arrests of the accused, the USSR Council of Ministers reprimanded the deputy minister of geology V. I. Smirnov "for relaxed vigilance and irresponsible attitude toward his duties that resulted in the breach of the state secret law by publications of the Ministry of Geology."* In fact, the data contained in these publications were neither secret nor listed in the Glavlit instructions.

* RGASPI, fond 17, opis' 118, ed. khr. 717, l. 60.

[28] At approximately the same time, paleontologist from New York J. Chronic also expressed concern about the lack of new scientific publications from the Soviet Union. He wrote to Professor B. K. Likharev, Leningrad, on 21.12.1948: "There are many publications from the USSR in the library of the American Museum of Natural History but no recent ones." This letter is now in the materials of Likharev's case, instituted after he was arrested by the USSR Ministry of State Security in April 1949.*

* Tsentral'ny arkhiv FSB RF, sledstvennoye delo P-24208, t. 1, l. 122.

[29] RGASPI, fond 17, opis' 118, ed. khr. 369, l. 131–132, 146–147. The thoughtless self-isolation could not but have a boomerang effect on Soviet science and its prestige. Here is an interesting document illustrating the undesirable (even for authorities) consequences of imposing secrecy on each scientific work. A. Ya. Vyshinsky advised Stalin (16.9.1951) not to respond to the invitation of Soviet scientists to the International Congress on marine meteorology in Genoa. He argued his recommendation as follows: "The majority of studies to be discussed at the Congress were carried out at Soviet research institutions in the framework of secret projects; hence, our representatives will not be able to deliver reports reflecting the high level of these investigations in our country." As a result, the Politbureau gave no permission to send the Soviet delegation to Genoa.*

* The resolution adopted on 18.9.1951 (protocol 83, p. 545) is published in: "Akademiya nauk v resheniyakh Politbyuro TsK RCP(b)-VKP(b)-KPSS 192–1952 (Academy of Sciences in the Decisions of Politbureau of the Communist Party Central Committee". M.: ROSSPEN, 2002, pp. 504–505.

[30] "Vavilov as a human being showed quiet and restrained dignity. He commanded deep respect by his sound opinions, straightforwardness and wholeness of his character. He died at his post, most likely of overwork, but what he has done for the benefit of his country surpasses the deeds of an ordinary person. He will be remembered as one of the great creators of science in Russia, along with Lomonosov."*

* *Nature*, 168, No. 4277, 1951, p. 679.

[31] RGASPI, fond 17, opis' 118, ed. khr. 627, l. 200. Statements about the maleficence and political indifference of scientific publications became commonplace during those years whenever an excuse was needed to persecute a researcher. Specifically, the booklet *World harvest of aquatic animals* by T. S. Rass, prohibited for circulation in 1949, was characterized exactly in these terms. The editor of the booklet was relieved of his duties and the director of the All-Union Institute of Fishery and Oceanography was sacked and lost his position as a member of the Board of the Ministry of Fisheries.* A lesson in ideological loyalty was taught to Professor Rass a year before his booklet was out, in November 1948. The secretary of the Communist Party organization of the Institute of Oceanography, the USSR Academy

of Sciences, where Rass worked at that time said: "Speaking about the transformation of marine nature, he states that aquatic organisms can be introduced only to proper water-bodies. If Professor Rass really thinks so, he contradicts the teaching of Michurin." (?!).**

* RGASPI, fond 17, opis' 118, ed. khr. 974, l. 88.

** CAODM, fond 2971 opis' 1, ed. khr. 1, l. 98. The meeting where this episode took place was held in connection with the resolution of the Presidium of the USSR Academy of Sciences about the situation in biological science following the notorious session of VASKhNIL (31 July-7 August 1948) that proclaimed the victory "of Michurin's teaching, of our Soviet Michurinian Darwinism".

[32] E. Schrödinger. *What is life? The physical aspects of the living cell.* 1944.

[33] RGASPI, fond 17, opis' 118, ed. khr. 129, l. 29.

[34] Accusations against Linus Pauling of "machist sins" date to 1947 when his fundamental work *The Nature of the Chemical Bond* was published in the Russian language and gained wide popularity among Soviet natural scientists. The Institute of Organic Chemistry, USSR Academy of Sciences, was at the head of the struggle against the resonance theory. The decision of the bureau of the Institute's party organization emphasized: "the majority of leading workers of the Institute have until recently credulously and uncritically believed in the idealistic 'resonance' concept."* Under such leading workers were meant the director of the Institute, a member of the USSR Academy of Sciences, his disciple and close co-worker R. Kh. Freidlina, future corresponding member of the Academy etc. Party functionaries of another academic organization, the Institute of Oil, regretted that they were not the first to combat "idealistic nonsense."** The annual report of the Party bureau for 1950–1951 contained the following phrase: "The Institute of Organic Chemistry criticized, even if incompletely, the theory of resonance and did it without our assistance."***

University students were also made to criticize the theory of resonance. The known publicist and diplomat A. E. Bovin, student of the law faculty, University of Rostov, at the time being considered, recollected that they also had to criticize the idealism of "resonance in chemistry."**** It is worthwhile to note that Pauling's resonance models so severely criticized by incompetent ideologists between the 1940s and 1950s eventually turned out to make a bridge between classical (qualitative rather than quantitative) methods of description of chemical compounds by structural formulas and modern quantitative theory of molecular orbitals.*****

* TsAODM, fond 299, opis' 1, ed. khr. 10, l. 111.

** Freidlina was criticized at a sitting of the bureau (1951) for theoretical "mistakes and support of the 'resonance' theory"; *ibid.*, l. 142.

*** TsAODM, fond 7332, opis' 1, ed. khr. 5, 65.

**** A. E. Bovin. XX vek kak zhizn' (The XXth Century as Life). M.: Zakharov, 2003, p. 38.

***** "Vestnik RAN", No. 9, 2003, p. 813.

[35] In fact, the criticism of Pauling was only a pretext to accuse Soviet physical chemists Ya. K. Syrkin and M. E. Dyatkina, who extensively used the theory of resonance in their studies. They were incriminated in "concealing and belittling the role of Russian and Soviet scientists", citing "the White Guard Kistyakovsky working for American monopolies." These charges were followed by a rhetoric question: "Where is the patriotism of Soviet scientists?"*

The ignorant criticism of the resonance theory gave rise to protests in the press that made V. M. Tatevsky, one of the authors of the above statements, apply for support to the Secretariat of the AUCP(b) Central Committee. He obtained it. A week and a half after the application (26.7.1950), the secretary of the Central Committee G. M. Malenkov instructed the "Department of Propaganda and Agitation to report its opinion within a week."** On 4 January 1951, a clerk made a note that "the journal *Bolshevik* (No. 3, 1951) published the article by Yu. Zhdanov 'Selected works of A. M. Butlerov', which touches on the questions mentioned in the application of comrade V. M. Tetaevsky."*** Ideological attacks on leading scientists affected their disciples too. One of them was Syrkin's postgraduate student, L. Ya. Blumenfeld, war veteran, who performed quantum calculations based on the resonance theory in the L. Ya. Karpov Physico-Chemical Institute, Moscow, for the Soviet atomic project. In 1951, he had to leave the Institute accused of idealistic views alien to dialectic materialism.****

Ideological attacks on supporters of the resonance theory weakened very soon after the death of Stalin. On 6.6.1953, Professor M. V. King, secretary of the chemical section, American National Council of Fine Arts, Sciences and Free Trades, invited N. D. Sokolov, professor at the Moscow Physico-Technical Institute and specialist in quantum chemistry, to write a paper for an international volume of collected articles devoted to the valent resonance theory and its role in modern natural sciences. Similar invitations were sent to the founder of the theory Professor L. K. Pauling and Professor C. A. Coulson, a known expert in the valency theory and quantum-chemical computation of the molecular structure. After preliminary discussions with the Department of Science and Culture, CPSU Central Committee, the chief scientific secretary of the USSR Academy of Sciences, A. V. Topchiev, forwarded an official application (25.9.1953) to the head of the Department, A. M. Rumyantsev, asking for permission to N. D. Sokolow to give his American correspondents consent to participate in the edition being planned.*****

* "Voprosy filisofii", No. 3, 1949, p. 191.

** RGASPI, fond 17, opis' 118, ed. khr. 999, l. 151.

*** *Ibid.*, l. 165. The article of the head of the Department of Science Yuri Zhdanov, educated as a chemist, told that "Soviet scientists exposed the inconsistency of the resonance theory"; Syrkin and Dyatkina were additionally charged that the American physico-chemist K. Fiance in the review of their book *Chemical bond and molecular structure* (published in 1946) did not find significant difference between their views and opinions of western scientists. *Bolshevik*, No. 3, 1951, pp. 70–71.

The ignorance of Tatevsky and other participants of attacks on the universally recognized theory of resonance structures in organic chemistry was strictly condemned by Roald Hoffmann, Nobel laureate in chemistry for 1981, who conducted studies in Moscow in 1960–1961. Knowing the Russian language, he could thoroughly read shorthand records of the "anti-resonance" discussion of 1951. See: I. Hargittai. Otkrovennaya nauka. Besedy so znamenitymi khimikami (Candid Science. Talks with Famous Chemists). M.: YRSS, 2003, T. 1, pp. 182–183.

**** "Izvestiya"-"Nauka", 22.11, 2002, p. 11.

***** RGANI, fond 5, opis' 17, ed. khr. 414, l. 118–120.

[36] TsAODM, fond 992, opis' 2, ed. khr. 3, l. 34.

[36a] In the context of this most important scientific breakthrough, it is interesting to avert to the observation made by Velimir Khlebnikov, son of the biologist and future decadent poet,

who was at that time (1903) a first-year student at the Physico-Mathematical Department, University of Kazan. The 18-year-old student formulated his poetical manifesto in the following words, prompted by certain inner vision: "Let them read on the tombstone [...] he linked time with space" a few years before Einstein published the special theory of relativity.*

 * See "Novy mir", No. 5, 1983, p. 252.

[37] RGASPI, fond 17, opis' 118, ed. khr. 503, l. 203.

[37a] The authoritative Marxist philosopher and historian of exact sciences E. Ya. Kol'man, connoisseur of philosophical problems of modern physics, described the attempts of his colleague to disprove what he could not conceive: "Maksimov considered himself specialist in the methodology of physics although his knowledge of this science was much poorer than his knowledge of philosophy and he had no talent."*

 * See V. I. Vernadsky. Dnevniki 1935–1941 v dvukh knigakh. Kn. 1. 1935–1938 (Diaries 1935–1941 in Two Books. Book 1. 1935–1938). M.: Nauka, 2006, p. 139 (from the comments of the compiler V. P. Volkov, D. Geol.-Mineral Sci.).

[38] V. A. Fok, member of the USSR Academy of Sciences, sent a letter to G. M. Malenkov, secretary of the AUCP(b) Central Committee, saying that such articles denying "the physical theory on which modern physics is based" do "great harm to the development of physics and technology." He asked the influential functionary for permission to publish his answer to Maksimov "in one of the authoritative organs of the Soviet press". After the meeting of the Central Committee Secretariat, Malenkov chose not to meddle in settling the conflict and left the scientist's letter unanswered. The lack of response of authorities made a group of nuclear scientists (future Nobel laureates L. D. Landau, A. D. Sakharov, and I. E. Tamm) to seek the support of L. P. Beria, the head of the Soviet atomic industry. They tried to explain that "the most important problems related to [...] elementary particles and nuclear forces can not be solved without Einstein's theory of relativity". Beria, personally responsible for the development of Soviet nuclear physics, had to seek support from Malenkov. As a result, the scientist's letter was soon (before Stalin's death) published in the journal *Voprosy filisofii** although Maksimov was a member of its editorial board.** In half a century, the eminent nuclear physicist N. S. Rabotnov published a vivid essay under the title "Iosif Vissarionovich changes profession" in which he described the situation in physical sciences during the last years of the dictatorship: "The coryphaues of all sciences*** was very close to do with physics what he did with biology — physical Lysenkos were ready to attack and only waited for his command.**** However, the ruin of nuclear physics would have too serious consequences. Fear for his own fate made him listen to clever people [...]"*****

 * *Voprosy filisofii*, No. 1, 1953, pp. 168–174. In fact, this issue of the journal was approved for printing on 7.3.1953, i.e. two days after Stalin's death. Fok's article was sent from the AUCP(b) Central Committee for immediate publication in early January 1953.

 ** See the author's article in *Nezavisimaya gazeta* (3.6.1997) for details; see also V. P. Vizigin. "The nuclear shield in the 'thirty-year war' of physicists against ignorant criticism of modern physical theories". (*Uspekhi fizicheskikh nauk*, 169, No. 12, 1999, pp, 1381–1383). It should be mentioned in addition that the USSR Academy of Sciences organized in 1951, i.e. a year before Maksimov's article was published, the workshop "The situation in the theory of chemical structure in organic chemistry". It was designed after a model of the VASKhNIL session of 1948 to discredit leading physical chemists for their support of Pauling's resonance theory. Young participant in the project L. Ya. Blumenfeld, future authoritative biophysicist, who then performed quantum calculations based on the resonance theory in the L. Ya. Karpov Physico-Chemical Institute, Moscow, had to leave the Institute; see also notes 34 and 35 to this chapter. The article by Maksimov doomed leading physicists of that time to the same fate were it not for the counteraction of the key figures of the atomic project led by its head I. V. Kurchatov.

*** The term coined by propaganda of those times to praise and eulogize the "boundless abilities" of the chief.

**** Stalin initiated the organization of an All-Union workshop of heads of physical departments of universities and higher education institutions in January 1949. It was designed to bring the same result as the VASKhNIL session in 1948 (see doc. 9.6). The idea met with strong opposition from top officials responsible for the development of nuclear weapons; it was given up in spring 1949.

***** *Znamya*, No. 8, 2002, p. 209.

[39] A. F. Shestakova, correspondent of *Pravda* in Krasnoyarksy *krai* (territory) made an absurd denunciation (being mentally unbalanced) that led to repressions against the USSR Ministry of Geology accompanied by numerous discharges of its staff and arrests of professors, practical geologists, and even students. They all were accused of deliberate concealment of information about the existence of uranium, copper, and other ore deposits in Krasnoyarsky *krai.** The arrests started during the night of 30–31 March, immediately after the meeting of Politbureau of the AUCP(b) Central Committee on 30 March. Stalin referred to Shestakova's information that copper ore samples from Minusinskaya Depression had been sent to US museums and asked the minister for geology I. I. Malyshev (since 1937) whether it represented real facts. Malyshev confirmed the information but explained that it dated from the period before the 1930s. Stalin said: "You are responsible for all years [...]"**

Such an absurd claim to subordinates was in the nature of Stalin's character. N. M. Lun'kov, a diplomat, recollects the grand meeting in the Bolshoy Theatre on the occasion of Stalin's 70th birthday in 1949. Lun'kov, acting as an interpreter for the DDR leader, Walter Ulbricht, stopped for a moment and was immediately rebuked by Stalin. "I am thinking how to better translate your important assertion." "You should have thought before" was the reply.***

* See *Poisk*, No. 31–32, 2000, p. 10.

** P. N. Pospelov, editor-in-chief of *Pravda* was present at the meeting of Politbureau on 30 March 1949 and put down this dialogue in his notebook (presently in his personal fund: RGASPI, fond 629, opis' 1, ed. khr. 95a, chast' 2. l. 241.

*** N. M. Lun'kov Russky diplomat v Europe (A Russian Diplomat in Europe). M.: OLMA-PRESS, 1999, p. 67.

[40] RGASPI, fond 17, opis' 118, ed. khr. 777, l.112. Other methods were used when jamming proved impracticable. Administration of the Grumant coal-mine, West Svalbard, leased by the Soviet Union from Norway, prohibited new communication wiring in miners' living apartments and reduced to a minimum the existing ones under the pretext of strengthening fire safety. Simultaneously, supply of radio-transmitters to the local shops was discontinued. The Central Administration of the Northern Sea Route reported these measures to the AUCP(b) Central Committee (20.1.1952).*

* RGASPI, fond 17, opis' 119, ed. khr. 758, l. 88–93.

[41] "29/3, 22 hr 42 min. Wife of M. R. Galaktionov phoned to the emergency care service (of Sanupr)* and said that her husband just attempted hang himself."** "Morbid suspiciousness. Talks about changed attitude to him after he returned from America [...]"***

* Central Health Care Department, USSR Ministry of Health, providing medical aid to officials of the AUCP(b) Central Committee.

** RGASPI, fond 629, opis' 1, ed. khr. 97, l. 72.

*** *Ibid.*, l. 72

[42] P. M. S. Blackett. *Military and political consequences of atomic energy.* L.: Turnstile Press, 1948, pp. 222.

[43] *Novoye vremya*, No. 5, 1949, p. 31.

[44] Documents of the AUCP(b) Central Committee contain dossiers of even such antifascist scientists as Nobel laureates N. Bohr, M. Born, and E. Schrödinger (RGASPI, fond 17, opis' 118, ed. khr. 796, l. 72, 73, 99–100). Dossiers of eminent foreign scientists were kept under Khrushchev's rule too. Archives of the Division of Physico-Mathematical Sciences, the USSR Academy of Sciences, have a characteristic of the Nobel laureate Louis de Broglie* dated from February 1958 with information that he "works on principal problems of quantum mechanics and develops ideas essentially different from the idealistic concepts of the Bohr-Heisenberg school prevailing in the West."**

* An unsigned and dateless document without indication of the purpose for which it has been drawn up.

** Arkhiv RAN, fond 471, opis' 1, ed. khr. 308, l. 61.

[45] The notebook of P. N. Pospelov, editor-in-chief of *Pravda*, contains an entry telling about his night-time meeting with Malenkov to discuss the forthcoming article. "With comrade Malenkov. 27 January 1949, 3 hr 55 min. Corrections to the article 'On one antipatriotic group of theatre critics'. Three terms should be used for the sake of variety: 'Hurrah-cosmopolitanism', where the word 'cosmopolitanism' occurs for the first time; second, 'unbridled cosmopolitanism'; third 'kinless cosmopolitanism'. These corrections inserted, the article may be published in tomorrow's issue."* However, the start of this anti-intellectual campaign actually dates to summer 1948, as evidenced by the letter from O. E. Kurganov, editor of the international sector of *Literaturnaya gazeta* to the secretary of the AUCP(b) Central Committee, G. M. Malenkov (21 June 1949). Kurganov had a quarrel with the editor-in-chief, V. V. Ermilov, and reported a wealth of discrediting information concerning him. According to Kurganov, the secretary of the Union of Soviet Writers, A. A. Fadeev, recommended Ermilov a year ago to inspect the work of the All-Union Theatre Society (VTO), "a seat of formalists alien to Soviet fine arts", with special reference to the activities of Yuzovsky, Gurvich etc. "A publication in *Literaturnaya gazeta* would have greatly contributed to unmasking the antipatriotic group of theatre critics in summer 1948. However, V. Ermilov disregarded this recommendation because at that time VTO was publishing a series of his books about Chekhov's plays."**

The above recommendation of Fadeev coincided with the closed Communist Party meeting at the Institute of Philosophy, USSR Academy of Sciences, held on 19 May 1948, i.e. 8 or 9 months before the nationwide campaign "against cosmopolitanism". The speaker at the meeting was secretary of the local Party bureau P. E. Vyshinsky, who presented the report "Class analysis and national problems in Marxism-Leninism."*** The author extensively used definitions that seemed exotic in 1948. For example: "Cosmopolitanism [...] is nationalism disguised as internationalism. Cosmopolitanism stems from shameful remnants of servility and toadyism characteristic of the most conservative fraction of Soviet intellectuals that has lost links to the people [etc.]."*** There is no doubt that such assertions were formulated by ideologists of the Central Committee to be spoken out by the secretary-general of the Union of Soviet Writers and the Party functionary at the Institute of Philosophy.

True, A. A. Fadeev was the first to apply the formally neutral term "cosmopolitanism" in a negative sense five years before 1948. He wrote a large article published in the Party press under the title "On Soviet patriotism and national pride of the peoples of the USSR", saying that "enemy agents may try [...] and undermine the feeling of national honour and pride in our peoples [...] by their hypocritical propaganda of groundless 'cosmopolitanism.'"*****

* RGASPI, fond 629, opis' 1, ed. khr. 98, l. 1.

** *Ibid.*, fond 17, opis' 118, ed. khr. 451, l. 23–24.

*** The text of this report was later used in the booklet "Cosmopolitanism, idealism, nationalism" published by P. E. Vyshinsky and Yu. P. Frantsev, his co-author, in Budapest (1949) in the Hugarian language: P. Vusinskij, J. Francev. "Kosmopolitizmus, idealizmus, nacionalizmus. Budapest, "Szikra", 1949.

**** TsAODM, fond 2502, opis' 1, ed. khr. 5, l. 61; see also note 3 to this Chapter.

***** "Pod znamenem marksisma", No. 11, 1943, p. 34.

[46] One of the main components of the 1949 ideological campaign was obtrusive anti-Semitic propaganda.* For example, an author subjected to criticism had his Russified pseudonym disclosed** in violation of the existing copyright law.*** It was not a mere coincidence that leading representatives of Soviet Jewish culture were arrested in the same January 1949. The classical Yiddish poet Peretz Markish was caught on the night of 28 January, i.e. when the aforementioned article was published in *Pravda.* Also, S. A. Lozovsky, a known statesman, was dismissed from all his posts "for association with leaders of the spying Jewish nationalist centre on the former Jewish Antifascist Committee."****

One more characteristic feature of the campaign "against cosmopolitanism" was unprecedented hypocrisy. Despite its undisguised anti-Semitic trend, public references to this fact were strongly discouraged. By way of example, a speaker at a meeting of the Communist Party organization at the Institute of World Literature, the USSR Academy of Sciences, stated: "There were attempts in certain Party organizations of Moscow to represent the struggle against cosmopolitanism and bourgeois nationalism as a manifestation of anti-Semitism. Such attempts are provocations of our enemies and must be denounced.***** The hypocrisy of the authorities was immediately responded to in the then popular expression: "Call a Jew cosmopolitan if you do not wish to pass for an anti-Semite"...

After half a century, S. S. Averintsev, a known Russian intellectual, recollected the 1949 campaign "against cosmopolitanism" with its undisguised anti-Jewish slant and reminded that "in the worst time of late Stalinist anti-Semitism [...] the Soviet press remained blameless in terms of political correctness [...] speaking only of *contemptible cosmopolitans*" [italicized in the original publication of *Averintsev*].****** An illustrative example of such "political correctness" was given at the end of Stalin's rule by ideological services of the CPSU Central Committee in response to the letter from I. G. Erenburg to the secretary of Central Committee M. A. Suslov dated 30.1.1953, i.e. at the peak of the "doctors' case" fabricated by Stalin (see Chapter 7). Erenburg told the addressee: "on 29 January, I read written notes of comrade Akshinsky (of the staff of the CPSU Central Committee) on my novels *Den'vtoroy*

(The Second Day) and *Ne perevodya dykhaniya* (Without Taking Breath) published in the early 1930s." The reviewer found that the novels "abound in characters of non-native nationalities", i.e. Jews. The situation further developed unfavourably for V. S. Akshinsky, instructor of the Department of Propaganda and Agitation and the author of more than 100 poems, one of which was included in the volume under the title "Songs About Stalin" (see his note "on the circumstances under which I wrote the review on the novels of I. G. Erenburg" dated 3.2.1953). He forwarded his review to Goslitizdat (State Publishing House) where the collected works of I. G. Erenburg had been prepared for publication since April 1952. Erenburg could read the review two days after he received the International Stalin Prize "For strengthening peace among peoples" (27.1.1953). The award made him "untouchable", at least in the near future, and he competently took advantage of the situation. V. S. Kruzhkov, head of the Department of Propaganda and Agitation drew up a note for Suslov (5.2.1952) informing the secretary of the Central Committee about "a talk with comrade Kotov, director of Goslitizdat, in which it was brought to his attention that the Central Committee did not share the opinion of Akshinsky concerning the novels of I. G. Erenburg [...] as erroneous and inessential; these remarks should not be taken into account when preparing collected works of I. Erenburg for publication."*******

* This formerly unused stylistic device designed to draw the attention of the reader became the sort of "know-how" of the 1949 campaign. It was introduced into routine practice three days after the article in *Pravda* was out by *Kul'tura i zhizn'*, another newspaper of the AUCP(b) Central Committee controlled by Shepilov. The large article published by the newspaper developed propositions of the preceding (*Pravda*'s) one, and for the first time disclosed the pseudonym of the theatre critic Kholodov (of the passport surname Meerovich).

** This practice gave rise to a discussion in the Soviet press of 1951 between Konstantin Simonov (passport name Kirill), the famous poet and publicist, editor-in-chief of the *Literaturnaya gazeta*, and his colleagues Mikhail Bubennov and Mikhail Sholokhov (a future Nobel Prize winner), who did not conceal their anti-Semitic views. M. S. Bubennov published a demagogic article in *Komsomol'skaya pravda* (27.2.1951) under the title "Do we need literary pseudonyms now?" listing the true names of certain writers. K. M. Simonow characterized this article in *Literaturnaya gazeta* (6.3.1951) as an "off-hand and forward" initiative. M. A. Sholokhov supported Bubennov, known in literary circles for his "drunkenness and hooliganism" (see Kul'tura i vlast' ot Stalina do Gorbacheva. Ideologicheskie komissii TsK CPSS 1958–1964. Dokumenty (Culture and Authorities from Stalin to Gorbachev. Ideological Commissions of CPSU Central Committee 1958–1964. Documents). M.: ROSSPEN, 1998, p. 253). Sholokhov stood up for Bubennov with rhetorical questions in his article published in *Komsomol'skaya pravda* on 8.3.1951: "Whom does Simonov defend? What does he defend? He is not easy to understand [...]" Simonov wound up the discussion on 10.3.1951 by reference to the "strange attempts" by Sholokhov to defend Bubennov after he had kept silence for five years, withstanding discussion of the most topical literature problems. Three years later (in 1951), Bubbenov repaid Sholokhov's support. The 2nd Congress of Soviet writers was held in December 1954, i.e. after Stalin's death. Propagandistic work was underway within half a year preceding the Congress. In the framework of this propagandistic campaign, Bubennov wrote to Malenkov, secretary of the AUCP(b) Central Committee, in a letter of 24.9.1954: "many Soviet people [...] hope that the favourite writer of our time

Mikhail Sholokhov will be the main speaker at the Congress, but the heads of the Union of Soviet Writers debar Sholokhov from literary and public activity." (RGANI, fond 5, opis' 17, ed. khr. 486, l. 133; see also Chapter 10 and note 3 therein.)

One episode characteristic of those times as told by Simonov to one of the interviewers took place "at a meeting of the Stalin Prize Committee [...]" Stalin noticed the passport name of an author written in brackets after his pseudonym. "I would ask the respected comrades, what do they mean by writing double names? Doesn't a writer have the right to use a pseudonym? I think he has. You wish to underscore that he is a Jew... Why?" This rhetorical question put by the key initiator of the shameful campaign "against cosmopolitanism" once again emphasizes his principal quality of boundless hypocrisy (V. O. Osipova. Tainaya zhizn' Mikhaila Sholokhova (Secret Life of Mikhail Sholokhov). M.: Libereya, Raritet, 1995, p. 341.

There are more examples of Stalin's pathological guile. The archive of composer D. R. Rogal'-Levitsky keeps a few remarks made by the chief written at a night-time government banquet at the Bolshoy Theatre on the occasion of the presentation of the new Soviet Anthem (1944). Asked about the conductor, N. S. Golovanov, the chief conductor A. M. Pazovsky answered that he will continue to work in the theatre. Stalin: "I don't like him [...] The Bolshoy Theatre is not a proper place for him. This anti-Semite will turn everything upside down [...]" After a short time Golovanov was appointed chief conductor, obviously by the order of Stalin. A. N. Vaksberg. *Stalin protiv evreev* (Stalin Against Jews). N. Y. : House, 1995, pp. 45–48.

One more episode that occurred at a meeting of the Stalin Prize Committee is borrowed from the memoirs of composer Tikhon Khrennikov: "[...] when the works nominated for the prize were for the last time discussed in the presence of Stalin at the end of November or early December 1951 [...] I suddenly perceived his monstrous duplicity and craftiness. It was a hard time when anti-Semitism reached its utmost degree in this country [...] All of a sudden, Stalin remarked: 'Some members of Politbureau turned out to be anti-Semites. It is scandalous!' [...] In fact, this statement was addressed to us intellectuals. We had to tell everywhere that Stalin had nothing to do with the anti-Semitic campaign. He had to be a saint in our eyes. '*Tak eto bylo. Tikhon Khrennikov o vremeni i sebe.*' (It was like this. Tikhon Khrennikov on Time and Himself). M.: Muzyka, 1994, p. 179.

Nevertheless shameless demagogic statements made by the guileful leader sometimes brought fruits. The prime minister of Israel, speaking to members of the government in connection with the "doctors' case" in the USSR (see Chapter 7) said: "Stalin is not an anti-Semite [...] He just had to slander Jews and the Jewish state to reach certain political goals." L. M. Mlechin. "Zachem Stalin sozdal Izrail?" (Why did Stalin create Israel?). M.: Izd-vo "Yauza-Eksmo", 2005, pp. 288–289.

*** *Izvestia*, 38.1.1999. p. 7. The secret trial of Markish and his comrades was held in 1952; they were sentenced to capital punishment and shot dead on the night of 12–13 August.

**** RGASPI, fond 17, opis' 118, ed. khr. 296, l. 63–64. Lozovsky was arrested a few days after he was expelled from the Communist Party (26.1.1949) and executed on 13.8.1952 after the aforementioned trial of the Jewish Antifascist Committee.

***** TsAODM, fond 2472, opis' 1, ed. khr. 5, l. 28.

****** S. S. Averintsev. Preodolenie totalitarizma kak problema (Overcoming Totalitarism as a Problem). "Novy mir", No. 10, 2001, p. 148.

******* "Kul'tura i vlast' ot Stalina do Gorbacheva. Apparat TsK CPSS i kul'tura. 1953–1957. Dokumenty (Culture and Authorities from Stalin till Gorbachev. Apparatus of CPSU Central Committee and Culture. 1953–1957. Documents)". M.: ROSSPEN, 2001, pp. 15–19).

[47] The well-known literary critic T. L. Motylyova was expelled from the Communist Party in 1949 for "systematic serious cosmopolitic mistakes"* by the decision of the Party meeting at the Institute of World Literature, the USSR Academy of Sciences (9.3.1949). She was accused of distorting the image of L. N. Tolstoy, who "Motylyova described as [...] a teacher of Maeterlinck and Shaw, decadents and proponents of imperialistic banditry."** Another anecdotic accusation ran: "she wrote the book *German literature in the struggle against fascism*, published in 1942. There is not a single word in it about Russian culture. The book is harmful."***

* TsAODM, fond 2472, opis' 1, ed. khr. 5, l. 51.

** *Ibid.*, l. 40–41.

*** *Ibid.*, l. 49.

[48] TsAODM, fond 992, opis' 2, ed. khr. 24, l. 20. Interestingly, the question of changing geographic names was raised in the Presidium of the USSR Academy of Sciences ten years after the said letter. It was proposed to rename the Frantz Josef Land in the Barents Sea after Russian scientists Lomonosov and Mendeleev. E. N. Pavlovsky, vice-president of the Russian Geographic Society tactfully drew the attention of the Academy (27.12.1961) to the fact that "it is too late to think of renaming" the archipelago because it received its name from the Austrian scientific expedition that discovered it in 1873. On 6 April, S. G. Korneev informed the Geographic Society on behalf of the USSR Academy of Sciences that specialists of the Institute of State and Law recommended to address the question of changing names given by foreigners "taking into consideration political aspects and international ethics".* This put an end to the attempts to change the historical name of the archipelago.

* Arkhiv RAN, fond 579, opis' 1, ed. khr. 643, l. 9–10.

[48a] The decision did not concern the variety art alone. Classical music was also affected. By way of example, the famous violin play *Interpretation* by the Swiss composer Ernest Bloch had been prohibited from performing since 1949 because some melodies were found to be reminiscent of Old Testament themes. It was excluded from the curriculum of musical schools and was not played until the "thaw period" after Stalin's death.

[49] RGASPI, fond 17, opis' 118, ed. khr. 303, l. 59. Ideologists of Central Committee thoroughly controlled all articles written for the second edition of the *Great Soviet Encyclopaedia*, publication of which started in 1950. The purpose was to exclude everything that was at variance with the current ideological trend. Here is an example of the absurd argumentation with respect to the article "Volya" (Will) by a certain Gurevich, prepared for volume 9 (published in 1951). The reviewers from the Department of Propaganda and Agitation wrote to G. M. Malenkov, secretary of the AUCP(b) Central Committee, that "the author failed to properly characterize the importance of I. P. Pavlov's teaching for the understanding of human actions and to show volitional qualities of Soviet people."*

* RGASPI, fond 17, opis' 119, ed. khr. 538, l. 65–70.

[50] Here is an extract from the expert conclusion of Professor A. Tiselius, who the Nobel Committee in chemistry entrusted with the consideration of nominations of A. N. Frumkin for

the 1946 Nobel Prize: "Frumkin is a very active and productive researcher especially in the field of surface chemistry. He is certainly one of the best-known physical chemists in the Soviet Union."*

* Archive of the Royal Academy of Sciences, Stockholm. Fund of the Nobel Committees in physics and chemistry. Materials of committees for 1946, l. 293.

[51] RGASPI, fond 17, opis' 118, ed. khr. 433, l. 34.

[52] See "Aleksandr Nikolaevich Frumkin. Ocherki. Vospominaniya Materialy (Aleksandr Nikolaevich Frumkin. Sketches. Memoirs. Materials). M.: Nauka, 1989, p. 23. GSE published the following text: The history of physical chemistry "begins from the works of Lavoisier, who formulated the law of conservation of mass in chemical reactions (1770). This law was discovered even earlier by Lomonosov (1745), but his work passed unnoticed."* In other words, there was no reason to speak about underestimation of Lomonosov's role...

* BSE, 1st ed., T. 57, 1936, column 319.

[53] The author wrote in the preface: "There are great theories that highlight the entire areas of complex and tangled labyrinths of nature. Such is the theory [...] of solutions suggested by van't Hoof and Arrhenius [...] The clarity and reliability of the theory [...] make it possible to write a book in strong, bright strokes rendering it understandable and simple to read."* This "mistake" was recalled fifteen years later by Semenov's ill-wishers who found support from ideological services of the AUCP(b) Central Committee. The USSR Council of Ministers took a decision (16.7.1949) to hold a regular general meeting of the USSR Academy of Sciences in December the same year to elect new members. In the course of preparation, the Presidium of the Academy submitted lists of members of expert commissions for each Division to the AUCP(b) Central Committee. Yu. A. Zhdanov, head of the sector of science, struck the name of Semenov off the list of experts for the Division of Chemistry.**

The abuse of foreign science became commonplace and an instrument by which to seek benefits from authorities. Some requests of this kind looked like nasty anecdotes. By way of example, the editor-in-chief of the newsletter *Meteorologia and gidrologia* (Meterorology and Hydrology) wished to turn the edition into a monthly journal and emphasized in the letter to the AUCP(b) Central Committee (11.3.1950) that it greatly contributed to "the struggle against cosmopolitanism and servility to foreigners, in support of the priority of Russian and Soviet science." This contribution was exemplified by the recently published article "On certain principal questions of short-term weather forecasts" that "revealed the mechanistic approach to resolving problems of weather forecast that originates from the works of foreign authors."***

*N. N. Semenov. *Tsepnye reaktsii* (Chain Reactions). Leningrad: Goskhimizdat, 1934, p. 4.

** RGASPI, fond 17, opis' 118, ed. khr. 673, l. 59.

*** *Ibid.*, ed. khr. 914, l. 194.

[54] TsAODM, fond 8099, opis' 1, ed. khr. 13, l. 1. Moscow University is mentioned here because the reactionary leadership of its physics department played a key role in defaming Semenov in 1949.

[55] "Nelinienye volny. Fizika i astronomiya (Non-Linear Waves. Physics and Astronomy)". M.: Nauka,1993, pp. 12–13.

[56] See the author's article for detail. *Poisk*, No. 25, 2000, p. 11.

[57] A similar error was made by the journal *Novoye vremya* (No. 41, 1949). Germanist I. I. Anisimov described Professor Barker Farley of Toronto, vice-president of the National Council of Canadian-Soviet Friendship, as an "obscurantist and pseudoscientist." The author imputed to Professor Farley participation in the Goethe Congress (USA) in July 1949, organized by the "reactionary H. Hoover" and of the authorship of a newspaper article where he "interpreted Goethe's ideas of the Utopian state as Socialist aspiration and predicted that in 50 years Marxists will recognize Goethe as a partisan of their doctrine."* The telegram from the Soviet embassy in Ottawa characterizing Farley as "a progressive scientist known for his sincere sympathy with the Soviet Union" gave reason to the Secretariat of the Central Committee to reprove Anisimov for his "irresponsible attitude toward publications in press" and impose penalty on him.**

* Tagesspiegel, 28.7.1949.

** RGASPI, fond 17, opis' 118, ed. khr. 717, l. 60.

[58] The journal *Voina i rabochiy klass* was founded by the trade-union newspaper *Trud* (Labour) in June 1943.* The announcement of the appearance of its first issue said that the journal is open "for wide public circles to express opinion on all topical questions of international relations and foreign policy of different states [...]"** The newspaper has been published under its present name since the second half of 1945.

*Hence the word combination "working class" in the name of the newspaper.

** *Voina i rabochiy klass*, No. 1, 1943, p. 4.

[59] Nobelprisen 50 ar. Forskare. Diktare. Fredskampar, Stockholm, 1950.

[60] The Russian translation of B. Russell's lectures "The impact of science on Society",* read at the University of Columbia in November 1950 after the author received the Nobel Prize in literature, was published in 1952 for "very restricted administrative use". In the foreword to this edition, Russell was called an "inveterate advocate of American imperialism"; see also note 30 in Chapter 12 and note 1 in Appendix 3.

* B. Russell. *The Impact of Science on Society*. New York, 1951.

[61] During the Great Patriotic War, secret radiograms from General Headquarters referred to I. V. Stalin as comrade Vasiliev and G. K. Zhukov as comrade Konstantinov, etc.

[62] There was one exception. A few months prior to this article, the journal *Novoye* vremya published one more paper signed "S. Maksimov."* Like the 1951 article, it was a sort of review of a new book, the biography of Trygve Lie, UNO Secretary-General and former minister for foreign affairs (Norway), published in Oslo in 1950. The two articles are similar in terms of style and scathing characterisation: Trygve Lie — "a henchman of American imperialism"; Andre Gide — "an accomplice of American imperialists"...

*S. Maksimov. Biografiya stavlennika amerikanskogo imperializma (Biography of a Henchman of American Imperialism.) *Novoye vremya*, No. 44, 1950, pp. 28–31.

[63] Such actions frequently took place in the post-Stalinist period too. For example, Soviet scientists could not participate in the prestigious scientific forum dedicated to the 100th anniversary of A. Einstein held at Princeton University, which brought together prominent representatives of modern natural sciences. In this way, the AUCP(b) Central Committee protested against the official position of the US National Academy of Sciences that expressed its support of A. D. Sakharov.

[64] TsAODM, fond 7334, opis' 1, ed. khr. 3, l. 37, 39. The author had a talk (22.3.2002) with S. M. Nikol'sky, member of the USSR Academy of Sciences, who had taken part in the work of the Congress in Budapest as a member of the Soviet delegation in 1950. He told that "many delegates from Eastern European countries collected signatures under various welcoming addresses. One of the Czech delegates asked for my signature, but I did not sign anything. On the whole, our foreign colleagues felt more at ease than the Soviet participants. Governments of socialist countries treated scientists far more favourably than the Soviet leadership. The key figure in our delegation was not its official head, I. M. Vinogradov, member of the USSR Academy of Sciences but his deputy appointed either by the AUCP(b) Central Committee or Ministry of State Security [...]"

[65] RGASPI, fond 17, opis' 132, ed. khr. 523 , l. 4. "The serious mistake" of the newspaper fell in the period of fierce air-battles during the Korean War of 1950–1953, provoked by Stalin.* Thousands of Soviet pilots and servicemen (not only regular military personnel but also specially mobilized ones) fought on the side of the North Korean and Chinese armed forces.

* N. S. Khrushchev said in a talk with Carlos-Rafael Rodrigez, member of the leadership of the Cuban People's Socialist Party, that the Korean War of 1950–1953 was unleashed by Stalin and Mao Tse-tung rather than the Americans. The talk took place on 11.12.1962, i.e. soon after the Caribbean crises when the whole world was put on the brink of a catastrophic clash between the USSR and the USA. *Novaya i noveishaya istoriya,* No. 5, 2003, p. 77.

[66] In 1964, A. M. Prokhorov and N. G. Basov shared the Nobel Prize in physics with the American scientist Ch. H. Townes (who independently resolved the same problems) for "fundamental work in the field of quantum electronics which has led to the construction of oscillators and amplifiers based on the maser-laser principle".

[67] In half a century, Professor A. S. Khoroshev summarized some data related to the discovery of the first birch bark writing (a promissory note of the fourteenth century). Further studies confirmed the great importance of the finding and introduced into scientific use a wealth of previously unknown materials that gave evidence of a high literacy rate of the population of north-east Europe (peasants and monks, men and women) in the first part of the second millennium. A total of 1002 documents have been found over 50 years, 915 of them in Novgorod, where soil properties are most favourable for bark preservation. People of those times extensively used birch bark as the writing material for correspondence not only on political and trade matters but also on everyday social contacts. By way of example, a birch bark note sent by a young man to a girl reads: "And I am waiting for you..."*

**Izvestiya*, 24.7.2001, p. 12.

[68] Yu. V. Knorozov. Kratkie itogi izucheniya drevnei pis'menosti maiya v Sovetskom Soyuze (Results of Investigations into the Ancient Written Language of Maya in the Soviet Union). M.: Izd-vo AN SSSR, 1955.*

* From: "Doklady Sovetskoy delegatsii na X Mezhdunarodnom kongresse istorikov v Rime." The reports were published in Russian and English.

[69] After the creation of the betatron in 1940, Kerst was nominated for the Nobel Prize in 1943 by Professor F. Watson (University of California) and Professor H. Alfven of Stockholm (future Nobel laureate, 1970). Thereafter, he was nominated every year (alone or together

with other candidates) from 1945 until at least 1953 (excepting 1950). Alfven was his nominator again in 1952.

[69a] The delayed recommendation of Bialobrzeski was considered by the Nobel Committee after his death in 1953...

[70] "Vospominaniya o V. I. Vekslere pod red. M. A. Markova i A. N. Gorbunova (Recollections About V. I. Veksler)". M.: Nauka, 1987, pp. 20–221.; Lawrence emphasized "logic in the development of science that leads to simultaneous discoveries in different parts of the world".

[71] Colleagues of Veksler fairly well understood the importance of his idea. The jury at the competition of scientific works at the Physical Institute, the USSR Academy of Sciences, in 1944 came to the following paradoxical conclusion concerning two of Veksler's papers: "If the work of V. I. Veksler is correct, it is not for us to award him; if it is wrong, it does not deserve the prize... Anyway, it is interesting and he should work further."*

* "Vospominaniya o V. I. Vekslere...", p. 20.

[72] *Ibid.*, p. 248.

[73] McMillan shared the Nobel Prize in chemistry with Seaborg in 1951 "for their discoveries in the chemistry of the transuranium elements" (not for the method of construction of high-power accelerators of elementary particles).

[74] Siegbahn formulated his nomination of 1948 as follows: "E. M. McMillan, Berkeley, with V. Veksler, Moscow, or D. V. Kerst, Urbana."

[75] The lack of information from the Soviet Union accounted for the following formulation of Siegbahn's proposal for the 1948 Nobel Prize: despite the incontestable priority of Veksler as the discoverer of the principle of autophasing, "he did not demonstrate, as far as I know, the possibility of its technical application in experiment."* Nevertheless, the authors of this discovery, Veksler and McMillan, received the international prize "Atom for Peace" instituted by the American "Ford Motors" Foundation for outstanding merits in the field of the peaceful use of atomic energy. The laureates received a gold medal and prize money ($70,000). The prize-giving ceremony was held at the Rockefeller University.

* Archive of Royal Academy of Sciences, Stockholm. Fund of Nobel Committees in physics and chemistry. Materials of committees for 1948, l. 468.

[76] In the same 1951, the Nobel Committee in physics received proposals to award the prize to McMillan and Kerst from Professor E. Hulten of the Stockholm Technological Institute. T. Svedberg nominated McMillan alone; he simultaneously sent his nomination of McMillan and Seaborg to the Nobel Committee in chemistry. The two nominees received the Nobel Prize the same year.

[77] Besides the nomination of McMillan and Seaborg for the prize in chemistry by three Nobel laureates in 1951, the candidacy of Seaborg alone was submitted by five other nominators including the 1936 Nobel prize winner E. Fermi, then professor of the University of Chicago. Two of the remaining four were professors of the Universities in Strasburg and Amsterdam, for whom it was their first nomination.

[78] V. P. Filatov, an eminent ophthalmologist, was born in the city of Odessa, famous for the inherent wittiness of its residents. In 1955, his 80th anniversary was celebrated at the

University of Odessa, where Filatov's portrait hung over the scene featuring him in the midst of a large eye. Filatov noted with a smile: "I am lucky indeed not to be a gynaecologist." Had Filatov known about the nomination from Rumania, his response to the lack of nominations from compatriots might have been equally striking even if hidden because he was well aware that the inactivity of Soviet scientists in this context resulted from the unannounced prohibition of contact with the Nobel organizations imposed by the Kremlin.

[79] The nickname of Timofeev-Resovsky given by writer Daniil Granin. Nikolai Vladimirovich graduated from Moscow University in 1925 and was sent by the Institute of Experimental Biology, People's Commissariat of Health, to continue scientific work in Germany. He carried out joint biophysical studies with German scientists. In 1937, Timofeev-Resovsky disregarded the demand of a Soviet representative in Berlin to immediately return to Russia. In 1945, he was found by agents of the intelligence service in the Soviet occupation zone, arrested on 13 September, and brought to Moscow to be sentenced for "high treason". The group of arrested persons also included several German geneticists working at the Institute of Genetics and Biophysics.* Timofeev-Resovsky was exiled to the Urals where he worked in the so-called "*sharashka*"**, together with his German colleagues, continuing investigations into the effects of radiation on human genomes started in Germany. He was released after Stalin's death.***

*Zh. A. Medvedev. Stalin i atomnaya bomba (Stalin and the Atomic Bomb). "Vestnik RAN", No. 1, 2002, p. 62. Interestingly, after the seizure of Berlin by the Soviet troops the clinical departments of this institute in Berlin-Buch were turned into a field hospital of the 3d Army. It is in this hospital that autopsies were performed to identify the burnt corpses of Adolf Hitler and Joseph Goebbels.

** Colloquial name of research institutions created in GULAG for the work of arrested scientists.

*** D. Vronskaya, V. Chugaev. Kto est' kto v Rossii i byvshem SSSR. Vydayushchiesya lichnosti byvshego Sivetskogo Soyuza, Rossii i emigratsii (Who is Who in Russia and the Former USSR. Outstanding Personalities of the Former Soviet Union, Russia, and Emigration). M.: Terra, 1994, p. 534.

[80] Isaiah Berlin called him "my friend [...] Bowra, expert in Russian affairs".*

* *Oktyabr'*, No. 11, 2000, p. 15.

[81] RGASPI, fond 17, opis' 118, ed. khr. 261, l. 129–130, 133. Only a few copies of the total 25.000 are known today, those transferred (under the existing regulations) to the Knizhnaya palata (Book Chamber) immediately after printing was over. The negative attitude of Party ideologists toward Pasternak's poetry was apparent before 1948. The decision of the AUCP(b) Central Committee Secretariat dated 20.9.1947 excluded *Poems by Pasternak* along with 80 other editions from the list of foreign publications ordered by Saltykov-Shchedrin Public Library, Leningrad.*

* RGASPI, fond 17, opis' 116, ed. khr. 320, l. 109; in all likelihood, it was the book *B. Pasterna: Selected Poems.* London, Drummond, 1946.

[82] Swedish Academy, Stockholm. Archive of Nobel Committee in literature for 1947.

[83] The words of Erenburg's politicized works are consonant to the following episode that took place in the USSR Ministry of Foreign Affairs in 1949. One of its workers informed the Communist Party organization of the Ministry that a colleague of his "expressed critical attitude to Erenburg's novel *Thunderstorm* saying that it did not deserve the Stalin Prize,*

that he would not award the Stalin Prize for this novel". This comment was classified as "anti-Soviet".**

* "Thunderstorm" was awarded the 1st grade Stalin Prize in 1948.

** RGASPI, fond 17, opis' 118, ed. khr. 872, l. 47.

[83a] B. Boyd. Vladimir Nabokov. Russkie Gody (Vladimir Nabokov. Russian Years). M.: Izd-va "Nezavisimaya gazeta" and "Symposium", 2001, p. 458.

[84] RGASPI, fond 17, opis' 118, ed. khr. 207, l. 12.

[85] Here are a few typical examples, for illustration. A "Memorandum" drawn up by the Staff Department of the AUCP(b) Central Committee in July 1946 noted that V. I. Pozdnyakov was expelled from the Communist Party in 1937 "for relations with the enemy of the people, Skorospeshkin". Later, Skorispeshin was released from arrest and Party membership was restored to Posdnyakov in 1940 as "expelled without grounds".* The characteristisation of B. V. Gruzdev presented for approval by the head of a department of the State Committee on New Technologies contained detailed data on his expulsion from the AUCP(b) "for psychology dominated by intellectual tradition" in 1921 and severe reprimand "for unwarranted trustfulness of subjects that subsequently proved to be the people's enemies" in 1949".** Documents of a serviceman who was to be appointed by the AUCP(b) Central Committee in as a division commander (May 1948) mentioned that he had been expelled from the Lenin's Young Communist League in 1932, when he was twenty, "for pornographic drawings on official papers".***

* GASPI, fond 17, opis' 118, ed. khr. 107, l. 203–204.

** *Ibid.*, ed. khr. 86, l. 175.

*** *Ibid.*, ed. khr. 45, l. 75.

[86] The mention of A. A. Akhmatova's name in the 1946 resolution of the AUCP(b) Central Committee was predetermined by the negative attitude toward her works of A. A. Zhdanov, the main organizer of the first post-war anti-intellectual campaign. His specific interpretation of Akhmatova's verses was autographically fixed in 1940 on the first page of a message under the title "About the collected poems of Anna Akhmatova" passed to him on 25.9.1940 (the signature is illegible). His instruction to subordinates in the Department of Propaganda and Agitation, handwritten with blue pencil ran as follows: "Attn: comrades Aleksandrov and Polikarpov. Publication of Churilov's collected verses* by Sovetskiy pisatel' is followed by 'poems' by Akhmatova.** I was told that the head of this publishing house also leads the Molodaya *gvardiya.* It is a shame to publish such books. How could Akhmatova's 'lechery with a prayer to the glory of God' appear at all? Who allowed it? What is the position of Glavlit? Please, clarify and submit proposals. Zhdanov." The task was diligently fulfilled, the results reported to Zhdanov,*** and the book was destroyed. Today, the sole copy of it is exhibited in the Book Museum of the Russian State Library.

It may be thought that Stalin also contributed to the aforementioned resolution. He was displeased by the news of the visit of I. Berlin, an emigrant and a member of the British mission, to Akhmatova's house in Leningrad in summer 1945. Even greater irritation was caused by the enthusiastic welcome given to the poetess at the public meeting with Leningrad writers in the Hall of Columns (Dom Soyuzov) on April 3, 1946. The audience twice stood up to welcome Akhmatova.**** At this news, Stalin asked angrily: "Who organized this rising?"*****

* *Stikhi Tikhona Churilina* (Poems of Tikhon Churilin). M.: Sov. Pisatel', 1940.

** *Iz shesti knig. Stikhotvoreniya Anny Akhmatovoy* (From Six Books. Poems of Anna Akhmatova).

*** RGASPI, fond 17, opis' 117, ed. khr. 192, l. 152, 153.

****From the memoirs of I. G. Erenburg: "Anna Akhmatova was among those reciting poetry. She was welcomed with great enthusiasm. Two days later, Anna Andreevna was my guest. When I mentioned the meeting, she said: 'I don't like it [...] And what is more important, those at the top do not like it [...]' I tried to reassure her saying that we do not live in 1937... I was still naive at 50." I. Erenburg. Sobranie sochineniy v 8 tomakh. T. 3 (Collected Works in 8 volumes. V.8). M.: Khudozhestvennaya literatura, 2000.

*****A dateless note by Akhmatova made in 1962: "A literary party in the Moscow writer's club [a slip of the pen: in the Column Hall] when the audience stood up twice. Stalin asked: "Who organized this rising?" *Zapisnye knizhki Anny Akhmatovoy* (1958–1966) (Notebooks of Anna Akhmatova (1958–1966)). Torino: Einaudi, 1996, p. 204.

[87] The resolution of the AUCP(b) Central Committee "About journals *Zvezda* and *Leningrad* was issued on 16.8.1946. The secret order of Glavlit No. 42/1629-f "On the withdrawal of books by M. M. Zoshchenko and A. A. Akhmatova from the book trade and public libraries"* dated 27.8.1946. Two books of Akhmatova's poems that had been printed by that time in the *Pravda* publishing house, Moscow, and the "Goslitizdat" affiliation in Leningrad, with a circulation of 100,000 and 10,000 respectively, were destroyed.**

*GARF, fond P-9425, opis' 1, ed. khr. 404, l. 57; see also "Ideologicheskie komissii TsK CPSS 1958–1964. Dokumenty (Ideological Commissions of CPSU Central Committee 1958–1964. Documents). M.: ROSSPEN, 1998, p.97.

** *Znamya,* No. 10, 2001, pp. 140–141.

[88] This first post-war anti-intellectual campaign was so close to the hearts of ideological heirs of Stalin and Zhdanov that the absurd accusations contained in the resolution of the AUCP(b) Central Committee remained in force and all attempts to revise them were severely suppressed. Yu. G. Burtin recollected: "I had just joined the 'Sovetskaya entsiklopedia' publishing house when a conflict broke out around the article 'Russian Soviet Literature', written for GSE.* It was provoked by my refusal to sign it as the editor after the heads of the publishing house crossed out from the article the critical appraisal of Zhdanov's campaigns against literature and demanded to insert negative characterisations of the poem "Terkin na tom svete [Terkin in the other world]" by A. Tvardovsky), the novels *Master and Margarita* by M. Bulgakov and *Ne khlebom edinym* (Not by Bread Alone) by V. Dudintsev or to not mention them at all. After long debates with the participation of top officials of the Communist Party and the government including the chief ideologist M. A. Suslov, the article appeared as it was designed by the heads of the publishing house,** and I was debarred from editing Soviet literature in encyclopaedic editions."***

*BSE, 3-e izdanie, T. 22, 1975, pp. 274–280.

** *Ibid.,* p. 278. "Resolution of the AUCP(b) Central Committee on ideological issues ('About journals *Zvezda* and *Leningrad* etc.') adopted in 1946–1948 had great importance for the development of post-war literature. They were directed to raise ideological work in the country to a higher level, emphasize the mission of Soviet literature and art [...]" etc.

*** Yu. G. Burtin. "Druzhba narodov", No. 2, 2001, p. 148.

[89] D. S. Danin. Bremya styda (The Burden of Shame). Izd-vo: Moskovskiy rabochiy", 1996, p. 266. This "passion" was recalled in three years at the meeting of the Communist Party organization of the Union of Soviet Writers (30.11.1949). The meeting was convened for the sole purpose of discussing "the personal case of the AUCP(b) member A. K. Tarasenkov". V. V. Vishnevsky, the editor-in-chief of the journal *Znamya* said: "In 1946, Tarasenkov brought me a large article glorifying Pasternak. I wrote a letter to him: "I still think of Pasternak's portrait as you see it. You are about to take an erroneous and scandalous step. It will have bad consequences [...] You forget what the word "great" means in the people's, the party's conscience [...]"* The article was rejected.** Pending the coming storm, Tarasenkov tried to keep himself afloat in the dirty ocean of anti-cosmopolitanism. Many years later A. T. Tvardovsky wrote in his notebook that Tarasenkov had "betrayed Pasternak"*** meaning in his scandalous *Critic's Notes*, published a month before the above Party meeting in the October issue of *Znamya*,**** where he was deputy editor-in-chief. The recent admirer of Pasternak characterized his works as "the most conspicuous manifestation of decadence", as "philosophy of a convinced enemy of sensible ideological poetry" for which "he is supported by enemies of the Soviet people". The fate of Tarasenkov shows how the Regime managed to subdue people.

* A warning. Stalin was the only "great" living person at those times.

** TsAODM, fond 8131, opis' 1, ed. khr. 26, l. 14.

*** *Znamya*, No. 11, p. 161; note of 3.4.1964.

**** *Znamya*, No. 10, 1949, pp. 167–168.

[90] It may be thought that the name of Akhmatova was first associated with Pasternak by Nukolai Gribachev, a functionary of the Union of Soviet Writers and one of the key initiators of persecutions in the period of "struggle against cosmopolitanism". In the article published in *Pravda* (16.2.1949), he described Daniil Granin as an heir of "infamous methods of cosmopolites who persecuted Mayakovsky and glorified B. Pasternak and A. Akhmatova". In other words Gribachev could borrow the association of the names of the two poets from his earlier publications mentioning Tarasenkov among others.

[91] Such additions to official documents were in the order of things even if the basic legal material was quite enough to condemn a "guilty" person. Here is my personal memory. Having graduated from a geological prospecting institute, I worked in a field team in Ryazanskaya *oblast'* (province). The "host" of the Skopinsky district was Z. S. Alekseev, first secretary of the local AUCP(b) committee, known as a strong leader. In summer 1952, his wife was caught when she sold meat from the meat-packing plant in Skopino. Alekseev's guilt was evident and he was sentenced to a seven-year imprisonment. But the sentence also mentioned that Alekseev had served as a policeman in the territory occupied by German troops during the Great Patriotic War. His service record found in AUCP(b) Central Committee archives reads: "Alekseev Zinoviy Stepanovich was in the armed forces since 1941 until 1947."* Interestingly, the secret services distributed similar slander about the future Nobel laureate A. I. Solzhenitsin thirteen years after the trial of Alekseev: "By that time [1965], the practice of "official slander" began to spread, impossible in an open society because the accused can always expose it, but in our closed society it is a sure business: the press keeps silent [...] and speakers at closed meetings can safely pronounce any lie about a disagreeable person at the command from the top [...] This practice was still in bud, had not yet acquired a definite form, but I was already accused of high treason, having been taken prisoner and served as a policeman.**

* RGASPI, fond 17, opis' 119, ed. khr. 266, l. 90.

** A. I. Solzhenitsin. Bodalsya telenok s dubom (The Oak and the Calf). M.: Soglasie, 1996, p. 110.

92 Correspondence of Gorky and Rolland, published and annotated by the A. M. Gorky Fund, contains 122 of the Russian writer's letters;* the editors emphasized that the published materials are "complete".** The first and last letters from Rolland to Gorky date from 27.1.1917 and 5.6.1936 respectively. Gorky died on 18.6.1936.

* M. Gorky i R. Rollan. Perepiska (1916–1936) (M. Gorky and R. Rolland. Correspondence (1916–1936). "Arkhiv Gor'kogo", T. 15. M.: Nasledie, 1995.

** *Ibid.*, p. 9.

93 *RGASPI, fond 17, opis' 118, ed. khr. 553, l. 61, 62. In his letter to the AUCP(b) Central Committee, Fadeev meant some messages from Rolland written in 1922–1923 after he had read Gorky's works translated into French. The letter of 4.4.1922 (after reading *Childhood*) ran: "[...] I did not plunge as deep as you did into the depths of the awful lot of this ignorant, agitated people devoid of moral principles, people that do not know themselves and deny themselves twenty times a day like the apostle Peter."* The letter of 12.10.192 (after reading *About Russian Peasantry*): "Throughout your life, you have borne in yourself they sufferings of this dreadful Russian people, who you wanted and could not save."** In the letter of 28.8.1923, the great French writer predicted the pseudopatriotic rhetoric of 1949 and warned of its danger for the society with specific Soviet mentality: "It has long been observed that the Slavs and the French are the best psychologists in Europe. Both are capable of introspection that can turn into greatness or dangerous mania: those who look into an abyss too long may have dizziness. Such risk is negligible in France where the 'abyss' is sensibly cleared, ploughed, and planted; if you fall down you find yourself on the soft bed of cultivated land. For you, Russians, the abyss is still pristine with all its terrible fauna hiding under stones. Beware dizziness!"***

*"Arkhiv Gor'kogo", T. 15, p. 31.

***Ibid.*, p. 44.

****Ibid.*, pp. 63–64.

Document 6.1

From: Swedish mission in Moscow.
To: VOKS.
31.1.1947

The Swedish mission has the honour to attach a letter from Professor Sven Johansson, chief physician of the Salgren Hospital, Stockholm, and asks to pass it to the addressee, Professor Klyueva. If Professor Klyueva answers the questions of Professor Johansson concerning cancer therapy,[1] the mission is always ready to transmit her replies to Professor Johansson. The mission thanks the All-Union Society of Cultural Relations with Foreign Countries for the assistance.[2]

In the top left corner: "Attn: comrade Shumilov. Please send the letter and follow up the reply. Zharikov. 5.2.1947." "For execution. 7.2.1947." (The signature is illegible.)

GARF, fond 5283, opis' 20, ed. khr. 273, l. 151.

[1]All international correspondence, including private letters, concerning scientific and cultural relations could be carried on only through the agency of VOKS.
[2]The letter carries the stamp of the Swedish mission in Moscow; the signature of the sender is absent.

Document 6.2

From: Professor Sven Johansson.
To: Professor H. G. Klyueva.

Dear colleague,

I have read your paper in *Le medicine de France*[1] and it interested me very much indeed. I have experimented with a variety of non-invasive methods for the treatment of cancer for many years; certainly without success. You work appears to open up quite new prospects. As soon as I have an assistant as I am a surgeon myself), I shall be able to experiment in the direction outlined in your work.

I should greatly appreciate your sending me a reprint of your paper that I would have translated for me [...] I would be indebted if you sent me a specimen of your preparation (for a certain consideration) via our envoy in Moscow [...][2]

GARF, fond 5283, opis' 20, ed. khr. 273, l. 153.

[1] Medicine de France, 7, No. 2, 1947.
[2] The original letter from Professor Johansson was not found; the translation was done by the translator of the Swedish mission in the USSR.

Document 6.3

From VOKS.
To: Swedish mission in Moscow.
No. 63.[1] 12.2.1947.

The All-Union Society of Cultural Relations with Foreign Countries has the honour to acknowledge the receipt of the letter from Professor Sven Johansson kindly dispatched to us by the mission. The letter was passed to Professor Klyueva N. G. [2]

[1] GARF, fond 5283, opis' 20, ed. khr. 273, l. 149.
[2] The reply of Klyueva is unknown.

Document 6.4

From: P. L. Kapitsa, member of the USSR Academy of Sciences.
To: V. M. Molotov, chairman of the Council of People's Commissars.
6.7.1936, in relation to the article against N. N. Luzin in the newspaper *Pravda* dated 5.7.1936.

[...] first Lazarev was arrested[1], then Speransky sacked[2], now Luzin has been oppressed. No wonder such scientists as Uspensky,[3] Chichbabin, Ipatiev[4] etc. escaped from this 'gentle' treatment. I know myself how heartless you can be when dealing with people."[5]

In the top right corner: "Return to Kapitsa as unwanted. Molotov."[6]

From the archive of the Kapitsa Museum; published by P. E. Rubinin in "Vestnik RAN", 64, No. 6, 1994, p. 510.

[1]P. P. Lazarev, member of the USSR Academy of Sciences, was arrested in 1931 and then exiled to Sverdlovsk.

[2] M. N. Speransky was excluded from membership of the USSR Academy of Sciences in December 1934, after the assassination of S. M. Kirov "for participation in counterrevolutionary actions".

[3] Ya. V. Uspensky, member of the USSR Academy of Sciences, left for the USA in summer 1929 and did not return to the Soviet Union; he was excluded from membership of the USSR Academy of Sciences on 29.11.1930.

[4] D. E. Chichibabin and V. N. Ipatiev, members of the USSR Academy of Sciences, were sent (30.6.1930) to participate in the 2nd International Congress of Energetics held in Berlin and did not come back; they were expelled from membership of the USSR Academy of Sciences on 29.12.1936.

[5] Kapitsa came to Leningrad to take part in the VII Mendeleev Congress in summer 1934 from the UK, where he worked in the laboratory of E. Rutherford; he was not allowed to return to Great Britain.

[6] Eight months before this angry reply from the prime minister, he was equally displeased with the scientist's letter to Stalin dated 1.12.1935 (the chief sent a copy of the letter to Molotov). Kapitsa complained of the attitude of Soviet bureaucracy. Molotov wrote on the typewritten copy: "We have more reasons to be angry with Kapitsa than he has [...]"*

* V. D. Esakov, P. E. Rubinin. Kapitsa, Kreml' i nauka Tom. 1. Sozdanie instituta fizicheskikh problem. 1934–1938 (Kapitsa, Kremlin, and Science. Foundation of the Institute of Physical Problems. 1934–1938). M.: Nauka, 2003, p. 364.

Document 6.5

From: G. P. Shumilov, head, Department of Scandinavian Countries, VOKS.
To: A. S. Kaplin, VOKS representative in Stockholm.
No. 421/occ, 2.7.1952

VOKS received a message from N. P. Suvorov, Cand. Phys.-Math. Sci., historian of discoveries and inventions in physico-mathematical sciences, technology, natural sciences, and geography. He asks for assistance in obtaining [...] "Lists of Nobel Prize Winners in Physics, Chemistry etc. Please collect these materials if you can and send them to VOKS.

In the top left corner is a note by Kaplin in pencil: "Fulfilled, file 800, 15.8.1952."

AVP RF, fond 200, opis' 38, papka 70, ed. khr. 37, l. 65.

Document 6.6

From: A. S. Kaplin, VOKS representative in Stockholm.
To: G. P. Shumilov, head, Department of Scandinavian Countries, VOKS.
No. 830/292 shv, 18.8.1952

With reference to your No. 421/oos, I am sending you the list of scientists awarded Nobel Prizes.[1]

AVP RF, fond 200, opis' 38, papka 70, ed. khr. 37, l. 85.

[1] Sending such a document via VOKS can be accounted for by the fact that lists of Nobel laureates up to 1951 inclusive were either absent in Soviet public libraries or kept in closed funds inaccessible to readers.* Regular information about awards of Nobel Prizes was not published in the Soviet mass press up until the early 1990s.

*The Nobel Foundation systematically supplied the Soviet embassy in Stockholm with its reference materials. For example, the VOKS representative in Sweden, F. M. Mal'gin, sent to E. P. Meleshko, head of the Department of Scandinavian Countries a "Calendar of the Nobel Foundation" for 1947 (30.9.1947); it was readdressed to the Presidium of the USSR Academy of Sciences (20.10.1947).
(GARF, fond 5283, opis' 20, ed. khr. 271, l. 134–135).

Document 6.7

From: F. M. Mal'gin, VOKS representative in Sweden.
To: V. S. Kemenov, head, VOKS Executive Board.
No. 810/267, 20.6.1947

Swedish physicist Manne Siegbahn, working together with physicist T. Svedberg over nuclear fission, addressed to the Soviet envoy[1] a request for assistance in obtaining copies of the journal *Nauka i Zhizn'*, No. 3 and 13, 1994. According to Siegbahn, these issues of the journal contain a paper by Soviet physicist Kapitsa about liquid helium[2] of great interest for the Swedish Academy of Sciences and important for its scientific research.

We do not know what sort of papers are those mentioned by the Swedish professor and whether we are entitled to pass such publications to foreigners. Please clarify this issue and inform us about the result.[3] In case of a positive decision, please send us the said issues of *Nauka i zhizn'* that will be given to Siegbahn.[4]

GARF, fond 5283, opis' 20, ed. khr. 270, l. 102.

[1] I. S. Chernyshev.

[2] Academician P. L. Kapitsa. Superfluidity of liquid helium II. *Nauka i zhizn'*, No. 3, 1944, pp. 113–17. The above document contains a mistake by the translator. It should be thought that

the original letter of Siegbahn (absent in the archive file) referred to *Nauka i zhizn'* 3, 13, 1944, where the first figure is the issue number and the second the number of the first page of the article of interest.

[3] The request is absurd because the paper was published in the popular science journal with a circulation of 35,000 copies (in 1945) that was distributed via subscription and sold freely by "Soyuzpechat'".

[4] Siegbahn wanted to have the article by Kapitsa published in a popular science journal because he knew the author's papers on liquid helium that appeared in 1941 in the Soviet *Journal of Physics*", printed in the English language: "The Study of Heat Transfer in Helium II" (V. 4, No. 3, 1941, p. 181) and "Heat Transfer and Superfluidity of Helium II" (V. 5. No. 1, 1941, p. 59. They were translations of Russian language publications in the *Zhurnal of eksperimental'noy i teoreticheskoy fiziki* (10.10.1940 and 16.5.1941 respectively). The article in *Nauka i zhizn'* appeared three years later, and Siegbahn might hope to find in it the most recent experimental results absent in the 1941 publications. He needed all these materials in the first half of 1947 for the nomination of the Soviet scientist for the 1948 Nobel Prize (see Table 6.1).

Document 6.8

From A. V. Karaganov, deputy head, VOKS Executive Board.
To: N. G. Bruevich, academician-secretary of the USSR Academy of Sciences.
No. 445, 24.7.1947

Please find enclosed for your consideration the following requests of Swedish researchers:

1. The known Swedish physicist Manne Siegbahn working on nuclear fission asks for copies of the journal *Nauka i Zhizn'*, No, 3 and 13[1] with the article by P. Kapitsa about liquid helium.
2. Torsten Ensson, a researcher from the Institute of Physics, Geteborg Higher Technical School, asks for a copy of *Astronomicheskiy zhurnal*, volume 23, No. 6 with a paper by I. Shklovsky about radiowaves emitted by the Milky Way and the Sun.[2]
3. Holger Arbman, professor of archeology of prehistoric and medieval times, director of the Historical Museum, University of Lund, asks for publications of Soviet archeologists and the journal *Sovetskaya arkheologiya* for the library of the museum. He offers in exchange the annual edition *Transactions of the Historical Museum of Lund University* starting from 1930 and other works of Swedish archeologists.

Please write directly to the above-named scientists if you find it appropriate to keep contact with them and to exchange literature; please inform VOKS about your decision.

GARF, fond 5283, opis' 20, ed. khr. 270, l. 98.

[1]See note 2 to doc. 6.7; Curiously, none of the authors of docs. 6.7 and 6.8 noticed that the well-known popular science monthly could not have No. 13.

[2] I. S. Shklovsky. On the emission of radiowaves by the Galaxy and upper layers of the solar atmosphere. *Astronomicheskiy zhurnal*, 23, vyp. 6, 1946, p. 333.

Document 6.9

From: S. V. Kaftanov, minister of Higher Education, and S. I. Vavilov, president of the USSR Academy of Sciences.
To: G. M. Malenkov, secretary of AUCP(b) Central Committee.
3.12.1948

[...] Physics is taught in many higher education institutions, ignoring fundamentals of dialectic materialism. The great Lenin's work *Materialism and empiriocriticism* is not used by lecturers in full measure [...] They neither expose nor criticize idealistic philosophical concepts making use of achievements of modern physics. Of special danger for students are idealistic corollaries from modern theoretical physics (quantum mechanics and the theory of relativity). Certain known bourgeois physicists (Bohr, Heisenberg, Schrödinger) have insistently preached idealistic philosophy during the last 20–25 years trying to base their doctrines on new discoveries in physics. Some of our scientists share these idealistic views alien to Marxism-Leninism instead of denouncing them and thereby facilitating their penetration into our institutes. Books and papers by bourgeois physicists are translated into Russian without critical comment and become widespread in our literature. An example is the popular book by Schrodinger *What Is Life. The Physical Aspect of the Living Cell*[1], etc. Certain books on quantum mechanics and relativity published in the Russian language expound these theories from bourgeois idealistic positions [...], there is great confusion in the interpretation of such key physical notions as mass and energy, space and time. Textbooks in physics do not properly highlight the role of Russian scientists in the development of this science; they are full of foreign names.

With this in mind, the USSR Ministry of Higher Education and the USSR Academy of Sciences ask for permission to convene (during the winter holidays) the All-Union workshop of heads of physical departments of universities and higher education institutions with the participation of the Division of Physico-Mathematical Sciences of the Academy.[2]

RGASPI, fond 17, opis' 118, ed. khr. 244, l. 237–239.

[1] See this chapter and notes 32–33.
[2] See: A. S. Sonin "Fizichesky idealism. Istoriya odnoy ideologicheskoy kampanii (Physical Idealism. The Story of One Ideological Campaign)". Izd-vo Fiziko-matematicheskoy literatury, 1994, pp. 114–115.

Document 6.10

Extract from the minutes of the Communist Party meeting at the Institute of Organic Chemistry, the USSR Academy of Sciences, held on 14.9.1951 to discuss the personal case of R. Kh. Freidlina, D.Chem.Sci., in connection with the arrest of her husband by the USSR Ministry of State Security.[1]

Question:[2] Why was G. E. Syroezhkin expelled from the party?

Reply: According to Syroezhkin, [...] for the deception of the Party [...] *Kuibyshevsky raikom* believes that he concealed his involvement in trotskyist opposition before he joined the Party [...]

Q: Why do you think that your husband was justly arrested?

R: Because the investigation is over, but he has not been released.

Q: Did he tell you about his former trotskyist activity?

R: Yes, he told me about his speech at a meeting of the Young Communist League in 1928 and regarded it as a serious mistake.

Q: When did you learn about the anonymous letter?

R: I knew about it but did not worry because the *raikom* was satisfied with the explanation given by Syroezhkin.

Q: Did you tell about the letter to the secretary of the Party organization?

R: No, even the organization where Syroezhkin was a member did not consider this question.

Q: Who carried out the investigation of this case?

R: The regional procurator's office of the Ministry of State Security.

Q: Am I right in understanding your words as an attempt to justify your husband?

R: No, I am just telling things as they are.

Q: Do you still maintain personal relations with your husband?

R. Yes, I sent a parcel to him (200 rubles and underclothes) on September 8.

Q: Did you ask the secretary of the Party organization about permissibility of such action?

R: Yes (confirmed by comrade S. S. Novikov, the secretary).

In the discussion that followed, the chairperson of the commission set up by the Party bureau said that "Freidlina deserves to be expelled from the AUCP(b), but the bureau takes into account some previous merits[3] of hers and decides to severely reprimand and warn her."

Q: What merits were taken into account?

R: The high productivity of labour and participation in public activities.

One of the three members of the commission stated that she did not trust Freidlina: "What you display is not a relaxing of vigilance, it is a total loss of vigilance [...] All this shows that it will be difficult to trust Freidlina in the future. I think that she must be expelled from the AUCP(b).

Results of the voting: 42 votes in support of the proposal of the bureau, twelve for the exclusion of Freidlina from the AUCP(b) membership, one abstained from voting.[4]

CAODM, fond 999, opis' 1, ed. khr. 10, l. 34–37.

[1] Abridged.
[2] Participants of the meeting asked questions after the secretary told about the application of Freidlina to the bureau.
[3] Characteristic of the results of Freidlina's studies of metallo-organic compounds. During the work at the Institute of Organic Chemistry since 1935, she developed a system of methods for the synthesis of organic compounds of mercury, lead, and antimony; moreover, she was the co-author (with Nesmeyanov) of the discovery of dual reactivity of a large group of chemical compounds.*

* Vydayushchiesya khimiki mira. Biograficheskiy spravochnik (Outstanding chemists of the world. Biographic Reference Book). M.: Vysshaya shkola, 1991, p. 463.

[4] Freidlina was expelled from the AUCP(b) in 1952. She was restored to Party membership in 1954 without record of the previous term.

Document 6.11

From: Petrov, head, and Sologubova, inspector, Staff Department, the AUCP(b) Central Committee.
To. A. A. Kuznetsov, secretary of Central Committee.
19.11.1947, in connection with the proposal of A. N. Sukhov, minister of social security, to sack V. D. Chaklin, director of Moscow Research Institute of Prosthetics.

[...] The facts of Chaklin's inappropriate conduct are established, such as the loss of vigilance and servility to foreigners. Professor Chaklin is in close contact with the Norwegian embassy. The ambassador Andwort, secretary Jakobsen, and physician Ustwed many times visited the Institute. Professor Chaklin told them about works carried out at the Institute and invited to them to participate at the conferences of its researchers. Certain members of the embassy including the ambassador were treated at the Institute's clinic [...] When in the clinic, Andwort showed interest in the way of life of Soviet people, put questions about agriculture etc. Professor Chaklin and his wife several times visited the ambassador in his residence and received him at home. They exchanged presents.[1] He was expelled from the AUCP(b) by the Communist Party organization of the Institute.[2]

RGASPI, fond 17, opis' 117, ed. khr. 969, l. 172.

[1] It is unbelievable that under Stalin's rule Chaklin maintained such close contact with Norwegian diplomats on his own initiative, without a sanction of top officials; otherwise these contacts would have been interrupted in the very beginning. This document dates to the period of "Klyueva–Roskin case". Professor Chaklin fell victim to the campaign against antipatriots launched in the Soviet Union at that

time when backdated charges were brought for actions that had previously been regarded as legal (see, for instance, doc. 6.38).

Actors in a similar situation were P. I. Egorov, head of the Kremlin Health Department, and his wife, who visited embassies of Egypt and Syria "in search of work". Both were arrested in the framework of the "doctors' case" in 1952 together with N. S. Vlasik, head of the Central Guards Department, USSR Ministry of State Security. Asked by the interrogator why he did not take measures to stop the actions of Egorova, Vlasik said that "the upper spheres had a plan with regard to Egorova" although "it was not completely realized until March 1952; I reported this to the minister [of state security] S. D. Ignatiev".*

* J. Brent, V. Naumov. Poslednee delo Stalina (The Last Action of Stalin). M.: Izd-vo "Prospekt", 2004, p. 243.

[2] When back in Sweden, Jacob Ustved published a paper about Soviet medicine in the Norwegian medical journal *Nordisk Medicine* and his observations during his stay in Moscow in the newspaper *Verdens Gang*. The Norwegian embassy referred these materials to VOKS with the request to pass them to several Soviet scientists including Professor Chaklin.* However, the parcel remained in VOKS until 20.9.1947, when it was passed to N. A. Vinogradov, deputy minister of health.** In 1947, V. M. Molotov, chairman of the USSR Council of Ministers, ordered to sack all foreign citizens employed in auxiliary positions in Soviet missions abroad or as family servants of their members.*** Taken together, these actions of top authorities in the second half of the 1940s were designed to maximally restrict personal contacts of Soviet people with foreigners.

*GARF, fond 5283, opis' 20 , ed. khr. 101, l. 47.

** *Ibid.*, l. 19.

*** RGASPI, fond 17, opis' 118, ed. khr. 990, l. 14.

Document 6.12

Materials to the case "On the journal *Vestnik Akademii meditsinskikh nauk SSSR*.

I

From: Maevsky, deputy head of Staff Department, AUCP(b) Central Committee.
To: A. A. Kuznetsov, secretary of Central Committee.
24.4.1948

"On the journal *Vestnik Akademii meditsinskikh nauk SSSR*.

Vestnik Akademii meditsinskikh nauk SSSR, No. 2, 1948, published the summary report of academician-secretary Sarkisov S. A. "About the work of the USSR Academy of Medical Sciences" without proper preliminary editing.

For example, the paragraph in the report mentioning political events in the life of the Academy contains the following assertion: "The court of honour considering the case of Klyueva, corresponding member of our Academy, and Professor Roskin, working in one of the academic institutes, demonstrated that some scientists are not free from servility and toadyism to foreigners." Another paragraph reads: "the courts of honour considering the 'Klyueva–Roskin case' and the case of the former minister Mitirev organized by the USSR Ministry of Health contributed to the improvement of ideological work." One more quotation: "Was not the 'Klyueva–Roskin case' a serious lesson for all of us?"

The editorial body of the journal (acting editor-in-chief P. G. Sergiev) neglected the instruction of the AUCP(b) Central Committee prohibiting publications of materials about courts of honour in the open press. Sergiev let even the following passage be published: "It should be emphasized that these conclusions are consistent with the analysis of the situation in medical science in light of the problems facing the health care system in the post-war period, made by the government commission that transferred authorities to the new minister E. I. Smirnov." Why did P. G. Sergiev permit mentioning the government commission and appraise its work? Also, the journal published the plan of research activity of academic institutions and data about their staff classified as secret.

Comrade Sergiev is totally unfit for the position of the editor-in-chief for political considerations. In the past, he was relieved of the duties of *narkomzdrav*.[1] There is evidence that he kept close relations with enemies of the people (Kaminsky, Rakovsky, Boldyrev) in 1937–1938. Recently, Sergiev himself displayed servility to foreigners and placed (together with comrades Gauze and Brazhnikova) gramicidin developed in this country at the disposal of specialists of the USA and Britain.[2]

[...] I think it necessary:

1. to cut out the report by academician-secretary Sarkisov S. A. "On the work of the USSR Academy of Medical Sciences" from the journal and reprint it only after thorough editing;
2. to relieve P. G. Sergiev from the duties of the acting editor-in-chief of the journal *Vestnik Akademii meditsinskikh nauk SSSR* for serious political mistakes and publication of secret data.

In the upper part of sheet 95 is a note in ink: "Attn: comrade Zhdanov. A. Kuznetsov. 24.4.1948."

Below, in pencil: "Attn: comrade Suslov. Please clarify at your earliest convenience. Zhdanov."

Left of the previous note, in red pencil: "Attn: comrade Morozov. M. Suslov. 28.4.1948.

RGASPI, fond 17, opis' 118, ed. khr. 58, l. 95–96.

II

Resolution of the Secretariat of the AUCP(b) Central Committee (minutes 350, p. 299, 20.5.1948) "About the divulgence of secret data[3] in the journal *Vestnik Akademii meditsinskikh nauk SSSR*.

1. to relieve P. G. Sergiev from the duties of the acting editor-in-chief of the journal *Vestnik Akademii meditsinskikh nauk SSSR* for publication of secret data;
2. to bring to the notice of academician-secretary Sarkisov S. A. that he made a mistake sending for publication in the journal *Vestnik Akademii meditsinskikh nauk SSSR* his report at the fourth session of the USSR Academy of Medical Sciences containing secret data without its proper preliminary editing.

RGASPI, fond 17, opis' 118, ed. khr. 58, l. 92.

[1]People's commissar of health.
[2] An absurd statement characteristic of the "patriotic" rhetoric of the late 1940s and the earlier 1950s. The placement of a new preparation synthesized in a given country at the disposal of specialists of other states* can not by definition be regarded as "servility to foreigners"; *vice versa*, it demonstrates the high level of science in such a country and inspires a feeling of pride for its intellectual and technical potential.

* Specimens of gramicidin could not be given by the head of the Soviet health care system to anyone without the sanction of higher authorities including departments of the AUCP(b) Central Committee.

[3] The final text of the document does not include the first paragraph or the closing phrase from item 1 of the resolution of the Central Committee Secretariat drawn up by M. Morozov on the order of M. A. Suslov, head of the Department of Propaganda and Agitation and secretary of the Central Committee, and endorsed by Morozov and L. Ilyichev, Suslov's deputy. They ran as follows: "The AUCP(b) Central Committee notes that secret data published in the journal can be used by the enemies of this country and inflict harm on the Soviet state"; "the cost of corrections and reprinting of this issue of the journal should be defrayed by comrade Sergiev."*

* RGASPI, fond 17, opis' 118, ed. khr. 58, l. 94.

Document 6.13

From: A. Komarov, *charge d'affaires* of the USSR in Sweden.
To: A. N. Abramov, head, 5th European Department, USSR Ministry of Foreign Affairs.
No. 481. 30.10.1949

Enclosed find please the letter of the Nobel Committee in physics and chemistry mailed to us together with letters to members of the USSR Academy of Sciences

Ioffe A. F., Kapitsa P. L., Semenov N. N., Frumkin A. N. who the Nobel Committee asks to propose candidates for the Nobel Prize on physics and chemistry in 1950.

A handwriiten note in pencil below reads: "Attn: comrade G. F. Aleksandrov; to be passed to the USSR Academy of Sciences. 1/XI. B.K."

AVP RF, fond 0140, opis' 40, papka 159, ed. khr. 48, l. 79.

Document 6.14

From: Arne Olander.
To: Embassy of the USSR in Stockholm.
October 1949.

The Nobel Committee in physics and chemistry of the Royal Academy of Sciences would greatly appreciate your sending the four attached letters to members of the USSR Academy of Sciences Ioffe A. F., Kapitsa P. L., Semenov N. N., Frumkin A. N. in which they are asked to propose candidates for the Nobel Prize on physics and chemistry for the next year. Thank you for your assistance.

Translated from English* by Boltunova.

AVP RF, fond 0140, opis' 40, papka 159, ed. khr. 48, l. 80.

* The original English letter has not been found.

Document 6.15

From: P. Orlov, acting head of the 5th European Department, USSR Ministry of Foreign Affairs.
To: A. V. Topchiev, chief scientific secretary, USSR Academy of Sciences.

The USSR embassy in Sweden received letters from the Nobel Committee in physics and chemistry of the Royal Academy of Sciences addressed to members of the USSR Academy of Sciences Ioffe A. F., Kapitsa P. L., Semenov N. N., Frumkin A. N., in which they are asked to propose candidates for the Nobel Prize on physics and chemistry for 1950. It is emphasized that the proposals must reach the Nobel Committee no later than 31 January 1950.[1]

Enclosed find please the said letters and advise us how to respond to them.[2]

A handwritten note in the bottom left corner reads: "comrade Svetailo[3] told comrade Orlov by phone that the USSR Academy of Sciences does not plan to nominate candidates."

AVP RF, fond 0140, opis' 40, papka 159, ed. khr. 48, l. 81.

[1] It means that the USSR Ministry of Foreign Affairs practiced perlustration.
[2] Unrestricted control of correspondence addressed to Soviet scientists from abroad by government authorities continued in the post-Stalinist period. Here is an example. A committee of the UNO planned to hold a meeting to discuss the effects of

scientific and technical progress on cultural and moral values of mankind. According to the previous agreement, P. L. Kapitsa was to participate in the work of the committee but the official invitation was delayed. Kapitsa wrote to L. I. Brezhnev (17.8.1975) saying that the invitation was shelved for two moths in the USSR Ministry of Foreign Affairs: "I was not informed about the receipt of the letter although it was sent personally to me and bears my address." He further emphasized the importance of the problem to be discussed in Geneva and concluded: "these problems may provoke an ideological struggle and we should not stand aside [...] I would like to assist to the best of my abilities in the promotion of our policy of collaboration but can not do this for the lack of proper attention [...]" The situation of the 1970s was by far more favourable than in the 1940s-1950s, and the CPSU Central Committee immediately sanctioned the trip of Kapitsa to the UNO.*

*RGANI, fond 5, opis' 68, ed. khr. 28, l. 47.

[3] Head of the Foreign Department of the USSR Academy of Sciences.

Document 6.16

From: M. S oloviev, head, Foreign Department, USSR Ministry of Higher Education.
To. 5th European Department, USSR Ministry of Foreign Affairs.
No. 435c. 2.2.1948

Enclosed find please a letter from the Latvian University, Riga, concerning the Nobel Prizes for 1948.[1]

In the top left corner, handwritten in ink: "To file. 2.2.48."(The signature is illegible).

AVP RF, fond 0140, opis'39, papka 154, ed. khr. 51, l. 42.

[1]Thus, the document 6.17 was delivered to Riga without passing by VOKS.

Document 6.17

From Augustus Shon, secretary of the Norwegian Nobel Committee.
To: Latvian University, Riga.
15.12.1947

You will receive a few copies of the letter concerning nomination for the Nobel Peace prizes of 1948. I dare to ask you to distribute these documents among interested scientists in the University of Riga.[1]

Translated from French.

AVP RF, fond 0140, opis'39, papka 154, ed. khr. 51, l. 43.

[1]See note 2 to doc. 4.42.

Document 6.18

From: V. A. Zorin, deputy minister of foreign affairs.
To: I. S. Chernyshev, USSR ambassador in Sweden.
No. 118/5-eo, 16.1.1948

Swedes Hans Euler and Nilsson Ehle[1] are to be excluded from membership in the USSR Academy of Sciences for their fascist activities hostile to the USSR. We therefore need more facts than contained in the previously sent records of talks held by comrades Mal'gin and Usov. In particular, we need texts of the anti-Soviet speeches of Euler and Ehle or at least data on where and when they were delivered, concrete information about their membership (and its duration) in Swedish organizations, press releases and news containing evidence of their fascist activities etc. We are looking forward to having these data at your earliest convenience because the lack of them delays resolving the issue in question.

AVP PF, fond 0140, opis'39, papka 154 ed. khr. 51, l. 19.

[1] Foreign members f the USSR Academy of Sciences.

Document 6.19

From: I. S. Chernyshev, USSR ambassador in Sweden.
To: V. A. Zorin, deputy minister of foreign affairs.
No. 119, 23.3.1948

[...] Hans von Euler-Chelpin, German by birth,[1] is professor of chemistry in the Highest School of Stockholm. He is 75. He served in the German Army during World War I. He worked in Stockholm before and during the last war. According to the executive secretary of the Soviet–Swedish Society, Eva Palmer (whose father knew Euler), he frequently visited Berlin during World War II as a member of the committee set up to purge Jews from research institutions. The newspaper *Ny Dag* reports his activity as the head of the German colony here in Stockholm during the last war (22 March 1948).[2] Euler is an old member of the Swedish–German Society in Stockholm. Archives of the newspaper *Dagens Nycheter* contain data that Euler was for the first time elected vice-president of this society on 3 November 1933; he has been deputy chairman since 29 October 1943. The other deputy chairman is the known Swedish fascist Sven Hedin.[3] Fascist leaders including Rudolf Hess spoke at the meetings of this society. It sent a welcoming telegram to Hitler on 12 March 1936. *Dagens Nycheter* and *Svenska Dagbladet* published the protest of a group of reactionary professors "against the persecution of Czechoslovakian scientists and students by the new government and against the Communist dictatorship" (19 March 1948). Euler signed the protest.[4]

These are a few facts concerning Euler's activities that we managed to collect. We think they fairly well characterize him as a reactionary person.

Professor Nilsson Ehle is a long-standing member of the Swedish–German Society known as an active actor in those Swedish circles that supported the Nazi newspaper *Dagposten.* Holger Carlsson, author of the book *Nazism in Sweden* (published in 1942), characterizes him as an active participant of the Nazi movement in this country. According to the author, a group of Swedes confidentially addressed certain Swedish organizations, calling them to establish the Swedish–German Society in order to promote "the understanding of new Germany". Ehle signed this address and became one of the first members of the new society. He was editor-in-chief of the newspaper published by the society [...]

Zorin's note on the first page of the letter: "Attn: comrade Abramov. It seems insufficient to raise the question. M.b.[1] to ask Kollontai: 'What does she know?'"
[1] May be

AVP RF, fond 0141, opis' 39, papka 154, ed. khr. 51, l. 21–22.

[1] Underscored with red pencil by the addressee.
[2] The appearance of the article mentioning the pro-fascist views of von Euler-Chelpin, Nobel Prize winner in chemistry for 1929, in the newspaper of the Swedish Communist party *Ny Dag* a day before the Soviet embassy dispatched its message to Moscow could not be a mere coincidence. It appears that the embassy asked journalists to look for materials compromising Euler.
[3] Sven Hedin had complicated relations with Russia. On the one hand, he was favourably treated by Russian authorities during his trips to China and Mongolia and back* across the country's territory. Also, he was a member of the Russian Imperial Geographic Society. On the other hand, he published the booklet *The Word of Warning* in 1912** that provoked a scandal in Sweden and indignation in Russia. He stated that Russia is a major threat to Sweden and called the government to be properly armed.*** He visited Russia again in 1924 and presented collections of his scientific works to five research and educational institutions in Moscow, Leningrad and Irkutsk.**** In May 1935, Hedin became a member of the Society of Cultural and Scientific Relations between Sweden and the USSR.***** In 1936, he accepted a personal invitation from Hitler to be a guest of honour at the Olympic Games in Berlin. Thereafter, he broke up all relations with the Soviet Union. A. M. Kollontai, Soviet ambassador plenipotentiary in Sweden, had a talk with Hedin six years before his contacts with Hitler and wrote in her diary (17.2.1930): "Sven Hedin is a suspicious personality."******

* Baron K. Stael von Holstein, *charge d'affaires* of Russia in Sweden, wrote to V. N. Lamzdorf (25.4 (8.5).1905), minister of foreign affairs, after the defeat of Russia in the Russian-Japanese War of 1904–1905 about Hedin's warning that "Russia, pushed aside from the Pacific, may try to occupy non-freezing Norwegian ports." Von Holstein concluded his letter by the characterisation of Hedin as "having the reputation of a supporter of Russia and an expert in Russian affairs". AVP RF, fond 133, opis' 470(1905), ed. khr. 111, l. 70.
** See: "The booklet *The Word of Warning* by Sven Hedin and a polemic aroused by it in Sweden", SPb: Gosudarstvennaya tipografiya, 1912. The original booklet *Ett Varnings Ord*

is kept in the archive of the office of the minister of foreign affairs of the Russian Empire. AVP RF, fond 133, opis' 470(1912), ed. khr. 175.
*** The Russian mission in Stockholm sent a note to the Swedish government in connection with this publication. The Swedish side stated that "the royal government deeply regrets the actions of Hedin" and that it "relies on the common sense of the Swedish people, the majority of whom do not share the views of Hedin" (doc. 6.40, 6.41).
**** GARF, fond 5283, opis' 5, ed. khr. 452, l 107.
***** *Ibid.*, ed. khr. 540, l. 40.
****** A. M. Kollontai. Diplomaticheskie dnevniki 1922–1940. T.1 (Diplomatic Diaries 1922–1940. V. 1). M.:Academia, 2001, p. 31.
[4] During the Communist *coup d'état* in Czechoslovakia in February 1948.

Document 6.20

Memorandum of the USSR Ministry of Foreign Affairs (from correspondence of 1947): "On the exclusion of Swedish scientists Hans Euler and Nilsson Ehle from the USSR Academy of Sciences", 25.12.1948.

The USSR Academy of Sciences (comrade Vavilov) was asked about who proposed and when to elect these scientists honorary members of the Academy (4 January 1947). The Academy revealed (6 January 1947) that the Swedish botanist[1] Hans Euler was elected its corresponding member in 1927 and the Swedish geneticist Niels Herman Nilsson Ehle in 1932, both on the proposal of the Biological Association, Division of Biological Sciences, the USSR Academy of Sciences [...] In April 1947, a letter to N. G. Bruevich was drafted requesting that the Academy take measures and expel these scientists from its membership. Comrade Novikov postponed the dispatch of this letter (16.4.1947) until "additional facts about the activities of these persons are collected". Comrade Chernyshev was asked to provide the necessary material (letter No. 680 of 30.4.1947). Bazarov sent records of conversations of comrade Mal'gin with J. Johansson containing some data about the two scientists (letter No. 413 of 31 July 1947). Extracts from the records were passed to the AUCP(b) Central Committee (L. S. Baranov) (letter No. 1399/5eo of 8.8.1947). Comrade Bazarov was informed by letter No. 138/5eo of 8 August 1947 that more data are needed about the Swedish scientists. Bazarov told the Department[2] that the embassy does not possess more detailed information (letter No. 567 of 26.12.1947); new data will be transmitted as soon as they are available.

Note: the file contains correspondence for 1948.

AVP RF, fond 0140, opis'39, papka 154, ed. khr. 51, l. 25.
[1] A mistake. H. Euler-Chelpin was a biochemist. He received the Nobel Prize in chemistry for "investigations on the fermentation of sugar and fermentative enzymes".
[2] 5th European Department of the USSR Ministry of Foreign Affairs.

Document 6.21

From: A. N. Abramov, head, 5th European Department, USSR Ministry of Foreign Affairs.
To: V. A. Zorin, deputy minister of foreign affairs.
No. 2349/5eo, 27.12.1948

Taking a decision on the exclusion of anti-Soviet Swedish scientists Hans von Euler-Chelpin and Herman Nilsson Ehle from membership in the USSR Academy of Sciences is delayed (it was to be discussed on 8 June 1948 by order of comrade Vyshinsky A. Y a.,). It is therefore advisable to ask comrade Chernyshev for additional information about the recent political activities of the said scientists [...] I am waiting for instructions.

AVP RF, fond 0140, opis' 39 , papka 154, ed. khr. 51, l. 26.

Document 6.22

From: A. Komarov, *charge d'affaires* of the USSR in Sweden.
To: V. A. Zorin, deputy minister of foreign affairs.
No. 100, 17.3.1949

This is to inform you, with reference to your letter No. 3/5eo of 3 January 1949, that Swedish scientists Euler-Chelpin and Nilsson Ehle have recently displayed no political activity, and we have no data about them other than that contained in our letter No. 119 of 23 March 1948.[1]

AVP RF, fond 0140, opis' 40, papka 159, ed. khr. 48, l. 3.

[1] See doc. 6.19.

Document 6.23

From: L. D. Kislova. member of VOKS Executive Board.
To: S. I. Nikishin, VOKS representative in Sweden.
No. 3195c, 13.10.1949

The available data indicate that Swedish botanist[1] H. Euler is involved in reactionary anti-Soviet activities. VOKS asks you to collect and transmit to us as soon as possible facts characterizing Euler's hostile actions (comprehensive biography, articles, speeches, etc.).

AVP RF, fond 0140, opis' 40, papka 159, ed. khr. 48, l. 48.

[1] See note to doc. 6.20.

Document 6.24

From: V. S. Kruzhkov, head, Department of Propaganda and Agitation, AUCP(b) Central Committee, and Yu. A. Zhdanov, head, Sector of Science.

To: G. M. Malenkov, secretary of Central Committee.
27.2.1950

In pursuance of your instructions, the Presidium of the USSR Academy of Sciences (comrades Vavilov and Topchiev) submitted materials and proposals concerning foreign honorary members and corresponding members of the Academy. The Presidium [...] recommends the expulsion of Hans Euler, a biochemist known for his fascist views, from membership in the Academy [...] The available materials indicate that the majority of foreign members of the USSR Academy of Sciences are loyal to the USSR [...] At the same time, some of them are involved in hostile activities, e.g. Swedish biochemist Hans Euler, [...] biochemist A. Szent-Gyorgyi (emigrated from Hungary to the USA),[1] Italian mathematician Francesco Severi, Slavonic scholar Andre Mason (France), and Alexander Belich, president of the Serbian Academy of Sciences. The Presidium asked for the opinion of A. Y u.Vyshinsky, minister of foreign affairs, concerning the membership of these scientists in the Academy. The minister did not object to the exclusion of Swedish biochemist Hans Euler from the Academy. As regards other persons mentioned in the letter, their exclusion from the Academy membership is currently undesirable.[2]

[...] The Department of Propaganda and Agitation, AUCP(b), supports the proposal of the Presidium of the USSR Academy of Sciences to expel Swedish biochemist Hans Euler from foreign membership in the Academy.

RGASPI, fond 17, opis' 118 , ed. khr. 796, l. 59–61.

[1] Szent-Gyorgyi's emigration was related to the political situation in Hungary in December 1946-May 1947 when the leaders of the party of petty farmers (that received 57% of the votes at the parliamentary elections of 1945) were arrested and communists came to power.
[2] In view of the inevitable negative international resonance. It was especially true of Szent-Gyorgi, Nobel Prize winner in physiology and medicine in 1937, known as a participant in the resistance to Nazi occupation in Hungary (see docs. 3.24–3.26); he was recommended for election to the USSR Academy of Sciences by the Politbureau of the AUCP(b) Central Committee in June 1947 (see doc. 4.37) shortly before his secret emigration to the United States.

Document 6.25

From: A. N. Abramov, head, 5th European Department, USSR Ministry of Foreign Affairs.
To: S. I. Vavilov, president of the USSR Academy of Sciences.
No. 645/5eo, 31.3.1949

[...] Erick Stensi is characterized as a non-Party apolitical bourgeois scientist who never made public political statements. His political views are known. He speaks positively of Soviet scientists. At the same time, he maintains close contact with British paleontologists and highly appreciates the values of English science and culture.

AVP RF, fond 0140, opis' 40, papka 159, ed. khr. 48, l. 5.

Document 6.26

Extract from the shorthand record of the interview of A. A. Zhdanov, secretary of the AUCP(b) Central Committee, with organizers of the 1st International Congress of Slavists[1] scheduled for 1946 (27.3.1946).

Zhdanov: Someone may regard the Congress as a step in the preparation of the Slavs for the confrontation with America. We certainly try to unite the Slavs, but this activity should not be advertized. Let the initiative of creating blocks come from the Americans.[2] The United States and Britain look for any pretext to accuse us of uniting the Slavs against them. We should not give them this trump card.[3]

[...] Who is to be invited? I don't think we should invite anyone from the USA or UK. Let these scientists first learn to control their rulers and deserve the honour to be invited to the Soviet Union. The same applies to France.

Voznesensky:[4] It will create an awkward situation...

Zhdanov: Let them in France feel awkward. It's time for us not to feel embarrassed when dealing with foreigners. Let them feel awkward for not being invited to the Soviet Union for their behaviour.[5]

Istoricheskiy arkhiv, No. 5, 2001, pp. 10–11.

[1] The Congress was not convened in 1948.

[2] Western states started the creation of the North-Atlantic Treaty Organization (NATO) in accordance with the agreement signed in Washington on 4.4.1949.

[3] A masterpiece of cynicism.

[4] AA. V oznesensky, RSFSR minister of education.

[5] The last phrase, especially the mention of "behaviour," shows that the secretary of the Central Committee had completely lost touch with reality...

Document 6.27

Extract from the minutes of the meeting of the Communist Party organization of the journal *Novoye vremya,* 15–19.3.1949

I

Extract from Sheidin's speech.

Last year, the journal published the letter from Soviet scientists to Professor Einstein,[1] who joined the movement for the establishment of the "world government."[2] As Professor Einstein continued to persist,[3] the journal again published materials demonstrating the absurdness and harm of the idea of the "world government." Nevertheless, none of the published documents contains the word "cosmopolitanism" although it is by no means new and has long been in wide usage after the exposure of a group of cosmopolitan theatre critics. This fact indicates that we do not know how to notice and characterize in due time selected features of international situation [...]

CAODM, fond 63, opis' 23, ed. khr. 216, l. 23.

II

Exract from N. S. Sergeeva's speech.

Comrade Sheidin said that we did not carry out propaganda against cosmopolitanism. I should think so! The thing is we did not expose cosmopolitanism. M. Rubinstein aptly criticized the idea of "world government" in his article,[4] but he never used the word "cosmopolitanism".[5]

We had an article by Parin submitted for publication. It did not appear. But in what position would he who offered this article have found himself if it did? It is not always possible to hide behind the editorial board. We must be vigilant [...][6]

Ibid., fond 63, opis' 23, ed. khr. 216, l. 32–33.

[1] "On Certain Errors of Professor Albert Einstein. The Open Letter by Soviet Scientists," signed by members of the USSR Academy of Sciences S. I. Vavilov, A. F. Ioffe, N. N. Semenov, and A. N. Frumkin. *Novoye vremya*, No. 48, 1947, pp. 14–15.

[2] In September 1945, very soon after the bombardment of Hiroshima and Nagasaki by the Americans, Einstein stated that the only way to save civilization and mankind is to establish a world government ensuring national safety based on the law applying to all member-states of the world community. Einstein was sceptical about the ability of the UNO to resolve the situation because its Charter did not confer such power. The scientist more than once tried to revive his idea.*

* See: A. P. Pais. The Science and the Life of Albert Einstein" (translated from English). M.: Nauka, 1989, p. 451.

[3] Einstein responded to the letter from Soviet scientists published in the bulletin of nuclear physicists. He described their disagreement with his arguments as an "escape to isolationism", especially dangerous for the Soviet Union where "the government controls not only military forces but also all educational and information channels as well as welfare of every citizen."

[4] "On Carelessness in Policy and Persistence in Mistakes"; the article was unsigned, i.e. published as an editorial. *Novoye vremya* No. 11, 1948, pp. 12–14.

[5] The formally neutral term "cosmopolitanism" was first used in the negative sense in the unsigned editorial in *Pravda* of 28.1.1949 (see also note 45 to this chapter). It was therefore absurd to demand that the author anticipated such a radical

transformation of the usage of this notion in the article published in *Novoye vremya* (No. 11, 10.3.1948) one year before the above editorial.
[6] V. V. Parin, academician-secretary of the USSR Academy of Medical Sciences, was arrested in February 1947 in connection with the "Klyueva–Roskin case" (see note 1 to doc. 4.34 and doc. 4.35). His article was certainly ordered when he was considered a leader of Soviet medical science. Therefore, the demand for "vigilance" in this case was akin to the demand to use the term "cosmopolitanism" a year before it appeared in the official newspaper of the AUCP(b) Central Committee.

Document 6.28

Information of G. S. Migirenko, secretary of the Communist Party organization, Mathematical Institute of the USSR Academy of Sciences, about the fulfillment of the decisions of Party meetings and the critical remarks of their participants, 27.3.1951.

In the Party organization, of which I was a member for about a year,[1] they severely criticized the cosmopolitan, anti-patriot, and thief of his colleagues' data Khanovich.[2] In the end, Khanovich was exposed as a pseudoscientist whose open publications included slightly modified unofficial data.[3] Concomitantly, a few other careerists, boasters, eye-washers, and dealers were unmasked [...][4] We must be vigilant, reveal such negative tendencies, and immediately rebuff them [...] There are such workers in our institute among both communists and non-Party bolsheviks.[5]

TsAODM, fond 7334, opis' 1, ed. khr. 3, l. 31.

[1] Naval Academy in Leningrad and its participation in the campaign "against cosmopolitanism" in 1949.
[2] I. G. Khanovich, professor of the Naval Academy before the dismissal in 1949, specialist in the theory of shipbuilding, author of the textbook *Hydrodynamics and Water Resistance* etc. In the 1950s, he filled the position of professor at the Electrotechnical Institute, Leningrad, and focused on jam-resistant electrical communication systems. The institute published a bibliography of his works on the occasion of his 70th birthday in 1969.
[3] Certain authors involved in secret research were allowed to publish their studies in the open press provided they eliminated all secret data from their papers. Here is an example from my personal experience. The existing regulations permitted to publish results of the surveys of uranium deposits without mentioning their geographic location.
[4] It was a typical story of reprisals against a disagreeable person during campaigns of the "struggle against cosmopolitanism."
[5] "Non-Party bolshevik" was a propagandistic term to describe those outside the Communist party who were "with heart and soul devoted to the cause of Lenin and

Stalin." Its usage in the given context was obviously out of place, which fact was beyond the understanding of the simple-minded Party leader.

Document 6.29

Extract from the minutes No. 503 of the meeting of the Secretariat of the AUCP(b) Central Committee, 5.5.1950. Strictly confidential.[1]

On additional elections of full members and corresponding members of the USSR Academy of Sciences.

1. The Secretariat accepts the proposal of the Presidium of the USSR Academy of Sciences to provide 4 additional vacancies of full members of the Academy (2 in physics, 1 in geophysics, 1 in history) and 4 vacancies of corresponding members (3 in physics, 1 in paleontology).
2. Secretariat accepts the following proposals of the president of the USSR Academy of Sciences comrade Vavilov:
   (a) to recommend candidates for the election of full members of the Academy (as per list No. 1) at its general meeting;
   (b) to recommend candidates for the election of corresponding members of the Academy (as per list No. 2) at its general meeting.[2]
3. This decision is subject to the approval of Politbureau.

Signed: G. M. Malenkov, secretary of the Central Committee.[3] Round stamp of the AUCP(b) Central Committee.

In the right margin below the text of the resolution are the signatures of the members of Politbureau: V. Molotov, L. Beria, A. Mikoyan, L. Kaganovich, A. Kosygin, K. Voroshilov, A. Andreev (all approved the above decision).

In the left margin: "The issue in question was not put to the vote as subjected to criticism. Sukhanov.[4] 2/VIII 1950".

RGASPI, fond 17, opis' 118, ed. khr. 854, l. 5.

[1] The extract is typewritten on an official form of the Politbureau of the AUCP(b) Central Committee, which had to approve this decision of the Secretariat.
[2] The attached lists No. 1 and 2 are not published here.
[3] Facsimile.
[4] D. N. Sukhanov was Malenkov's assistant at the Secretariat.

Document 6.30

From: L. Slepov and V. Lebedev, Department of Propaganda and Agitation, AUCP(b) Central Committee.
To: G. M. Malenkov, secretary of Central Committee.
24.1.1951

The editorial board of *Literaturnaya gazeta* has recently made two serious mistakes [...][1] On 28 December [...] it published the article by V. Druzhinin "Prizes Stuffed with the Dynamite of War"[2] in which Nobel Prize winners for 1950 are called "pseudoscientists" and "favourites of imperialism"; one of them is Cecil Powell, a prominent English scientist, vice-president of the British Peace Committee and member of the Bureau of the World Peace Council.[3]

These mistakes stem from the vicious practice alien to the bolshevist press that made possible the publication of important materials without their previous verification and sometimes based on tendentious and dubious facts.

[...] The mistake was due to the irresponsible attitude towards official duties on the part of comrade Rozentsveig, head of the department "Life and morals abroad" (he had to control the contents of the article), and comrade Krivitsky, editor of materials highlighting international events. Krivitsky was well aware of the resolution of the AUCP(b) Central Committee (4 July 1950) that obliges editorial boards to thoroughly check information about foreign scientists and art workers to avoid groundless criticism and alienation from the peace movement.[4] Nevertheless, he did not take measures to ensure the practical implementation of the resolution, which led to the publication[5] of the article "Prizes Stuffed with the Dynamite of War" with a serious mistake...

RGASPI, fond 17, opis' 119, ed. khr. 216, l. 58.

[1] The first "mistake" was unfounded criticism of a specialist in photoelectronic technique.
[2] See note 3.
[3] International non-governmental organization established in November 1950 with the active participation of Soviet propagandistic services.
[4] The resolution was adopted to smooth the consequences of the campaign for the "struggle against cosmopolitanism and servility to the West" that seriously affected the prestige of the Soviet Union in the eyes of world public opinion.
[5] In other words, mention of Powell's name was the sole cause for criticism.

Document 6.31

Resolution of the Secretariat of the AUCP(b) Central Committee; 26.2.1951, minutes No. 546, p. 1c.

[...] These mistakes stem from the vicious practice alien to the bolshevist press that made possible the publication of important materials without their previous verification and sometimes based on tendentious and dubious facts. The editorial board does not properly select responsible authors able to understand the material they submit for publication.

The AUCP(b) Central Committee resolves:

1. to reprove comrade Ryurikov B. S., deputy editor-in-chief and acting editor-in-chief of *Literaturnaya gazeta*, for failure to take measures necessary to exclude publication of unverified facts and materials.
2. to rebuke comrade Krivitsky Z. Yu., editor of materials highlighting international events, for inadequate control of the facts included in the article "Prizes Stuffed with the Dynamite of War" that led to a serious mistake.
3. to oblige comrade Simonov, editor-in-chief of the *Literaturnaya gazeta* to ensure thorough preparation and verification of materials submitted to publication in the newspaper and to strengthen the editorial staff with qualified and responsible workers.

RGASPI, fond 17, opis' 119, ed. khr. 216, l. 1.

Document 6.32

Resolution of the Secretariat of the AUCP(b) Central Committee; 26.1.1951, minutes No. 546, p. 14c.

The Secretariat of the AUCP(b) Central Committee commissions comrades Grigoriyan, Kruzhkov, and Sleptsov to submit to the consideration of the Central Committee (within a week) candidates for the positions of editors of sectors of international life (deputy editor-in-chief) and science, and the executive secretary of the editorial board of *Literaturnaya gazeta.*

RGASPI, fond 17, opis' 119, ed. khr. 216, l. 147.

Document 6.33

Extract from the decision of the editorial board of *Literaturnaya gazeta*, 18.1.1951[1]

All members of the editorial board guilty of the serious mistake in the article by V. Druzhinin, "Prizes Stuffed with the Dynamite of War" (defamation of Professor Powell, a prominent peace champion) are subject to severe penalty [...]

Comrade Rozentsveig, head of the department "Life and morals abroad," is relieved of his duties for an irresponsible attitude to the control of materials included in the said article. Comrade Yartseva of the same department is reprimanded and warned [...]

Comrade Krivitsky, editor of the department of international life, did not take necessary measures to ensure adequate control of materials submitted for publication which led to the serious mistake [...] Comrade Krivitsky is committed to comply a comprehensive dossier with materials concerning peace movement and its champions occupying elective offices and appearing in mass media. The writer V. D. Druzhinin must be called out from Leningrad to the next meeting of the editorial board to give an explanation.

RGASPI, fond 17, opis' 119, ed. khr. 216, l. 13–16.

[1] In accordance with the existing practice, the AUCP(b) Central Committee adopted a decision after the meeting of the editorial board of *Literaturnaya gazeta* with the participation of representatives of the Department of Propaganda and Agitation, at which the work of rank-and-file editorial staff was discussed.

Document 6.34

From: I. P. Bardin, vice-president, and A. V. Topchiev, chief scientific secretary, the USSR Academy of Sciences.
To: V. M. Molotov, deputy chairman of the USSR Council of Ministers.
No. IO-5621c, 30.8.1950

I

The International Council of Theoretical and Applied Physics organizes an International Conference on radiospectroscopy to be held in Amsterdam (Holland) on 18–23 September 1950. Leading scientists from Holland, England, France, USA, and other countries including the Soviet Union are invited to participate in the conference. Its program includes discussion of theoretical and applied problems of primary importance for in-depth studies of matter. Reports by foreign scientists cover the entire spectrum of topics pertinent to radiospectroscopy [...] The Presidium of the USSR Academy of Sciences considers it highly desirable for our representatives to participate in the work of the conference because it would enable Soviet scientists to more effectively organize research in this field, where we have not yet enough experience.

The Presidium of the USSR Academy of Sciences asks for permission to send to the Netherlands for 12 days a delegation of Soviet scientists comprising:

1. Andronov A. A., member of the USSR Academy of Sciences, head of the delegation.
2. Prokhorov A. M., deputy head of laboratory, Physical Institute, USSR Academy of Sciences.
3. Vladimirsky K. V., senior research scientist, Physical Institute, USSR Academy of Sciences.
4. Oraevsky P. A., senior assistant for the Foreign Department, USSR Academy of Sciences, secretary of the delegation.

RGASPI, fond 17, opis' 119, ed. khr. 65, l. 59–61.

II

From: M. A. Suslov, secretary of the AUCP(b) Central Committee, and V. Grigoryan, head, Department of Foreign Policy.
To: G. M. Malenkov, secretary of Central Committee.
No. 25-c-1633, 14.9.1950[1]

The International Council of Physicists and the Dutch Physical Society organizes the 2nd International Conference on radiospectroscopy to be held in Amsterdam on 18–23 September 1950. The Soviet Union did not participate in the first conference. The USSR Academy of Sciences is entitled to send one representative.[2]

Leading scientists from Holland, England, France, USA, and other countries including the Soviet Union are invited to participate in the conference. Its program includes discussion of theoretical and applied problems of primary importance for in-depth studies of matter. Reports by foreign scientists cover the entire spectrum of topics pertinent to radiospectroscopy [...] The Presidium of the USSR Academy of Sciences considers it highly desirable for our representatives to participate in the work of the conference because it would enable Soviet scientists to more effectively organize research in this field where we have not yet enough experience.

The Presidium of the USSR Academy of Sciences (comrades Bardin and Topchiev) asks the AUCP(b) Central Committee for permission to send to Amsterdam in September 1950 a delegation of Soviet scientists comprising Vvedensky B. A., member of the USSR Academy of Sciences, head of the delegation; Prokhorov A. M. and Vladimirsky K. V., Cand.Phys.-Math.Sci.[3] The composition of the delegation does not meet the requirements of the organizers of the conference. The USSR Academy of Sciences can not propose other candidates. I[4] therefore consider it inadvisable to send a delegation to the International Conference on radiospectroscopy.

RGASPI, fond 17, opis' 119, ed. khr. 65, l. 57.

[1] Because materials concerning the scientific mission were considered by the Department of Foreign Policy of the AUCP(b) Central Committee, the Council of Ministers did not object to sending the delegation.
[2] The application of the USSR Academy of Sciences does not mention any restriction on the composition of the delegation; in all probability, this argument is invented as a pretext for the negative decision.
[3] Thus, the head of the delegation was changed and the officer of the Academy administration excluded from it within two weeks after the application of the USSR Academy of Sciences was forwarded to the USSR Council of Ministers.
[4] It appears that the letter was initially written to be signed by the head of the department alone; later, it was decided to support the refusal by the signature of the secretary of the Central Committee.

Document 6.35

Parallel Mathematical Congress in Budapest in 1950.

I

From: N. V. Svetailo, head of the Foreign Department, USSR Academy of Sciences.
To: I. M. Vinogradov, member of the USSR Academy of Sciences, director of Mathematical Institute, USSR Academy of Sciences.
13.6.1950

Attached find please a copy of the letter received from organizers of the Hungarian Congress of Mathematicians. I would appreciate your consideration concerning issues touched upon in the letter and concrete proposals of Soviet scientists for participation in the work of the congress.[1]

Arkhiv RAN, fond 383 opis' 1, ed. khr. 236, l. 9.

II

From: The Janos Bolyai Mathematical Society.
To: C. I. Vavilov, president of the USSR Academy of Sciences.
26.5.1950[2]

The Janos Bolyai Mathematical Society of Hungary organizes the 1st Hungarian Mathematical Congress to be held in Budapest from 27 August to 3 September this year. We invite mathematicians of the Soviet Union and countries of the people's democracy to participate in our congress and are determined to make it an instrument for establishing close relations between scientists of our countries [...]

We ask you to form the delegation and confirm your participation before 15 June because the opening of the congress is very close. We also ask members of the delegation to advise us about topics of their reports that we need to draw up the program...[3]

We attach special importance to this congress, the first in the history of mathematics in Hungary, because two honorary chairmen of our society, Leopold Fejer and Friedes Rees, will be 70 this year.[4]

The time of the congress has been specially chosen such that it coincide with the so-called "international" congress[5] held in the USA. Hungarian mathematicians will not go to that congress because the American government closed the Hungarian consulate and pursues a hostile policy toward Hungary and other countries of the people's democracy; participation in the American congress is inconsistent with our patriotic feelings.[6]

We are sure that the American administration will use the "international" congress in the USA as an instrument of its imperialist policy.[7] We would like to emphasize, however, that we do not organize a counter-congress because it is designed to be a Hungarian but not international forum. We invite scientists from friendly countries who can contribute to its success and promote further development of Socialist science.

Ibid., l. 10, 11.

III

From: M. V. Keldysh, member of the USSR Academy of Sciences, deputy director of the Mathematical Institute, USSR Academy of Sciences.
To: A. N. Kolmogorov, member of the USSR Academy of Sciences.
15.8.1950

With reference to the forthcoming celebrations of Rees and Fejer, I ask you and P. S. Aleksandrov, corresponding member of the USSR Academy of Sciences, to prepare welcoming addresses to be handed to these scientists by our delegation at the 1st Hungarian Mathematical Congress.

Ibid., l. 15.

[1] The delegation of the Mathematical Institute, USSR Academy of Sciences, constituted A. N. Kolmogorov, its full member, P. S. Aleksandrov and A. O. Gel'fond, corresponding members, N. M. Korobov, A. A. Markov, S. M. Nikol'sky, and I. R. Shafarevich (all doctors of Physico.-Math. Sci.). I. M. Vinogradov, member of the USSR Academy of Sciences, was appointed to lead the delegation.* All these researchers presented scientific reports at the Congress.**

* *Ibid.*, ed. khr. 228, l. 3.

** The texts of the reports by the members of the Soviet delegation were previously submitted to the administration of the Mathematical Institute. *Ibid.*, ed. khr. 233.

[2] The letter was translated from Hungarian into Russian in the Foreign Department of the USSR Academy of Sciences.

[3] Successful organization of such an ambitious international congress (it was called the 1st Hungarian Congress for political considerations) for only three months has no precedent in the history of big international scientific forums. This example gives one more confirmation of the potential of totalitarian regimes determined to demonstrate "the triumph of will." It would be absolutely impossible to do in a democratic society...

[4] This circumstance may account for the choice of Hungary as the host state of the congress.

[5] Inverted commas were used to emphasize the slighting attitude of the initiators of the congress toward the international status of the scientific forum in American Cambridge.

[6] The authors of the letter talk about discriminative measures of the American administration taken in response to the Communist *coup d'état* in Hungary in 1947.

[7] Literal confirmation of the contents of note 5.

Document 6.36

Recollections of I. M. Frank, member of the USSR Academy of Sciences, Nobel Prize winner in physics.

[...] Nuclear physics was at the threshold of mastering nuclear energy. [...][1] His [Veksler's] works created a new field of research, high-energy accelerators. He came to the Institute[2] in 1944 inspired by fresh ideas.[3] I did not immediately understand the principle of autophasing underlying the theory of modern accelerators. However, Vladimir Iosifovich knew perfectly well the importance of his investigations. He spent some time in the rest home "Uzkoye" near Moscow and came back with the manuscripts of the presently famous papers. S. I. Vavilov submitted them for publication in *Doklady Akademii Nauk.*[4] Happily, they appeared just in time to secure priority of Soviet science, and American specialists had to recognize it.[5] After a short time, it would be impossible to publish these papers because everything directly or indirectly related to nuclear physics was classified as secret and could not appear in open press.

"Vospominaniya o V. I. Vekslere pod red. M. A. Markova i A. N. Gorbunova (Recollections About V. I. Veksler)." M.: Nauka, 1987, pp. 10–11.

[1] After returning to Moscow from evacuation in 1943.
[2] P. N. Lebedev Physical Institute, the USSR Academy of Sciences.
[3] According to E. L. Feinberg, corresponding member of the USSR Academy of Sciences, the discovery dates to February 1943 "when [...] V. I. Veksler (thinking over acceleration regardless of what he had to do in other fields) cut the Gordian knot; he found the way to surpass many times the relativistic barrier and jump across it."* In a few pages, Feinberg again emphasizes: "I am absolutely sure [...] that it took place in February."**

* In: Sergei Ivanovich Vavilov. Ocherki i vospominaniya (Sergei Ivanovich Vavilov. Essays and Memoirs) (pod red. I. M. Franka). M.: "Nauka" 1991, 162, 281.

** *Ibid.*, p. 285.

[4] "Doklady Akademii Nauk," T. 43, No 8, 1944, p. 346; T. 44, No. 9, 1944, p. 393. S. I. Vavilov submitted only the first paper; the second was submitted by N. D. Papaleksi.
[5] In the first place, E. Lawrence and E. McMillan (see this chapter and doc. 6.37).

Document 6.37

Comments on the publication of the discovery of V. I. Veksler in the open press at the Communist Party meeting at the Physical Institute, USSR Academy of Sciences, held 25.9.1947 to discuss the closed letter of the AUCP(b) Central Committee concerning the "Klyueva–Roskin case."

S. E. Khaikin: We already had similar cases and should have been cautious not to tolerate new ones, such as the publication of the principle of synchrotron by V. I.

Veksler [...] I do not mean that new ideas of our scientists should not be published at all. But they can be published only when we simultaneously start their implementation.

D. I. Blokhintsev: I would like to speak of a few facts, in the first place those related to the work of V. I. Veksler. Semyon Emmanuilovich said that it should not have been published. I think that Vladimir Iosifovich was right to publish his results. The commission set up at our Institute was sceptical about his idea and did not recommend to award the prize to the author despite all the good words addressed to him [...][1] Vladimir Iosifovich was left alone for some time and decided to publish his study [...] What was the result? Lawrence sent a telegram that changed the situation. In one respectful organization [...] they read the Lawrence's telegram[2] before the report of Vladimir Iosifovich.

TsAODM, fond 6862, opis' 1, ed. khr. 7, l. 47–48, 55–56.

[1] See note 71 to this chapter.
[2] Nobel Prize winner in physics in 1939, awarded "for the invention and development of the cyclotron and the results obtained with it, especially with regard to artificial radioactive elements." Carl Manne Siegbahn, member of the Royal Academy of Sciences and Nobel laureate in physics for 1924, stated after Lawrence received the prize that the invention of the cyclotron "was a breakthrough in nuclear research."* See also note 70 to this chapter.

* "Laureaty Nibelevskikh premiy. T. 1 (Nobel Prize laureates. V. 1)," M.: Progress, 1992, p. 271.

Document 6.38[1]

I

From: S. N. Polipchuk, executive secretary of Antifascist Committee of Soviet Scientists.
To: professor Zhebrak, Timiryazev Agricultural Academy.
11.11.1943[2]

Dear Anton Romanovich,

You kindly consented to write an article about genetics in the USSR and planned to complete it by the 1 November. Please advise when I can have the article if it is ready or when I should expect to receive it if it is not.

II

Copy of the certificate signed by the executive secretary of the Antifascist Committee of Soviet Scientists, 12.9.1947, for the administration of Timiryazev Agricultural Academy in Moscow.

The paper by Professor Zhebrak referred to in the article by Professor Laptev in *Pravda* of 2 September 1947[3] was ordered by S. N. Polipchuk, the former executive secretary of the Antifascist Committee of Soviet Scientists, and sent by him to *Science* (USA) on 25 April 1945.[4]

From the personal archive of E. A. Zebrak, son of Professor A. R. Zhebrak.

III

Extract from the article "Antipatriotic actions under the flag of 'scientific' criticism" signed by Professor I. Laptev.

[…] In this context, the feeling of deep indignation is aroused by the antipatriotic paper "Soviet biology" published by A. Zhebrak, member of the Byelorussian Academy of Sciences, in the American journal *Science* […] Zhebrak fully shares the views of Professor Sax under the guise of a polemic with this author. He joined reactionary foreign scientists running down our progressive biological science and its outstanding contemporary representative academician, T. D. Lysenko […] Zhebrak's solidarity with the reactionary Sax is unworthy of a Soviet scientist. A. Zhebrak writes: "Unfortunately, Professor Sax […] identifies modern biology, in the first place genetics, with the name of Lysenko. He criticizes the erroneous views of Lysenko as if they were fundamentals of Soviet biology and the attitude of the Soviet government to science".

One should totally lose the feeling of patriotism and scientific dignity to state that the world-renowned scientist Lysenko, an innovative geneticist, stands apart from Soviet biology […] Such a statement arouses the feeling of deep indignation in each Soviet farm-worker, each person engaged in Socialist agriculture, the high productivity of which has been for many years based on the scientific methods developed by T. D. Lysenko!

The aim of Zhebrak was to give T. D. Lysenko away for delectation of the capitalist press and represent himself as a devoted ally […] of Saxes and the like who criticize Lysenko and defend reactionary idealism against dialectic materialism […]

Zhebrak undertook to defame his countryman and colleague in a hostile foreign journal to satisfy the aspirations of reactionaries of the world. Only he who lost any sense of civic duty could do it! […]

Pravda, September 2, 1947, p. 2.

[1] This document illustrates the common practice of fabricating charges under which a person could be punished for the actions that were initially sanctioned by authorities or even initiated by them (as in the present case).

[2] A reminder of Zhebrak's promise was probably prompted by the information about the meeting of the American–Soviet Society held in New York on 7.11.1943 to celebrate the anniversary of the October Revolution. Professor L. Denn, University of Columbia, made a report about biological science in the Soviet Union that was soon

published in the journal *Science.** The publication was criticized in the same journal by Professor K. Sax of Harvard University,** who described it as "a panegyric."

* *Science*, 99, No. 2561, 1944, pp.5–67.

** *Science*, 99, No. 2572,1944, p. 298.

[3]The article entitled "Antipatriotic actions under the flag of 'scientific' criticism" was obviously inspired and preceded the closed Communist Party meeting in the Timiryazev Agricultural Academy (9.10.1947) and the court of honour at the Ministry of Higher Education (22–23.11.147), at which Zebrak was accused of "anti-Party actions." The decision to organize the court of honour was taken by the Orgbureau of the AUCP(b) Central Committee on 27.8.1947. P. N. Pospelov, editor-in-chief of *Pravda*, present at the sitting of the Orgbureau wrote in his diary: "*A propos* the court of honour: Zhebrak opposes Lysenko in the American journal *Science.** The article in *Pravda*, signed by a certain Professor I. Laptev (most probably a pseudonym) is concluded by the following words fully consistent with the decision of the Orgbureau: 'Those who try to slow down the fulfillment of this task,** who discredit Soviet science by their antipatriotic actions must be brought to the public court!'"***

One month after the appearance of this article, the *Literaturnaya gazeta* published a twin article in which the scientist was subjected to even more severe criticism.**** Moreover, it touched on an entirely new topic: the recent publications of physicist B. L. Ginzburg described as the manifestation of servility to foreigners and mistreatment of the achievements of Soviet science.**** This public delation was concluded by the allusion to "shameful attempts to conceal discoveries of Soviet scientists and discredit Soviet authors." In other words, initiators of the article pointed to one more potential candidate for prosecution.***** Only the involvement of Ginzburg in the "Atomic project" saved him from a sweeping denunciation at a "court of honour."

* RGASPI, fond 629, opis' 1, ed. khr. 95a, chast' 2, l. 75. This note is characteristic of the experienced bureaucrat Pospelov, who went through all the stages of being an ideological carrier, from a provincial rank-and-file Party member in the first years of Soviet power to secretary of the CPSU Central Committee under Khrushchev and member of the Presidium of Central Committee under Brezhnev. He could allow to show his true self writing the diary and tell what he subconsciously felt about the so-called "Zhebrak's case." He understood perfectly well that Zhebrak was persecuted not because he adhered to Mendel and Morgan's views but for his opposition to the ignorant "people's academic" Lysenko, encouraged by Stalin [...]

** In a preceding paragraph, the author formulated this task in the spirit of Stalin's call to "surpass scientific achievements of foreign countries in the near future".

*** *Pravda*, 2.9.1947, p. 2 (see also doc. 6–38-III).

**** *Literaturnaya gazeta*, 4.10.1947, p. 4.

***** Ginzburg described this episode in his memoirs. V. L. Ginzburg. "O nauke, o sebe i o drugikh (About Science, Myself, and Others)." M.: Fizmatlit, 20003 (3d ed.), p. 381.

[4] Zhebrak was received by V. M. Molotov, deputy chairman of the USSR Council of Ministers, a week and a half before his article was sent to *Science* (16.4.1945) by the Antifascist Committee of Soviet Scientists.* He was allowed to publish his work in the USA. In the letter sent to G. M. Malenkov on the eve of this meeting, Zhebrak asked for permission to reply to Sax because his paper "contained wrong political inferences." The Central Committee had no objections to what Zhebrak wrote in his article in 1945, as evidenced by his appointment to the position of the head of the Sector of Science, Department of Propaganda and Agitation, which he occupied until April 1945.**

* RGASPI, fond 17, opis' 125, ed. khr. 360, l. 5.

** E. S. Levina "Vavilov, Lysenko, Timofeev-Resovsky...." M.: Izd-vo "AIRO-XX," 1995, p. 92.

Document 6.39

Exposure of the pseudoscientific theory of the nature of viruses and microbes of G. M. Bosh'yan in 1950–1951.

I

From: Professor V. N. Orekhovich, director of the Institute of Biological and Medical Chemistry, the USSR Academy of Medical Sciences.
To: G. M. Malenkov, secretary of the AUCP(b) Central Committee.
27.6.1950

The letter states that the appearance of the monograph *On the nature of viruses and microbes*[1] by G. M. Bosh'yan gave rise to "rumours that the book was approved by the AUCP(b) Central Committee, that the author is the nephew of a known statesman who supports him,[2] and that some ministers (e.g. of public health) were instructed to debar its criticism[3], etc. The mass media launched a noisy campaign to advertize the book as if there were actually grounds for such rumours [...] although it is a pseudoscientific rather than scientific work [...] Nobody took the risk to criticize it and it would be simply impossible to do." Neither *Meditsinsky rabotnik* nor *Literaturnaya gazeta* consented to publish the critique of the book. The author of this letter read his review at a closed meeting of the Scientific Council of his Institute, and its members unanimously supported it. But most of them did not attend the open scientific conference at which the review was again presented. "I came to know that these professors read the editorial in the journal *Bolshevik* No. 11 in which Bosh'yan was described as an innovator[4] and decided not to meddle in the discussion [...] To be frank, I also anticipate negative consequences of my interference. Still, I dare criticize Bosh'yan as a devoted member of our Party understanding that his book may inflict serious harm to Soviet science and the interests of our great country [...]"[5]

GASPI, fond 17, opis' 118, ed. khr. 975, l. 145–148.

II

Extract from the attached copy of the review of Professor V. N. Oreckhovich.

Bosh'yan tries to prove the following theses: (1) microbes are immortal; (2) living nature supports spontaneous development of antibiotics, antibodies, toxins, antigens, etc.; (3) organic matter is eternal and organic substances are involved in a continuous turnover; (4) life on the Earth is continuously self-generated; (5) "living" non-protein bodies really exist; (6) microorganisms have a "common" progenitor, etc.

All experimental data were obtained by Bosh'yan from 12 May to 8 July 1949. "Such a pace is amazing taking into account the painstaking and time-consuming work needed to isolate, identify, and cultivate individual microorganisms." In addition, the author managed to obtain 40 species of microbes and viruses in crystalline form during the same period. He started writing the book in July and had it published in the autumn. "This case has no precedent in the history of science [...]," but the reader is left unaware of experimental conditions and has no ways to objectively verify the reported data.[6]

Ibid., l. 149–159.

III

From: E. I. Smirnov, USSR minister of health.
To. G. M. Malenkov, secretary of the AUCP(b) Central Committee.
5.1.1951

The letter confirms the importance of Bosh'yan's work and the value of the studies of O. B. Lepeshinskaya[7] that "proved the existence of life in a simpler form at the pre-cellular level, brought biology to the borderline between the animate and the inanimate, and created prerequisites for the conversion of the inanimate into the animate." In conclusion, the minister notes that "Professor Orekhovich does not understand it, as a scientist unfamiliar with biology."

Ibid., l. 160–168.

IV

From: Professor S. R. Mardashev, editor of *Voprosy meditsinskoy khimii.*
To: M. A. Suslov, secretary of the AUCP(b) Central Committee.
7.1.1951

"The volume of *Voprosy meditsinskoy khimii* (No. 2) published by the USSR Academy of Medical Sciences contains the review of the book of G. M. Bosh'yan by Professor V. N. Orekhovich [...] The editorial board [...] decided to publish it and thereby open discussion of the said monograph. The volume was ready for printing and passed to Glavlit on 3 November 1950. Glavlit kept it for two months on the grounds that it contains the critical review of Bosh'yan's book." After many attempts to overcome the difficulty, the problem was submitted to the consideration of the Presidium of the USSR Academy of Medical Sciences. The Presidium had already approved the publication of the review in August 1950. In November 1950, the reelected Presidium and the new president of the Academy took an opposite position but "at the same time abstain from taking an official decision [...] There are rumours that the book of Bosh'yan is nominated for the Stalin Prize, that he will receive the prize, and that his critics "will be punished."(!)

Glavlit and the USSR Academy of Medical Sciences delay the decision in order to make Professor Orekhovich take his review away or the editorial board refuse to publish it [...] Such methods remind the Arakcheevshchina of Ivan the Terrible times and do not cease to be such because they are performed in "kid gloves" [...] I am writing to you because I am seriously concerned about stopping criticism in medical science. Please, help us."

Resolution of Suslov: "Attn: comrade Zhdanov. Please clarify. 15.1.1951."

Ibid., l. 169–171.

V

From: N. I. Glagolev, head, and A. N. Cherkashin, instructor, Sector of Natural Sciences, Department of Science and Higher Education, AUCP(b) Central Committee.
To: M. A. Suslov, secretary of the Central Committee.
10.2.1951

In response to the letter of Professor Mardashev, the Presidium of the USSR Academy of Medical Sciences took the decision to publish the review by Professor Orekhovich and organize discussion of the problem of speciation in microorganisms in the near future.[8]

Ibid., l. 187.

[1] G. M. Bosh'yan. O prirode virusov i mikrobov (On the Nature of Viruses and Microbes). M.: Medgiz, 1949, circulation 25,000; 2nd edition. (revised), 1950, circulation 100,000.

[2] There were rumours at that time that Bosh'yan was a close relative of A. I. Mikoyan, member of Politbureau of the AUCP(b) Central Committee; they were not confirmed.

[3] Rumours that the minister of health supported the works of Bosh'yan, either voluntarily or not, were confirmed by his later letter to the AUCP(b) Central Committee; see doc. 6.39-III.

[4] The author of the article "In support of creative development of progressive science" wrote: "The discussion of the theory of heredity [at the 1948 session of

VASKhNIL] posed serious problems pertaining to the struggle of true materialistic science against reactionary idealistic remnants in research work. [...] The recent studies by O. B. Lepeshinskaya contribute to the development of materialistic biology, break up old traditional concepts that slow down further progress in life sciences, and give up an instrument to control biological laws. Of special interest in this context is the innovative work of G. M. Bosh'yan, "On the Nature of Viruses and Microbes."*

* *Bolshevik*, No. 11 (June), 1950, p. 10.

[5] Orekhovich occupying a high administrative position demonstrated fidelity to scientific principles and personal courage. Criticism of an official press organ of the AUCP(b) Central Committee was severely prosecuted at that time. Therefore, his move was as unique as the final positive response of Party ideologists to the arguments of both professors.

[6] A. Kozlov, head of the Agricultural Department, AUCP(b) Central Committee, proposed (in a letter of 4.7.1950 to Malenkov, secretary of the Central Committee) to instruct the ministers of health and agriculture to "investigate the case and report the results to the AUCP(b) Central Committee". The Secretariat of the Central Committee adopted the pertinent resolution on 14.7.1950 (minutes of the sitting No. 517, p. 326).

[7] O. B. Lepeshinskaya. Kletka, ee zhizn' i proiskhozhvenie (The Cell, Its Life and Origin). M.: Sel'khozgiz, 1950, circulation 10,000; 2nd edition, 1951, circulation 100,000; 3d edition, 1952, circulation 50,000.

[8] This document closes the story of the "untouchable" Bosh'yan. Three weeks before this note was sent to M. A. Suslov (12 January) volume 2 of *Voprosy meditsinskoy khimii* was approved for printing after all necessary permissions, including that of Glavlit, were obtained. The decision to convene a workshop of microbiologists lifted the ban on the criticism of pseudoscientific theories developed by Bosh'yan.

Document 6.40

From: Myasoedov, *charge d'affaires* of Russia in Sweden.
To: S. D. Sazonov, minister of foreign affaires of Russian Empire.
31.1(13.2).1912

[...] One can hardly believe, reading the booklet, that it was written by Sven Hedin who (according to himself) visited Russia 16 times and always enjoyed our hospitality and assistance in his journeys to Central Asia; he had the audience of our monarch in the beginning of this month.[1]

The famous traveller decided to warn his compatriots about the necessity to give more attention to self-defense in view of the inevitable danger from the East [...] Your Excellency can easily see what influence this booklet may have on the people still frightened by "Russian danger." The easiness to fight the long-standing enemy promised by Hedin will further strengthen chauvinistic trends in Swedish society and the martial spirit of these people.

The other day, I had a talk with the deputy minister of foreign affairs and drew his attention to the unfavourable impression this action of Hedin will produce on public opinion in Russia and its harmful effect on the attitude of the Swedes to our country. Baron Ramel answered that he shares my opinion and that the royal government deeply regrets this incident and would have never allowed publication of Hedin's booklet if it had tools to control the press [...][2]

AVP RI, fond 133, opis' 470 (1912), ed. khr. 118, l. 22–25.

[1] Sven Hedin more than once had an audience with Nikolai II when he visited St. Petersburg. For example, he was received by the tsar on his way to the next journey to Central Asia (11(24).6.1902)* and on the way back to Sweden (15(28).11.1904).**

*AVP RI, fond 133, opis' 470(1902), ed. khr. 96, l. 15.

***Ibid.*, fond 133, opis' 470(1904), ed. khr. 119, l. 30.

[2] See also note 6 to doc. 2.8 and note 1 to doc. 2.9.

Document 6.41

From: Myasoedov, *charge d'affaires* of Russia in Sweden.
To: S. D. Sazonov, minister of foreign affaires of Russian Empire.
4.(17).2.1912

The Minister of foreign affairs asked me to convey to Your Excellency his deep regret for Hedin's actions. The freedom of the press does not allow the government to take measures against the distribution of the booklet,[1] but the minister relies on the common sense of the Swedish people, the majority of whom do not share Hedin's views. Count Erensverd informed me confidentially that he is going to take an opportunity to make a relevant statement.

AVP RI, fond 133, opis' 470(1912), ed. khr. 161, l. 4.

[1] See also note 2 to doc. 6.40.

Chapter 7 • 1953–1957

From the Peak of the Clinch to the Thaw

The First Nobel Prize for the USSR

The peak of the ideological clinch fell on the first three months of 1953 and coincided with the last weeks of Stalin's life.[1]

On the 13th of January, newspapers published an official communication under the heading "News by TASS" that shocked not only this country but also the entire world. It announced the arrest of a group of prominent representatives of the medical profession, the majority of whom were Jews.[2] The editorial in "Pravda" entitled "Vicious spies and killers under the mask of academic physicians" and reprinted by the mass media of the country stated that most members of "the terrorist group"... were bought by the American intelligence service... recruited by its branchoffice — the international Jewish bourgeois-nationalist organization called "Joint". "Pravda" revealed that "the filthy face of this Zionist spy organization, covering up their vicious actions under the mask of charity, is now completely revealed". The text of the communication was approved by Bureau of the Presidium of CPSU Central Committee*18 four days before its publication (i.e. on January 9).[3]

After three days (16 January), "Pravda" published the article "Against subjective perversion in natural sciences" signed by Yuri Zhdanov, head of the Sector of Science, CPSU Central Committee, son of the deceased member of the Politbureau and Stalin's son-in law. The author severely criticized L. A. Orbeli, member of the USSR Academy of Sciences, a nominator of the Nobel Committee in physiology and medicine and nominee for the Nobel Prize in 1934 (see Table 3.1).

The first attacks on the prominent Soviet scientist, I. P. Pavlov's disciple and colleague, dated back to 1948, i.e. immediately after the notorious session of VASKhNIL (see Chapter 6). Orbeli was academician- secretary of the Division of Biological Sciences, USSR Academy of Sciences. In this capacity, he tried to adhere to neutrality and did not lay down administrative obstacles to opposing groups of researchers. However, the aggressive policy of Lysenko actively supported by ideological services of the AUCP(b) Central Committee made Orbeli retire following the sitting of the Presidium of the USSR Academy of Sciences three weeks after the session of VASKhNIL (26 August 1948).[4] In 1950, he lost all other administrative posts.[5]

Supporters of Lysenko criticized Orbeli in the first place for the alleged opposition to the physiological doctrine of Pavlov. Driven to despair by absurd accusations,

Orbeli wrote a letter to Stalin (late 1952) to protest against statements that his studies "are at variance with the concepts of my great teacher Pavlov".[6] Zhdanov's article was a sort of response to this letter. The author stated that Orbeli "falsified the teachings of Pavlov, declared Pavlov's objective method as limited, and [...] claimed for its supplementation by the subjective method" etc.[7]

Publication of the article of the high-ranking official incidentally fell on the onset of the "doctors' affair". However, the two events were not a mere coincidence for all the differences of their consequences. They were natural consequences of the anti-intellectual policy that prevailed in the last years of Stalin's life.

One more feature of this policy was rigorous state control over the system of post-diploma education based on research institutions and higher schools. The campaign for the struggle "against cosmopolitanism" launched in the early 1949 virtually deprived Jews of the right to continue education as postgraduate students.[8] The incompetent interference of authorities in the well-established system of continuous education very soon (even during Stalin's life) resulted in the deficit of young scientists for research institutions. The situation is exemplified by the discussion that took place at the Institute of Physical Problems, one of the leading research centres of the USSR Academy of Sciences.

A meeting of the Communist party organization of the Institute was held on 26 February 1953, three days before Stalin died of a stroke, at the heat of hysteria that engulfed the country after the announcement of the "doctors' plot". The agenda included a single item concerning the situation with recruitment of young specialists. Non-Party researchers invited to participate in the discussion included L. D. Landau, A. I. Shal'nikov (members of the USSR Academy of Sciences), and other scientists. The speakers emphasized that the enrolment of post-graduate students substantially decreased in the recent years, and those selected by the Presidium of the USSR Academy of Sciences and staff department of the Institute were not infrequently unfit for carrying out research at the high level characteristic of the Institute. The participants of the meeting condemned, even if in euphemistic terms, the situation in which leading scientists were devoid of the possibility to choose graduate students meeting their high requirements (doc. 7.1).

In the first months of 1953, any document concerning workers of science and culture was met with suspicion by officials faced with a threat to be accused of relations with the next victim of repressions. One such document was the letter from A. N. Abramov, head of the 5th European Department, the USSR Ministry of Foreign Affairs, to his immediate superior G. M. Pushkin, member of the Board of the ministry (doc. 7.2). It was written after the department received (via the USSR embassy in Sweden) the proposal of L. G. Sillen, expert for the Nobel Committee in chemistry, to nominate a Soviet scientist for the 1953 Nobel Prize. Sillen asked to recommend a candidate and present the list of his scientific works.[9]

The letter of the top official of the USSR Ministry of Foreign Affairs responsible for the development of relations with Scandinavian states is remarkable for its confrontational and contradictory tone. On the one hand, he observes that the Nobel organizations never awarded a single prize to Soviet scientists. On the other hand, he reminds the addressee that participation of Soviet scientists in the nomination of candidates for the Nobel Prizes "was deemed unadvisable" despite numerous attempts of the Swedish side to restore contacts with colleagues in the USSR. In conclusion, Abramov asks for permission "not to reply to the proposal of Sillen."

Attached to the letter were four pages of an article cut from the journal "Novoye vremya" for 1951 with Abramov's note that "activities of the Nobel Committee were exposed in our press". The tendentious article abounds in doubtful facts (see Appendix 4). There is reason to believe that the article had been written by someone of the USSR Ministry of Foreign Affairs and that Abramov was well aware who was its author and why he signed it with a pseudonym.[10]

In conclusion, it is worth noting the date of the document, 9 February 1953. In other words, it had no sense at all taking into account that Sillen asked to submit a candidate for the Nobel Prize of the same year and that the closing date for nominations was 31 January. The professional diplomat Abramov could not help knowing this requirement of the Nobel Statutes.

In other words, officials of the USSR Ministry of Foreign Affairs played the same way as their counterparts from the Department of Propaganda and Agitation, AUCP(b) Central Committee, who had similarly treated letters of the secretary of the Nobel Committee in physiology and medicine two years ago (see Chapter 4). True, it is not known when Sillen resorted to the help of Eva Palmer, chairperson of the Swedish-Soviet Society (educated as chemist); there is no doubt, however, that they had a talk before spring 1952. This inference is confirmed by the second letter of Sillen (doc. 7.4) who again tried to obtain the necessary materials, this time for the nomination for the 1955 prize.

...Nobody could think that in a month after the appearance of doc. 7.2, the coffin with the corpse of the deceased dictator would be placed in the mausoleum at Red Square. This, however, did not bring an immediate change into the political and ideological atmosphere of the country. The press continued propaganda of xenophobia and spying mania alternated with swearing fidelity to the cause of "the chief of the peoples". Large cities, in the first place Moscow, were engulfed by rumours about the forthcoming mass deportation of Jews to Siberia and the Far East.[11]

In March 1953, when General M. D. Ryumin, one of the main organizers of the "doctors' case" was already arrested (although it was not announced in press; see footnote* to note 12), two key figures in the Union of Soviet Writers, A. A. Fadeev and K. M. Simonov, sent a letter to Khrushchev, secretary of the CPSU Central Committee, dated March 24. Their message reflected traces of anti-Semitic hysteria of the two first months of 1953. The authors recognized that "the number of Jewish members in the Union was really very high" and attributed this fact to "national nepotism". They took on themselves "responsibility for such situation" and promised to change it.

Whether the letter was written on the initiative of the two authors or prompted by ideologists of CPSU Central Committee is difficult to say for the lack of relevant documents. But there was another letter of Fadeev to the CPSU Central Committee written on the day of his suicide, May 13, 1956, in which he plainly characterized the years of Stalin's rule in the following words: "After the death of Lenin, we were reduced to the position of small boys, annihilated and frightened; all this was called "Party spirit" in ideology... No motivation is left for creative work."[11a]

The situation changed abruptly on the morning of April 4 when the mass media published "Communication of the Ministry of Internal Affairs". It announced the release of the arrested professors and dismissal of the criminal lawsuit against them. The document read: "Inspection demonstrated that these people were falsely accused and that the investigators used entirely invented evidence."[12] This day may be regarded as a threshold after which the situation in the country radically changed towards cautious liberalization referred to as "the thaw" (*ottepel*).

One of the first results of this change was the gradual restoration of lost contacts with the world community. It proceeded along all lines and with practically all states excepting politically unacceptable regimes (e.g. Franco's Spain). Sweden was among the countries where this process developed faster than elsewhere.[13]

An early response to the changes in the Soviet Union was the first nomination of L. D. Landau for the 1954 Nobel Prize (30.1.1954). His candidacy was submitted by Robert Marshak, professor for the University of Rochester, USA. He proposed to share the 1954 prize between Landau and Fritz London, and emigrant from Nazi Germany. The Soviet scientist was nominated for his pioneering theory of quantum nature of helium superfluidity suggested in 1941 to explain the phenomenon discovered earlier by P. L. Kapitsa.

Two other Russian scientists (emigrants) were nominated for the Nobel Prize in physics in the same year. V. K. Zvorykin left Russia in 1917 and lived in the USA. I. R. Prigogine (Brussles) was taken abroad by his parents when he was only four (see Chapter 12). The candidacy of Zvorykin, who had invented iconoscope, the first transmitting TV tube, in 1931 was proposed by Professor V. Pavlovich, Belgrade. Prigogine was nominated for the theory of non-equilibrium (far from equilibrium) thermodynamics, together with L. Onsager, Yale University, by the Dutch professor E. F. M. van der Held. After a time, the two latter scientists actually received the Nobel Prizes in chemistry, Onsager in 1968 and Prigogine in 1977.

No nomination came to Stockholm from the Soviet Union. An undeclared boycott of the Nobel organizations continued despite great progress in scientific relations with Swedish research institutions since 1953. An important event was the participation of the delegation of Soviet physico-chemists in the XIII International Congress on general and applied chemistry held in Stockholm in August 1953. Such a visit would have been unfeasible half a year earlier (see Chapters 4 and 6). Professor Sillen actively participated in the organization of the Congress as a secretary of its physico-chemical section. In this capacity he had an opportunity to directly address himself to the future participants of the Congress, bypassing the Soviet embassy or VOKS (since May 1953).

The failure to obtain the necessary information from Moscow in 1952 was not the first for Sillen. He had already applied to the Soviet embassy between 1950 and 1951 for assistance in search of the author of a scientific paper that interested him. He received no answer even though his letter reached VOKS (doc. 7.3). Therefore, he could appreciate the deep changes that occurred in the Soviet Union and wrote his second letter in the hope that it will bring the desired reply.

Direct contacts with Soviet colleagues participating in the Congress gave him an opportunity to better see the level of physical chemistry in the USSR and name concrete pretenders for the Nobel Prizes in his second message. One of them was N. N. Semenov, the discoverer of chain reactions in chemistry,[14] the other, M. M. Dubinin, specialist in solid adsorbents, who participated in the Congress (doc. 7.4).[15]

Sillen's second letter contained another proposal very unusual for the practical work of Nobel organizations but certainly approved by the Nobel Committee in chemistry and the Royal Academy of Sciences. It was intended to help Soviet colleagues to pass from the ban on contacts with the Nobel organizations to cooperation with them. Sillen told about the possibility to address "all physicists and chemists of the Soviet Academy of Sciences and its leading figures like Topcheiv"[16] in order to enable them to submit proposals "personally or in letters *signed by more than one scientist*" [underlined by the author].[17]

The initiative of Sillen was supported in late 1954 by members of the Nobel Committee in chemistry Arne Tiselius and Arne Fredga. They met A. I. Oparin, vice-president of the USSR Academy of Sciences, visiting Stockholm for participation in the session of the World Peace Council, and expressed regret about the absence of contacts between Soviet scientists and the Nobel organizations. When back to Moscow, Oparin immediately informed A. N. Nesmeyanov, president of the USSR Academy of Sciences, about this meeting and was instructed to send a letter to the CPSU Central Committee (doc. 7.5).[18]

The postscript in doc. 7.5 about the forthcoming discussion of the problem raised by Oparin at the next sitting of the Presidium of the USSR Academy of Sciences proved to be misinformation. The minutes of the Presidium sittings for December 1954 and 1955 contain neither mention about such discussion nor any trace of correspondence with the CPSU Central Committee on this issue. It appears that the question was never considered officially but was solved at informal meetings; the results were verbally reported to the CPSU Central Committee.[19]

This suggestion is confirmed by documents.

The letter from Sillen reached the USSR Academy of Sciences via the USSR Ministry of Foreign Affairs long before Oparin met Tiselius and Fredga, i.e. in early June 1954.[20] In July, A. P. Vinogradov, acting academician-secretary of the Division of Chemical Sciences, informed the Soviet embassy in Stockholm that the USSR Academy of Sciences was ready to send the materials requested by Sillen (doc. 7.6). This message was certainly forwarded to Sweden; its archival copy is dateless[21] but has an outgoing number. Moreover, Sillen himself reported to Arne Westgren, chairman of the Nobel Committee in chemistry, (doc. 7.7) that he received positive information from S. N. Istomin, secretary of the Soviet mission.

The Swedish side also proposed, by way of exception, to send to the USSR Academy of Sciences a certain number of blank nomination forms instead of personal invitations to selected scientists, thereby giving the Academy the right to chose nominators. A parcel of such forms was sent to the Academy by the Nobel Committees in physics and chemistry as soon as the reply from Moscow was received (doc. 7.6). They were distributed among 29 members of the Academy in October 1954 (doc. 7.15).

However, the documents of these committees for 1955 that became available for researchers in 2006 mention only one nominator from the USSR, Professor V. A. Fok who proposed the candidacy of P. L. Kapitsa. The nomination lists of both committees contain no other name of the 28 Soviet scientists who received the invitation as per doc. 7.15

It does not mean, however, that none of them used an opportunity to propose his candidate. In the early post-Stalinistic period they could send correspondence abroad only via the Presidium of the USSR Academy of Sciences that acted on the advice of the Sector of Science, CPSU Central Committee. Therefore, many recommendations of the nominators could be disapproved and never reached the addressees.

Alternatively, the packet of proposals from Soviet nominators could reach Stockholm too late for consideration for the 1955 Nobel Prizes, and they were taken into account in the selection process of 1956. It will be possible to know after the Nobel archives open their files in 2007.

V. A. Fok sent his letter to Stockholm, bypassing the Foreign Department of the USSR Academy of Sciences. The envelope presently kept in an archive file of the Nobel Committee in physics bears the stamp of a Leningrad post office dated 9 October 1954; the letter was received in Stockholm on October 22. Whether Fok acted at his own risk or obtained permission from the Foreign Department is immaterial. Anyway, his nomination was among the first to reach the Nobel Committee; the overwhelming majority of proposals begin to come in late November and early December.

The text of this letter is published in the newspaper "Poisk" (No 34–35, 2006, p. 22). It was very unusual for the nomination addressed to the Nobel Committee. The author did not attach great importance to scientific arguments in favour of the epoch-making discovery of helium superfluidity by Kapitsa. Instead, he was focused on what he regarded as the discrimination policy of the Nobel organizations towards Soviet candidates for the prestigious awards.

Nevertheless, Fok's proposal was duly considered and included in the nomination list of the Nobel Committee in physics. Had such nomination come from another country it would have been turned down in the very beginning. But it was the first that the Nobel Committee had received from the Soviet Union for a decade and a half.

As mentioned above, the efforts of Sillen laid a basis for the establishment of contacts between the Soviet side and the Nobel organizations. A. P. Vinogradov, academician-secretary of the Division of Chemical Sciences, USSR Academy of Sciences, sent an official letter to the USSR Ministry of Foreign Affairs for the Soviet embassy in Stockholm (doc. 7.6).

However, it proved difficult to send the promised materials to Sweden. In all likelihood, bureaucrats in the Foreign Department of the USSR Academy of Sciences did not take the risk to dispatch printed matter (see doc. 7.9) without sanction of higher authorities. Sillen received only lists of publications of Semenov and Dubinin in August or early September (doc. 7.8), and their collation against catalogues of Swedish libraries "took more time than he expected to spend."

Leaders of the USSR Academy of Sciences applied to CPSU Central Committee (September 3, 1954) for permission to transmit the materials of interest to Professor Sillen via the USSR Ministry of Foreign Affairs (doc. 7.10).[22] In addition, A. V. Topchiev, chief scientific secretary of the USSR Academy of Sciences, presented to the Central Committee (September 21) information about the scientific work of N. N. Semenov and reprints of his publications on chemical kinetics (doc. 7.11), probably after a verbal consent from the Department of Science and Culture, CPSU Central Committee.

On September 30, A. M. Rumyantsev, head of the Department of Science and Culture, CPSU Central Committee, sent to the Central Committee Secretariat his conclusion concerning the two letters from the USSR Academy of Sciences (doc.7.12). Rumyantsev referred to Sillen's mention of two potential candidates for the 1955 Nobel Prize and recommended to send to Stockholm materials characterizing both scientists. He confirmed his opinion in a few days in doc. 7.13. On 10 October, the head of the Foreign Department, the USSR Academy of Sciences, informed the USSR Ministry of Foreign Affairs that these materials "will be dispatched in the near future" (doc. 7.14).

This "near future" came only in December 1954 when the Soviet embassy in Stockholm at last received materials concerning Semenov alone.[23]

The absence of documents on Dubinin can hardly be attributed to intrigues at the Presidium of the USSR Academy of Sciences or his participation in some secret research referred to by Topchiev (see doc. 7.13). In all probability, Dubinin withdrew his candidacy himself, understanding that he could not compete with Semenov in terms of significance of their research. Sillen eventually came to a similar conclusion (doc. 7.7).

The two-month delay in dispatching materials on Semenov to Stockholm was not a mere result of bureaucratic slovenliness. The negligence of rank-and-file executives would have had serious consequences for them taking into consideration how many high-ranking organizations were involved in the matter (the Soviet embassy, the Presidium of the USSR Academy of Sciences, the USSR Ministry of Foreign

Affairs, the Department of Science and Culture, CPSU Central Committee)... The true cause was the policy pursued by the Central Committee.

The remnants of the Cold War were still alive in the minds of top officials. Here are two examples of the deliberate hindrance of contacts between Soviet scientists and their colleagues abroad by the Department of Science and Culture, CPSU Central Committee, and its head, A. M. Rumyantsev, dated from December 1954 when the materials on N. N. Semenov lay in the Foreign Department of the USSR Academy of Sciences.

The first example is the story of invitation to a group of Swedish scientists including L. Sillen and A. Tiselius to visit the Soviet Union. The invitation was initiated by the Soviet embassy in Stockholm, influenced by the activity of Professor Sillen, who managed to receive support from Eva Palmer, chairperson of the Swedish–Soviet Society. The Presidium of the USSR Academy of Sciences asked the CPSU Central Committee for permission to send the invitation (doc. 7.16) but did not receive it (December 3) under the ridiculous pretext that "Swedish scientists did not apply [...] with a request to visit the Soviet Union" (doc. 7.17). Curiously, the refusal was formulated the day after A. N. Cherkashin of the Department of Science and Culture made a note on the letter from A. I. Oparin testifying to the positive attitude of the Central Committee toward restoration of contacts between the USSR Academy of Sciences and the Nobel organizations (see doc. 7.5). In other words, the right hand did not know what the left was doing (a sort of unstable equilibrium during the transitional period).

One more example of inconsistent decisions made by high-ranking Party officials is related to the request of the USSR Academy of Sciences to invite a group of leading foreign scientists for the participation in the All-Union Conference on quantum electrodynamics and theory of elementary particles (November 3 1954). The Conference was to be held in the second half of March 1955. The decision of the CPSU Central Committee taken on 25 December 1954 allowed to invite four of the nine scientists from capitalist countries including future Nobel laureates Richard Feynman and Andus Salam. Five others had to represent all states of people's democracy (barring Albania) instead of Wolfgang Pauli, Nobel Prize winner of 1945, and future laureate Sin-Itiro Tomonaga (who shared the 1965 Nobel Prize with Feynman).

President of the USSR Academy of Sciences A. N. Nesmeyanov replied on December 30: "The invitation of representatives from the countries of people's democracy is impracticable because there are no specialists in the field of the Conference in most of them. He suggested that only "Professor L. Infeld (Poland) and Ning-Chu (China) be invited" and insisted on the invitation of "S. F. Edwards (UK), M. Levi (France), M. Chini (Italy), W. Pauli (Switzerland) and S. Tomonaga (Japan) mentioned in the letter of 3 November." Rumyantsev refused again and emphasized, to strengthen his arguments, that "the last five candidates were considered and declined by the Department of Science and Culture."[24]

The CPSU Central Committee approved this decision of the Department of Science and Culture at the same time when it allowed to send the selection of

Semenov's works and related materials to Stockholm. This example once again demonstrates the inconsistency of the behaviour of authorities in the early post-Stalinistic period that reflected the struggle between retrogrades and those understanding the necessity of changes.

One more initiative of the USSR Academy of Sciences dated to the last week of 1954 found positive response from the CPSU Central Committee. A. N. Nesmeyanov and A. V. Topchiev asked the Department of Science and Culture for permission to resume (after the seven-year ban, see Chapter 4) publication of English abstracts in academic journals (24 December, i.e. the same day when the CPSU Central Committee Secretariat refused to invite five leading physicists from Western states and Japan to the conference on quantum electrodynamics). Also, they asked to permit Soviet scientists to send reprints of their publications abroad by mail without the agency of the Foreign Department. The CPSU Central Committee gave the long-awaited permission after a month of discussions and hesitations (doc. 7.18).

In the end, Sillen received documentation on Semenov only a month before the 1st of February 1955 and found that he had not enough time to prepare a well-reasoned nomination. Therefore, he decided to add it to the nominations for the next year. It turned out later that this delay, the responsibility for which lay totally at the Soviet door, gave rise to another collision in academic circles.

A regular sitting of the Bureau of the Division of Physico-Mathematical sciences, the USSR Academy of Sciences, was held on 1 November 1955 (minutes No. 13, signed by the academician-secretary M. A. Lavrentiev and secretary A. N. Lobachev). Item 19 of the agenda under the heading "On the nomination of Soviet scientists for the Nobel Prize" was in the focus of the discussion.[25] Its participants stated that the Nobel Prizes "can not be regarded as international" for which reason the Bureau refused to propose candidates for the 1956 prize (doc. 7.19). A similar statement on the same issue was made in two months (22 December). It was signed by M. M. Dubinin, academician-secretary of the Division of Chemical Sciences and addressed to S. G. Korneev, head of the Foreign Department of the Academy (doc. 7.20). Dubinin informed that the Division was not going to nominate its candidate for the Nobel Prize of 1956.

Both statements were a response to the absence of N. N. Semenov among the Nobel Prize winners of 1955. Today, we know that Sillen received the relevant documents too late, not in August as was his wish (doc. 7.4) but after the Christmas holidays of 1955, and simply had no time enough to prepare the nomination.

The question is whether item 19 of the agenda was actually discussed at the sitting of the Bureau on November 1 in the presence of its authoritative members.[26] Textological analysis of the minutes for 1955 from archive files of the Academy[27] gives evidence that the item was inserted in the official text *post factum* on the order of organs having little to do with science (not later than the last week of November). This inference is documented below.

The Bureau of the Division had a total of 15 sittings during 1955. The registry contains dates of dispatch of the copies of each minutes to the Presidium of the

USSR Academy of Sciences and extracts from them to other interested divisions of the Academy (doc. 7.21).

Normally, the copies and the extracts were sent within 6–8 days after the respective sitting. In case of delay for objective reasons, the dates of sending copies and extracts differed by only 2–3 days. For example, the dispatch of the copies of minutes dated 28 June and the relevant extracts was delayed by 11 and 13 days respectively. The delay was caused by the death of G. A. Gamburtsev, director of Geophysical Institute, the USSR Academy of Sciences, who made a report of its work at the previous sitting of the Bureau.

Minutes No. 13 are essentially different from the remaining fourteen. Extracts of items 1–17 were forwarded to the addresses between 9 and 14 November, the delay being due to the celebration of the anniversary of the October Revolution (7–8 November and the rest-day of 6 November). The extract from item 18 (development of scientific relations with Great Britain) was dispatched on November 18 and that from p.19 only on the 19th of November.[27] It may be deduced that original minutes No. 13 did not contain p.19 and the question was not discussed at the sitting of 1 November 1955. Lavrentiev occupying a high position in the Presidium of USSR Academy of Sciences could not resist the pressure from the outside and had to sign the minutes with the inserted item 19.

There is one more fact in support of the aforesaid. Document 7.19, like many others mentioned above (including 7.2), used the word combination "Nobel Committee" in the singular. If members of the Bureau participating in the sitting actually discussed item 19 (see-note 26), this mistake could not be repeated in the minutes because all of them and especially P. L. Kapitsa fairly well knew the structure of the Nobel Foundation.

A later note of M. M. Dubinin, academician-secretary of the Division of Chemical Sciences, the USSR Academy of Sciences, to the head of the Foreign Department of the Academy (doc. 7.20) also informed him that the Division refused to nominate candidates for the 1955 Nobel Prize. However, the dry, lapidary style of this note, the lack of examples and characteristics made it markedly different from official decisions of the Bureau.

It should be noted, for the sake of justice, that the positions of the two academician-secretaries in that situation were quite different. Lavrentiev was actually a detached onlooker in the conflict whereas it was a matter of serious concern for Dubinin. He twice visited Stockholm (the last time in the same 1955; see note 15 to this chapter) and could not sign the falsified document; it would be tantamount to moral suicide. He managed to resist and confined himself to writing a formal memorandum almost two months (22 December) after the notorious "decision" ascribed to his colleagues.

All these actions were certainly initiated by the Department of Science and Culture, CPSU Central Committee. A. M. Rumyantsev filled the position of its head in 1952, that is at the peak of ideological confrontation with the West. Like his predecessors, he could not help feeling puzzlement towards the international collaboration of scientists (doc. 7.35). The first signs of the thaw in the end of 1954 made him reconcile with the necessity to present documentation expected by Professor Sillen. When the Soviet scientist did not receive the prize in 1955, Rumyantsev interpreted the fact as incontrovertible evidence of his righteousness in regarding Nobel Committees as "anti-Soviet" organizations.

The episode with the minutes of the Bureau of Division of Physico-Mathematical Sciences was confirmed by one of its participants. A. S. Monin, member of the USSR Academy of Sciences, was an instructor at the Department of Science and Culture in 1955 and had to participate in the sittings of the Bureau. He told the author that he would have certainly remembered such an "unnatural" discussion if it had taken place in his presence (doc. 7.23).

Asked about the situation in academic circles after December 22 1955, i.e. after Dubinin sent his note to the Foreign Department of the USSR Academy of Sciences (doc. 7.20), A. S. Monin suggested his version agreeing with later documentary evidence. He correctly guessed the incentives to radically change relations between the USSR Academy of Sciences and the Nobel organizations in the end of 1955.

Dubinin repudiated the refusal to nominate candidates for the 1956 Nobel Prize and urgently began to prepare documentation on A. E. and B. A. Arbuzovs for the Nobel Committee in chemistry between 23 and 30 December 1955. He passed five pages of the handwritten nomination translated into English to the Foreign Department on January 3 1956 (see note 3 to doc. 7.27) and informed the president of the USSR Academy of Sciences about this on the next day (doc. 7.22). Also, he suggested that the Presidium of the Academy "informally" discussed the nomination.

The incentive for changes at the border of 1955 and 1956 was given by the CPSU Central Committee although it was again initiated by Sillen. He met A. V. Topchiev, chief scientific secretary of the USSR Academy of Sciences, during his visit to Stockholm in the second half of December 1955 and raised the old question. Topchiev informed the Department of Science and Culture, CPSU Central Committee, about his talks and obtained permission to further develop relations with the Nobel organizations.[28] It became possible after the dismissal of Rumyantsev[29] from his position in November 1955* that was occupied by V. A. Kirillin, corresponding member of the USSR Academy of Sciences and member of its Division of Technical Sciences. According to S. G. Korneev, deputy head of the Foreign Department of the Academy at that time, Kirillin understood the importance of the development of contacts between the USSR Academy of Sciences and foreign institutions (doc. 7.35). This position of the new head of the Department of Science and Culture encouraged leaders of the USSR Academy of Sciences to send him two letters in January 1956 (docs. 7.24 and 7.25).

However, all these good intentions were mired in bureaucratic procedures. Topchiev informed the CPSU Central Committee about his contacts in Stockholm in December 1955 and was allowed to immediately start preparing documentation on Soviet scientists for the Nobel Prize of 1956. However, the official permission was not received before January 27 (doc. 7.26) Accordingly, nominations of A. E. and B. A. Arbuzovs, V. N. Kondratiev, and N. N. Semenov (docs. 7.27 and 7.28) were dispatched to the Soviet embassy in Sweden for their transmission to Sillen only on January 28 (doc. 7.29), i.e. three days before the 1 February deadline!

However, it could not hamper the nomination of Semenov. His nominator was L. Sillen, the key figure in the process that eventually brought the first Nobel Prize to Soviet science in 1956. It might have been lost for the Soviet Union without his efforts, just because of sheer bureaucratic irresponsibility even though the question has already became a matter of concern for high authorities.

I was told about it by N. M. Lun'kov, *charge d'affaires* in Stockholm who accompanied Tage Erlander, Swedish prime minister, during his official visit to Moscow in summer 1956. Looking through his notebook, Lun'kov found the following entry: "Khrushchev, Erlander-Semenov." Lun'kov had made it to remind Khrushchev that he had to speak about the Nobel Prize for the Soviet scientist during his talks with the Swedish leader...[30]

Another remarkable date in the history of "rehabilitation" of the Nobel organizations in the Soviet Union was 31 January 1956. This day, the Presidium of the CPSU Central Committee approved the proposal of the Ministry of Higher Education to nominate for the Nobel Peace Prize of 1956 two members of the USSR Academy of Sciences, D. V. Skobeltsin, chairman of the Committee for International Lenin Prizes "For the Strengthening of Peace Between Nations," and the known writer M. A. Sholokhov.[31] The nominations did not reach Oslo in time and were disregarded by the Noble Peace Prize Committee. However, the importance of the decision taken by the CPSU Central Committee was in that it opened up ways to contacts of Soviet institutions not only with the Nobel Committees in sciences but also with the Nobel Peace Prize Committee so severely blamed by Communist Party ideologists during the past two decades.

One more episode is worth mentioning that coincided with the events preceding the dispatch of documents to Stockholm in the end of January 1956.

P. L. Kapitsa, member of the USSR Academy of Sciences and future Nobel Prize winner in physics, sent a letter describing the situation in Soviet science to N. S. Khrushchev (15 December 1955). The letter was published in 1989.[32] The reply was given in the form of a note signed by V. A. Kirillin, head of the Department of Science and Education, CPSU Central Committee, and N. I. Glagolev, his deputy. The authors recognized that criticism of Kapitsa was on the whole justified and proposed measures to improve the situation...[33]

N. N. Semenov was announced to win the Nobel Prize in chemistry for investigations into mechanisms of chemical reactions on November 1, 1956. He shared the award with Cyril Hinshelwood of Oxford University. Hinshelwood was the first scientist who nominated Semenov for the Nobel Prize in 1946.

By that time, the contribution of Hinshelwood to the theory of chemical reactions was great enough to rank him among the contenders for the prestigious award. His first book on kinetics of chemical conversion in gaseous systems was published in

1926[34] and attracted attention of specialists all over the world. He was well aware that he would never receive the Nobel Prize if it were first awarded to Semenov alone (the Nobel Prizes for similar results are given to two or three scientists simultaneously but not in different years). Nevertheless, he used his right as a nominator to draw attention of the Nobel Committee to the brilliant studies of Moscow scientist and proposed his candidacy again in 1950.

The public opinion in the Soviet Union welcomed the news from Stockholm. The President of the USSR Academy of Sciences A. N. Nesmeyanov congratulated Semenov at the sitting of the Presidium a day after the Nobel Committee announced the names of the winners (2 November 1956). He emphasized in a brief speech that "the award of the Nobel Prize to Semenov reflects the worldwide recognition of success of chemical sciences in the USSR".[35] In half a month (10 December), N. N. Semenov received the diploma and the gold medal from the hands of the king of Sweden.

However, the presentation day was saddened by the tragic events in Hungary between 23 October and 4 November 1956. They were triggered by the Soviet Army invasion and aroused indignation in most countries of the world. The presence of the Soviet Nobel laureate created *force-majeur* circumstances for the Nobel Foundation, which had to change the traditional format of the ceremony of 10 December.

The formal dinner in honour of the Nobel Prize winners traditionally given at the Stockholm City Hall was transferred to a smaller reception hall of the Swedish Academy (doc. 7.30). It was attended by the king and queen, top figures of the Nobel Foundation, and Nobel laureates of the past years staying in Stockholm at that time. The diplomatic corps including ambassadors of the countries whose citizens had received the Nobel Prize in 1956 were absent. Also absent were representatives of students' organizations, customary participants of all Nobel celebrations. The usual concert was cancelled.[36] All newspapers reported that ladies present at the banquet including the queen were attired in dresses of grey colours.[37]

Semenov was accompanied only by his wife, in contrast to other laureates who arrived in Stockholm with their families and friends (Werner Forssmann receiving the Nobel Prize in physiology and medicine brought with him seven children). Semenov's daughter was not allowed to join her parents by Soviet authorities (doc. 7.36). Other Nobel laureates from the Soviet Union invariably found themselves in the same situation. The authorities were afraid that once abroad, the entire family might refuse to return to the USSR. Sholokhov, Nobel Prize winner in literature and member of CPSU Central Committee, was the first to go to Stockholm with his family and close friends (see Chapter 10).

The positive reaction to the award of the Nobel Prize to the Soviet scientist was certainly sincere. It influenced other aspects of relations between Soviet researchers and their foreign colleagues especially in Sweden. One example is presented below.

Kai Siegbahn, professor at the University of Uppsala and future Nobel laureate in physics (1981), became editor-in-chief of the new international journal *Nuclear*

Instruments in 1956. He decided to have several Soviet scientists on the editorial board and applied to the USSR Academy of Sciences for advice (three letters dated 31 May, 4 August, and 3 October 1956 respectively). The first two letters remaining unanswered, the third one contained the request to tell "as soon as possible if Soviet scientists are ready to participate in the work of the editorial board". The last letter was sent shortly before the Royal Academy of Sciences announced the award of Semenov.

This time, I. P. Bardin, vice-president of the USSR Academy of Sciences, and A. V. Topchiev, its chief scientific secretary, asked the CPSU Central Committee for permission to give a positive answer (8 February 1957 or nine months after the reception of Siegbahn's first letter). The Department of Science, Higher Educational Institutions and Schools recommended to the Central Committee Secretariat to give the permission (14 February); two weeks later, A. S. Monin phoned to S. G. Korneev, head of the Foreign Department of the Academy, and informed him about the positive decision of the Secretariat.[38]

For all that, contacts of Soviet scientists with foreign organizations and especially with the mass media of Western states continued to be limited from above even after the invitation of Siegbahn was favourably accepted.

The largest conservative newspaper *Svenska Dagbladet* asked several leading natural scientists in different countries to answer four questions concerning the future of basic science. A similar poll was conducted in 1937; then, the replies were published in the newspaper and thereafter in a separate volume.

Kapitsa was one of the researchers chosen by the newspaper. Allan Gernelius, editor-in-chief, sent a letter to the Presidium of the USSR Academy of Sciences (22.9.1956 asking Kapitsa to answer the questions. Whether Kapitsa knew about it or not is unclear. Anyway, Gernelius had to send another letter (3.11.1956) saying that the replies were expected before mid-December.

The Presidium informed the CPSU Central Committee about the request of the Swedish newspaper (22.11.1956) and expressed the opinion that "P. L. Kapitsa should not answer the questions." It was exactly what high-ranking Party officials wanted to hear because it gave them an opportunity to demonstrate that Soviet organizations can make decisions and act independently.

N. I. Glagolev, deputy head of the Department of Science, Higher Educational Institutions and Schools, CPSU Central Committee, and A. S. Monin, instructor of this department, informed leaders of Central Committee that they agree with the opinion of the USSR Academy of Sciences and asked for permission to bring it to the notice of the Presidium. The final document was signed by half a dozen members of the Central Committee Secretariat and bore a note dated 8.12.1956: "comrade Kapitsa P. L. is notified via the USSR Academy of Sciences."[38a]

The award of the Nobel Prize to Semenov convinced sceptics that Soviet scientists may receive the worthiest tributes. In 1957, the USSR Academy of Sciences recommended four its members for all categories of the Nobel Prizes in sciences: P. L. Kapitsa and A. P. Cherenkov in physics, A. E. Arbuzov in chemistry, and

E. M. Pavlovsky in physiology and medicine (doc. 7.37). However, the bureaucratic machine continued to work as if it had learnt no lesson from the past negative experience.

As before, nominations of Soviet scientists for the Nobel Prize had to be approved by the CPSU Central Committee. The USSR Academy of Sciences transmitted them to the Central Committee only on 23 January 1957. Each of the four nominations was signed by I. P. Bardin, vice-president of the Academy, and A. V. Topchiev, its chief scientific secretary. They were registered in the Central Committee on the same day under No. 01906 and 01909 for A. E. Arbuzov and P. L. Kapitsa respectively.

The Department of Science, Higher Educational Institutions and Schools prepared and passed the relevant document to the Secretariat on the same day, and the secretaries of the Central Committee signed the permission to dispatch the materials to Stockholm on January 25 (doc. 7.31).[39] N. I. Glagolev confirmed (January 29) that Topchiev was informed about the decision of the Secretariat. In other words, all these decisions were again taken close to the deadline for presenting the nominations.

No one knows how they managed to reach the Nobel Committees in time, but Professor A. F. Westgren, permanent secretary of the Royal Academy of Sciences later confirmed that "all the four works [...] were received" (doc. 7.34).

A remarkable event was the meeting of a representative of the Soviet embassy in Stockholm with Professor Arne Westgren, a prominent actor of the Nobel movement, member of the Nobel Committee in chemistry and thereafter its chairman for almost 40 years (1926–1965). It put an end to the ideological confrontation of Soviet authorities at different levels with the Nobel organizations. Also, it was the first official contact of a Soviet diplomat with the Nobel organizations since the time when Kollontai had represented the USSR in Sweden.

It is appropriate to once again recall Lars Gunnar Sillen, whose unselfish insistence and sincere fidelity to the ideals of science played a key role in overcoming the inertness of the bureaucratic machine confronting progress and love of freedom, and eventually brought the first Nobel Prize to the Soviet scientist, Nikolai Nikolaevich Semenov.

Scandals accompanying the awards of the Nobel Prizes to Boris Pasternak, Aleksandr Solzhenitsin, and Andrei Sakharov, attempts to reanimate absurd pretensions of the late Stalin times were still ahead.[40] But they were just echoes of the former supremacy of ideology in the life of the country that still led to diametrically different appraisals of the Nobel movement. Suffice it to mention its characteristic as a phenomenon of world culture on the one hand and an instrument of anti-Soviet propaganda on the other hand in 1958 (when the Nobel Prizes were awarded to a group of Soviet physics and poet Pasternak) and in later years (Nobel Prizes to Sholokhov and Solzhenitsin).

Of course, the contribution of Soviet science and literature to world culture would be sooner or later recognized at the level of the Nobel movement. But it is equally clear that it would not have happened in 1956 without selfless efforts of L. Sillen.

The conversation between Professor A. F. Westgren and I. I. Murzin, representative of the Soviet embassy in Stockholm, (doc. 7.34) once again demonstrated the ignorance of Soviet diplomats as regards the inner structure of the Nobel Foundation. Murzin kept speaking about one Nobel Committee even though the Soviet scientists were nominated for three categories of the Nobel Prize and the relevant documents were submitted by the USSR Academy of Sciences to three different Nobel Committees (doc. 7.37). Also, he wrongly interpreted the words of Westgren about the announcement of awards on 31 October 1957. The Royal Academy of Sciences, Karolinska Institutet, and Swedish Academy always announce the names of Nobel laureates at different times. Speaking about October 31, Westgren certainly meant nominees of the Royal Academy but not of the Karolinska Institutet, which considered the candidacy of E. M. Pavlovsky.

To conclude, P. L. Kapitsa was not the sole Soviet scientist whose name was included in the nomination lists of two Nobel Committees of the Royal Academy of Sciences in 1955. Also listed were P. A. Cherenkov (in physics) and N. N. Semenov (in chemistry). Like Kapitsa, they received the Nobel Prize in later years. Cherenkov was nominated by Lamek Hulthen, professor of the Institute of Mathematical Physics at the Higher Polytechnical School of Stockholm, and Semenov (together with Hinshelwood) by Lars Gunnar Sillen.

Unable to recommend Semenov in time for the failure to receive his works form Moscow, Sillen decided to postpone the formal nomination till the next year and confined himself in 1955 to submitting a brief characteristic of the two candidates.

This tactics brought success in 1956 when Semenov and Hinshelwood shared the Nobel Prize in chemistry. The 18-line nomination presented by Sillen in 1955 mentioned only Semenov's monograph *Chain reactions* published in Oxford in English in 1935 and the equally well known work by Hinshelwood *Kinetics of Chain Reactions* published by the same publishing house five years later...

Notes

[1] Stalin died on 5 March 1953. In half a century, N. I. Dedkov, deputy editor-in-chief of the journal *Svobodnaya mysl'*, legal successor of "Kommunist", organ of CPSU Central Committee,* summarized the role of the dictator in the life of the country in the following words: "He betrayed [...] true victors [in the Great Patriotic War], taking their glory for his own,** leaving the fallen unburied, filling new concentration camps by inmates of Hitler's. He betrayed them as he had betrayed those who won the Civil War, his Bolshevik companions, ideals of freedom, equality, and fraternity. He was a traitor and criminal, and the incredible scale of his murderous deeds can hardly serve as evidence of his greatness. If Supreme Justice does exist in the world, he already gets his just deserves. Let his name be cursed for ever!"***

One more act of villainy should be added to this list. Stalin left as bloody a trace in history as the Grand Inquisitor Thomas Torquemada who was the first to burn people not only for the open demonstration of dissidence but also for the suspicion of it. This practice gave birth to a crowd of informants both in Spain of those times and in the Soviet Union of the 1930s to the early 1950s. No concrete facts were required to confirm their evidence...

* The journal has been published under this title since 1991; its first issue had No. 14 of its predecessor.
** Those trying to moderate flattery of Stalin were doomed to imprisonment in the Gulag. Historian Elena Zubkova told in the program *Documents of History* broadcast by Radio Svoboda about an officer of the Soviet Army who proposed to give the rank of generalissimo to marshal G. K. Zhukov in a letter to the Presidium of the USSR Supreme Council. He was sentenced to a term in a labour camp as a counter-revolutionary...
*** *Svobodnaya* mysl', No. 3, 2003, p. 3.

[2] V. A. Malyshev, a leading organizer of Soviet industry was present at the sitting of the Presidium of the CPSU Central Committee 1.12.1952, where it was decided to publish in the press information about arrests of medical professors in the framework of the "doctors' case". He put down in his diary the following words of Stalin: "The more success we achieve the more enemies we have. People forgot about it, impressed by our success. They were lulled into complacency, thoughtlessness, and conceit. Every Jewish nationalist is an agent of American intelligence services. Jewish nationalists think that their nation was saved by the USA (where you can become rich, bourgeois etc.). They think they are indebted to the Americans. There are many Jewish nationalists among doctors."...* Stalin proposed the title of the article: "Arrest of a group of saboteur-doctors". He also changed the original title of the editorial in *Pravda* from "Spies and killers under the mask of physicians" to "Vicious spies and killers under the mask of academic physicians" and corrected the text.** The article was written by Malenkov and published under the heading "Khroniki TASS".

This muddle of words and ideas reflects the inadequate mental condition of Stalin in the last months of his life that was apparent to those having an opportunity to observe him. Fourteen years later (27.3.1967), V. S. Semenov, a known diplomat, recollected the words of B. V. Petrovsky, minister of public health: "[...] he [Stalin] was a schizoid person and was close to mental disorder."***

* "Istochnik", No 5, 1997, pp. 140–141.
** In: "Politbureau CK AUCP(b) i Sovet Ministrov SSSR. 1945–1953. (Politbureau of AUCP(b) Central Committee and the USSR Council of Ministers. 1945–1953). M.: ROSSPEN, 2002, pp 396–397.
*** "Novaya i noveishaya istoriya", No 3, 2004, p. 129.

[3] R. G. Pikhoya. Sovetsky Soyuz: istoriya vlasti, 1941–1991 (the Soviet Union: History of the Power, 1941–1991). Novosibirsk: Izd-vo "Sibirsky khronograf", 2000, pp. 76. . The list of killer-doctors included Vovsi M. S., Vinogradov V. N., Kogan M. B., Kogan B. B., Egorov P. I., Fel'dman A. I., Etinger Ya. G., , Grinshtein A. M., Maiorov G. I. It was said that "the majority of the members of the terrorist group (Vovsi M. S., Kogan B. B., Fel'dman A. I., Etinger Ya. G., Grinshtein A. M.) were associated with the international Jewish bourgeois-nationalist organization "Joint" created by American intelligence services under the guise of providing material help to Jews in different countries". Ethnic Russians were included in the list together with M. B. Kogan, B. B. Kogan's brother, as "long-standing agents of British intelligence service". The first arrested were doctors who treated A. A. Zhdanov, secretary of the AUCP(b) Central Committee, of whom Stalin knew from the letter of L. F. Timashuk as long ago as 1948 (see note 29 to Chapter 5). Zhdanov's attending physician G. I. Maiorov was arrested in September 1952 together with pathologist anatomist A. N. Fedorov* who allegedly reported the results of the autopsy so as to confirm clinical diagnosis of V. N. Vinogradov, V. Kh. Vasilenko, and

P. I. Egorov excluding myocardial infarction. Egorov was arrested on October 18; his wife V. Ya. Egorova was arrested three weeks earlier (September 27).** Vinogradov was arrested on November 4, Vovsi ("head of the Zionist gang of killer-doctors") and B. B. Kogan on November 11, A. M. Girnshtein, A. I. Fel'dman, and L. C. Temkin*** on December 6...

As usual in those years, arrests of intellectuals were accompanied by prohibiting measures of Glavlit (central censorship agency). A note by Glavlit on the withdrawal from circulation of works of saboteur-doctors and all works of S. Mikhoels"**** appeared four days after the publication of information about the arrests. The note was replicated and distributed in all organizations of the country. It was received in the Skopinskaya geological team, Ryazanskaya region, where I worked at that time.

> * Stalin crossed the name of Fedorov from the initial text of "Khroniki TASS".
> ** V. Ya. Egorova was arrested for attending embassies of Egypt and Syria in search of work. She did it on the instruction of the USSR Ministry of State Security for which she served as a secret agent (see note 1 to doc. 6.11).
> *** G. V. Kostyrchenko. Tainaya politika Stalina. Vlast' i anti-Semitism (Secret Policy of Stalin. Power and Anti-Semitism). M.: Mezhdunar. otnosheniya, 2001, pp. 643, 645, 652.
> **** GARF, fond P94–25, opis' 1, ed. khr. 1603, l. 24; see also Kul'tura i vlast' ot Stalina do Gorbacheva. Ideologicheskie komissii TsK CPSS 1958–1964. Dokumenty (Culture and Authorities from Stalin till Gorbachev. Ideological Commissions of CPSU Central Committee 1958–1964. Documents). M.: ROSSPEN, 1998, p.97. S. M. Mikhoels (1890–1948), a known public figure and great actor, was killed by secret agents on 13 January 1948 on the order of Stalin. (see, for instance, *Argumenty i fakty*, No. 19, 1992, p. 5). He was the first cousin of Professor M. S. Vovsi accused of leading the group of "terrorist-doctors" five years later. Mikhoels was officially announced to have died in a car accident and was given state funerals as "a distinguished representative of Soviet culture". He was for the first time called the "known Jewish bourgeois nationalist" in "Khroniki TASS" five years after his death. Hence inclusion of his name together with "saboteur-doctors" in one list by Glavlit. This coincidence is another example of the cynicism and designing abilities of the dictator...

[4] "Vestnik AN SSSR, No 9, 1948, pp. 21–24. L. A. Orbeli was dismissed from the position of academician-secretary "in connection with the discussion of the situation in biological science".* Simultaneously, he was relieved of his duties as the editor-in-chief of "*Zhurnal obshchei biologii*".

> *RGASPI, fond 17, opis' 118, ed. khr. 247, l. 7.

[5] A. V. Topchiev, chief scientific secretary of the USSR Academy of Sciences, spoke at the meeting of the Communist Party organization of its Presidium (17.7.1950) and explained actions against Orbeli in the following way: "The central organ of our Party* has for a long time given much attention to the teaching of I. P. Pavlov. Hence the conclusion that science can not develop without conflict of opinions and free criticism. We must put an end to the monopoly of individual scientists. Orbeli has developed Pavlov's ideas in the wrong way for 15 years; he misinterpreted it and did great harm."** Under Stalin's rule, such characteristics could not help affecting people close to the persecuted person. I. A. Orbeli, member of the USSR Academy of Sciences, director of the Hermitage Museum and brother of L. A. Orbeli, was also attacked by authorities in 1950. He was accused of incompetence and organization of an exhibition of French impressionist painters*** whose "formalistic" works were regarded as

incompatible with the criteria of the Russian realistic school. I. A. Orbeli had to retire from his post in 1951.

* Newspaper *Pravda.*
** Cited from the shorthand record of the meeting. TsAODM, fond 992, opis' 2, ed. khr. 24, l. 84.
*** RGASPI, fond 17, opis' 118, ed. khr. 969, l. 15–21.

[6] L. G. Leibson. Akademik L. A. Orbeli. Neopublikovannye glavy biografii (L. A. Orbeli. Unpublished Chapters of Biography). Leningrad: Izd-vo "Nauka", 1990, p. 155. Addressing his letter to Stalin, Orbeli had no idea that the addressee was one of the inspirers of persecution and the author of statements like: "he did everything he could to slander and defame Pavlov, to discredit his teaching by cowardly and dishonest* attacks." (see doc. 7.39)

* Sic!

[7] "Pravda", 16.1, 1953, p. 3. A. V. Voino-Yasenetsky, Orbeli's disciple, noted that "Orbeli found himself in a most unfavourable situation in January-February 1953."* A. V. Voino-Yasenetsky and Z. I. Barbashova, co-workers of L. A. Orbeli, wrote a letter to Stalin (October 1952) asking "to interfere in the fate of academician Orbeli and ease conditions around him". The letter did not reach Stalin and was readdressed by his secretariat to Yuri Zhdanov.**

* L. A. Orbeli v vospominaniyakh sovremennikov (L. A. Orbeli in Recollections of His Contemporaries). Leningrad: Izd-vo "Nauka", 1983, p. 60.
** RGASPI, fond 558, opis' 11, ed. khr. 882, l. 59.

[8] These secret instructions met with hidden resistance of the heads of many research and educational institutions. According to A. I. Vaksberg, no Jew was admitted to the internal postgraduate courses of the prestigious All-Union Institute of Juridical Sciences in 1953, but all those who wished (including himself) could enter the extra-mural postgraduate study.* The author of this book, a graduate from the Moscow Geological Prospecting Institute (1951), was enlisted in the extra-mural postgraduate courses of the Geological Institute, the USSR Academy of Sciences, in 1955. Professor A. A. Gapeev, head of the Chair of Coal Geology, tried to have me admitted for the internal study but failed.

* Ark.Vaksberg. *Stalin protiv evreev* (Stalin Against Jews). Liberty Publishing House, New York, 1995, pp. 354–355.

[9] Characteristically, the request of Sillen objectively related to the lack of reliable information about research activities of Soviet chemists is reminiscent of an earlier statement by Professor Liljestrand, secretary of the Nobel Committee in physiology and medicine, who complained of the absence of Russian scientific literature in Sweden published during the last 10–20 years. In both cases, the situation was created by the actions of Soviet authorities aimed at curtailing scientific and cultural relations with foreign states since the mid-1930s.

[10] Professional diplomat N. M. Lun'kov occupied key positions in the Soviet embassy in Stockholm in 1954–1957. He is sure that the article in "Novoye vremya" was written by A. M. Aleksandrov, a member of the mission (see doc. 7.37). In 1952, he was appointed first secretary of the 5th European Department, the USSR Ministry of Foreign Affairs. In other words, he was directly subordinate to A. N. Abramov, head of the Department. Naturally, the letter by Sillen received from the embassy in 1952 was passed to him; hence, the four pages cut from "Novoye vremya" and attached by Abramov to his letter.

[11] From recollections of the literary critic E. G. Gershtein: "The most terrible time for me was the late 1940s and the early 1950s during the struggle against cosmopolitanism. They dismissed me from any position I happened to occupy, everything seemed unsettled [...] my university mate Feinberg, a brilliant Pushkin scholar, told me that they started to build up labour camps for Jews."*

* Emma Gerstein. Memuary (Memoirs). M.: Zakharov, 2002, p. 747.

[11a] See "Voprosy literatury", No. 2, 2005, pp. 60–62.

[12] "Pravda", 4.4.1953, p. 2. It was decided to close the "doctors' case" very soon after Stalin's funeral. General M. D. Ryumin, deputy minister of state security and head of the investigation body of the USSR Ministry of State Security, was arrested as the principal initiator of this case* on 17.3.1953**, i.e. less than two weeks after Stalin's death and almost three weeks before the mass media reported the release of the accused doctors. The arrested doctors were set free on the night of 3–4 April. When at home, they could hear an official report about it broadcast in a news program.

* The name of the general was disclosed in the editorial in "Pravda" of 6.4.1953. The former lieutenant colonel obtained quick promotion after he wrote a denunciation of B. S. Abakumov, minister of state security, in the personal office of S. D. Ignatiev, head of the Staff Department, AUCP(b) Central Committee (in mid-1951). Abakumov was arrested on 13.7.1951 and subsequently executed. Ignatiev occupied his post. Ryumin initiated the "Zionist conspiracy" case when he was deputy minister.
** "Istoricheskiy arkhiv", No 3, 2002, p. 3. Ryumin was arrested on the order of L. P. Beria; after the death of Stalin, Beria was appointed head of the USSR Ministry of Internal Affairs that incorporated the former Ministry of State Security. He occupied this position till 26.6.1953 when he was in turn arrested and thereafter executed. The Military Board of the USSR Supreme Court sentenced Ryumin to capital punishment on 7.7.1954, i.e. more than half a year after Beria had been shot.

[13] Here is an example of the rapid restoration of contacts with Swedish partners. Sweden took active part in the development of modern telephone and postal systems in Russia since the mid-19th century. In June 1953, the minister of communication N. D. Psurtsev, one of the most progressive Soviet leaders, wrote to the USSR Ministry of Foreign Affairs for information about "the structure and operation of the Swedish postal service in the post-war period [...] its qualitative characteristics, mechanization, labour conditions for postal personnel [...] relations between postal authorities and dealers providing postal services to the population [...]"*

* AVP RG, fond 0140, opis' 44, papka 117, ed. khr. 35, l. 50.

[14] The candidacy of N. N. Semenov, member of the USSR Academy of Sciences, was considered by the Nobel Committee in chemistry in 1946–1948 and 1950 (tables 4.1 and 6.1).

[15] Two years after the 1955 Congress, M. M. Dubinin, academician-secretary of the USSR Academy of Sciences, was again invited to read lectures on solid adsorbents in Sweden. In his report to the USSR Academy of Sciences, he told about the "great consideration shown to us by the Sillens. They invited us to have dinner at their country house on the day of arrival [...] they also invited us to a ballet performance at the Opera House, organized a sightseeing tour in Stockholm; we many times had dinners and suppers with them in restaurants, etc. We

could see from their talks and behaviour that they sympathize with our country. They could hardly give us so much care and attention from mere politeness."*

* RGANI, fond 5, opis'35, ed. khr. 31, l. 43.

[16] Sillen and A. Tiselius, president of the International Union of general and applied chemistry, maintained business correspondence with A. V. Topchiev, chief scientific secretary of the USSR Academy of Sciences, during organization of the XIII Congress of the Union in Stockholm. Tiselius, Nobel laureate of 1954, was vice-chairman of the Nobel Committee in chemistry (see also doc. 7.5).

[17] An official response to this proposal was not found in Russian archives. Probably, verbal consent was given. Such a mode of distributing invitations persisted for a long time. M. Sulman, executive director of the Nobel Foundation told the author in February 1998 that "parcels" of blank invitation forms were sent to the USSR Academy of Sciences and its Presidium distributed them among members of the Academy. Also, the Nobel Committee in physiology and medicine regularly sent such "parcels" to the USSR Ministry of Health that passed them to the USSR Academy of Medical Sciences.

[18] The All-Union Communist Party (bolsheviks) was renamed as the Communist Party of the Soviet Union at its XIXth Congress in October 1952.

[19] There was a serious reason why leaders of the USSR Academy of Sciences did not want to include this issue in the official agenda of the sittings of the Presidium. Contacts of members of the Academy with the Nobel organizations were not banned by an official document of higher organs. They were limited by verbal recommendations. The Presidium could not officially object to what was not officially forbidden.
The Presidium of the USSR Academy of Sciences usually resorted to verbal "understanding" when it came to settling "delicate" issues, without entering them into official minutes. For example, the director of the Physical Institute, the USSR Academy of Sciences, demanded that the chief accountant stop paying a salary to A. D. Sakharov, Nobel Prize winner, after he was exiled to the city of Gorky in 1980 (see Chapter 11). The accountant requested an extract from the decision of the Presidium but was not given it for the lack of relevant written documents. As a result, Sakharov received salary during his seven-year absence from the Institute. The Czech physicist F. Janouch told this story and summarized: "in the jungles of Brezhnev's time lawlessness, the law was all of a sudden defended at the lowest level; an accountant saved the honor of the Academy [...]"*

* In: "On mezhdu nami zhil... Vospomimaniya o Sakharove (He Lived Among Us. Recollection About Sakharov)". M.: Praktika, 1996, p. 828.

[20] The cover note with which the Foreign Department of the USSR Academy of Sciences passed Sillen's letter to M. M. Dubinin,* academician-secretary of the Division of Chemical Sciences, (29.6.1954) bears the outgoing number and the date of receipt (9.6.1954).** In other words, the letter dated by Sillen from 10.5.1954 was referred by the USSR Ministry of Foreign Affairs*** to the Presidium of the USSR Academy of Sciences one month after it had been received by the Soviet embassy in Stockholm.

* The head of the Foreign Department showed surprising tactlessness and broke a condition of confidentiality by passing the letter of Sillen to Dubinin, mentioned in it as a potential nominee for the Nobel Prize of 1955.

** Arkhiv RAN, fond 579, opis' 1, ed. khr. 639, l. 17.
*** The letter was sent from the USSR Ministry of Foreign Affairs on 8.6.1954.

[21] K. A. Vlasov, acting chief scientific secretary of the USSR Academy of Sciences, passed the letter of Vinogradov to the deputy head of the 3d European department, the USSR Ministry of Foreign Affairs, S. A. Afanasiev on 6.8.1954; the cover note No. 1–46–398* refers to the outgoing No. 1410/3eo (8.6.1954) on the cover note with which the USSR Ministry of Foreign Affairs forwarded the letter of Sillen.

* Arkhiv RAN, fond 579, opis' 1, ed. khr. 639, l. 24.

[22] Supervising organs were traditionally afraid of sending printed matter abroad. Suffice it to recall the anecdotic correspondence of 1947 *a propos* the request of the known Swedish physicist and Nobel laureate M. Siegbahn to pass to him the journal "Nauka i zhizn'" with the article by P. L. Kapitsa about liquid helium (docs. 6.7–6.8).

[23] It may be thought that the final decision (verbal, see note 19) to dispatch the scientific works of Semenov to Sillen was taken at the sitting of the Presidium of the USSR Academy of Sciences held on 25.12.1954 after the telephone talk of A. V. Topchiev, chief scientific secretary of the Academy, with officials of the Sector of Science, AUCP(b) Central Committee, on 2.12.1954 (doc. 7.5).

[24] RGANI, fond 5, opis' 17, ed. khr. 458, l. 180–183. It is difficult to explain the decision of the CPSU Central Committee Secretariat to forbid the visit of these five physicists to the Soviet Union and simultaneously allow the visit of four others (Feynman and Dayson (USA), Andus Salam (Pakistan), and Callen (Switzerland)). M. Chini (Italy) and S. Tomonaga (Japan) could be regarded as representatives of hostile countries during World War II but the antifascist views of Pauli were no secret to Soviet authorities. If the United States was the main antagonist in the Cold War, why was the invitation of Feynman and Dayson approved while Edwards (UK) and Levi (France) were not invited. No wonder Rumyantsev provided no rationale for his decision.

[25] Arkhiv RAN, fond 471, opis' 1, ed. khr. 144a, l. 47–56. This archive file contains all minutes of the sittings of Bureau of the Division of Physico-Mathematical Sciences for 1955. An extract from the minutes of the sitting on 1.11.1955 (with item 19 (doc. 7.19) is also kept in the RAN archive among documents of the Foreign Department (Arkhiv RAN, fond 579, opis' 1, ed. khr. 639, l. 32).

[26] The following members of the Academy were present at the meting of the Division of Physico-Mathematical Sciences on 1.11.1955 (as recorded in minutes No 13): M. A. Lavrentiev, academician-secretary of the Division, A. N. Lobachev, its secretary, L. A. Artsimovich, G. V. Kurdyumov, P. L. Kapitsa, I. M. Vekua, M. G. Meshcheryakov, E. R. Mustel', and D.Phys.-Math. Sci. I. I. Novikov (all members of the Bureau). Invited participants were P. S. Aleksandrov, full member of the USSR Academy of Sciences, M. S. Molodensky, corresponding member, S. V. Fomin, D.Phys.-Math.Sci., A. S. Fomin,* Cand.Phys.-Math.Sci (representative of CPSU Central Committee).

* Arkhiv RAN, fond 471, opis' 1, ed. khr. 144a, l. 178.

[27] Arkhiv RAN, fond 471, opis' 1, ed. khr. 144a, l.188. It appears that item 18 was also a backdated entry. Such practice was very common in Soviet organizations.

[28] Topchiev was educated as an organic chemist, which certainly facilitated his personal contacts with Sillen. Moreover, Sillen maintained correspondence with him in the capacity of secretary of the physico-chemical section of the XIII International Congress of general and applied chemistry during its organization in 1953. It may be supposed that Sillen explained to Topchiev why he could not present nomination of Semenov for the Nobel Prize of 1955 when the two scientists met in December that year.

[29] Rumyantsev was moved from the position of the head of department of the CPSU Central Committee to the editorial board of the journal "Communist" and thereafter to "Problemy mira and sotsializma" (theoretical journal of Communist and Workers' Parties of several countries printed in Prague). Three weeks after Khrushchev's downfall (November 1964), he was appointed editor-in-chief of the newspaper "Pravda". This appointment was welcomed by many journalists because Rumyantsev proved himself to be a liberal during his previous work in press organs. V. Ya. Lakshin, member of the editorial board of the journal "Novy mir", wrote in his diary (9.11.1964): "Some liberal trends are emerging. Rumyantsev is appointed editor-in-chief of "Pravda" instead of Satyukov. Karyakin* worked with him in Prague and gives high praise to him. Tvardovsky said: '*Oktyabrists* (partisans of the fundamentalist communist literary journal *Oktyabr*') will hardly drink champagne.'"** Such ambivalence is a characteristic feature of the actors of totalitarian regimes that make them perform deeds at variance with their moral principles.

> * Yu. F. Karyakin, publicist and literary critic, one of the prominent public figures of the sixties, irreconcilable opponent of the creeping restoration of Stalinism undertaken by M. A. Suslov's assistants after Brezhnev came to power.
> ** V. Ya. Lakshin "Novy mir" vo vremena Khrushcheva: Dnevnik i poputnoye ("Novy mir" in Khrushchev' Time: Diary and Incidental Notes)." M.: Knizhnaya palata, 1991, p. 255.

[30] The valuable information of Lun'kov needs to be commented on. To begin with, it suggests attempts of L. Sillen to get in touch with the Soviet embassy; Lun'kov had no other source from which to learn that Semenov was included in the nomination lists of 1956. Secondly, the combination of names "Khrushchev–Erlander–Semenov" reflects the old (since Kollontai's time) disease of Soviet diplomacy — the inability to understand that the Nobel organizations are independent of state authorities, and that neither the king nor the prime minister has any influence on the selection of candidates for the next award.

[31] See V. O. Osipov. Tainaya zhizn' Mikhaila Sholokhova (Secret Life of Mikhail Sholokhov). M.: Lubereya, Raritet, 1995, p. 321.

[32] P. L. Kapitsa. Pis'ma o nauke (Letters About Science). Izd-vo "Moskovsky rabochiy", 1980, pp. 314–319.

[33] RGANI, fond 5, opis' 35, ed. khr. 16, l. 34–36; see also newspaper "Kuranty", 29.7.1994, p. 9.

[34] C. N. Hinshelwood. The Kinetics of Chemical Chains in Gaseous Systems, 1926.

[35] Arkhiv RAN, fond 2, opis' 6a, ed. khr. 136, l. 7.

[36] The official explanation for holding the banquet in the reception hall of the Swedish Academy instead of the Stockholm City Hall was formulated in the following statement: "The current international situation and difficulties encountered by many states necessitate the restriction of the number of guests at the official dinner following the prize-giving

ceremony. The dinner will be held in the large hall of the Swedish Academy known as the "Stock-Exchange Hall."* This urgent decision was taken by the Nobel Foundation without convening the meeting of its Executive Board.**

* Les prix Nobel, 1956, p. 46, Stockholm, 1957.
** The last sitting of the Executive Board (before the ceremony of 10 December 1956) was held in September.

[37] L. N. Semenova related that her parents went to Stockholm by plane together with the Swedish ambassador and his wife, who told Nataliya Nikolaevna that female participants in the ceremony were expected to wear grey dresses rather than the dark evening gown she had with her; Semenova bought the new one in Stockholm.*

The Sulmans officially accompanied the Semenovs to Nobel festivities. In accordance with protocol practices of the Swedish Ministry of Foreign Affairs, only one of the ambassadors accredited in the countries whose representatives received the Nobel Prizes in a given year could participate in the official ceremony. Sulman was chosen because the 1956 prize was the first for the Soviet Union.

[38] RGANI, fond 5, opis' 35, ed. khr. 49, l. 34–37, 41.

[38a] *Ibid*, ed. khr. 31, l. 110–113.

[39] Doc.7.31 mentions only three of the four nominees (Arbuzov's name is absent). The copy of original nomination of Arbuzov has the following note: "CPSU Central Committee permitted the USSR Academy of Sciences (26.1.1956) to nominate A. E. Arbuzov, member of the Academy, for the Nobel Prize in chemistry. This decision was brought to the notice of comrade Topchiev* (24.1.57). V. Smirnov, instructor of the Department of Science and Education, CPSU Central Committee. 25.1.57. N. Glagolev."**

* See doc. 7.26.
** RGANI, fond 5, opis' 35, ed. khr. 51, l. 24.

[40] One of such attempts initiated by security services was the resolution of the CPSU Central Committee "On measures for the protection of state secrets" adopted in autumn 1958. This resolution was not published in the open press but was discussed in all Communist Party organizations of the country. The Secretary of the Party bureau at the Presidium of the USSR Academy of Sciences emphasized (30 October 1958) that "previously classified books are sent abroad without consent of the Divisions of the Academy." The speaker referred to a huge amount of published materials classified as secret in the late 1940s including those printed in the 1920s-1930s and readily available in foreign libraries (see note 22 to Chapter 6). Another example illustrating the absurdity of this campaign in the last years of Stalin's rule was the accusation of sending abroad the booklet *Geochronological scale of the USSR** brought against radiochemist I. E. Starik, corresponding member of the USSR Academy of Sciences. This edition was classified in the late 1940s although it contained purely scientific data on the age of geological formations. Starik was also charged with transmitting abroad the recently published jubilee volume *35 years of the Radium Institute*** readily available through retail.

In half a month (13 November), a speaker at the Party meeting in the Joint Institute of Nuclear Research, Dubna, drew attention to the publication of the report of a scientist from the

Laboratory of High Energies in the local newspaper "Za kommunizm" that "gave a good idea of the conditions of the synchrophasotron and its operation for the work with deutrons."***

* TsAODM, fond 992, opis' 2, ed. khr. 69, l. 123.

** "35 let Radievogo instituta im. V. G. Khlopina AN SSSR (pod red. I. E. Starika). Leningrad: Izd-vo AN SSSR, 1957

*** TsAODM, fond 7913, opis'1, ed. khr. 2, l. 60.

Document 7.1

Extract from the minutes of the open meeting of the Communist Party organization held at the Institute of Physical Problems, the USSR Academy of Sciences[1] 26.2.1953.

I

Replies to the questions after report on the work with the personnel by M. P. Malkov, deputy director.

Question: What measures are taken to organize the discussion of Landau's lectures in the laboratories where they were not yet discussed?[2]

Reply: Administration had talks with the heads of these laboratories. I believe that such discussion is absolutely necessary.

Question: Why does the Theoretical Department have no postgraduate students?[3]

Reply: It is very difficult to find postgraduate students for theoretical disciplines. It needs much work. There is one candidate for whom we are waiting for a year and a half. We do our best to have him referred to this Institute, but it remains to be seen whether the problem will be positively solved.[4]

TsAODM, fond 995, opis' 1, ed. khr. 10, l. 12.

II

Statement of L. D. Landau, member of the USSR Academy of Sciences.

I am rebuked for having no young researchers. This question has two aspects. On the one hand, it should be recalled that more than half of the leading theoretical physicists are my former pupils.[5] On the other hand, training theoretical physicists encounters serious difficulties that we are trying to overcome. It is difficult to recruit new postgraduate students. All those admitted to the Institute were recruited by me. But I do not teach at educational institutions[6] and therefore can not have enough postgraduate students. The best of them are taken by their lecturers. Now Lifshits became a lecturer[7] and probably will be able to help. However, it is impossible to surmount all obstacles because we can not influence many things. By way of example, we selected two candidates; one we shall have, but there is no guarantee that we shall have the other.[4]

Ibid.

III

From the statement of A. I. Shal'nikov, corresponding member of the USSR Academy of Sciences.

[...] About the work of the seminars.[8] The reports were bad, confined to retelling published papers. There were good examples but on the whole the preparation of seminars was unsatisfactory, and they had to be substituted by the lectures of Landau. The low scientific level of junior researchers is confirmed by the fact that they have to additionally study Landau's lectures by themselves.

Ibid., l. 13.

IV

From the statement of V. T. Khozyainov, a recent post-graduate student of Landau.[9]

Lev Davidivich has an unusual method of teaching that differs from the routine one in that it gives an opportunity for independent work. Post-graduate students taught by Landau are self-sufficient researchers unlike pupils of other scientists. Naturally, one has to overcome difficulties posed by Landau and meet his high requirements.[10]

Ibid., l. 20.

[1] An open Party meeting could be attended by everybody regardless of membership in the AUCP(b)-CPSU.

[2] The cycle of Landau's lectures on fundamentals of theoretical physics was organized by the administration of the Institute for postgraduate students and junior research scientists. After each lecture, the listeners had to discuss it in the respective departments of the Institute.

[3] The department headed by Landau.

[4] I. E. Dzyaloshinsky, a graduate from the physical faculty of Moscow State University. He was a gifted third-year student when he drew the attention of Landau, who read lectures at the faculty. Being a Jew, he could not be admitted to the Institute as a postgraduate student. He passed "theoretical minimum"* after the death of Stalin in 1953 and next year was invited to the Theoretical Department; he obtained the candidate of science (PhD) degree in 1957.

* A comprehensive examination on a special program covering all aspects of theoretical physics developed by Landau that he invited anyone to pass before admission to his school. Only 43 people passed the exam during the period from 1933 to 1961 (A. M. Livanova. Landau. M.: Znanie, 1983, p. 78).

[5] Those who passed "theoretical minimum".

[6] Landau had to leave Moscow University on September 1, 1952 as a result of the anti-Semitic campaign.

[7] E. M. Lifshits, later member of the USSR Academy of Sciences and co-author with Landau of *The Course of Theoretical Physics.* He was sacked from the Moscow State University together with Landau and thereafter taught physics in the V. I. Lenin 1st Pedagogical Institute, Moscow, for four years.* At that time, several graduates

from this Institute worked as junior research scientists at the Institute of Physical problems.

* Data included in notes 6 and 7 were reported to the author by Lifshits' widow Z. I. Gorobets.

[8] The seminar was organized at the Institute of Physical Problems, the USSR Academy of Sciences, to update postgraduate students and junior research scientists on recent progress in world physics.

[9] V. T. Khozyainov defended his candidate thesis on 10.1.1953.

[10] Khozyainov spoke in response to the remark of a participant at the meeting that Landau pays little attention to his postgraduate students.

Document 7.2

From: A. N. Abramov, head of the 5th European Department, USSR Ministry of Foreign Affairs.
To: G. M. Pushkin, member of the Board of USSR Ministry of Foreign Affairs.
No. 238/5eo, 9.2.1953

Palmer, member of the Executive Board of the "Sweden–USSR" Society, informed the Embassy of the USSR in Sweden that professor of inorganic chemistry Sillen of the Higher Technical School in Stockholm is going to nominate a Soviet scientist to the Nobel Committee[1] for the award of the Nobel Prize in 1953.
Sillen would like to have a candidate to be proposed by the Soviet Union and a list of his works with data on the time and place of their publication.
The Swedish Nobel Committee is reactionary and hostile to the Soviet Union as appears from its choice of candidates for the Nobel Prizes. It never awarded the prize to a Soviet scientist during the period of its existence. The reactionary character of the Nobel Committee was exposed in our press.[2] In the post-war years (1947, 1948, 1949), the Swedes invited Soviet scientists to take part in nomination of candidates for the Nobel Prize.[3] However, participation of Soviet scientists in the nomination was deemed unadvisable.[4]
I therefore believe that we should not respond to the proposal of Sillen and inform the embassy accordingly.
Given that Sillen directly addresses the embassy, the proposal should be additionally discussed depending on the form of its presentation.
I am waiting for instructions.[5]

AVP RF, fond 0140, opis' 44, papka 177, ed. khr. 35, l. 1.

[1] As before (see the preceding chapters), the author of the letter refers to prize-awarding organizations, in this case to the Royal Academy of Sciences and the Karolinska Institutet.

[2] The addressee underlined what he regarded as the most important phrases.

[3] See Chapter 4.

[4] Recognition of the fact that ideological services of the CPSU Central Committee during two years deliberately delayed the answer to G. Liljestrand, secretary of the

Nobel Committee in physiology and medicine, who asked to advise of the names of medical professors to be invited to nominate candidates for the Nobel Prizes (see Chapter 4 and docs. 4.20–4.25 and 4.28–4.35).

[5] In the top right corner: the stamp of Pushkin's secretariat with the date of receipt (9.2.1953) and incoming No. 887-GP. The addressee made no note on the letter. The superscription: "Inform the embassy. 24/IV.53" appeared two and a half months later, most probably on the verbal instruction of either the author or the addressee.

Document 7.3

From: L. G. Sillen, professor of inorganic chemistry, Higher Technical School of Stockholm.
To: S. S. Nikol'sky, 2nd secretary of the USSR embassy in Sweden.
Dateless[1]

During recent years, I have read a few abstracts from the works of A. K. Babko (the Ukrainian Academy of Sciences).[2] They report research in the field of my interests (equilibrium between ions in solutions); therefore, I would like to write to Professor Babko and exchange reprints of our papers. I would appreciate your advising me via VOKS a) to what address should I send letters and reprints to Prof. Babko to be sure that he receives them?; b) what language should I use in my letters? Prof. Babko may write to me in the Russian language which I can read tolerably well.[3] I could write in broken Russian too, but I have no Cyrillic typewriter. Unfortunately, my handwriting is illegible in both Russian and Swedish; c) it is very likely that in the future I will need to get in touch with another Soviet researcher of whose name and whereabouts I am not quite sure. May I rely on your assistance or shall I apply to other authorities?

Translated from the English by S. S. Nikol'sky.

AVP RF, fond 995, opis' 37, papka 64, ed. khr. 37, l. 2.

[1] The original letter in English was not found. The translation made by the secretary of the embassy is dateless. The cover letter under which the translation was sent to G. P. Shumilov, head of the Department of Scandinavian Countries, VOKS, is dated 9.1.1951, No. 830/6 shv.* It means, that Sillen wrote his letter in the second half of December 1950 or early January 1951.

* *Ibid.*, l. 1.

[2] Head of Department of Analytical Chemistry, Institute of General and Inorganic Chemistry, the Ukrainian Academy of Sciences; specialist in chemistry of coordination compounds and analytical chemistry.

[3] In 1998, Professor I. Grente, head of the Department of Inorganic Chemistry, Technological Institute of Stockholm, showed the author a shelf of books on chemistry in the Russian language that belonged to Sillen.

Document 7.4

From: Professor L. G. Sillen.
To: G. A. Zaraisky, attaché of the USSR embassy in Sweden.
10.5.1954
I am writing to you for assistance and advice in an important affair. You may be aware that I have the honour to nominate candidates for the Nobel Prize. I consider it an unpardonable omission that none of the Russian scientists received the Nobel Prize after Pavlov despite a large number of valuable works done in the Soviet Union. Nevertheless, the responsibility for this omission lies not at the Swedish door alone.
To begin with, you know that only limited contacts exist between Swedish and Soviet researchers. It makes it difficult for the Swedish side to judge whether a given Soviet scientist is actually a creative personality worthy of the Nobel Prize or merely a clever administrator knowing how to choose gifted assistants. Moreover, the majority of Soviet works were published in recent years in the Russian language and remain unavailable to most members of the Nobel Committees.
The Nobel Committee many times addressed Soviet researchers via the USSR Academy of Sciences with the request for help in selecting candidates for the Nobel Prizes. Such letters were annually sent after World War II, one year via VOKS, thereafter via the embassy, mostly by mail. However, they remained unanswered.[1]
1. Now to the issue in which I need your help and advice. Can you obtain for me reprints of scientific papers of N. N. Semenov and M. M. Dubinin and their CVs? I mean that one of them could be proposed as a candidate for the Nobel Prize in chemistry in 1955, Semenov for his studies on kinetic chain reactions and Dubinin for the works on solid adsorbents. If I propose one of them, the Nobel Committee would like to have the aforementioned information.
Dubinin and Semenov are physical chemists, like me. Therefore, it will be natural if I nominate one of them. I could hardly propose anyone else beyond my sphere of interests, e.g. organic chemist or physicist, although I know that there are worthy researchers in these fields too in the Soviet Union. It is desirable that the scientists of which I am speaking be unaware of the purpose for which the required information is collected because only one of them can receive the Nobel Prize in a given year. It would spare them the disappointment.[2]
I am acting on my own initiative but have reason to think that my well-substantiated nomination would be favourably met by the Nobel Committee.[3]
Preparation of decisions of the Nobel Committee always takes much time because discussion of scientific achievements of each candidate and their comparison is rather a long procedure.
Consideration of candidates for the 1955 Nobel Prizes begins only in January 1955,[4] but I must prepare in advance reviews, translations, etc. and distribute them among members of the Nobel Committee. It seems I already told you that I am very busy in teaching and research programs.[5] I would greatly appreciate it if you answered my

first question within a month or two; it would give me enough time for preparation and increase the probability of success.
2. One more question: "How should I advise the Nobel Committee as regards next years? Who of the Soviet scientists must they [members of the Committee — note of the translator] address invitations to nominate candidates for the Nobel Prizes. I should probably tell you that the Nobel Committee addresses only individuals but not academies or other institutions.
However, the Nobel Committee can certainly write to all physicists and chemists of the USSR Academy of Sciences and its leaders, such as Topchiev[6]; of course, Soviet scientists are free to present individual or collective nominations.
It is desirable that your reply to this second question reach me no later than mid-August 1954 because the first recommendations are expected in September.[7]

RGANI, fond 5, opis' 17, ed. khr. 460, l. 65–68.[8]

[1] See docs. 4.1, 4.2, 4.10, 4.20, 4.32, 7.2.
[2] See footnote* in note 20 to this chapter.
[3] This statement by Sillen, an expert of the Nobel Committee, is worthy of credit. The candidacies of Semenov and Hinshelwood he proposed in 1953 were favorably met by the members of the Nobel Committee in chemistry (see Table 6 and the accompanying text). The main obstacle was bureaucratic barriers deliberately created by Moscow in the last years of Stalin's rule...
[4] Preliminary selection of nominations by the Nobel Committee starting before the 1 February deadline.
[5] Before writing this letter Sillen had a preliminary talk with the addressee of the letter.
[6] In 1954, A. V. Topchiev was the chief scientific secretary for the USSR Academy of Sciences.
[7] See note 17 to this chapter.
[8] Translated from English; the original in the English language is in the archives of the Russian Academy of Sciences, fond 579, opis' 1, ed. khr. 639, l. 23 (both sides).

Document 7.5

From: A. I. Oparin, academician-secretary of the Division of Biological Sciences, the USSR Academy of Sciences.
To: CPSU Central Committee.
No. 252c, 30.11.1954

When in Sweden (to participate in the session of the World Peace Council, 18–25 November 1954), I was invited by Professor A. Tiselius to visit his Institute of Biochemistry in Uppsala. A. Tiselius is vice-president of the Nobel Prize committee.[1] After the visit, he introduced to me another member of this committee, Professor Fredga, an organic chemist. We talked about Nobel Prizes. Tiselius told us about his talks with members of the Nobel Committee who consider it unnatural that the Soviet Union is not involved in the Nobel movement although the Nobel Prize had been awarded to Pavlov. The Nobel Prize is an international award and the committee is interested in that Soviet scientists nominated their compatriots or researchers from other countries (if they deem it appropriate) for the prizes in physics, chemistry, biology, and medicine.[2] Moreover, there are prizes to writers and peace champions, but the latter is awarded by the parliament.[3] The Nobel Committee[4] many times sent letters of invitation to individual scientists and Soviet organizations directly or via our embassy but never received replies. At present, it again officially raises this question and asks for nominations from the Soviet side.

I said that I shall transmit this information to the USSR Academy of Sciences and asked to send their proposals in the written form together with the Nobel Statutes necessary to prepare nominations in accordance with the existing regulations. Professors Fredga and Tiselius agreed with me and informed that the deadline for nominations is 31 December.[5]

I am waiting for instructions in connection with the aforesaid. I presented a detailed report of my visits to research institutes of the Swedish Academy of Sciences to the Presidium of the USSR Academy of Sciences. I would like to emphasize that I met with a favourable attitude from Swedish scientists wherever I had contacts with them: they all wish to have close collaboration with their Soviet colleagues.

I reported information expounded in this letter to A. N. Nesmeyanov, president of the USSR Academy of Sciences, on November 29. He told me that Yu. B. Khariton (physicist) has recently received an invitation to nominate a candidate for the Nobel Prize and asked him (Nesmeyanov) for advice.[6]

The note by A. N. Cherkashin, head of the Sector of Science, CPSU Central Committee (2,12.1954) at the bottom of page 1: "Comrade Topchiev, chief scientific secretary of the USSR Academy of Sciences, informed that the Presidium will consider the letter of Oparin and report the proposals to the CPSU Central Committee.

RGANI, fond 5, opis' 17, ed. khr. 463, l. 209–210.

[1] A mistake: Vice-chairman of the Nobel Committee in chemistry.

[2] Nobel Prize in physiology and medicine.

[3] Peace prizes are also awarded by the Nobel Committee.

[4] As before, the author of the letter refers to prize-awarding organizations of the Nobel Foundation.

[5] A mistake. The nomination deadline is 31 January.

[6] Yu. Khariton, member of the USSR Academy of Sciences and an atomic scientist, was one of the 29 persons included by the Presidium of the Academy in the list of nominators for the Nobel Committees in physics and chemistry of the Royal Academy of Sciences (see doc. 7.15).

Document 7.6

From: A. P. Vinogradov, acting academician-secretary of the Division of Chemical Sciences, USSR Academy of Sciences.

To: G. A. Zaraisky, attache of the Soviet embassy in Stockholm.

No. 141–64,[1] July 1954

I have read the letter from Professor L. G. Sillen of May 10 addressed to you. Please tell him that we consider it a great honour for any scientist to participate in the competition for the Nobel Prize. The Nobel Statutes give the right to nominate candidates to Swedish scientists.[2] We are ready to assist them in the selection of worthy candidates. In what follows, I answer the questions of Professor Sillen.[3]

Professor N. N. Semenov is presently in Moscow, and we are collecting his publications as requested by Professor Sillen. We shall soon send them to you for dispatch to Professor Sillen.[4]

The second question is how to invite Soviet scientists to recommend their candidates for the Nobel Prizes. I believe that the Nobel Committee should first better inform them about the nomination procedure. It would allow the scientific community in the Soviet Union to have a better idea of the Nobel movement and facilitate assistance to Swedish scientists in the choice of worthy candidates. Certainly, this does not exclude the possibility for the Nobel Committee to directly address individual scientists in the Soviet Union.

Please thank Professor Sillen on my behalf for his attention and interest.

RGANI, fond 5, opis' 17, ed. khr. 460, l. 69.

[1] The letter was sent to the USSR Ministry of Foreign Affairs 6.8.1954 (see note 21 to this chapter).
[2] The author seems to have believed that only Swedish scientists had the right to nominate candidates for the Nobel Prizes while their foreign colleagues played the role of consultants-advisers.
[3] The materials could be sent to Sillen only at the end of 1954 because the CPSU Central Committee did not give its consent. The letter of Oparin (doc. 7.5) promoted a positive decision.
[4] See note 2 to this document.

Document 7.7

From: Professor L. G. Sillen.
To: Professor A. F. Westgren, chairman of the Nobel Committee in chemistry.
Dateless[1]

The 1st secretary of the Soviet embassy told me by phone the other day that the Soviet side recommends for the Nobel Prize of 1955 Professors Semenov and Dubinin. The official notification will come later. Because formalities there[2] take much time, I would like to inform you about their plans just now. I am ready to assist if I can be of any use, e.g. as a translator.[3]

I am going to nominate Semenov and Hinshelwood.[4] Of course, Dubinin made innovative sorbent studies and proposed a better multidisciplinary approach to the work with porous sorbents (activated carbon, silicagel, etc.) than Western scientists whom I know. However, his studies are quite new and known to few researchers working in this field in the West. Also, the language barrier hampers their wider knowledge. More time is needed to make these results generally recognized.

Archive of L. G. Sillen in the Technological Institute, Stockholm (translated from Swedish).[5]

[1] The letter was most likely written in August or in the first half of September 1954 because the talk with the secretary of the Soviet embassy could take place only during this period (see also doc 7.8).
[2] In the Soviet Union.
[3] Sillen read and spoke broken Russian.
[4] Quite a reasonable decision taking into account the benevolent attitude of the Nobel Committee in chemistry toward both scientists when they were first nominated in 1946 and 1947 (see Chapter 4).
[5] Photocopies of docs. 7.7 and 7.8 were received by the author from Professor Ingmar Grente, head of the Department of Physical Chemistry, who succeeded Sillen in this capacity after his premature death in 1970.

Document 7.8

From: Professor L. G. Sillen.
To: S. N. Istomin, 1st secretary of the Soviet embassy in Sweden.
19.9.1954

Attached find please two photocopies of the lists of publications of Professors Semenov and Dubinin. It took more time than expected to check which of their works could be found in Swedish libraries. Titles in brackets are not needed for our purpose or can be found translated into other languages. It is desirable to have the works indicated by arrows as original publications and in several copies (photocopies or microfilms). Minor works would probably be most convenient for Soviet librarians to present as microfilms. Also, it would be good to have Dubinin's[1] works published after 1949 and Semenov's after 1946 (the lists do not include later publications).[2]
Thank you for your assistance.[3]

Archive of L. G. Sillen in the Technological Institute, Stockholm (translated from Swedish).

[1] Sillen considered it incorrect to omit the name of Dubinin as he had decided to nominate only Semenov by that time.*
* See doc. 7.7.
[2] It means that the list of Semenov's works did not include his recently published fundamental monograph *On selected problems of chemical kinetics and reactionary capacity.* M.: Izd-vo AN SSSR, 1954.
[3] Subsequently, Sillen maintained good business relations with Istomin. The letter of 23.11.1954 in his archive was written after the diplomat visited the Technological Institute. It informs Istomin about the dispatch of certain statistical and other materials about this Institute to the Soviet embassy including "a copy

of the album *Technological Institute in photographs* by Gullers. I have two copies and it is a pleasure to give one to you. Unfortunately, it bears my name scribbled on the title-page. You may keep it if you find it interesting or transmit it to Moscow."

Document 7.9

From: A. P. Vinogradov, acting academician-secretary of the Division of Chemical Sciences, USSR Academy of Sciences.
To: G. A. Zaraisky, attache of the Soviet embassy in Stockholm.
July 1954[1]

In addition to my previous letter, find enclosed materials concerning N. N. Semenov, member of the USSR Academy of Sciences: 1) biography of N. N. Semenov; 2) selected works of N. V. Semenov (a book and 31 reprints).[2] Please pass all these materials to Professor L. G. Sillen with a request to return them after they are no longer needed because we have only a few volumes of the book[3] and an almost equally small number of certain reprints.

RGANI, fond 5, opis' 17, ed. khr. 460, l. 70.

[1] The letter was not dispatched.
[2] The list of the attached works is not published here.
[3] V. Smirnov, an instructor of the Department of Science and Education, CPSU Central Committee, made the following note on doc. 7.9 (27.1.1956): "The book of N. N. Semenov "On selected problems of chemical kinetics and reactionary capacity" is returned to the USSR Academy of Sciences by mail.*

*RGANI, fond 5, opis' 35, ed. khr. 16, l. 733.

Document 7.10

From: K. V. Ostrovityanov, vice-president of the USSR Academy of Sciences, and A. V. Topchiev, chief scientific secretary.
To: CPSU Central Committee.
No. IO-2/0253, 3.9.1954

Professor Sillen of the Royal Technological Institute of Sweden, member of the Nobel Committee in chemistry,[1] applied to the USSR Academy of Sciences (via the USSR Ministry of Foreign Affairs) with a request to nominate Soviet scientists for the Nobel Prize of 1955. He informs in his letter that Soviet chemists N. N. Semenov and M. M. Dubinin could be proposed as candidates for the Nobel Prize. The USSR Academy of Sciences prepared materials concerning scientific works of the known Soviet chemist N. N. Semenov.[2]
The Presidium of the USSR Academy of Sciences asks for permission to pass these materials to Prof. Sillen (via the USSR Ministry of Foreign Affairs).

RGANI, fond 5, opis' 17, ed. khr. 460, l. 64.

[1] Professor Sillen was not a member of the Nobel Committee in chemistry.
[2] A brief review of scientific works of Semenov on chemical kinetics was prepared by his colleagues V. N. Kondratiev, N. M. Emanuel, and V. V. Voevodin (Institute of Chemical Physics, USSR Academy of Sciences) and sent to the USSR Ministry of Foreign Affairs on 10.9.1954.*

* Arkhiv RAN, fond 579, opis' 1, ed. khr. 639, l. 27.

Document 7.11

A. V. Topchiev, chief scientific secretary of the USSR Academy of Sciences.
To: CPSU Central Committee.
No. IO-2/02506, 21.9.1954 in addition to No. IO-2/02353 of 3.9.1954

Enclosed find please a brief review of the scientific works of N. N. Semenov, member of the USSR Academy of Sciences. The Academy plans to send this document to the Nobel Committee in response to its proposal to nominate a Soviet chemist as a candidate for the Nobel Prize in 1955.[1]
Please return the attached materials once they are no longer needed.[2]

RGANI, fond 5, opis' 17, ed. khr. 460, l. 73.

[1] Inaccuracy. This proposal was made by Professor Sillen, an expert of the Nobel Committee, but not the committee itself.
[2] The materials were returned as evidenced by the note on doc. 7.13.

Document 7.12

From: A. M. Rumyantsev, head of the Department of Science and Culture, CPSU Central Committee.
To: CPSU Central Committee.
30.0.1954

The Presidium of the USSR Academy of Sciences (comrades Ostrovityanov, Topchiev)[1] informs that Professor Sillen of the Royal Technological Institute, Sweden, member of the Nobel Committee in chemistry,[2] wants to have a selection of scientific works and biographies of N. N. Semenov and M. M. Dubinin, members of the USSR Academy of Sciences.
Professor Sillen is going to nominate these scientists for the 1955 Nobel Prize. Sillen believes that a properly substantiated nomination will be favourably met by the Nobel Committee. He also plans to compile and distribute among members of the committee a review of the works of his nominee.[3] The Presidium of the USSR Academy of Sciences asks for permission to pass this information and some works by Prof. Semenov on chemical kinetics to Prof. Sillen via the USSR Ministry of Foreign Affairs.
With reference to the request of Prof. Sillen to have materials about scientific works of Semenov and Dubinin, members of the USSR Academy of Sciences, we consider

it possible to give the Presidium of the Academy permission to send them to Sillen. The dispatch of materials concerning N. N. Semenov alone would be taken as an official support of his candidacy and would bring harm to the prestige of Soviet science in case the Nobel Prize is not awarded to him.[4]
We ask for permission to inform comrade Nesmeyanov about this decision.

RGANI, fond 5, opis' 17, ed. khr. 460, l. 80.

[1] See doc. 7.10.
[2] An inaccuracy. Professor Sillen was not a member of the Nobel Committee in chemistry.
[3] See doc. 7.4.
[4] This statement should be regarded as a forerunner of the events in the USSR Academy of Sciences that followed the announcement of the decision of the Royal Academy of Sciences to award the Nobel Prize in chemistry for 1955 to foreign nominees and accounted for the appearance of docs. 7.19 and 7.20.

Document 7.13

From: A. M. Rumyantsev, head of the Department of Science and Culture, CPSU Central Committee.
To: CPSU Central Committee.
Dateless[1]

The Presidium of the USSR Academy of Sciences (comrades Ostrovityanov, Topchiev) informs [...][2]

Comrade Topchiev told, in addition, that it is undesirable to send the works of Dubinin because most of them are secret. It should be emphasized that Prof. Semenov also does secret work but it did not obstruct preparation of materials concerning his scientific activity and collection of publications for the dispatch to Prof. Sillen.[3] Professor Dubinin is the author of almost 100 works on gas adsorption by porous substances published in the open press. Adsorption processes are used in industry and of importance for chemical defense. The book by Dubinin *The Physico-Chemical Basis of Sorption Technique* deals with theoretical and experimental studies of adsorption (two editions).[4] Professor N. N. Semenov was awarded the Stalin Prize for investigations into adsorption processes and sorbent structure published in journals "Doklady akademii nauk SSSR" and "Zhurnal fizicheskoy khimii."[5]

With reference to the request of Prof. Sillen to have materials about scientific works of Semenov and Dubinin, members of the USSR Academy of Sciences, we consider it possible to give the Presidium of the Academy permission to send them to Sillen. The dispatch of material concerning N. N. Semenov alone would be taken as official support of his candidacy and would bring harm to the prestige of Soviet science in case his is not awarded the Nobel Prize.[4]

We ask for permission to inform comrade Topchiev about this decision.

The note above the text on the first page of the document says: "Agree" (the signature is illegible). The note at the bottom: "To be shelved (for 1 year).[6] Brought to the notice of comrade Topchiev. One book and 35 booklets[7] are returned to the Presidium of the USSR Academy of Sciences by A. Cherkashin." The signature is illigible.

RGANI, fond 5, opis' 17, ed. khr. 460, l. 81–82.

[1] According to the note on the letter, it was read on 12.10.1954 (signature is illegible), i.e. the document was prepared before this date.
[2] The following text is identical with the first paragraph of doc. 7.12.
[3] A sound observation doing credit to the author. In half a year (18.3.1957), however, the top officials of the same department of the CPSU Central Committee recommended to refuse permission to N. N. Semenov to visit Great Britain for the participation in the international symposium on radicals in the gaseous phase (9–12 April). The pretext was that "he is widely informed about the work of strictly secret research and industrial facilities"* as if he were not equally well-informed in December 1956 when he went to Stockholm to receive the Nobel Prize. The allusion to secrecy was a standing means employed by Soviet authorities to prohibit a person from doing something or going somewhere at his (or her) free will. In September 1957, Semenov and his wife were allowed to visit Czechoslovakia to participate in the work of international symposium on macromolecular chemistry.**

* RGANI, fond 5, opis' 35, ed. khr. 51, l. 38.

** *Ibid.*, l. 118.

[4] First edition: Goskhimtekhizdat, 1932, 381 pp; second edition (revised and extended). Goskhimizdat, 1935, 536 pp.
[5] The official formulation of the award of the grade 2 Stalin Prize for 1949 read: "for investigations into adsorption phenomena and sorbent structure published in journals "Doklady akademii nauk SSSR" and "Zhurnal fizicheskoy khimii" in 1947–1949."*

* "Pravda, 4.3.1950, p. 1.

[6] Clerks on the CPSU Central Committee appear to have forgotten about the one-year term for the storage of the document and thus preserved it for history.
[7] Reprints of papers published in scientific journals.

Document 7.14

From: V. D. Bobkov, head of the Foreign Department, USSR Academy of Sciences.
To: S. A. Afanasiev, head of the 3d European Department, USSR Ministry of Foreign Affairs.
No. 46–1862, 19.10.1954, in addition to No. 1-46-398, 6.8.1954[1]

The USSR Academy of Sciences asks to inform Professor Gunnar Sillen,[2] member of Nobel Prizes International Committee, that the Soviet side proposes for the award of 1955 two scientists, N. N. Semenov and M. M. Dubinin, members of the Academy. All materials necessary for their nomination will soon be dispatched to you.

Arkhiv RAN, fond 579, opis' 1, ed. khr. 639, l. 25.

[1] Draft letter dated to August and signed by A. P. Vinogradov where Semenov alone was mentioned as the candidate.
[2] See note 1 to doc. 7.10.

Document 7.15

List of members of the USSR Academy of Sciences to whom the Presidium of the Academy sent (in October 1954) invitations of the Nobel Committees in physics and chemistry to nominate candidates for the Nobel Prizes of 1955.

1. Semenov N. N.
2. Vvedensky B. A.
3. Tamm I. E.
4. Linnik V. P.
5. Kikoin I. K.
6. Lebedev S. A.
7. Alikhanov A. I.
8. Landay L. D.
9. Landsberg G. S.
10. Kurchatov I. V.
11. Khariton Yu. B.
12. Dorodnitsin A. A.
13. Aleksandrov A. P.
14. Kurdyumov G. V.
15. Lebedev A. A.
16. Frumkin A. I.
17. Ioffe A. F.
18. Sakharov A. D.
19. Artsimovich L. A.
20. Fok V. A.
21. Lavrentiev M. A.
22. Skobeltsyn D. V.
23. Petrovsky I. G.
24. Leontovich M. A.
25. Andreev N. N.
26. Lukirsky P. I.
27. Terenin A. N.
28. Shuleikin V. V.
29. Krylov A. P.

Arkhiv RAN, fond 579, opis' 1, ed. khr. 639, l. 28.[1]

[1]The archives of the Nobel committees in Stockholm related to the prizes of 1955, were opened in 2006 and we were able to find out at this point how actively Soviet nominators used their rights. Due to the publication of archive document by Pr. Prokhorov from the Physics Institute at Saint Petersburg University, it became widely known that one of the scientists listed, namely Academician V. A. Fok used the newly-created possibility and sent to the Nobel Committee in Physics his recommendation. His candidature was P.L. Kapitsa for "the discovery of the phenomena of superfluidity of helium."*

*"Poisk" N 4, 2004, page 12; see also note 38 in Chapter 1.

Document 7.16

From: A. N. Nesmeyanov, president of the USSR Academy of Sciences, and A. V. Topchiev, chief scientific secretary.
To: CPSU Central Committee.
"On the invitation of delegation of Swedish scientists to the USSR."
No. IO-2/022989, 25.11.1954

In 1954, the USSR Academy of Sciences received proposals of the Soviet embassy in Sweden, VOKS, and Eva Palmer, secretary-general of the "Sweden–USSR" Society, to invite a delegation of Swedish scientists to acquaint them with the work of the USSR Academy of Sciences and its research institutions. However, the Soviet embassy deems it inconvenient to invite eminent Swedish scientists as members of a delegation and recommends sending them personal invitations to visit the Soviet Union on the pretext that it is impossible to draw all of them together at a time. Also, the embassy holds that eminent Swedish scientists would hardly accept an invitation to come as members of a delegation because they usually receive personal invitations from other foreign institutions.

The Presidium of the USSR Academy of Sciences considers it advisable to invite a group of Swedish scientists to the USSR for the above purpose. The Presidium insists on their invitation in a small group because individual invitations would require the program of identical visits to be repeated many times and researchers of different institutes to be unnecessarily distracted from their everyday work.[1]

The Presidium asks for permission to invite to the Soviet Union for two weeks in 1954–1955 a group of Swedish scientists constituted by:
1. Lars Gunnar Sillen, professor of the Higher Technical School, Stockholm;
2. Tiselius, professor of the University of Uppsala, Nobel Prize winner;
3. Sven Brahualt, lecturer in chemistry;
4. Sven Ulrich Palme, historian;
5. Tamm, professor of the Higher Forestry School, forestry expert.

These candidates are recommended by Eva Palmer, secretary-general of the "Sweden–USSR" Society. The embassy of the USSR in Sweden considers it necessary to invite these scientists through the "Sweden–USSR" Society to promote

its authority in the country. The Swedish specialist could read lectures on different topics for the Soviet scientific community. In case of your consent, please give instruction to the USSR Ministry of Finance to approve the budget to cover expenses on the forthcoming visit including travelling expenses from Stockholm to Moscow and back.
Appendix: data about Swedish scientists on eight pages (non-secret).[2]

Lars Gunnar Sillen

Lars Gunnar Sillen, professor of the University of Stockholm[3] was born in 1916 in Stockholm in the family of a professor. He obtained his M.Phil. in 1936, PhL. in 1937,[4] and doctor of inorganic chemistry degree in 1941. Sillen published over 1000 works on crystal structure in solutions.[5]
He is loyal to the Soviet Union, actively helped the Soviet delegation in settling problems and promoting proposals.[6] Sillen maintains correspondence with the USSR Academy of Sciences and certain Soviet scientists (Professors Dubinin, Poroshin, etc.).
In 1954, Sillen forwarded a proposal to the USSR Academy of Sciences to nominate a known Soviet scientist as a candidate for the Nobel Prize in chemistry for 1955.[7]

Arne Tiselius

Arne Wilhelm Tiselius, president of the international union of Theoretical and Applied chemistry, was born in Stockholm in 1902. He graduated from the University of Uppsala, obtained MPh. in 1924, PhD in 1930, and became a professor of biochemistry in 1936. He headed the first department of biochemistry in Sweden. In 1934–1935, Tiselius worked in the laboratory of G. S. Taylor, Princeton University, USA. His further studies in Sweden were focused on electrophoresis. The results of his research were published in the transactions of the Faraday Society in 1937, most of them concerned with protein chemistry. In 1944–1946, Tiselius was a member of the Swedish Scientific Committee in Natural Sciences, in 1946 and 1947 he chaired the Committee. Simultaneously, he became a member of the Administrative Council of the Research National Defense Committee and of the Nuclear Power Committee.

A. Tiselius has been a member of the Nobel Committee in chemistry of the Royal Academy of Sciences since 1.1.1949 and vice-president of the Administrative Council of the Nobel Foundation since October 1949.[8] Tiselius was awarded the 1948 Nobel Prize for his work on electrophoresis and adsorption analysis, especially for the discoveries related to heterogeneous nature of serum proteins.[9]

Tiselius has published 75 works since 1926,[10] most of which are considered to be of paramount importance for the development of electrophoresis and adsorption theory and application of these phenomena to biochemistry. Later studies concern the use of paper chromatography for separation of amino acids and determination of the molecular weight of proteins.

Tiselius sympathizes with the Soviet Union and highly appreciates achievements of Soviet scientists.

The above data are borrowed from the materials of the USSR Academy of Sciences.

Signed: Agayants. 13.11.1954.[11]

RGANI, fond 5, opis' 17, ed. khr. 457, l. 226–230.

[1] A groundless statement most likely dictated by security services that find it easier to supervise groups rather than individuals.
[2] Data on Brahualt, Palme, and Tamm are not published.
[3] A mistake; Sillen was professor at Technological Institute, Stockholm.
[4] Licentiate of philosophy, scientific degree intermediate between master and doctor of philosophy.
[5] Sillen was a specialist in structural chemistry of solutions.
[6] See also note 15 to this chapter.
[7] The first proposal of Sillen dates to 1952 (see doc. 7.2).
[8] Vice-president of the Nobel Foundation.
[9] Official formulation: "for his research on electrophoresis and adsorption analysis, especially for his discoveries concerning the complex nature of serum proteins."
[10] Sic!
[11] In all probability, KGB general I. I. Agayants, curator of the Presidium of the USSR Academy of Sciences. The Presidium could not send to the CPSU Central Committee proposals to invite foreigners if not endorsed by Agayants. In 1959, I. I. Agayants was appointed head of KGB Department D (disinformation) and organized a series of successful provocations abroad.* There was no person with this name among the personnel of the Presidium of the USSR Academy of Sciences.**

* J. Barron. "KGB. The work of Soviet secret agents" (translated from English into Russian. Tel Aviv, "Effect Publications," 1998, pp. 225–227, 407; zhurnal "Zvezda," No. 4, 2003, p. 134.

** Information received from S. G. Korneev working in the Foreign Department of the USSR Academy of Sciences in 1955.

Document 7.17

From: A. M. Rumyantsev, head, and I. Vinogradov, instructor, Department of Science and Culture, CPSU Central Committee.
To: CPSU Central Committee.
No. 25-C-2625, 3.12.1954

The Presidium of the USSR Academy of Sciences (comrades Nesmeyanov and Topchiev) forwarded a proposal to invite a delegation of Swedish scientists to the USSR for two weeks to acquaint them with the work of the Academy and its research institutions [...] The proposal is the initiative of the Soviet embassy in Sweden and heads of the "Sweden–USSR" Society. The Soviet embassy deems it inconvenient to invite Swedish scientists as members of a delegation and recommends sending them personal invitations to visit the Soviet Union because it is impossible to draw all of them together at a time. Moreover, the Swedish scientists would hardly accept an invitation to come as members of a delegation because they usually receive personal invitations from other foreign institutions [...]

The Presidium of the USSR Academy of Sciences considers it inadvisable to send personal invitations to Swedish scientists because it would require the program of identical visits to be repeated many times and researchers of different institutes to be unnecessarily distracted from their everyday work. Swedish scientists themselves did not apply to the Soviet embassy in Stockholm, VOKS, and the USSR Academy of Sciences with a request to visit the Soviet Union and attend its research institutions; therefore, we consider it unadvisable to invite them.[1]
The Presidium of the USSR Academy of Sciences (comrade Topchiev) does not insist on its invitation.

There are three illegible signatures (most likely of secretaries of the Central Committee) in support of the opinion expressed in the letter in the left margin of sheet 236. The handwritten note below the text reads: "Comrdae Topchiev is informed. M. Yakovlev, V. Gorbunov. 7/XII."

RGANI, fond 5, opis' 17, ed. khr. 457, l. 236–237.

[1] A ridiculous reason for refusal. It may be thought, from the mixed composition of the group of Swedish specialists comprising representatives of different disciplines (from chemistry to history and forestry), that the Soviet embassy in Stockholm had already received permission for their visit to the Soviet Union; otherwise it would not have taken the initiative.

Document 7.18

Resuming publication of English abstracts in Soviet scientific press.

I

From: A. N. Nesmeyanov, president of the USSR Academy of Sciences, and A. V. Topchiev, chief scientific secretary.
To: Department of Science and Culture, CPSU Central Committee.
24.12.1954

With reference to the considerable extension of foreign relations of the USSR Academy of Sciences and the importance of their further development, the Presidium of the Academy asks to consider the following issues.

1. Many foreign scientists express their wish to be better acquainted with the work of the Academy and ask the Presidium to publish abstracts of scientific papers in foreign languages in academic journals. As many as 15 such requests from different foreign organizations were delivered to the library of the Academy in 1954. According to V. A. Ambartsumyan, member of the Academy, who participated in the sittings of the Executive Board of International Astronomical Union in Liege (Belgium) in July 1954, certain foreign journals are ready to publish Russian abstracts of scientific works in exchange of the publication of their abstracts in foreign languages in Soviet journals. Moreover, most researchers in the Chinese People's Republic know English. Therefore, publication of English abstracts in some Soviet journals would make it easier for them to know the achievements of Soviet science.

2. In recent years our scientists send increasingly more scientific literature to their foreign colleagues and scientific institutions. This trend would be greatly promoted if they were allowed to dispatch their publications directly to the addresses, bypassing the Foreign Department. The Presidium of the USSR Academy of Sciences asks for such permission taking into account that the analysis of scientific materials prior to publication includes assessment of the feasibility to transmit them abroad [...]

RGANI, fond 5, opis' 17, ed. khr. 507, l. 86–87.

II

From: A. M. Rumyantsev, head, and I. Vinogradov, head of the sector, Department of Science and Culture, CPSU Central Committee.
To: CPSU Central Committee.
21.1.1955

[...] According to the decision of the Presidium of the USSR Academy of Sciences No. 015 of 8.2.1954, all scientific literature published by the Academy is looked through before it appears in the open press or is transmitted abroad. Current practice of its additional reviewing and dispatching to foreign scientists and research institutions via the Foreign Department of the Academy leads to delays and hampers further development of foreign relations of the Academy. Also, it appears desirable to consider the possibility to publish English abstracts of Russian papers in academic journals, in the first place to make achievements of Soviet science known to Chinese and Indian researchers.[1] The Presidium of the USSR Academy of Sciences should be instructed to present a list of journals where such abstracts need to be published in the first place[2] and to clear up what reciprocal steps the respective foreign journals are prepared to take.

We deem it necessary to support the proposal of the USSR Academy of Sciences to revise regulations concerning correspondence of Soviet scientists with foreign colleagues and organizations with a view to their simplification.
We ask for permission to inform comrade Nesmeyanov about these considerations.

Ibid, l. 84–85.

[1] Officials on the CPSU Central Committee took up the idea of the Presidium of the USSR Academy of Sciences to lay emphasis on the necessity to satisfy the needs of researchers in friendly China and India and only marginally mention the interests of Western scientists.
[2] English abstracts appeared again in Russian scientific publications in the second half of 1955, first in 20 academic journals, the list of which was agreed with the CPSU Central Committee (see Chapter 4 and note 21 therein).

Document 7.19

Extract from minutes No. 13 of the sitting of Bureau of the Division of Physico-Mathematical Sciences, the USSR Academy of Sciences, (1 November 1955) passed to the Foreign Department of the Academy.
No. 101–11/1145, 21.11.1955[1]

XIX. On nomination of Soviet scientists for the Nobel Prize.
The Bureau of the Division considers it unadvisable to nominate Soviet scientists for the Nobel Prize because it can not be regarded as an international award; in the past, the Nobel Committee did not give this prize to the distinguished representatives of our science, such as D. I. Mendeleev, L. N. Tolstoy, A. P. Chekhov,[2] M. Gorky.
Chairman: M. A. Lavrentiev, member of the USSR Academy of Sciences, secretary: A. N. Lobachev, Cand. Chem. Sci.[3]

Arkhiv RAN, fond 579, opis' 1, ed. khr. 639, l. 32.[4]

[1] The entry in the registry gives an earlier date of the transmission of minutes No. 13 to the Foreign Department (19.11.1955).* However, the date on the document seems more reliable since it is confirmed by the clerk's note indicating that it was received and registered on 22.1.1955.

* Arkhiv RAN, fond 471, opis' 1, ed. khr. 144a, l. 188.

[2] A similar reproach was cast on Swedish "humanitarian scientists" of the Swedish Academy by *Literaturnaya gazeta* almost 40 years later.* However, the statements were irrelevant in both cases. Chekhov had never been nominated for the Nobel Prize. He became popular in the West after he died on 17.7.1904; for this reason he could not attract the attention of foreign nominators. None of the countrymen entitled to propose candidates recommended the candidacy of Chekhov to the Nobel Committee either.** Some of them did not regard Chekhov as worthy

of the Nobel Prize. Great Prince Kostantin Konstantinovich, president of the Academy of Sciences and honorary member of its Division of Russian Language and Literature, who had the right to nominate candidates annually, wrote in his diary (18.6 (1.7).1905): "I have just read the foolish article by Kuprin 'In memory of Chekhov'; Kuprin calls the deceased writer 'great', almost a man of genius having an unusually beautiful style and cites some of his unwitty bywords."*** This comment has much in common with the attitude of the conservative press towards Chekhov. The newspaper *Novoye vremya* reviewed the developments of the past 19th century and alluded to the arising decadent trends in literature and art. It counted Chekhov and Gorky among proponents of the new ideas since "they depicted in their works neurasthenics and vagabonds."**** A. M. Skabichevsky, a known historian of literature of the second half of the 19th and the early 20th centuries, treated Chekov in the condescending manner ("shallowness and pointlessness"; "talent most suitable for such ephemerous works" etc.). Skabichevsky did credit to Chekhov as "a favourite of the public," who won "a leading place in modern literature" in volume 6 of his *History of the New Russian Literature*, published after Chekhov's death.***** The ambivalent treatment of Chekhov by Russian intellectuals persisted until the 1920s. Prince Vladimir Golitsin recollected in his diary the words of P. D. Korin, an outstanding artist: "I don't like Chekhov, he smells of bugs."******

* *Literaturnaya gazeta.*

** The right to nominate candidates for the Nobel Prize was given to full and honorary members of the Division of Russian Language and Literature, Imperial Academy of Sciences.

*** "Novoye vremya," 1.1.1901, p. 2.

**** "Krasny arkhiv," t.1 (44), 1931, p. 1333. The obituary was published in a volume of "Znanie" for 1904, No. 3, SPb., 1905, pp. 5–14.

***** A. M. Skabichevsky. History of the New Russian Literature, 1st ed., SPb., 1891, p. 416; 6th ed., *ibid*, 1906, p. 380.

****** "Yunost", No. 3, 2003, p. 90.

[3] Candidate of Chemical Sciences.

[4] A similar text can be found in the documents of the Bureau of the Division of Chemical Sciences, the USSR Academy of Sciences.

Arkhiv RAN, fond 471, opis' 1, ed. khr. 144a, l. 187.

Document 7.20

From: M. M. Dubinin, academician-secretary of the Division of Chemical Sciences, USSR Academy of Sciences.
To: S. G. Korneev, head, Foreign Department, the USSR Academy of Sciences.
No. 141–640, 22.12.1955

No of minutes and dates of meetings	Dates of dispatch	
	Copies	Extracts
Minutes No 1, 27 January	No information	No information
Minutes No 2, 10 February	No information	No information
Minutes No 3, 15 February	18 February	18 February
Minutes No 4, 1 March	4 March	4–5 March
Minutes No 5, 29 March	2 April	1–5 April
Minutes No 6, 3 May	9 May	11 May
Minutes No 7, 17 May	25 May	25–26 May
Minutes No 8, 7 June	14 June	11–16 June
Minutes No 9, 28 June	9 July	11 July
Minutes No 10, 27 July	30 July	1–2 August
Minutes No 11, 6 September	27 September	24–26 September
Minutes No 12, 11 October	19 October	12–20 October
Minutes No 13, 1 November	12 November	12–19 November*
Minutes No 14, 15 November	19 November	16–20 November
Minutes No 15, 29 November	3 December	6 December

* Arkhiv RAN, fond 471, opis' 1, ed. khr. 144a, l. 47–56.
The date of dispatch indicated on the extract including item 19 of minutes No. 13 (see doc. 7.19) is 21 November.

This is to inform you that the Division of Chemical Sciences will not nominate a candidate for the Nobel Prize of 1956.

Arkhiv RAN, fond 579, opis' 1, ed. khr. 639, l. 31.

Document 7.21

List of minutes of the sittings of Bureau of the Division of Physico-Mathematical Sciences, the USSR Academy of Sciences for 1955, with dates of dispatch of copies to the Presidium of the Academy and extracts from the minutes to other divisions.

Document 7.22

From: M. M. Dubinin, academician-secretary of the Division of Chemical Sciences, USSR Academy of Sciences.
To: A. N. Nesmeyanov, president of the USSR Academy of Sciences.
4.1.1956

I recommend A. E. and B. A. Arbuzovs for informal discussion[1] of candidates for the Nobel Prize; their investigations into chemistry of organophosphorous compounds

made an outstanding contribution to world science. The history of the Nobel Committee knows of examples of nomination of two candidates for one prize (e.g. Pierre and Mari Curie).[2]
Attached find please a brief description of scientific achievements of A. E. and B. A. Arbuzovs.[3]

Arkhiv RAN, fond 579, opis' 1, ed. khr. 639, l. 40.

[1] The author suggests an informal consideration of the proposed candidates taking into account that collective discussion is in conflict with the Nobel Statutes requiring strict confidentiality of nomination.
[2] The Nobel Prize for 1903 was shared between P. and M. Curie and A. Becquerel.
[3] The handwritten description was translated into English and passed to the Foreign Department of the Academy (3.1.1956).

Document 7.23

Interview with A. S. Monin,[1] corresponding member of the Russian Academy of Sciences (18 February 1999).[2]

Question: Andrei Sergeevich, you were present at the sitting of the Bureau of the Division of Physico-Mathematical Sciences held on 1 November 1955 as a representative of the CPSU Central Committee. It follows from the minutes of the sitting that the participants refused to collaborate with the Nobel organizations because the Nobel Prize "can not be regarded as an international award."[3] Could you recall the discussion of this issue at the meeting?

Reply: I am surprised at this decision. I could hardly forget it if it had been actually adopted at that sitting. To begin with, the Academy and its divisions never held collective discussions of candidates for the Nobel Prizes. I can't remember such discussions at any meeting in which I participated.[4] They would contradict the requirements of the Nobel Committee that consider only personal nominations and disregard collective proposals. Secondly, I represented the Department of Science, CPSU Central Committee, and had nothing to do with questions concerning the work of the Nobel Committees; they were covered by the terms of reference of the International Department and the USSR Ministry of Foreign Affairs. Thirdly, the above decision looks quite unnatural. I am convinced (and it is the common opinion of all members of the USSR Academy of Sciences) that the Nobel Prizes in natural sciences were always regarded as prestigious international awards. Therefore, any honest scientist would have refused to support such a resolution. I guess the resolution included in the minutes of 1 November 1955 was never brought to the attention of the members of the Division of Physico-Mathematical Sciences and was known only to those who backdated it.

Question: M. M. Dubinin, academician-secretary of the Division of Chemical Sciences, also stated (22 December 1955) that the Division was not going to propose candidates for the Nobel Prize. In a few days, however, (3 January 1956) he passed the English translation of the nomination of A. E. and B. A. Arbuzovs to the typing office. What happened during the last week of 1955? What made Dubinin so quickly change his position?

Reply: I think that all you are talking of was the doing of V. D. Bobkov, head of the Foreign Department of the Academy. However, he could not take such an initiative by himself; in all probability he acted on the instruction of his superiors; Bobkob was directly subordinate to A. V. Topcheiv, chief scientific secretary of the Academy. Topchiev, in his turn, received orders from CPSU Central Committee; that is, from one of the secretaries or members of the Presidium because such issues were beyond the competence of apparatus of the Central Committee.
[5] Topchiev fulfilled the order, he could not do otherwise.

[1] The general meeting of the USSR Academy of Sciences elected A. S. Monin its full member (26.5.2000).
[2] Answers to the two questions are related to docs. 7.19–7.21.
[3] See doc. 7.19.
[4] Monin means that such questions were not included in the official agenda. For this reason, Dubinin recommended the candidacies of the Arbuzovs for "informal discussion by the Presidium of the USSR Academy of Sciences" (see doc. 7.22).
[5] An inaccurate statement. The Department of Science could not take major initiatives with respect to the USSR Academy of Sciences without their preliminary approval by the Secretariat of CPSU Central Committee. Nevertheless, it maintained everyday contact with divisions of the Academy including those pertinent to relations with the Nobel organizations. An example is a telephone conversation between A. V. Topchiev, chief scientific secretary of the Academy, and A. N. Cherkashin, head of the Department of Science, (2.12.1954) concerning information on Oparin (doc. 7.5).

Document 7.24

From: A. N. Nesmeyanov, president of the USSR Academy of Sciences, and N. M. Sisakyan, acting chief scientific secretary.
To: CPSU Central Committee.
No. IO-3/075, 16.1.1956

The Nobel Committee in chemistry invites Soviet scientists to propose their candidates for the Nobel Prize in chemistry for 1956. The Presidium of the USSR Academy of Sciences considered this invitation and deems it advisable that those invited nominate N. N. Semenov and his last work *Some problems of chemical kinetics and reactivity*[1] as well as A. E. and B. A. Arbuzovs for their works on organophosphorous compounds.

Enclosed find please materials about the scientific activity of these members of the Academy.
Also, we ask for permission for the scientists invited by the Nobel Committee to address their recommendations of N. N. Semenov, A. E. and B. A. Arbuzovs and all related documents to members of the Nobel Committee in Sweden.[2] The proposals must reach the Nobel Committee before the 1st of February this year.

RGANI, fond 5, opis' 33, ed. khr. 16, l. 5.

[1] The monograph was issued in 1954 by the publishing house of the USSR Academy of Sciences, with a circulation of 3,000.
[2] The CPSU Central Committee approved the request of the USSR Academy of Sciences. The Nobel Committee in physics and chemistry received a large number of twin nominations. B. Malmstrom, member of the Nobel Committee in chemistry and its chairman in 1977–1986, condemned this practice when he recalled (in 1998) the flow of nominations almost identical with one another (see doc. 8.53).

Document 7.25

From: A. N. Nesmeyanov, president of the USSR Academy of Sciences, and N. M. Sisakyan, acting chief scientific secretary.
To: CPSU Central Committee.
No. IO-3/0104, 20.1.1956

In addition to No. IO-3/075 of 16 January 1956.[1]
In 1954, the USSR Academy of Sciences received an invitation to recommend Soviet chemists as candidates for the Nobel Prize of 1955.[2] The Presidium of the USSR Academy of Sciences presented to the CPSU Central Committee (3 September 1954) proposals to nominate N. N. Semenov and M. M. Dubinin, members of the Academy.[3]
All necessary materials were forwarded to Professor Sillen, member of the Nobel Committee, at the end of 1954.[4] However, the Committee took no decision concerning the award of the prizes to Soviet scientists.[5]
Professor Sillen again addressed A. V. Topchiev when he visited Sweden in late December 1955 and asked to recommend Soviet scientists for the Nobel Prize in chemistry for 1956.
The Presidium of the USSR Academy of Sciences considered this invitation and deems it advisable that N. N. Semenov [...] A. E. and B. A. Arbuzovs be nominated for this prize and asks for permission to inform the Nobel Committee about it and send all necessary materials to Sweden.

RGANI, fond 5, opis' 35, ed. khr. 16, l. 6, 7.

[1] Probably upon the verbal request of the CPSU Central Committee after the receipt of doc. 7.24.
[2] Letter from Sillen dated 10.5.1954 (doc. 7.4).

3 See doc. 7.10.
4 See doc. 7.14.
5 Because the Soviet side did not present the necessary documents in time.

Document 7.26

From: N. I. Glagolev, deputy head, and V. Smirnov, instructor, Department of Science and Education, CPSU Central Committee.
To: CPSU Central Committee.
24.1.1956. Urgent.

The Presidium of the USSR Academy of Sciences (comrades A. N. Nesmeyanov and N. M. Sisakyan) informs that the Nobel Committee in chemistry invites Soviet scientists to propose their candidate for the Nobel Prize in chemistry for 1956. The Nobel Prize is annually awarded for the most important discovery or best improvement in this scientific discipline. The Presidium deems it advisable that N. N. Semenov be nominated for the Nobel Prize for his monograph *Some problems of chemical kinetics and reactivity* and A. E. and B. A. Arbuzovs for their works on organophosphorous compounds [...]
According to A. N. Nesmeyanov, president of the USSR Academy of Sciences, the candidacies of N. N. Semenov, A. E. and B. A. Arbuzovs for the Nobel Prize in chemistry for 1956 will be favourably met by Swedish chemists.[1]
N. N. Semenov, A. E. and B. A. Arbuzovs were nominated for the Nobel Prize by eminent Soviet scientists in conformity with the Nobel Statutes.
We agree with the opinion of the Presidium of the USSR Academy of Sciences and ask for permission to inform comrade A. N. Nesmeyanov about it.[2]

V. Smirnov made a note below this text (27.1.1956): "Comrade A. N. Nesmeyanov is informed. The book by N. N. Semenov *Some problems of chemical kinetics and reactivity* was sent back to the USSR Academy of Sciences by mail."
The note is signed by Smirnov and endorsed by Glagolev (30.1.1956). The document was shelved on 30.1.1956.

RGANI, fond 5, opis' 35, ed. khr. 16, l. 32, 33.

1 Neither document (7.24 or 7.25) contains the assertion that all three candidacies will be "favourably met by Swedish chemists."
2 The handwritten word "Agree" on the left side of doc. 7.26 (l. 32) is followed by five signatures of secretaries of the CPSU Central Committee and the date, 26.1.1956.

Document 7.27

From: M. M. Dubinin, academician-secretary of the Division of Chemical Sciences, the USSR Academy of Sciences.
To: Professor L. G. Sillen.
28.1.1956

I nominate Soviet scientists Aleksandr Arbuzov and Boris Arbuzov, members of the USSR Academy of Sciences, for the Nobel Prize; they both work in the field of chemistry of organophosphorous compounds.
Enclosed find please characteristics of scientific achievements of the candidates, a list of their publications, and reprints of their papers. The description of their research will be forwarded to you in the near future translated into English.[1]

The note in pencil below the text reads: "Dispatched on the instruction of Oraevsky."[2]

Arkhiv RAN, fond 579, opis' 1, ed. khr. 639, l. 34.[3]

[1] The letter was addressed to Sillen as a member of the Nobel Committee in chemistry although he was not such. It could hardly reach the committee in time being sent three days before the deadline for nominations for the 1956 Nobel Prizes. It means that the proposed candidates were not considered by the Nobel Committee.*

* Archives of the Nobel Committees are still closed at the time of writing and will be opened on 2 January 2007.

[2] Deputy head of the Foreign Department, the USSR Academy of Sciences.

[3] Sheet 34 is followed by 35–39 containing handwritten text of the English translation of the nomination. A handwritten note in a left corner of sheet 35 reads: "MB-INO,* 4 copies. E. Lebedeva. 3.1.1956."

* Typing office of the Foreign Department.

Document 7.28

From: V. N. Kondratiev, member of the USSR Academy of Sciences.
To: Professor L. G. Sillen.
Telegram No. 346, dated 28.01.1956.

In response to your invitation, I recommend N. N. Semenov, member of the USSR Academy of Sciences, for the Nobel Prize in chemistry.[1] All necessary materials were sent to you at the end of 1954 via the USSR embassy in Sweden. I simultaneously dispatch new materials and details by airmail.

Arkhiv RAN, fond 579, opis' 17, ed. khr. 639, l. 7.

[1] Recommendation of the author was immaterial. The main purpose was to inform Sillen about the dispatch of new materials concerning the scientific work of N. N. Semenov.

Document 7.29

From: N. M. Sisakyan, acting chief scientific secretary of the USSR Academy of Sciences.
To: M. G. Gribanov, head of the department of Scandinavian States, the USSR Ministry of Foreign Affairs.
No. 1–46–398, 28.1.1956.

This is to inform you that A. E. and B. A. Arbuzovs as well as N. N. Semenov, members of the USSR Academy of Sciences, are nominated for the Nobel Prize of 1956. The USSR Academy of Sciences sends materials on A. E. and B. A. Arbuzovs by airmail. Materials on Semenov were dispatched via the USSR Ministry of Foreign Affairs in the end of 1954.

Arkhiv RAN, fond 579, opis' 1, ed. khr. 60, l. 41.

Document 7.30

Program of visit of N. N. Semenov to Stockholm.[1]

9.12.1956 Cocktail in the office of Ekeberg, President of the Nobel Committee[2] (Sturegatan 14). Evening: Opera House.

10.12.1956 11.30 A rehearsal in the Concert Hall.[3]
4.30 Prize-awarding ceremony in the Concert Hall.
7.00 Dinner at the Nobel Committee (Banquet Hall of the Stock Exchange, old town).[4]

11.12.1956 11.00 Lecture of Professor Semenov in the Technological Institute of the Higher Technical School.
7.30 Dinner with the King.
12.12.1956 10.00 Visit to the Nobel Committee to receive the check (Sturegatan, 4).[5]
11.0 Visit to the Nobel Institute of Chemistry
20.00 Meeting with the Executive Board of the "Sweden— USSR" Society
13.12.1956 8.00 Departure to Uppsala. Visit to the University.
19.30 Saint Lucia Day[6] at the Higher Technical School.

14.12.1956 7.30 Lecture of Professor Semenov in the Swedish Chemical Society.
15.12.1956 Sightseeing, attendance of museums.
16.12.1956 Free time.
17.12.1956 Free time.
18.12.1956 Departure to the USSR.

AVP RF, fond 200, opis' 42, papka 88, ed. khr. 30, l. 32.

[1] Time is indicated in accordance with the traditional (in Sweden) division of the day into periods from midnight to afternoon and from afternoon to midnight.
[2] President of the Nobel Foundation
[3] The rehearsal of the appearance of the laureates at the scene of the Concert Hall, their seating, meeting with the king in the centre of a white circle with the letter N on a blue carpet, making bows to the king, members of the Nobel Committees, and

the public. The role of the king during the rehearsal is usually played by the executive director of the Nobel Foundation.

[4] In 1956, the formal dinner traditionally given after the prize-awarding ceremony at the Stockholm City Hall was transferred to the smaller reception hall of the Swedish Academy in the building of the former Stock Exchange (see note 36 to this chapter). The reason was the tragic events in Hungary and the presence of the Soviet citizen among the Nobel laureates. Only a small number of guests were invited. Mentioning dinner at the "Nobel Committee" was a mistake.

[5] These days, the Nobel laureates receive cheques of prize money from the Nobel Foundation.

[6] 13 December is Saint Lucia Day in Sweden (since 1927) in memory of a Syracuse girl executed in the year 304 during the mass persecution of Christians. She was canonized in the 15th century and the day of 13 December is celebrated as Saint Lucia Day. The festivities in Sweden include a procession of girls attired in white dresses and crowns with five lighted candles playing the role of Saint Lucia and her entourage. People exchange presents, watch beauty competitions, merry students' shows, etc. Nobel laureates have been participants of Saint Lucia festivities for several decades. Saint Lucia and her entourage wake them in the morning in the rooms of the Grand Hotel, their traditional whereabouts in Stockholm.

Document 7.31

From: D. M. Kukin, head, and N. I. Glagolev, deputy head, Department of Science, Higher Educational Institutions and Schools[1], CPSU Central Committee.
To: CPSU Central Committee.
24.1.1957. Urgent.

The Presidium of the USSR Academy of Sciences (comrades Bardin and Topchiev) asks for permission to nominate P. A. Cherenkov, D.Phys.-Math.Sci. and P. L. Kapitsa, member of the USSR Academy of Sciences, for the 1957 Nobel Prize in physics, and E. N. Pavlovsky, member of the USSR Academy of Sciences, in physiology and medicine.

Cherenkov P. A., born in 1904 of Russian descent, member of the CPSU, doctor of physico-mathematical sciences, head of a sector in the Physical Institute, the USSR Academy of Sciences, professor of Moscow Engineering Physics Institute, specialist in experimental atomic physics, author of almost 30 scientific publications. In 1934, Cherenkov made an outstanding discovery: he found that charged particles travelling at a speed higher than the phase velocity of light in a given medium emit waves of hard electromagnetic radiation. This phenomenon is called the Cherenkov effect throughout the world and is used in many atomic physical studies. Cherenkov's discovery laid down the foundation for the new physical discipline, superlight velocity optics.

Kapitsa P. L., born in 1894 of Russian descent, non-Party man, doctor of physico-mathematical sciences, member of the USSR Academy of Sciences, director of the Institute of Physical Problems, the USSR Academy of Sciences, editor-in-chief of the journal *Experimental and Theoretical Physics*, specialist in physics of low temperatures and magnetism, Hero of Socialist Labor. In 1938, Kapitsa discovered superfluidity of the so-called liquid helium II, i.e. quantum state of a liquid. In 1940, Kapitsa published the results of the measurement of thermodynamic constants of helium II. Kapitsa and his co-workers devised an original apparatus for the production of large amounts of liquid air (turbodetander).

Pavlovsky E. N., born in 1884 of Russian descent, member of the CPSU, member of the USSR Academy of Sciences and Academy of Medical Sciences, doctor of biological and medical sciences, director of the Zoological Institute, USSR Academy of Sciences, head of department at S. M. Kirov Military Medical Academy, specialist in zoology and parasitology, author of more than 1000 scientific publications. Pavlovsky discovered the origin of many serious diseases (tick-borne encephalitis, tick-borne relapsing fever, tularemia, desert variant of skin leishmaniasis) and developed the theory of their natural foci underlying modern epidemiological surveys and practical measures to control these diseases.

N. N. Semenov, visiting Sweden in late 1956 to receive the Nobel Prize, reveals that representatives of the Nobel Committee would positively consider Cherenkov[2] and Kapitsa as candidates for the Nobel Prize in physics. Also, Kapitsa may be nominated for the Nobel Prize by foreign scientists.[3]

We consider it advisable to approve the proposal of the USSR Academy of Sciences and ask for permission to inform them accordingly.

Note at the top of page 1: "Agree", the signature of P. N. Pospelov, secretary of the CPSU Central Committee, (25.1.1957) and five signatures of the members of the Secretariat.

Note at the bottom of page 2: "CPSU Central Committee. Comrade Topchiev A. V. is informed. N. Glagolev. 29.1.1957."

RGANI, fond 5, opis' 35, ed. khr. 49, l. 11–12.

[1] Department of Science and Higher Education was redesignated as the Department of Science, Higher Educational Institutions and Schools on 1.1.1957.

[2] The opened files of the Nobel Committees in physics (dated to 1955 and earlier) indicate that Cherenkov was first nominated for the Nobel Prize in 1952 (Table 6.1) by the English physicist L. Rosenfeld of Manchester University. In 1955, he was nominated by Professor L. Hulten, Stockholm.

[3] P. L. Kapitsa was first nominated in 1946 (see Chapter 4). In the second half of the 1950s, his nominators were foreign scientists. The facts of the nomination of Kapitsa for the Nobel Prize were widely known despite the rigorous secrecy surrounding the nomination process. P. Dirak, his nominator in 1946 and 1950, openly expressed

indignation that his recommendations were disregarded by the Nobel Committee. According to P. E. Rubinin, assistant to Kapitsa, Dirac himself told Kapitsa about this when they met in Moscow in 1956.

Document 7.32

From: A. V. Topchiev, chief scientific secretary of the USSR Academy of Sciences.
To: K. K. Rodionov, head of Department of Scandinavian States, the USSR Ministry of Foreign Affairs.
Copy: to N. M. Lun'kov, *charge d'affaires* of the USSR in Sweden
January 1956[1]

This is to inform you that the following Soviet scientists were nominated for the Nobel Prize in 1957:
P. L. Kapitsa, in physics (nominators A. V. Topchiev and N. N. Semenov, members of the USSR Academy of Sciences);
P. A. Cherenkov, in physics (nominator V. N. Kondratiev, member of the USSR Academy of Sciences);
A. E. Arbuzov, in chemistry (nominator A. V. Topchiev, member of the USSR Academy of Sciences);
E. N. Pavlovsky, in physiology and medicine (nominators A. N. Bakulev and I. A. Sarkisov, members of the USSR Academy of Medical Sciences).
The USSR Academy of Sciences dispatched all relevant materials (nominations, lists of scientific publications, and the most important works) directly to the Swedish Royal Academy of Sciences; they are addressed to the Nobel Committees in physics, chemistry, physiology and medicine.

Please get in touch with the said committees and inform the USSR Academy of Sciences about their response to our proposals.[2]

Arkhiv RAN, fond 579, opis' 1, ed. khr. 639, l. 90–91.

[1] The date of signing is absent, the dispatch date: 2.2.1957, No. 1-6-643.
[2] The author of the letter ignores the requirement of confidentiality set by the Nobel Committees. Also, he is unaware that the Nobel Committee in physiology and medicine is attached to the Karolinska Institutet rather than the Royal Academy of Sciences.

Document 7.33

From: P. Kosharin, scientific secretary of the Soviet part of the Soviet–Finnish Commission on scientific and technical collaboration .
To: S. G. Korneev, head, Foreign Department of the USSR Academy of Sciences.
15.11.1957.

Enclosed find please the record of conversation between comrade Murzin I. I., councillor of the Soviet embassy in Sweden, and Professor Westgren, scientific secretary of the Swedish Academy of Sciences.[1]

Notes in ink: "Attn: comrade Shchetinin. Please repeat nominations.[2] Korneev. 19.11";
"Reply cabled 26.11. Shchetinin";
"Cable returned 2.12. Shchetinin".

Arkhiv RAN, fond 579, opis' 1, ed. khr. 639, l. 92.

[1] Permanent secretary of the Royal Academy of Sciences, chairman of the Nobel Committee in chemistry.
[2] Because Soviet scientists were not awarded the Nobel Prizes in 1957.

Document 7.34

Extracts from the diary of Murzin I. I., councillor of the Soviet embassy in Sweden. They were sent to the addressees in Moscow on 28.10.1957 (No. 1293 shv.), copy No. 3.[1] Record of the conversation with Professor Westgren, scientific secretary of the Swedish Academy of Sciences.

I visited Professor Westgren, scientific secretary of the Swedish Academy of Sciences, in his office on 25 October 1957 and had a talk on the procedure of awarding Nobel Prizes.
I referred to the publications in Swedish newspapers about the beginning of work of the Nobel Committees and said that the Soviet embassy would like to clarify:
1. whether the Nobel Committee received nominations of Kapitsa P. L, Arbuzov A. E, Pavlovsky E. N, members of the USSR Academy of Sciences, and Cherenkov P. A., D.Phys.-Math.Sci. sent by the USSR Academy of Sciences.
2. whether the presented materials satisfy the Nobel Committee and whether it has additional questions;
3. when these materials will be considered by the Nobel Committee.
Professor Westgren replied that all four nominations have been received by the Nobel Committee,[2] they satisfy the requirements, and the Nobel Committees have no additional questions. Westgren revealed that the Nobel Committee will take the decision about the award of Nobel Prizes[3] for the best presented works on Thursday next week (31 October 1957). In accordance with protocol practices of the Nobel Committee, it will announce the Nobel winners and their works via the press-bureau of the Swedish Ministry of Foreign Affairs. The Soviet embassy will know the decision on Thursday if it applies to the press-bureau.

Professor Westgren spoke about the work of the Nobel Committee. The Nobel Committee is constituted by distinctive scientists and specialists and works for seven months every year. The final decision is taken in September, but preparation of the last sitting begins in January. It is therefore important that nominations for the next Nobel Prize reach the Nobel Committee at least in January.

Westgren asked to bring this information to the notice of the president of the USSR Academy of Sciences even though the Nobel Committee had already forwarded it to the president.[4]

Westgren showed me a list of the most well-known Soviet scientists (33 physicists and 22 chemists) whose published works could be recommended to the Nobel Committee.[5] I promised Professor Westgren to transmit his words to the interested people.

The conversation lasted 20 minutes, in the presence of V. V. Veselko, secretary of the USSR embassy in Sweden.

Arkhiv RAN, fond 579, opis' 1, ed. khr. 639, l. 93–94.

[1] The extract was sent to the USSR State Committee of External Economic Relations. The first two copies were sent to the USSR State Technical Committee and the USSR Ministry of Foreign Affairs, respectively.
[2] Westgren could not speak about a Nobel Committee in the singular because nominations of the Soviet scientists were forwarded to three different Committees (see doc. 7.32). It is an obvious mistake made by the interpreter.
[3] Another mistake. Speaking about the date of the announcement of the Nobel winners, Westgren meant only the nominees of the Royal Academy of Sciences.
[4] Westrgen could write to the president of the USSR Academy of Sciences in the capacity of chairman of the Nobel Committee in chemistry in connection with the delay in the receipt of nominations from Moscow for the 1956 Nobel Prize (see note 1 to doc. 7.27).
[5] The author asked Professor Barani, secretary of the Nobel Committee in physics, to search for this document in the archives; it was found in November 2000. Information on the diplomat from the Soviet embassy in Stockholm proved incorrect. The list contained names of potential nominators (not nominees) for the Nobel Prize, who were selected by the Nobel Committees in physics and chemistry based on their scientific works. The list included a total of 33 physicists and 20 chemists plus N. N. Semenov, who had the life-long right to nominate candidates as a Nobel laureate.* The names of the remaining 53 scientists mentioned in the list will be known only after a lapse of 50 years when the Nobel Archives open the respective files.

* "Poisk", No 50m, 2000, p. 14.

Document 7.35

Interview with S. G. Korneev, deputy head of Foreign Department, USSR Academy of Sciences (7 April 1999).

Question: What was the cause of difficulties in the organization of contacts between the Nobel Committee and the Presidium of the USSR Academy of Sciences in the mid-1950s? Negligence, bureaucracy, or deliberate delays?
Reply: I joined the Foreign Department on May 5, 1955. The same day, the USSR Academy of Sciences held a meeting of its Communist Party organization. Half of the speakers at the meeting criticized the work of the Foreign Department that slowed down the development of international relations,[1] in the first place contacts with the Nobel organizations. Leaders of the Academy came to the correct conclusions and the attitude towards visits of Soviet scientists abroad and foreign specialists to the USSR changed dramatically. The Department of Science, CPSU Central Committee, was of great help in this respect. At that time, it was headed by V. A. Kirillin, corresponding member of the USSR Academy of Sciences.[2] I did not personally know his predecessor, A. M. Rumyantsev, but the two men had different views on the international relations of our scientists. Kirillin was a broad-minded person whereas Rumyantsev saw the same problem from a narrower perspective;[3] he came to understand its importance later. Some leaders of the USSR Academy of Sciences actually believed that the Nobel Committees underestimated the achievements of Soviet science. When they gave us an encouraging signal, we selected a number of candidates among physicists and chemists whom the Academy could recommend for Nobel Prizes.[4]
Question: Did the Presidium and Foreign Department give preference to certain candidates over others? Specifically, why did not the Academy fulfil the request of Professor Sillen to provide information on both N. N. Semenov and M. M. Dubinin?
Reply: Yes, they did. Leaders of the Academy (in the first place, its chief scientific secretary A. V. Topchiev) were afraid that the prize could be awarded to Dubinin instead of Semenov, which was considered undesirable. The same refers to the choice of Kapitsa and Landau. The Nobel Committee might choose Landau if both were nominated simultaneously. The candidacy of Cherenkov also seemed doubtful because he was not member of the USSR Academy of Sciences at that time.

[1] Those allowed to go abroad had to fulfil a number of humiliating conditions. Specifically, one could not meet foreign colleagues unless their names were included in the program of the visit. P. S. Aleksandrov, a world-renowned mathematician invited to read lectures on topology in Sweden reported after returning to the Soviet Union: "After the lecture I read on 27.9.1955, Professor Muntz, living in Stockholm (professor of Leningrad University in the late 1920s to early 1930s), invited me to be his guest. I refused pleading busyness and ilness. I had no other meetings with Professor Muntz."*

* Arkhiv RAN, fond 417, opis' 1, ed. khr. 277, l. 2–3.

[2] V. A. Kirillin, specialist in energetics and thermal physics, professor of the Moscow Energy Institute, was elected corresponding member of the USSR Academy of Sciences in 1953 before he enrolled in public service. He was appointed head of the Department of Science, Higher Educational Institutions and Schools in November 1955.

[3] A. M. Rumyantsev started to work in Young Communist League and Party organs very soon after graduation from the Institute of National Economy, Khar'kov. He was appointed head of the Department of Science, CPSU Central Committee, in 1952.
[4] See doc. 7.13.

Document 7.36

Interview with Lyudmila Semenova, daughter of N. N. Semenov, Nobel Prize winner (6 November 1999).

Question: Why were you not allowed to join your father during his visit to Sweden?
Reply: After father was awarded the Nobel Prize, we certainly discussed who would accompany him to Sweden. He said that he could take mother and one of the children with him. I had an elder brother. I studied in Gnesins Musical Institute at that time. Father chose me. I was very happy. Before the award of the Nobel Prize, I felt shy to speak about my father being a member of the USSR Academy of Sciences, laureate of the Stalin Prize, etc. and was ill at ease when people congratulated me on the award of the Stalin Prize to my father. Now it was quite different. It was pleasing to receive congratulations. I sometimes told myself that my father was a Nobel laureate, especially when I was abroad.
Of course, I felt upset when thinking of my brother but was really happy all the same. I began to prepare. Mother had a black dress made for her in a fashion house. True, she did not need it after all and had to buy a light one of proper length in Stockholm. Father thought about having a tail-coat made for himself in Moscow,[1] but after a thought decided to hire it in Stockholm. I planned what to take with me but did not yet order anything. The Swedish side was informed that Father would be accompanied by his wife and daughter.
One evening, Father told us that I was not allowed to go. I was greatly disappointed that father could not convince the decision-makers. He explained nothing. Maybe he had something to tell, but he did not.[2]
Question: How did they take it in Stockholm that Semenov came with his wife alone, without you? How did he explain it to the organizers?
Reply: They were not alone. They were accompanied by an old friend, Kondratiev, member of the USSR Academy of Sciences and one of Father's first pupils in the Physico-Technical Institute, Leningrad. Another accompanying person was P. S. Kostikov, "secretary" to Father but actually his bodyguard. At that time, all members of the USSR Academy of Sciences involved in secret research had such "secretaries".[3] My parents received three invitations to all celebrations (i.e. including one for myself). There is a photograph in our family album taken at the prize-giving ceremony in the Concert Hall and featuring mother sitting among relatives of other Nobel laureates and an empty chair for me. The parents told everybody that their daughter studied at a musical institute and had to take exams in December.

Question: Did your parents tell you anything about why the formal dinner was transferred from the City Hall to the building of the Swedish Academy? Did they feel any inconvenience in connection with the events in Hungary?
Reply: Mother said that holding the dinner in an unusual place was somehow related to the uprising in Hungary. It means that participants of the Nobel festivities discussed the theme. But my parents did not feel that there was any unfriendly attitude to them, nor were they asked improper questions.[4] My mother was a very sociable person, she knew three foreign languages and spoke with other guests either in English or in German. She had a talk with members of the royal family, told them about a kindergarten where she taught the nuts and bolts of music. The Swedish press published well-disposed articles about the Russian visitors. There was a funny report about a Russian academic in a long overcoat and overshoes (they gave up this style long ago in Sweden). Were there any talks about Hungarian events? My parents never told me about them. They returned very pleased. I was after all happy too. Some prize money was spent on a gift for me. They bought a Steinway piano that was delivered by sea in three months. I still play it [...]

[1] See note 37 to this chapter for the true reason for buying the light dress in Stockholm.

[2] N. N. Semenov did his best to obtain the permission for his daughter to join him and applied to A. V. Topchiev, chief scientific secretary of the USSR Academy of Sciences, for help. Also involved was N. M. Lun'kov, *charge d'affaires* of the USSR in Stockholm. He told the author that he had a telephone call with Topchiev who asked to inquire via the USSR Ministry of Foreign Affairs whether it was possible to reverse the decision to ban the visit of Semenov's daughter to Stockholm. Lun'kov fulfilled the request but his superiors in Moscow recommended that he not be concerned with things beyond his competence. "The prohibition was imposed by state security services."

[3] The decree of the USSR Council of Ministers (26.3.1949) established the institute of bodyguards for leading figures of the super-secret atomic project. Semenov had three of them, officers of the USSR Ministry of State Security.* This practice was eliminated after Stalin's death at the request of the concerned scientists. However, many officers were very close to their chiefs and were almost members of their families. Therefore, they continued to work for them as secretaries or assistants. P. S. Kostikov was among them.

> * "Atomny proekt SSSR. Dokumenty i materialy (pod red. L. D. Ryabev) Tom 2. Atomnaya bomba 1945–1954. Kniga 1 (The USSR atomic project. Documents and materials. Volume 2. Atomic bomb of 1945–1954. Book 1)".M.-Sarov: Nauka, 1999, p. 512.

[4] Minutes of a meeting of the Communist Party organization of the USSR Academy of Sciences held three and a half weeks after the Nobel festivities (4.1.1957) give evidence that Hungarian events had nevertheless some repercussion for Semenov's visit. The document was found one year after the interview in the archive of the

Moscow City Party organization. I. I. Mints, member of the USSR Academy of Sciences, referred to Semenov's narration about the meeting with the executive board of the "Sweden–USSR Society" included in the program of his visit (see doc. 7.30). According to Semenov, the question of "our relations with other countries" rose but the discussion was nipped in the bud (doc. 7.38). In other words, Semenov behaved like Sienkiewicz, who had to respond to questions about Russian–Polish relations when he received the Nobel Prize half a century ago (1905) (see doc. 1.20).

Document 7.37

Interview with N. M. Lun'kov, *charge d'affaires* of the USSR in Sweden in 1956[1] (3 February 2002).

Question: Nikolai Mitrofanovich, you filled the top position in the Soviet embassy in Stockholm in late 1956. Your appointment coincided with the events in Hungary from 23 October to 4 November. How did these events influence the work of the mission and the prize-giving ceremony with the participation of the Soviet scientist?
Reply: The work of the embassy was seriously affected. From ten to twenty thousand people, mostly emigrants from Baltic states, gathered day and night near our building for about two weeks, up to mid-November. One day, I was invited to the Swedish Ministry of Foreign Affairs for a talk with minister B. Unden. As soon as the car left the embassy, the crowd rushed to it with threatening cries. I thought that they might turn the car over. Fortunately, the mounted police acted very efficaciously and restored order.
N. N. Semenov, who arrived on 10 December, was worried about the situation. He was especially nervous during the first half of 10 December, in anticipation of demonstrations in front of the Concert Hall where the prizes had to be awarded. However, the ceremony went off smoothly. The police blocked local Hungarian demonstrators far from the place where participants of the festivities arrived.[2] The audience of the Hall welcomed Semenov as heartily as other laureates.
Question: But the Nobel Foundation changed the format of the traditional banquet in response to the reaction of Swedish public opinion to the invasion of Hungary by the Soviet Army (see note 36 to this chapter). Did the Soviet side protest against this decision?
Reply: No, it did not. Our Ministry of Foreign Affairs was prepared for the worst. They were afraid that the ceremony would be shifted to a different place and no guests would be invited.[3]
Question: Did Semenov know the reason for which the banquet was held in the smaller reception hall of the Swedish Academy instead of the City Hall?
Reply: When he asked me about the reason, I told that it might be related to Hungarian events.
Question: Did they hoist the Soviet flag atop the Grand Hotel where Nobel laureates and the invited guests traditionally stay during festivities?

Reply: Yes, they did, together with the flags of USA, UK, Federal Republic of Germany, and Spain, whose citizens received the Nobel Prizes in 1956.
Question: Do you have an idea who was "S. Maksimov", the author of the article in the journal *Novoye vremya* published in 1951? [See Appendix 4.]
Reply: Based on the style and the word usage, it may be conjectured that the article was written by A. M. Aleksandrov, the councillor for the Soviet embassy in Sweden who knew the Swedish language. Later, he became one of Brezhnev's assistants. Suffice it to read an article criticizing Swedish social democrats in the journal *Bolshevik* for 1951. You will notice the same word stock and the same use of epithets. I know for sure that this article was written by Aleksandrov [...][4]
It was in the order of time that officials of the USSR Ministry of Foreign Affairs signed their publications by pseudonyms. I chose to use my patronymic and signed "N. Mitrofanov".[5]

[1] Lun'kov was subsequently the USSR ambassador in Norway, Great Britain and Italy.
[2] One side of the Concert Hall faces the central street Sveavagen and its facade looks out on a large square. On weekdays, the square is busy with a small market of flowers, fruit, berries, dairy produce etc. On December 10 each year, the police encircle the entire square, and people gathering behind their backs could watch the arrival of the participants of the prize-awarding ceremony. For this reason, even the opposition could not protest against the actions of police in 1956. The Soviet embassy was situated in an ordinary street, and any attempt to keep people far from it would be regarded as an encroachment on the freedom of speech and demonstrations.
[3] The huge Concert Hall seats 2,000 people including almost two hundred members of the Nobel Committees, laureates, and members of the Royal family sitting at the scene.
[4] The negative attitude to Aleksandrov is in line with later impressions of A. N. Yakovlev, member of the USSR Academy of Sciences, who happened to deal with the man. Yakovlev frequently talked with the assistant of Brezhnev when he worked in the CPSU Central Committee and left the following characterisation: "[...] Aleksandr Mikhailovich had un awful character. Many were afraid of him and all tried to keep away."*

* A. N. Yakovlev. Omut pamyati (The Depths of Memory). M.: Vagrius, 2001, p. 140.

[5] The choice of the pseudonym was probably prompted by the first letter of A. M. Aleksandrov's patronymic. The aforementioned article in the journal *Bolshevik* entitled "Right-wing social democrats of Sweden — servitors of American imperialism" (see Appendix 5) was signed by a different pseudonym, "A. Vasiliev". In this case, the author used the first letter of his passport name, "A". Vasiliev had been the pseudonym of I. V. Stalin, supreme commander-in-chief during the Great Patriotic War. It was probably used by the author out of mere vanity, since Stalin was sill alive in 1951.

Document 7.38

Extract from the speech of I. I. Mints, member of the USSR Academy of Sciences, at the meeting of the Communist Party organization in the Presidium of Academy, 4.1.1957.[1]

[...] N. N. Semenov told of the following episode. He was invited to the Society of Relations[2] after all official receptions in Stockholm were over. He was prepared to listen quietly how the participants of the meeting would praise Soviet power. Instead, the chairman addressed[3] Semenov and started to talk about our relations with other countries[4] to diminish the value of Soviet achievements in the person of Professor Semenov.[5] Certainly, Semenov successfully repulsed the attack [...]

TsAODM, fond 992, opis' 2, ed. khr. 58, l. 10.

[1] It is an extract from the shorthand record of Mints' speech. Hence fragmentary character and clumsiness of the text.
[2] "Sweden–USSR Society".
[3] The executive board of the Society was chaired by Eva Palmer. The speaker did not know that and therefore spoke of the chair*man*.
[4] About Soviet oppression of the Hungarian uprising from 23 October to 4 November 1956.
[5] Nonsense, probably the fault of the stenographer. It appears that the speaker wanted to represent an allusion to the Hungarian events as an attempt to compromise scientific results of Semenov.

Document 7.39

From: I. V. Stalin.
To: Yu. A. Zhdanov, head, Sector of Science, Department of Propaganda and Agitation.
6.10.1949.

I have received your letter about Professor Pavlov and his scientific legacy.
I am glad that you gave attention to this matter. I share your opinion on all aspects touched upon in the letter. You absolutely correctly assessed both the theory of the great Russian scientist and the position of his opponents.
To my mind, Professor Orbeli inflicted much harm to Pavlov's teaching. Orbeli hypocritically presents himself as the main disciple of Pavlov and at the same time does his best to discredit him by ambiguous statements, dishonest concealment,[1] and cowardly camouflaged attacks. The sooner Orbeli is denounced and the more convincingly his monopoly is eliminated, the better.[2]
Beritov[3] and Shtern[4] are not as dangerous because they openly oppose Pavlov and thereby make it easier to expose them.

Professor Bykov[5] is the most true and consistent successor of Pavlov, only he looks timid and does not like "to be spoiling for a fight". It is necessary to support him in all possible ways and explain to him that he must be brave enough to interfere in the fight because there is no other way to defend the great teachings of Pavlov.
I agree with your conclusions, I am even ready to cube them.[6]
A few words about the tactics of the struggle against opponents of Pavlov's teaching. First it is necessary to stealthily gather its supporters, organize them, and distribute roles. This done, you may convene the workshop of physiologists of which you write[7] to give the decisive battle to the opponents. You may be defeated without such preparation. We must beat the opponents[8] to ensure complete success.
It is necessary to secure the support of Vavilov and other members of the USSR Academy of Sciences.[9] The same refers to Smirnov, minister of health. I had a talk with him two weeks ago and it seems you may rely on him.

RGASPI, fond 558, opis' 11, ed. khr. 762, l. 24–25.

[1] Sic!
[2] The style of this paragraph is very close to that of the speech by A. V. Topchiev, chief scientific secretary of the USSR Academy of Sciences, made at a meeting of the Communist party organization of its Presidium nine months later (see note 5 to this chapter). Certainly Zhdanov did not show Stalin's message to Topchiev but instructed him as the head of the Sector of Science, CPSU Central Committee, in the spirit of the rhetoric of his father-in-law.
[3] I. C. Beritashvili (Beritov), member of the USSR Academy of Sciences, was the founder of the Georgian school of physiology. In the late 1940s, he was accused of having "views at variance with Pavlov's materialistic theory of conditioned reflexes".*

* BSE, 2-izd., t. 5, 1950, p. 21.

[4] L. S. Shtern was the first woman elected full member of the USSR Academy of Sciences in 1939. Mentioning her name in this message as a candidate for persecution in the future once again illustrates the features of paranoia in Stalin's character. A year before he wrote the letter, Shtern was dismissed from the positions of director of the Institute of Physiology, the USSR Academy of Sciences, and professor of the 2nd Moscow Medical Institute. She was arrested in January 1949 and accused in the framework of the trumped-up "case of the Jewish Antifascist Committee".* The indictment brought on 28.3.1952 incriminated her with "servility to the bourgeois West, propagation of cosmopolitanism in science, and affirmation that Soviet science must be apolitical."**

* See: A. M. Borshchagovsky. "Obvinyaetsya krov' (The Accused Blood)". M.: Progress-Kul'tura, 1994.

** *Ibid*, p. 116.

[5] K. M. Bykov succeeded L. A. Orbeli as director of I. P. Pavlov Institute of Physiology, USSR Academy of Sciences, in 1950.

[6] When proof-reading the first typewritten version of the letter, Stalin substituted the word "to cube" for "to square" at the end of the phrase to make it more emphatic.*

* RGASPI, fond 558, opis' 11, ed. khr. 762, l. 26.

[7] Stalin means the organization of the scientific session devoted to I. P. Pavlov's teachings. The session was held in Moscow from 28 June to 4 July 1950 as an openly political/propagandistic demonstration. Shorthand records of the participants' speeches opportunely appeared in *Pravda*, and the complete shorthand report was published as a separate volume in the same 1950.*

* Nauchnaya sessia posvyashchyonnaya problemam fiziologicheskogo ucheniya akademika I. P. Pavlova. Stenograficheskiy otchyot. M.: Izd-vo AN SSSR, 1950, 736 pp.

[8] See also Chapter 6 and note 14 therein.

[9] In the first variant of the letter (see note 6), Vavilov was mentioned together with A. V. Topchiev, chief scientific secretary of the USSR Academy of Sciences. Stalin crossed off his name when he re-read the text.*

* RGASPI, fond 558, opis' 11, ed. khr. 762, l. 27.

Chapter 8 • 1958–1965

The Academy of Sciences on the Way to Civilized Nomination

The awarding of the 1956 Nobel Prize to N. N. Semenov substantially changed the visible aspect of the relationship between the Presidium of the USSR Academy of Sciences and the Department of Science, CPSU Central Committee. There is practically no documented evidence of the Department's meddling in the recommendation of Soviet scientists for the Nobel Prizes since the 1958 nomination campaign. The Academy of Sciences no longer asked the CPSU Central Committee for permission to nominate concrete candidates.

Of course, it did not mean complete independence of the Academy; only the exchange of letters regarding each candidate stopped. However, verbal contacts on the issue in question were maintained for at least ten more years (see, for instance, a note oby S. G. Korneev, head of the Foreign Department, USSR Academy of Sciences, on doc. 8.14). True, the nomination of Professor N. N. Bogolyubov in 1959–1960 was an exception, but the correspondence was initiated neither by the Presidium of the USSR Academy of Sciences nor by the Department of Science, CPSU Central Committee (see docs. 8.16–8.20 below).

Abandonment of the practice of petty surveillance by the CPSU Central Committee was a corollary to the general liberalization in this country most prominent after the open denunciation of Stalin's personality cult.[1] The new political situation made Party functionaries look for "softer" coercive measures. The rude threats and arm-twisting of Stalin's time became ineffective. Soviet intellectuals already had a taste of freedom and refused to blindly obey ignorant authorities. A latent upheaval ripened in the Academy of Sciences in 1957 in the form of nomination of several Soviet scientists for the Nobel Prizes of that year secretly prepared by the Presidium (doc. 7.31). Professor V. L. Ginzburg, one of the principal actors of that affair, recounted its details (doc. 8.1).

Despite the secret character of doc. 7.31 signed by A. V. Topchiev, chief scientific secretary of the USSR Academy of Sciences, leakage of information gave rise to rumours that only two candidates were approved by the Presidium for the Nobel Prize in physics, viz. N. P. Cherenkov for the discovery of the new physical phenomenon (called the Vavilov–Cherenkov effect) and P. L. Kapitsa for his works on liquid helium superfluidity. The final list included neither I. E. Tamm nor I.

M. Frank, recommended by nominators of the Academy together with Cherenkov,[2] nor L. D. Landau nominated together with Kapitsa. The reason is perfectly clear today. As follows from docs 8.3 and 8.4, L. D. Landau and I. E. Tamm[3] actively opposed to the policy of the Presidium in 1955–1956 and its curators in the CPSU Central Committee. S. G. Korneev, one of the leading and well-informed figures in the Foreign Department of the Academy, confirmed that strong selection bias actually impacted the decision of the heads of the Academy concerning the approval of candidates for the Nobel Prize in physics for 1957 (see doc. 7.35).

S. G. Korneev certainly attended the portback election meeting (6.12.1957) as a member of the Communist Party organization of the Presidium of the USSR Academy of Sciences. He heard the report of the secretary of the Party committee and probably participated in its preparation. Summarizing the work of the committee for 1956–1957, the speaker commented on the actions of certain leading members of the Academy: "They (Artsimovich, Tamm) suggested the abolishment of the Staff Department, positions of vice-directors of staff in academic institutes etc.). These wrong proposals were met with approval by scientists affiliated with the Divisions of Physico-Mathematical and Biological Sciences [...] Such sentiments seem to imply attempts to escape the Party's control and the guidance of government bodies."[4] No wonder the nomination of I. E. Tamm in 1957 was so actively opposed by the functionaries of the Presidium.

Did the Department of Science, CPSU Central Committee, make any attempt to influence the situation? There are no documents to this effect dated from 1957 because written permission was no longer needed to send documents to Stockholm. There is little doubt, however, that the Department was directly involved in the preparation of doc. 7.31 bearing in mind that functionaries of the AUCP(b)/CPSU Central Committee always interfered in the distribution of awards, e.g. Stalin Prizes in science (doc. 8.2). The Communist Party continued to play the leading role despite the denunciation of Stalin's cult of personality, and a private talk or a telephone call remained as efficient tools of influence as ever before...[5]

It appears from doc. 8.1 that in 1956 I. V. Kurchatov, member of the USSR Academy of Sciences having the right to nominate candidates for the Nobel Prizes, entrusted V. L. Ginzburg and E. K. Zavoisky, experts in Cherenkov radiation, to prepare a well-substantiated nomination of P. A. Cherenkov, I. E. Tamm, and I. M. Frank for the Nobel Prize for the discovery and theoretical explanation of the Vavilov–Cherenkov effect.

This previously unknown type of radiation was first observed by Cherenkov, Vavilov's postgraduate student, in 1934 when he worked with luminescent uranium salt solutions treated with gamma rays. The researchers immediately understood that the observed radiation was not the ordinary luminescence. Vavilov hypothesized that it was exited by the motion of electrons generated by gamma irradiation of the solution and that its mechanism was related to the so-called *bremsstrahlung* resulting from interactions of moving electrons with the medium. However, calculations showed that the contribution of *bremsstrahlung* radiation was negligibly small, which suggested a different mechanism of the new phenomenon.

It was explained after two years by I. E. Tamm and I. M. Frank of the P. N. Lebedev Physical Institute, USSR Academy of Sciences. The researchers presented theoretical evidence that the new radiation was emitted as fast electrons moved through a medium at superluminal velocity. The authors submitted their manuscript with relevant calculations to *Doklady AN SSSR* on January 2, 1937; it was publised in Russian and English on 21 Januray. Only then could the scientific community appraise the epoch-making discovery in full measure, and the American physicists G. Collins and V. Reiling coined the term "Cherenkov radiation" for what was initially referred to as the "blue glow."[6]

V. L. Ginzburg and E. K. Zavoisky prepared the nomination of the Soviet physicists and handed it out to Kurchatov, who signed and passed it to the Academy Secretariat. Nobody could directly address the Nobel Committees in those years.

It is known today that the materials dispatched by Topchiev to the USSR Ministry of Foreign Affairs did not include Kurchatov's nomination. The nomination of Cherenkov alone was signed by V. N. Kondratiev.[7] It is this document that gave rise to rumours and made V. L. Ginzburg and E. L. Feinberg undertake active actions, the result of which was the nomination of Cherenkov, Tamm, and Frank signed by N. N. Andreev, A. I. Alikhanov, and L. D. Landau. There is reason to believe that this nomination provided a basis for the decision of the Nobel Committee to award the prize to the three Soviet scientists. At any rate, no other evidence of nominations for the 1958 Nobel Prize was found in the archives of the Foreign Department, USSR Academy of Sciences.[8]

The truth will be known soon, that is when the Nobel archives open their files in 2009. Then, after 50 years of secrecy, the world will learn whose nominations contributed to the recognition of one more triumph of Soviet science in 1958. If the prize was actually awarded based on the nomination of the above three persons, we shall have to bow down before them and those "rebels" whose courage and fidelity to principle made this recognition possible.[9]

To conclude the discussion of the situation preceding the award of the Nobel Prize in physics for 1958, it should be emphasized that the nomination of Cherenkov alone by an uninformed specialist representing a different field of knowledge ensued from a political game undertaken by the heads of the USSR Academy of Sciences in 1957. The nomination was *a priori* doomed to be rejected by experts of the Nobel Committee. The outstanding discovery of new radiation brought together two groups of scientists, S. I. Vavilov and P. A. Cherenkov who first described it and I. E. Tamm[10] and I. M. Frank who offered theoretical explanation of its physical nature. Their separate nomination had no sense, only together could they receive the prestigious international prize.[11] One can easily imagine what a difficult task it would have been for the Nobel Committee and the Royal Academy of Sciences to choose three of the four researchers if S. I. Vavilov had been still alive in 1958.[12]

The news of the award of the Nobel Prize to Soviet scientists in 1958 was again the focus of political collisions as two years before.

The award of the Nobel Prize to N. N. Semenov in 1956 coincided with the Hungarian revolt that made the Nobel Foundation modify the traditional prize-giving ceremony (see Chapter 7 and note 36 therein). In 1958, when the Nobel Prize in physics was given to the three Soviet scientists, a scandalous situation was provoked by the October 23 announcement of the Nobel Prize in literature awarded to Boris Pasternak. It may be thought that the leakage of information (doc. 8.5) four days before the Royal Academy of Sciences promulgated its decision was largely designed to quell the rising tide of hysteria in the Soviet Union, the first signs of which were apparent as early as 24 October (see Chapter 9).

The date doc. 8.3 was registered in the CPSU Central Committee indicates that the chief Party ideologist M. A. Suslov could discuss the situation with V. A. Kirillin, head of the Department of Science, on 27 October, i.e. the day before the formal announcement of the names of the new Nobel laureates in physics. Due to this, the two officials had time to take proactive measures. As a result, the day after the announcement central press organs published brief communication about the event, and the newspaper *Pravda* appeared with a large article signed by authoritative members of the USSR Academy of Sciences.

Of special interest in this article is a paragraph in which the authors referred to V. I. Lenin and set off "bourgeois natural scientists" against those "seeing public events [...] entirely from the standpoint of ruling class ideology." It appears unlikely that the scientists who signed the article formulated this thesis. In all probability, it was included by the editorial staff on the order of ideologists from the CPSU Central Committee.[13] Such practice was common for the Soviet press and frequently resorted to whenever deemed necessary.[14]

It was long unclear whether the laureates were allowed to receive the prize in Stockholm or made to turn it down as "a token of protest". The first document related to the arising conflict of interests dates to 17 November 1958 (doc. 8.7). It informed the Soviet embassy in Sweden about the situation in Moscow around the newly awarded Nobel Prize in physics. The USSR Academy of Sciences organized the first press conference of the new laureates only a week after the announcement of this important occurrence (4 November). It appears that the note on doc. 8.5 dated November 5 refers to this event. After two more days, the Prize winners sent a cable of gratitude to Stockholm, thereby giving consent to receive the prize. Next week (13 November), they informed the Swedish side about their resolution to attend the prize-giving ceremony.

In two weeks, however, the Swedish ambassador again addressed the USSR Ministry of Foreign Affairs asking to confirm the arrival of Soviet scientists to Stockholm (doc. 8.9) as if the organizers of the ceremony were not quite sure about the intention stated half a month ago. The USSR Academy of Sciences issued definitive information about the date of departure of the Nobel laureates to Stockholm and application to the CPSU Central Committee for permission for them to go abroad only on November 26 (doc. 8.11). On the same day, the business manager of the Academy got himself in trouble about dress coats for the laureates and their companions (doc. 8.12).

The cause of delays and ambiguities is quite clear. Ideologists of the CPSU Central Committee could not reconcile with the necessity to bring together in Stockholm the "slanderer" Pasternak and the academic scientists. Indeed, the possibility of their visit to Sweden began to be seriously considered only after Pasternak was made to refuse to receive the Nobel Prize (his decision was announced in the mass media on November 2). Only then was the USSR Academy of Sciences permitted to organize the first press conference. However, it gave no guarantee that the laureates would be able to attend the prize-giving ceremony.

Unfortunately, archive materials of the CPSU Central Committee for the period of interest remain unavailable, and one can only conjecture what obstacles had to be overcome before November 26. It appears that the most tough ideologists in the Central Committee were not satisfied with Pasternak's refusal alone and thought that the appearance of Soviet physicists in Stockholm would be regarded as a token that the Communist rulers swallowed "the insult"...[15] This is only a hypothesis, however. The real state of affairs will be possible to establish only after the CPSU Central Committee opens its archive files.

A characteristic episode was recorded in the stenographic report of the meeting in the Polytechnical Museum at which the three Nobel laureates told the Moscow public about their visit to Stockholm. I. E. Tamm was the main narrator. After he finished his story, someone shouted from the audience: "Long live Soviet science governed by the Communist Party, leading mankind to peace in the world!" The prearranged demonstration was followed by applause.[16]

This stylistically awkward slogan habitual at those times sounds like a satire today...

As follows from documents of the Foreign Department, USSR Academy of Sciences, only one nomination was sent to Stockholm in 1959. It was signed by two leading members of the Academy, A. N. Nesmeyanov, its president, and A. V. Topchiev, chief scientific secretary. The two scientists nominated V. I. Veksler, full member of the Academy, and E. K. Zavoisky, its corresponding member, for the Nobel Prize for "their outstanding discoveries in the field of physics" (doc. 8.13).[17] The text of the nomination is traditional bureaucratic phrases and cliché listing ethnic affiliation, place of birth, work status, and other biographic data of the candidate irrelevant for expert appraisal of his contribution to science. Each line of the document exposes the professional style of the bureaucrat who prepared it.

It was the first nomination of Veksler by his compatriots; previously (since 1947), he had been nominated only by foreigners (see Chapters 4 and 6). The new nomination became possible because information about Soviet accelerators of charged particles was declassified by that time, after the sensational report of I. V. Kurchatov at the Atomic Energy Research Establishment at Harwell, Oxfordshire, (25 April 1956) in the framework of the official visit of the Soviet government delegation to Great Britain led by N. S. Khrushchev.

In 1965, P. A. Cherenkov dispatched one more nomination of Veksler via the Foreign Department, USSR Academy of Sciences (doc. 8.14). There is no doubt that it contained a much more sound substantiation of the nominee's services than the previous one although this document was not found in archive materials. Suffice it to say that Cherenkov simultaneously nominated American physicists E. McMillan and N. Kristofilos. The former had been nominated together with Veksler by foreign nominators in 1947, 1948, and 1951; the latter was known for his principle of charge-exchange focusing suggested in 1950 to substantially raise the upper energy limit attainable in accelerators. In other words, recommendations of Cherenkov were based on his thorough knowledge of the problem and contribution of other scientists to its development.

E. K. Zavoisky was for the first time nominated for the Nobel Prize in 1959 (doc. 8.13).[18] Zavoisky was known among physicists as "a wizard of experiment". He deserves special mention in the history of Nobel Prizes for Soviet scientists. Virtually all of them were convinced that he was an indisputable candidate for the prize and considered his absence among the laureates as an inexcusable omission of the Nobel Committee.

In 1944, Zavoisky discovered electron paramagnetic resonance (EPR), of paramount importance for experimental physics. The method provided a valuable tool for in-depth studies of substance structure at the molecular and atomic level. Nominating Zavoisky for the Lenin Prize in 1957, the Presidium of the Kazan Division of the USSR Academy of Sciences emphasized that the discovery of EPR "gave impetus to the description of a number of related effects, such as nuclear paramagnetic resonance (Purcell, Bloch, and co-worker, USA, 1947), ferromagnetic resonance (Griffiths, UK, 1946; Zavoisky, 1947), antiferromagnetic resonance (Gorter, the Netherlands, 1951), and diamagnetic (cyclotron) resonance (Dorfman, USSR, 1952; Dingl, UK, 1952). Detailed analysis of these phenomena created a new large area of physics known as magnetic radiospectroscopy." Also, the nominators ascertained that the discovery of nuclear magnetic resonance (NMR) by Felix Bloch and Edward Purcell, awarded the Nobel Prize in 1952,[19] was actually "an extension of the fundamental discovery of Zavoisky to nuclear paramagnetics."[20]

Soviet scientists several times nominated Zavoisky for the Nobel Prize after 1959. By way of example, his candidacy was recommended by A. P. Aleksandrov[21] and S. V. Vonsovsky (doc. 8.15). V. L. Ginzburg, elected full member of the USSR Academy of Sciences in 1966, twice nominated Zavoisky, in 1974 and 1976.[22] P. L. Kapitsa was of the opinion that the discovery of EPR "deserves the Nobel Prize".[23] Even L. A. Artsimovich, who never had good personal relations with Zavoisky, said in the late 1960s that "Evgeniy Konstantinovich is the most worthy candidate for the Nobel Prize among our scientists."[24]

There is little doubt that foreign researchers also nominated Zavoisky for the Nobel Prize. In the early 1960s, M. A. Markova, member of the USSR Academy of

Sciences, had talks with Professor Ivar Waller, member of the Nobel Committee in physics, in Trieste, to which he was invited by Abdus Salam, Nobel Prize winner and director of the International Institute of Theoretical Physics. One day, Waller started to talk about Zavoisky. "I was first surprised to see him so much excited. Little by little I came to understand that he was greatly disappointed that the works of Zavoisky were not awarded the Nobel Prize."[25] This episode suggests that Professor Waller, an expert in resonance phenomena, was one of the nominators of the Soviet physicist and supported his candidate at the sittings of the Nobel Committee.

However, it was too late to discuss the candidacy of Zavoisky after Bloch and Purcell had received the Nobel Prize for the discovery of NMR in 1952. The Nobel organizations do not give prizes for discoveries essentially similar to the previously awarded works. Had the candidacy of Zavoisky been presented in the mid-1940s, i.e. immediately after his publications on EPR, he would certainly have received the Nobel Prize either alone or (after the discovery of NMR in 1947) together with the two American scientists.

However, such a scenario was absolutely out of the question in the mid-1940s in the post-war Soviet Union, where the author of the outstanding scientific discovery became a hostage of political situation. In the beginning of 1946, Soviet authorities broke all contact with the Nobel organizations; the boycott continued until Stalin's death in March 1953.[26]

Had the nomination of Zavoisky been proposed by one of his countrymen under Stalin's rule, it would all the same have never reached the Nobel Committee. The campaign of mass arrests in the 1930s affected Zavoisky's family. According to his daughter, N. E. Zavoiskaya, the husband of the scientist's sister was arrested in July 1937. His brother was arrested and shot on December 26. The arrests took place in Kiev while Zavoisky lived in Kazan.

N. E. Zavoiskaya was given access to some documents from NKVD archives that however did not mention the name of her father. She found it in the list of university researchers dated to early 1938 and stored in the State Economics Archive. Zavoisky was mentioned as "having a brother who was arrested by NKVD". It means that administration of the Kazan University was aware of the "black spot" in the biography of its professor; it automatically referred him to the category of "politically suspect persons." Thus, nomination of Zavoisky for the Nobel Prize by any Soviet scientist would have never been approved by the Presidium of the USSR Academy of Sciences and the CPSU Central Committee. Of course, the nomination could be transferred to Stockholm illegally, but there was hardly a single associate of the scientist brave enough to take such a risky step.

The reality of the above situation is confirmed by documents concerning another person dated to the time when colleagues of Zavoisky ought to have nominated him for the Nobel Prize.

In the beginning of 1948, D. V. Skobeltsin, member of the USSR Academy of Sciences, was dismissed from the position of the councillor representing the USSR in the Atomic Energy Control Commission, UN Security Council.[27] The

Staff Department of the AUCP(b) Central Committee proposed to replace him with P. A. Cherenkov of the Physical Institute, USSR Academy of Sciences, a future Nobel laureate (3 February 1948). A. A. Kuznetsov, secretary of the Central Committee responsible for its staff policy, presented the proposal for the consideration of the USSR Ministry of Foreign Affairs. V. M. Molotov, deputy prime minister and minister of foreign affairs gave his consent on the condition that L. P. Beria also approved the candidacy. Beria, another deputy prime minister, heading the USSR atomic project, informed Molotov (March 15) that "according to the Ministry of State Security, Cherenkov's father was sentenced to capital punishment and executed in 1938. Therefore, Cherenkov can not be appointed to fill the position in question." The final resolution read: "Cherenkov can not be sent [to Geneva]." V. Molotov. 15/III."[28]

This document shows that all attempts to nominate Zavoisky for the Nobel Prize for his breakthrough into previously unknown depths of matter would have been in vain, even if short-sighted officials had not broken all contact with the Nobel organizations.

To conclude, it seems appropriate to emphasize one more feature of Zavoisky as a natural scientist that has a direct bearing on his anticipatory research. Being of advanced age, not long before his death, Zavoisky took an interest in high-temperature superconductivity when this problem was not yet in the focus of physical research. In 1976, he wrote a review for the journal *Uspekhi Fizicheskikh Nauk* but did not see it published. Zavoisky died on 9 October 1976. The review appeared in the journal together with the obituary...[28a]

For many years, millions of law-abiding Soviet people lived under the burden of the cruel totalitarian regime. Hundreds and thousands of them were brilliant representatives of the scientific elite, "the gold reserve" of the nation. Here are a few more names added to the list of researchers whose relatives were repressed in the years of Stalin's terror to illustrate the scale of the global calamity that befell the country.

D. V. Skobeltsin's brother was repressed too (see note 4 to doc. 8.1). Even two presidents of the USSR Academy of Sciences could not avoid being affected. S. I. Vavilov's brother, the great geneticist N. I. Vavilov, died of hunger in a Saratov prison. The brother of M. V. Keldysh, a young graduate historian, was shot in the late 1930s. The martyrology is endless...

At the end of 1959, certain scientists proposed to nominate Professor N. N. Bogolyubov for the Nobel Prize in physics. In this case, the Department of Science, CPSU Central Committee, was again involved as years ago. This time, the affair was not confined to verbal instructions that remained in force until at least the late 1960s and the early 1970s.[76] The Communist Party of Armenia initiated a formal application to the CPSU Central Committee (doc. 8.19). It was preceded by correspondence (doc. 8.16) started by the secretary of the Communist Party committee of the city of Dubna, where Bogolyubov headed the Laboratory of Theoretical Physics at the Joint Institute of Nuclear Research.

The Department of Science took on itself to coordinate the nomination procedure. Its active efforts brought to life a letter by the secretary of the Communist Party committee of the Mathematical Institute, USSR Academy of Sciences (doc. 8.17) that appeared five and a half weeks after the message from Dubna (26 November 1958). The next day (November 27), heads of the Department sent their own document (doc. 8.18) to the Secretariat of CPSU Central Committee that included information obtained from both Dubna and Mathematical Institute. There is no doubt that doc. 8.17 was compiled on instruction from CPSU Central Committee...

The cause of the trouble was the microscopic theory of superconductivity developed by the American physicists J. Bardeen, L. Cooper, and J. Schrieffer in 1957. The theory was underlain by the fundamental idea of Leon Cooper that under certain conditions conduction electrons can pair together by weak attraction under the action of crystal lattice. His brief communication[29] refuted the then current concept that postulated the absence of attraction between electrons and gave a secondary role to lattice effect in the analysis of superconductivity. This fruitful idea brought together into a strong intellectual alliance its author, his coeval Schrieffer, and venerable Bardeen recently (1956) awarded the Nobel Prize for the invention of the first transistor. Moreover, it greatly impressed many scientists in different countries by its simplicity and cogency. One of them was the brilliant Soviet physicist and mathematician N. N. Bogolyubov.

When a new scientific concept motivates independent researchers to study a formerly unknown physical phenomenon, a positive result may be expected to come any time and at any place. Cooper's idea was exactly this sort of thing. That so many researchers took interest in it promised a tense race for priority in the development of a new theory within a few weeks or months.

The Americans won the race. They published their short communication on the development of the theory of superconductivity in April 1957. In June, they submitted to a journal a large article with detailed analysis of the results and took care to ensure priority by simultaneously distributing its reprint.

Bogolyubov developed a similar theory independently and came to virtually the same conclusions as the American authors. It occurred in autumn 1957,[30] when the above reprint was already known to many foreign scientists but not to Bogolyubov.

The Soviet physicists A. A. Abrikosov and I. M. Khalatnikov recounted the story of the creation of the superconductivity theory in their papers published in 1958. They naturally gave up the palm of supremacy to the Americans and described Bogolyubov as the developer of "a few original methods permitting to obtain the results of Bardeen, Cooper, and Schrieffer." Also, they "used Bogolyubov's methods themselves to carry out calculations similar to those of Bardeen, Cooper, and Schrieffer."[31]

These papers by the young scientists aroused indignation among Party functionaries, who viewed their objective analysis based on the succession of publications as a manifestation of "narrow group interests" etc. This brings us back to the events preceding the dispatch of the letter from the Mathematical Institute to the CPSU Central Committee.

Participants of the portback election Party meeting held at the Institute on 21 November 1957 unanimously gave priority to the development of the microscopic theory of superconductivity as the most remarkable result of research done in the past year.[30] Another meeting held a year later (13 November 1958, i.e. after Abrikosov and Khalatnikov had published their papers), again discussed the works of Professor Bogolyubov, who had just received the Lenin Prize, the highest scientific award in this country.[32]

Evidently, it was high time for the secretary of the Party organization to criticize the articles of the above authors and charge them with "scientific unconscientiousness", belittling "the leading role of Soviet science in the world" etc. as he did 13 days later in the letter addressed to the CPSU Central Committee (doc. 8.17). However, nothing of the kind was said at the meeting. It would be naïve to assign his silence to mere forgetfullness and the said letter to a sudden recovery of sight.

The behaviour of the man after the meeting was obviously dictated by a signal he received from the CPSU Central Committee. It was probably an urgent signal since the minutes of the Party bureau sittings contain no trace of discussion of the hastily written letter.

Later notes on the first page of doc. 8.18 indicate that secretaries of the CPSU Central Committee readily approved the proposals of the Department of Science. Fortunately, the affair was not set going and the document was shelved. There is only one rational explanation of the situation. It appears that some sober-minded people managed to persuade the functionaries of the Central Committee that the new "witch-hunt" campaign would be totally out of time so soon after the triumph of the Soviet Nobel laureates in Stockholm (see note 3 to doc. 8.18).

In 1959, it was decided to nominate N. N. Bogolyubov for the Nobel Prize for his work published in January 1958.[33] It appears that the nomination signed by V. A. Ambartsumyan, member of the USSR Academy of Sciences, was sent to Stockholm (doc. 8.19) unlike that of I. I. Artobolevsky. Notes on doc. 8.20 indicate that A. V. Topchiev, chief scientific secretary, instructed the Foreign Department (6 January 1960) to prepare a relevant letter to the CPSU Central Committee. However, the issue was closed in a month and a half, taking into consideration the objections of the Division of Physico-Mathematical Sciences. No properly argued document was found but the objections seem self-evident. In his nomination, Artobolevsky totally disregarded the generally accepted priority of the American scientists and recommended awarding the prize to Bugolyubov alone. He did not even mention Cooper, whose theoretically predicted electron pairing underlies the theory of superconductivity including its variant suggested by Bogolyubov himself.[34]

One more aspect needs to be considered in connection with the debates of 1958. As expected, the world scientific community met with great interest the original publications of Bogolyubov on the theory of superconductivity. He received numerous invitations from research centres of Europe and USA to attend conferences, universities, schools, etc. He was asked to read lectures for leading theoretical physicists including the developers of BCS theory (Bardeen, Cooper, Schrieffer). Personal contacts with these people would have given Bogolyubov a possibility to further develop his original ideas and correct certain inexact propositions of the theory of his American colleagues, that is to directly participate in what is traditionally considered to be a normal scientific process of search for the final truth.

L. A. Artsimovich, the then acting academician-secretary of the Division of Physico-Mathematical Sciences, insistently asked the Presidium of the Academy to permit Bogolyubov to go abroad and thereby enable him to demonstrate the "priority of Soviet science in a most important field of physics" (docs. 8.21, 8.22).[35] The request was in vain. The visit was described as "unpractical".[36] Meanwhile, personal contacts of Bogolyubov with his foreign colleagues, in the first place with the authors of the BCS theory, would have given him place among its generally recognized developers and posed a difficult task for the Nobel Committee to choose three of the four pretenders for the 1972 Nobel Prize.

The immediate cause of refusal to grant Bogolyubov permission for foreign trips was the notorious problem of secrecy. In the period from 1950 until 1953, Bogolyubov was engaged in calculations for the Soviet atomic project. True, he participated in the work of the International Congress of theoretical physicists in Seattle, USA, in September 1956. In the same year of 1958 when the scientist was prohibited to accept invitations from Benilux, France, and the USA, he took part in three international forums. Two of them were held in Geneva (conference on the problems of high energy in June and July, conference on the peaceful use of nuclear energy in September) and the third one in Edinburgh (International Mathematical Congress in August).[37] This seeming inconsistency is easy to explain.

In all these cases, Bogolyubov was a member of large groups of scientists. By way of example, the Soviet delegation at the congress in Edinburgh comprised 32 specialists.[38] Secret agents always included in such groups could better control the behaviour of these people knowing, each other fairly well and joined by common interests. Also, it was easier to introduce them as members of large delegations (doc. 8.23 and note 2 therein).

The situation was quite different when foreigners invited Bogolyubov to come alone. In such cases, the supervising "organs" did not give consent.[39]

The episode with the nomination of Bogolyubov was the last in the history of correspondence between the USSR Academy of Sciences and the CPSU Central Committee on Nobel matters. The procedure of nominating Soviet and foreign scientists was gradually becoming more confidential, as required by the Nobel Statutes. V. A. Ambartsumyan and I. I. Artobolevsky still applied to the Central Committee for permission to send documents to Stockholm (docs. 8.19, 8.20), but their colleagues preferred to act independently.

To conclude the story of Bogolyubov's Nobeliana, it is opportune to ask whether he had any chance whatsoever to receive the Nobel Prize for the development of the superconductivity theory. The answer should be sought in the probable outcome of his foreign trips, had they not been prohibited by the Soviet secret services.

At the dawn of the sensational discoveries of the American physicists and Bogolyubov himself (January–February 1958), he could meet and talk with specialists from Belgium and Holland. In half a year, he could see John Schrieffer in France and convince the audience of theoretical physicists from all over the world that his superconductivity model had some advantages over that of Bardeen's group. The next visit would have been to Princeton for the meeting with Robert Oppenheimer, the legendary leader of the American atomic project.

Doubtless these trips and discussions would have greatly contributed to the image of the Soviet scientist as a key figure in the discovery of physical nature of superconductivity and superfluidity. These discussions and the resulting publications in the world scientific press would not have passed unnoticed by the Nobel Committee and the Royal Academy of Sciences when they considered candidates for the 1972 prize to be awarded for the development of the theory of superconductivity.

As the Nobel organizations open their files for researchers after 50 years of secrecy, increasingly more information will be available on how the unannounced upheaval of Soviet scientists in the late 1950s and the early 1960s against the humiliating procedure of applying for permission to send nominations for the Nobel Prizes influenced their dynamics. Meanwhile, there is no doubt that this upheaval brought about full freedom of contact with the Nobel organizations by the mid-1960s. This small victory over bureaucrats was one of the main achievements of the romantic times of the Khrushchev Thaw. It survived the new "glacial period" dated from October 1964 of Brezhnev's era.

Despite separate cases of interference from ideologists of the CPSU Central Committee in the life of the Academy, their influence gradually lessened unless they faced outright opposition, as in the situation with A. D. Sakharov. An example is the above story of the theory of superconductivity and the related decision of the CPSU Central Committee to "take measures to strengthen the editorial board of the journal *Uspekhi fyzicheskihk nauk* (doc. 8.18). In the language of Party ideologists, "to strengthen" meant "to break up". In Stalin's time such a decision would have been immediately fulfilled, but in 1958 it remained on paper. E. V. Shpolsky, an eminent experimental physicist and one of the founders of the journal, remained its editor-in-chief until his death in 1975 despite strict orders of top officials.

In the first years after the restoration of contacts between the USSR Academy of Sciences and the Nobel organizations, it was planned to nominate only Soviet scientists as was stated by an official of the Foreign Department, USSR Academy of Sciences (see doc. 8.29). In 1962, however, the same department sent to Stockholm the nomination of Peter Poisson, an English chemist studying metallo-organic compounds (doc. 8.24). Poisson was nominated by A. N. Nesmeyanov, president of the USSR Academy of Sciences, for the discovery of ferrocene from a group of metallocenes with the so-called "sandwich" structure. Today, this compound is widely used in catalytic processes and as a medicinal agent.[40]

Two years later, D. V. Skobeltsin nominated N. G. Basov and A. M. Prokhorov together with their American colleague Ch. Townes (doc. 8.39). In 1964, the Academy nominated A. N. Frumkin together with M. Volmer (DDR) and A. P. Vinogradov together with A. Holmes (USA).[41] Frumkin and Volmer were nominated again by the Nobel laureate N. N. Semenov in 1965.[42] Certain nominators still asked their superiors in the Academy for permission to send their proposals to Stockholm and cautiously prepared draft letters to the same effect to the CPSU Central Committee. However, such letters were usually not sent to the addressee because a verbal consent was deemed sufficient (see note 1 to doc. 8.38). The Foreign Department of the Academy sometimes did not register nominations and even did not store their copies.

The situation is illustrated by two letters from the secretary of the Nobel Committee in chemistry dated 26.1.1962 addressed to A. V. Topchiev and A. E. Arbuzov. The letters informed the members of the USSR Academy of Sciences about the receipt of their recommendation of geochemist A. P. Vinogradov for the 1962 Nobel Prize.[43] The letters are found in the archive of the Foreign Department, but there are no copies of the nominations in the respective files.

Nor does the archive contain a copy of the nomination of V. I. Veksler and two his American colleagues by P. A. Cherenkov for the prize of 1965. However, the cover letter with a request to send the nomination to Stockholm "as soon as possible" addressed to S. G. Kornev, head of the Foreign Department, bears the note in the latter's hand testifying to the verbal consent of L. F. Ilyichev, secretary of the CPSU Central Committee (doc. 8.14).

Interesting information is derived from the correspondence between the USSR Ministry of Foreign Affairs and the USSR Academy of Sciences related to the request of the Rumanian embassy in Moscow that wished to know who of the Soviet scientists and for which works were nominated by their countrymen for the Nobel Prize of 1963 (doc. 8.25). After a month of inquiries, the Foreign Department of the Academy had to notify the USSR Ministry of Foreign Affairs that they do not posses such information (docs. 8.26, 8.27). Incidentally, officials of the Department found a long-forgotten invitation from Stockholm to recommend candidates for the Nobel Prize of 1962. The invitation was certainly received in due time, i.e. not later than September 1962, but never reached nominators for mere negligence of the departmental staff.

A year before, the Nobel Prize was awarded to L. D. Landau. According to V. L. Ginzburg (doc. 8.1), the scandalous story of the approval of P. A. Cherenkov by the Presidium of the USSR Academy of Sciences as the sole candidate for the Nobel Prize for the discovery of Cherenkov radiation was paralleled by the exclusion of Landau from his joint nomination with Kapitsa for the discovery and explanation of liquid helium superfluidity (doc. 8.1). There was one cause behind both cases: I. E. Tamm in the former and L. D. Landau in the latter case were politically unacceptable for ideologists of the CPSU Central Committee. Their dislike of Landau was clearly expressed in the letter from the Department of Science dated 16 January 1956.

To recall, P. L. Kapitsa for the first time experimentally observed superfluidity in the end of 1937 (doc. 8.36). Liquid helium cooled to below 2 K acquired a fantastic ability to instantaneously pass through such capillaries that would take water weeks to penetrate. Viscosity of liquid helium at 2.17 K dropped jump-like by 10–11 orders of magnitude and became unmeasurable. The new phenomenon could not be explained in terms of traditional concepts.

L. D. Landau resolved the problem in 1941. He was the first to apply quantum mechanics for explaining events in the macroworld which enabled him to create the theory of quantum liquids composed of two components, normal and superfluid ones. The normal component is a set of elementary excitations in the form of quasiparticles in a range from absolute zero to 1.7–1.8 K; it is involved in the interaction of helium with the walls of the container. The superfluid component (presently called helium II) has no viscosity and can easily flow through narrow holes and capillary tubes.

It was totally absurd to nominate the outstanding discovery of superfluidity disregarding the related breakthrough into the depth of matter. Any competent nominator giving credit to Kapitsa as a worthy candidate for the Nobel Prize could not overlook the equally impressive achievement of another contributor, Landau.

Until the 50 year-long term for the storage of documents of the Nobel organizations is expired, there is only very scarce information about nomination of the discovery of superfluidity for the Nobel Prize in the 1950s. Documents from the personal archive of Niels Bohr in Copenhagen indicate that he twice nominated both Kapitsa and Landau for the prizes of 1956 and 1960.[44] There is reason to think that he was not the sole nominator of the two scientists.[44a] Time will show whether Soviet researchers nominated them too. If they did, the nominations were sent to Stockholm either by mail or taking a convenient opportunity, i.e. bypassing the Presidium of the USSR Academy of Sciences, whose leading figures had to take into account opinions of ideologists of the CPSU Central Committee until the collapse of the Communist regime in this country.

The aforementioned nominations were two before the last of the total 29 ones sent by Bohr to the Nobel Committee during the 40-year-long period, from 1920 till 1962. His last nomination dated to 1962, a fatal year for himself and Landau.

On the 7th of January 1962, Landau was in a car accident on the way from Moscow to Dubna. His car hit the lorry head on. V. L. Ginzburg, a close friend and co-worker of Landau, recollected at the meeting in memory of the scientist held in the Polytechnical Museum 16 years after the accident that "his delicate constitution of no consequence under routine conditions proved fatal in that accident [...] Eggs in a bag remained intact, but Landau was severely injured."[45]

The skill of the doctors and heroic efforts of Landau's colleagues all over the world including Niels Bohr prolonged his life for six more years but could not restore the unique intellect of the great physicist. Bohr and other friends of Landau were well aware that recovery was impossible. They urgently (in January) sent to the

Nobel Committee a new nomination of Landau formulated in quite different terms compared with the previous ones.

They proposed to award the Nobel Prize in physics for 1962 to Landau alone "taking into account the decisive influence of his original ideas and remarkable works on modern atomic physics."[46]

The Royal Academy of Sciences listened to the voice of the renowned nominator whose opinion was many times before shared by the prize-giving institutions[47] and awarded the 1962 Nobel Prize to his candidate. It was very unusual. Normally, the Nobel organizations avoid giving prizes for theoretical works apart from experimental findings.[48]

Bohr witnessed the success of his nomination; he died on the 18th November 1962. Landau was awarded for "his fundamental theories of condensed matter, especially liquid helium."

This award was quite unexpected for the USSR Academy of Sciences. Moreover, the physical condition of the laureate made him unable to attend the prize-giving ceremony. Hence an extensive correspondence between the Department of Scandinavian States, the USSR Ministry of Foreign Affairs, the USSR Academy of Sciences, and the Soviet embassy in Stockholm in the first days after the announcement of the name of the new Prize winner. The result was "a tentative variant of the speech of the Soviet ambassador in Sweden on the occasion of awarding the Nobel Prizes for 1962" prepared by the USSR Academy of Sciences on December 1. A later note on the document says that it was drafted in the Foreign Department of the Academy. It did not occur to its officials to apply for assistance to a competent member of the Division of Physico-Mathematical Sciences.

On the 10th of December, the ambassador N. D. Belokhvostikov read the text standing on the steps of the huge Blue Hall of the Stockholm city hall, the traditional place of formal Nobel dinners. He introduced minor corrections in the document received from Moscow (see note 4 to doc. 8.31). They however did not substantially change its overall bureaucratic style.

The prize-giving ceremony took place in the conference hall of the Central Hospital, the USSR Academy of Sciences, no the morning of December 10 (doc. 8.32). The attending doctors insisted that only a few persons be invited to participate. Even the Nobel Prize winners P. A. Cherenkov and I. M. Frank were cut off the list of the guests. Rolf Sulman, Swedish ambassador in the Soviet Union, represented King Gustav VI Adolf. He made a short welcoming speech in good Russian and handed over the Nobel gold medal and diploma to Landau sitting at the table.

The protocol of the ceremony did not envisage a response speech. Nevertheless, the laureate found strength to thank the Nobel organizations in English.

On the evening of the same day, the Swedish embassy in Moscow gave a reception simultaneous with the similar reception in Stockholm. The laureate was represented by his wife, K. T. Drobantseva-Landau. Cherenkov and Frank were among the invited guests. Sulman and K. N. Rudnev, deputy chairman of the USSR Council of Ministers, exchanged speeches.[49]

P. L. Kapitsa, director of the Institute of Physical Problems with which Landau was affiliated, attended both ceremonies, at the hospital and at the embassy. Kapitsa did much good for Landau, who never forgot it. One of Kapitsa's most remarkable actions was the liberation of Landau, arrested by NKVD, which required both courage and quick wit (doc. 8.34). When Kapitsa himself was dismissed from all his positions and had to live unemployed in his country house,[50] Landau and very few other persons regularly visited him although such visits were fraught with much trouble with Soviet authorities.

Collisions associated with the exclusion of Landau from the joint nomination with Kapitsa for the Nobel Prize in 1957 narrated by V. L. Ginzburg (doc. 8.1) had nothing to do with the personal relations of the two eminent representatives of Soviet science. They ensued from the political game undertaken by the Department of Science, CPSU Central Committee, and their accomplices in the Presidium of USSR Academy of Sciences.

Foreign specialists many times nominated P. L. Kapitsa alone for the Nobel Prize since 1946 (see Tables 4.1 and 6.1). His nominators included P. Dirac,[51] W. Pauli, T. Swedberg, and other scientists. They proposed to award him for the totality of experimental studies on the behaviour of matter in strong magnetic fields and at superlow temperatures rather than the discovery of superfluidity. For example, Bohr nominated him in 1947 (doc. 8.35). In 1956 and 1960, Bohr decided to confine his nomination to the brilliant discovery of helium superfluidity; therefore, he proposed to share the prize between Kapitsa (the discoverer) and Landau (the developer of the superfluidity theory).

Kapitsa received the Nobel Prize only in 1978 and shared it with the American physicists R. Wilson and A. Penzias. The prize was awarded to him for the inventions and discoveries in the field of low temperatures and to the Americans for the discovery of relict cosmic radiation also made with reference to superlow temperatures.

The cause of such a long delay and the overcautious attitude of the Nobel Committee will be clear only in 25 years, when archives of the Nobel organizations open their files for historians. At present, one can only speculate on the subject although some information is available from sources other than the Nobel archives. To begin with, here are some personal recollections dated from the early 1970s.

At that time, the author had a talk with I. M. Lifshits, member of the USSR Academy of Sciences and the closest co-worker of Kapitsa, and asked him about the cause for which so few Soviet scientists received the Nobel Prizes. Lifshits said that he annually nominated Kapitsa and could not understand why Petr Leonidovich did not yet receive it. Then, he added as an afterthought that the reason might be the appearance of two similar papers on formerly unknown effects in liquid helium in one issue of *Nature*.[52] He referred to the paper by Kapitsa himself and the other by John Allen and Ostin Meisner working in the Mond Laboratory, Cambridge, that Kapitsa had headed in 1934. Lifshits talked about "the difficult situation facing the Nobel Committee that did not wish to give the prize to Allen and Meisner, considering them unworthy of it, and was afraid to award Kapitsa alone because the papers of these authors were published simultaneously."[53]

P. E. Rubinin, assistant to Kapitsa for many years, tried to analyse this situation. He published correspondence of Kapitsa with his English colleagues that may give a clue to the strange case of the simultaneous publication of two similar papers.[54]

Science knows many cases when independent researchers practically simultaneously came up with an epoch-making discovery. It is unclear from the aforementioned correspondence whether Kapitsa knew about the works carried out in the Mond Laboratory. But Allen was totally unaware of the studies in Moscow as appears from his letter written in 1977 (note 2 to doc. 8.36). It should be emphasized that Kapitsa immediately understood the nature of the new phenomenon and described it as superfluidity in his very first publication. After a time, Allen called this conjecture "prophetic".[55]

All further events were nothing else but the fatal coincidence of random circumstances. The editorial board of *Nature* asked Webster to look through the galley proofs of Kapitsa's paper. Had not Webster owed a visit to France (doc. 8.36 I–III), he would have done it and the paper appeared at least a week earlier. On the other hand, Allen would have read Kapitsa's paper published in the journal rather than in galley proofs if Cockroft had not proved an interested person unexpectedly to Kapitsa. Anyway, the two papers would not have been received by the editorial board in one envelope,[56] and the English authors would have had to refer to the pioneer article by Kapitsa, as G. S. Landsberg and L. I. Mandel'shtam had done years ago after they read the short communication of Raman in the same journal (see Chapter 2)...

It is presently unclear whether this situation had anything to do with the position of the Nobel Committee, which had exhibited a negative attitude towards the candidacy of Kapitsa for so many years.[57] If it had not, then his nomination by such an authoritative scientist as Theodor Swedberg, Nobel laureate and member of the Royal Academy of Sciences, in 1947 and 1948, must have considerably increased the probability of awarding the Nobel Prize to Kapitsa. Swedberg nominated the Soviet scientist together with Willem Keesom, of the Netherlands. The co-worker of the famous Kamerling-Onnes, professor of the University of Leiden, first observed phase transition in helium at 2.19 K in 1932. It was later shown that below this temperature helium converts to a quantum liquid (helium II). Kapitsa discovered the ability of helium II to flow without viscosity, i.e. its superfluidity, in 1938.

Keesom was for the first time nominated for the Nobel Prize in physics in 1940 together with Johannes de Haas, his colleague at the University of Leiden, for the discovery of jump-like changes in magnetic susceptibility of metals at superlow temperatures. Thereafter, de Haas and other scientists thrice nominated him till 1945. All four nominators were Keesom's countrymen. In 1948, Swedberg nominated him together with Kapitsa. Seeing the Nobel Committee reluctant to award Kapitsa, Swedberg refused to support the nomination of Keesom in 1949, considering it unjust to give the prize to only one of the two discoverers equally deserving it.

Kapitsa made a hint at the lateness of his award in the Nobel lecture of 1978 (doc. 8.37). The audience met the hint with understanding. Kapitsa demonstratively devoted his lecture to a subject unrelated to that for which he was given the prize and thereby for the first time broke up the long-standing tradition. The lecture was called "Plasma and controlled thermonuclear reactions."

N. G. Basov and A. M. Prokhorov, corresponding members of the USSR Academy of Sciences, were the last Soviet Nobel laureates in physics whose nomination was sent to Stockholm via the Presidium of the Academy and its Foreign Department. They were nominated together with the American physicist C. Townes by D. V. Skobeltsin, director of the Physical Institute, USSR Academy of Sciences (doc. 8.38–8.39). All three received the prize in 1964.

Skobeltsin paid much attention to the work of Basov and Prokhorov from the early stages of the development of optical quantum generators. He understood that it could lead to a revolutionary breakthrough in the physics of electromagnetic radiation and actively supported it despite the opposition of the conservative members of the Scientific Council of his own Institute and of the Presidium of the Academy. The creation of laser generators was a remarkable event in physics of the mid-1990s, the importance of which was immediately recognized by the Nobel Committee and the Royal Academy of Sciences.[58]

It is worthwhile to mention an episode that coincided with the beginning of the consideration of nominations for the Nobel Prizes by the Nobel organizations in September–October 1964. President of the Royal Academy of Sciences Rolf Granit and his wife visited the Soviet Union from late August to early September. Granit expressed his wish to visit the USSR in a talk with the Soviet ambassador in Sweden, N. D. Belokhvostikov, and was soon invited by the USSR Academy of Sciences to come as its guest. He described the invitation as "a great honour to him" (doc. 8.45).

There is no documented evidence that he had talks about the awards of the 1964 Nobel Prizes in Moscow. Clearly, Granit could easily avoid discussing this subject since nominations had not yet reached prize-giving organizations and were not considered by them. There is no doubt, however, that he supported the candidacies of the Soviet scientists when they were voted for in the Nobel Committee on 29 October 1964.

In 1964, the new Soviet laureates found themselves in a much more comfortable political situation than their predecessors. To recall, N. N. Semenov received his award in 1956 very soon after the events in Hungary. A. Cherenkov, I. E. Tamm, and I. M. Frank were awarded after the scandalous attempts of the Presidium of the USSR Academy of Sciences to debar the two latter scientists from participation in the nomination process and simultaneously with the award of the Nobel Prize in literature to Boris Pasternak (this decision of the Swedish Academy infuriated Soviet authorities and compelled them to delay the visit of the Soviet physicists to Stockholm). Finally, the USSR Academy of Sciences did nothing to promote the nomination of Landau. A complicated situation arose around the nomination of L. V. Kantorovich, one of the first Nobel laureates in economics, awarded in 1975.

In 1964, the nomination procedure proceeded in a civilized form. The candidates were recommended by D. V. Skobeltsin, director of their Institute, without an objection from the Department of Sciences, CPSU Central Committee (doc. 8.40). Their scientific lecture was dispatched to Stockholm in July 1964 as requested by the Nobel Committee (8.41). Even instructions of "higher organs" to "propagandize the policy of the Communist Part and Soviet government" were nothing else but routine formality. Curiously enough, such documents were usually drafted by those for whom they were intended and thereafter submitted for the approval of the respective officials.

The only failure was the refusal to grant A. M. Prokhorov permission to take his son, a university student, to Stockholm (note 3 to doc. 8.43). The refusal was, however, predictable from the previous experience with the children of N. N. Semenov, I. E. Tamm, and I. M. Frank[59] (note 2 to doc. 8.12). Even diplomats and foreign journalists were not allowed to have with them children above 12 years of age when abroad. The adolescents were sent back to the Soviet Union to live with relatives or in a boarding house; sometimes it ended in a family drama.[60]

This unnatural practice that separated parents and children for many years was said to be dictated by the necessity to give the children a proper education in the Russian language that was really difficult to organize abroad. However, the true reason was to have them as hostages thereby reducing the risk of the parents' refusal to return to the USSR. The same rule extended to the scientists going to Stockholm to receive the awards. The protocol of the Nobel festivities required that they should be accompanied by their wives but did not regulate participation of other close relatives. Therefore, Soviet authorities were free to act at their own discretion.[61]

It should be noted in conclusion that the Soviet inventors of the molecular generator were lucky to create it in due time. Had they done it two years earlier or had Stalin lived a couple of years longer, they would have never received the Nobel Prize, as was the case for V. I. Veksler a decade or so before.

All works on quantum electronics in the Soviet Union were regarded as strictly secret till the second half of 1953. Suffice it to say that Prokhorov defended his thesis for the degree of doctor of sciences (summarizing results of his studies on coherent radiation in a synchrotron in the centimetre wavelength range) in 1951 but was able to publish it only in 1956.[62] The first maser was almost simultaneously created in FIAN and Columbia University in late 1954 to early 1955. It is not difficult to guess what would have happened if this event had occurred at the peak of ideological clinch in 1952 and the beginning of 1953, when so many research projects in the Soviet Union were considered secret. C. Townes alone or with his close co-workers J. Gordon and H. Zeiger would have been given the Nobel Prize and Soviet science would have lost another prestigious award.

Chronologically, the last documents filed in the archive of the Foreign Department, USSR Academy of Sciences, were nominations of A. N. Frumkin[63] and the German chemist M. Volmer by N. N. Semenov, of P. L. Kapitsa by A. L. Mints, V. I. Veksler and his American colleagues E. McMillan and N. Kristofilos by P. A. Cherenkov.[64]

The nomination process at last acquired a civilized form of the direct submission of proposals to the Nobel Committees.

The possibility of free contact with foreign scientists demonstrated the decay of the totalitarian regime rather than its strength. The loss of ability to exercise overwhelming control led to unexpected *force majeure* situations, such as faced ideologists of the CPSU Central Committee and members of the Presidium of the USSR Academy of Sciences in October 1975.

The Norwegian Nobel Committee awarded the Nobel Peace Prize for 1975 to A. D. Sakharov, member of the USSR Academy of Sciences. It was not unexpected because the world press discussed such a possibility long before the Nobel Committee made its formal announcement. Therefore, the ideological apparatus of the CPSU Central Committee was fairly well prepared to respond.[65] Very soon, however, the Royal Academy of Sciences announced the award of the Nobel Prize in economics to L. V. Kantorovich, member of the USSR Academy of Sciences, and Professor T. Koopmans, USA. It was quite unexpected because the confidentiality level of discussions in the Nobel Committees in science was traditionally much higher than in Oslo.

The Soviet authorities disliked Kantorovich even in the pre-war years when he was a professor of Leningrad University. Kantorovich was the first to develop the linear programming method and apply it to practical purposes in 1939. The method proved valuable for optimal planning of production processes, from the use of raw materials to the manufacture of the finished product. As a man of independent thinking, Kantorovich not infrequently criticized dogmatic principles of Soviet political economy and therefore was a target of demagogic attacks (doc. 8.47).

His opponents pointed out what they regarded as the unreasonably extensive use of mathematical formulas that overshadowed the economic essence of the problems under consideration. The principles of linear programming that allowed to substantially reduce unproductive consumption of natural resources seemed inconsistent with traditional socialist economy in which gross output was the main criterion of efficiency. High-performance economics based on these principles was primarily oriented toward maximum cost effectiveness, reduced labour requirement, and markedly decreased production-related waste. They were in manifest conflict with the existing economic system[66] (by way of example, Soviet industrial enterprises had to annually give up a preplanned amount of scrap metal; departure from this practice would lead to the loss of rises in the low wages normally paid to workers). Socialist economy was extensive and costly; it could not objectively be made more efficient without radical transformation. All of this made Kantorovich to give up his applied projects and focus on "pure mathematics" pending a more favourable turn of events. In the meantime, the totalitarian regime wallowed in numerous outdated dogmas and as unable to assimilate fresh ideas and proposals of its most educated and gifted citizens, kept afloat only by virtue of the country's enormous natural resources and infinite patience of its people...

Kantorovich found himself in a somewhat more advantageous situation during the thaw period in the second half of the 1950s and early 1960s. He owed it in the first place to some of his colleagues, who understood perfectly well the scale of the losses suffered by the economy of the country that ignored the scientist's proposals. One of them was A. A. Markov, corresponding member of the USSR Academy of Sciences, an expert in the theory of fundamental systems and one of the developers of the theory of algorithms.

In his speech at a meeting of the Communist Party organization at the Mathematical Institute (22 April 1957), Markov emphasized the great potential of mathematical methods for scientifically grounded planning of national economy. "We have good prerequisites for the development of this section of mathematics in the works of Kantorovich. Unfortunately, practical application of his results is hampered by hide-bound attitudes of economists. Meanwhile, such work is being done in the United States [...]" M. A. Lavrentiev, member of the USSR Academy of Sciences who chaired the meeting, immediately proposed: "Maybe comrade Markov will submit a note on this issue to the Presidium of the Academy [...]"[67]

Markov used the opportunity and thereby made the first breach in the Chinese Wall around Kantorovich.[68] In 1958, he was elected a corresponding member of the USSR Academy of Sciences and joined the newly organized Siberian Division of the Academy in Novosibirsk where he had all conditions for fruitful creative work. His fundamental monograph *Economic calculation of optimal utilization of resources*[69] was published in 1960 although it had been written in 1942. The USSR Academy of Sciences elected him its full member in 1964. His innovative ideas long ago applied abroad began to be used in national economy.

However, a group of his permanent opponents formed during Stalin's rule and led by K. V. Ostrovityanov, academician-secretary of the Division of Economic Sciences and thereafter vice-president of the USSR Academy of Sciences, continued to attack him. Ostrovityanov was an orthodox proponent of the "radiant future" for socialist economics as it developed under Stalin. He was virtually unable to conceive of the new ideas of Kantorovich. This group, even after the dismissal of its leader from his official posts, did a lot to prevent awarding the Lenin Prize to the prominent Soviet economist (doc. 8.47).

As mentioned above, the award of the Nobel Prize to A. D. Sakharov was not a surprise for Soviet authorities. The award of Kantorovich was quite unexpected. Leaders of the USSR Academy of Sciences and their inspirers in the Department of Science, CPSU Central Committee, found nothing better than to try and make Kantorovich sign a letter against Sakharov and the decision of the Nobel Committee to give him the Nobel Peace Prize, in exchange for permission to go to Stockholm (doc. 8.46).

The scientist did not sign the letter (see also note 7 to doc. 8.46) although he had to insert a few rhetorical phrases in his message to the President of the USSR Academy of Sciences (doc. 8.48). Certainly, these passages by no means reflected his personal opinion, nor did they ensue from his scientific publications.

Kantorovich came to Stockholm with his wife. He could not even think of taking his son to accompany them in the situation created around him. In 1964, Prokhorov's son was not permitted to accompany his father. Since then, there was a precedent to which Kantorovich could refer under more favourable conditions. Sholokhov visited Stockholm to receive his prize in literature for 1965 accompanied by numerous relatives (see Chapter 10). However, he was a member of the CPSU Central Committee, one of the top figures in the official Soviet hierarchy...

The Nobel festivities of 1975 were organized in a big way. The day of June 29, 1975 was a jubilee date; the king of Sweden signed the Nobel Statutes and the Nobel Committees were established in Stockholm 75 years ago. Therefore, the Nobel Foundation invited all living Prize winners in the categories of physics, chemistry, physiology and medicine, literature, and economics. Soviet scientists invited to Stockholm included N. N. Semenov, P. A. Cherenkov, I. M. Frank, N. G. Basov, and A. M. Prokhorov.

When the Soviet participants of the festivities returned to Moscow, they were invited for a talk to the Department of Science and Higher Education (22.12.1975). They were hosted by S. P. Trapeznikov, head of the department, V. L. Barsukov, head of its sector, and A. P. Aleksandrov, the newly elected President of the USSR Academy of Sciences (doc. 8.50). V. L. Kantorovich was absent,[70] but his name was twice mentioned during the meeting. N. N. Semenov spoke about publications in the mass media concerning the Nobel Prize for Kantorovich and sometimes wrong interpretation of his replies to interviewers. N. G. Basov commented on speculations in Swedish newspapers preceding their visit about chances for Kantorovich to have permission to come (doc. 8.51).

Of special interest in the shorthand report of the conversation is a remark by Aleksandrov at variance with the widespread opinion about the lack of objective attitude of the Nobel organizations toward the achievements of Soviet science. According to the speaker, "[...] although our scientists are awarded less frequently than Americans [...] the awards are generally given according to deserves."

In a month after the above meeting, the Presidium of the USSR Academy of Sciences sent to the CPSU Central Committee a letter listing concrete measures necessary for the strengthening of its contacts with the Nobel organizations in Stockholm (doc. 8.50). Some proposals suggested poor knowledge of the Nobel Statutes requiring strict confidentiality of nominations.[71] Nevertheless, this document marked the complete reconciliation of the USSR Academy of Sciences with the Swedish institutions awarding prizes in sciences.

It is confirmed by more, even if indirect, evidence. The year of 1978 was the final year of the publication of the *Great Soviet Encyclopaedia* in 30 volumes. The first volume of this third edition appeared in 1969. The editor-in-chief was A. M. Prokhorov, a Nobel Prize winner and participant in the meeting at the CPSU Central Committee, mentioned in a preceding paragraph. The last volume contained a list of almost 500 biographies not published for a variety of reasons in the previous books. It included 43 Nobel laureates, 20 of whom had already received the award by the time of appearance of the respective volumes. They were the first Nobel laureate in physiology and medicine Emil Behring (developer of antidiphtheritic and antitetanic serum), Robert Barani, who received the prize in the same category in 1914, Julius Wagner-Jauregg (1927), Selman Waksman (1952) etc. Some of them could not be included in the word-list compiled in the mid-1960s for purely ideological reasons. For example, Waksman was an emigré from the Russian Empire, and his visit to Moscow in 1946 by invitation of the USSR Academy of Sciences gave much trouble to the Department of Agitation and Propaganda, AUCP(b) Central Committee (see note 24 to Chapter 4, and Chapter 12). But the majority of the scientists on the additional list had not been included in the published volumes just because the authors' slighting attitude towards decisions of the Nobel organizations. These omissions were somewhat corrected after the December meeting in the CPSU Central Committee...

True, some prejudices developed in the past and their bearers persisted. But now their views were a matter of private opinion. Soviet scientists received freedom to maintain contacts with their foreign colleagues, the right to which they started to vindicate in the late 1950s and succeeded by the mid-1970s. The most prominent of them are now equitable members of the international corps of nominators for the Nobel Prizes and annually function in this capacity.

To conclude, there is one more story of the nomination of a Soviet scientist for the Nobel Prize. It is based on personal evidence since there are yet no relevant documents in the available archive files of the Karolinska Institutet. The man in question is I. A. Rapoport, an eminent physiologist and in a sense a legendary personality.

Rapoport graduated from the University of Leningrad in 1935 and focused on chemical mutations. Just before the Great Patriotic War, the young scientist obtained very important data, basis for the development of radiation genetics, and started to work on his doctoral thesis. The next day after the Germans attacked the Soviet Union (23 June 1941), he came to a military enlistment office to volunteer to the front. He remained there till the end of the war despite a few serious wounds. Sent for short-term training to M. I. Frunze Military Academy, Rapoport managed to defend the previously prepared doctoral thesis and was given the scientific degree of doctor of biological sciences.

Authors of the obituary of Rapoport published by the Presidium and the Division of General Biology, Russian Academy of Sciences, emphasized "the outstanding courage and heroism" shown by Rapoport in military operations. He needed still more bravery in the post-war period to speak against Lysenko, from whose dominance he "fearlessly and straightforwardly defended biological science."[72] Rapoport was the only one in the audience at the notorious session of the All-Union Academy of Agricultural Sciences (1948) who cried out "Obscurants!"[73] to hold up to shame ignorant supporters of Lysenko taking the floor.

After the session, Rapoport was sacked from the academic institute and expelled from AUCP(b),[74] which he had joined in 1943. In 1957, N. N. Semenov gave him an employment at the Institute of Chemical Physics, the USSR Academy of Sciences, on the advice of S. Hinshelwood (doc. 8.55). When the era of Lysenko came to an end in 1964, Semenov appointed Rapoport head of the newly organized Department of Chemical Genetics.

R. I. Rapoport, son of the scientist, told the author that the geneticist Oke Gustaffson, professor at the University of Lund, had nominated his father for the Nobel Prize in physiology and medicine together with the English physiologist Scharlotta Auerbach, who received similar results in 1942. As in the case of Semenov (see Chapter 7), the Nobel Committee needed additional information about the nominee and his works and sent a confidential message to this effect to Moscow.

According to R. I. Rapoport (doc. 8.55), his father was permitted to send the necessary materials to Stockholm in exchange of application for restoring his Party membership. Rapoport refused. The scientist and the warrior, he did not wish to give up his moral principles.

Here is one more closing remark...

Paradoxically, Soviet scientists won the freedom of action in their relations with foreign colleagues and the Nobel organizations under very unfavourable political conditions in this country. It was the period when Khrushchev was displaced from power by the Brezhnev–Suslov group in October 1964, who did not care to conceal their plans of a creeping restoration of Stalinism. They started the campaign from gradual elimination of references to "the personality cult" in the mass media and related repressions of millions of guiltless people. The work of rehabilitation commissions established during the Khrushchev Thaw was little by little reduced to nought.

In response, opponents of the deceased chief worked out an Aesopian language, allowing undercover thoughts to appear in censored publications. This activity peaked in the late 1960s and mid-1970s. The writer V. N. Voinovich narrated one of the most remarkable examples of such mimicry.

In 1975, mathematician Yu. A. Gastev published a purely scientific work on mathematical logic. In the preface to the book, he expressed "special thanks to doctors Cheyne and Stockes for their assistance in his work." Neither John Cheyne nor William Stockes was an expert in mathematics or logic... They belonged to the medical profession and gave their names to the Cheyne–Stockes respiration that develops in man in death agony. This term was used in the official medical bulletin issued after Stalin's death (it was exactly what Gastev hinted at)... Gastev had been arrested when Stalin was still alive, and only the death of the chief enabled him to continue education.[75]

Notes

[1] It is appropriate to mention a poorly known episode unrelated directly to the subject in question but vividly characterizing the situation and expectations in the first years of the Khrushchev Thaw. A new election campaign of candidates for the USSR Supreme Council was launched in January 1958. Yu. G. Burtin, a schoolteacher in the provincial town of Bui,

Kostroma region, nominated in his constituency and on his own initiative A. T. Tvardovsky, a known poet and editor-in-chief of the journal *Novy Mir.* The initiative was readily supported by his former pupils, local railway workers. The authorities regarded this move as a demonstrative challenge to the existing system of non-alternative elections to the highest legislative body of the country. The unusual attempt was nipped in the bud, and Burtin was expelled from the membership in the AUCP(b).* This initiative of a provincial school teacher was really a reckless act, but it objectively reflected hopes of Soviet intellectuals that the death of the dictator would bring about the liberalization of public life and put an end to the pathological fear of the cruel regime.

* *Znamya,* No. 10, 2002, p. 179.

[2] S. I. Vavilov died in 1951; therefore he could not be nominated for the Nobel Prize in 1957 although he himself had designed the experiments with gamma-luminescent solutions during which his postgraduate student P. A. Cherenkov discovered an unusual luminescence (in 1934), later called by his name.

[3] I. M. Frank was cut off the list of nominees for the 1957 Nobel Prize in physics together with I. E. Tamm. The two scientists could not be nominated one without the other because the theory of Cherernkov radiation was their joint creation. Recommendation of Cherenkov alone looked logical since he was officially nominated for the discovery of new radiation.

[4] TsAODM, fond 992, opis' 2, ed. khr. 58, l. 191.

[5] Telephone talks and visits to the Department of Science, CPSU Central Committee, were part of the everyday activities of the chief scientific secretary of the USSR Academy of Sciences and his subordinate, head of its Foreign Department.

[6] G. B. Collins, V. G. Reiling. Phys. Rev. 54, 1938, p. 499. There is an interesting story related to the foreign origin of the term "Cherenkov radiation". It was narrated by I. M. Frank: "The results obtained by Cherenkov were first noticed and theoretically interpreted only by Soviet physicists. It appears that foreign researchers did not read much Russian literature although *Doklady AN SSSR* was published in two languages [...] It is after the new phenomenon was fairly well understood by 1937* that Vavilov sent a small paper of Cherenkov summarizing the experimental data and the theory to *Nature.* The paper was turned down on a plausible pretext.** The true reason was the serious journal did not want to publish data that seemed at least debatable. The "Physical review" proved more liberal and published the paper.*** Soon, Collins and Reiling obtained another experimental confirmation of the theory."****

* The paper of I. E. Tamm and I. M. Frank "Coherent radiation of a fast electron in the medium." *Doklady AN SSSR,* 14, No. 3, 1937, p. 107.

** The letter of the editor-in-chief accompanying returned Cherenkov's paper was dated 26.6.1937.

*** Phys. Rev., 52, 1937, p. 378.

**** I. M. Frank. Izluchenie Vavilova-Cherenkova. Voprosy teorii (Vavilov-Chererkov radiation. The Questions of Theory). M.: Nauka, 1988, p. 34.

[7] Physico-chemist Kondratiev was member of the Division of Chemical Sciences, USSR Academy of Sciences, whereas Cherenkov was nominated for the prize in experimental physics. Kondratiev's works were largely devoted to chemical kinetics including the theory of chain reactions, molecular spectroscopy, photochemistry, etc. All these problems are rather

far from the areas of physics where the Vavilov–Cherenkov effect finds application. This circumstance clearly suggests that Kondratiev's nomination was designed to counterbalance that of Kurchatov, unacceptable for the leadership of the Academy and the Department of Science, CPSU. The choice of the member of a different Division had to ensure confidentiality and prevent the nomination from being known to physicists. Nevertheless, it came to light through information leakage and caused protests of members of the Division of Physico-Mathematical Sciences.

[8] The same is confirmed by a very ambiguous phrase in the newspaper *Pravda* concerning the participation of the Academy in this affair (see note 6 to doc. 8.1).

[9] The nomination could be sent to Stockholm only by mail. Such a step was fraught with most serious trouble for those who signed it if the letter were opened and inspected.

[10] Tamm was somewhat disappointed to receive the prize for the work that he did not regard as his most important. He believed his best result to be the analysis of nuclear forces underlying the hypothesis of the primary role of the so-called beta-forces or electropositron pairs suggested in 1934.* The fundamental significance of this hypothesis was appreciated by Hideki Yukawa, a future Nobel laureate, who described (in 1935) nuclear force genesis as a result of hypothetical particle exchange. He also computed the mass of the particles. In 1937, Carl David Anderson discovered these particles in cosmic rays and called them mesons.

* "Kapitsa, Tamm, Semenov v ocherkakh i pis'makh (Kapitsa, Tamm, Semenov in essays and letters)." M.: "Vagrius-Priroda", 1998, p. 319.

[11] As mentioned above, foreign nominators recommended Cherenkov alone in 1952 and 1955, but his candidacy did not attract the attention of the Nobel Committee.

[12] The four scientists received the grade I Stalin Prize in January 1946 for "the discovery and investigation of radiation of electrons during their motion in matter with a superluminal speed."* However, the choice of the awardees also encountered difficulty. Indeed, there were proposals to award the prize only to the discoverers of the effect. Initially, all the four were mentioned in the resolution of the Stalin Prize Committee in science dated 28.3.1944 (doc. 8.54-I). However, S. V. Kaftanov, deputy chairman of the Committee, objected to awarding I. E. Tamm and I. M. Frank in a letter sent to the USSR Council of People's Commissars. Kaftanov wrote that they "participated only in theoretical interpretation of the phenomenon" (doc. 8.54-II). It seems unlikely that Kaftanov could write this letter of his own accord. He was educated as a chemist technologist and could not appreciate in full measure the fundamental importance of the theoretical work of the two scientists. The true authors were Party ideologists for whom Tamm was a "politically suspect" person. Suffice it to note that Kaftanov wrote his letter when Tamm was not yet involved in the supersecret atomic project initiated in 1943.**

* *Pravda.* 27.1.1946, p. 1.

** "Kapitsa, Tamm, Semenov [...]", p. 240.

[13] Pasternak's son confirmed this in his memoirs. After the article appeared in *Pravda,* M. A. Leontovich, member of the USSR Academy of Sciences, asked him to go together to Peredelkino where the poet stayed at that time since "he considered it his duty to assure Pasternak that the physicists did not share opinions of the newspaper and that the tendentious phrases were added without their knowledge."*

Half a century ago, B. M. Boltovsky, D.Ph–M.Sc., recollected that, being a young research scientist in FIAN, he was interviewed about Cherenkov radiation for a radio program. The broadcast text contained a paragraph inserted without his consent. It told how justly the Soviet physicists were awarded and how worthless and provocative was awarding the prize to Pasternak.**

Such forgeries were common at those times. The everyday practice of adaptation of original articles without the knowledge and consent of their authors was fairly widespread when editorial boards prepared for publication regular issues of journals and newspapers. A scandalous example of this practice was the story of a manuscript submitted to the journal *Bolshevik* in 1948 "half of which the editorial board substituted by a new text."***

* Pasternak E. B. Boris Pasternak. Biographia. M.: Tsitadel', 1997, p. 706.

** *Priroda*, No. 7, 2004, pp. 33–34.

*** In: "Politbyuro TsK VKP(b) i Sovet Ministrov SSSR, 1945–1953 (Politbureau of AUCP(b) Central Committee and the USSR Council of Ministers, 1945–1953)." M.: ROSSPEN, 2002, p. 290.

[14] Such insertions in the text unauthorized by the author being unlawful, the executors acted on verbal instructions of their superiors. This circumstance was sometimes used against them if deemed necessary. Here is an example. Politbureau of the CPSU issued a resolution (13.7.1949) after the arrest of its member N. A. Voznesensky in which the editorial board of the journal *Bolshevik* was charged with "publishing an article praising the book of N. Voznesensky *Military Economics of the USSR During the Great Patriotic War* [...] and inserting extracts from this book into publications of other authors without their consent."*

* RGASPI, fond 17, opis' 118, ed. khr. 449, l. 4–5.

[15] I. M. Frank cautiously mentioned the unusual situation surrounding the forthcoming visit of Soviet physicists to Stockholm in his memoirs: "[a] suspicious attitude toward our group (of course not from our colleagues) was stirred up by awarding the Nobel Prize in literature to B. L. Pasternak in the same 1958)."*

* "Vestnik AN SSSR", No. 7, 1978, p. 108.

[16] In: "Kapitsa, Tamm, Semenov v ocherkakh i pis'makh (Kapitsa, Tamm, Semenov in essays and letters)." M.: "Vagrius-Priroda", 1998, p. 322.

[17] The nomination looked senseless. The nominators were chemists although the Academy had the Division of Physico-Mathematical Sciences. This intrigue developed as a continuation of the events around the nomination for the Vavilov–Cherenkov effect for the prize in 1957–1958. Then, the nomination signed by N. N. Andreev, A. I. Alikhanov, and L. D. Landau, members of the USSR Academy of Sciences, was sent to Stockholm by mail (see this chapter and note 9). Probably, physicists used the same route in 1959, to bypass the bureaucratic apparatus of the Foreign Department.

[18] Very likely, some foreign nominators also recommended Zavoisky between 1954 and 1959. But relevant documents of the Nobel Committee were still unavailable at the time when this monograph was published.

[19] F. Bloch was for the first time nominated for the Nobel Prize in physics by Professor G. Wentzel (Zurich) in 1948. In the period from 1949 to 1952, both scientists were nominated. In 1951, their nominator was the Nobel laureate M. von Laue of Gettingen; in 1952,

K. Siegban (see Table 6.2 and the accompanying text in Chapter 6), a future (1977) Nobel laureate J. van Vleck (Harvard), and Professor D. Williams, Columbus, Ohio. Professor K. Gorter (Leiden)* was one more nominator of F. Bloch in 1952, but his nomination reached the Nobel Committee after 31.1.1952 and was officially included in the 1953 nomination list.

* S. V. Vonsovsky, member of the USSR Academy of Sciences, several times (since 1967) nominated C. Gorter for the Nobel Prize together with E. K. Zavoisky and I. Weller (Sweden) for their contribution to the use of resonance techniques in experimental physics (see doc. 8.15).

[20] "Charodei eksperimenta, Sbornik vospominaniy ob akademike E. K. Zavoiskom (The wizard of experiment. Collected recollections about academician E. K. Zavoisky)." M.: Nauka, 1993, p. 141.

[21] *Ibid.*, p. 59. Nomination of Zavoisky in 1964 was signed by Nobel laureates N. N. Semenov and I. E. Tamm, besides Aleksandrov.*

* Arkhiv RAN, fond 579, opis' 1, ed. khr. 639, l. 129–132.

[22] V. L. Ginzburg. "O nauke, o sebe i o drugikh (About Science, Myself, and Others)." M.: Fizmatlit, 2001, p. 393.

[23] "*Charodei eksperimenta*"..., p. 161.

[24] *Ibid.*, p. 99.

[25] *Ibid.*, p. 166.

[26] Zavoisky was awarded an international prize for his outstanding experimental work of the mid-1990s. The International Magnetic Resonance Society conferred it on May 21, 1977, i.e. one year after the scientist's death. The president of the Society was Alfred Kastler, a leading expert in magnetic resonance and Nobel Prize winner in 1966.

[27] It is not known why Skobeltsin was recalled from New York, where he stayed since 1946. Probably, the reason was the appearance of materials concerning the arrest of his brother in 1937 (see note 4 to doc. 8.1). Also, there were rumours that his sister was married to Rolf Sulman, a Swedish diplomat. Skobeltsin was on friendly terms with P. L. Kapitsa, whose wife, Anna Alekseevna, told the author that Petr Leonidovich used to frighten him in jest with this relationship.* Characteristically, the recall of Skobeltsin coincided with the appointment of Sulman Swedish ambassador in Moscow.**

* P. E. Rubinin, the closest assistant of Kapitsa for 30 years and thereafter curator of the scientist's museum, told the author that he found allusions to such a relationship in the diary of Anna Alekseevna. However, they are absent in the two volumes of this diary published in 2005. Zinaida Aleksandrovna, wife of Sulman and mother of Mikael Sulman, Director of the Nobel Foundation, is mentioned only once in this publication (20.2.1979), in connection with news from Stockholm about her death. (Dvadtsaty vek Anny Alekseevny. Dnevniki. Zametki. T. 2 (The twentieth century of Anna Alekseevna. Diaries. Notes. V. 2)). M.: Izd-vo "Agraf", 2005, p. 162.

** Mikael Sulman, son of the diplomat and Executive Director of the Nobel Foundation, is unaware of such a relationship.

[28] RGASPI, fond 17, opis' 118, ed. khr. 291, l61. The clerk of the Staff Department, AUCP(b) Central Committee, who prepared the document was sacked "for neglect of his duties"*

because he "relied on the information from the Ministry of State Security and secretary of the Communist Party organization of FIAN containing no discreditable data and did not study the personal file of Cherenkov."** It is opportune to mention that reputation of Cherenkov's wife was also "soiled". She was the daughter of Professor A. M. Putintsev (Voronezh), philologist accused of "counter-revolutionary activity" in the framework of the so-called "case of local lore students" dated to 1930 and sentenced to confinement in a labour camp. Putintsev's brother was arrested and shot in 1937.***

* RGASPI, fond 17, opis' 118, ed. khr. 29, l. 160.

** *Ibid.*, ed. khr. 291, l. 148.

*** In: "P. A. Cherenkov. Chelovek i otkrytie (P. A. Cherenkov. Man and discovery)." M.: Nauka, 1999, pp. 132–134.

[28a] "Uspekhi fizicheskikh nauk", V 121, No. 4, 1997, pp. 735–736 (Obituary) and 737–743 (paper: "In search of high-temperature superconductivity").

[29] L. Cooper. Bound electron pairs in a degenerate Fermi gas. "Phys. Rev.", 104, No. 6, 1956, p. 1189.

[30] The importance of Bogolyubov's findings was specially emphasized at the Communist Party meeting at the Mathematical Institute, USSR Academy of Sciences, on 21.11.1957 as a contribution to the solution of "one of the problems that for several decades defied the efforts of the greatest scientists."*

* TsAODM, fond 7334, opis' 1, ed. khr. 9, l. 67.

[31] A. A. Abrikosov, I. M. Khalatnikov. Modern theory of superconductivity. Uspekhi fizicheskikh nauk", 65, vyp. 4, 1958, p. 552.

[32] CAODM, fond 7334, opis' 1, ed. khr. 167, l. 82.

[33] "Zhurnal eksperimental'noy i teoreticheskoy fiziki, 34, vyp. 1, 1958, pp. 73–79; received 17.10.1957.

[34] Bardeen, Cooper, and Schriffer received the Nobel Prize in 1972 for "the development of the theory of superconductivity, referred to as BCS theory."*

* First letters of the laureates' names.

[35] Professor Schenberg (Cambridge, UK) invited several Soviet physicists including N. N. Bogolyubov to take part in the International Conference on Superconductivity. The invitation addressed to A. V. Topchiev, chief scientific secretary of the USSR Academy of Sciences, emphasized that their participation will make "our conference truly international."* L. A. Artsimovich twice submitted the question to the consideration of the Presidium of the Academy but "received no reply".**

* Arkhiv RAN, fond 471, opis' 1, ed. khr. 345, l. 228.

** *Ibid.*, l. 227

[36] Arkhiv RAN, fond 471, opis' 1, ed. khr. 311, l. 200.

[37] *Ibid.*, fond 383, opis' 1, ed. khr. 407, l. 1.

[38] *Ibid.*, ed. khr. 405, l. 33.

[39] Scientists somehow involved in the Soviet atomic project were not granted permission to go abroad until the 1970s. The known mathematician I. M. Gelfand, member of the USSR

Academy of Sciences, made computations for the project, similar to Bogolyubov. According to Professor A. D. Gvishiani, deputy director of the Institute of Physics of the Earth and grandson of A. N. Kosygin, prime minister, he once told his grandfather that Gelfand was not allowed to visit Great Britain to receive the mantle and cap of doctor *honoris causa* at Oxford University. Kosygin "was lost in thought for a minute and said: 'It is wrong.' He raised the question at a sitting of Politbureau. Gelfand was given the foreign passport and went to England." (Interview for the newspaper *Izvestiya.*)* Thus a precedent was created for individual foreign trips of Soviet scientists invited by research institutions of different countries.

* *Izvestiya,* 8.7.2002, p. 4.

[40] Arkhiv RAN, fond 579, opis' 1, ed. khr. 639, l. 147.

[41] *Ibid.*, ed. khr. 640, l. 69–96 and 98–122.

[42] *Ibid.*, p. 141.

[43] *Ibid.*, ed. khr. 639, l. 142–145.

[44] U. Rosenberg. Niels Bohr. Leben und Werk eines Atomphysikers. Stuttgart, 1985, S. 327.

[44a] It appears from archive files of the Nobel Committee in physics opened in 2007 that Kapitsa was nominated not only by Niels Bohr and his three colleagues but also by the 1954 Nobel laureates Max Born and L. D. Landau (Moscow). Even earlier (1955), the Nobel Committee registered the nomination of Kapitsa by the Soviet physicist V. A. Fok.

[45] V. L. Ginzburg. O fizike i astrofizike (On physics and astrophysics). M.: Byuro-Kvantum, 1995, p. 365.

[46] See "Vospominaniya o Landau (Recollections about Landau)." M.: Nauka, 1988, p. 111.

[47] N. Bohr nominated a total of 23 scientists for the Nobel Prize from 1920 to 1962, when he died. Sixteen of them eventually received the prize: A. Einstein, J. Franck, O. Richardson, Ch. Raman, I. Lengmuir, W. Heisenberg, E. Schrödinger, F. and I. Joliot-Curie, E. Lawrence, O. Stern, G. Hevesi, M. Born, L. Landau, H. Bethe, P. Kapitsa.* Nine of them were awarded the prize in the year of their nomination by N. Bohr.** It is an example of unparalleled scientific intuition.

* The laureates are listed in the order of prize-winning years.

** U. Rosenberg, S. 326–327.

[48] For example, the authors of the microscopic theory of superconductivity mentioned earlier in this chapter published it in 1957 but received the Nobel Prize only after fifteen years, in 1972. An even more striking case is the discovery of M. Planck's fundamental physical constant, the quantum of action. It was made in 1900, nominated for the Nobel Prize in 1907, and awarded only in 1919.*

* The Royal Academy of Sciences awarded two prizes in physics in 1919, one to M. Planck for 1918, the other to J. Stark for 1919.

[49] *Les prix Nobel, 1962,* Stockholm, 1963, pp. 58–59.

[50] By the efforts of Beria sanctioned by Stalin,* Kapitsa fell into disfavour for a long seven years. After the death of the latter and the arrest of the former, the Presidium of the USSR Academy of Sciences decided to take measures to help the scientist, who lived in exile in his country house near Moscow. The resolution dated 28.8.1953 stipulated the "creation of

the most favourable working and living conditions for Kapitsa including the refurbishing of his country house, its adequate heat and water supply, installment of telephone** etc., in compliance with the existing project."*** On 28.1.1953, Kapitsa was reappointed director of the Institute of Physical Problems, USSR Academy of Sciences, which he had created in 1935.

* According to A. A. Kapitsa, the scientist's wife, "he had to work closely with Beria in post-war time although they had for a long time had tense relations." General A. V. Khrulev, an intimate friend of the family, told her how "he once happened to be present in the private office of Stalin when Beria entered and started to persuade Stalin that Kapitsa should be arrested. Stalin said after a minute's silence: 'I shall dismiss him, but you don't touch him.'" See Dvadtsaty vek Anny Alekseevny. Dnevniki. Zametki. T 1, str. 346–347 (The twentieth century of Anna Alekseevna. Diaries. Notes. V 1, pp. 346–347)). M.: Izd-vo "Agraf," 2005.

** P. E. Rubinin, the assistant of Kapitsa for many years, told how one day in summer 1953, G. M. Malenkov, chairman of the USSR Council of Ministers, needed to have a talk with Kapitsa. He had to phone to a neighbouring settlement and wait till the scientist was found and had gotten access to the telephone set.

*** RGANI, fond 5, opis' 17, ed. khr. 412, l. 110.

[51] See note 2 to doc. 7.31 for nomination of Kapitsa by P. Dirac.

[52] *Nature*, 141, N0 3558, p. 74 (P. L. Kapitsa), p. 75 (J. Allen and A. D. Misener).

[53] I. M. Lifshits correctly guessed the source of difficulties related to awarding the prize to Kapitsa, as follows from the expert appraisal of the Nobel Committee for 1946 opened to the public in 1997 (see Chapter 4 and note 38 therein).

[54] P. E. Rubinin. History of one discovery in letters and document (on the 60th anniversary of the discovery of superfluidity). "Uspekhi fizichrskikh nauk", 167, No 12, 1997, pp. 1349–1360. Rubinin told the author that he had heard Kapitsa saying many times in private talks touching on the paper of Allen and Misener: "Let them give the prize to all the three [...]" The Georgian member of the USSR Academy of Sciences E. L. Andronikashvili evidenced that in the 1960s the candidacy of Kapiza (along with Allen) was supported by such an authoritative nominator as J. Bardeen (USA), winner of two Nobel Prizes in physics.*

* E. L. Andronikashvili. Nachinayu s El'brusa (I start from Mount Elbrus), Tbilisi, Izd-vo "Metsnieraba", 1982, p. 297.

[55] J. F. Allen. Early superfluidity in Cambridge, 1936 to 1939. "Advances in cryogenic engineering," V. 35, 1990, p. 35. See also P. E. Rubinin, p. 1357, note 36.

[56] Dirac emphasized in his nomination of Kapitsa for the 1946 Nobel Prize that Allen and Misener "were much less accurate in measuring very low viscosity of liquid helium II" while "Kapitsa made measurements in fixed conditions, and his data on zero entropy [...] were used to formulate principles underlying all the above anomalies."*

* Archive of the Royal Academy of Sciences. Fund of Nobel Committees in Physics and Chemistry. Materials of Committees for 1946, sheet 384.

[57] There were many talks about the negative role of N. S. Khrushchev, who allegedly stood against awarding the Nobel Prize to Kapitsa. It was said that during the visit of Tage Erlander to Moscow in spring 1956 Khrushchev recommended to award the prize to I. V. Kurchatov

instead of Kapitsa in a conversation with the Swedish prime minister. Such talks have no grounds whatever. Neither Erlander nor other statesmen had any chance to influence the Nobel organizations. The executive director of the Nobel Foundation Mikael Sulman once stated that not a single Nobel organization "ever received instructions from any side; least of all could it receive them from the Swedish government" (see doc. 10.16-II).

[58] The creation of masers and lasers, the first instruments that laid a basis for the development of the new field of physics and technology called quantum electronics, dates to 1955 and 1960 respectively.

[59] See the author's article "Zalozhniki (Hostages)" in the newspaper *Izvestiya,* 3.2.2000, p. 9 and doc. 7.36.

[60] Here is one such story. The elder son of a foreign correspondent for the newspaper *Izvestiya* was accommodated in a boarding house in Moscow for a few years. He graduated from a secondary school, was recruited to the Army, deserted with a stolen pistol, and killed a man to have his documents. After the younger son reached 12 years and could not be taken abroad, his mother refused to leave Moscow. The father had to retire.*
When the assistant naval attache in Great Britain went through formalities prior to leaving for the country of residence, one of the documents confirmed that his son was to stay in the boarding house of the Military Ministry and that he had mother, father, and sister living in the city of Odessa.** Initially, children could join their parents abroad during school holidays. Politbureau of the CPSU Central Committee cancelled this practice by its decision of 8.6.1950.***

    * RGASPI, fond 17, opis' 118, ed. khr. 384, l. 4.
    ** *Ibid.*, ed. khr. 595, l. 134.
    *** *Ibid.*, ed. khr. 907, l. 146.

[61] The existence of the institute of hostages was plainly recognized in secret office correspondence between Party functionaries. An employee of the Scientific and Technical Department, USSR Ministry of Cinematography, emigrated from the Soviet occupation zone in Germany to the British sector of Berlin in November 1949. The deputy minister N. I. Sakontikov was dismissed from his position and accused of disregarding the future runaway's statement that she "lived alone having neither permanent lodging nor relatives in the Soviet Union."*

    * RGASPI, fond 17, opis' 118, ed. khr. 948, l. 3–4.

[62] "Vestnik RAN", No 1, 2003, p. 69.

[63] Co-workers of Frumkin recollected that J. O'MacBokris (USA) nominated him for the Nobel Prize in 1966* and a group of scientists from DDR in 1977.**

    * In: "A. N. Frumkin. Ocherki, Vospominaniya. Materialy (A. N. Frumkin. Sketches. Memoirs. Documents)." M.: Nauka, 1989, p. 287.
    ** *Ibid.*, p. 251.

[64] Arkhiv RAN, fond 579, opis' 1, ed. khr. 640, l. 141–145.

[65] It appeared in the first place as launching a campaign in the mass media, inspiration of collective letters, etc.

[66] When a young specialist affiliated with a geological prospecting team working in Ryazanskaya *oblast'* (province), I encountered the most absurd situation. The staff of the

team expected a substantial rise in wages for successful results of work in 1952. We only needed to collect a certain amount of scrap metal preplanned by the superior organization. To fulfil the plan, we gave up part of the copper-core cable just received from the manufacturer for field power stations. Such economic absurdity was commonplace all over the country...*

* Samuel Beckett, Nobel Prize winner in literature and creator of the so-called theatre of the absurd once remarked: "Russia can't have theatre of the absurd for life there is absurdity."

[67] TsAODM, fond 7334, opis' 1, ed. khr. 9, l. 22.

[68] Soon after the meeting in April 1957, Kantorovich delivered a lecture at the philosophical and methodological seminar at the Mathematical Institute. V. S. Nemchinov, another eminent economist who shared the 1965 Lenin Prize with Kantorovich, was invited to participate. Nemchinov said that "the questions raised in the lecture are of interest for economists." It was decided that "Kantorovich would hand over his lecture to Nemchinov for a more detailed consideration."*

* *Ibid.*, l.63.

[69] L. V. Kantorovich. "Economic calculation of optimal utilization of resources." M.: Izd-vo AN SSSR, 1960 (preface by V. S. Nemchinov).

[70] In early January 1976, Kanorovich returned to Moscow from Sweden where he lectured at university centres.

[71] Item 6 in doc. 8.52 suggested annual workshops of Soviet nominators under the auspices of the Presidium of the USSR Academy of Sciences for "the development of agreed recommendations of Soviet scientists." The Swedish side was aware of this "know-how." See doc. 8.53 for its reaction.

[72] *Izvestiya*, 2.1.1991, p. 8.

[73] "O polozhenii v biologicheskoy nauke. Stenografichesky otchyot sessii VASKhNIL (Situation in biological science. Stenographic report of VASKhNIL session)." M.: Sel'khozgiz, 1948, . 410.

[74] Practically all opponents of the "people's academic" were dismissed from their positions in research institutions after the session of VASKhNIL and moved to work in other fields having little to do with science. None of them was arrested, however. At the same time, the critical attitude toward the defeat of biological science was used as an additional pretext to accuse some people imprisoned by the KGB in the framework of other political cases of the late 1940s. Geologists from Krasnoyarsk V. V. Bogatsky and V. N. Vereshchagin were arrested and accused under "the geologists' case" of 1949. When they were confronted, the interrogator demanded that Bogatsky "gave concrete evidence of his hostile activity." Bogatsky stated in reply: "I once said to Vereshchagin in a talk about Lysenko's report on the situation in biological science that researchers must have the right to freely express different views and choose by themselves areas of scientific works." Vereshchagin confirmed this evidence and added: "I don't conceal my anti-Soviet views."*

* Central archive of Federal Security Service of the Russian Federation. Sledstvennoye delo P-3, l. 230–231. It should be borne in mind that phrases in the minutes of

examinations were formulated by interrogators and are not necessarily authentic sayings of those on trial. This fully refers to the statement of Vereshchagin indicating that he regarded the critical attitude towards Lysenko and his actions as "anti-Soviet views".

[75] V. N. Voinovich. Antisovetsky Sovetsky Soyuz (Anti-Soviet Soviet Union). M.: Materik, 2002, p. 166. To add, Yu. A. Gastev mentioned in the preface seven names of those who "had the greatest influence on the author as far as he can judge."* Three of them had nothing to do with the subject matter of the book. One was the author's father, A. K. Gastev, expert in scientifically grounded organization of labour, two others, Cheyne and Stockes who lived and died in the 19th century. Gastev referred to their joint article in his monograph although both the article and the journal *Mind* in which it was published were contrived by the author. The reference gives March 1953 as the date of publication,** that is the month and the year of Stalin's death. Gastev's book was approved for printing by the Council on Cybernetics, USSR Academy of Sciences. It is unknown whether the reviewer or any member of the Council participated in the mistification...

* Yu. A. Gastev. Gomomorfizmy i modeli. Logiko-algebraicheskie aspekty modelirovaniya (Homomorphisms and models. Logico-algebraic aspects of simulation). M.: Nauka, 1975, p. 4.

** J. Cheyne and W. Stockes. "The breath and the death marks the rebirth of spirit." *Mind,* March 1953. Reference 55 on page 15 of Gastev's monograph pertains to the article related to the author's discourses on biological and psychological issues. In fact, all works of J. Cheyne (1777–1836) and W. Stockes (1804–1878) date from the 19th century; they had no joint publications (catalogue of the Library of Congress, Washington).

[76] Ideological structures of the CPSU Central Committee had to smooth pressure not only in science but also in the Arts. For example, publication of the novel *Rakovy korpus* by Solzhenitsin in the journal *Novy mir* in 1967 (see Chapter 11 and note 19 therein) was prohibited on the initiative of his brother — writers, in the first place K. A. Fedin known as a participant in the persecution of Pasternak (see Chapter 9). A. T. Tvardovsky, editor-in-chief of the journal wrote in his diary on 6.1.1968: "[...] now leading figures in the Party and government no longer meddle in such specific and delicate matters [...] they are happy to make no decision and even happier to see others taking negative decisions ("deside yourselves")."*

* *Znamya,* No. 8, 2003, p. 137.

Document 8.1

Interview of V. L. Ginzbirg, member of the USSR Academy of Sciences, 11 May 1999.

Question: You wrote about discussions in the USSR Academy of Sciences on the choice of nominees for the Nobel Prize in physics for 1957.[1] You argued that P. A. Cherenkov could not be nominated apart from I. E. Tamm and I. M. Frank and P. A. Kapitsa without L. D. Landau. Would you dwell at greater length on the subject?

Answer: In the mid-1950s, the leadership of the country decided to join the Nobel movement and nominate candidates for the forthcoming Nobel Prizes.[2] I. V. Kurchatov, member of the Presidium of the USSR Academy of Sciences, entrusted E. K. Zavoisky and me to prepare materials for nomination of Soviet scientists who discovered and interpreted the Vavilov–Cherenkov effect for the 1957 Nobel Prize in physics. The choice of Kurchatov was quite logical. I. E. Tamm was my teacher; also, I was the deputy head of his theoretical department at FIAN and one of his group involved in the development of the hydrogen bomb. Zavoisky was known as a brilliant experimenter. Nomination for the discovery of the new physical effect and its theoretical explanation required the knowledge of both a theorist and an experimenter.

In 1957, we came to know of rumours that higher organs decided to nominate only Kapitsa for the discovery of superfluidity and Cherenkov alone for the description of the above effect. I was not well aware of the situation around Kapitsa. What I knew was that some of his associates were dissatisfied and started preparation of a certain document. Probably, it was signed by M. A. Leontovich, member of the USSR Academy of Sciences having the right to nominate pretenders for the Nobel Prizes. Anyway, there were people resolute enough to express their attitude toward the actions of the leadership of the Academy.

E. L. Feinberg and I took the initiative in the case of nomination for the Vavilov–Cherernkov effect. Feinberg was also a pupil of I. M. Frank. We decided that we should not allow such crying injustice and prepared a draft letter to the Nobel Committee. The difficulty was to find a nominator since neither of us had the right to propose candidates, me being the corresponding member of the Academy and Feinberg doctor of sciences. We undertook to act in different directions. I applied to Landau although he did not appreciate Cherenkov radiation as a prominent scientific discovery, unlike I myself.[3] Landau consented to sign the nomination provided I change its closing portion in a sense that the prize should be awarded to all three scientists.

We corrected the text and Landau signed it. Feinberg managed to obtain signatures of N. N. Andreev and A. I. Alikhanov

We also thought about having the signature of the highly influential D. V. Skobeltsin. I explained the situation to him, and he agreed that it would be incorrect to nominate Cherenkov alone. However, he refused to sign the nomination, probably out of fear.[4] I do not blame him. He was a decent person. Very few were as brave in those times as to openly oppose the authorities [...][5]

I am unaware whether our nomination contributed in any way to the decision of the Nobel Committee to award the 1958 prize to Cherenkov, Tamm and Frank for the discovery and interpretation of the new effect.[6] I once asked Frank about it but did not receive an explicit answer. He seemed to be displeased with my question, and I never talked with him on this subject again.

Question: It appears from the declassified archive documents for 1956–1957 that the Department of Science and Education, CPSU Central Committee, treated Tamm and Landau as unacceptable opponents after they spoke at the meeting of the Division of Physico-Mathematical Sciences dedicated to the 50th anniversary of the theory of relativity (December 1955) and at the general session of the USSR Academy of Sciences (December 1956).[7] Was there any hidden political motive for their exclusion from the list of nominees for the 1957 Nobel Prize signed by A. V. Topchiev in January 1957?[8]

Answer: I do not know concrete facts. What I do know is both were never trusted by the authorities. Tamm was considered the first pretender for the vacancy of the full member of the Academy even in 1946. However, he was slandered in a letter to the AUCP(b) CPSU Central Committee, and Zhdanov crossed his name off the list of candidates recommended for election. He was elected only in 1953. He was then supported by Kurchatov, who referred to his contribution to the development of the hydrogen bomb. It is really a miracle that Tamm was not repressed and survived in Stalin's time. He had been affiliated with Mensheviks before the 1917 revolution, and the probability of his arrest was always very high. He told me that he had had at home a bag with essentials in anticipation of arrest throughout the period of "great terror."[9]

[1] V. L. Ginzburg. O fizike i astrofizike (On physics and astrophysics). M.: Byuro-Kvantum, 1995, p. 327–373.
[2] See doc. 7.6.
[3] Ginzburg confirmed Tamm and Frank's theoretical model of Cherenkov radiation suggested in 1937 from the standpoint of the quantum theory (1940).*
 * V. A. Kirillin. Vstrechi s interesnymi lyud'mi (Meetings of interesting people). M.: Nauka, 1994, p. 45; V. L. Ginzburg. "O nauke, o sebe i o drugikh (About Science, Myself, and Others)." M.: Fizmatlit, 2001, p. 76–77.
[4] Skobeltsin had more than one reason to be afraid. His brother, Professor Yu. V. Skobeltsin, was exiled from Leningrad to Saratov as "a socially dangerous element" in 1935 and then sentenced to eight years in a labour camp "for participation in a counter-revolutionary group and anti-Soviet propaganda." After release, he was twice refused rehabilitation by the special commission of the USSR Ministry of State Security. He was granted it only after his brother personally applied to G. M. Malenkov, secretary of the AUCP(b) CPSU Central Committee.* Hence, D. V. Skobeltsin's caution in situations that might bring him into conflict with higher authorities.
 * RGASPI, fond 17, opis' 119, ed. khr. 634 l. 89, 95
[5] Ginzburg was one of them. He demonstratively refused to sign the letter "against awarding the Nobel Peace Prize to A. D. Sakharov" drawn up by the Presidium of the USSR Academy of Sciences and signed by its 72 members. He unequivocally motivated his refusal by "personal considerations" (doc. 8.49).
[6] The newspaper *Pravda* commented on the decision of the Royal Academy of Sciences in the following words: "Vavilov–Cherenkov luminescence was long ago recognized in physics. The USSR Academy of Sciences recommended this study for the Nobel Prize."* The phrase sounds ambiguous because the Nobel Prizes are awarded to scientists and not to "studies". In all likelihood, the newspaper meant the nomination of Cherenkov alone in 1957. Ginzburg and Feinberg should be

given credit to for their initiative that resulted in the nomination of Tamm and Frank signed by Alikhanov, Andreev, and Landau.

* *Pravda*, 29.10.1958, p. 5.

[7] See docs. 8.3 and 8.4.

[8] See doc. 7.31.

[9] A similar denunciation was scribbled against Landau. I. A. Serov, KGB chairman, sent to V. A. Kirillin, head of the Department of Science, CPSU Central Committee, "a document containing data on anti-Soviet views and sentiments of Landau obtained from informants and secret agents."*

* "Arkhivy Kremlya i Staroy Ploshchadi (Archives of the Kremin and CPSU Central Committee", No. 3, 1993, p. 15.

Document 8.2

Materials on awarding the Stalin Prize to A. I. Shal'nikov, corresponding member of the USSR Academy of Sciences, in 1948.

I

Annotation of A. I. Shal'nikov's works submitted for the Stalin Prize in physico-mathematical sciences; from the document prepared by the Stalin Prize Committee for the AUCP(b) Central Committee and the USSR Council of Ministers.

A. I. Shal'nikov developed a micromethod for measuring magnetic fields in very narrow spaces. The method was used to demonstrate experimentally the layered structure of superconductors in an intermediate state, its formation and disappearance after transition of the specimen into the superconducting state upon a change of temperature and magnetic field... Favourable reports of A. F. Ioffe, member of the USSR Academy of Sciences, and Professor B. G. Lazarev.

RGASPI, fond 17, opis' 125, ed. khr. 586, l. 148.

II

Extract from the letter of A. A. Zhdanov, secretary of the AUCP(b) Central Committee, D. T. Shepilov, head of the Department of Propaganda and Agitation thereof, Yu. A. Zhdanov, head of the Department of Science, to I. V. Stalin; 27.5.1948.

The expert opinion about the works submitted for the 1947 Stalin Prize in sciences recommends that A. I. Shal'nikov, corresponding member of the USSR Academy of Sciences and head of laboratory at the Institute of Physical Problems, the USSR Academy of Sciences, "be awarded for the experimental study of superconductivity, the results of which are published in his papers "Structure of superconductors in an intermediate state" and "Surface phenomena in superconductors in an intermediate state" in 1946–1947. The work is submitted by the Institute of Physical Problems, Physical Institute, and Physico-Technical Institute, the USSR Academy of Sciences.

We think that the proposal of the Stalin Prize Committee should be turned down because the work of Shal'nikov continues the low-temperature studies for which the prize has already been awarded to Kapitsa (1943), Landau (1946), and Peshkov (1947). Shal'nikov did not make any further important contribution to low-temperature physics and only verified experimentally the calculations of Landau.[1] His work does not deserve the prize as having no significant theoretical and practical value.[2]

Ibid., 48.

III

Extract from the article of A. N. Nesmeyanov, member of the USSR Academy of Sciences and chairman of the Stalin Prize Committee in science and invention, the USSR Council of Ministers.

[...] The prize in physics is also awarded to Professor A. I. Shal'nikov, whose intricate experiments at temperatures close to absolute zero completely resolved the problem of a metal state intermediate between ordinary electroconductivity and superconductivity at very low temperatures.[3]

Pravda, 30.5.1948, p. 2.

[1] The phrase is ridiculous. Experimental confirmation of a theoretically predicted important physical phenomenon was always regarded as an important breakthrough into the nature of matter.

[2] This conclusion is altogether at variance with what was specially underscored by A. N. Nesmeyanov (doc. 8.2-III).

[3] That the USSR Council of Ministers, the prize-giving organization, acted contrary to the recommendation of the secretary and top officials of the CPSU Central Committee would be an unparalleled event were it not for certain accompanying circumstances. Yu. A. Zhdanov, heading the Department of Science, was educated as chemist. Speaking at the All-Union Seminar for lecturers of the regional Communist Party organizations, he severely criticized the anti-scientific views of Lysenko (10.4.1948). Stalin was infuriated and ordered (at the sitting of Politbureau on 12.4.1948) to set up a commission headed by Malenkov to investigate the case.* In the end, Malenkov replaced A. A. Zhdanov in the position of secretary of the CPSU Central Committee and Shepilov had to retire a few months later. Probably, the shaky situation of all three authors of the present document in May 1948 encouraged the government to include Shal'nikov in the resolution awarding the Stalin Prizes for 1947 (29.5.1948). Nevertheless, the refusal to award the prize to I. I. Shmal'gauzen (an opponent of Lysenko) mentioned in the same document** remained in force. It could not be otherwise at the time when Lysenko supported by Stalin was preparing the ruin of Soviet genetics. T. D. Lysenko was deputy chairman of the Stalin Prize Committee and presented his "separate opinion". He stated that the book *Factors of Evolution* by Shmal'gauzen "does not deserve the prize named after comrade Stalin" because "Shmal'gauzen based it on the theories of Mendel,

Morgan, Dobrzhansky and the like and did not criticize bourgeois outlooks in biological science."***

* D. T. Shepilov recollected that Stalin said at that sitting: "The guilty should be punished. Yuri Zhdanov is still young and inexperienced. 'Fathers' must be punished: Zhdanov (he pointed to Andrei Aleksandrovich with the tip of his pipe) and Shepilov." "Voprosy istorii", No 8, 1998, p. 10.

** RGASPI, fond 17, opis' 125, ed. khr. 586, l. 57.

*** *Ibid.*, l. 137–138.

Document 8.3

From: V. A. Kirillin, head, Department of Science, Higher Educational Institutions and Schools, CPSU Central Committee; N. I. Glagolev, deputy head; A. S. Monin, instructor.
To: Secretariat of the CPSU Central Committee.
16.1.1956

Division of Physico-Mathematical Sciences held a scientific session dedicated to the 50th anniversary of the theory of relativity (30 November-1 December 1956). Serious mistakes were made during the organization and holding of the session.
The program of the session was drawn up by the organizing committee comprising Tamm and Landau, full members of the USSR Academy of Sciences, Ginzburg, its corresponding member, and Professor Lifshits[1] who do not work in the field of relativity and are known for their nihilistic attitude toward methodological aspects of this theory.[2] The program contained only one report of the specialist in the theory of relativity (corresponding member Mikhailov). The lecture by Fok, full member of the academy and the most authoritative expert in methodological problems was not included, same as reports of researchers from the Moscow State University with their many years of experience in the field of relativity.
The Department of Science, Higher Educational Institutions and Schools, CPSU Central Committee, pointed out these drawbacks to comrade Nesmeyanov, president of the USSR Academy of Sciences. The program of the session was extended. Reports of Landau, Ginzburg, Mikhailov, and Aleksandrov were scheduled for the first day, reports of Fok, Subbotin, Lifshits and Shirokov for the second. However, this plan failed. Aleksandrov, living in Leningrad, was informed only in the evening preceding his lecture and could not arrive in time. Tamm, chairing the session, suggested that Lifshits spoke instead of Aleksandrov. Therefore, the program of the first day was virtually that initially proposed by the organizing committee for the entire session. Many scientists including Landau and his pupils did not come on the second day.
The organizing committee collected abstracts of reports from certain speakers but not from Lifshits. There are no stenographic records of the session. These are serious omissions, the more so that the report of Lifshits contained ideological mistakes. The author overviewed publications on relativistic cosmology and popularized the

idealistic "theory of expanding universe."[3] This "theory" stems from the unjustified extension of non-stationary solutions of Einstein's gravitational field equations to the entire universe; it was developed by abbot Lemetre on the initiative of the Pope. According to the "theory", the universe has a finite age; it had a negligibly small volume at the moment of birth but started to expand thereafter; the expansion still continues.

Lifshits did not even mention the numerous studies of Soviet scientists criticizing the "expanding universe theory" and offering a correct interpretation of cosmological problems (works of Fesenkov, Bogorodsky, Zel'manov, Ogorodnikov, Fok, and other authors). Lifshits represented this theory as a sound scientific concept allegedly confirmed by experimental data. He ignored the existing publications on hierarchical distribution of matter in the universe (specifically, on the structure of metagalaxies) and argued that matter is spread throughout the universe with a certain mean density as postulated by the above "theory". He stated that this density may serve as an indicator of whether the "expanding universe" has a finite or infinite volume. His computations suggest that our universe is in the intermediate situation between the two options.

Also, Lifshits calculated the age of the universe that he called "characteristic time". He argued that correlation of this age with the age of the Earth's geological rocks determined by the radioactive method is not a mere coincidence and confirms the validity of the "expanding universe theory". As a matter of fact, it demonstrates nothing but the inanity of such estimation not only for the metagalaxy but also for individual galaxies. Nevertheless, the statement of Lifshits was supported by Zel'dovich, corresponding member of the USSR Academy of Sciences, according to whom not only geological rocks on the Earth but also all chemical elements are that old. The expression "the time from the moment when the trigger was pulled" was used. Lifshits did not mention restricted application on the "models" in question nor did he clarify their ideological meaning.

The program of the session did not stipulate discussions. However Professor Fok considered it necessary to comment on the report of Lifshits and indicated that his "models" should not be understood in a literal sense and can be used only for the description of the "background" gravity field in certain finite parts of the universe. Other specialists (astronomer Parenago, corresponding member of the USSR Academy of Sciences, philosopher professor Kuznetsov) confirmed the idealistic contents of Lifshits' lecture.

Reports by Landau and Ginzburg were not free from drawbacks either. Landau expressed his personal negative attitude toward the works of Einstein, his unified field theory, and criticism of the logical basis of quantum mechanics. The perfunctory report of comrade Ginzburg was devoted to experimental verification of the theory of relativity; its scientific level was low[4] which became especially obvious after the lectures of comrades Mikhailov and Subbotin on the same subject. After comrade Fok finished his lecture, Ginzburg started an argument with him to criticize his conclusion about advantages of Copernicus' system over that of Ptolemy.

This is not the only case when the Presidium of the USSR Academy of Sciences did not pay proper attention to the organization and holding of workshops by the Division of Physico-Mathematical Sciences. Serious mistakes were made during the All-Union conference on quantum electronics and the theory of elementary particles in spring 1955. These drawbacks were discussed and analysed by Bureau of the Division of Physico-Mathematical Sciences but the Presidium disregarded the results of the discussion and took no measures to prevent them in the future.
We deem it necessary to point out to the Presidium of the USSR Academy of Sciences that it should give more attention to the preparation of workshops by the Division of Physico-Mathematical Sciences and suggest that the Presidium organize discussion of the ideological mistakes contained in the report of Lifshits at a sitting of the Bureau.
We are asking for your consent.

Note in the left part of page 1: "Agree. Attn. comrade Kirillin: please do it in one form or another [...] (signature illegible). D. Shepilov. 29.1.58." Signatures of P. Pospelov, M. Suslov, A. Aristov, N. Mukhitdinov below.

RGANI, fond 5, opis' 35, ed. khr. 19, l. 70–73

[1] E. M. Lifshits, future member of the USSR Academy of Sciences and brother of Lifshits mentioned earlier.
[2] "Nihilistic attitude of Landau and E. M. Lifshits toward methodological aspects" was recalled in five years, when their world-renowned *Course of Theoretical Physics* was nominated for the Lenin Prize of 1962. V. A. Kirillin, head of the Department of Science, Higher Educational Institutions and Schools, CPSU Central Committee, wrote to its Secretariat (13.4.1961): "Methodologically, the 'Course' of Landau and Lifshits has serious drawbacks. It fully ignores philosophical aspects of modern physics, formally treats physical phenomena, etc. [...]"* That time, these primitive arguments were disregarded, probably because Landau was in a car accident. The two authors received the highest scientific award of this country.

* RGANI, fond 5, opis' 35, ed. khr. 157, l. 38.

[3] Ideologists started to attack the theory of expanding universe some twenty years before the appearance of this document. The cited passage almost literally reproduces a phrase from the article published in the theoretical journal of the AUCP(b) Central Committee "Pod znamenem marksizma" in 1938: "The theory of the expanding universe [...] gives food for ideological reactionaries." Even the small letter in the word "universe" is retained...*

* "Pod znamenem marksizma", No 7, 1938, p. 126.

[4] Ginzburg recollected events of that day in a quarter of a century: "[...] I am recalling the sitting of the Division of Physico-Mathematical Sciences, USSR Academy of Sciences, held on 30 November 1955 in the conference hall of the Presidium. It was dedicated to the 50th anniversary of the theory of relativity and the memory of Einstein, who died on 18 April 1955. I. E. Tamm opened the sitting with a short

welcoming address. A few reports followed, including mine (on experimental verification of the general theory of relativity). Landau's concluding lecture was on the life and works of Einstein. It impressed the audience, but I remembered especially well what Landau said about "the tragedy of Einstein" in the last period of his life. First, he did not "accept" quantum mechanics, did not understand it as many believed. Second, he spent much time developing the unified field theory but did not succeed either."*

* V. L. Ginzburg. O fizike i astrofizike (On physics and astrophysics). M.: Byuro-Kvantum, 1995, p. 365.

Document 8.4

From: V. A. Kirillin, head, Department of Science, Higher Educational Institutions and Schools, CPSU Central Committee.
To. CPSU Central Committee.
10.1.1957

The general session of the USSR Academy of Sciences was held on 28 December 1956. The Presidium of the Academy limited participation to its full and corresponding members, presidents of the Academies of the Union Republics.[1]
Comrade Nesmeyanov, president of the Academy, made a report on the results and prospects of its activity.
Twenty-eight participants took part in the discussion of the report... The letter of the CPSU Central Committee emphasizing the necessity to strengthen political work of local Party organizations and cut short attacks of hostile anti-Soviet elements was read to the members of the Communist Party participating in the session prior to its opening...[2] Most speakers raised interesting and important questions pertinent to the further development of Soviet science. Some told about a lag in our biological science. Comrades Engel'gardt and Tamm emphasized that biology acquires the key role in modern natural sciences pending fundamental discoveries in such fields as photosynthesis, genetics, protein biochemistry, and biophysics. According to the speakers, biology will be a leading scientific discipline in the second half of the 20th century as atomic physics was during its first half. Meanwhile, some branches of biology, biophysics, and biochemistry are still poorly developed in this country.[3]
[...] It is worthwhile to note that Professors Artsimovich, Tamm, and Ryzhkov advanced some wrong ideas. Artsimovich misquoted Engels in his speech and said: "Someone remarked that materialism must acquire a new form with each new discovery in natural sciences. It might be a metaphor, but it is the right view in essence." According to him, the theory of relativity and quantum mechanics "radically changed fundamentals of scientific outlook" and it is necessary now "to formulate new notions and philosophical concepts (e.g. in application to physics), to offer new interpretation of such problems as causality, limitation of physical experiments,

and relationship between natural laws and logical construction of relevant theories, etc." Also, Artsimovich stated that mutual understanding between physicists and philosophers was hampered by "conciliatory attitudes" of the latter ("they are too often guided by timeserving considerations").
Comrade Tamm argued that "guiding ideas and fundamental discoveries in physics are still coming from abroad."[4] He supported Artsimovich in finding faults with Soviet philosophers ("our philosophers have only a compass but have no means of travel"). Speaking on the teaching of Lysenko [...], Tamm stated: "What academician Lysenko calls Michurinian materialistic biology is to my mind a dismal caricature of dialectic materialism." The audience applauded his statement. Finally, Tamm demanded that the USSR Academy of Sciences condemn the decisions of the VASHNIL session: "We must honestly and straightforwardly say that what was spoken at the session of VASHNIL in 1948 has nothing to do with reality."[5] Ryzhkov, corresponding member of the Academy, supported this statement saying that the decisions of the session of VASHNIL "cancelled all real biological knowledge accumulated by thousands of researchers [...]"
It should be noted that comrade Artsimovich said in a conversation held at the Department of Science, Higher Educational Institutions and Schools, CPSU Central Committee, to discuss the work of the Division of Physico-Mathematical Sciences that his speech at the general session of the USSR Academy of Sciences needs clarification and that he regards dialectic materialism as the basis of the scientific outlook.

RGANI, fond 5, opis' 35, ed. khr. 48, l. 2–5.

[1] Hungarian events (23 October-4 November 1956) with the invasion of Soviet troops caused latent protest of Soviet intellectuals. The CPSU Central Committee expected criticism of the government's actions by members of the USSR Academy of Sciences and recommended to hold its general session "behind closed doors"; no information about it could be found in the central press.
[2] This "closed letter" of the CPSU Central Committee was read only to members of the Communist Party; it concerned public reaction to the participation of the Soviet Army in the Hungarian events.
[3] As a result of the crush of genetics in this country in 1948 and the resulting dominance of Lysenko and his associates.
[4] Here is an extract from the stenographic record of Tamm's speech: "It was correctly noted that neither a single elementary particle nor a transuranium element was discovered [in the Soviet Union]. Our science actually develops at the level of the world standards. But unfortunately, we must recognize that major discoveries and physical ideas determining the tendency of further research are still coming from abroad, fundamental discoveries are coming from abroad. (I mean even such ones that are reported every two or three years and determine the tendency of work for the near future, to say nothing about fundamental ideas determining the development of science for a century). We assimilate them very quickly [...] but the creative initiative takes its origin abroad."*

* RGANI, fond 5, opis' 35, ed. khr. 48, l. 134

[5] Tamm always remained a man of independent thinking. He did not hesitate to express his civil sentiments when Warsaw Pact countries oppressed the democratic movement in Czechoslovakia in August 1968. When film director M. I. Romm drew up a collective letter in protest against these actions, "thorough and emotional", Tamm visited him on the third day of the invasion (evening of 23 August) and signed the letter.*

"Svobodnaya mysl'", No 8, 2001, p. 62.

Document 8.5

TASS information message "A Swedish newspaper about the forthcoming award of the Nobel Prize) dated 25.10.1958 for M. A. Suslov, secretary of the CPSU Central Committee; typewritten document, sheets 7–10.[1]

New York, 25 October. Associated Press correspondent reports from Stockholm:
It is said that the Nobel Prize in physics for 1958 will be awarded to three eminent Soviet atomic scientists for their discovery made use of on the third Soviet artificial Earth satellite.[2] According to the newspaper *Aftonbladet* (Stockholm), the first Nobel Prize in physics ever given to Soviet scientists will be shared by Cherenkov, Tamm, and Frank. They all are outstanding representative of Soviet nuclear physics. They receive the Nobel Prize for their pioneer investigations into the nature of the "blue glow" produced by a nuclear reactor submerged in water.[3] If this report is confirmed by the Swedish Academy of Sciences at its sitting of 28 October, it may somewhat quell the rage of Soviet authorities provoked by awarding the Nobel Prize in literature to the writer Boris Pasternak earlier this week.
The discovery of the blue glow, called the Cherenkov effect by the three Russian atomic physicists, turned out to be of greatest importance for this profession throughout the world. The Russians used the Cherenkov effect in a new type of cosmic ray counter installed on their third artificial satellite still orbiting the Earth.

The document was registered in the General Department, CPSU Central Committee, on 27.10.1958. The note above the text reads: "Attn. Comrade Kirillin. I need a talk with you. M. Suslov"; dateless.[4]
The note below: "Fulfilled, 5.XI.58" (signature illegible).[5]

RGANI, fond 5, opis' 35, ed. khr. 70, l.205.

[1] Such messages (the so-called "tassovki") were prepared for administrative use by decision-making authorities and never appeared in the open press.
[2] The third Soviet artificial Earth satellite was launched 15.5.1958.
[3] Cherenkov luminescence discovered and explained long before the appearance of nuclear reactors.
[4] The talk might have taken place on October 27, the day before the Royal Academy of Sciences officially announced its decision.

[5] The note of fulfillment probably referred to the press conference of the Nobel laureates organized by the USSR Academy of Sciences for Soviet and foreign journalists on November 4.

Document 8.6

Extract from the article in *Pravda* of 29.20.1958 signed by I. V. Kurchatov, N. N. Semenov, A. V. Topchiev, A. P. Aleksanadrov, A. F. Ioffe, V. A. Fok, full members of the USSR Academy of Sciences, and B. M. Vul, corresponding member; p. 5.

Vavilov–Cherenkov luminescence was long ago recognized in physics. The USSR Academy of Sciences nominated this work for the Nobel Prize...[1] The Swedish Academy of sciences awarded the Nobel Prize in physics for 1958 to Soviet scientists and thereby reflected the opinion of scientific community that recognized the primary importance of the discovery of our physicists [...] In light of this fact [...] the decision to give the Nobel Prize in literature to Pasternak for his slanderous representation of life in this country looks especially tendentious [...] This decision is dictated by political considerations. In connection with this, it is in order to recall V. I. Lenin's utterance that bourgeois scientists can be objective so far as factual[2] special studies are concerned but see public events including literary works entirely from the standpoint of ruling class ideology [...] It was the scientific value of the work that influenced its choice for awarding in the first case whereas reactionary political considerations prevailed in the latter [...]

[1] The expression "this work" substituting the names of concrete scientists stems from the fact that no nomination for the Nobel Prize in physics for 1958 passed via the Foreign Department, USSR Academy of Sciences. The Academy nominated Cherenkov alone for his study in 1957 (see docs. 7.31 and 8.1).
[2] "factual" instead of "factologic." The scientists who signed the article could not make such a mistake (if it is not a mere misprint) which suggests that the phrase was included in the text by someone else (see also note 13 to this chapter).

Document 8.7

"Memorandum" signed by A. F. Sokolov, Department of Scandinavian States, USSR Ministry of Foreign Affairs; 17.11.1958

In connection with the award of the Nobel Prize to Soviet scientists Cherenkov, Tamm and Frank, comrade Vinogradov, head of the Foreign Department, the USSR Academy of Sciences, informed me by phone on November 17 about the following:

1. The USSR Academy of Sciences organized a press conference of Soviet Nobel Prize winners for Soviet and foreign journalists (November 4).

2. Soviet scientists sent a cable to the Swedish Academy[1] thanking it for awarding the Nobel Prize (November 6).
3. Soviet scientists informed the Swedish Academy that they would arrive in Stockholm to receive the prize (November 13).

Below: the note in pencil by Sokolov dated 17.11.1958: "The embassy is informed".
AVP, fond 0140, opis' 47, papka 70, ed. khr. 37, l. 88.

[1] Royal Academy of Sciences

Document 8.8

From: Rolf Sulman, Ambassador Extraordinary and Plenipotentiary of Sweden to the USSR.
To: Laureates of the Nobel Prize in physics for 1958.
18.11.158

Dear Mr. Academician,

Being aware that you were awarded the Nobel Prize in physics for 1958 and that you intend to personally receive it in Stockholm, I have the honour to congratulate you on this high award and inform you that the Swedish embassy is naturally prepared to render you practical assistance that you may need in connection with your forthcoming visit and the stay in Stockholm.
Arkhiv RAN, fond 579, opis' 1, ed. khr. 639, l. 98.

Document 8.9

From: Rolf Sulman, Ambassador Extraordinary and Plenipotentiary of Sweden to the USSR.
To: K. K. Rodionov, head, Department of Scandinavian States, the USSR Ministry of Foreign Affairs.
25.11.1958

I should appreciate your transferring the attached letter to Mr. Ambassador Gusev[1]; also find please a copy for yourself.[2]
I would like to bring it to your attention that the director of the Nobel Committee[3] emphasized in a telephone conversation with me that it is highly desirable that the Nobel laureates inform him at their earliest convenience whether they all intend to arrive in Stockholm to personally receive their prizes, also whether their wives or some of their associates plan to accompany them.
AVP RF, fond 0140, opis' 47, papka 70, ed. khr. 37, l. 83.

[1] Ambassador Extraordinary and Plenipotentiary of the USSR to Sweden F. T. Gusev.
[2] See doc. 8.10.
[3] See note 2 to doc. 8.10.

Document 8.10

From: K. K. Rodionov, head, Department of Scandinavian States, the USSR Ministry of Foreign Affairs.
To: S. G. Korneev, deputy head, Foreign Department, the USSR Academy of Sciences.
28.11.1958

Attached find please the two-page letter of R. Sulman dated 25.11.1958 and addressed to F. T. Gusev, ambassador of the USSR in Sweden, that was received by the Department of Scandinavian States, USSR Ministry of Foreign Affairs. It concerns the prize-giving ceremony for the Soviet scientists Cherenkov, Frank, and Tamm.

Copy.
Royal embassy of Sweden
Moscow, 25.11.58

Dear Mr.Gusev,

In connection with my letter of yesterday to you,[1] I have the honour to bring to your notice that I received additional information from the Nobel Committee concerning the program of Nobel festivities.

Tuesday, 9 December
15.30–17.00 Reception by riks-marshal Ekeberg, chairman of the Nobel Committee.[2]
Evening. Probably dinner given by Professor Ivar Waller for the laureates and their wives.

Wednesday, 10 December
11.30 Preparation for the prize-giving ceremony.[3]
16.00 Awarding of the Nobel Prizes in the concert hall.

Thursday, 11 December
16.00 "Nobel lecture" of Nobel Prize laureates in physics.
Evening. Dinner for the laureates with His Majesty the King of Sweden.

Saturday, 12 December
Evening. Professor Manne Siegbahn invites the laureates and their wives for dinner.

To my knowledge, heads of respective embassies usually invite laureates from their countries for breakfast, sometimes dinner, at a time convenient for them.
Swedish Ministry of Foreign Affairs deputes its following officers to accompany the Soviet laureates: attache Chellen for Professor Cherenkov, attaché Cornell for Professor Frank, attaché Willen for Professor Tamm.
To conclude I would like to inform you that, according to Professor Rutbek, Chalmers Technical University, Goteborg, and University of Goteborg invite the laureates in physics for a visit and delivery of lectures on 14–15 or 17–18 December, after the

prize-awarding ceremony in Stockholm. Finally, the dinner with Professor Westgren, president of the Swedish Academy of Sciences, scheduled for Tuesday, 9 December, will most probably be held at Professor Waller's, because madam Westgren is ill.
Yours sincerely
Rolf Sulman
Ambassador of Sweden in the USSR

Note: The Sweden–USSR Society schedules its meeting with the participation of Soviet scientists on December 14 at 20.00. The USSR embassy in Sweden deems advisable participation of the Soviet scientists.

Arkhiv RAN, fond 579, opis' 1, ed. khr. 639, l. 102–104.

[1] See doc. 8.9 also dated 25.11.1958. The two dates are most likely the dates of dispatch of both letters with the morning mail. The former letter was probably written in the evening 24.11.1958.
[2] Chairman of the executive board of the Nobel Foundation.
[3] Rehearsal of the ceremony, see note 3 to doc. 7.29.

Document 8.11

Memorandum of A. Ryl'nikov, Department of Scandinavian States, the USSR Ministry of Foreign Affairs; 26.11.1958

S. G. Korneev, deputy head of the Foreign Department, USSR Academy of Sciences, told me by phone with reference to the letter of Sulman addressed to comrade Rodionov K. K that Professors Cherenkov, Frank and Tamm are going to Stockholm by plane on December 7. Cherenkov goes with his wife, but the wives of Tamm and Frank are ill and will not accompany them. Comrade Korneev asks to inform Sulman about it. He also told that the USSR Academy of Sciences applied to the higher instance in connection with this visit.[1]

The note below: "The Soviet scientists informed the Swedish Academy[2] by telegraph about their intention to arrive in Stockholm to receive the prize. Rodionov."

AVP RF, fond 0140, opis' 47, papka 70, ed. khr. 37, l.86.

[1] To the CPSU Central Committee. The visit was sanctioned between November 26 and December 3, 1958 (see note 2 to doc. 8.1. 20).
[2] Royal Academy of Sciences.

Document 8.12

From: M. V. Nikitin, business-manager of the USSR Academy of Sciences.
To: K. K. Rodionov, head, Department of Scandinavian States, the USSR Ministry of Foreign Affairs.
26.11.1958

Members of the USSR Academy of Sciences Tamm I. E with his son, Frank I. M. with his son,[1] and doctor of sciences P. A. Cherenkov accompanied by his wife leave Moscow for Stockholm on the 7th of December to receive the Nobel Prize awarded to them. The Management Department, the USSR Academy of Sciences, asks you to instruct the Soviet embassy in Sweden to assist in hiring five dress coats (two of size 50 and three of size 46, 52, and 54 respectively). It proved impossible to buy the dress coats in Moscow. I would appreciate you informing the embassy about this request before the scientists leave Moscow.

Note in the top left corner: "Attn. comrade Murzina. The USSR Academy of Sciences asks for dress coats of size 54 (for Cherenkov), 52 (for Tamm), and 50 (for Frank)."[2] 3.12.1958 (signature illegible).

Note below the text: "It is agreed with Nesmeyanov that no telegram needs to be sent."[3] 4.12.1958 (signature illegible).

AVP RF, fond 0140, opis' 47, papka 70, ed. khr. 37, l. 87.

[1] The "higher instance" did not grant permission to the sons of Tamm and Frank to accompany their fathers at the prize-giving ceremony (see note 2).

[2] It appears that the visit of Tamm and Frank to Stockholm together with their sons was initially approved by the CPSU Central Committee (probably, at the level of the Department of Science), otherwise they would not be mentioned in this document. However, the Secretariat of the Central Committee did not grant permission at one of its sittings between November 28 and December 3. In 1956, the CPSU Central Committee already refused such permission to the daughter of N. N. Semenov, who was accompanied only by his wife.*

*See doc. 7.36

[3] E. P. Cherenkova, daughter of the Nobel laureate, told the author that her mother was acquainted with the wife of the former Soviet ambassador in Sweden who told her that the dress coats were easy to hire in Stockholm; therefore, there was no need to send a telegram to the embassy.

Document 8.13

Nomination signed by A. N. Nesmeyanov, president of the USSR Academy of Sciences, and A. V. Topchiev, its chief scientific secretary, that was sent to the Nobel Committee in physics on 27.1.1959.

The USSR Academy of Sciences proposes eminent Soviet physicists Vladimir Iosifovich Veksler, full member of the Academy, and Evgeniy Konstantinovich Zavoisky, its corresponding member, for the Nobel Prize in physics in 1959 for their outstanding contribution to science.[1]

Professor Veksler born 1907 into a Jewish family in the city of Zhitomir, Ukraine, graduated from the Moscow Energy Institute, doctor of physico-mathematical sciences, professor, head of the Laboratory of High Energies at the International Institute of Nuclear Research.[2] V. I. Veksler is a known Soviet specialist in nuclear physics. He suggested the principle of autophasing of paramount importance for the development of accelerators of protons and electrons.

All high-energy accelerators currently operated and built in the world are based on the principle of autophasing discovered by Veksler. His priority in this discovery is recognized by scientists working in the field of high energies in the USA and UK. The outstanding discovery of V. I. Veksler gives grounds to nominate him for the Nobel Prize in physics in 1959.

Professor Zavoisky was born in 1907 in the city of Mogilev. He is Russian, graduated from the University of Kazan, doctor of physico-mathematical sciences, professor, head of a sector in the laboratory of Measuring Devices, Institute of Atomic Energy, the USSR Academy of Sciences. He was awarded the Lenin Prize in 1957 for the discovery and investigation of paramagnetic resonance. Professor Zavoisky is a highly experienced experimenter, doing research for over 20 years.[3] His prominent discovery of paramagnetic resonance substantially extends the possibilities for investigations into the structure of atoms, molecules, liquids, and solids. Recently, phenomenon of paramagnetic resonance has found practical application for generation and acceleration of radio waves. International significance of the discovery of Professor Zavoisky gives reason to nominate him for the Nobel Prize in 1959.

Arkhiv RAN, fond 579, opis' 1, ed. khr. 639, l.108–109.

[1] Each prize is given for one study or two studies of similar design or using common methods. For example, the Nobel Prize in physics for 1978 was shared between P. L. Kapitsa and the American scientists A. Penzias and R. Wilson who studied the behaviour of material particles at temperatures close to absolute zero. The above nomination concerned two unrelated discoveries.

[2] Joint Institute of Nuclear Research in Dubna, Moscow region.

[3] This information in the nomination for the Nobel Prize could provoke nothing but bewilderment because the award is given for concrete scientific results rather than the length of the nominee's service.

Document 8.14

From: P. A. Cherenkov, corresponding member of the USSR Academy of Sciences and Nobel Prize winner in physics.

To: S. G. Korneev, head, Foreign Department, the USSR Academy of Sciences.

January 1965

Please instruct your staff to send the attached materials to the Nobel Committee in physics. The materials contain the nomination of V. I. Veksler, E. McMillan, and N. Kristofilos for the Nobel Prize in physics for 1965. Nomination of these scientists was approved by the president of the USSR Academy of Sciences M. V. Keldysh. The Nobel Committee does not consider proposals received after the 31st of January.

It is therefore important that the attached materials are dispatched as soon as possible. I also ask for your permission to send the attached telegram to the Nobel Committee.[1]

Below the text: "Attn. Comrade Pochitalin. Please send these materials urgently by airmail. President is informed and comrade Ilyichev L. F. gave his consent.[2] 22.1. Korneev."

Arkhiv RAN, fond 579, opis' 1, ed. khr. 640, l. 144.

[1] The text of the telegram is not published here.
[2] Secretary of CPSU Central Committee.

Document 8.15

From recollections of S. V. Vonsovsky, member of the USSR Academy of Sciences, about E. K. Zavoisky (1993).

[...] After I was elected full member of the USSR Academy of Sciences in 1966, I began to annually receive letters from the Nobel Committee in physics, Stockholm, with invitations to nominate candidates for the Nobel Prize for the next year. I used to regularly send nominations of Evgeniy Konstantinovich for the discovery of EPR from 1967 till 1976[1]. I always nominated all three pioneers of the discovery: E. K. Zavoisky, I. Waller (Sweden), and K. Gorter (Netherlands). This is my first nomination of E. K. Zavoisky dated from 1967: "In response to your kind invitation to propose a candidate for the Nobel Prize in physics for 1968, I would like to recommend the world-known Soviet physicist E. K. Zavoisky, member of the USSR Academy of Sciences, the key author of the experimental discovery of the very important physical phenomenon, electron paramagnetic resonance (EPR) made in 1944. The main reports of this discovery were published in the *Zhurnal fiziki SSSR* [...] in 1944 and 1945;[2] they are referred to in monographs where all possible applications of EPR are discussed. Today, EPR techniques are widely employed in all natural sciences, namely in theoretical and applied physics, chemistry, and molecular biology. In many cases, information obtained by these methods provides the unique possibility to study electron structure, bonding forces, and physical mechanisms of various processes at the atomic and molecular levels. As EPR is becoming a routine method, we tend to forget pioneers who discovered this secret of nature and thereby opened up a possibility to use it for the benefit and welfare of people. I therefore believe that awarding the Nobel Prize to E. K. Zavoisky would be a just and objective appraisal of his contribution to the development of modern science.
[...] Unfortunately, my efforts did not produce any positive effect. When I tried to clarify the reason, I was told that the Nobel Prizes for resonance studies had already been awarded to I. Rabi in 1944, F. Bloch and E. Purcell in 1952. This argument does not convince me because those prizes were given for nuclear physics and nuclear resonance whereas EPR is an original efficient method for physics, chemistry, and

biophysics of electron systems in the condensed phase applicable to quite a different area of research than the methods of the above authors.

"Charodei eksperimenta, Sbornik vospominaniy ob akademike E. K. Zavoiskom (The wizard of experiment. Collected recollections about academician E. K. Zavoisky)." M.: Nauka, 1993, pp. 22–23.

[1] The year of Zavoisky's death.
[2] Journal of Physics USSR, 8, No. 6, 1944, pp. 377–380; 9, No. 3, 1945, pp. 211–216.

Document 8.16

From: A. G. Skvortsov, secretary of the city Communist Party committee, Dubna.[1]
To: CPSU Central Committee.
17.10.1958

In April 1958, N. N. Bogolyubov, director of the Laboratory of Theoretical Physics, Joint Institute of Nuclear Research, was awarded the Lenin Prize for the development of methods in the quantum field theory and statistical physics that helped to explain mechanisms of superfluidity and superconductivity. President of the USSR Academy of Sciences Nesmeyanov characterized N. N. Bogolyubov as the developer of the theory of superconductivity in his article about the new laureates published on 22 April 1958. Scientific Council of the Joint Institute of Nuclear Research highly appraised these works of Bogolyubov. Moscow State University awarded the Lomonosov Prize to him in 1958.

These facts indicate that Bogolyubov made an important contribution to the study of superconductivity. Certainly, the solution of this complicated problem required efforts of scientists from many countries who studied some of its aspects in parallel and independently. In the light of the aforesaid, it looks strange that the editorial board of the journal *Uspekhi fizicheskikh nauk* published the article of A. Abrikosov and I. Khalatnikov "Modern theory of superconductivity",[2] in which they give American researchers alone credit of developing the superconductivity theory.[3] Scientists of the Joint Institute of Nuclear Research fairly well knowing the problem are convinced that publication of the paper by Abrikosov and Khalatnikov is a manifestation of narrow group interests and does much harm to the prestige of Soviet science.

RGANI, fond 5, opis' 35, ed. khr. 70, l. 194–195.

[1] The city of Dubna, Moscow region, location of the Joint Institute of Nuclear Research.
[2] *Uspekhi fizicheskikh nauk*, 65, vyp. 4, 1958, pp. 551–591.
[3] Deliberate misinterpretation of facts; the authors of the article only emphasized that J. Bardeen, L. Cooper, and J. Schrieffer independently developed the complete theory of superconductivity and published it earlier than the Soviet scientist. The history of science is rich in such cases. Suffice it to recall a dramatic collision around the discovery of the molecular structure of nucleic acids (the famous "double

helix"). It was made in the situation of strong competition between L. Pauling and the Anglo-American group comprising F. Crick, J. Watson, and M. Wilkins. The latter won the race and received the Nobel Prize in 1962. The interval between publications of the final results was much shorter* than in the case of Bogolyubov and Bardeen's group. In the end, the Nobel Prize for 1962 was awarded to the joint team.

* J. D. Watson. The double helix. A personal account of the discovery of the structure of DNA. New York: "Atheneum." 1968.

Document 8.17

From: E. F. Mishchenko, secretary of the Communist Party organization, V. A. Steklov Mathematical Institute, the USSR Academy of Sciences.
To: V. A. Kirillin, head, Department of Science, Higher Educational Institutions and Schools, CPSU Central Committee.
26.11.1958

[...] As known, N. N. Bogolyubov, member of the USSR Academy of Sciences, resolved (in 1957) a most important and complicated problem, creation of the consistent microscopic theory of superconductivity [...] In this context, publication of an article by A. Abrikosov and I. Khalatnikov in the August issue of the journal "Uspekhi fizicheskikh nauk" raises indignation [...] The authors write (p. 552): "The complete and consistent theory of superconductivity was developed by Bardeen, Cooper, and Schriffer [...] N. N. Bogolyubov proposed a few original methods permitting to obtain the results of Bardeen, Cooper, and Schrieffer." In other words, [...] the authors give credit to the Americans Bardeen, Cooper, and Schriffer for the development of the superconductivity theory while the Soviet scientist is given the role of the developer of a method leading to the known results. This assertion contradicting public opinion in this country (and other countries too) evidences in the first place scientific inconscientiousness of Abrikosov and Khalatnikov. What is worse, it perverts and belittles the role of Soviet science in the world.[1] Publication of the article [...] is not incidental. An earlier paper by Abrikosov in "Vestnik Akademii Nauk", No. 4, 1958[2] was virtually to the same effect. It appeared together with the article of Bogolyubov expounding the theory of superconductivity.[3]
Responsibility for the publication of both articles lies at the door of "Uspekhi fizicheskikh nauk" and "Vestnik Akademii Nauk." E. V. Shpolsky editor-in-chief of "Uspekhi fizicheskikh nauk" [...] deserves blame more than anyone because the paper of Abrikosov and Khalatnikov was published in the journal after N. N. Bogolyubov had received the Lenin Prize.

RGANI, fond 5, opis' 35, ed. khr. 70, l. 201–202.

[1] The document is formulated in the same words that were used to accuse V. L. Ginzburg, the 2003 Nobel Prize co-laureate (together with A. A. Abrikosov), in the period of persecution of Klyueva and Roskin initiated by Stalin and Zhdanov. *Literaturnaya gazeta* wrote at that time about the then last paper of Ginzburg in

Izvestiya Akademii Nauk, N 2, fiz. ser., as showing "absurd reverence for American science. Speaking of the effect of interest [physical effect observed by D. D. Ivanenko and I. Ya Pomeranchuk], Doctor Ginzburg conceals the authorship of Ivanenko. When he has to refer to earlier publications, he cites the review of Schiff (USA) [...]" Furthermore: "[...] in the recent paper by Doctor Ginzburg 'The theory of mesotron and nuclear forces' [...] the author deliberately conceals this achievement of Soviet science."*

* *Literaturnaya gazeta*, 4.10.1947, p. 4.

[2] "Vestnik Akademii Nauk", 4, 1958, pp. 30–36.
[3] *Ibid.*, pp. 25–29.

Document 8.18

From: V. A. Kirillin, head, Department of Science, Higher Educational Institutions and Schools, CPSU Central Committee; A. S. Monin, instructor.
To: Secretariat of CPSU Central Committee.
27.11.1958

The city Communist Party committee, Dubna (comrade Skvortsov), informs us about the inadequate interpretation of the works of N. N. Bogolyubov, member of the USSR Academy of Sciences awarded the Lenin Prize in 1958 in journals of the Academy and asks the CPSU Central Committee to take necessary measures [...] A. Abrikosov and I. Khalatnikov are disciples of Professor Landau and authors of several publications on the theory of superconductivity that contain no fundamental results. They unscrupulously advertise their works in the two papers mentioned in the letter of Skvortsov. The city Communist Party committee characterizes publication of the papers by Abrikosov and Khalatnikov as a manifestation of narrow group interests doing serious harm to the prestige of Soviet science.

We believe that K. V. Ostrovityanov,[1] editor-in-chief of "Vestnik Akademii Nauk" and E. V. Shpolsky, editor-in-chief of "Uspekhi fizicheskikh nauk", made a mistake [...] It should be emphasized that Professor Shpolsky made similar mistakes in the past, and they led to the wrong interpretation of the works of certain scientists. These facts were brought to the notice of the Presidium of the USSR Academy of Sciences, but it has not thus far taken any steps to introduce order in the work of the editorial board of "Uspekhi fizicheskikh nauk." We deem it advisable to point out to the Presidium of the USSR Academy of Sciences that some of its editions inadequately interpreted works awarded with the Lenin Prize. Also, measures should be taken to strengthen the editorial board of the journal "Uspekhi fizicheskikh nauk."[2]
We are looking forward to knowing your opinion [...]

Note in the upper left corner of sheet 203: "Agree. P. Pospelov [...]"; the other signature is illegible.

Note in the bottom of sheet 203: "CPSU Central Committee. Presidium of the USSR Academy of Sciences (comrade Topchiev) is informed. A. Monin."
Below: "Agreed with comrade Mukhitdinov. N. A. V. Kirillin. N. Glagolev."
Left of these two notes: "To be shelved. V. Gorbunov. 19.12.1958."[3]

[1] That Party functionaries took issue with the paper by Abrikosov "Problems of superconductivity" published in the "Vestnik Akademii Nauk" seems especially absurd today. This paper proved a forerunner of the discovery of the formerly unknown class of superconductors. The author for the first time emphasized that the available observations prompted "investigation of the behaviour of superconductors in a strong magnetic field."* The scientist soon resolved the problem and put into use such notions as type II superconductivity and "Abrikosov vortex lattice." It allowed to extend the generally accepted Ginzburg–Landau theory (describing the behaviour of type I superconductors able to completely displace magnetic fluxes at a given low temperature) to superconductivity in strong magnetic fields. This outstanding discovery brought to Abrikosov the Nobel Prize of 2003 that he shared with V. L. Ginzburg (Russia) and the English professor A. Legget working in the United States; the scientists were awarded for "pioneering contribution to the theory of superconductors and superfluids."

* "Vestnik Akademii Nauk", No. 4, 1958, p. 35.

[2] "Strengthening the editorial board" implied the dismissal of Professor Shpolsky. However, the recommendation remained on paper, and Shpolsky occupied his position till his sudden death on 21.8.1975; he had time to approve publication the September issue of the journal.* The happy end of another persecution of scientists planned by Party ideologists was due to the "revolt of academics" in early 1958. Then, several members of the USSR Academy of Sciences prevented incorrect nomination of Soviet scientists for the Nobel Prize in physics (see doc. 8.1). As a result, it was awarded to P. A. Cherenkov, I. E. Tamm, and I. M. Frank (28 October 1058). New repressions against scientists would be totally out of time so soon after the triumph of the Soviet Nobel laureates.

* *Uspekhi fizicheskikh nauk.*

[3] The date of shelving the document (meaning that no further repressive actions will be undertaken) is interesting in itself. That very day, the three Nobel Prize winners returned from Stockholm to Moscow.

Document 8.19

From: V. A. Kirillin, head, Department of Science, Higher Educational Institutions and Schools, CPSU Central Committee; A. S. Monin, instructor.
To: Secretariat of CPSU Central Committee.
No. 32-A-749; 21.12.1959; Urgent.

Comrade Tovmasyan, secretary of the Central Committee of the Communist Party of Armenia informs that president of the Armenian Academy of Sciences,

Ambar-tsumyan, received the invitation of the Nobel Committee[1] to recommend a candidate for the prize in physics for 1960. The nomination deadline is December this year.[2] Professor Ambartsumyan suggests the candidacy of N. N. Bogolyubov, member of the USSR Academy of Sciences, for his outstanding works in theoretical physics.

N. N. Bogolyubov obtained fundamental results of importance for the theory of non-linear oscillations, statistical mechanics, quantum field theory, theory of superfluidity, and other fields of theoretical physics. One of his recent prominent achievements is the exhaustive explanation of the phenomenon of superconductivity resulting from a thorough analysis and strict generalization of the hypotheses simultaneously developed by Cooper (USA). This work of Bogolyubov is widely known abroad. It may be hoped that the Nobel Committee will award the Nobel Prize to Bogolyubov (probably jointly with Cooper) for this outstanding study.

We deem it desirable that comrade Ambartsumyan nominated comrade Bogolyubov for the Nobel Prize.[3] The president of the USSR Academy of Sciences Nesmeyanov supports this proposal.

We are asking for your consent.

In the top left corner: "Agree. N. Mukhitdinov. 2.XII.59."

Signatures of P. Pospelov, O. Kuusinen, M. Suslov, secretaries of the CPSU Central Committee, right of this note.

In the bottom right part of the document, below the text: "CPSU Central Committee. Central Committee of the Communist Party of Armenia (comrade Tumanyan S. D.) is informed today. N. Glagolev. 28.XII.59."

Below this note: "To the cipher communication from comrade Tovmasyan No. 971sh, 17.XII.59." (unsigned). Shelving date: 28.12.1959."

RGANI, fond 5, opis' 35, ed. khr. 114, l. 71.

[1] The Nobel Committee in physics.

[2] A mistake. The deadline is 31 January of the current year.

[3] V. A. Ambartsumyan later wrote that he was "uninformed in many scientific disciplines in which versatile talents of N. N. Bogolyubov shone so brightly (mechanics, quantum field theory, etc.)."*

* Nikolai Nikolaevich Bogolyubov. Matematik, mekhanik, fizik (Nikolai Nikolaevich Bogolyubov. Mathematician, mechanic, physicist. Dubna, 1994, p. 37.

Document 8.20

From: I. I. Artobolevsky, member of the USSR Academy of Sciences.
To: A. V. Topchiev, vice-president of the USSR Academy of Sciences.
28.12.1959

I have received the invitation to recommend a candidate for the Nobel Prize in physics for 1960 from the Swedish Academy of Science, the Nobel Committee. I think that N. N. Bogolyubov, creator of the theory of superconductivity and superfluidity, is the worthiest candidate for this prize [...] Bogolyubov developed a new mathematical method for the solution of this problem based on special canonical transformations [...] His studies [...] on the construction of the microscopic theory of superconductivity are methodologically and conceptually related to his works in the field of superfluidity and ensure their further development. Generalization of the canonical transformation method enabled Bogolyubov to propose the consistent microscopic theory of superconductivity [...]
Bardeen, Cooper, and Schriffer published a detailed paper on superconductivity almost simultaneously with Bogolyubov;[1] however, they treat the problem in a narrower sense.[2] Specifically, they substituted electrophoton interaction by certain equivalent electron-electron interaction.
In the first approximation the results of Bogolyubov on the one hand and Bardeen, Cooper, and Schriffer on the other hand coincide. However, the method of Bogolyubov provides a much more detailed description of the phenomenon of interest, e.g. as regards a branch of the spectrum of elementary excitations corresponding to collective oscillations of the system [...]
I beg you to review my considerations and grant permission to send nomination to the Nobel Committee (the draft is attached).[3]

Notes on the margins of sheet 134: "Attn. comrade Oraevsky P. S. Comrade A. V. Topchiev ordered to prepare a letter to the CPSU Central Committee. V. Vinogradov, 6.1.1960"; "Attn. Belyaev. For preparation according instructions. P. Oraevsky. 9.1.1960."
"The decision is tentatively postponed since DPMS objects.[4] V. Vinogradov, 23.2.1960."[5]

Arkhiv RAN, fond 579, opis' 1, ed. khr 639, l. 134–137.

[1] Error. The preprint of the three American physicists appeared much earlier that the publication of Bogolyubov (see this chapter).
[2] N. N. Bogolyubov treated both superconductivity and superfluidity whereas BCS theory focuses on superconductivity alone. According to V. L. Ginzburg, Nobel Prize winner of 2003 for "pioneering contribution to the theory of superconductors and superfluids", Bardeen certainly understood the analogy between superconductivity and superfluidity long ago [...]* It appears that superfluidity is not mentioned in the paper of BCS by mere accident."**

* "Handbuch der Physik." Heidelberg, 1956, B. 15, S. 274.

** V. L. Ginzburg. O fizike i astrofizike (On physics and astrophysics). M.: Byuro-Kvantum, 1995, p. 439.

[3] Not published.
[4] It may be thought that objections of the Division of Physico-Mathematical Sciences arose from inconsistencies contained in the nomination by Artobolevsky.

[5] The sitting of Bureau of the Division of Physico-Mathematical Sciences closest to this date was held 17.2.1960;* however, the question is not mentioned in the minutes. Probably, it was discussed verbally.

* Arkhiv RAN, fond 471, opis' 1, ed. khr. 419.

Document 8.21

From: L. A. Artsimovich, acting academician-secretary, Division of Physico-Mathematical Sciences, the USSR Academy of Sciences.
To: A. V. Topchiev, chief scientific secretary.
13.12.1957

Division of Physico-Mathematical Sciences intercedes to obtain permission for N. N. Bogolyubov's trip to Belgium and Holland[1] for lecturing on the superconductivity theory he developed. The fact is, the American scientists Bardeen, Cooper, and Schriffer has recently proposed a theory that also offers some explanation of the phenomenon of superconductivity. The authors recognize that the theory is methodologically inconsistent and makes use of certain arbitrary assumptions [...] The theory of Bogolyubov is based on quite different ideas and ensues from generalization of his superfluidity theory. Hence,the necessity to acquaint foreign physicists with ideas and results of Bogolyubov. The question at issue is the priority of this country in a most important filed of modern physics. Bogolyubov should expound his theory by himself in order to be able to answer questions and clarify debatable items.

Arkhiv RAN, fond 471, opis' 1, ed. khr. 307, l. 119.

[1] "Invitation of N. N. Bogolyubov to the University of Brussels (letter of Professor Prigogine) and University of Utrecht (letter of Professor Van Gova). Delivering lectures and reports on the theory of superconductivity in research centres of low temperatures and statistical physics. The time is not specified, hopefully as soon as possible but not later than January-February. Duration: 2 weeks."*

* From a handwritten dateless note of Bogolyubov; *Ibid.*, l. 121.

Document 8.22

From: L. A. Artsimovich, acting academician-secretary, Division of Physico-Mathematical Sciences, USSR Academy of Sciences.
To: S. G. Korneev, head, Foreign Department, the USSR Academy of Sciences.
22.1.1958

Division of Physico-Mathematical Sciences intercedes to obtain permission for N. N. Bogolyubov's trip to France for 1 month in July-August for lecturing on the superfluidity and superconductivity theory in the Summer School of theoretical physics.[1] Other lecturers will include the most authoritative theoreticians from the United States invited to report their recent results in related areas of statistical physics [...][2]

Simultaneously, the Division asks to grant Bogolyubov permission to visit the Institute of Advanced Research, Princeton,[3] inviting him as its member in 1958–1959 academic year [...]

Arkhiv RAN, fond 471, opis' 1, ed. khr. 307, l. 124.

[1] N. N. Bogolyubov received an invitation to participate in the summer school of theoretical physics at the University of Grenoble in early autumn 1957. In a letter of 7.11.1957, Professor Philipp Nesier, organizer of the School, thanked Bogolyubov for his consent to read ten lectures and informed him about financial aspects of the visit.* At those time, a Soviet scientist could not accept an invitation from abroad without approval of the Presidium of the USSR Academy of Sciences because even regular letters were dispatched via its Foreign Department. The Presidium, in its turn, needed the approval of the CPSU Central Committee. Later, all previous agreements were disavowed by its Secretariat, supposedly on the recommendation of state security organs.

Ibid., l. 115–116.

[2] Invitation to read lectures on my theory of superfluidity (poorly known abroad) at the summer school of theoretical physics in "Les Honches" in July–August 1958.* They pay the expenses. Simultaneous lecturing on statistical physics by the most authoritative specialists (mostly the Americans).** Invitation of Professor Oppenheimer to visit Princeton from September 1958 to April 1959. Duration is unlimited. The expenses will be covered. Delivering lectures on the subject of my research (hopefully with the wife).***

* The official program of the summer school included a cycle of Bogolyubov's lectures under the general title "Superfluidity of Bose and Fermi systems." *Ibid*, l. 10.

** The program envisaged seven seminars on the topical problems of theoretical physics, besides six cycles of lectures by different scientists including B. R. Mottelson (Denmark), a future Nobel laureate. R. Schriffer, one of the authors of the complete theory of superconductivity was to chair a seminar.

*** From a handwritten dateless note of Bogolyubov; *ibid.*, l. 120–121.

[3] The letter of Oppenheimer dated 14.11.1957; *ibid.*, l. 120–121. In a year, Oppenheimer again invited Bogolyubov to visit Princeton for the 1959/1960 academic year. Bogolyubov wrote in the letter of 23.3.2959: "Dear Professor Oppenheimer, unfortunately it has just become clear that we [with his wife] shall not be able to visit your Institute. I thank you once again for the most kind invitation. Yours, Bogolyubov."*

*Arkhiv RAN, fond 383, opis' 1, ed. khr. 444, l. 21; translated from English.

Document 8.23

From: I. M. Vinogradov, head of the Soviet delegation at the International Mathematical Congress, Edinburgh.

To: Department of Science, Higher Educational Institutions and Schools, CPSU Central Committee.
5.9.1958

[...] National Committee of Soviet mathematicians approved the membership of the Soviet delegation at the International Mathematical Congress in Edinburgh and of the incorporated tourist group.[1] However, eight persons were crossed off the list of participants without the knowledge of the National Committee. Six others were included by an authority unknown to us. They did not present reports at the Congress.[2]
All communications delivered by the members of the delegation were met with great interest and demonstrated the impressive success of Soviet mathematical science. Participants of the Congress were especially interested in the works of M. M. Postnikov, A. V. Pogorelov, V. S. Vladimirsky, and some others included in the delegation by the National Committee but not allowed to join it for unknown reason. On the other hand, the absence of N. I. Akhiezer, M. G. Krein, and I. M. Gel'fand passed virtually unnoticed although they had been personally invited by the organizing committee of the congress[3]

Arkhiv RAN, fond 383, opis' 1, ed. khr. 420, l. 11–13.

[1] These people paid their own travelling and accommodation expenses.
[2] The Soviet delegation comprised 32 members. Twenty-six presented reports* while those six "included by an authority unknown to us" did not. They were obviously concerned about things far from higher mathematics. Inclusion of secret agents in delegation of scientific and cultural workers was a common practice in the Soviet Union. For example, the minister of culture N. A. Mikhailov interceded for the permission for the famous Soviet pianist S. T. Rikhter to go to DDR on the occasion of the 100th anniversary of R. Schuman' death in 1956. The Secretariat of the CPSU Central Committee granted the permission (6.7.1956) on the condition that Rikhter was accompanied by an official of the Department of Foreign Relations, USSR Ministry of Culture.**

* Arkhiv RAN, fond 383, opis' 1, ed. khr. 405, l. 39.

** Kul'tura i vlast' ot Stalina do Gorbacheva. Apparat TsK CPSS i kul'tura. 1953–1957. Dokumenty (Culture and Authorities from Stalin till Gorbachev. Apparatus of CPSU Central Committee and Culture. 1953–1957. Documents). M.: ROSSPEN, 2001, pp. 505, 512.

[3] The closing phrase of the letter exposes well-known psychopathic personality traits of the author of the letter, such as his life-long morbid anti-Semitism apparent from the list of the names of world-renowned Soviet mathematicians not allowed to join the delegation. I. R. Shafarevich, a pupil and close associate of Vinogradov in the Mathematical Institute, USSR Academy of Sciences, characterized him in the following words: "[...] his unusual personality made itself felt at first sight. For example, his immediate reaction to any request was always negative even if he should seemingly approve of it [...] Such "negativism" is a well-known morbid symptom; I could

see many signs of Vinogradov's unhealthy mentality."* Furthermore: "He was an acrimonious person [...] his negativism was paralleled by misanthropic irony [...] It led to frequent scandals [at the Institute] that Vinogradov observed with great interest."**

* Rossiyskaya nauka v litsakh. Kniga 2 (Russian science and its actors. Book 2). M.: Academia, 2003, p. 104.

** *Ibid.*, p. 106.

Document 8.24

From: R. N. Nikolaeva, assistant of A. N. Nesmeyanov, president of the USSR Academy of Sciences.
To: S. G. Korneev, head, Foreign Department, the USSR Academy of Sciences.

Stepan Gavrilovich, Aleksandr Nikolaevich asks you to send the attached letter to Stockholm (Academy of Sciences) so that it reach the addressee before 1 February 1962.[1]
Note below: "Attn. Comrade Gromovoy. Do please. 23.1. Korneev." "Fulfilled, No. 55, 23.1.62."

Arkhiv RAN, fond 579, opis' 1, ed. khr. 639, l. 146.

[1] The attached letter is the nomination of P. L. Poisson, the discoverer of ferrocene, typewritten in the French language.* Nesmeyanov emphasized that "the discovery has already opened up wide prospects for chemical research in different countries although it was made only 10 years ago."**

* *Ibid.*, l. 147.

** Ferrocene was synthesized in 1951.

Document 8.25

From: S. T. Astavin, head, 5th European Department, USSR Ministry of Foreign Affairs.
To: M. N. Agoshkov, acting chief scientific secretary, the USSR Academy of Sciences.
No. 434/5eo/ns, 27.12.1962

The embassy of Rumania addressed the 5th European Department, USSR Ministry of Foreign Affairs, for assistance in obtaining information about the works of Soviet scientists that will be nominated for the Nobel Prize in 1962. If such data are yet unavailable, the embassy would appreciate unofficial information about the character of works used to be nominated in the USSR. Please advise what could be said in response to the request of the Rumanian embassy.[1]

Arkhiv RAN, fond 579, opis' 1, ed. khr. 640, l. 25.

[1] It remains unknown why the Rumanian embassy in Moscow showed interest in nomination of Soviet scientists for 1963. The aforementioned letter was not found

in AVP RF. Probably, it was a byproduct of unofficial contacts between Academies of Sciences of socialist countries with a view to joint nomination of candidates for the Nobel Prizes that existed in the early 1960s and were planned "to be extended" in 1976 (see point 5 in doc. 8.52).

Document 8.26

"Memorandum" signed by Y. G. Belyaev, Foreign Department, the USSR Academy of Sciences, 17.1.1963.[1]

1. The right to nominate candidates for the Nobel Prizes is granted to the winners of theses prizes and other world-renowned scientists invited by the Nobel Committee.
2. No Soviet scientists were nominated for the Nobel Prizes in 1963, at least via the USSR Academy of Sciences (nominations must reach Stockholm before 1 February of the current year; otherwise, they are not considered).[2]
   *Arkhiv RAN, fond 579, opis' 1, ed. khr. 640, l. 24.

[1] A handwritten copy.
[2] This handwritten "memorandum" was drawn up at least half a month before dateless doc. 8.29 also signed by Belyaev. He did not know then about a parcel from Stockholm containing invitations to Soviet nominators to recommend candidates for the 1963 prizes (the parcel was lost and found later in the chancellery of the Foreign Department; see note 2 to doc. 8.28).

Document 8.27

From: S. G. Korneev, head, Foreign Department, the USSR Academy of Sciences.
To: S. T. Astavin, head, 5th European department, the USSR Ministry of Foreign Affairs.
21.1.1963

With reference to you No. 1434/5eo/ns dated 27 December 1962, I have to inform you that no Soviet scientist was officially nominated for the Nobel Prizes here in the USSR.[1] It should be borne in mind, however, that any Nobel laureate or other world-renowned scientist abroad may nominate either Soviet or foreign specialists whose names are known only to the members of the Nobel Committees.
Arkhiv RAN, fond 579, opis' 1, ed. khr. 640, l. 23.

[1] See note 2 to doc. 28.

Document 8.28

From: N. M. Sisakyan, chief scientific secretary, USSR Academy of Sciences.
To: S. G. Korneev, head, Foreign Department, the USSR Academy of Sciences.

30.1.1963

Attached find please the materials I received from the Nobel Committee. Please clarify what we should do and report the result.[1]

Note below the text: "Attn. Comrade Zhigalin N. M. Please prepare the reply. P. Oraevsky. 31.1.63." "Attn. Belyaev Yu. G. Please prepare. 2.2.63. Zhigalin." "It should have been done [candidates nominated] before 1 February. Yu. Belyaev." (dateless)[2]

Arkhiv RAN, fond 579, opis' 1, ed. khr. 640, l. 19.

[1] Sisakyan, member of the USSR Academy of Sciences, was appointed to serve as chief scientific secretary from January 1, 1963. He previously occupied the position of academician secretary of the Division of Biological Sciences.
[2] This document evidences that invitations to Soviet nominators to recommend their candidates for the Nobel Prizes in physics and chemistry for 1963 (received by the Presidium of the USSR Academy of Sciences not later than autumn 1962) never reached the addresses, most likely for mere negligence of the departmental staff.

Document 8.29

"Memorandum" signed by Y. G. Belyaev, Foreign Department, the USSR Academy of Sciences, dateless.[1]

This is the proposal of the Nobel Committee in chemistry to nominate a Soviet scientist[2] for the Nobel Prize in 1963.
According to the Nobel Statutes, the nomination deadline is 31st January of a current year. Therefore, nothing can be done in this case. If a similar proposal comes again, it will be necessary to send to the Nobel Committee the following materials before the 1st of February of a current year:
(a) detailed nomination,
(b) characteristic of scientific activity of the nominee,
(c) his biographic note,
(d) bibliography of his works, and
(e) reprints of main publications.

Arkhiv RAN, fond 579, opis' 1, ed. khr. 640, l. 20.

[1] This is a reply to the letter of Sisakyan (see doc. 8.28).
[2] Perhaps a slip of pen. It was not officially prohibited to nominate foreign scientists. For example, documents of the Foreign Department, USSR Academy of Sciences, contain the English text of the nomination of Professor P. L. Poisson by A. N. Nesmeyanov for the Nobel Prize in chemistry for 1962 (20.1.1962).* In 1964, nominations of foreign candidates were still sent to Stockholm via the Foreign Department (doc. 8.40).

* Arkhiv RAN, fond 579, opis' 1, ed. khr. 639, l. 147

Document 8.30

From: M. N. Agoshkov, acting chief scientific secretary, the USSR Academy of Sciences.
To: F. A. Mikhailov, acting head, Department of Scandinavian States, the USSR Ministry of Foreign Affairs.

The Presidium of the USSR Academy of Sciences supports the opinion of N. D. Belokhvostikov, USSR ambassador in Sweden, that he should speak on behalf of L. D. Landau at the prize-giving ceremony in Stockholm scheduled for December 10. Please send to comrade Belokhvostikov the draft text of the speech.[1]

The note below: "The announcement of awarding the prize to L. D. Landau appeared in the central press on November 2, 1962.[2] Yu. Belyaev. 3.12.1962."

Arkhiv RAN, fond 579, opis' 1, ed. khr. 640, l. 1

[1] See doc. 8.31.
[2] The decision of the Royal Academy of Sciences to award the Nobel Prize in physics to L. D. Landau was announced on November 1, 1962.

Document 8.31

Draft speech of the ambassador of the USSR to Sweden at the prize-giving ceremony of 1962.

Your Majesty. Mr. Chairman. Ladies and gentlemen.
I have the great honour to represent here Lev Davidovich Landau,[1] member of the USSR Academy of Sciences, who is unfortunately unable to attend this solemn ceremony, being ill. He entrusted me to express his cordial thanks to the Nobel Committee for the recognition of his scientific works in the form of the Nobel Prize in physics for 1962.[2] Professor Landau asked me to underline that he considers this award as a sign of recognition of not only his personal contribution but also of the achievements of other Soviet scientists working with him and of Soviet science at large. He believes that this award will promote further development of close scientific collaboration between Soviet scientists and their foreign colleagues both here in Sweden and in other countries all over the Globe.[3] Thank you for attention.[4]

Note below: "Attn. Comrade Belyaev Yu. J. Agreed with Agoshkov. He has no objections. 1.12.1962. P. Oraevsky."

Arkhiv RAN, fond 579, opis' 1, ed. khr. 639, l. 3.

[1] It is unlikely that Landau could have given any instructions bearing in mind his health condition in November 1962.
[2] The authors of the document from the Foreign Department, the USSR Academy of Sciences, appear to have been unaware that the Nobel Prizes are awarded by prize-giving organizations, not by the Nobel Committees; in this case, it was the Royal Academy of Sciences that should be thanked.

[3] An example of bureaucratic style including the word "Globe" spelled with the initial capital letter.
[4] "The draft speech" was slightly modified (specifically, the Russian Nobel Prize winners preceding Landau were enumerated, viz. I. P. Pavlov, I. I. Mechnikov, N. N. Semenov, P. A. Cherenkov, I. E. Tamm, and I. M. Frank). It was presented by Belokhvostikov at the official dinner in the City Hall in the evening 10 December 1962. The ambassador corrected the mistake ("cordial thanks" to the Nobel Committee) and thanked the Royal Academy of Sciences. His speech translated, from Russian into French, was published in *Les prix Nobel. 1962.* Stockholm, 1963, pp. 36–37.

Document 8.32

Ceremony of awarding the Nobel Prize in physics for 1962 to L. D. Landau.

Presenting the award: Rolf Sulman, ambassador of Sweden to the USSR.
Location: Hospital of the USSR Academy of Sciences, conference hall.
Time: 10 December 1962. 11.00.
Duration of ceremony: 15–20 min.
Present at the ceremony:
(a) from the Soviet side
— M. V. Keldysh, president of the USSR Academy of Sciences,
— M. I. Agoshkov, acting chief scientific secretary of the Academy,
— L. A. Artsimovich, academician-secretary of the Division of Physico-Mathematical Sciences,
— N. N. Semenov, G. G. Chakhmachev[1]
— I. E. Tamm,
— I. M. Frank,
— P. A. Cherenkov,
— P. A. Kapitsa,
— P. S. Oraevsky,
— K. T. Landau, laureate's wife,
— 2–3 journalists;
(b) from the Swedish side
— ambassador R. Sulman (hands out the prize and makes a short speech; no return speech is envisaged),[2]
— 3–4 representatives of the Swedish embassy,
— 2–3 journalists.

This text is endorsed by P. S. Oraevsky, acting head of the Foreign Department, the USSR Academy of Sciences, and Zhigalin I. M., deputy head.

Arkhiv RAN, fond 579, opis' 1, ed. khr. 640, l. 9.

[1] Written in ink, with one letter missing. G. G. Chakhmachev held the position of business manager for the USSR Academy of Sciences from 1960. His name was most likely inserted on the personal initiative of some of the officials. In order to comply

with the quota on the number of participants assigned by the doctors, they simultaneously crossed off the Nobel laureates I. M. Frank and P. A. Cherenkov, who had personally known Landau for a long time.
[2] Landau briefly gave thanks in English for the high praise of his scientific works.

Document 8.33

From: Embassy of Sweden in Moscow.
To: State Committee of Cultural Relations with Foreign Countries, the USSR Council of Ministers.
13.2.1963

The embassy of Sweden has the honour to ask you, on behalf of the Nobel Committee, for assistance in obtaining a facsimile or photo of the signature of the Nobel Prize winner L. D. Landau. It will be reproduced in the yearbook of the Nobel Committee *Les prix Nobel* together with the laureate's portrait.

Arkhiv RAN, fond 579, opis' 1, ed. khr. 640, l. 8.

N. M. Zhigalin, deputy head of the Foreign Department, the USSR Academy of Sciences, informed (16.3.1963) K. S. Proskurnikova, head of the Department of Western Countries, State Committee of Cultural Relations with Foreign Countries, that the photograph of Landau's signature was passed to the Swedish embassy (*ibid.*, l. 6).

Document 8.34

From the Decision of the USSR NKVD.
28 April 1939

[...] Landau Lev Davydovich, born in 1908 in Baku, prior to arrest professor of physics, n/p, and c-n of the USSR,[1] has been convincingly exposed as a member of an anti-Soviet group, is guilty of sabotage and attempted to publish and disseminate an anti-Soviet leaflet. However, taking into account that:

1. Landau L. D. is a major specialist in the field of theoretical physics, and may be useful in future for the Soviet science;
2. Academician Kapitsa P. L. has consented to pledge his word for Landau L. D.;
3. Acting on orders from the People's Commissar for Internal Affairs of the Union of Soviet Socialist Republics, Commissar of State Security of the 1st Rank Comrade L. P. Beria, to release Landau in the trust of Academician Kapitsa, we hereby order that the detainee Landau L. D. be discharged from custody, the investigation discontinued, and case files sent to the archive.[2]
   "Izvestiya TsK KPSS", No. 3, 1991, p. 155–156.

[1] No-Party citizen.

[2] Here is the story about the liberation of Landau told by P. L. Kapitsa: "I was summoned at about 1 a.m [...] They escorted me to a large room where two men were already present. They turned out to be Beria's deputies, Kobulov and Merkulov (both were shot dead afterwards).* 'Do you understand whom your are pleading for? He is a most dangerous man, a spy who confessed everything. Please read this.' And they handed me a huge volume. I did not read it. I said: 'May I ask just one question?' 'If you please', they laughed. 'Tell me what could Landau gain by all this, what is the motive of those crimes you claim he has committed?' They replied that nobody cares about the motives. I hung on, quoted examples from the literature. We thus talked till four in the morning [...] At last, one of them said: 'All right, Kapitsa, if you pledge your word for Landau, then give us a written guarantee. If anything happens, you will be held responsible.' I wrote a letter,** and Landau turned up at the Institute in a couple of days."***

* After arrest and execution of Beria in 1953.

** The letter of Kapitsa to Beria dated 26.4.1939 was published in a volume of "Vospominaniya o L. D. Landau (Reminiscences of L. D. Landau). M.: Nauka, 1988, p. 346.

*** Pyotr Leonidovich Kapitsa. Vospominaniya. Pis'ma. Dokumenty. (Pyotr Leonidovich Kapitsa. Recollections. Letters. Documents). M.: Nauka, 1994, p. 323.

Document 8.35

Nomination for the Nobel Prize of 1947 sent by N. Bohr to the Nobel Committee in physics from Copenhagen on 29.1.1947.

In reply to the Committee's request to send in proposals for awarding the Nobel Prize for Physics for 1947, I shall propose that this prize be awarded to Professor Pyotr Kapitsa for the development of new ingenious methods for investigating the properties of matter in very strong magnetic fields and at extremely low temperatures, as well as for the important results that his investigations using these methods have given and which, especially as regards the behavior of helium at low temperatures, have led to the discovery of new surprising phenomena, that in the most striking fashion have illuminated problems of the greatest importance for the development of our knowledge and understanding of the properties of matter under the most different conditions. In connection with this proposal, I would like to state that the reason that I have not repeated this year the proposal that I sent last year to the Nobel Committee for Physics to award the Nobel Prize to Professor Lise Meitner and Professor Otto Robert Frisch, is that after further consideration I am sending this proposal to the Nobel Committee for Chemistry.

Yours sincerely,

N. Bohr

Archive of N. Bohr in Copenhagen[1] (translated from English).

[1] Published in the journal "Uspekhi fizicheskikh nauk", 167, No. 12, 1997, p. 1360.

Document 8.36

Materials related to the publication of Kapitsa's paper in the journal *Nature* in 1938.

I

Kapitsa to the Editor of *Nature*, P. A. Gregori.[1]

I am sending herewith a short note: "Viscosity of liquid helium below the #-point", which I hope you will kindly publish in your "letters to the editor." I think this is an important note and I should be glad if you could arrange it to be published as soon as possible, and with the day of dispatch. Please do not bother to send the proofs to me here in Moscow, it takes too much time. If necessary please send them either to Prof. P. A. M. Dirac, St. John's College, Cambridge, Dr. J. D. Cockcroft, The Royal Society Mond Laboratory, Free School Lane, Cambridge, or to Dr. W. L. Webster, 46 Brook St., London. The all are my good friends and sufficiently competent to make the necessary corrections [...]

II

P. L. Kapitsa to N. Bohr, 10.12.1937.

[...] All this time I was very busy working on the viscosity of helium below the #-point. Maybe you remember what I was telling you during your visit here about the idea of the work.[2] Experiments are underway, and the preliminary results are very interesting. It appears that viscosity of helium below the #-point actually drops more than 1000 times.

I expected the viscosity to be 10-fold lower, but it proved 10,000 times lower than that of hydrogen at the lowest temperature that viscosity measurements were made. It is rather difficult to imagine a `viscousless' fluid, and its properties are quite amusing [...] The low viscosity phenomenon is obvious and easy to observe. I repeated experiments almost 20 times varying the conditions and looking for some possible errors but could not find any. I am sending herewith a copy of my preliminary note to "Nature" [...]

III

J. D. Cockcroft to P. L. Kapitsa, 23.12.1937.

[...] During the last few weeks, we experimented on viscosity of liquid helium by measuring its flow through capillary tubes with a diameter down to 1/50 mm. We have found it impossible to produce a laminar flow and estimated viscosity at less than 10 -9. We [...] observed many curious things which may mean that the theory of ordinary liquid flow is inapplicable.
Yesterday, just as these experiments were being completed, Webster came and told us that you had done similar work. We were glad to hear that you obtained similar results [...] Webster asked me to look through your proofs prior to sending them for publication since he is going away to France. I shall be glad to do this. We shall send a letter to "Nature" at the end of this week [...]"

IV

John Cockcroft to P. L. Kapitsa, 23.12.1937.

I have just looked through your proofs for "Nature" and sent them back. The paper will appear on January 8 [...] I asked Allen to send you a copy of our letter to "Nature."[2]

V

N. Bohr to P. L. Kapitsa, 26.12.1937.

I thank you heartily for your kind letter with the copy of the note you have sent to "Nature" about your most beautiful and important discovery of the superfluidity of helium below the transition point. It is really a most remarkable and unexpected contribution to our knowledge of the properties of matter which will surely have great consequences for the development of theoretical ideas [...]

Translated from English by P. E. Rubinin.

[1] The original letters cited above are stored in the archive of P. L. Kapitsa at the Institute of Physical Problems, USSR Academy of Sciences. They were published by P. E. Rubinin in the journal *Uspekhi fizichrskikh nauk*, 167, No. 12, 1997, pp. 1349–1360.

[2] From the notes of P. E. Rubinin:* "I am grateful to Professor Allen for permission to publish an extract from his letter to Schoenberg dated 22 October 1997. I first learned about the works of K.** on superfluidity from his letter to 'Nature' [...] When he sent his 'superfluidity' letter to "Nature", he simultaneously asked them to pass the proofs to Cockroft. Cockroft asked me to look through the proofs. When I returned them to 'Nature', he put a letter to this journal prepared by Misener and myself in the same envelope [...] He asked the editorial body to publish both letters in one issue, Kapitsa's first (as it was received earlier) then ours." David Shoenberg then worked in IPP, Moscow.*** He recalls that Kapitsa asked him to translate his letter to *Nature* into good English.

* Note 36; "Uspekhi fizichrskikh nauk", 167, No. 12, 1997, p. 1357
** Kapitsa.
*** Institute of Physical Problems, USSR Academy of Sciences, headed by P. L. Kapitsa.

Document 8.37

Extract from the Nobel lecture of P. L. Kapitsa presented in English in Stockholm on 8.12.1978.[1]

The choice of theme for my Nobel lecture presents some difficulty for me. Usually, such lectures are concerned with a work recognized by the prize. In my case the prize was awarded for the work in low temperature physics (temperature of helium liquifaction, i.e. a few degrees above absolute zero). It so happened that I left this field some 30 years ago,[2] although at the institute under my directorship they still carry out low temperature studies. Personally, I am now studying plasma phenomena at those very high temperatures that are necessary for a thermonuclear reaction to take place. These investigations have led to interesting results and opened up new possibilities. I think they are more interesting as a subject for the lecture than my low-temperature works that I have long forgotten [...]"[3]

"Les prix Nobel 1978", Stockholm, 1979, p. 73.

[1] Translated from English into Russian by P. E. Rubinin; the above extract is borrowed from the publication of Rubinin in "Uspekhi fizichrskikh nauk", 167, No. 12, 1997, p. 1360.
[2] Kapitsa was sacked from the Institute of Physical Problems, USSR Academy of Sciences, on 17.8.1946 after he refused to participate in the project of the development of the atomic bomb lead by L. P. Beria till his arrest in June 1953. As a result, he could not continue his low-temperature studies using the unique equipment of the Institute.
[3] A note of P. E. Rubinin to this extract reads: "The publisher of these letters was among the listeners of Kapitsa's lecture and remembers fairly well how this phrase livened up the audience. (By the way, neither the original manuscript nor the text of the lecture published in 1978 (both in English) contained the words 'works that I have long forgotten')... They were 'my past low-temperature work...' But when Kapitsa spoke in English in the conference hall of the Swedish Academy of Sciences filled to capacity by young people, largely undergraduate and postgraduate students, he did not resist the temptation and said: 'works that I have long forgotten,') The audience appreciated this good-natured although ironical note to the address of the Nobel Committee [...]"

Document 8.38

From: D. V. Skobeltsin, director, Physical Institute, USSR Academy of Sciences.

To: V. A. Kirillin, vice-president of the USSR Academy of Sciences.
23.12.1963

As previously agreed, find please attached the draft letter[1] concerning nomination of N. G. Basov and A. M. Prokhorov, corresponding members of the USSR Academy of Sciences, for the Nobel Prize of 1964.

Arkhiv RAN, fond 579, opis' 1, ed. khr. 640, 40.

[1] The addressee of the draft letter with nomination of Basov and Prokhorov (*ibid.*, l. 41–45) is not specified, but there is no doubt that it was intended for the Department of Science, CPSU Central Committee. Such inference follows from the opening and closing phrases: "The Presidium of the USSR Academy of Sciences asks to consider nomination of Soviet physicists N. G. Basov and A. M. Prokhorov for the Nobel Prize in physics in 1964 [...] The aforesaid gives grounds to nominate the works of N. G. Basov and A. M. Prokhorov as well as Townes for the Nobel Prize in physics. These candidacies might be proposed by the USSR Academy of Sciences or someone of those Soviet scientists invited by the Nobel Committee to forward their proposals. The Presidium of the USSR Academy of Sciences asks to take a decision on this issue." The relevant archive document is not found, which suggests that the letter of Skobeltsin was not sent to the CPSU Central Committee; both the request and the reply (see doc. 8.40) had a verbal form.

Document 8.39

From: D. V. Skobeltsin, director, Physical Institute, USSR Academy of Sciences.
To: S. G. Korneev, head, Foreign Department, the USSR Academy of Sciences.
21.1.1964

Attached find please materials for nomination for the Nobel Prize of N. G. Basov and A. M. Prokhorov, corresponding members of the USSR Academy of Sciences (jointly with C. Townes, USA).[1]
I nominate them on the invitation of Mr. Olender, secretary of the Nobel Committee in physics, on behalf of the Royal Academy of Science, Sweden.[2]

Arkhiv RAN, fond 579, opis' 1, ed. khr. 640, 40.

[1] Full text of the nomination of Basov and Prokhorov in Russian and English is in the same file.
[2] It is quite natural that V. D. Skobeltsin nominated for the Nobel Prize three scientists including two from his Institute (FIAN). He paid special attention to investigations of Basov and Prokhorov in quantum electronics and the totally new method they developed for generating electromagnetic waves. This sometimes excited jealousy in heads of other laboratories. One of them complained in the address at the general meeting of the Institute's staff that Skobeltsin could rarely be seen at the Institute.* "When he comes, he calls a few researchers with whom he likes to converse, e.g. comrade Prokhorov [...]"** This remark was not caused by jealousy alone.

Prokhorov's colleagues were skeptical about his ideas, which seemed exotic at that time. Prokhorov recollected at the end of his life: "[...] many believed that Basov and ! went crazy. When Nikolai [Basov] visited Landau in his country house, the man of genius said that it cannot be so because it can never be so. But Kapitsa being a very clever person, stated that the ideas are very interesting even if impracticable and are worth supporting [...] After many years, I repaid the "debt" to science by supporting Alferov's idea*** about heterotransitions, in which nobody believed."****

* Skobeltsin simultaneously headed the Research Physical Institute of Moscow State University (NIPI-2). A. V. Topchiev, chief scientific secretary of the USSR Academy of Sciences, reported to the Department of Science, CPSU Central Committee, in May 1953 that Skobeltsin categorically refused to give up his job in NIFI-2." RGANI, fond 5, opis' 17, ed. khr. 412, l. 41.

** TsAODM, fond 6862, opis' 1, ed. khr. 47,17.

*** See note 1 to Chapter 12.

**** "Izvestiya-Nauka", 11.1.2002. p. 1.

Document 8.40

"Memorandum" signed by S. G. Korneev, head, Foreign Department, the USSR Academy of Sciences, 247.1.1964.[1]

V. A. Kirillin[2] passed for me information from A. S. Monin[3] about permission to nominate Sov[iet] scientists for the Nobel Prize (those who were proposed).[4]

Arkhiv RAN, fond 579, opis' 1, ed. khr. 639, 149.

[1] Filed as a handwritten copy.
[2] Vice-president of the USSR Academy of Sciences in 1964.
[3] Deputy head of Department of Science, Higher Educational Institutions and Schools in 1964.
[4] A total of four nominations by members of the USSR Academy of Sciences were sent via its Foreign Department in 1964. These were nomination of N. G. Basov, A. M. Prokhorov, and Ch.Townes in physics (docs. 8.38, 8.39); E. K. Zavoisky in physics (ed. khr. 639, l. 129–130); A. N. Frumkin and M. Volmer (DDR) in chemistry (ed. khr. 640, l. 69–96), A. P. Vinogradov and A. Holmes (UK) (*ibid.*, l. 98–122).

Document 8.41

"Memorandum" signed by the scientific secretary of the Division of Nuclear Physics, the USSR Academy of Sciences.
No. 13–648/189, 28.7.1964

P. N. Lebedev Physical Institute, the USSR Academy of Sciences, is permitted (28 July 1964) to send to Sweden the report of N. G. Basov and A. M. Prokhorov "Application of molecular beams in radiospectroscopy" (15 pages).

Arkhiv RAN, fond 579, opis' 1, ed. khr. 640, l. 62.

Document 8.42

From: Executive director of the Nobel Foundation.
To: Nobel Prize winner N. G. Basov, corresponding member of the USSR Academy of Sciences; dateless (translated into Russian).[1]

The Nobel Foundation is aware of the decision of the Royal Academy of Sciences taken at its session of 29 October 1964 to award you the Nobel Prize in physics for this year divided into two halves, one being due to Professor Charles H. Townes of the Massachusetts Technological Institute (Cambridge, USA) the other to Professor A. M. Prokhorov (Moscow) and yourself. The Foundation extends congratulations and invites you to visit Stockholm as its guest to personally receive your prize. We do hope that you will be able to come and would be grateful for a telegram of consent because we have very short time to be prepared for this event. Attached find please information for your notice about the prize-giving ceremony. We could transfer your share of the prize via your bank if you wished so. Also, we ask you to fill and send to our address by airmail (at your earliest convenience) the attached questionnaire concerning your wishes and intentions. We also attach information about materials needed for publication in our edition *The Nobel prize*[2] together with the copyright assignment form concerning your Nobel lecture as stipulated by the Nobel Statutes. You will find more details on this issue in the attached memorandum. We should appreciate your bringing the filled form to Stockholm; it will allow avoiding a delay of the publication.

In case, you can not come to Stockholm, please send at your earliest convenience the materials listed in the memorandum (barring your Nobel lecture, which can be presented later). Please address these materials to the Nobel Foundation, publisher of *The Nobel prize.*

You will receive by mail under a separate cover the following materials: last year's issue of *The Nobel prize*; calendar of the Nobel Foundation; the code of Nobel Statutes; the booklet *Alfred Nobel and the Nobel prize. The Nobel prize* may be useful to you in that it contains the Nobel lectures. The calendar contains the names and addresses of the Nobel organizations and certain individuals.

You will receive all invitation cards and other information upon arrival in Stockholm. We hope that you will find answers to the questions you may have in this letter and the attached materials. Do not hesitate to contact us for any additional information.

Arkhiv RAN, fond 579, opis' 1, ed. khr. 640, l. 67–68.

[1] The original letter is absent; the Russian translation does not give the name of the sender.

[2] *Les prix Nobel.*

Document 8.43

From: D. V. Skobeltsin, director, Physical Institute, USSR Academy of Sciences.
To: S. G. Korneev, head, Foreign Department, the USSR Academy of Sciences.
3.11.1964

On October 30, 1964, The Swedish Academy of Sciences officially informed H. G. Basov and A. M. Prokhorov, corresponding members of the USSR Academy of Sciences affiliated with P. N. Lebedev Physical Institute, that the Nobel Committee awarded them the Nobel Prize in physics for "fundamental work in the area of the quantum radiophysics, which has led to the construction of oscillators and amplifiers in radio and optical wavelength ranges (lasers and masers)."[1] Mr. Yarring, the ambassador of Sweden to the USSR, invited N. G. Basov and A. M. Prokhorov as well as their close relatives to arrive in Stockholm by December 10 and participate in the prize-giving ceremony. Simultaneously, Basov and Prokhorov received many invitations to read traditional lectures at Swedish research institutions during their visit to the country. The P. N. Lebedev Physical Institute asks you to apply to the Presidium of the USSR Academy of Sciences for permission to visit Sweden for two weeks in December (2.XII-16.XII.64)[2] for the following persons: N. G. Basov with his wife, Basova K. T, and A. M. Prokhorov with his wife, Prokhorova G. A. and son Prokhorov K. A (undergraduate student of MGU).[3]

Arkhiv RAN, fond 579, opis' 1, ed. khr. 640, l. 65.

[1] The Royal Academy of Sciences made the decision on 29.10.1964. The Nobel Prize in physics for 1964 was officially awarded for "fundamental work in the area of the quantum electronics, which has led to the construction of oscillators and amplifiers based on the maser-lasers principle."*
[2] 22 is inserted in ink instead of 16.
[3] The son of the Nobel laureate was refused permission to accompany father (see note 2 to doc, 8.12), as before on such occasions. Characteristically, Prokhorov was advised to refer to his son having exams at the university, i.e. the same excuse as in the case with N. N. Semenov's daughter (doc. 7.36) although no exams were ever scheduled for early December... An interesting fact related to this episode emerged almost 40 years later when the former press-attaché of the Soviet embassy in Stockholm E. P. Rymko wrote in a newspaper article: "the 1 and 10 year-old sons of the Basovs remained with their grandmother in Moscow and Prokhorov's son, aged 20, took exams at the university."*

* "Obshchaya gazeta", No. 42, 2001, p. 5.

Document 8.44

Directives to H. G. Basov and A. M. Prokhorov, corresponding members of the USSR Academy of Sciences, visiting Sweden (for three weeks) signed by Professor

N. A. Dobrotin, deputy director of the Physical Institute, the USSR Academy of Sciences (1.12.1964); approved by N. M. Sisakyan, chief scientific secretary of the Academy.

The main purpose of the visit is to receive the Nobel Prize in physics for 1964. N. G. Basov and A. M. Prokhorov are advised:

1. To take part in official ceremonies and receptions in the framework of Nobel festivities;
2. To deliver the attached lectures based on their own and other materials allowed for publication ...];[1]
3. To familiarize themselves with investigations into quantum radiophysics carried out at the Universities of Stockholm, Uppsala, Goteborg, and Umea.
4. To popularize the policy of the Communist Party and Soviet government, success of building Communism in our country in their talks with foreign scientists.
   Arkhiv RAN, fond 579, opis' 1, ed. khr. 640, l. 63–64.

[1] The subject matter of the lectures allowed for delivering in Swedish universities is omitted.

Document 8.45

Materials related to the visit of Ragnar Granit, president of the Royal Academy of Sciences, Nobel Prize winner in physiology and medicine for 1967, to the Soviet Union in 1964.

I

From: N. D. Belokhvostikov, Ambasador Extraordinary and Plenipotentiary of the USSR to Sweden.
To: V. A. Kirillin, vice-president of the USSR Academy of Sciences.
6.3.1964

President of the Royal Academy of Sciences Ragnar Granit shows great interest in the development of contacts between Soviet and Swedish scientists. He tells about plans of the Swedish Academy to repeat the invitation to P. A. Kapitsa to visit Sweden[1] and also to invite M. V. Keldysh. On his part, Granit would like to visit the USSR in summer 1964 (with his wife). He is prepared to pay the related expenses himself but would like to receive the author's royalties due to him from the sales of his book *Electrophysiological investigations of reception* published in the USSR by "Inostrannaya literatura" publishing house, Moscow, in 1957 [...]
The embassy believes that such a visit of R. Granit, an eminent physiologist and president of the Royal Academy of Sciences, would be useful for the further development of contacts with Sweden. It is therefore desirable that the question of royalties

in the ruble equivalent be positively resolved[2] and Granit could receive an invitation from the USSR Academy of Sciences.

Arkhiv RAN, fond 679, opis' 1, ed. khr. 646, l. 88.

II

From: N. M. Sisakyan, chief scientific secretary of the USSR Academy of Sciences.
To: N. D. Belokhvostikov, Ambassador Extraordinary and Plenipotentiary of the USSR.
15.6.1964

Please forward the attached letter to Ragnar Granit, president of the Royal Academy of Sciences, Sweden. It contains an invitation from the USSR Academy of Sciences for him and his wife to visit the Soviet Union for 2 weeks in summer 1964.[3] It would be desirable to know Professor Granit's considerations about the program of the visit and the date of his arrival.
The question of royalties for the publication of his book in the Soviet Union has been positively settled. The royalties (around 900 Soviet rubles) will be paid to him on arrival in this country.

Ibid., l. 89.

III

Extract from the diary of A. A. Zhitkov, councilor for the embassy of the USSR in Sweden, 20.7.1964.

I received the president of the Swedish Academy of Sciences Prof. G. Granit at the embassy on July 4, 1964 and handed over to him the invitation to be a guest of the USSR Academy of Sciences for 14 days at any time in summer 1964 [...] Prof. Granit thanked me for the invitation and said that it is a great honour to him and that he is very interested in such a visit. He expressed the desire to see research institutions and meet with scientists of his profession, neurophysiology. To his knowledge, the most interesting institutes are situated in Moscow, Leningrad, and Kiev. He also said that, being president of the Swedish Academy of Sciences, he would find it interesting to be present at one of the sittings of the USSR Academy of Sciences or its Presidium if such are held during his visit. At the same time, he is not going to focus on business matters alone as he and his wife would be happy to go sightseeing, attend theatres and museums. Prof. Granit asks for the opportunity to discuss the final program with his wife taking into account her health conditions. Prof. Granit's wife is a daughter of Baron Theodor Brun, former member of the State Council, Finland. The maiden name of her mother was Henley. Dacy Granit, the professor's wife, lived in Russia till 1917 [...] Prof. Granit tentatively plans his visit for late August and early September because he will chair the sitting of the Swedish Academy of Sciences on the 9th of September and will thereafter go to Rome (mid-September) [...]

Ibid., l. 95–96.

[1] Kapitsa arrived in Stockholm as a guest of the royal Academy of Sciences in May 1965 after he received the Niels Bohr International Gold Medal in Copenhagen. It was his first foreign trip after 1934 when he returned to the USSR to take part in the Mendeleev Congress and was not allowed to go back to Great Britain (on the order of Stalin) where he had worked with Rutherford since 1921. The son of P. L. Kapitsa, S. P. Kapitsa, accompanied him to Denmark and Sweden while his wife, A. A. Kapitsa, remained in the USSR as a hostage. Kapitsa could not accept earlier invitations from abroad including Sweden because Khrushchev personally refused permissions although regretted his actions later.

[2] In 1964, the Soviet Union had not yet joined the Universal Copyright Convention; therefore, the government discussed paying royalty fees to foreign authors whose works were published in the country on an individual basis.

[3] The letter of invitation from the USSR Academy of Sciences is not published.

Document 8.46

Interview of V. L. Kantotovich, son of the Nobel laureate in economics L. V. Kantorovich; 17 June 1999.

Question: Please tell how your father learned that he had been awarded the Nobel Prize.

Reply: To be frank, Father expected the award. He came to know from some sources that the Nobel Committee was going to share the prize between him and Vasiliy Leontovich (who emigrated in 1925) as early as 1973. A representative of the Nobel Committee visited Moscow and met with N. K. Fedorenko, academician-secretary of the Division of Economic Sciences, USSR Academy of Sciences. He asked Fedorenko how the award of one prize to a Soviet scientist and an emigrant would be taken in the USSR. Fedorenko said that such a situation would be undesirable, and the prize was given to Leontovich alone in 1973.[1] Father learnt that he was awarded the 1975 prize from the Swedish ambassador, who found him at some conference.[2] The Soviet press kept silent for almost a week before TASS made a very short announcement.[3] It was due to the fact that the Norwegian Nobel Committee awarded the Nobel Peace Prize to A. D. Sakharov in the same year.

Question: Did Leonid Vitalievich receive congratulations from his compatriots in the period between the announcement of the award by the Royal Academy of Sciences and in the Soviet press?

Reply: Yes, of course. Foreign radio stations immediately broadcast the announcement. Father received telegrams, telephone calls etc. Curiously, V. M. Lagutkin, deputy chairman of the State Logistics Committee, USSR Council of Ministers, who had business contacts with Father sent a telegram marked "government message." He congratulated Father on winning "the high government award and the prize." "The high government award" was the Red Banner Order given to Father among many other members of the Academy a month before on the occasion of its 250th anniversary;[4] under "the prize" was meant the Nobel Prize.

Question: It should be thought that Leonid Vitalievich was a nominator for the Nobel Committee in economics. Do you know whom he nominated for the 1975 Nobel Prize and who nominated him?

Reply: Neither he nor I knew who nominated him. That year, Father nominated two American economists, Tjalling Koopmans who eventually shared the prize with him and George Dantzig, one of the developers of the linear programming method for which Father and Koopmans received the Nobel Prize.

Question: How did the leaders of the USSR Academy of Sciences take the announcement of award to one of its members?

Reply: They reacted frostily and looked ill-disposed. Certainly, the simultaneous award of the Nobel Peace Prize to Sakharov had some effect. The authorities initiated propagandistic campaigns and inspired collective letters all over the country in protest against the latter award. Members of the Academy underwent especially strong pressure, many of them had to sign prefabricated documents.[5] In connection with this, Father had a very unpleasant talk with G. K. Skryabin, chief scientific secretary of the USSR Academy of Sciences, who demanded that Father sign aprotest against the decision of the Norwegian Nobel Committee in exchange for permission to go to Stockholm.

The threat was quite real. Father could not go abroad since 1969 and even earlier. He did not participate in a single scientific conference. True, he was permitted to receive diplomas of doctor *honoris causa* from the Universities of Glasgow (1966) and Grenoble (1967). In 1969, the American Academy of Sciences and Arts elected him foreign member along with Sakharov and Solzhenitsin. It put an end to all foreign trips. When the Foreign Department of the USSR Academy of Sciences received information from Finland that Father was elected honorary doctor of the University of Helsinki (1971), they did not even tell him about it. He learned the news incidentally, from a letter sent to the home address by a tailor who had to make a top hat for the ceremony and asked for the size.

Thus, the threat of Skryabin was not mere words. After that talk, Father applied to V. A. Kotel'nikov, acting president of the USSR Academy of Sciences in the period between the death of Keldysh and the election of Aleksandrov. He explained in his letter that he could not sign any documents against the Nobel organizations for moral considerations.[6] In the end, the dispute was settled.[7]

[1] Such combination of a Soviet citizen and an emigrant was absolutely unacceptable, regardless of its form. Here is another example also related to Sweden but dated

to the early 1930s. The head of the sector of Central Europe and Scandinavian States, VOKS, informed its representative in Sweden (28.3.1931): "the appearance of a Soviet singer in one concert with Shalyapin is out of the question and [...] we have to categorically refuse such an invitation." It is really a pity that a concert tour by Barsova suggested by the Swedish side encountered serious obstacles in the very beginning...and ended in a proposal unacceptable for a Soviet actress."*

* AVP, fond 0140, papka 115, ed. khr. 278, l. 12.

[2] See doc. 8.47.

[3] The decision of the Royal Academy of Sciences was announced on the 12th of October 1975, the first information in the Soviet press appeared almost in a week (18 October). Here is its complete text: "Decision of the Swedish Academy of Sciences. Stockholm. 17 (TASS). It is announced here that the Academy of Sciences of Sweden awarded the Nobel Prize in economic sciences to the Soviet mathematician Professor* L. V. Kantorovich and the American professor T. Koopmans. The prize is awarded for their 'contribution to the theory of optimal allocation of resources.'"**

* L. V. Kantorovich was elected corresponding member of the USSR Academy of Sciences in 1964.

** *Pravda.* 18.10.1975, p. 5.

[4] The decree of the Presidium of the USSR Supreme Council was signed on 17.9.1975 and published in the press next day.

[5] See doc. 8.49.

[6] See doc. 8.48.

[7] It is worth mentioning that P. A. Cherenkov and I. M. Frank refused to sign this letter too. The two Nobel laureates were invited to participate in the festivities in Stockholm on the occasion of the 75th anniversary of the Nobel Foundation in 1975 (see doc. 8.51). It appears that the Presidium of the USSR Academy of Sciences learnt the lesson of Kantorovich; therefore, Cherenkov and Tamm could accept invitation without problem.*

* The widow of I. M. Frank accompanying her husband to Stockholm in 1975 told the author that she never heard Ilyia Mikhailovich talking about demands to sign the letter against awarding the Nobel Peace Prize to Sakharov drawn up in the Presidium of USSR Academy of Sciences.

Document 8.47

From recollections of A. G. Aganbegyan, member of the USSR Academy of Sciences, communicated to V. L. Kantorovich, son of L. V. Kantorovich, Nobel Prize winner, in July 1997.

[...] I accidentally witnessed how L. V. came to know that he was awarded the Nobel Prize. We both were members of a commission that inspected the ASU[1] system in a department of Minpribor.[2] We participated in a business meeting somewhere in Moscow when the Swedish ambassador arrived and broke up the news.

I should say that L. V. encountered much hardship upholding his principles.[3] I do not know how matters stood in the 1940s and 1950s but I remember fairly well all later events. Official Marxist science took up arms against L. V. His opponents were led by K. V. Ostrovityanov, who was an indisputable authority in political economy [...] and occupied the position of vice-president of the Academy [...] The struggle for the Nobel Prize lasted 3 years, and it was eventually awarded to L. V.[4] It was a real war, and L. V. was at the forefront of it. He was an honest and frank man unable to scheme [...] He always acted regardless of obstacles and expressed his thoughts straightforwardly, giving his opponents an opportunity to take him on his word [...] An adult man having passed all ordeals of difficult years, he had something of a child's naivety in him. He remained true to what he regarded as the main criterion, that is to have a clear conscience [...]

He suffered too much injustice [...], he could not go abroad for many years [...] An Econometric Congress was held in Rome.[5] He was elected vice-president and me a member of the organizing committee. He had to read a lecture at the first plenary session and me to chair a section and present a report [...] I was had not been abroad for many years before and decided to learn whether I could accept the invitation. We were assured that everything would be OK. In the end, we were not allowed to go. Instead, they formed a delegation of junior research scientists for fear of a scandal if the Soviet Union were not represented at all. However, they were refused to be registered as participants of the congress, it was a great scandal, the president of the Econometric Society read the lecture of L. V. himself, the audience applauded it standing up and demanded to adopt a resolution, etc. [...]

The prize-giving ceremony was not announced [...] although it was an important event in our science. A. P. Aleksandrov came to Akademgorodok[6] for the banquet. He proposed a toast and said: "Gathered here are many members of the Academy, its Presidium, heads of Divisions, big scientists. But let us honestly speak the truth: there is only one Nobel laureate here, and he is Kantorovich." I was really happy to hear that at least one man understood what it all meant, especially for a Russian scientist winning the prize in a situation of severe competition [...]

From the private archive of L. V. Kantorovich stored at the Academy of People's Economy, Moscow.

[1] Automatic management system.

[2] USSR Ministry of Instrument-Making Industry.

[3] Linear programming method.

[4] The Lenin Prize for 1965 was awarded to L. V. Kantorovich and V. S. Nemchinov, members of the USSR Academy of Sciences, and Professor V. V. Novozhilov "for the development of linear programming methods and economic models."*

* *Pravda*, 22.4.1965, p. 2.

[5] The congress was held in 1970.

[6] Siberian Division of the USSR Academy of Sciences, near Novosibirsk.

Document 8.48

From: L. V Kantotovich, member of the USSR Academy of Sciences.
To: V. A. Kotel'nikov, acting president of the USSR Academy of Sciences.
25.10.1975

I do not appreciate many proposals and actions of A. D. Sakharov but consider my involvement in a campaign against him irrelevant for the following reasons:

1. The other day I sent abroad a paper by the authority of the Academy with reference to the Nobel Statutes that do not envisage any appeal or modification of the decisions of the Nobel Committees.[1] In these conditions, signing a document to the opposite effect[2] within a few days after the preceding one would seem strange.
2. Several days ago, I already made a few statements in which I disproved some assertions in the Western press with an object to uphold my position as a Soviet scientist of orthodox views. Moreover, now (see *Literaturnaya gazeta* 21 of October) and many times before in my books and articles, I expressed (on my own initiative) my ideological credo with an economic slant underlying the advantages of Socialist economy and the drawbacks of capitalist system. I believe that such well-grounded arguments of a scientist having international authority in the field of economics may be regarded as a contribution to ideological struggle. Their effect would be lessened if I extended them beyond the scope of my special competence. It is another reason why I consider it irrelevant to join the above campaign.
3. Participants in such demonstrations are usually leading figures of the Academy of Sciences, such as members of its Presidium, Heroes of Socialist Labour, directors of big institutes. I do not belong to this category and never before participated in such actions. Therefore, the absence of my signature will pose no questions[3] especially bearing in mind my peculiar situation this year.

From the private archive of L. V. Kantorovich stored at the Academy of People's Economy, Moscow.

[1] Kantorovich alludes to the telegram of his American colleague T. Koopmans, co-laureate of the 1975 Nobel Prize, which he received after the Royal Academy of Sciences announced its decision. Koopmans suggested that they should protest against excluding the American economist G. Dantzig, one of the developers of the linear programming method, from the 1975 nomination list. In the return telegram, Kantorovich referred to p. 10 of the Nobel Statutes according to which the prize-awarding organizations consider no appeal as regards an adopted decision.
[2] The letter of protest against awarding the 1975 Nobel Peace Prize to A. D. Sakharov drawn up in the Presidium of the USSR Academy of Sciences to be signed by its members (see doc. 8.49).

[3] The author foresaw an inquiry to be made by the CPSU Central Committee to the leadership of the USSR Academy of Sciences. Doc. 8.49 shows that his apprehensions had strong grounds.

Document 8.49

From: V. A. Kotel'nikov, acting president of the USSR Academy of Sciences; G. K. Skryabin, chief scientific secretary.
To: CPSU Central Committee.
Dateless, but no later than 24.10.1975.

This is to inform you that 72 members of the Academy signed the statement in protest against awarding the Nobel Peace Prize to A. D. Sakharov.[1] Those who did not include:

— Ya. B. Zel'dovich, who states that the letter should be formulated differently and that he is going to draw up a separate letter;
— Khariton Yu. B. believes that the letter is of no current importance because members of the Academy including himself already protested against Sakharov's actions;[2]
— Kapitsa P. L. suggests that Sakharov should be called to a sitting of the Presidium of the Academy to explain his actions; only after that the Academy should respond as appropriate;[3]
— Kantorovich L. V. cannot sign a collective letter since he has recently received the Nobel Prize and is going to protest individually;[4]
— Ginzburg V. L. refuses to sign for personal considerations.[5]

The USSR Academy of Sciences asks for permission to publish the statement on October 25, this year.[6]

From the private archive of V. Bukovsky placed on the Internet.[7]

[1] Signatures were collected till the last moment prior to dispatch of the letter to the CPSU Central Committee. V. I. Gol'dansky, then a corresponding member of the Academy (1975), recollected that A. N. Frumkin, who celebrated his 80th birthday on 24.10.1975, was urgently summoned to the Presidium for signing the letter just when "tens of people swarmed at his office door waiting their turn to embrace and kiss the hero of the day."*

* See "Aleksandr Nikolaevich Frumkin. Ocherki. Vospominaniya Materialy (Aleksandr Nikolaevich Frumkin. Sketches. Memoirs. Materials). M.: Nauka, 1989, p. 182.

[2] In August 1973, Khariton was forced to sign a collective letter against the remedial actions of Sakharov which "caused a family discord and made him deplore his conduct later."* Khariton did not support Sakharov, his subordinate in the framework of the Arzamas-16 based secret project, in his fight with authorities foreseeing the consequences. Sakharov recollected that Yu. V. Andropov, KGB chief, once

asked Khariton to have a talk with him and try to obtain his manuscript "on the problems of war and peace, ecology and freedom of convictions." The manuscript was readily provided. "In the morning, we met again. I asked: 'What will you say?' 'It's awful.' 'What, the form?' Khariton smiled: 'I don't care for the form. The content is awful.'"**

* In: "Yuliy Borisovich Khariton. Put' dlinoyu v vek" (Yuliy Borisovich Khariton. A Century-Long Road)." M., 1999, p. 471; from recollections A. Yu. Semenov, grandson of Khariton and Semenov.

** A. D. Sakharov. Vospominaniya. In: "Vegetarian epokha (Memoirs. In: The Vegetarian Epoch)." M.: Pik, 2003, p. 231.

[3] A. A. Kapitsa wrote on 27.10.1975: "The Presidium tried to persuade P. L. to sign the letter against Sakharov, but he refused after a heated debate [...] We went to the countryside. A. V. Kot phoned [A. V. Zaitseva, assistant to V. A. Kotel'nikov; note by P. E. Rubinin], but I apologized and said that P. L. is unwell and cannot speak [...] P. L. is alarmed and excited, says that Sakharov is a prominent scientist [...] he must not be persecuted. P. L. looks unsettled."*

* (Dvadtsaty vek Anny Alekseevny. Dnevniki. Zametki. T 2 (The twentieth century of Anna Alekseevna. Diaries. Notes. V 2). M.: Izd-vo "Agraf", 2005, p. 116.

[4] The intention of Kantorovich "to protest individually" was most likely invented by the authors of the document since it is at variance with doc. 8.48.

[5] The day after the 2003 Nobel Prize in physics was awarded to Ginzburg, he explained in an interview to S. L. Leskov (*Izvestiya*) why he did not sign the letter: "Because I did not like it. Many did not like it, but people were afraid. They are not to blame. I took some risk and lost a medal due to me from the government. Should I not give up a medal and act according to my convictions?"*

Eight years before this interview, Vitaliy Lazarevich described this episode in the following words: "The Vice-president of USSR Academy of Sciences V. A. Kotel'nikov invited me to sign the letter [...] We had rather a peaceful talk. V. A. neither frightened or threatened nor persuaded me. In a word, he did his duty without much enthusiasm [...] I refused to sign and was very calm because I had firmly resolved to behave like that beforehand."**

* *Izvestiya*, 9.10.2003, p. 12.

** V. L. Ginzburg. O fizike i astrofizike (On physics and astrophysics). M.: Byuro-Kvantum, 1995, p. 482.

[6] The statement of Soviet scientists was published in *Izvestiya*. 26.10.1975, p. 3.

[7] This document is one of more than one thousand units of the issue that Bukovsky managed to secretly copy in AP RF using a mini-computer and scanner. He consented to act as an expert at the sessions of the Supreme Court held in 1993* to investigate "the case of KPSS" on condition of having access to archive documents of the "Special File" still closed for researchers.

* *Izvestiya*, 23.8.2000, p. 7.

Document 8.50

From: S. P. Trapeznikov, head, Department of Science and Higher Education, CPSU Central Committee.
To: Secretariat of CPSU Central Committee.
25.12.1975

The Soviet Nobel laureates participating in the festivities on the occasion of the jubilee of the Nobel Foundation in Sweden[1] were invited for a talk to the Department of Science and Higher Education (22.12.1975).
Members of USSR Academy of Sciences Aleksandrov A. P.,[2] Semenov N. N., Frank I. M., Prokhorov A. M., and Basov N. G. described their visit to Sweden as very useful since it allowed the establishment of new and the strengthening of old contacts with foreign researchers, Nobel Prize winners. Swedish scientists welcomed the visit of Soviet representatives. They and the Nobel Committees gave special attention to the Soviet delegation. There were no malevolent actions. Swedish Nobel Committees in natural and technical sciences[3] totally distanced themselves from the actions of the Norwegian Nobel Committee that awarded the prize to Sakharov.[4]
The USSR Academy of Sciences plans to take additional steps with a view to further strengthening contacts with foreign Nobel laureates and Swedish Nobel Committees in natural and technical sciences.[5]
This information is reported for your attention. The shorthand record of the conversation is attached.[6]

RGANI, fond 5, opis' 17, ed. khr. 492, 1. 2.

[1] The 75th anniversary of signing the Nobel Statutes by King Oskar II of Sweden.
[2] President of USSR Academy of Sciences A. P. Aleksandrov did not go to Sweden.
[3] An obvious mistake. The Nobel Prizes are not awarded in technical sciences. Probably, the author meant the prize in economics as distinguished from natural sciences.
[4] A classical example of rhetoric in conflict with the real state of affairs. Certainly, none of the Nobel Committees made any statement regarding the award of the Nobel Peace Prize to A. D. Sakharov.
[5] See doc. 8.52.
[6] See doc. 8.51.

Document 8.51

Stenographic report of the meeting with the Soviet participants of the Nobel festivities in Stockholm held at the Department of Science and Higher Education on 22.12.1975; the Department was represented by S. P. Trapeznikov, its head, and V. L. Barsukov, head of a sector.

Trapeznikov: Comrades, we would like to hear your impressions about the trip to Sweden.

Semenov: They treated us decently, without exaggerated attention but quite tolerably. We had talks with scientists, but on the whole it was a show. There were no unpleasant incidents because neither Sakharov nor Solzhenitsin attended the festivities. They tried to interrupt Kantorovich.[1] Newspapers published incorrect information. But we could not feel any hostility. They cut short our reports. In fact, we only had an opportunity of free talks. Prokhorov alone made a report. Some talks were simply funny. Wilson (Nobel laureate in chemistry from England making very good catalysts) offered his products for a certain price. No ideological considerations. We expected it all to be more interesting.

Frank: Nikolai Nikolaevich was right describing the festivities as a show. I was interested in meetings with scientists; some of them I knew before, others I did not, especially biologists and medical men. I met Bohr[2] and Messbauer. Those were useful meetings. As regards the general situation, our hosts took care to avoid undesirable situations. Was it a mere accident? I think not [...] Those receiving prizes were guests. The prize in literature was given to an Italian poet whom nobody knows.[3] My impression is, the four Nobel Committees[4] are at odds. The Swedish Academy[5] and the Karolinska Institutet[6] are more or less in contact with them. They treat the Nobel Committee in literature with some humour and the Norwegian Committee with irritation and sarcasm. None of those I spoke to ever mentioned the names of Sakharov and Solzhenitsin. There was no such situation. We were welcomed on arrival at the airport by Bernhardt, secretary of the Swedish Academy, but he did not come to see us off. My feeling is the Swedes tried to make our visit pleasant. We (and our wives) were always given much attention. As far as benefits are concerned, one advantage is that a large Soviet delegation visited Sweden and held a worthy place among this big gathering of scholars including some 80 Nobel laureates.

Prokhorov: The meeting was arranged to raise the prestige of the Nobel Prizes in physics, chemistry, and biology. Our delegation enjoyed a bit more attention than others. Our wives and comrades were seated at the best places both in the City Hall and at the table. Many Swedes and Nobel laureates addressed me. It was good that the Russian delegation was so large. Western propaganda would use an opportunity to speak against us if the delegation did not visit the festivities. It would make a racket, and we would lose contacts. But we established good contacts. It is very important. We received letters, some of them from Lvoff. He introduced himself to me pretending to see me for the first time. I said: "You write a lot."[7] One scientist came up to me and said that he knew Sakharov. "You are wrong to persecute him." I answered that one should know the reason. He did not discuss the subject any longer and moved off. At the dinner with the king, Lvoff spoke with me only about champagne. It should be noted that scientists in the so-called "free" world are under strong pressure. Those who try to resist are included in black lists. I mean that scientists are forced to sign letters etc. They speak openly about this in private talks. Our propaganda should use these facts, of course without naming particular scientists.

Basov: I have almost nothing to add. Everybody seemed satisfied that the delegation came in full strength. It would be bad if we did not come. The Swedes treated us very well, they seemed proud to receive us, we were seated in the best places. We met Lvoff, an emigrant and Nobel laureate. He was born in France but is of Russian descent.[8] Kastler was present and wrote letters to us. He was in a dismal mood, spoke about the coming death of civilization.[9] However, nobody supported such talks and they quickly ended. There was another man of his sort. They were always seen together at receptions. The Swedish Academy and Engineering Academy showed great interest in our delegation. We had a very interesting discussion with representatives of five departments and the Engineering Academy. The Director of the Academy is a more influential person than the president. His deputy was also present. They wished to clarify the situation with basic science in our country. The Americans cited some figures that we knew before. But the Swedes took an opportunity to turn them to advantage and emphasize that they (Americans) do not have enough equipment and should raise funds to purchase it. Townes ridiculed them. He said that the laser would have never been invented if funds had been allocated for science in general. Rabi argued that we should work in those areas in which money is invested. But investments must be increased and scientists supported more lavishly. Both capitalists and socialists support science, there is no difference between them from this standpoint. The Swedes show great interest in collaboration, and we should develop it. The Physical Institute has good contacts with them. There is much reason to invite big scientists, Nobel Prize winners, to visit our institutes for a week or two. Our press should better highlight the Nobel movement. One interview was published.[10] The situation was better at the time when we received the Nobel Prize.

Trapeznikov: Could you visit their institutes?

Basov: We did not have enough time. It took us a long time to search for dress coats. We had invitations from several institutes but had no time to use them.

Aleksandrov: I was shown many institutes when I visited Sweden.

Prokhorov: We did not emphasize this point when we drew up the program of our visit; therefore, the Swedes did not plan taking us to their institutes. I could see their thermonuclear institute, and it was interesting. We must estimate beforehand what use can be drawn from cooperation, and we must cooperate if it is to our advantage. We really did not have enough time. The Swedes invited too many people and were very busy. We did not give radio or TV interviews. They had asked about our intentions in this respect before we came, but we disregarded the proposal. So, we were not interviewed. I made many acquaintances. For example, I had a very good talk with Bardeen from America, who won the Nobel Prize twice. They would be glad to have an invitation to visit our country. Also, I had a ten minute talk with the Queen of Denmark.

Basov: The embassy held a reception in honour of its member who was leaving the country [...] We invited everybody, and they all came excepting one scientist who had signed a letter in support of Sakharov and therefore found it improper to attend a reception at the Soviet embassy. They were sure that Kantorovich would not be allowed to visit Sweden and prepared unfriendly publications in the mass media. However, he arrived and the local scribblers were greatly disappointed. Their plans of an anti-Soviet campaign were frustrated.

Frank: There were people (not Swedes) who would have been happy if Kantorovich had not received permission to come.

Basov: The Swedes are friendly to us. Lvoff asked me whether I received his letter. I answered that I did but it was not to my liking. He moved off. It seems that he gave up science and fully engaged in political activity.

Prokhorov: Many Nobel Prize winners from abroad participated in the jubilee session celebrating the 250th anniversary of our Academy.[11]

Aleksandrov: After all the talks that we had here in this country, I looked through the lists of Nobel laureates: maybe our scientists are awarded less frequently than Americans [...] the awards are generally given according to deserts.[12]

Frank: The Academy of Stockholm[13] is a highly competent organization.

Prokhorov: The Italian ambassador could not say who was the Italian poet given the Nobel Prize this year.[3]

Semenov: They become flustered when asked why Tolstoy did not receive the prize.[14]

Basov: In chemistry, they disregarded Nesmeyanov, who deserved the award for his metallo-organic studies.

Prokhorov: Zavoisky and Kapitsa deserved the prize too, but never had it.[15] I told them about the discrimination of our scientists.

Basov: Prize-awarding activities are strictly confidential.

Prokhorov: The Americans are much more influential in the Nobel Committee than our country.

Aleksandrov: We must act in concert when it comes to the nomination of Soviet scientists.

Basov: Too few Nobel laureates are members of our Academy. The Committee[16] is constituted by highly reputed physicists and chemists. We should award some of them, e.g. with nominal medals. They speak sarcastically about prizes in literature, make fun of them.

Trapeznikov: Did you notice any attempts at provocation?[17]

Prokhorov: No, they were simply afraid.

Trapeznikov: May I report[18] that the mission was accomplished honoroubly.

Prokhorov: Our visit prevented a slanderous campaign that would have been launched if we had not come. The Swedes were glad to receive our delegation.

Trapeznikov: Thank you for the interesting talk and useful information.

RGANI, fond 5, opis' 68, ed. khr. 492, l. 3–9.

[1] L. V. Kantorovich awarded the Nobel Prize in economics for 1975 had difficulties in obtaining permission to go to Stockholm for participation in the prize-giving ceremony (see docs. 8.46–8.49). He remained in Sweden after the festivities were over, on the invitation of the Swedish side to read lectures at several universities. He came back to Moscow in January 1976 and therefore was not present at the meeting in the Department of Science, CPSU Central Committee.

[2] Oge Bohr, son of Niels Bohr and Nobel Prize winner in physics for 1975.

[3] Eugenio Montale. The dominating motives of his poetry were personal solitude and disharmony with the world. During World War II, however, he wrote protest verses against the inhumanity of Nazism. His brief biographic note had been published in the 3rd edition of the *Great Soviet Encyclopaedia*, V 16, two years before he received the Nobel Prize.* Therefore, the words of the Italian ambassador that he did not know his name look surprising.

* By the way, Prokhorov, making this remark, was editor-in-chief of the 3rd edition of the *Great Soviet Encyclopaedia.*

[4] In order to make himself more understandable, Frank regarded the Nobel Committees in physics and chemistry of the Royal Academy of Sciences as a single organization; hence, "four committees" instead of five as stipulated by the Nobel Statutes.*

* The sixth committee (in economics) already existed at the Royal Academy of Sciences in 1975.

[5] Royal Academy of Sciences.

[6] Another slip of the tongue. What is meant is the Karolinska Medico-Surgical Institute.

[7] Andre Lvoff, Nobel Prize winner in physiology and medicine for 1965, was a son of physician Solomon Lvoff, who left Russia in 1879 for political circumstances, and Maria Simanovich, cousin of the famous artist Valentin Serov.* Lvoff actively participated in the international campaign against persecution of geneticists in the Soviet Union and in defense of Soviet biologists against attacks by Lysenko and his associates.

* She is portraited in the painting *Girl Lit by the Sun* (1988), now in the Tretyakov Gallery, Moscow.

[8] Lvoff's parents originated from Russia. He was born in 1902 and lived in France all his life.
[9] The French scientist Alfred Kastler, Nobel Prize winner in physics for 1966, was a convinced opponent of nuclear weapons. Hence his impression of the international situation in the 1970s characterized by the arms race between the two superstates.
[10] Interview of the members of the Soviet delegation to Swedish pressmen during the celebration of the 75th anniversary of the Nobel Foundation.
[11] The following Nobel Prize winners and members of the USSR Academy of Sciences visited Moscow and Leningrad as the participants in the jubilee session celebrating its 250th anniversary: Abdus Salam (Pakistan, prize in physics for 1979), Hannes Alfven (Sweden, prize in physics for 1970), Severo Ochoa (USA, prize in physiology and medicine for 1966), Linus Pauling (USA, prize in chemistry for 1954 and Nobel Peace Prize in 1962).
[12] Zh. I. Alferov, Nobel Prize winner in physics for 2000, almost literally repeated this thesis a quarter of a century after Aleksandrov: "[...] there are hardly two or three physical works on the 'Nobel list' for the past hundred years that do not meet the criteria of the Prize. These rare exceptions only confirm the rule: the Nobel Foundation brilliantly organizes both nomination and expertise of the proposed works. It is one of the factors accounting for the high prestige of the Nobel Prize."*

* *Poisk*, No. 3, 2002, p. 4.

[13] Royal Academy of Sciences.
[14] See Chapter 1.
[15] See this chapter for materials about Zavoisky. Kapitsa received the Nobel Prize in physics for 1978.
[16] Two Nobel Committees of the Royal Academy of Sciences, in physics and chemistry.
[17] This unambiguous question confirms the characterisation of Trapeznikov as the most "zealous defender of militant dogmatism"* given many years later by A. N. Yakovlev, a leading politician in the period of Gorbachev's *perestroika*. Yakovlev had to leave service in the CPSU Central Committee in early 1973 after he published an article in the *Literaturnaya gazeta* found unacceptable by conservative members of the Central Committee.

* A. N. Yakovlev. Sumerki (Twighlight). M.: Materik, 2003, p. 570.

[18] To the Secretariat of the CPSU Central Committee.

Document 8.52

From: A. P. Aleksandrov, president of the USSR Academy of Sciences, and G. K. Skryabin, acting chief scientific secretary.
To: CPSU Central Committee.
28.1.1976

A group of Soviet scientists, Nobel laureates, visited Sweden in December 1975 to participate in festivities on the occasion of the 75th anniversary of the Nobel

Foundation. The Presidium of the USSR Academy of Sciences considered the results of the visit and the relevant information from the Soviet embassy in Sweden and brings to your notice the following measures planned by the Academy to further develop scientific contacts with the Royal Academy of Sciences, Sweden. The objective of these measures is to strengthen the influence of the USSR Academy of Sciences on the nomination of Soviet scientists for international Nobel Prizes in natural and economic sciences and awarding them to researchers in this country. The Presidium of the USSR Academy of Sciences proposes:

1. To elect a few foreign Nobel laureates in natural and economic sciences members of the USSR Academy of Sciences. The selection of such candidates is underway.[1]
2. To strengthen scientific contacts between the two academies by means of exchange of researchers and delegations, personal contacts between co-investigators in the framework of agreements between the USSR Academy of Sciences, Royal Academy of Sciences, and Royal Academy of Engineering Sciences.
3. To invite the most influential members of the Nobel Committees and other organizations as well as foreign Nobel laureates to visit this country as guests of the USSR Academy of Sciences for establishing business contacts with the Presidium and authoritative Soviet scientists.
4. To ensure more active participation of Soviet scientists in scientific symposia organized by the Nobel Foundation.
5. To extend unofficial contacts with the Academies of Sciences and individual researchers in socialist countries with a view to joint nomination of candidates for the Nobel Prizes.[2]
6. To hold annually (in October[3]) special meetings of Soviet Nobel laureates and foreign members of the Royal Academy of Sciences (having the right to nominate candidates in accordance with the Nobel Statutes) for the development of agreed proposals of Soviet scientists as pretenders to the Nobel Prizes.

RGANI, fond 5, opis' 68, ed. khr. 492, l. 10–11.

[1] The general session of the USSR Academy of Sciences (31 May–1 June 1976, i.e. half a year after the meeting in the Department of Science, CPSU Central Committee, see doc. 8.50) elected 35 new foreign members including 18 scientists from the socialist camp and the third world and 17 from Western countries. The latter were represented by Sune Bergstrom, rector of the Karolinska Institutet and president of the Nobel Foundation*; members of the Royal Academy of Sciences Trugve Gustafsson (professor of zoophysiology, University of Stockholm), Stieg Classen (professor of physical chemistry, University of Uppsala); and five Nobel laureates including Robert Woodworth (USA, prize in chemistry for 1965), Alan and Dorothy Hodgkin (UK, prizes in physiology and medicine for 1963 and in chemistry for 1964 respectively), Ernst Chain UK, prize in physiology and medicine for 1945), Manfred Eigen (Germany, prize in chemistry for 1967).

* In 1977, Bergstrom retired due to old age from the position of rector of the Karolinska Institutet, a prize-awarding organization in physiology and medicine. In 1982, he was awarded the Nobel Prize for his 1956–1965 studies on isolation and investigation of prostaglandins, essential mammalian hormones with a broad spectrum of physiological activities. When rector, he blocked his nomination for the prize, giving a striking example of self-respect...

[2] See note 1 to doc. 8.25.

[3] This proposal violates the strict confidentiality of nominations stipulated by the Nobel Statutes. It appears that such meetings were actually held after 1975, which accounts for the appearance of twin nominations formulated practically in the same words (see doc. 8.53).

Document 8.53

From an interview of Professor B. G. Malmstrom of Goteborg University, chairman of the Nobel Committee in chemistry from 1977 to 1986, given to the author on 24 February 1998.[1]

Question: What else could you say about the problem raised ten years ago?[2]

Answer: The nomination procedure for the Nobel Prizes has not changed substantially since that interview in "Science." It should be noted that nominators give preference to candidates of their own country not only in the United States. Russian nominators behave in the same way. Nominators in Germany, France, and Great Britain adopt a more correct approach. Russian nominators not infrequently ignore the principle of confidentiality according to which candidacies can not be discussed with colleagues or in mass media. The Nobel Committees sometimes receive identical proposals from different Russian nominators formulated practically in the same words. It unequivocally means an organized campaign.[3]

[1] The full text of the interview is published in the newspaper *Poisk*, No. 10, 1998, p. 14.

[2] In 1989, Professor Malmstrom gave an interview to the journal *Science* that was published under the heading "Chauvinism in Nobel nominations." He complained of the behaviour of American nominators who frequently agreed between themselves candidacies of their compatriots for the Nobel Prizes.*

* C. Holden. Chauvinism in Nobel nominations." *Science*, 243, No. 890, 1989, p. 471.

[3] It is exactly what follows from doc. 8.52, especially p. 6.

Document 8.54

Materials related to the nomination for the Stalin Prize for the discovery and interpretation of the Cherenkov effect.

Extract from the memorandum to the resolution No. 1 of the Stalin Prize Committee in science and invention on prize-awarding for outstanding scientific achievements in 1943. 23.3.1944.

a) Physico-mathematical sciences. Grade 1 Stalin Prize (200,000 rubles).
[...] S. I. Vavilov and P. A. Cherenkov discovered and investigated a totally new physical phenomenon — luminescence of pure liquids induced by electron movements with superluminal velocity. This luminescence is characterized by a number of specific properties, such as speed-dependent direction, peculiar polarization, etc. This phenomenon proved especially interesting after I. M. Frank and I. E. Tamm managed to elucidate its mechanism. Their theory is important in that all previously described forms of luminescence were underlain by accelerated electron motion whereas uniform motion was believed to produce no radiation. The authors showed that the new phenomenon differs from all other forms of radiation as resulting from uniform electron motion at a velocity in excess of phase velocity of visible light in the environment but lower than the velocity of light in a vacuum. The existence of the newly discovered phenomenon was confirmed by experiments of American scientists and was highly appreciated by Soviet and foreign specialists. It is an outstanding scientific discovery.

It is nominated for the prize by P. N. Lebedev Physical Institute, USSR Academy of Sciences, and L. I. Mandel'shtam, full member of the Academy. Recommended to be awarded by an expert panel constituted by Ioffe A. F., Sobolev S. L., Kolmogorov A. N., Alikhanov A. I., Kapitsa P. l., Kurchatov I. V. full members of the USSR Academy of Sciences; Leontovich, corresponding member; Professor P. I. Popov.[1]

Voting results: 20 and 12[2] votes of the 33 in support of awarding grade 1 and grade 2 prize respectively.

RGASPI, fond 17, opis' 125, ed. khr. 233, l. 130–131.

[1] A. I. Alikhanov and I. V. Kurchatov nominated the authors of the discovery for the Nobel Prize in physics 13 years later (1957) (see doc. 8.1).
[2] The document does not explain the absence of the 33rd member of the Committee. The possibility can not be excluded that he abstained from voting or voted against the proposal as suggested by discontent of S. V. Kaftanov, deputy chairman of the Committee, with the list of candidates proposed for further consideration (see doc. 8.54-II).

II

Extract from the letter of S. V. Kaftanov, deputy chairman of the Stalin Prize Committee in science and invention, to I. V. Stalin and V. M. Molotov entitled "On the Stalin Prizes for outstanding works in science, outstanding inventions, and radical improvements of the production methods for 1943."
No. SP-278ss. 4.4.1944.

[...] This is to draw your attention to the works undeservedly nominated for the Stalin Prize of 1943.
1. Vavilov S. I., Tamm I. E, Frank I. M., Cherenkov P. A. "The discovery and investigation of luminescence induced by electron movements in matter with superluminal velocity." This phenomenon was discovered by S. I. Vavilov and P. A. Cherenkov while I. M. Frank and I. E. Tamm participated only in its theoretical interpretation. It is my opinion that the prize should be awarded only to the principal authors of the discovery, i.e. Vavilov S. I and Cherenkov P. A.

Ibid., l. 19.

Document 8.55

Interview of R. I. Rapoport, 23 July 2001.

Question: Tell me please, how did your father come to know about his nomination for the Nobel Prize?

Answer: In summer 1962, Father told me about a telephone call from the Presidium of the USSR Academy of Sciences that had received information from Stockholm about his nomination for the Nobel Prize. He had a talk with N. N. Semenov soon after that. Semenov explained to him that the instance[1] would not object to the support of his candidacy by the USSR Academy of Sciences if Father asked to restore his membership in the Communist Party. It was a long talk. It was very difficult for Father to say "No" to Nikolai Nikolaevich, who had helped him to recommence research work a few years ago. Later, Semenov several times resumed the talk. Also, Father was summoned to the Presidium of the Academy and the State Committee on Science and Technology. The chairman of the Committee, Kirillin, and his first deputy (I don't remember his name) had a 3 hour conversation with Father.[2] All talks came down to the restoration of the Party membership. It was easier for Father to say to them "No" than in his talks with Semenov. Father used well-known facts as his arguments. For example, the recently adopted program of the CPSU contained a thesis to the effect that the Michurinian teaching was the only true one. He also alluded to the fact that the pseudoscience developed by Lysenko became an element of state ideology and that he was expelled from the Party for disagreement with these false concepts. He refused to apply to restore his membership in the Party till organizers of the 1948 session are punished and the erroneous provisions in the Party program corrected. Otherwise, his behaviour would look unethical in his own eyes and in the eyes of Party members.

Father hoped that the leaders of the state would have enough reason not to neglect another opportunity to have the Nobel Prize awarded to a Soviet scientist, regardless of his affiliation or non-affiliation with the Party. One day in autumn, Father came home and said that our leadership had informed the Nobel Committee that it considers awarding him premature.[3] Such a decision might be dictated by apprehension of officials (not groundless) that Father would certainly comment on the events of 1948 in his Nobel lecture if he received the prize. True, Father did not altogether lose hope and spent much time studying colloquial Swedish (he had read scientific literature in this language before). In the end, he said with a smile that he was all the same satisfied that his contribution to science was so highly appreciated in Sweden.

Question: Did Iosif Abramovich know who nominated him?

Answer: He several times told me that Professor Gustafsson of the Forest Institute, University of Lund, nominated Charlotte Auerbach and him.[4] But I do not know how he found that out and never asked him. Gustafsson and his wife visited Moscow and were our guests. Gustafsson mentioned his nomination in the table talk.

Question: How did it happen that N. N. Semenov invited your father to his Institute?

Answer: Hinshelwood, Nobel co-laureate with Semenov in 1956, studied (unlike Semenov) chain reactions including those in biological objects. It turned out that he heard about Father's works and took them seriously enough. Meeting Semenov during Nobel festivities in Stockholm, he asked him about the state of research on chemical mutagenesis. Semenov knew nothing but promised to inquire into the matter. He and my father had no common acquaintances, that is why they met only after a few months.

[1] CPSU Central Committee.
[2] An inaccuracy. V. A. Kirillin headed the Department of Science, CPSU Central Committee, in 1962, and the meeting took place in the premises of the Central Committee (see doc. 8.56).
[3] I. A. Rapoport was certainly wrong thinking that the Nobel Committee needed the consent of the Soviet leadership to award the Nobel Prize to him.* In all probability, the letter contained a request for additional biographic data and reprints of his recent publications, as was the case with the request of L. G. Sillen concerning N. N. Semenov (see doc. 7.4).
* B. L. Pasternak was equally misled in the mid-1950s (see doc. 9.1-I).
[4] Oke Gustafsson (1908–1988), geneticist, professor of the State Forest Institute, Lund.

Document 8.56

Extract from stenographic record of the speech of I. A. Rapoport at Moscow State University in early 1988.

[...] To begin with, I was given a flat. I did not ask for it [...] Quite unexpectedly, the Academy allocated a flat. In a few days, I came to know that the Nobel Commission [sic!] recommended my candidacy for the Nobel Prize. I was summoned to various organizations and asked to restore my membership in the Party. I refused on the grounds that they had expelled me for my convictions. I joined the Party during the war and pursued no special goals by this act. But they expelled me. I had a two-hour talk with Kirillin, then head of the Department of Science, CPSU Central Committee, in his office. He and his deputy persuaded me to restore membership in the Party. But I refused. I said I would not have my membership restored for 60,000 dollars.[1]

Iosif Abramovich Rapoport. Uchyony, voin, grazhdanin. Ocherki. Vospominaniya. Materialy (Iosif Abramovich Rapoport. Scientist. Warrior. Citizen. Essays. Recollec-tions. Materials). M.: Nauka, 2001, p. 281.

[1] Half of the Nobel Prize in the early 1960s.*

* The other half would go to Charlotte Auerbach.

Chapter 9 • 1958

The Nobel Prize and Boris Pasternak
"Police Measures" During the Khrushchev Thaw

In 1983, in connection with a wide-ranging discussion among literary critics trying to ascertain the years in which Boris Pasternak was nominated for the Nobel Prize, the permanent secretary of the Swedish Academy Lars Gyllensten stated that there were definitely seven such nominations — five annual, in the period between 1946 and 1950, and two more, after the interval of six years, in 1957 and 1958.[1] At the present time the documents of the Nobel Committees are available for researchers up to 1955, which allows us to form an objective judgment about their nominations and nominees in the first five years after the end of the World War II.

Three times, in 1946, 1947 and 1949, Pasternak as a nominee was offered to the Nobel Committee by Cecil M. Bowra — a universally recognized expert on Russian poetry of the Silver Age. Besides his nominee's poems, he also translated Alexander Bloc, Osip Mandelshtam and a number of other Russian poets. In 1948 and 1950, it was Martin Lamm (Table 6.3), eminent scholar of literature and drama, since 1928 one of the eighteen "immortals" of the Swedish Academy who became the nominator of Pasternak.[2]

As it is clear from Pasternak's correspondence (doc. 9.1), the information about his nominations being considered in Stockholm sometimes reached the poet, although it was not always authentic. For example, it concerns the 1954 Prize, about which he wrote to his cousin to Leningrad (doc. 9.1-1), although that year the Nobel Committee did not receive any nominations for him.

An interesting thing about this correspondence is Pasternak's conviction that Nobel Prizes for Literature are awarded only in coordination with the government of the country of residence of the potential award-winner. It is all the more surprising because in the statutory requirements of the Nobel Foundation there is not a hint of the possibility of such coordination; on the contrary, they invariably emphasize the necessity of complete confidentiality at all stages of consideration of candidates, from the nomination to the moment when the name of the next prize-winner is published.

The roots of delusions like that are clearly seen. Soviet society, including the intellectuals, was for many decades deprived of information on the activities of the Nobel institutions. Even reports about new prize-winners did not appear in the

Soviet press — of course, with the exception of cases when there were our compatriots among them. A spectacular example of such ignorance is given in Pasternak's already mentioned letter to Olga Freidenberg, which shows that on November 22, 1974 he did not yet know about the Nobel Prize having been given, on October 28, to Ernest Hemingway (doc. 9.1–1). The Soviet media did not mention this event, not even the *Literaturnaya Gazeta*, the organ of the Soviet Writers Union, although the award-winner was a writer quite well-known in the country.[3]

Such ill-intentioned suppressing of information on the activities of the Nobel institutions did not in the least apply to the alert readers from the propaganda department of the Communist Party's Central Committee, where reports from the international press were carefully traced and analyzed correspondingly. Of course, that also involved leaks of information from the Nobel committees.

Ideological workers from the Central Committee also did not overlook the articles in the foreign press discussing Pasternak's chances of winning the prestigious award, the 1947 Nobel Prize, after his being nominated for the second time by the authority nominator from the UK, Cecil M. Bowra (Table 4.2). The Central Committee's organ, newspaper *Kultura i Zhizn* published, as early as March 21, the article by the poet A. A. Surkov under the significant heading "On the poetry of Pasternak". The article, obviously made to order, abounded with expressions like "reactionary backward outlook", "lives in discord with the new reality", "downright slander", "the Soviet literature cannot put up with his poetry", and so on. And when, in 1949, the same Bowra repeated his recommendation (Table 6.4), rumors that the poet had been arrested began to circulate among the Soviet reading public[3a]. There were more than enough reasons for such fears, as the flame of the "struggle against cosmopolitanism" was sweeping the country.

Coming back to Pasternak's later nominations, it should be noted that he was well enough informed of him being considered for the prize in Stockholm; among other things, it is clear from his letters to France and Germany (doc. 9.1-II, III). However, the communist party's ideological structures proved to be much better informed of the preferences of the eighteen voting members of the Swedish Academy. Reports on the possible awarding the Nobel Prize to Pasternak began to be delivered to the CPSU Central Committee several months before the decision was made public (doc. 9.2–9.3), and two days before, on October 21, the Secretariat of the Central Committee received a note signed by two heads of ideological departments, with a detailed list of measures to be taken if such a decision took place (doc. 4).

The cause for such nervous expectation in Moscow of the verdict of the Swedish Academy was given by Pasternak's novel *Doctor Zhivago*, the manuscript of which in the summer of 1956 had been handed over, with the author's consent, to the Milanese publisher, member of the Italian communist party Giangiacomo Feltrinelli. The story of this handover and the insistent efforts made by the Central Committee and the USSR Union of Writers in order to prevent publishing the novel abroad and to bring the manuscript back is well known,[4] and there is no need to repeat it here. We shall only say that the first copies of the novel in Italian appeared in Milan on November 23, 1957. Soon a French translation came out, then an English, and so on.[5]

The novel at once gained a position of an international bestseller.[6] As it was released simultaneously with the beginning of the 1958 Nobel Prize nominations, the press started to discuss the prospects for Pasternak to be awarded it. One of the echoes of these discussions was Pasternak's name being mentioned among the "great" names in the report of the French existentialist Albert Camus, 1957 prize-winner, which he read at Uppsala University on December 14 the same year.[7] It was also Camus who, shortly after that, nominated Pasternak for the 1958 prize.

So, owing to the press, a story with deliberately shifted emphases, which alleged that Pasternak got the prize for *Doctor Zhivago*, started to spread in the world. In reality such a definition did not correspond to the exact wording of the award-giving formulation which sounded in the report of the Swedish Academy and was repeated, on October 23, by the Academy's Secretary Anders Esterling in his telegram to Pasternak. It said that the prize was awarded "for his important achievement both in contemporary lyrical poetry and in the field of the great Russian epic tradition". That is, put in first place are the merits of the prize-winner as an outstanding lyrical poet of the modern time, which also was the object of nominating him in the previous years; whereas his prose, which confirms the versatility and depth of his talent, occupies in this formulation the second, certainly less important place.[8]

In spite of that, the very first phrase of the resolution of the Presidium of the CPSU Central Committee, which was passed on October 23, on the day when the decision of the Swedish Academy was published, stated, as though nothing were the matter, that the prize was awarded to the novel "which gave a slanderous portrayal of the October Revolution," and so this awarding "is a hostile act to our country and an instrument of international reaction directed at the kindling of cold war" (doc. 9.5). Already a day later such unsubstantiated assertions started to sound like the main motif in the media, in speeches at numerous meetings and rallies and in "letters of working people" written on instruction from the authorities.[9]

The campaign was started by the newspaper *Literaturnaya Gazeta*, which, in accordance with the decision of the Presidium of the CPSU Central Committee, published, on October 25, together with the editorial, the text of a negative review of the novel, which had been composed and sent to the author from the *Novy Mir* literary journal in 1956. The next day these articles were reprinted by many other major newspapers. But preparations for arranging "an outburst of indignation" had already started on the evening of October 24. Their centre was the Literary Institute, where the lists of students were being made for participation in a "spontaneous" demonstration planned for October 25. Posters with demands to strip Pasternak of the Soviet citizenship were being prepared, his caricature portraits against the background of sacks with dollars were being drawn, and so on.

Half a century later, the well-known poet Bella Akhmadulina, then a student of the Literary Institute, remembered that not only were the students being made to walk down Moscow streets, but also to put their signatures under the already written "stigmatizing" letter: "Two young men (they studied with me and visited B. L. many times, he treated them with affection) ran to Pasternak, in fact, to ask for his permission to sign. And the most scrupulous B. L. said: 'Of course, sign. I don't want to increase my worries.'" Akhmadulina herself refused pointedly to put her signature under the libel, which resulted in her being expelled from the institute.[9a]

The first signal of the possible awarding of the prize to Pasternak sounded in Moscow, however, seven months before the decision of the Swedish Academy, and the ideological divisions of the Central Committee started at once to prepare measures which, in their opinion, could prevent developments undesirable for them (doc. 9.2). And a month before the award giving, vilification of Pasternak took public form. Olga Ivinskaya wrote that on September 13, at the meeting with Italian poets that was taking place at the Polytechnic Museum, the chairman of the meeting, head of the Soviet Writers Union, A. A. Surkov, answering a note from the audience why Pasternak was not present, announced for all to hear that Pasternak "wrote an anti-Soviet novel, against the heart of the Russian revolution, and gave it to be published abroad".[10] Four days after Surkov's revelations the Central Committee got a letter from Boris Polevoi informing on Pasternak (doc. 3). That is, preparations for the expected decision of the Swedish Academy were fundamental and forestalling.

The Writers Union and the Cultural Department of the Central Committee were also well informed of the fact that among the participants of the approaching clash there was Albert Camus, Nobel prize-winner of the previous year. Not only did they know that he nominated Pasternak for the 1958 prize, but also known to them was the content of his report "The Artist and the Time" read to the audience of Uppsala University four days after the award-giving ceremony in Stockholm. Here is what Camus meant when mentioning Pasternak's name in his report:

> [...] artists rejecting the bourgeois society and its formal art, but aspiring to describe the reality, and nothing but reality, find themselves in a tragic impasse. They must be realists but they cannot become them. They want to submit their art to the reality, but it is impossible to describe it without resorting to the choice which submits it to the peculiarity of art. The beautiful and tragic creative work of the first years of the Russian revolution graphically demonstrates this torture to us. Everything that Russia showed in this epoch–Blok and the great Pasternak, Mayakovsky and Esenin, Eisenstein and the first singers of cement and steel[11] — is a brilliant laboratory of forms and subject matters, of fruitful and troubled quests. However, all that had to be abandoned on having come to the conclusion that it is impossible to be a realist when realism itself is impossible. Dictatorship there, like everywhere, cut things which are extremely painful to cut: realism, according to its principles, was necessary... providing that it would agree to be socialist.[12]

No wonder that the name of the French existentialist who touched at once two points painful for the ideological bosses in Moscow — Pasternak and the problem of realism in the Soviet literature — was repeatedly mentioned in October and November, 1958 in public speeches and newspaper articles having a deliberately negative color. The following passage by the writer S. S. Smirnov, who chaired the meeting of writers on "Pasternak's case", is worth quoting: "Last year the Nobel Prize for Literature was awarded to the pro-fascist French writer Camus, who is very little known in France and in terms of morality is a person to whom no decent French writer will hold out his hand. By the way, this Camus has now sent Pasternak a friendly telegram."[13]

After receiving the news of the decision of the Swedish Academy[14] and the emergency meeting of the Presidium of the CPSU Central Committee (doc. 5), the Central Committee's Cultural Department started to carry out the measures planned in doc. 9.4 and approved by the Presidium. The attempts to enlist the help of Vs. V. Ivanov, Pasternak's neighbor at his *dacha* in Peredelkino, near Moscow, and his friend, had to be abandoned at once, but with his other neighbor and friend, K. A. Fedin, one of the leaders of the Soviet Writers Union, they managed to come to an agreement. As early as the morning of October 24 Fedin came to Pasternak and demanded that he refuse the prize (doc. 9.6). It is interesting that the head of the Cultural Department D. A. Polikarpov, when informing the Secretary of the Central Committee M. A. Suslov of what was going on in Peredelkino, blatantly misinforms his immediate superior about the duration of Fedin's talk with his friend, who had fallen foul of the authorities (note 2 to doc. 9.60). Their five-minute meeting could have appeared to their bosses as the meeting having not been prepared well enough by the department and therefore could have caused anger; so Polikarpov preferred to increase the duration of his emissary's contact with Pasternak by twelve times.

In spite of the threats which he heard from Fedin, Pasternak sent, on this very day, October 24, a telegram to Stockholm expressing his gratitude for the award. A major role in keeping up the poet's spirits was played by Vs. Ivanov, to whom Pasternak ran for advice right after Fedin's visit. Ivanov told Pasternak that he deserved any prize.[15] Paradoxically, Fedin himself, although having turned into a minion of the Central Committee's Cultural Department, secretly revered Pasternak's great talent.[16] This permanent duplicity was a shameful sign of the époque, overwhelming the greater part of figures of culture, who very often had to immorally compromise with the cruel power (see also note 48 to Chapter 4).

In the evening of the same eventful day, October 24, Pasternak received a summons to appear at the meeting at the Writers Union. Actually, for him the perilous situation did not arise out of the blue. He was able to be prepared for the torrent of abuse in advance. That can be seen from his letter to his long-standing French correspondent, Jacqueline de Proyart, sent on September 6, 1958. Already informed that his candidacy for the prize was being considered in Stockholm (see doc. 9.1-II), he wrote: "You should work out your own attitude to the changes beyond our control, to which our plans are sometimes exposed...With each change like that a cry rises again that I have committed a terrible crime, a shabby treachery, that I should be expelled from the Writers Union and outlawed [...] I am only afraid that sooner or later I will be drawn into something that, perhaps, I would be able to stand if I were allotted some five or six years of healthy life."[17]

However, the poet's troubled, impulsive soul was not capable of using in practice what was imagined so distinctly in theory ("[...] perhaps, I would be able to stand [...]"). Kornei Chukovsky happened to be at Pasternak's place when a messenger from the Writers Union brought the summons. "His face darkened," he wrote in his diary three days later, "he grabbed at his heart and trudged up the stairs to his study. I realized that they would not spare him, that they are preparing a public punishment for him, that they will trample him until they kill him, like they had killed Zoshchenko, Mandelshtam [...]."[18] And according to the memories of Vsevolod. Ivanov's wife, who saw him that evening, after receiving the summons his arm became paralyzed for some time (see note 3 to doc. 9.7).

But he was not the only potential victim of the meeting of the commanding writers, which took place on October 27 and expelled Pasternak from the Union. Shortly after the visit of the messenger with the summons Ivanov had a spasm of brain vessels, which put him to bed for a long time (see note 3 to doc. 9.7). Chukovsky after these shocks was "ill with Pasternak" the whole November and could not sleep a single night without sleeping pills.[18] However, the fate was especially cruel to the poet Boris Slutsky. Like many of his colleagues, he was not able to resist the pressure of the ruthless authorities and spoke, on October 31, at the all-Moscow meeting of writers organized to discuss "the case of Pasternak." He made a speech which had such passages as: "The prize has been given to Pasternak out of hatred for us [...] Pasternak is a winner of the Nobel Prize against communism."[19] He was not able to forgive himself for this forced self-betrayal. The coercion plunged him into a mental illness, from which he did not recover until the end of his days.[20]

Pasternak, pleading illness, did not go to the meeting on October 27. He also refused to be present at the meeting on October 31. Instead, he sent a letter to the Presidium of the Writers Union. Having come to his senses after October 24, which was a dramatically hard day for him, he spelled out in his letter eight points leading to the conclusion which had the same impact on the electrified audience as a red rag on an enraged bull. "Circumstances," he wrote, "may make you go very far in dealing with me so as to rehabilitate me, again under the pressure of similar circumstances, when it is already too late. I ask you: do not hurry. It will not bring you more happiness, nor more glory." (doc. 9.7)

In the memo written by D. A. Polikarpov, head of the Cultural Department of the CPSU Central Committee, Pasternak's letter is described as "outrageously impudent and cynical, [...] with vile slander on our reality and foul accusations against Soviet writers." (doc. 9.8) One more phrase which attracts attention is the phrase about Vs. Ivanov. Several days before, this writer disappointed his hopes by refusing to twist Pasternak's arm together with Fedin; now he is trying to get even with him by setting high authorities on him. In the manner of reporting on Ivanov, he hints that the latter might have simulated the disease. However, the residents of Peredelkino, including Fedin, knew only too well about Ivanov's swoon which put him to bed for a long time; Polikarpov certainly knew it too.

Also revealing is the fact that the renowned writers L. M. Leonov[21] and N. F. Pogodin were absent at the meeting, without explaining the reasons. Their absence was certainly meant to be a demonstration. In the Soviet Union "voting with your feet" was the most common way of evading participation and support of some unwanted political event. Marked by the signs of passive resistance was also the silence, fraught with meaning, of the poet S. M. Kirsanov, although at the voting for expelling Pasternak he did not break the unanimity of the audience.[22]

But still, there was a member of the Writers' Union who, in response to the invitation, came to the meeting in order to upset the unanimous expression of obedience. That was A. S. Allilueva, sister of Stalin's wife, who had made away with herself. In January, 1948, at the instruction of her brother-in-law, she had been arrested and spent more than five years in the cells of the Vladimir prison. Her hand, raised in answer to the question from the Presidium "Who is against?" turned out to be the only one at the whole honorable meeting.

We should specially mention the position of A. T. Tvardovsky, who, as well as Leonov and Pogodin, was absent in the auditorium and did not take part in the vote. Nevertheless, he was in the building of the Writers Union — preferring, however, to be in the corridors of the Writers House, where he even managed to recommend to Polikarpov, who turned to him for advice, giving up the idea to punish Pasternak by expelling him from the Union.[22] It will be useful to add that in his now published notebooks of 1958 and 1959 there proved to be not a single mention of the turmoil around Pasternak's Nobel Prize, and between some routine notes dated October 21 (that is, two days before the decision of the Swedish Academy) and November 23, 1958,[23] there are no entries whatsoever. In such a way Tvardovsky deliberately held himself aloof from the dramatic developments — perhaps, consciously escaping into a drinking bout,[24] as if demonstrating his disgust over the anti-Pasternak hysteria provoked by the authorities.

Pasternak learnt about the decision taken at the meeting that very evening. The news again made him lose his equanimity, although it would have been naïve to expect a different outcome. According to Ivinskaya, he met with her, the following day, in Peredelkino and suggested committing suicide together. There are different versions as to the seriousness of this proposal. Specifically, E. B. Pasternak maintains that his father did not at all tend to be suicidal, although talks about suicide had also taken place earlier — for example, about his readiness to make away with himself if "cares about family and children become a hindrance to his artistic occupations," — but at the same time he "assured us that he was quite far from thinking anything of the kind."[25] However, Ivinskaya took these words quite seriously and at once, without telling Pasternak, went to Fedin's *dacha* to tell him what she heard (doc. 9.9).

The information of the possible tragedy, which Fedin immediately conveyed to Polikarpov by telephone, had due effect. Its consequences for the authorities were evident. It was easy to foresee the violent reaction of international public opinion to the demonstrative suicide of the poet, whom the Nobel Prize followed by the campaign of abuse made the man of the year. Naturally, no one in the Central

Committee wanted such a shock. They wanted to humiliate the poet, to force him to think, if at least outwardly, in their own way, that is, to turn him into a tame drummer of ideological services. But not more.

Measures were taken without delay. On the evening of the same day an ambulance with a doctor from the Writers Union clinic drew up at the *dacha* in Peredelkino. As the poet's wife Zinaida Nikolaevna later remembered, the doctor told her that "she was attached to Borya on instruction from the Central Committee and would live with us for a whole month."[26]

However, the fly-wheel of persecution was revolving faster and faster. The next day, October 29, turned out to be filled with still more serious events. The first of them was created by the persecuted poet himself. The phantasmagoria of the previous six days had broken his will. The last straw was the news brought by Ivinskaya that the Goslitizdat (state literary publishing house) refused to commission her to do literary translations, which were her only source of earnings.[27] There was no doubt about the reason for this initiative, nor about the names of those who ordered it.

The next morning, having come to Moscow from Peredelkino, Pasternak, without warning anybody, sent a telegram to Stockholm about his declining the Nobel Prize (see note 1 to doc. 9.11-I). It was followed by a message to the Central Committee addressed to Polikarpov: "Thank you for sending a doctor twice. I have declined the prize. I am requesting you to restore to Ivinskaya her sources of earnings in the Goslitizdat."[28]

But the main event of the day happened in its second half. The grand meeting at the Kremlin on the occasion of the 40th anniversary of the Young Communist League was addressed by the first secretary of the YCL Central Committee V. E. Semichastny. Under the arrangement with the CPSU Central Committee he included in his report a section dealing with Pasternak (doc. 9.10). Besides boorish attacks like comparing the poet with a pig,[29] it also contained threats to strip him of his citizenship and exile him from the country. That was the most terrible blow for Pasternak. When two days later he was signing the letter to N. S. Khrushchev with the words "To leave my country would be for me the equivalent of death" (doc. 9.11-I), there should not be any doubt about his sincerity.[30]

The prospect of becoming an exile completely broke Pasternak down. He had not foreseen such a move from the authorities, nor had he doubted that before the beginning of the ceremony in the Kremlin Polikarpov would inform the YCL leader about his having rejected the Nobel Prize. Some evidence that after Semichastny's speech he discussed with his near and dear ones the possibility of exile and which of them would agree to accompany him[31] only reflected the confusion of a man who was at a loss what to do. Such a finale of the drama directed by the authorities was unacceptable for him.

That is why he agreed without much hesitation to the recommendations given by third party persons to address Khrushchev and ask him not to take any extreme measures against him. The letter was not composed by him, he signed the text already prepared (doc. 9.11-I and note 2 to it). For him it was not the first case of such a procedure humiliating human dignity. He had already had to grudgingly agree to similar concessions when the authorities were trying to prevent publication of the novel abroad.

Before the Nobel story he had made an attempt to explain the reasons making him agree to such coercion to Jacqueline de Proyart, his agent in the matters of publishing *Doctor Zhivago* abroad. In his letter of November 3, 1957, touching upon the telegrams under his signature addressed to the Dutch and the British publishers of the novel, Gallimard and Collins, where he insisted that they hold back the publication, Pasternak confessed that he was happy about the fact that they "did not let government officials fool themselves with false telegrams, which they made me sign threatening to arrest, to outlaw and to deprive me of means of subsistence and which I signed only because I was sure that not a single person in the world would believe these false texts written not by me but by officials and imposed on me."[32]

The letter to Khrushchev, along with the corresponding TASS (Telegraph Agency of the Soviet Union) Statement (doc. 9.11-II) was published in newspapers on November 2. That was the first information on Pasternak's refusal of the Nobel Prize published in the Soviet press. Nobody said a word about the decision taken by him at the all-Moscow meeting of writers on October 31 — although all Russian language foreign radio stations had been broadcasting about that already for two days and, besides, the leaders of the writers organization had been informed through their own channels. Simply passing over in silence was one of the many inquisition ways and means for humiliating the poet "at fault".

At the meeting itself, the atmosphere made you think of a witches' Sabbath. The literary critic L. I. Lazarev, who was present there as a correspondent of the *Literaturnaya Gazeta*, where he was working at that time, vividly portrayed what was going on. "Even more terrible than the speakers", he remembered half a century later, "was the hooting, hysterically aggressive audience. Comments in the shorthand report — 'noise, laughter, heckling' — do not in the least give that incandescence of hatred, thirst for reprisal, desire to crush, to exterminate..."[33]

In different ideological institutions the public vilification of the poet stirred up the most reactionary forces, which in the previous years of the Khrushchev Thaw had tried to keep back their overflowing emotions. The insultingly demagogical rhetoric of the times of the "struggle with cosmopolitanism and obsequiousness before the West" started to sound again. On October 30, the next day after the memorable speech by Semichastny, at a meeting of members of the Communist party in the Presidium of the USSR Academy of Sciences, a secretary of the party organization made the following statement: "Recently we had a congress of Slavonic scholars. [...] Present at this congress in the capacity of a member of the American delegation was a certain Yakobson Roman Osipovich, vice-president of the Committee of Slavonic Scholars, traitor of our Motherland, defector, now professor of Harvard

University. Everybody knows that he is a scoundrel, a spy — but he was met with such obsequiousness [...]," and so on.[34]

Meanwhile, protests against the Moscow orgy were spreading acorss the world. Telegrams came from different parts, they were signed by intellectuals of most different political views. On October 31, the day of the all-Moscow meeting of writers on the "case of Pasternak," the leaders of the Writers Union cabled a selection of these messages to the Cultural Department of the CPSU Central Committee; at the meeting, however, receiving these protests was not mentioned. On November 3 the selection received by the Cultural Department was forwarded to the Secretariat of the CPSU Central Committee, from where the secretary of the Central Committee E. A. Furtseva directed it for familiarization to the Ideological Commission, where the document was enriched by the signatures of the secretaries of the Central Committee N. A. Mukhitdinov and M. A. Suslov. On November 5 it was directed to the archive (doc. 9.13).

It is interesting that the dates of the document going around within the Central Committee correspond well to the events that preceded the appearance of a letter signed by Pasternak in the *Pravda* newspaper, on November 6 (doc. 9.12-II). First of all, revealing is the date when Polikarpov sent doc. 9.13 to the Secretariat of the Central Committee. After he received the selection of telegrams from the Writers Union they lay for three days on his table. They were set going on November 3, the day after the poet's letter to Khrushchev appeared in *Pravda*. According to the memories of Ivinskaya, Polikarpov promised Pasternak not to bother him before the forthcoming November holiday, and only "later, in about a fortnight, he would invite him [...] to compose some new letter." However, on November 4 he changed plans dramatically and suggested a meeting in order "to ask Boris Leonidovich to write an appeal to the people".[35]

There is an evident connection between the date when the document was signed by members of the Ideological commission, about November 3 or 4 (doc. 9.13) and the summons of Pasternak, on November 4, for creating the next epistolary message. Less evident is the reason which made the Central Committee's ideologists initiate active moves practically right after the letter to Khrushchev, and on the eve of the holiday at that.

Of course, writers' protests could hardly have roused the Central Committee into such prompt action. It had to be something more substantial. One can suppose that such an impetus came from India's prime minister J. Nehru, whose reaction to the campaign of the persecution of Pasternak was definitely negative. It is alleged that the latter phoned Khrushchev in person to express his attitude to what was going on (see comment 4 to doc. 9.14). However, the crucial role was rather played by the note of G. A. Zhukov about his talk with K. P. S. Menon, the ambassador for India in the USSR, which Zhukov handed over to the Central Committee on November 4 (doc. 9.14), the very day, or, to be more exact, the very morning when Pasternak was urgently invited to the cultural department.[36] For the Soviet authorities Nehru was a key figure, and they were not going to quarrel with him. Perhaps, it was the factor which made the Party officials take a turn for stopping the hysteria they had unleashed, at least in its public form.

But the campaign was to be finished in a spectacular way — by compelling Pasternak to repent once again, which would also convince the international public opinion that, actually, there was no persecution whatsoever.

Pasternak accepted the rules of the game and wrote, in his own hand, a draft of a letter to the *Pravda* (doc. 9.12-I), which Ivinskaya took to the Cultural Department of the Central Committee the next morning, on November 5. However, Polikarpov was not satisfied with the text. Further on let us give the floor to Ivinskaya.

"Polikarpov said that he and I would 'work on this letter ourselves'. It was a job of downright falsifiers. We took B. L.'s separate phrases written or said at different times and on different occasions and put them together. Taken out of context, they did not reflect the general course of B. L.'s thinking. White was becoming black [...] When I immediately came to Borya with a new version of the letter, where almost all words were his, but his thought was not there — he only waved his hand. He was tired. He wanted to put an end to this extraordinary situation [...] And B. L., committing an irremediable violence on himself, signed this second letter."[37] (doc. 9.12-II).

True, in the finishing passage Ivinskaya is not quite accurate. The personality of Pasternak was present even in the newspaper text, with its original unceremoniously edited by high-positioned censors. First of all, his presence was felt in the words about his connection with Russia through his "birth, life and work," which made "leaving it and going abroad into exile impossible" for him. At that time, seen on the pages of the Party's leading newspaper, this phrase was taken by many as quasi-patriotic blather forcibly put into the poet's mouth. But in fact, these words are one hundred per cent sincere and quite adequate to his vulnerable inner world and his open "unprotectedness" before the face of the cruel power and its assistants, devoid of moral principles. It was the power that he had always been ready to tolerate as an insuperable element, but from which he had always kept his distance with his soul, his ego and innate loyalty and love for everything that a normal person invariably associates with the idea of home and homeland.

Pasternak's words recorded by the British intellectual Isaiah Berlin present a documentary evidence of that. These words were said in a talk with Berlin two years before the drama in the fall of 1958, during Berlin's visit to the Soviet Union in 1956, when the manuscript of the novel was not even handed over yet to Feltrinelli's messenger. We shall quote the poet's utterance as it was given by Berlin: "Now we both are in Russia; wherever you look, everywhere it is disgusting, scaring and loathsome — swinishness overall; however, I seem to be in real ecstasy from all that, I wander and look at everything with enchanted eyes."[38]

Against the background of massive support for the persecuted poet in the civilized world, quite different voices were sometimes heard from the side of Western intellectuals — voices in support of the Moscow anti-Pasternak hysteria.

Among such critics were Elsa Triolet,[39] French writer of Russian origin, and the Italian poet Salvatore Quasimodo, who was awarded the Nobel Prize the year after Pasternak. Quasimodo had also been a member of the Communist Party for a while, in the 1920s, but soon resigned his membership, as he did not agree to the proposal to concentrate his talent on poetical support of the Party's ideology and policy. In his letter to Jacqueline de Proyart Pasternak touched upon the Italian's response — "half a page against me in the the spirit of *Pravda* and the *Literaturnaya Gazeta* "(doc. 9.15). He also mentioned that when he was summoned to the Central Committee's Cultural Department in connection with the preparation of the letter to Khrushchev, Polikarpov, whom Khrushchev's assistant, literary critic V. S. Lebedev will call, in 1964, "Pasternak's killer,"[40] demonstrated this "scathing review" as evidence of "the opinion of the West."[41]

It is quite obvious that the Cultural Department had taken part in the prompt obtaining of this "scathing review", so desirable for the Central Committee, from the poet from the West, who turned out to be at hand because of his unexpected illness.[42] It is also not difficult to understand the ways of approaching him. He was certainly grateful for the first-class medical aid granted to him, the more so that it was free, contrary to Italian medicine; on returning home he expressed his gratitude in the poem dedicated to the nurse Varvara Alexandrovna Kruchinina, who took care of him. This poem was published in *Pravda* newspaper.[43]

The Soviet authorities did not leave this service unanswered. The *Literaturnaya Gazeta* marked Quasimodo's coming out of the hospital and leaving for his homeland, on March 28, 1959, by publishing, this very day, a warm article and his two poems about his native land, Sicily. Two years later the first collection of his poems came out in Russian, and in 1967 it was followed by the second one.[44] The first edition had a detailed preface written by the poet Alexei Surkov, in the recent past the head of the Writers Union. In it he lavishly praised both the poet's works and his utterances at the theoretical discussion of the subject "Poetry and Modern Times" organized by the "Italy–USSR" society in Rome, in October 1957. Then, however, speaking at the discussion itself Surkov never mentioned the name of the orator whom he liked so much two years later. In the verbose writings of the prominent critic Kornelii Zelinsky there are even wider discrepancies. In 1958, before the scandal with Pasternak, in an article by which the *Inostrannaya Literatura* (Foreign Literature) journal preceded a publication of speeches of Italian and Russian poets in Rome, he gave a rather adverse response to the theoretical positions of the hosts of the discussion, including Quasimodo. Whereas in November 1959, when the *Literaturnaya Gazeta* published his article in connection with the news of the Nobel Prize having been given to Pasternak, his voice sounded in quite a different register.[45]

The gratitude to Quasimodo for his support in the campaign against Pasternak made the authorities, for the first time in many years, break the taboo against positive information on decisions of the Swedish Academy. On October 29 the *Literaturnaya Gazeta* published a congratulatory telegram to Quasimodo from the

Board of the Writers Union: "We are sending our brotherly congratulations on the award you have received, Nobel Prize, which this time worthily marked one of the major poets, singer of Italy [...]" But this became the only mention in the mass media of the Nobel Prize as such.[46] Already in the article by K. Zelinsky which appeared on November 10, there is not a single hint at the event which had lead to its publication. In the same way, *Pravda*'s postscript to the mentioned poem "*Varvara Alexandrovna*" characterizes the author as "the well-known Italian poet." The newspaper kept silence on the fact that he was also a Nobel Prize winner. That is, the long-lasting ban on mentioning the Nobel Prizes for Literature was promptly restored and was not removed until 1965 when the Swedish Academy announced awarding the Nobel Prize to Mikhail Sholokhov.

Quasimodo was not the only Western intellectual who joined the campaign of abusing Pasternak. The man who could well compete with him in that was the American writer of Russian origin, Vladimir Nabokov, Pasternak's long-standing ill-wisher. As far back as in 1927, in his review of the poems written by two third-rate poets living in emigration, like him, he also caustically touched upon the works of the young Pasternak. "In Russia there is a rather gifted poet, Pasternak," he wrote at the beginning of this publication. "His verse is convex, goitrous, goggle-eyed, as if his muse is suffering from a disease of the thyroid gland [...]" and so on.[47]

His attitude to *Doctor Zhivago* was as intolerant, to a large extent because he was jealous of the success of the novel, which fell exactly on the times when his *Lolita* was published. Further on we shall speak about the competition of the two bestsellers in more detail. Here we shall only say that, according to the literary critic N. A. Anastasiev, in his private correspondence Nabokov, for whom the success of *Doctor Zhivago* was a painful experience, "from the very beginning spoke of *Doctor Zivago* extremely harshly. And it has to be admitted that after the unquestionable success of this novel with American readers, and especially after the decision of the Nobel Committee [he means the Swedish Academy], these utterances became especially intolerant." For illustration Anastasiev quoted an excerpt from Nabokov's epistle to one of his correspondents, of January 12, 1958: "I think that the phrase 'Nabokov is a second Pasternak' is a journal absurdity. Maybe it will be right [...] to maintain that Pasternak is the best Soviet poet, like Nabokov is the best Russian prose writer, but here the parallels end [...]".[48]

In spite of the wrung-out public repentance, the authorities went on keeping Pasternak in the grip of a tight blockade. Paying royalties was stopped — not only for the translations ordered earlier but also for the already published books. Theatres closed the performances of dramas by Schiller and Shakespeare in his translations, and if they continued to be played, the censorship cut out his name as translator in the posters. His correspondence was controlled and sometimes withdrawn (doc. 9.21).

Not only did all that aggravate his financial situation and everyday problems,[49] but it also gave him unbearable pain. Things became especially hard for him in the winter.[50] On January 16, 1959 he sent Polikarpov a desperate letter (doc. 9.17). In it he reproached the Party leader of the Soviet writers that, in spite of the obedience that he had showed ("I shamelessly lied at your dictation") and repeated promises to lift the imposed penance, "after that I was left out in the cold" and got nothing. About a week later, seized with a feeling of complete hopelessness, he materialized it in the poem *The Nobel Prize.*[51] This cry of a soul in torment was to bring about a completely new situation in Pasternak's opposition to the authorities, to a large extent due to his own imprudence and, on the other hand, due to the obvious unscrupulousness of the Paris correspondent of the *Daily Mail* newspaper Anthony Brown.

Being in Moscow as a tourist, Brown visited Pasternak in Peredelkino, on January 30, 1959. Since he was going to return to Paris the next day, the poet asked him to pass to Jacqueline de Proyart the autographs of his three new poems, to be added to the collection of verse being prepared for publication, *The Nobel Prize* among them. According to E. B. Pasternak, his father did not give the journalist permission to publish his poems in the newspaper.[52] This is confirmed by the information from an official of the Ministry for Foreign Affairs, who happened, by chance, to witness Pasternak handing over his poems (doc. 9.19).

Nevertheless, on February 11 the poem *The Nobel Prize* appeared in the newspaper. The publication is accompanied by Brown's imprudent comment, in which Pasternak's works of the latest years, first of all, *Doctor Zhivago,* are characterized as "a symbol of one man's struggle against the dictates of a rigid, unforgiving totalitarian order". (doc. 9.18)

It was difficult to imagine a better present for the authorities. In the same February, 1959, says E. B. Pasternak, the country was expecting an official visit from a British governmental delegation headed by Prime Minister Harold Macmillan. Such events are always accompanied by large groups of journalists coming from different countries. In order to prevent undesirable contacts of representatives of the Western press with the dishonored poet the authorities ordered Pasternak to leave Moscow and its environs for the time of the visit. "Although choking with indignation", he submitted, once more, to coercion and on February 20, accompanied by his wife, went to Tbilisi, to stay with the widow of his long-standing friend, poet Titsian Tabidze, killed in the years of the Great Terror.[52a]

It was on the day of his leaving for Georgia that two major Soviet leaders started to prepare coordinated anti-Pasternak actions. The chairman of the KGB (the State Security Committee) A. N. Shelepin sent to the CPSU Central Committee detailed information on Pasternak, his contacts, his correspondence, etc. (doc. 9.21). Meanwhile, the procurator-general of the USSR R. A. Rudenko, on the same day, spelled out his proposals, including the option of stripping Pasternak of his citizenship and sending him into exile (doc. 9.22). At the meeting of the Presidium of the CPSU Central Committee (doc. 9.23) these initiatives were slightly corrected. The proposal of expulsion seems to have found no support; anyway, at Pasternak's meeting with the procurator, which followed soon (doc. 9.24), this question was not raised.

Pasternak came back home on March 13 and already the next day was brought to the procurator-general in a not quite usual way. Instead of sending him an official summons, he was offered to get into a car passing by when he went out for his traditional walk in Peredelkino. The main topic at the interrogation was the story of the publication of the poem in the British newspaper. In conclusion Rudenko threatened the possibility of criminal prosecution and warned about the responsibility for divulging the topic of the interrogation.

However, the authorities did draw certain conclusions from the content of the poem made public by the *Daily Mail.* The fact that Pasternak agreed to refuse the expected invitation to visit a reception in the Swedish embassy also played a significant part.[53] In exchange they promised to renew the payment of royalties and to restart preparing for publication the collection of poems stopped by the Goslitizdat publishing house in connection with the story of passing the novel abroad.[54] His name as translator of dramas by Schiller and Shakespeare reappeared on theatre posters.

Nevertheless, his forceful isolation continued. Rather characteristic is the story how the prominent Swedish Slavonic scholar from Stockholm University Oke Nilsson, expert on Chekhov, Mandelshtam and Pasternak, made an attempt to meet the poet in the spring of 1959. A year before, when the Nobel Prize had not yet been awarded, he had had a meeting with Pasternak in Peredelkino and published in the press information on his talks with him.[55] In ideological circles these publications were regarded as "hostile." When he came to Moscow again, with a group of Swedish journalists, and again wanted to go to Peredelkino, the information received from the USSR embassy in Stockholm (doc. 9.25) was used to overtake him at the railway platform and make him return to Moscow without being given any explanations (note 1 to doc. 9.25).

The publication of the poem *The Nobel Prize* in the West attracted unkind attention not only in the Party and power structures of the Soviet Union. The above mentioned Pasternak's adversary, Vladimir Nabokov, also did not fail to make his contribution. On December 27, 1959, less than a year after Pasternak described in his verses the suicidal state of the man driven into a corner by the authorities, Nabokov, staying in blissful San Remo, where he came together with his family and friends for the Christmas vacation, reacted to them by his poem. The implied meaning of this poem, which came into being in a luxurious suite at the Excelsior-Bellevue hotel, where the company was staying, can only be taken as a cruel parody of the wail of a tortured soul.[56]

Here it is appropriate to add that Nabokov met Pasternak in person during the latter's stay in Berlin, from the autumn of 1922 to February 1923. As this author was told by the poet's son E. B. Pasternak, at that time Nabokov had great respect for Pasternak's works, and his personal dislike of his colleague started to appear only in later years.

The reason why Nabokov made such an unjust attack against the dishonored poet is obvious. As already mentioned, Pasternak and Nabokov chanced to be rivals in rating races. Pasternak's *Doctor Zhivago* and Nabokov's *Lolita,* both key works of their authors, were begun in 1948 and both appeared as bestsellers in the books market practically simultaneously, in 1958. As Stacy Schiff, a friend and talented biographer of the Nabokov family, expressively put it, at the end of the 1950s these novels were locked in a mortal combat for the top position in the list of bestsellers. She wrote that *Zhivago* for the first time appeared in this list at the beginning of October, 1958, when *Lolita* was occupying the first place there, but six weeks later *Zhivago* left *Lolita* behind. At first Nabokov tried to control himself, but when he learnt that in a library the list of orders for *Zhivago* was longer than the one for *Lolita*, he became indignant. And Vera Nabokova, supporting her spouse in his righteous anger, in those days set down in her diary the following inference of truly unparalleled absurdity: "The communists succeeded in shoving their shoddy concoction into the club of "Nobel Prize Laureates" — by way of sheer pretense that it had been "illegally taken out" from the USSR! Mass psychosis of idiots, which is conducted by scoundrels of procommunist views [...]" Summing up, the biographer said that "the Nabokovs were both convinced that in the USSR the novel was approved and the condemnation was merely pretended."[57]

In conclusion of this theme let us come back for a while to the place where Nabokov's poem appeared. It is worth mentioning that San Remo is the town where Alfred Nobel died. According to the well-established tradition, every year by December 10, the date of his death, from this town on the shore of the Ligurian Sea a cargo of fresh cut flowers is delivered to Stockholm. They are used for decoration of the stage of the Concert hall, on which the king gives the awards to the new award-winners, and of the Blue hall of the Town hall, where the Nobel banquet is given the same evening. In the same 1959, when enchanting landscapes of San Remo provided an inspiration for Nabokov's another exquisite piece of poetry (see note 56), it was only three weeks after a consignment of flowers expected in the Nobel capital was shipped to it from there. Among those who were surrounded with these flowers when receiving the Nobel Prize was the above mentioned Italian poet Salvatore Quasimodo, one of the most noticeable participants of the campaign of humiliating Boris Pastermak. Was it accidental that this award was given the year after the prize given to the poet who was persecuted in the Soviet Union? Or was there in this decision of the Swedish Academy a secret wish to please the Party bosses from the Central Committee and their yes-men from the Writers Union so as to try in such a way to protect Pasternak against new persecution?

Boris Pasternak died on May 30, 1960. The cause of death was a cancerous tumor in his lungs. Undoubtedly, the total torment that he had been subjected to over the four last years of his life precipitated the decease. The authorities realized it only too well and took the funeral of the great son of his homeland as a continuation of their opposition to him. A short report of the death appeared only in the *Literaturnaya Gazeta* on the day of the funeral, without mentioning the time and place of the burial;[58] the radio and television did not get permission for publication of this sad information.[59] Handwritten advertisements about the funeral, which the poet's admirers put up on Kievski railway station, from which commuter trains go to Peredelkino, were torn off by the people specially charged with the task, and the procedure of the burial became the subject of a special report to the Central Committee's Secretariat (doc. 9.26).

On the second day after the funeral, as if in a state of catharsis, a poem burst out from the poet German Plisetsky: "[...] poetry that scumbags have put in the ground [...]" (see note 64 to this chapter).

The indelible shame of the dirty anti-Pasternak campaign is entirely on the conscience of the Central Committee's ideological structures. The personal responsibility of the country's chief person, Nikita Khrushchev is also beyond any doubt. After his forced resignation he realized it and expressed regret that when he had had in his hands all levers of power he had not used them. The definition "police measures" offered by him to characterize the actions of the authorities gave an exhaustive diagnosis of what had been going on (doc. 9.27).[60]

However, those who had executed the "police measures" did not feel any twinges of regret, nor did they have any need to repent before the memory of the poet whom they had ruined. Nearly two years after his death, on February 26, 1962, the chairman of the USSR Council of Ministers' State Committee on Cultural Connections with Foreign Countries G. A. Zhukov informed the CPSU Central Committee that one of the Italian newspapers published an interview with the young Moscow writer Vasily Aksenov, who dared to call Pasternak "a great poet" and "also made some other things totally unacceptable from our point of view."[61]

On May 5, 1962, owing to the head of the Glavlit (the state censorship committee) P. K. Romanov, Pasternak's name again becomes the subject of consideration in the Central Committee. Now it is connected with the "apologetic assessment" of the poet's works by the writer Ilya Ehrenburg in his memoirs *Lyudi, Gody, Zhizn'* (People, Years, Life) (doc. 9.29).

In 1967, in the framework of preparations for celebrating the 50th anniversary of the October Revolution, the poetry evening "From the History of the Soviet Poetry: Block, Esenin, Mayakovsky, Pasternak" took place in the Central House of Writers. A special memo from the Cultural Department to the Secretariat of the Central Committee described what was said in the auditorium. Among other things, the memo accused the organizers of the evening of exaggerating the significance of Pasternak's works. It was also noted, with condemnation, that "B. Akhmadulina read two enthusiastic poems dedicated to the memory of Pasternak", and "in conclusion pieces of music by B. Pasternak were performed."[62]

Shortly after this memo, as E. B. Pasternak wrote in his preface to the collection of documents *Pasternak i Vlast'* (Pasternak and Power), publishing a collection of Pasternak's works at the Iskusstvo publishing house was stopped, and the typeset of the book was undone.[63]

L. K. Chukovskaya in her diaries told about one more episode of the combat with the shadow of the great poet. It deals with Anna Akhmatova's poetic response to the decease of the poet:

A voice ineffable's no more. He's gone.
He'll never talk to coppices again.
He must have turned now into ears of corn
Or one of his old friends, the finest rain.

And every single flower in this world
Has blossomed out in honor of that death.
And one can hear a hush on this globe fall —
The globe that we call modestly the earth.[64]
Translated by Sergei Roy

The poem was written in the Botkin hospital, where the poetess was staying at that time with the diagnosis of intercostals neuralgia. According to the entry in the diary, Chukovskaya first heard the poem on June 6, 1960, when she came to visit the sick Akhmatova.[65] However, when publishing this poem called by Akhmatova *In Memory of Boris Pasternak* in the October, 1964 issue of the *Znamya* journal, she had, at the demand of the censorship, to change the title to a less specific one, *The Death of a Poet*, and to agree to remove the date when it was written. Pointedly trivial, repeatedly used by other poets, this title must have lulled the censor's vigilance, and he did not pay attention to the fact that nearly 35 years before that Pasternak responded to the death of Mayakovsky with a poem with the same title.[66] Outwardly submitting to the coercion, Akhmatova, in fact, outwitted the censorship. Calling up the past, the verse so entitled gave all the more strength to the secret sounding of her poetic wreath laid on the grave of the poet killed by the unenlightened power.

But still it was the censor that had the last word. When her collection *Beg Vremeni* (The Flight of Time), the last published during her lifetime, came out in 1965, she found that "already without asking me the scoundrels changed the dates: instead of 1960 [...] they put 1957, in hope for the lack of memory of the readers [...]".[67]

Those who hated Boris Pasternak did not abandon their attempts to remove the name of their great compatriot from the people's memory.

And here a pertinent question comes to mind. Among the powers that be was there anybody who — not after the resignation, like Khrushchev, but when they were firmly holding the reins of power — got alarmed when the authorities and their innumerable assistants were killing, in sight of the whole world, a great poet who found himself to be a guiltless victim of the totalitarian regime? Anybody who felt something like an inner need of self-purification, in order to analyze *post factum* what had happened and to tell of their idea how to prevent something like that in future?

It turned out that there was such a man. He showed himself over eleven years after the death of Pasternak, when the Central Committee's ideologists, who had learned nothing, surprised the world by another, no less clumsy propaganda action. This time chosen as a victim was Alexander Solzhenitsyn, who was awarded the Nobel Prize, like Pasternak (see Chapter 11). The man in question was the USSR

Minister for Internal Affairs N. A. Shchelokov. He did not leave the best memories of himself,[68] but in this case, bringing out his better nature, he managed to analyze the lessons of the anti-Pasternak orgy, although with a good deal of overt cynicism characteristic of the top Soviet bureaucracy.

In the report "Regarding the question of Solzhenitsyn" that he sent, in the fall of 1971, to the General Secretary of the CPSU Central Committee L. I. Brezhnev, he explicitly formulated his opinion of what was going on in the late 1950s around Pasternak and his Nobel Prize. He recommended not to repeat "the same grave errors that were committed in dealing with Pasternak" (doc. 9.30). Alas, it was no good making these warnings. The authorities once again carelessly made the same regrettable blunder.

The early hours of November 10, 1989 were marked by an event which became a turning point for the second half of the 20th century. For the first time in history the gate of the Berlin wall swung open for everybody, and a week later the wall itself collapsed. The destruction of this symbol of enslavement of the man, which was, indeed, a cancerous tumor of the post-war Europe, was being accompanied by the sounds of a cello suite by Johann Sebastian Bach, performed by the great musician Mstislav Rostropovich, Pasternak's compatriot. A month after the materialized stronghold of totalitarian regime in Berlin started to show signs of cracking, in Stockholm, on December 9, one of its metastases was solemnly amputated. Then at the reception, traditionally organized during the Nobel festivities on the eve of the day of the award-giving ceremony in the great hall of the Swedish Academy, an end was put to the 31-year-old injustice. The permanent secretary of the Swedish Academy Sture Allen gave the gold medal of the Nobel Prize laureate belonging to Boris Pasternak to his son.

This ceremony was preceded by the initiative of the Swedish ambassador in Moscow, Erian Berner, who turned to the Swedish Academy with the suggestion that Pasternak's rejection of the prize he was awarded should be declared invalid as made under duress, with explicit threats on the part of the power. Such a situation was not new for the Nobel institutions. In 1939 the German chemists Adolf Butenandt and Richard Kuhn and biologist Gerhard Domagk were forced to decline the prize, as after the Nobel Peace Prize was awarded, in 1936, to the prisoner of a Nazi concentration camp Carl Ossietzky, Hitler declared the Nobel awards illegal and forbade the citizens of the Reich to receive them. That is why the Royal Academy of Sciences did not have any difficulty stating obvious signs of coercion in this affair[69] and giving to all three their laureate diplomas and gold medals.[70] In 1989 the Swedish Academy came to the same decision.

On December 1, 1989, at the opening of the exhibition "The World of Pasternak" at the Moscow museum of Fine Arts, the ambassador Berner passed to the organizers the diploma of the Nobel Prize winner to be exhibited.[71]

The gold medal was to be given by the Swedish Academy personally to the son of the deceased laureate, within the framework of the Nobel ceremonial events of the same year. At the ceremony of giving the medal Sture Allen in his speech distinctly formulated the signs of coercion used in that case, which, in fact, did not at all differ from what had taken place exactly half a century before in Nazi Germany.

And here again, like half a month before at the Berlin wall, sounded the divine cello tunes of Mstislav Rostropovich, who specially came to Stockholm. And again these were the solemn and spirited chords of Bach's suite. It was as if the musician made a visible sound path between the symbolism of the two events inwardly connected with each other: the destruction of the Wall, identified with the inhumanity of the totalitarian power, and the restoration of the honor and dignity of the great poet, a guiltless victim of totalitarianism.[72]

Before performing the suite, on December 10, from the landing of the famous staircase at the Blue Hall of Stockholm's City Hall, Rostropovich said a few words at the traditional Nobel banquet. "At this magnificent festival," he said, "I would like to make you think of the great Russian poet Boris Pasternak, who during his lifetime was devoid of the right to receive the award he won and enjoy the happiness and honor of being the Nobel Prize laureate."[73]

In conclusion of this narrative about the fate of the great Russian poet I shall allow myself to mention one more fact concerning him. This time it is of quite the opposite, truly phantasmagorical sort — something that was said nearly half a century after the poet's death and that during his lifetime could not have appeared to him even in his most terrifying dreams.

The prolific writer C. G. Kara-Murza in one of his chaotically verbose pieces of writing tried, like some authors before him, to substantiate the idea that the Jews are historically to blame for all the troubles of Russia. For the role of one of these Judaic plotters he decided to offer Boris Pasternak, making him personally at fault for the national sin of our homeland, for which each of us is basically responsible — that is, for Stalinism.[74] He stunned the reader by the unsubstantiated statement that "the first conceptual creators of the cult of Stalin were, in 1934–1935, Isaak Babel and Boris Pasternak."[75] Without giving any reason, even if debatable, as if in passing, he kicked both these victims of the totalitarian regime. He was not concerned about the memory of these writers, like a voluptuous womanizer is not concerned about the fates of the girls he seduced. He did not even consider it necessary to give at least a brief explanation why he tied up this gibberish with the mentioned years, already very close to the approaching époque of the Great Terror, and not with the other years more corresponding to the true time when Stalin's bloody cult originated.

In conclusion I would like to support my impulsive reaction to Kara-Murza's vicious slander by what the great poet himself said about Stalin. This was said in 1956: "For so long we were reigned by a madman and murderer, and now, by a fool and swine; the murderer had some impulses, he felt something intuitively, despite his abysmal obscurantism." Giving dues to Stalin's inborn intuition Pasternak concludes his thought by the following tough statement: "[...] now we are seized by the reign of mediocrity."[76]

Notes

[1] "Artes," No. 1, 1983, p. 112.

[1a] The information about Pasternak being nominated for the 1946 prize somehow reached the Department of Propaganda and Agitation of the Communist Party Central Committee and caused angry reaction there. On March 21, 1946 *Kultura i Zhizn',* organ of the Central Committee, published an article by the secretary of the Soviet Writers Union A. A. Surkov entitled "On the poetry of Pasternak" and certainly written on the orders from above. Here are some of the expressions used there: "reactionary backward outlook," "lives in discord with the new reality", "Soviet literature cannot put up with his poetry," and so on.

[2] Now we have the possibility to see the documents of the Nobel Committee for Literature up to 1953 inclusive, which gives food for thought concerning Pasternak's nominations in the first five post-war years. Bowra, with his comprehensive knowledge of his works, is the most appropriate figure for a nominator of the Russian poet.* Recommending his protégé for two years running, he must have expected that his proposals would rapidly be a success. Lamm, although not a Slavonic scholar, might have made an opinion of Pasternak's poetry through translations of his verse and, as a voting member of the Swedish Academy, the institution endowing the literary prizes, might have supported the initiative of the renowned British Slavonic scholar in the discussions of 1946 and 1947. In 1948, probably being informed that the Committee did not receive the nomination from Bowra,** before the end of the nominations, he decided to nominate Pasternak in his own name.*** In 1950, because of the similar circumstances, he introduced the poet from Moscow to the Nobel Committee for the second time.**** In the same 1950 Lamm died, while Bowra, who must have made sure that his further insistence would have been useless, stopped sending his proposals to Stockholm. Hence the seven-year break, until 1957, when the candidature of Pasternak was absent in the lists of the Nobel Committee's nominees.***** But when, after awarding the prize to the poet, in 1958, the latter turned out to be in the center of a scandal, Bowra was the first among British intellectuals who signed a telegram condemning the persecution of the poet respected by the whole civilized world (see doc. 9.13-III).

* It was Bowra, the first nominator of Pasternak in 1946, who was mentioned, in 1949, in the scandalous article by A. K. Tarasenkov. Having stated that "Pasternak is supported by enemies of the Soviet people," he introduced as such the renowned poet and Slavonic scholar in the following passage: "The English professor O. M. Bowr, when speaking about Pasternak's poetry is just beside himself with delight. 'Remarkable impressionability,' 'a great talent of giving emotions,' 'dynamic perception of life', 'a powerful poet of the Russian world' — such expressions overflow in the article by the honorable professor in the *British Ally* No. 5 1946." (A weekly of the British information ministry published in Moscow in the period of the war and in the first post-war years.) *Znamya* No. 10, 1949, p. 168.

** According to the information received from the Executive Director of the Nobel Foundation M. Sohlman, M. Lamm, from 1948 till his death in 1950, held the position of official inspector of the Nobel library at the Swedish Academy. So it was not difficult for him to be in the know of the nominations received and, if there were no proposals on the persons he was interested in, to manage to prepare his introductions in time.

*** An indirect confirmation of the suggestion described is given by the following curious coincidence. In 1948 Lamm, along with Pasternak, nominated one more candidate, the Israeli writer Schmuel Agnon, who was also nominated, in 1947 and then in 1949, by Hugo Bergman, professor of philosophy at the Jewish University in Jerusalem. In 1948 there was no nomination of him in the Nobel Committee, and Lamm decided, like in the case with Pasternak, to hedge the candidature of Agnon, who eventually got the Nobel award in 1966.

**** It is worth adding here that such hedging was also done in the committees dealing with scientific awards. One example of such kind is the nomination of the German scientist Carl.Bosch for the 1930 Nobel Prize by the chairman of the Committee for Chemistry V. Palmer. Previously, starting from 1916, his candidature had been repeatedly nominated by his German colleagues, but in 1930 none of them nominated him (see Chapter 3).

***** See also comment 1 to doc. 9.1.

[3] E. B. Pasternak. *Boris Pasternak. Biography*, Moscow, Citadel, 1997; O. V. Ivinskaya. *Gody s Borisom Pasternakom. V Plenu Vremeni* (The Years with Boris Pasternak. In Captivity of the Time), Moscow, Libris, 1992.

[3a] *Novy Mir* No. 6, 1988, p. 235.

[4] Actually, this strange silence is not all that surprising. Hemingway was given the Nobel Prize, in 1954, not just for his "narrative mastery" but also for "showing this mastery once again," specifically, in the book *The Old Man and the Sea.* This formulation by the Swedish Academy was at variance, once again, with the tastes of officials from the CPSU Central Committee's Cultural Department, who found this book "naturalistic and uninspired" (see doc. 3.33).

[5] The French translation of the novel *Doctor Zhivago* came out in June, 1958; the English in August; the Danish, German and Swedish in the fall of the same year.* The Russian translation was published in the Mutona publishing house (Holland) in August 1958, just before the end of the consideration of the 1958 nominations at the Nobel Committee. "That," wrote E. B. Pasternak, "was the resolution of the unnatural situation — world popularity with no publication of the text in Russian — which made it difficult for the Nobel Committee to take a decision [...] The book was not on sale. Part of the edition was handed out to visitors of the World Exhibition in Brussels.** As rumor had it, Soviet representatives brought out some books from the exhibition, and these books were later given 'for familiarization of members of the Central Committee' during the Nobel scandal. The total number of copies printed was slightly over 500."***

* *Continent* 95, No. 1, 1998, p. 237.

** B. Pasternak wrote to Paris to P. P. Suvchinsky, on September 14, 1958; "Is it true that D[octor] Zh[ivago] has come out in the original? It was allegedly seen by travelers at the exhibition in Brussels"; see: Vadim Kozovoi. *Poet v Katastrophe* (The Poet in Catastrophe), Moscow, Gnozis, 1994, p. 250.

*** *Novy Mir* No. 1, 1992, p. 151.

[6] In Moscow many people knew, as early as in 1956, that Pasternak finished the novel, and some already managed to read the manuscript. Among the latter was the recent Gulag prisoner, writer V. T. Shalamov. This is how he described his impressions from the novel in his letter to the recent fellow prisoner A. Z. Dobrovolsky, at that time living in exile in the Kolyma area in the north of Siberia: "[...] for him everything that happened to us did not

pass without leaving a trace, and he managed to keep flying the flag of the great Russian literature. If you read his novel, his work of genius, you would see that all these issues are raised and the answers are asserted with the force of Tolstoy [...]"*

The following information, among other things, shows how popular the novel was in the world. The well-known American publisher Andre Shiffrin, who was director of the Panteon publishing house for many years, wrote in his book that in 1957 they bought the rights to one Russian novel rather difficult for perception. The first edition made four thousand copies. But awarding the Nobel Prize to its author, Boris Pasternak, turned the novel into a world bestseller. As a result, it sold more than a million copies in hardback and five million more in paperback.**

* *Znamya* No. 6, 2001, p. 142.

** A. Shiffrin. *Legko li Byt' Izdatelem* (Is it Easy to Be a Publisher), Moscow, *Novoe Literaturnoe Obozrenie* (New Literary Review), 2002, p. 44.

[7] A. Camus. *Selected Works,* Moscow, Panorama, 1993, pp. 401–420.

[8] Eric Mesterton, Slavonic scholar from Goteborg, one of the experts who discussed Pasternak's nomination in 1958, said in his telephone interview, on December 9, 1999, that when spelling out the reasons for awarding the prize, the members of the Nobel Committee deliberately avoided mentioning the title of the novel, "in order not to irritate Moscow," since they were well aware of the possible consequences (see also note 14 to this chapter).

[9] Among other newspapers, the *Literaturnaya Gazeta* gave a whole page, on November 1, 1958, to the publication of extracts from "letters of working people." As the novel was forbidden in the Soviet Union and only very few people could read it, a scathing slogan came into being at that time in the country: "I haven't read Pasternak, but I'll speak my mind about him."

[9a] *Izvestiya,* July 28, 2006, p. 4.

[10] Ivinskaya, p. 260.

[11] He means the books *Cement* by F. V. Gladkov and *How the Steel Was Tempered* by N. A. Ostrovsky (1932–34).

[12] A. Camus, p. 410. In the summer of 1958 Camus sent Pasternak an autographed copy of his *Swedish Speeches.**

* E. B. Pasternak. *Boris Pasternak. Biography,* Moscow, Citadel, 1997, p. 701.

[13] *Aguitator* No. 22, 1958, p. 63. It will be useful to add to this quotation that during the Nazi occupation of France the "pro-fascist" Camus was an active participant of the Resistance movement; he was widely known among French intellectuals, and his being awarded the Nobel Prize was welcomed by the press of most different political shades.

[14] The prehistory of the Swedish Academy's decision was told to this author in answer to his questions, on December 12, 1999 by Eric Mesterton, one of the active witnesses of this decision being taken. In the summer of 1958 he met with Pasternak in Moscow* and recorded his voice on the tape.** On his return to Stockholm, Esterling, secretary of the Swedish Academy, asked his advice if, given the political situation in Moscow, it was worth awarding the Nobel Prize to a Russian poet, to which Mesterton answered in the affirmative. Later he was long tortured by the thought that he might have unintentionally become one of those to blame for Pasternak's untimely death. Being in Leningrad, in 1962, he asked Akhmatova if Pasternak wanted to get the Nobel Prize. The answer was unambiguous: "Yes, he desired it most of all."***

* Mesterton might have met with Pasternak as a secret emissary of the Nobel Committee. In the same way as he visited Akhmatova, in 1962, and Arthur Lundkvist visited Paustovsky (see Chapter 10).
** According to Evgeny Pasternak, Mesterton's recording of the voice of Boris Leonidovoch is the only one remaining today.
*** In his talk with this author Mesterton said this phrase in Russian.

[15] The wife of Vsevolod Ivanov, Tatyana Vladimirovna, remembered: "Pasternak asked our advice — Fedin gave him two hours to think about whether to decline the prize. Vsevolod said: 'Do what you yourself think you should do. Don't listen to anybody. Yesterday I told you many times [when the Ivanovs came to Pasternak in the late evening of October 23 to congratulate him on winning the prize] and I'm repeating it today: You are the best poet of the époque. You deserve any prize.'* 'Then,' Pasternak exclaimed, 'I'll send a telegram of gratitude.' 'That'll be fine!' Vsevolod said with a smile."**

* The members of the Ivanov family were acutely aware that their friend and neighbor in Peredelkino was a genius. The quoted words of the head of the family adequately reflected his genuine worship of Pasternak's verse. Nearly half a century later, the son of the Ivanovs, Vyacheslav Vsevolodovich, now a prominent linguist and philologist, member of the Russian Academy of Sciences, told a correspondent of the *Izvestiya* newspaper what he remembered about his talk with the poet shortly before his death. Pasternak made an attempt to figuratively explain his own vision of his verse being so complicated. He said that, in his opinion, "he did in art what science does: it does not describe reality as it is, since it is covered by a curtain, but it describes the movement of this curtain, which is in some way connected with what is hidden behind it [...]" On his own behalf Vyacheslav Vsevolodovich added that what was said by Pasternak can be interpreted as a spontaneous analogy of the probabilistic understanding of the picture of the world, that is, as a deep correspondence between his poetics and fundamental notions of the quantum physics.
** In the book *Vospominaniya o Borise Pasternake* (Memories of Boris Pasternak), Moscow, Slovo, 1993, p. 255.

[16] From the diary of K. I. Chukovsky: "September 1, 1956. Yesterday I visited Fedin. He told me as a great secret that Pasternak handed over his novel *Doctor Zhivago* to some Italian, who is going to publish it abroad [...] The novel, as Fedin says, is 'a work of genius.' Extremely egocentric, proud, satanically supercilious, exquisitely simple and at the same time bookish all throughout — autobiography of the great Pasternak. (Fedin was speaking about the novel with inspiration, walking about the room and gesticulating — showing great subtlety and deep insight — I enjoyed watching him...) He also admired Pasternak's translation of *Faustus*, the simplicity of speech in this translation, its flexible and rich phraseology, 'as if he learned by heart the whole dictionary of the Russian language.'* We went for a walk — and Fedin left me with such a radiant impression as I had not long had."**

* The same duplicity of the attitude to Pasternak — one to be officially expressed in the presence of the authorities and completely opposite when alone or with someone who could be trusted — was characteristic of the general secretary of the Soviet Writers Union, A. A. Fadeev; see note 48 to Chapter 4 and footnote*** to it.
** In the book *Vospominaniya...*, p. 270–271.

[17] Published in the *Novy Mir* journal No. 2, 1990, p. 149. In connection with the last phrase quoted it is worth recalling that Pasternak died less than two years after the letter was sent.

[18] In the book *Vospominaniya...* p. 277. K. I. Chukovsky had long admired Pasternak as a brilliant poet and a man of high culture. As early as in 1924 he wrote to his son Nikolai, who at that time was keen on writing poems: "Now only a man who is broadly educated (like Block or Gumilev, Pasternak or Khodasevich) can be a big poet."* The company in which Pasternak, young at that time, was mentioned speaks for itself.

* *Znamya* No. 1, 2004, p. 143.

[19] What was said by the renowned literary critic D. D. Blagoi more than a year before the mentioned writers' meeting can be seen as involuntary testimony that for Slutsky his forced participation in anti-Pasternak actions was highly unnatural. At the Communist Party meeting at the Institute of World Literature of the USSR Academy of Sciences, on June 20, 1957, he touched upon both these poets in close connection with each other. Stigmatizing young people from the university for their taste for poetry beyond the framework of socialist realism he mentioned Pasternak,* and finished this idea in the following way: "...and also Slutsky, whose poems I loathe."**

* See note 4 to doc. 10.30.

** The Central Archive of the Social and Political History of Moscow (CASPHM), fond 2472, opis' 1, ed. khr. 18, l.74.

[20] Almost at once after that meeting, Slutsky expressed what was torturing him in the following poetic lines:

For pleading illness I have no skill,
Nor talent to be not at home, like some of you.
That's why I have to plunge, against my will,
Into the dirt nobody else will ever get into...*

He returned to this shameful experience in 1964:

... I was a coward. And this experience,
Whatever it might be called,
Has been eating me with perseverance
From inside, like some caustic salt.

Salting all my actions and thinking,
Every minute it is with me,
Slightly trembling, tapping and tinkling,
No escape from it can I see.**

The poet Grigorii Pozhenyan, a friend of Slutsky, wrote later that for many years he had been "tortured by his sin" and "the doctors suspected that his speech at that infamous meeting became the cause of his incurable mental illness." The last straw for the poet was the death, in February 1977, of his wife, the closest person to him by the cast of mind; as wrote the literary critic Y. L. Boldyrev, his long-standing friend and commentator of his works: "in two and a half months he spoke out, in several thick notebooks, everything that he still wanted to tell people, and went away into the illness, into silence, into darkness — for nine terribly long years."*** By the way, Pozhenyan in the same newspaper article mentioned the chairman of that "infamous meeting": "Sergei Sergeevich Smirnov always suffered and many times when he met me, especially at table, blamed himself, making no secret of the fact that he was ashamed of what he had done."****

* Boris Slutsky. *Collected Works*. Vol. 1 — Moscow: Khudozhestvennaya Literatura, 1991, p. 306; the first posthumous publication appeared in the *Znamya* journal, No. 1, 1988.

** *ibid*, Vol. 2, p. 40.

*** the *Ogonek* magazine, N020, 1989, p. 26.

**** the *Sovetskaya Kultura* newspaper, October 13, 1988, p. 6.

[21] Three and a half decades after these memorable events the information appeared in the press that the secretary of the Soviet Writers Union K. V. Voronkov in person had had unsuccessful talks with Leonov. In answer to Voronkov's attempts to persuade him to speak against Pasternak the 60-year-old writer said: "I already do not have enough time to atone for my sins by prayer."*

* *Izvestiya*, January 20, 1993, p. 7.

[22] In spite of the pointedly asserted style of Polikarpov's memo, at the meeting he himself did not look certain that the decision approved in advance was correct. The silent opposition was not represented by Kirsanov alone. That was also the attitude of many people who were present at the meeting but preferred to spend the time in the bar and in corridors rather than in the auditorium. As the poet K. V. Vanshenkin remembered, Polikarpov himself "at some moment doubted whether the action was expedient." Having suddenly left his seat in the presidium of the meeting, he caught some participants wandering outside the auditorium, brought them into a separate study and "asked at once, without an invitation to sit down: 'So is it necessary to expel him or not?' 'He himself,' continued Vanshenkin telling this story to A. T. Tvardovsky, 'could not even halt the developments for some time and went to call Suslov,* who had sent him, asking on the way the opinion of some more writers, to be more certain. Suslov was out, and Polikarpov came back to the auditorium, where the meeting was drawing to a close.'"**

* A secretary of the CPSU Central Committee, Polikarpov's immediate superior.

** the *Voprosy Literatury* journal, No. 2, 1990. pp. 162–163.

[23] *Znamya*, No. 8, 1989, pp. 179–180

[24] As many other representatives of the Soviet creative intelligentsia, Tvardovsky suffered from this disease for many years, which eventually was one of the causes of his early death.

[25] E. B. Pasternak, p. 705.

[26] Pasternak's wife and son have similar memories of the poet's reaction to sending a doctor. The former says: "I told her: your precautions are unnecessary, he is not going to put an end to his life, quite the contrary."* The latter adds: "He spoke ironically about the doctor sent to him on the instruction from the Central Committee and staying on duty to avoid a suicide attempt."** The doctor stayed at Pasternak's house for three weeks.

* In the book *Vospominaniya...*, p. 223.

** E. B. Pasternak, p. 705.

[27] "A few years later [...] Olga Ivinskaya told us that that day she came down on Pasternak, reproaching him for his light-mindedness and selfishness. 'They won't do anything to you, but I will be ruined completely,' she told him over the phone. At that time she was very frightened by the refusal of the publishing house to give work to her. This reproach exhausted his patience."*

* E. B. Pasternak, p. 704.

[28] *Continent*, 95, No. 1, 1998, p. 242, the original is in the Russian State Archive of Literature and Art (RSALA), fond 379.

[29] Such an epithet used by Semichastny might have been not accidental and having the underlying cause. In 1956, in response to the suicide of A. A. Fadeev, Pasternak wrote a poem in three stanzas. Naturally, it was not published anywhere, but was widely read in manuscript.* This poem might have proved to be known to Semichastny or to his speechwriters who composed the text of his speech. Thus the comparisons which sounded from the tribune would have been sort of answers to that poem written two years before. Although taking into account the 1999 revelations of Semichastny, then living, (see note 3 to doc. 9.10) it is no less probable that they are entirely on his conscience. However, the boorish figurativeness of Semichastny's words had a predecessor, about whom the Comsomol leader might have had no notion. In 1949, on the wave of unbridled anti-cosmopolitan hysteria, in the report "Problems of Soviet Poetry (results of 1948)" read at the meeting of the poetic section of the Soviet Writers Union, the beginner poet M. K. Lukonin allowed himself the following analysis of the work of his older colleague: "Pasternak was only satisfied with and valued the fact that he was recognized by the degenerate rubbish abroad. He was always picked up by our enemies in order to oppose him to us. All his life in our poetry he was a pig under the oak."**

Touching upon the poem about the death of Fadeev, the literary critic N. B. Ivanova found in it, nearly half a century later, some parallels which previously had not attracted any attention. "Their rhythm and meter and their pattern of rhyming", she comments, "completely correspond to the 1936 poem 'I like the mulishness of a writer...'"*** In her opinion, these stanzas are directly connected with the period of Pasternak's contacts with Stalin, more specifically, with the moment of his preparing and sending, at the beginning of December, 1935, a letter addressed to the leader, where he asked him to facilitate the release from prison of A. A. Akhmatova's husband and son, who had been arrested. The positive result of his protection did not keep him waiting too long. As Ivanova supposes, "The letter is composed when the poem, later published in *Izvestiya*, most probably had already been written, or is already on the tip of his pen, [...] or is hovering in his conscience — in any case, there is hardly ten days between the letter and the poem..."**** All that research was necessary for her in order to give grounds to the poet's feeling of outward "unprotectedness" in the late 1950s. This time, "which not only brought him joy, but in many things warned and worried him — as it turned out, not in vain — dealt him a mortal blow."***** So, he opposed the Khrushchev Thaw times to the 1930s when Stalin himself did not shun contacts with him and listened to his requests.

* The cult of personality
is clogged with dirt, forty years on,
but the feral cult of conformity
is still in demand.

Even Pravda is unbearable:
truth a group photograph
of piggy faces. Every day.
Still in demand, the cult
of denunciation by philistines,
so people shoot themselves
because they are drunk,
drunk because they cannot bear it.

Translated by Ann Pasternak Slater and Craig Raine

Published: *Boris Pasternak. The Tragic Years.* 1930–1960. by *His Son Evgeny Pasternak.* — London: Collins Harvill, 1990, p. 218.

** Zvevda , No. 3, 1949, pp. 184–185.

*** I like the mulishness of a writer
at the top of his bent. He's forgotten
to phrase-monger, forgotten his audience,
would like to forget his previous books...
Translated by Ann Pasternak Slater and Craig Raine

Published: *Boris Pasternak. The Tragic Years...* p. 94.

**** N. B. Ivanova. *Pasternak I Drugie* (Pasternak and Others). — Moscow: EKSMO, 2003, p. 417

***** *ibid*, p. 434.

[30] According to the information found by N. B. Ivanova, the idea of the forceful expulsion of Pasternak belonged to N. M. Gribachev and S. V. Mikhalkov.* Pasternak's French correspondent, the young Slavonic scholar Jacqueline de Proyart, whom the poet sometimes called Jacqueline Yakovlevna, in the Russian manner, wrote, in 1991, in the preface to the collection of letters received from him, that he attached great importance to getting his novel *Doctor Zhivago* published as soon as possible: "He considered it his duty; in order to fulfill it he was ready to make any sacrifice — except expulsion from his home country."**

* N. B. Ivanova, p. 413.

** *Novy Mir*, No. 1, 1992, p. 127

[31] For example, *Vospominaniya...*, pp. 223–224; E. B. Pasternak, p. 797; O. V. Ivinskaya, p. 281. The fact that such talk on the part of Pasternak was not very serious is confirmed, among other things, by the words of Ivinskaya that forceful separation from his homeland, "with his love of the style of life of many years' standing would have certainly killed him."*

* O. V. Ivinskaya, p. 283.

[32] *Novy Mir* No. 1, 1992, p. 137.

[33] *Znamya*, No. 7, 2003, p. 132. Nevertheless, even in such an atmosphere, extremely difficult for a creative person, Pasternak did not in the least break his habitual working rhythm of many years. It was in those weeks that he completed, in a short space of time, the translation ordered to him of a play by the Polish poet and playwright of the first half of the 19th century, Juliusz Slowacki.*

* *Literaturnaya Gazeta*, No. 5, 1990, p. 5.

[34] CAODM, fond 992, opis' 2, ed. khr. 69, l. 129. As an antithesis to this irresponsibly slanderous statement it is worth adding the words of one more opponent, but already from "the other shore". Brian Boyd in his comprehensive biography of Nabokov, drawing attention to the groundless suspiciousness about other people, inherent in the writer's nature, mentioned, among other examples, R. O. Yakobson, whom Nabokov considered a Soviet agent.*

* B. Boyd. *Vladimir Nabokov. Americanskie Gody (The American Years).* Moscow, publishing houses Nezavisimaya Gazeta and Symposium, 2004, p. 374.

[35] O. V. Ivinskaya, p. 326.

[36] On the cover letter from Zhukov (doc. 9.14) there is an office note saying that Polikarpov read the document on November 10, 1958, that is, six days after it was received at the Central Committee and signed by Suslov. There cannot be any doubt that the head of the Cultural Department read the memo about the talk with the Indian ambassador on the same day, that is, on November 4, and the note appeared on the document already later, before handing it in to the archive.

[37] Ivinskaya, p. 326–327.

[38] Isaiah Berlin. *Istoriya Svobody. Rossiya* (The History of Freedom. Russia), Moscow, *Novoe Literaturnoe Obozrenie*, 2001, p. 450.

[39] Elsa Triolet, sister of Lilya Brick and wife of the well-known French writer Luis Aragon, wrote in her article about Mayakovsky and Pasternak published in the Paris weekly *Les Lettres Françaises* that she had not read the novel in a foreign language edition and would read it only in the original. In conclusion she summed up that the publication of *Doctor Zhivago* "will probably harm its author but not the author's homeland."*

* See Vadim Kozovoi. *Poet v Katastrophe,* Moscow, Gnozis, 1994, p. 237.

[40] *Znamya* No. 7, 2001, p. 134. It is worth adding to the quoted phrase that Polikarpov had shown an inclination for the role of a killer long before the anti-Pasternak campaign, which he conducted as the head of the Cultural Department of the Central Committee. He had managed to make himself famous in this field when, in July 1951, he occupied the position of director of the Lenin Moscow Pedagogical Institute. He marked his starting to manage the people who were new to him by prompt expelling 15 leading teachers from the institute. As grounds for dismissal he used such exotic formulations as, for instance, "for secretly bringing bourgeois nationalism and Jewish chauvinism into the teaching work." Even at those times, which were hard for creative intelligentsia, such large-scale measures directed against highly qualified workers of an educational institution were unanimously condemned. The Ministry of Education of the Russian Federation, having admitted that the actions of the newly appointed director were excessive, turned for help to the Chairman of the Council of Ministers of the Russian Federation, A. M. Puzanov. He forwarded the letter to the Central Committee of the Communist Party, from where it was sent to the Moscow city party committee. Its leaders, however, stated in their explanatory note that the new director "is following the right course for strengthening the staff of the pedagogical institute". With such support Polikarpov got rid of 23 more teachers, having found out that they had complained about him..."*

* *Kentavr* No. 11–12, 1992, pp. 69–70.

[41] Pasternak and Ivinskaya were at the Cultural Department of the CPSU Central Committee on November 1, 1958, when Polikarpov already had a selection of protest telegrams (doc. 9.13), but of course, he kept silent about them, confining himself to demonstrating Quasimodo's "negative opinion."

[42] On September 15, 1958 a second meeting of Soviet and Italian poets took place in Moscow (the first one was held, in October 1957, in Rome). However, Quasimodo, who came to the meeting among other guests, could not take part in it. In a grave condition he got into Botkin Hospital, where he stayed for more than half a year.*

* *Literaturnaya Gazeta,* March 28, 1959, p. 4.

[43] *Pravda*, February 15, 1960, p. 2. Here is how Quasimodo himself described his condition in a poem:

I'm seriously ill. Yes, it's me
Who can die here at any moment.
Translated from Italian into Russian by G. Breitburd.

[44] S. Quasimodo. *Moya Strana — Italiya* (My Country, Italy), Moscow, IL, 1961; by the same author: *Izbrannaya Lirika* (Selected Poems), Moscow, Molodaya Gvardiya, 1967. Soon after coming out of his first collection of poems in Russian, on March 11–15, 1962, a congress of the European Association of Writers took place in Florence. Among its participants was the translator of the Italian poet into Russian, G. Breitburd. There he met his colleague from Stockholm Anders Esterling, permanent secretary of the Swedish Academy, who translated Quasimodo into Swedish, and they exchanged books. This seemingly immaterial episode was favored with a special mention in the official report of the Soviet delegation at the congress to the CPSU Central Committee* — which emphasizes once again that after his assistance in "the case of Pasternak" the figure of this poet became very important for the party officials from the Central Committee's Cultural Department.

* RGANI, fond 5, opis' 36, ed. khr. 142, l. 46.

[45] Here are two mutually exclusive utterances of the critic about one and the same person, with an interval of a year and a half. April, 1958: "Reading the speeches [...] of Quasimodo and other participants of the discussion, [...] you have an impression that they want to speak about poetry as an occupation of the few, to whom the usual standards and criteria cannot be applied, which should enjoy a kind of diplomatic extraterritoriality with regard to politics and society requirements."* November, 1959: "A remarkable feature that has always been characteristic of Quasimodo as a poet is his acute feeling of the contemporary époque [...] I repeat, acute feeling of the contemporary époque with its struggle — that is what has always been inherent in Quasimodo, a passionate polemicist in his poems and in his articles."** Such an abrupt change of orientation in his utterances about the works of the Italian poet was not accidental for Zelinsky. A year before he, like Quasimodo, found himself in the avant-garde of the anti-Pasternak campaign and now was paying his debts to his unexpected external ally. And in the fall of 1958 the critic distinguished himself by ferocious attacks not only on Pasternak, but also on those who supported him in the most difficult days. Speaking, on November 20, 1958, at the meeting of the Party leaders of the Institute of the World Literature of the USSR Academy of Sciences, who were discussing his complaint about his colleagues ostracizing him, he declared: "I am told that I shouldn't have mentioned the people surrounding Pasternak. But I spoke about them with good reason. The Ivanov family*** are directly to blame for Pasternak's fall. These people talked Pasternak out of rejecting the prize. That is, they are responsible for Pasternak's unscrupulous behavior. And conclusions should be drawn from that."****

* *Inostrannaya Literatura* No. 4, 1958, p. 201.

** *Literaturnaya Gazeta,* November 10, 1959, p. 4.

*** The writer Vsevolod Ivanov and his son, literary critic Vyacheslav Ivanov.

**** CASPHM, fond 2472, opis' 1, ed. khr. 19, l. 115.

[46] True, the poetess Vera Inber in her preface to the selection of Quasimodo's poems in the January, 1960 issue of the *Inostrannaya Literatura* journal timed for the news of the Nobel Prize, also duplicated the text of the Board of the Writers' Union: "This time, the laurels crowned someone really worthy."*

* *Inostrannaya Literatura* No. 1, 1960, p. 78.

[47] This passage appeared in the 1920s in the daily newspaper *Rus* coming out in Russian in Berlin. The complete text is published in the book of collected works of Nabokov which came out in Russia at the end of the 1990s: V. V. Nabokov. *Sobranie Sochinenii Russkovo Perioda* (Collected Works of the Russian Period), St. Petersburg, Simposium, Vol. 2, 1999, p. 638.

[48] N. A. Anastasiev. *Vladimir Nabokov. Odinokii Korol'* (The Lonely King), Moscow, Centrpoligraph, 2002, p. 416.

[49] Twelve years before "the case of Pasternak" the authorities successfully tried the method of deprivation of literary earnings on Anna Akhmatova and Mikhail Zoshchenko. After the memorable party resolution* they were crossed out from literature, and Zoshchenko was driven, through faints from hunger, to an untimely death. The same methods of "taming" were applied to the poet G. B. Plisetsky, the author of a heartfelt poem in memory of Pasternak (see note 64).

* See Chapter 4.

[50] "I crossed the threshold of the new year with suicidal mood and wrath."*

* See doc. 9.17.

[51] The Nobel Prize

Like a beast in a pen, I'm cut off
From my friends, freedom, the sun,
But the hunters are gaining ground.
I've nowhere else to run.

Dark wood and the bank of a pond,
Trunk of a fallen tree.
There's no way forward, no way back.
It's all up with me.

Am I a gangster or murderer?
Of what crime do I stand
Condemned? I made the whole world weep
At the beauty of my land.

Even so, one step from my grave,
I believe that cruelty, spite,
The powers of darkness will in time
Be crushed by the spirit of light.

The beaters in a ring close in
With the wrong prey in view.
I've nobody at my right hand,
Nobody faithful and true.

And with such a noose on my throat
I should like for one second
My tears to be wiped away
By someone at my right hand.

Translated by Jon Stallworthy and Peter Frans.

Published: *Boris Pasternak. Selected Poems,* London, Allen Lane, 1982, p. 154. According to O. V. Ivinskaya, the poem was written on January 20, 1959.*

* O. V. Ivinskaya, p. 345–347.

[52] E. B. Pasternak, p. 713, and his oral information to this author.

[52a] *Znamya* No. 3, 2005, p. 156.

[53] The search for some written evidence of Pasternak having been invited to the Swedish embassy, undertaken at this author's request, did not give any positive results. There can be only one explanation for the failure of the search — the embassy had not sent Pasternak such an invitation. The ambassador of Sweden in the Soviet Union Rolf Sohlman understood perfectly well the situation around the poet. Not willing to aggravate it, he must have sounded out, in oral form, the possible reaction of the USSR Ministry for Foreign Affairs to such an initiative. This reaction had most likely been evasive — something like saying that after what had happened Pasternak would hardly agree to such a visit, and so on. Simultaneously, Pasternak was informed that if he did receive such an invitation, he should decline it under any pretext, to be rewarded by improvement of his financial situation.*

* In the Soviet times it was common practice to sound out power structures in advance about their attitude to inviting one or another Soviet citizen to events organized in embassies. As an example we can remember the story of the invitation of the composer S. S. Prokofiev, in 1947, to a reception in his honor at the Swedish embassy (see note 8 to Chapter 4).

[54] E. B. Pasternak, p. 716; the mentioned collection *Stikhotvoreniya i Poemy* (Poems and Big Poems) was published in Goslitizdat in 1961 after Pasternak's death. The text of the book, which was to come out in 1958, had already been edited, but after the beginning of the Nobel scandal its type was decomposed.*

* L. K. Chukovskaya. *Zapiski ob Anne Akhmatovoi* (Memories of Anna Akhmatova), Vol. 2, Moscow, Soglasie, 1997, p. 224.

[55] In the newspapers *Le Monde* of October 10, 1958 and *The Reporter* of November 26, 1958 Nilsson told the story that had been told to him by Pasternak about his refusal to sign the letter where writers demanded that Marshal Tukhachevsky and other people who were tried in the same case should be sentenced to be shot. According to Pasternak, this letter aroused in him intense hatred of murder but eventually his signature did appear in newspapers — since the leaders of the Writers Union, who prepared the letter, simply were afraid to inform the authorities of the poet's refusal.*

* *Continent,* 95, No. 1, 1998, p. 250.

[56]

Have I committed any evil or crime,
and am I a corrupter or a fiend,
I, who has made the whole world dream
of poor girl of mine.

I know that I'm frightening for people,
they burn for magic poets like me
and, like from poison in a hollow ring

die from my art.

But isn't it funny that at the end of story,
at odds with the proof-reader and the age,
a shadow of a Russian branch will sway on
the marble of my hand.
December 27, 1959, San Remo.

The poem was first published in the journal *Vozdushnye Puti* (Aerial Ways) New-York, No. 2, 1961. It was accompanied by the author's note: "Line 1–4: the first stanza of this novel imitates the beginning of Boris Pasternak's poem, the first line of which being was completely borrowed by me"* This note is inaccurate; probably, when writing the poem Nabokov did not have Pasternak's *The Nobel Prize* at hand. Comparing the given poem with the poem by Pasternak, it is easy to make sure that their first stanza, in fact, parodies the third stanza of Pasternak's poem, and the second one, the first (see note 51 to this chapter). The words that the "first line [...] was completely borrowed" may show that at the moment of writing this poem Nabokov did not yet read Pasternak's *The Nobel Prize* in the original and used its English translation.

By the way, for Nabokov that was not the first experience of mockingly disparaging parodying of works of his colleagues, poets who remained in Russia. Somewhat earlier he devoted his unkind attention to Anna Akhmatova. In the novel *Pnin* (1957), implying her in the character named Lisa, the writer included in the episode with Lisa's participation a caustic parody of the wording and melody of the poetess's early verse:

'Ya nadela tyomnoe plat'e,
I monashenki ya skromney;
Iz slonovoy kosti raspyat'e
Nad holodnoy postel'yu moey.

No ogni nebivalih orgiy
Prozhigayut moyo zabityo...

(I have put on a dark dress
And am more modest than a nun;
An ivory crucifix
Is over my cold bed.

But the lights of fabulous orgies
Burn through my oblivion...)
and so on.**

According to L. K. Chukovskaya, Akhmatova, on familiarizing herself with the novel, called it "slanderous."

N. N. Berberova, one of the first wave emigrants, in her letter from Princeton addressed to L. K. Chukovskaya, on March 17, 1981, gave an enthusiastic response to her *Memories of Anna Akhmatova*. At the same time, she noted that "Nabokov did not slander Anna Andreevna. He was parodying poems of her imitators."*** That is obvious cunningness on the part of Berberova, who was an ardent admirer of Nabokov and knew him in person. As is known, in the 1910s Akhmatova was one of those who created *akmeism*, the trend in poetry which opposed symbolism. Nabokov at that time was creating his own trend, close to *akmeism* but having some substantial nuances ("let tangible objects remain as they are, without implying anything else, but let the picture of the world itself hint at some mysteries from the beyond").**** But don't we know from our own historical experience what cataclysms could be brought about by some immaterial nuances of these or other dogmas — not only in the sphere of political fights but also in sheer creative differences.

Parodying, a talent as almost everything with Nabokov, became an essential point of his writings and was used repeatedly. In April 1945, he wrote a brilliant parody on the rhymes and intonations of Mayakovsky, by that time already dead for a decade and a half. As it was noted by Brian Boyd, this poem is seething with contempt for all those who idolized Stalin and other leaders. The first work by a Soviet poet which gained not contempt but wholehearted acceptance of Nabokov was *Sentimental March* by Bulat Okudzhava. At the beginning of 1966 he made a wonderful translation of this poem into English.*****

* See also: V. Nabokov. *Stikhotvoreniya i Poemy* (Poems and Big Poems), Kharkov-Moscow, 1997, p. 297 and 563.

** Vladimir Nabokov. *Selected Prose and Verse,* Moscow, Raduga Publishers, 1990, p. 409.

*** *Znamya* No. 8, 2005, p. 161.

**** B. Boyd. *Vladimir Nabokov. Russkiye Gody* (The Russian Years), Moscow, Nezavisimaya Gazeta and Symposium, 2001, p. 116.

***** B. Boyd. *Vladimir Nabokov. Americanskiye Gody,* Moscow, Nezavisimaya Gazeta and Symposium, 2004, p. 104, 604.

[56a] The period mentioned here is described more precisely by Nabokov's biographer, Brian Boyd. In mid-September, 1958, right at the moment when the English translation of *Doctor Zhivago* appeared on the book counters, *Lolita* was holding fourth place in the list of bestsellers, and by the end of the month it had moved up to the first line. Seven weeks later, after it was announced, on October 23, that Pasternak had been awarded the Nobel Prize, his novel ousted *Lolita* into the second position.*

* B. Boyd. *Vladimir Nabokov. Americanskiye Gody,* Moscow, Nezavisimaya Gazeta and Symposium, 2004, p. 445.

[57] S. Schiff. *Mrs. Vladimir Nabokov. Vera,* Moscow, Nezavisimaya Gazeta, 2002, pp. 325–326.

[58] From the memoirs of the writer B. Y. Yampolsky: "I learned of his death ("just this night") by chance. From a man next door, when I was going to work. The man is from *The Voice of America.* Our people kept silent. A brief mention in the *Literaturnaya Gazeta* flashed bashfully only already on the third day, the day of the funeral. Neither earlier, nor later! I say 'flashed' because it was no bigger than a tram ticket, typed in brevier. And it informed of the death of merely 'a member of the Literary Fund', in the name of the Literary Fund."*

* Voprosy Literatury No. 5, 2001, p. 284.

[59] L. K. Chukovskaya wrote in her diary about the rumors that the writer Voronkov sent by the Writers' Union came to Pasternak's wife Zinaida Nikolaevna to tell her that "the Union will do everything to organize the funeral if she allows them to put the coffin in the CDL.* "Zinaida Nikolaevna, to her honor, refused."** The poet's son E. B. Pasternak told this author that from her he had not heard anything about that proposal, but what was written by Chukovskaya could be trusted. Thus, if this rumor corresponded to the reality, the authorities expected in such a way to take under their control both the list of the speeches allowed for the official ceremony and the possible spontaneous actions at the freshly-dug grave.

* The Central House of Writers on Povarskaya *ulitsa* in Moscow.

** L. K. Ckukovskaya, p. 394, entry of May 31, 1960. In the entry made in the evening of June 1, 1960 she repeated: "... I heard again what I had already heard before [...] about Voronkov and the proposal to put the coffin in the CDL. If he [Pasternak] is merely "a member of the Literary Fund, why should he be honored in such a way? But the ways of the bosses are inscrutable." (p. 397)

[60] Six years later, after the death of Anna Akhmatova, the leadership of the Soviet Writers Union tried to organize her funeral according to a similar scenario (see doc. 10.32).

[61] RSAMH, fond 5, opis' 36, ed. khr. 140, l. 52.

[62] "*A za Mnoyu Shum Pogoni*" *Boris Pasternak i Vlast'. Dokumenty. 1956–1972* ("But the Hunters Are Gaining Ground..." Boris Pasternak and the Power. Documents), (edited by V. Y. Afiani and N. G. Tomilina), Moscow, ROSSPEN, 2001, p. 334. The pieces of music mentioned in the quoted document were written by the future poet in his youth. In 1979 his piano sonata was published for professional performance by the Sovetskii Kompositor publishing house; that and a number of other works of his were repeatedly performed at concerts.*

* E. B. Pasternak. *Boris Pasternak. Biography*, Moscow, Citadel, 1997, pp. 92–97.

[63] E. B. Pasternak. *V Osade* (Under Siege); the preface to the collection of documents *A za Mnoyu Shum Pogoni* (But the Hunters Are Gaining Ground) already mentioned in note 62. The story told here took place in the Iskusstvo publishing house.

[64] Anna Akhmatova was not the only one who gave a poetic response to the death of Pasternak. In the early 1960s people passed from hand to hand a manuscript copy of a poem by an unknown author, which sounded as if it belonged to Russian classical poetry. It was highly valued by S. Y. Marshak and K. G. Paustovsky, and A. A. Akhmatova, comparing it with Lermontov's poem *To the Death of the Poet* considered it "the best of everything created in memory of Boris."* The poem was first published in 1989.** It turned out to be that the author of the poem was German Plisetsky, a poet and translator from Persian, who had not been published for twenty years in revenge for his taking part in Pasternak's funeral. The quarantine ended in 1981, when his translations were published, but his own collection of poems *Prigorod* (Suburb) came out only in 1990.

In Memory of Pasternak

The poets of Russia, her natural children! When buried
through back door your bodies have always been carried.

....

I'm crying today, unabashed by my tears,
the tears of shame for what's happening here.

I've suffered from pain, and I've often been hungry,
And now I'm dying of shame for my country.

I don't care what rumor will go the rounds.
It's poetry that scumbags have put in the ground.

We give our glory away, we don't store it,
It probably means that we do not deserve it.

Indeed, we're enormously wealthy in glory,
So, if we loose some, why should we feel sorry?

It's only from pines that will poetry get what is fair —
Their roots grasping it not to give it to anyone ever.

June 4, 1960

* See *Mir Paustovskogo* (Paustovsky's World) No. 17, 2000, p. 179.
** *Druzba Narodov* No. 6, 1989, p. 11.

65 L. K. Chukovskaya, Vol. 2, p. 402.

66 First published: *Novy Mir* No. 1, 1931, p. 117.

67 L. K. Chukovskaya. *Zapiski ob Anne Akhmatovoi* (Memories of Anna Akhmatova), Vol 3, Moscow, Soglasie, 1997, p. 298. In the collection of poems by Akhmatova the poem *The Death of a Poet** is followed by the poem *I Snova Osen' Valit Tamerlanom* (Like Tamerlan, autumn is thronging again...), which was really written in 1957; probably, that is the reason why 1957 was chosen by the editor of the book when faking the year.
* A. A. Akhmatova. *Beg Vremeni* (The Flight of Time), Moscow, Sovetskii Pisatel, 1965, p. 432.

68 In 1939–1941, Shchelokov worked together with Brezhnev in Dnepropetrovsk and knew him well. In 1951 their friendly relations brought him to Kishsnev, where Brezhnev became First Secretary of the Communist Party of Moldavia; and after the latter was elected, in 1964, First Secretary of the CPSU Central Committee, Shchelokov moved to Moscow, in 1966, to take up the key position of minister for internal affairs. His 16 years' tenure in office was accompanied by a number of corruption scandals and numerous conflicts with the KGB chairman, Y. V. Andropov. As a result, a month after Brezhnev's death on November 9, 1982, he was dismissed. On June 14, 1983 he was expelled from the members of the Central Committee. Expecting another search and imminent arrest, he preferred to commit suicide.

The Moscow years of Shchelokov's life and his inglorious end "in his own luxurious apartment, [...] clad in a full-dress general uniform, with orders and medals, in a white shirt and trousers with wide stripes" were featured in the memoirs of V. V. Pribytkov, long-standing assistant of K. U. Chernenko, secretary and, in 1984–85, General Secretary of the CPSU Central Committee.*
* V. Pribytkov. *Apparat* (The Apparatus), St. Petersburg, VIS, 1995, pp. 191–200.

69 After it became known that he had been awarded the Nobel Prize, Domagk even found himself behind bars for a while and was released only after he declared that he declined the award.

70 Domagk received his awards in 1947, and the other two, who repudiated their former statements, some time later, in 1949. The monetary part of the prize, according to the charter regulations, is transferred to the reserve fund of the Nobel institution immediately after the declaration of rejection and is not subject to return.

71 Instead of the original the museum exhibited an exact copy of the Nobel diploma made at the request of Ambassador Berner. E. B. Pasternak told this author about the reasons which made this fake necessary. After the Swedish academicians made the decision to disavow Boris Pasternak's rejection of the prize, the permanent secretary of the award-giving institution Sture Allen asked the poet A. A. Voznesensky, who happened to be in Stockholm, to pass the diploma to the laureate's son. Due to a number of circumstances connected with Voznesensky's trips around the world, E. B. Pasternak was able to receive the document only several months after the ceremony of giving him his father's gold medal of the Nobel Prize winner.

[72] Rostropovich had his own reasons for taking part in the 1989 Stockholm ceremony. On Monday, October 10, 1958, prepared simultaneously with the writers' meeting, there was to be a similar meeting of representatives of creative intelligentsia. As Rostropovich told this author, the organizers of this event included his name into the list of those who were to publicly condemn Pasternak. For him such a speech was morally unacceptable, and in order to avoid it he decided to take advantage of the forthcoming trip to Ivanovo with a planned concert program over the weekend. Under a fake pretext he stayed in Ivanovo until Tuesday and came back to Moscow when the meeting had already taken place. By the way, in August, 1973, the renowned actors Yuliya Borisova and Mikhail Ulyanov used a similar method so as not to find themselves among those who signed the letter against academician Sakharov (see Chapter 11). Those days the Vakhtangov Theater was giving guest performances in Leningrad. Having learned that their bosses would be phoning them from Moscow, for a few days they did not come back to their hotel rooms until late, and when a phone call reached them there they did not answer it.

[73] Novy Mir No. 2, 1990, pp. 193–194.

[74] Developing this theme, it will be relevant to mention the great composer Sergei Prokofiev, who passed away, incidentally, on the same day as the dictator, March 5, 1953. The equally great pianist of the century, Svyatoslav Rikhter, wrote in his memoirs that "he was at odds with principles, being quite capable of composing music to order, for example, *Zdravitsa* (a toast for somebody's health) — a laudatory ode he was ordered for Stalin's next jubilee. These days performing these lyrics is absolutely unthinkable: "Glory to Stalin, the father of all peoples!"; the words should be changed. But the music is a work of genius...* He even did it with a certain effrontery, with some noble immorality: "Stalin? What Stalin? Oh, yes, of course! Well, why not? I can do everything, even such things."**

> * Given Rikhters' appraisal of *Zdravitsa* as a musical work, it is appropriate, from the viewpoint of today, to give Rikhter credit for the professional accuracy of his foreseeing. On May 4, 2006, on the occasion of the 75th anniversary of the outstanding conductor Gennady Rozhdestvensky, who most subtly interpreted and propagated the work of Prokofiev, the very same *Zdravitsa* was performed at the Grand Hall of Moscow Conservatory. As the musical observer of the *Izvestiya* newspaper did not fail to mention, "old residents remember whose birthday was on December 21." Further on he concluded his own "*zdravitsa*" to Rozhdestvensky by saying that the cantata is quite worthy of the present-day "musical leader" — with the corresponding transformation of its text, though, which was done, as Rikhter had foretold.
>
> ** B. Monsaingeon. *Rikhter. Dialogui. Dnevniki.* (Dialogues. Diaries), Moscow,: Klassika-XXI, 2002, p. 51.

[75] C. G. Kara-Murza. *Evrei, Dissidenty i Evrokommunizm* (The Jews, Dissidents and Eurocommunism), Moscow, Algoritm, 2003, p. 113.

[76] D. L. Bykov. *Boris Pasternak*, Moscow, Molodaya Gvardiya, 2006, p. 536.

Document 9.1

Boris Pasternak: about the rumors on the occasion of his being nominated for the Nobel Prize

I

From the letter to his cousin Olga Freidenberg,
of November 12, 1954

... Here the same rumors are going around.[1] I am the last person whom they reach... But you cannot imagine how strained are my relations with the official reality and how afraid I am to remind of myself. At the first move of mine I might be asked questions about my most essential views, and there is no force in the world which would make me answer these questions in the way they are answered by everybody else without exception. And all that is becoming the more aggravated and more terrible, the stronger, happier, healthier and more fruitful has my life been lately becoming.[2] And I have to live voicelessly and mysteriously.

I feared that this gossip might become true rather than wished it, although this award entails an obligatory trip for receiving the prize, going out into the broad world, exchanging ideas — but then again, I would not be able to make this trip like an ordinary clockwork doll, as others do, and I have the life of my family and the novel that I have to finish,[3] and how all that would be aggravated![4] Here's a Babylonian captivity! Perhaps God was merciful — this danger has passed. Probably, some candidature was offered that had long-standing and broad support. Belgian, French and West-German newspapers wrote about it.

Readers saw it, so it is said. Then people heard on the BBC[5] that allegedly (I heard it on the grapevine, so I don't know whether it's true) I was nominated, but, knowing our ways, they asked the consent of the government,[6] which petitioned them to replace me by the candidature of Sholokhov; and after he was rejected the commission nominated Hemihgway, who is most likely to be awarded the prize.[7]

But it was a joy for me even to have been supposed to get into the category in which Gamsun and Bunin were,[8] and, even if only by misunderstanding, to find myself next to Hemingway...

Boris Pasternak. *Collected Works*, Vol. 5, Moscow, Khudozhestvennaya Literatura, 1992, pp. 534–535.

[1] The rumors of Pasternak's candidature being considered for the 1954 prize appeared not only in the Soviet Union. The same can be seen, with the reference to "an anonymous authoritative source", in the book about Pasternak written by the literary critic from Oklahoma, Guy de Mallac.* The factual part of the book was severely criticized by Lars Gyllensten, permanent secretary of the Swedish Academy and a member of the Nobel Committee for Literature** — among other things, for the assertions mentioned. He emphasized that Pasternak was nominated for the Nobel Prize seven times, from 1946 to 1950 annually and also in 1957 and 1958, so in 1954 the Swedish Academy could not consider his candidature.

* Guy de Mallac. *Boris Pasternak. His Life and Art* — Univ. of Oklahoma Press, 1981.

** L. Gyllensten. *Some notes on Pasternak' Nobel Prize in 1958, Artes* No. 1, 1983, pp. 112–113.

[2] As E. B. Pasternak told this author, his father's optimism in this phrase was connected with the fact that his work over the novel *Doctor Zhivago* was drawing to a close.

[3] *Doctor Zhivago.*

[4] Pasternak was able to foresee the future, indeed.

[5] The Russian programs of the British Broadcasting Corporation.

[6] The story of the Swedish Academy's inquiry directed to the Soviet government through diplomatic channels, in order to sound out their attitude to the possible awarding of the 1954 Nobel Prize to Pasternak, was also included in Guy de Mallac's book mentioned in note 1, and was also rejected by Gyllensten.

[7] There are two enigmas in the given phrase. First, Hemingway got the Nobel Prize in 1954, when, judging by Gyllensten's mentioned publication, Pasternak was not nominated. On the other hand, from this publication it become clear that in mid-Nobember, 1954 Pasternak did not yet know of the Swedish Academy's decision, which was made public on October 24, 1954. True, none of the Soviet newspapers reported this news, but it became widely known from radio programs in Russian of foreign broadcasting companies, as Hemingway's name was quite popular in the USSR. The fact that Pasternak at that time was rather estranged from current developments was mentioned by A. A. Ahmatova in her talk with L. K. Chukovskaya, which took place half a year later, in May, 1955: "I have seen Boris Leonidovich. He makes a sad sight — growing old and week. Looks very bad. Is finishing his novel. [...] He lives a terrible life. Just imagine, he still has not heard of Lozinsky's death!* So where is he living, whom is he seeing?"**

* M. L. Lozinsky, a poet and outstanding translator of classical West-European and Oriental poetry, died in Leningrad on February 1, 1955. Pasternak knew him well.

** L. K. Chukovskaya. *Zapiski ob Anne Akhmatovoi,* Vol. 2. — Moscow: Soglasie, 1997, p. 116.

[8] In the context of the quoted letter the mention of the name of the two award-winning writers has an touch of the poet's unconcealed free-thinking, obviously reflecting the atmosphere of romantic expectations of the near beginning of purifying changes in the country. In 1954 Bunin, who was remembered or known by only a few in the USSR after several decades of his name having been ignored here, was still banned.* Gamsun, again in the opinion of the overwhelming majority of Soviet people, is merely a collaborationist, who cast a stain on his reputation in the years of the Nazi occupation of Norway. But the winds of liberation from Stalin's gloomy despotism, which began to be felt and brought Pasternak the publication of ten poems from *Doctor Zhivago* in the *Znamya* journal,** accompanied by his own announcement about the near completion of the novel,*** made him yield to the illusions that the times most difficult for creative self-expression are forever behind.

* The first Soviet edition of Bunin, after publication of the short story *The Gentleman from San-Francisco,* in 1930, appeared only a quarter of a century later:

Rasskazy (Short Stories) (with a foreword by Lev Nikulin), Moscow, Goslitizdat, 1955.

** The publication in the *Znamya* journal was preceded, in 1953, by an unsuccessful attempt to publish these poems in the *Literaturnaya Gazeta.* According to the literary critic and journalist V. F. Ognev, to whom the poet gave them for publishing, he then managed to persuade the editor-in chief B. S. Ryurikov "to break the period of ignoring Pasternak," But "when I brought the type (the whole page!) from the printing house," Ryurikov, on reading the selection, "crossed out at first *Rassvet* [Dawn], and then the whole page." *Znamya* got the notebook with the poems from the poetess Vera Inber, who assured Ognev that "Vadim listens to her." As it turned out, she did not overestimate her authority. *Literaturnaya Gazeta* No. 16, 2004, p. 13.

*** "Supposedly, the novel will be finished this summer. It is set in the period from 1903 to 1929, with the epilogue relating to the Great Patriotic War. The protagonist, Yuri Andreevich Zhivago, a doctor, thinking, with a quest, of creative and artistic nature, dies in 1929. He leaves notes and, among other papers, separate poems written in his younger years; these poems, some of which are published here, will in the aggregate make the last, conclusive part of the novel." (*Znamya* No. 4, 1954, p. 92).

II

From the letter to the French Slavonic scholar Helen Peltier-Zamoiska, of July 30,1958

... Where the Nobel Prize is concerned,[1] I am sure that this danger will not come true, because the actions of the committee usually include an inquiry to the government, to whose citizenship the personality being discussed belongs, about their agreement to this candidature.[2]

And in my case it will never be like that...

Znamya No. 1, 1997, p. 125

[1] Implied here is the leakage of information from the Nobel Committee for Literature, where in the summer of 1958 the predominant opinion was in favor of awarding that year's prize to Pasternak (see also doc. 9.3).

[2] See note 6 to doc. 9.1–1.

III

From the letter to the German writer Renate Schweitzer, of October 9, 1958

Some people assume that the N[obel] Prize of this year may be awarded to me.[1] I am firmly convinced that it will evade me and will be given to Alb[erto] Moravia.

You cannot imagine that even a doubtful supposition of such a possibility creates in my mind a lot of difficulties, fears and torments. There are situations in life when only a certain motionlessness keeps up the equilibrium for the people who surround you. One step aside, and your nearest and dearest are doomed to suffering...[2]

Renate Schweitzer. *Freundschaft mit Boris Pasternak.*
— Wien-Munchen-Basel:
Verlag Kurt Desch, 1963, p. 53.

[1] True, some foreign correspondents of Pasternak considered it more possible that he would be awarded the prize in 1959. P. P. Suvchinsky wrote to him, on June 8, 1958, from Paris:* "It is an absolute must that next year you get the highest European prize."* The words "an absolute must" were probably caused by the fact that when the letter was being sent there was no published Russian original of the novel, which could have created obstacles for the positive decision of the Swedish Academy. However, in 1958 such an edition appeared.**

* See: Vadim Kozovoi. *Poet v Katastrofe* (Poet in Katastrophe), Moscow, Gnozis, 1994, p. 230.

** See note 5 to this chapter.

[2] In the concluding remarks to Pasternak's letter of October 9, 1958 Schweitzer wrote: "I sent the last page of Pasternak's letter to Stockholm, to the Nobel Committee, with the request, in case the prize was to be awarded to Pasternak., to move the award presentation, if possible, to the following year. When I posted the letter, I started to doubt whether I was right to do it. But that had already happened..."*

* Schweitzer, p. 54. Interestingly, something like that happened in 1964. According to reports in the world press, when the decision of the Swedish Academy had not yet been announced, a letter addressed to the permanent secretary of the Academy came from Jean-Paul Sartre. In the letter he informed that he did not want to be awarded the Nobel Prize and was asking not to deprive of it some other nominee claiming it. However, in this case the Academy also did not tolerate any interference from the outside and announced Sartre the 1964 laureate.

Document 9.2

Resolution of the Commission of the Central Committee of the CPSU[1] "About the measures to be taken by the Soviet Ambassador in Sweden in connection with the possible Nomination of Boris Pasternak to the Nobel Prize," of April 7, 1958 (minutes 7, point 33c).[2]

To approve the text of the telegram to the Soviet ambassador in Stockholm (see the supplement).

Supplement to p[oint]33c, m[inute]7. Stockholm

There is information of the intention of certain circles to nominate Pasternak for the Nobel Prize.[3]

It would be desirable to give Swedish public opinion to understand, through culture personalities close to us,[4] that highly appreciated in the Soviet Union would be awarding the Nobel Prize to Sholokhov. Here you should emphasize the positive significance of Sholokhov as an outstanding writer and public figure, using, in particular, his last year trip to Scandinavia.

It is also important to give to understand that Pasternak as a writer has not obtained recognition from the Soviet writers and progressive writers of other countries. Nominating Pasternak for the Nobel Prize would be taken as an unfriendly act with regard to Soviet public opinion.[5]

RSAMH, fond 11, opis' 1, ed. khr. 7, p.p. 66–67; see also the collection of documents *Kultura I Vlast' ot Stalina do Gorbacheva. Ideologicheskie komissii CK KPSS, 1958–1964. Dokumenty* (Culture and Power from Stalin to Gorbachev. Ideological commissions of the Central Committee of the CPSU, 1958–1964. Documents, Moscow, ROSSPEN, 1998, p. 48.

[1] Ideological Commission of the Secretariat of the Central Committee of the CPSU.

[2] The grounds for the resolution were given by the letter from the Central Committee' s Cultural Department of April 5, 1958.*

* RSAMH, fond 11, opis'1, ed. khr. 242, l. 125; see also Ideological Commissions... p. 49–50.

[3] The formulation is inexact: in April 1958 there could have been information only of the intention to award the prize, as the campaign of nominating for the 1958 prize ended on January 31.

[4] The authors of the document probably had no notion that the activities of the Nobel institutions are confidential and their decisions are completely independent of any wishes, including those of "Swedish public opinion."

[5] It was the secretary of the Soviet Writers Union K. M. Simonov who was the first to inform the CPSU Central Committee of the possibility of the 1958 Nobel Prize being awarded to Boris Pasternak. In his letter of March 31, 1958 he wrote that, according to the information received by him from the Department of Scandinavian countries of the Ministry for Foreign Affairs, in the discussions which shortly before had taken place at the Swedish P. E. N.-Club about the most probable candidates for the Nobel award named as such were Sholokhov, Pasternak, the American of Italian origin Ezra Pound and Alberto Moravia from Italy.*

* RSAMH, fond 11, opis' 1, ed. khr. 242, l. 124, see also Ideological Commissions.... p. 48–49.

Document 9.3

From the Chairman of the Foreign Commission of the Soviet Writers Union B. N. Polevoi to the Central Committee of the CPSU; urgent, secret (no date)[1]

The Soviet Writers Union has long been informed, through our friends, writers from the West, that reactionary forces connected with the United States, which operate in the Nobel Prizes Committee, are planning, in order to spite the Soviet Union, to give the Nobel Prize for literature to Boris Leonidovich Pasternak for his creative work, and also for his novel *Doctor Zhivago*, which has not been published in the Soviet Union, but is being widely and sensationally published in the Western world. Our friends are warning that this fact may become material for creating a new anti-Soviet sensation, based on the talk about the lack of creative freedom in the USSR, about suppressing a number of writers for political reasons, etc.[2]

Informing the Central Committee of that, we would like to receive instructions as to what position on that question we should adopt in advance and what measures we should take.[3]

Beneath the text of the letter there are two notes: "CPSU Central Committee. The necessary measures in connection with the campaign of foreign reactionary press around the novel by B. Pasternak and its being awarded the Nobel Prize have been taken by the Cultural Department of the CPSU Central Committee.[4] Head of division of the Cultural Department of the Central Committee B. Yarustovsky. 11.10.1958." "Archive. Comrade Furtseva E. A. has familiarized herself. Agrees with the measures taken by the Department of the Central Committee and the Soviet Writers Union. 11.12.58. N. Kalinin."

RSAMH, fond 5, opis' 36, ed. khr. 61, l. 40

[1] The letter was registered in the office of the CPSU Central Committee on September 17, 1958. Judging by the note "urgent" it was written on the same day and passed to the Central Committee by the author in person.

[2] B. Polevoi had access to information coming "through our friends, writers from the West", as he was a member of the Bureau of the World Peace Council — organization created in 1950, at the height of the Cold War, with the purpose of propaganda support of the foreign policy of the Soviet Union in its opposition to the countries of NATO.

[3] A written answer to the author, in connection with the information he supplied, has not been found among the archive documents of the Central Committee. The author himself, speaking at the all-Moscow meeting of writers on October 31, 1958 expressed his position as to "what measures we should take" in the following way. Having called Pasternak "a literary Vlasov," he suggested "telling him in the name of Soviet public opinion: 'Get out of our country, Mr. Pasternak. We do not want to breathe the same air with you.'"*

* *Continent* 83, 1995, p. 194.

[4] In connection with B. Polevoi's letter the Writers Union suggested to the Central Committee's Cultural Department that, to avoid scandal, the novel should be published "in a limited edition, with an announcement in the press, in order to rule out the possibility of Western press making a fuss" about its publication being forbidden in the Soviet Union, but the Cultural Department rejected this suggestion

as "inexpedient".* Actually, Khrushchev said the same thing four years later, on December 17, 1962, at a meeting with creative intelligentsia: "If *Zhivago* had been published, he would not have got any prize."**

* *Continent*, 95, No.1, 1998, p. 238.

** A. I. Solzhenitsyn. *Bodalsya Telenok s Dubom* (The Oak and the Calf), Moscow, Soglasie, 1996, p. 80.

Document 9.4

From the Heads of the Cultural Department D. A. Polikarpov and of Propaganda and Agitation L. F. Ilyichev to the CPSU Central Committee, of October 21, 1958.

At the beginning of 1956[1] the writer B. Pasternak passed the manuscript of his anti-Soviet novel *Doctor Zhivago* to an Italian publishing house.[2] In 1957 the novel was published in Italy[3] (8 editions have already come out), then in France, England and some other capitalist countries. Lately the foreign circles hostile to the USSR have launched a campaign for awarding the Nobel Prize to Pasternak with the aim to use this act against our country. In case such an action hostile to us is carried out, we would consider it necessary to take the following reciprocal measures on our part:

1. To organize publishing articles on this occasion from the editorial board of the *Novy Mir* journal. To publish an abridged version of the letter of the members of the editorial board of this journal to Pasternak, sent in September 1956, which gives a detailed criticism of the novel and explains why the journal has not published this slanderous writing.[4]

2. To publish an article collectively written by major Soviet writers, in which awarding the prize to Pasternak should be assessed as an act hostile to our country.[5]

3. In the circles of representatives of the Western press a supposition was made that the Nobel Prize may be divided between Pasternak and Sholokhov. If comrade Sholokhov is awarded this year's prize along with Pasternak, it would be expedient that in protest he pointedly reject it and announce his unwillingness to be the winner of an award which is used to anti-Soviet ends. Such an action from Sholokhov seems all the more necessary if the prize is divided between him and Pasternak.

4. If the prize is awarded to Pasternak, it should be suggested to him that he should decline the award, which is clearly directed against the interests of our homeland. For explaining this position Pasternak could use his claims addressed to the publisher Feltrinelli when he objected to publishing an uncorrected version of the novel. To this end it would be expedient to use the influence on Pasternak of the oldest non-party member writers K. Fedin (see doc. 9.6) and Vs. Ivanov,[6] with whom Pasternak has been connected for many years and whose opinion he takes into consideration. In this case the impact of Fedin and Vs. Ivanov could be of crucial importance. It is desirable that the writers K. A. Fedin and V. V. Ivanov should be

invited to meet secretaries of the Central Committee of the CPSU for a discussion about this question.[7]

5. In case of the Nobel Prize being awarded to Pasternak we also consider it expedient to use in the Soviet press his correspondence with the Italian publisher Feltrinelli.[8]

RSAMH, fond 5, opis' 36, ed. khr. 61, l. 52–53.

[1] Error: it was in the middle of 1956.

[2] According to Pasternak's son, in May 1956 "the Moscow radio broadcast a program in Italian on the forthcoming publication of Pasternak's novel *Doctor Zhivago.* Soon after that a representative of the Soviet embassy in Rome brought Sergio d'Angelo, member of the Italian Communist Party and worker of the broadcasting company of the USSR Ministry of Culture. The manuscript of the novel was given to Angelo for familiarization. But Angelo also acted as a literary agent of the Milanese communist, publisher Giangiacomo Feltrinelli. Pasternak was fully aware of the danger that he would be exposed to by publishing the novel abroad. When seeing d'Angelo out he said: "By that I am inviting you to see how I will be facing my execution."*

* E. B. Pasternak. *Boris Pasternak. Biography,* Moscow, Tsitadel, 1997, p. 677.

[3] The first copies of the novel appeared in Milan on November 21, 1957.*

* *ibid,* p. 694.

[4] Published in the *Literaturnaya Gazeta* on October 25, 1958. There is evidence that the refusal of the *Novy Mir* journal to publish the novel was inspired by the Cultural Department of the CPSU Central Committee.*

* *Continent,* 95, No. 1, 1998, p. 226.

[5] On October 29, 1958 the *Pravda* published a resolution of the presidium of the Board of the Soviet Writers Union, bureau of the organizational committee of the Russian Federation Writers Union, presidium of the Board of the Moscow branch of the Russian Federation Writers Union entitled "On the actions of the member of the Soviet Writers Union B. L. Pasternak incompatible with the title of the Soviet writer."

[6] See note 15 to this chapter.

[7] The invitation of both writers to the Secretariat of the Central Committee did not take place, probably, for lack of time (the decision of awarding the Nobel Prize to Pasternak was taken a day after the date of the document in question).

[8] The writer Victor Nekrasov happened to be in Milan at Feltrinelli's at the moment when a telegram came from Moscow from Pasternak with the request to suspend printing the novel. Afterwards he told the writer Lev Ozerov: "The first thing I did was make a call to Moscow to the Writers Union to ask them to get this telegram published in newspapers. I asked Surkov to do that. In person. As you know, the telegram wasn't published. It was shelved. In such a way he was put in a spot. Deliberately... They were glad to take advantage of the opportunity."*

* Lev Ozerov. *Dver' v Masterskuyu.* (The Door to the Workshop), Paris-Moscow-New York, Tretya Volna, 1996, p. 88.

Document 9.5

Resolution of the Presidium of the Central Committee of the CPSU of October 23, 1958 (minutes No. 144, point XLII): "On the slanderous novel by B. Pasternak"[1]

1. To admit that awarding the Nobel Prize to the novel by Pasternak, which slanderously portrays the October socialist revolution, the Soviet people who carried out this revolution, and building socialism in the USSR, is a hostile act with respect to our country and an instrument of the international reaction directed at kindling the Cold War.

2. To publish in the *Novy Mir* journal and the *Literaturnaya Gazeta* the letter of the editorial board of the *Novy Mir* sent to Pasternak in September 1956.[2]

3. To prepare and to publish in *Pravda* a satirical article[3] giving a harsh critical assessment of Pasternak's novel itself and also revealing the sense of the hostile campaign conducted by the bourgeois press in connection with awarding the Nobel Prize to Pasternak.[4]

4. To organize and to publish an article by major Soviet writers, which will assess awarding the Nobel Prize to Pasternak as an aspiration to kindle the cold war.[5]

Published in the journal *Istochnik* No. 4, 1993, p. 104; quoted from the original, which is in the Archive of the President of the Russian Federation (APRF), fond 3, opis' 34, ed. khr. 269, l. 8).

[1] The published resolution of the Presidium of the CPSU Central Committee was preceded by the memo from the secretary and member of the Presidium of the Central Committee, M. A. Suslov, of the same date, entitled "On measures in connection with awarding the Nobel Prize to B. Pasternak".* The content of the memo repeats the main ideas of doc. 9.4, with the exception of the part concerning M. A. Sholokhov; and the five points of the draft resolution of the Presidium of the Central Committee are repeated word for word in the present document, with the exception of point 2, which is as follows: "An attempt should be made through the writer K. Fedin** to explain to Pasternak the situation which has arisen as a result of awarding him the Nobel Prize and to advise Pasternak to decline the prize and to make a corresponding statement for the press."

* *Istochnik* No. 4, 1993; in APRF — the same ed. khr., l. 10–11.

** See doc. 9.6.

[2] See note 4 to doc. 9.4.

[3] The article *Shumikha Reaktsionnoi Propagandy Vokrug Literaturnogo Sornyka* (The Fuss of the Reactionary Propaganda around a Literary Weed) by the journalist D. I. Zaslavsky was published in *Pravda* on October 26, 1958. Pasternak was completely aware of the fact that the article full of insinuations about "the payment for the treachery" and so on could not have gone without the sanctions from the highest level of the Central Committee. That is why afterwards he tried to be careful,

first of all in financial questions, so as not to give any cause for speculation. Evgeni Pasternak remembers that after he was informed by the Foreign Legal Board that large sums of money had been transferred in his name to banks of Norway and Sweden as the royalty for publication of the novel in different countries, "understanding that his consent to receive the money can be assessed as the pay for the treachery, he preferred the greater part of the money to remain abroad and the smaller part to be transferred here and divided equally between Zinaida Nikolaevna and Olga Ivinskaya."*

* See: E. B. Pasternak. Boris Pasternak. Biography, p. 715; Zinaida Nikolaevna was Pasternak's wife, O. Ivinskaya was the poet's last love and his proxy in the post-war years.

[4] On the same day Lidia Chukovskaya gave in her diary the following characteristic to the publication of the article by Zaslavsky: "Today *Pravda* let Zaslavsky loose on Pasternak. This "literary streetwalker", using the terminology of Herzen, one of those who sells the beauty of his style, is supposed to [...] remind Pasternak (Zaslavsky reminds Pasternak!) of his conscience, of his duty to the people [...] Looking at me from the newspaper pages are quotation marks — the favorite sign of the jargon used by our journalists of the executioner trend; I clearly see across two decades:* 'activities' — in quotes, 'small group' — in quotes, 'school of thought' — in quotes; and now it is 'award' — in quotes, 'martyr' — in quotes."**

* She means the 1930s, the time of the Great Terror.

** In the collection *Memories of Boris Pasternak*, p. 426.

[5] See note 5 to doc. 9.4.

Document 9.6

From the Head of the Cultural Department of the CPSU Central Committee D. A. Polikarpov, to the secretary of the Central Committee M. A. Suslov; no date.[1]

I

Mikhail Andreevich,

K. A. Fedin has carried out the conversation with Pasternak. They head a one-hour meeting.[2] At first Pasternak behaved belligerently, saying flatly that he would not make any statement about his rejection of the prize and they could do with him whatever they wanted. Then he asked to be given several hours to think over his position. After the meeting with Fedin Pasternak went to Vsevolod Ivanov to ask him for advice.[3] K. A. Fedin himself understands that it will be necessary to take drastic action against Pasternak unless the latter changes his behavior.

RGANI, fond 5, opis' 36, ed. khr. 61, l. 65.

II

Mikhail Andreevich,

K. A. Fedin has just told me by phone that in the time agreed with him Pasternak did not come to continue the conversation. It should mean that Pasternak will not make a statement about his rejection of the prize.[4]

RGANI, fond 5, opis'36, ed.khr.61, l.64.

[1] Both memos quoted were sent on October 24, 1958, after Polikarpov came back from Peredelkino, where he had come on the morning of the same day and waited at Fedin's place for the results of his talks with Pasternak; the talks were conducted on instruction from Suslov (see note 1 to doc. 9.5).

[2] This information is at variance with the memories of T. V. Ivanova: "The next morning [that is, on October 24] I was called to the phone by Zinaida Kapitonovna Ulina (then a secretary in the office of the Soviet Writers Union) and asked to inform Fedin (neither Fedin nor Pasternak had telephones at their *dachas* and they used our phone) that D. A. Polikarpov (from the Central Committee) had left for his place. I immediately sent a note to Fedin and went to Vsevolod's study. Discussing the events we looked out of the window from time to time, and soon we saw Fedin striding quickly along Pasternak's road. Barely five minutes later, Fedin strode back..."* Such considerable difference in the information about how long Fedin stayed at Pasternak's *dacha* can most probably be explained by Polikarpov's unwillingness to present the result of his trip as a five-minute talk of his emissary with the troublemaker. It was not accidental that Polikarpov drove away to Staraya Ploshchad** "infuriated", as Fedin put it, according to Chukovsky."***

* In the collection *Memories about Boris Pasternak*, p. 255.

** The square in Moscow where the Central Committee was situated.

*** *ibid*, p. 277.

[3] Further on Ivanova continued: "... in another five minutes Pasternak ran into Vsevolod's study out of breath and said that Fedin came to him with an "ultimatum", and added: "For the first time he came not as a friend but as an official."... Pasternak stayed for a short while, but he repeated several times how astounded, and more than that, painfully hurt he was by the official talk from Fedin..."*

**ibid*, p. 255.

[4] Most probably, it was after his meeting with Vs.Ivanov that Pasternak decided not to meet with Fedin and to answer the telegram from the secretary of the Swedish Academy, Anders Esterling, by a "telegram of thanks".

Its text was as follows: "Infinitely grateful, moved, proud, amazed and confused."*

* E. B. Pasternak. *Boris Pasternak. Biography*, p. 702. See also note 15 to this chapter.

Document 9.7

From the letter of B. L. Pasternak to the Presidium of the Soviet Writers Union, of October 27, 1958[1]

1. I was sincerely willing to come to the meeting[2] and came to town for that but unexpectedly felt ill.[3] The comrades should not take my absence as a mark of inattention. I am writing this note in haste, and perhaps it is not as smooth and convincing as I would like it to be.
2. Even now, after all the fuss made and the articles, I go on thinking that it is possible to be a Soviet man and write books like *Doctor Zhivago.* I merely have a broader understanding of the rights and possibilities of a Soviet writer and I do not lower this title by such a notion.
3. … Now printed in huge newspaper circulation are solely its [the novel's] unacceptable parts that had impeded its publication and which I had agreed to omit, and nothing happened but misfortunes striking me in person. Why couldn't it have been published three years ago, with the corresponding cuts.
4. I do not consider myself a sponger in literature. I have done something for it, cross my heart.[4]
[…]
6. I thought that I would not be alone in my joy on the occasion of being awarded the Nobel Prize, that this joy would concern the society which I am part of. As I see it, the honor given to me, a contemporary writer living in Russia, and therefore Soviet, is also given to the whole of Soviet literature. I am upset about having been so blind and deluded.
7. As for the prize itself, nothing can make me admit that this award is a disgrace and thank for the honor I have been given by being rude. Where the monetary side of the matter is concerned, I can ask the Swedish Academy to put the money in the Peace Council fund, not to go to Stockholm to receive it or leave it altogether at the disposal of the Swedish authorities. I would like to speak about it with one of our officials, maybe, with D. A. Polikarpov, a week and a half or two weeks later, when I have come to myself after the shock I have already got and still am expecting to get.[5]
8. I am expecting everything for myself, comrades, and I do not accuse you. Circumstances may make you go very far in dealing with me, so as to, again under the pressure of similar circumstances, rehabilitate me when it is already too late.[6] But in the past there already were so many such things!! I ask you: do not hurry. It will not bring you more happiness or more glory." (doc. 9.7)

Published in the *Continent* journal, 83, 1995, p. 196–197, by a copy from AP RF (fond 3, opis' 34, ed. khr. 269, l. 58–59).

[1] The letter was read, on October 27, 1958, at a joint sitting of the presidium of the Board of the Soviet Writers Union, the bureau of the organizing committee of the

Russian Federation Writers Union and the presidium of the Board of its Moscow branch and also, with the comments from the chairing writer S. A. Smirnov, at the all-Moscow writers meeting.

[2] Pasternak means the joint sitting on October 27,1958 mentioned above.

[3] According to the memories of T. V. Ivanova, on the day when he received the summons Pasternak "had his left arm on a sling. He said that when he was reading the summons from the Writers Union, his arm became 'sort of paralysed', now it is all right, but hurts." For Vs.Ivanov receiving the same summons turned out to be still more painful. "In the evening," continues Tamara Vladimirovna Ivanova, "when I [...] was at my sister's, the nurse Maria Egorovna called me from Peredelkino and asked me to come as soon as possible, because Vsevolod Vyacheslavovich feels very bad [...] Maria Egorovna said that at first a summons from the Writers Union had been brought (it was an invitation to the meeting of the presidium of the board on 'the unworthy behavior of Pasternak') [...] And suddenly she heard the sound of a body falling and an outcry [...] (the doctor diagnosed a spasm of the brain blood vessels, a possibility of a stroke)."*

* In the collection *Memories of Boris Pasternak*, p. 256.

[4] An indirect answer to the passages from the article by D. Zaslavsky in *Pravda* (see note 3 to doc. 9.5).

[5] The authorities were not in the least planning to reject the possibility of currency transfers at the expense of Pasternak. Having received from the Foreign Legal Board the notification of large sums accumulated in banks of Norway and Switzerland from the royalties for foreign editions of *Doctor Zhivago*, he submitted a request to Polikarpov to let him receive some part of them in order to give ten thousand dollars to the Literary Fund for assistance to aged writers. "In response he got a categorical demand to completely reject whatever money but to transfer all his royalties to Moscow, giving them to the World Peace Fund."*

* E. B. Pasternak. *Boris Pasternak. Biography*, p. 715.

[6] A hint at the recent exposure of the Stalin regime and the posthumous rehabilitation of its victims.

Document 9.8

From the memo of the Head of the Cultural department of the CPSU Central Committee D. A. Polikarpov to the Central Committee, of October 28, 1958

... On October 27 a joint sitting of the Presidium of the Board of the Soviet Writers Union, the bureau of the Organizing Committee of the Russian Federation Writers Union and the Presidium of the Board of the Writers Union, Moscow branch took place. It discussed the question of the behavior of Pasternak, which is incompatible with the title of a Soviet writer. Forty-two writers were present at the sitting...

26 writers did not come. Among those who did not come [...] without explaining the reasons were comrades Leonov and Pogodin. Pasternak's personal friend, writer Vsevolod Ivanov also did not come, reporting himself sick. Pasternak himself did not appear at the sitting, pleading illness. He sent to the Presidium of the Soviet Writers Union a letter, outrageously impudent and cynical.[1] In this letter Pasternak, transported with delight over being awarded the prize, viciously slanders our reality and makes vile accusations against Soviet writers. This letter was read at the sitting and was received with anger and indignation.

29 writers spoke at the sitting [...] It is worth mentioning that there were only a few speeches pointing out that for a long time some writers had extolled the significance of Pasternak for Soviet poetry. A participant of the sitting, the poet C. M. Kirsanov, who had once extolled Pasternak, did not express his attitude to the question being discussed. The writers who were present at the sitting took a unanimous decision to expel Pasternak from the membership of the Soviet Writers Union...

Published in the *Continent* journal, 83, 1995, p. 198–202,

by the original from AP RF (fond 3, opis'34, ed. khr. 269, l. 53–57).

[1] See doc. 9.7.

Document 9.9

One episode in the eyes of two interlocutors

I

From the Secretary of the Soviet Writers Union K. A. Fedin to the Head of the Cultural Department of the CPSU Central Committee D. A. Polikarpov, of October 28, 1958.

Dear Dmitry Alexeevich, today, at 4 o'clock Olga Vsevolodovna (I do not remember her name, she is Pasternak's friend[1]) came to me and told me — in tears — that this morning Pasternak declared to her that he and she had only "Lann's way out"[2] left. According to her, Pasternak allegedly asked her if she agreed "to go away together", and she, allegedly, agreed. The purpose of her coming to me was to learn if "it is still possible (in my opinion) to save P., or if it is already late, and if it is not late, what I could advise — what (in my opinion) should be done to save him.

I answered that such a declaration is a threat, in this case, to me, and in all other cases, to the person to whom it is made, and under such a threat it is impossible to ask for any advice. At the same time I said that the only thing that I consider absolutely necessary to advise her is that she must talk P. out of his crazy intention. I also said that I do not know if it is thinkable now, after everything that has happened, to "save" P., who refused point-blank to be "saved" when it was accessible. O. V. said that she is ready to compose "any" letter, to whoever possible, and to persuade P. to sign it. I answered that I cannot imagine what the content of the letter could be now and to whom it could be addressed.

I am not able to say for sure whether I should regard O. V.'s visit as Pasternak himself turning to me (she swore that she had not told him anything about it, although some time later she told me that he did not want her to come to me).[3] But I think you should know of P.'s intention, whether real or imaginary, serious or theatrical, of the existence of the threat or the attempt to maneuver it. In my long talk with O. V. she repeatedly asked me to whom it is "better" to address the letter or to whom "to go". I could not answer her anything and only promised that I would inform you that she came over and, of course, you would act in the way you would consider necessary, and maybe you would have a wish to summon her or Pasternak.[4] Just in case, I have taken her phone number in order to give it to you (B-7–33–70).

In the lower part of the sheet there are two notes:
"comrade Kirichenko A. I. has familiarized himself." (no date, signature illegible; "To keep in the archive. V. Gorbunov.5/XI."
RGANI, fond 5, opis' 36, ed. khr. 61, list 45 (s oborotom).

[1] O. V. Ivinskaya.

[2] According to the memoirs of O. V. Ivinskaya*, several days before October 28, 1958 she told Pasternak the news about the suicide of the writer and translator of the British and American literature E. L. Lann and his wife, which took place on October 3, 1958.

* O. V. Ivinskaya. *Gody s Borisom Pasternakom. V Plenu Vremeni* (The Years with Boris Pasternak (Captured by the Time), Moscow,Libris, 1992, p. 276.

[3] In the description given by Ivinskaya herself (see below, she does not say anything about Fedin having been mentioned in her conversation with Pasternak. It was only on the next day that she told Pasternak of her visit to him.*

**ibid*, p. 279.

[4] A doctor from the clinic of the Literary Fund coming to Pasternak's *dacha* in Peredelkino was one of the consequences of Fedin's information. Here is how it was remembered by E. B. Pasternak, the poet's son: "He spoke ironically about the doctor sent to him on the instruction from the Central Committee, who was on duty in the house to avoid a suicide attempt. Such a possibility became known through Ivinskaya, who came to Fedin to tell about Pasternak being driven to extremes. Fedin informed Polikarpov of that, and on the same day the Literary Fund clinic sent a lady with a set of first-aid medicines. She was put up in a small living room."*

* E. B. Pasternak. *Boris Pasternak. Biography,* Moscow, Tsitadel, 1997, p. 705.

II

From the memories of O. V. Ivinskaya

...I told him that B. L. is on the verge of suicide, that he has just now suggested this outcome to me:

"B. L. doesn't know that I'm here... You are an old friend of his, a cultured person, you understand that amid all that hue and cry your word will be important for him. So tell me — what else do they want from him now?[1] Can they be really expecting him to take his own life?"

Fedin came up to the window, and it seemed to me that I saw tears in his eyes.[2] But at last he turned round:

"Boris Leonidovich has dug such an abyss between himself and us that it cannot be crossed", he said with some theatrical gesture. And after a short pause added in a completely different tone:

"You've told me a terrible thing: will you be able to repeat it elsewhere?"

"Anywhere, even in hell", I answered. "...But it is you yourselves who are driving him to suicide."

"Will you please wait. Now I'll make a phone call and you'll meet with somebody to whom you'll tell everything that you have just told me." And he started to phone the same ill-fated Polikarpov.[3] "Will you be able to come to the Union[4] tomorrow at three o'clock? Dmitry Alexeevich will receive you — not in the Central Committee but as a writer, in the House of Writers [...] You certainly understand it yourself", he instructed me when saying goodbye "that you must restrain him, so that there would be no second blow to his motherland."

I understood that they did not want this suicide...

O. V. Ivinskaya. *Gody s Borison Pasternakom* (The Years with Boris Pasternak), Moscow, Libris, 1992, pp. 278–279.

[1] It is not quite clear why the author says "what else do they want from him now?" Her meeting with Fedin took place on October 28, while Pasternak's telegram to Stockholm with his refusal to accept the prize is dated October 29, that is, at the moment of the meeting there was no refusal yet and what the authorities wanted from the poet was absolutely clear. Most likely, it was some aberration of the author's memory, as she wrote the book many years after the events she described, in 1976.
[2] It is quite possible that it was really like that. K. I. Chukovsky wrote in his diary, on October 27, about Fedin's situation between the devil and the deep blue sea: "... Fedin was sad and irritated. "Pasternak will do a lot of harm to all of us. Now they will start a fierce campaign against intelligentsia." And he told me that Polikarpov was infuriated when leaving Peredelkino... And here at once Fedin started to say how sorry he was for Pasternak. "Because Polikarpov did not come on his own behalf. The answer is expected there. He will be most cruelly punished. And what shall I do? Because I am not a nominal chairman but really the head of the Uinon.* I have to speak against him..."**

* In the next year, 1959, Fedin was elected First Secretary of the Soviet Writers Union.

** *Vospominaniya o Borise Pasternake* (Memories about Boris Pasternak), p. 277.

[3] Probably, during his telephone conversation with Fedin Polikarpov suggested repeating what he had been told in the written form. After that the letter appeared, composed on that very day, October 28,1958. The next day Ivinskaya really visited Polikarpov, who told her that "this scandal should be settled [...] You can help him turn to his people [...] Don't pay attention to unnecessary cries, stay by his side, don't let him get any absurd ideas..."*

* Ivinskaya, pp. 279–280.

[4] The Soviet Writers Union.

Document 9.10

From the report of the first secretary of the Central Committee of the Young Communists League V. E. Semichastny at the solemn plenary session of the Central Committee in celebration of the Komsomol's 40th anniversary, on October 29, 1958.[1]

...Pasternak has lived 41 years in a socialist country; for 41 years he has fed on the bread of the people that was building the new on the debris of the old, that has experienced cold and hunger, raised the former Russia to a new life, created out of the former Russia a powerful state, which shakes the minds of all progressive people[2] and causes fear of the enemies of socialism; of the people that has gone through wars and smashed the fascist hydra. And this man lived among us and was better provided for than an average worker, who worked, did his job and was at war. And now this man suddenly spits in the people's face. What can we call it?

We sometimes say — by the by, quite undeservedly — all sorts of nasty things about the pig. I must say that it is slander against the pig. The pig — all the people who deal with this animal know its characteristic features — never fouls the place where it has meals[3], it never fouls the place where it sleeps. So, if we compare Pasternak with the pig, the pig will never do what he has done.(Applause). And Pasternak — this man ranks himself among the best representatives of our society — has done it. He has fouled the place where he ate, fouled it for those by whose labor he lives and breathes.(Applause).[4]

I would like to express my opinion on this question. Why shouldn't this internal emigrant try the capitalistic air, which he misses so much and to which he expressed his attitude in his work. (Applause). I am sure that our public opinion would welcome it.(Applause). Let him become a real emigrant and go to his capitalistic paradise. I am sure that neither public opinion, nor the government would pose any obstacles to him and, on the contrary, everybody would think that his leaving our country would make the air fresher.(Applause).

Komsomolskaya Pravda, October 30, 1958, p. 3.

[1] From the diary of Lidiya Chukovskaya: "October 29, 1958, evening. There are rumors in the city — about some speech of Semichastny at the 40th anniversary of the Komsomol, where he allegedly called Pasternak a pig.... Some Ciceros! And they speak as if their words should be remembered for ages..."*

* *Vospominaniya o Borise Pasternake*, p. 443.

[2] The evident ambiguity of these words certainly remained beyond the understanding of the Komsomol leader.

[3] From the diary of Lidiya Chukovskaya: "October 30, 1958, daytime [...] I have read Semichactny's speech in the *Komsomolskaya Pravda.* I am rewriting it here so as to reread it from time to time and never forget. At first there is comparison with the sheep. A bad sheep in a flock.* Well, that's quite ordinary. Then — an inconsistency of the image! — the sheep turns into a pig...The most interesting expression here is 'has meals'. 'The pig has meals.' Here is what he is, comrade Semichastny. He reckons that the word 'eats' is a rude word, and saying 'has meals' about a pig looks more cultured to him."**

* A paragraph from Semichastny's speech mentioned by the author of the diary is omitted in doc. 9.10.

** *Vospominaniya o Borise Pasternake*, p. 434.

[4] Having lived up to the age of 75, on his birthday, January 15,1999, Semichastny was favored by an interview on the NTV television channel. When asked by the interviewer what he thinks, 40 years after, of his utterances about Pasternak, the latter answered energetically: "Yes, that's the way it was. I called him a pig, and so on. I uttered it and said it in such a way with support from the whole audience.* And he reacted to that.** So, was he put into prison after that? No. Was he subjected to repression? No. Was something arrested from him? No.*** So, what's the matter?"****

* Here the former secretary of the YCL Central Committee and then the KGB chairman told the truth. As we can see from the text of doc. 9.10, his reflections on the author of *Doctor Zhivago* were five times interrupted by enthusiastic applause of those present at the meeting. I can personally testify to the fact that the applause was enthusiastic and long; it so happened that at the moment of the Komsomol leader's speech the radio at my home was turned on.

** He means Pasternak's rejection of the Nobel Prize (see doc. 9.11 and note 1 to it).

*** And here is what the poet's son says about these days: "His own correspondence [by mail] was completely blocked." (E. B. Pasternak. *Boris Pasternak. Biography*, p. 705).

**** A week later, on January 22, 1999, the *Izvestiya* newspaper did not miss the chance to immortalize this dialogue in a small article entitled *Krivaya Rozha v Krivom Zerkale* (A Crooked Face in a Distorting Mirror).

Document 9.11

The letter of B. L. Pasternak to the first secretary of the CPSU Central Committee N. S. Khrushchev of October 31,1958 and the official reaction to it.

I

Dear Nikita Sergeevich,

I am addressing myself to you personally, to the Central Committee of the Soviet Communist Party, and to the Soviet Government. I have learnt from the speech made by Semichastny that the government would not "put any obstacle to my departure from the USSR". For me this is impossible. I am linked to Russia by my birth, my life and my work. I cannot imagine my fate separate from it and outside Russia. Whatever my mistakes and errors, I could never have imagined that I should find myself the center of such a political campaign as has been kindled around my name in the West. Having realized what was happening, I informed the Swedish Academy of my voluntary renunciation of the Nobel Prize.[1] To leave my country would be for me the equivalent of death; and for that reason I ask you not to take this extreme measure. With my hand on my heart I can say that I have done something for Soviet literature, and I can still be useful to it.[2]

Pravda, November 2, 1958, p. 2.

[1] "Due to the resonance caused by my award in the society I belong to, I have to decline the prize. Don't consider my voluntary refusal an insult." The telegram was sent to Stockholm in the morning on October 29, 1958*, that is, before Semichastny's speech (doc. 9.10)

* *Continent,* 95, No.1,1998, pp. 241–242.

[2] According to Ivinskaya,* the draft of the letter to Khrushchev was initially written by an acquaintance of hers, whom she called Zoren'ka. Further on she wrote: "We rewrote the text provided by Zoren'ka, trying to follow the style of Pasternak. Ira and Koma** went to Peredelkino to get B. L.'s signature [...] B. L. signed several more blank sheets,*** so that I could correct something else if necessary. There was also a note in red pencil: 'Lelyusha, I'm leaving everything as it is, only, if possible, write that I was born not in the Soviet Union but in Russia.'"

* O. V. Ivinskaya, p. 283.

** Ivinskaya's daughter Irina Emelyanova and the son of the writer Vs.Ivanov, Vyacheslav Vsevolodovich.

*** That is, blank sheets of paper with Pasternak's printed letterhead.

II

TASS Statement

In connection with the letter of B. L. Pasternak to comrade N. S. Khrushchev being published today, TASS is authorized to declare that on the part of the Soviet authorities there will be no obstacle if B. L. Pasternak expresses his wish to go abroad to receive the prize he has been awarded.[1] The stories now being spread by the bourgeois press about Pasternak having been allegedly denied the right to go abroad are a blatant fabrication. As it has become known, up to the present moment Pasternak has not turned to any Soviet governmental institution with an application for a visa for going abroad, and on the part of these institutions there has not been, and will not be any objections to giving him an exit visa.

In case Pasternak has a wish to completely leave the Soviet Union, the social system and people of which he has slandered in his anti-Soviet writing *Doctor Zhivago*, the official institutions will not put any obstacle to it. He will be given an opportunity to leave the territory of the Soviet Union and to experience in person all "delights of the capitalistic paradise".

Pravda, November 2, 1958, p. 2.

[1] Thus, the official statement ignores the fact that Pasternak had already refused to accept the Nobel Prize. That can hardly be an accidental stylistic inaccuracy. Most likely, it seemed to the authorities that the telegram about the refusal and the letter to Khrushchev were not enough. They needed a more detailed repentance, and they achieved their goal — in the form of Pasternak's letter to *Pravda* (see doc. 9.12).

Document 9.12

Two variants of the letter of B. L. Pasternak to the Pravda newspaper

I

A handwritten text prepared by B. L. Pasternak before the meeting with the head of the Cultural Department of the Central Committee D. A. Polikarpov[1]

Over the last eventful week I was not legally persecuted, I did not risk my life, nor my freedom, nor anything whatsoever. If owing to the ordeal sent to me I put anything at stake, it was only my health, which I was able to preserve not because of my robustness, but because of my good spirits and human sympathy. Among a great lot of people who condemned me there might have been a few of those, unknown to me, who abstained. Rumor has it (it may be wrong) that Hemingway and Priestley stood up for me, maybe, so did the Trappist writer Thomas Merton and Albert Camus, my friends. Let them use their influence to put a stop to the fuss made around my name. Probably, there also were well-wishers in my homeland, maybe, even amid the highest government.[2] I would like to express my heartfelt gratitude to all of them.

There is no hopelessness in my situation. We shall live further on, actively believing in the force of beauty, good and truth. The Soviet government offered me a free exit abroad, but I did not take advantage of it, because my occupation is too closely connected with my home ground and cannot bear transplantation anywhere else.

Published in the *Continent* journal, 95, No.1, 1998, p. 24.

[1] The original is in RGALI, fond 379. According to Ivinskaya, the letter was written on November 4, 1958, after Polikarpov's phone call from the Central Committee.*

[2] This phrase certainly has some concealed meaning. Most probably, Pasternak meant the chairman of the Presidium of the Supreme Soviet of the USSR K. E. Voroshilov, who, after his official visit to India, at the beginning of 1958, brought from there a copy of *Doctor Zhivago* translated into *oria,** the language of a small nationality, and "passed it through someone to B. L."** Already after the end of the story of the publication of the letter to *Pravda* with his signature he learnt about one more possible high-ranking well-wisher. Jacqueline de Proyart told him, in January 1959, about some details of the visit to the United States of the deputy chairman of the Council of Ministers of the USSR A. I. Mikoyan. In his talk with the US State Secretary J. F. Dalles he said that the Soviet government had nothing to do with the scandal surrounding Pasternak's Nobel Prize and that it had been provoked by "intellectuals". American newspapers published photos of Mikoyan at a bookstore with copies of the novel *Doctor Zhivago* displayed in its window.***

* As said E. B. Pasternak, his father's correspondence with the Indian publisher in the *oria* language Prafulla Chandra Das, which resulted in the translation of *Doctor Zhivago* into that language, started with the writer I. G. Ehrenburg acting as intermediary. (*Novy Mir* No. 1, 1992, p. 181).

** O. V. Ivinskaya, p. 253.

*** *Novy Mir* No. 1, 1992, p. 154.

II

The text published in the Pravda with B. L. Pasternak's signature[1]

I ask the editorial board of *Pravda* newspaper to publish my statement. It is my respect for the truth that makes me do it.

Like everything that happened to me was a natural consequence of the actions performed by me, everything I did on the occasion of being awarded the Nobel Prize was free and voluntary. I took being awarded the Nobel Prize as a literary distinction, I was happy and expressed it in the telegram to the secretary of the Swedish Academy Anders Esterling. But I was mistaken. I had reason to be mistaken in such a way, as I had been nominated for this prize before, for example, five years ago,[2] when my novel did not exist yet. After a week passed, when I saw the increasing scope of the campaign around my novel and made sure that this awarding was a political step which has now led to monstrous consequences, I, of my own free will, not being forced by anybody, sent my voluntary rejection.

In my letter to Nikita Sergeevich Khrushchev[3] I said that I am linked with Russia by my birth, my life and my work and that leaving it and going into exile to a foreign land is unthinkable for me. Speaking about these links I meant not only being related to its land and nature, but, of course, also to its people, its past, its glorious present and its future. But insurmountable obstacles set up by the novel through my own fault emerged between me and these links.

I have never had any intention to do harm to my country and my people. The board of the *Novy Mir* warned me that the novel could be taken by readers as a work directed against the October Revolution and the basis of the Soviet system.[4] I was not aware of that, which I now regret. Indeed, if the conclusions resulting from the criticism of the novel are taken into consideration, it appears that in the novel I support the following erroneous ideas. It is as if I maintain that any revolution is a historically unlawful phenomenon, that one of these unlawful developments was the October Revolution, that is has brought Russia misfortunes and lead to the ruin of the successive Russian intelligentsia. It is clear to me that I am unable to put my signature under such assertions brought to absurdity. Meanwhile, my work awarded the Nobel Prize gave grounds for such a deplorable interpretation, and it is for this reason that I eventually declined the prize...

Over this eventful week I was not persecuted, I did not risk my life, nor my freedom, nothing whatsoever.[5] I want to emphasize once again that all my actions are taken voluntarily. People who are my close acquaintances know well that nothing in the world is able to make me be insincere or act against my conscience. This time it was also like that. There is no need to assure that nobody forced me to do anything and that I am making this statement with a free soul, with radiant belief in the common and my own future, with pride in the time in which I live and in the people who surround me.[6] I believe that I will have enough strength to restore my good name and the undermined trust of my comrades. B. Pasternak, November 5, 1958.[7]

Pravda, November 6, 1958, p. 4.

[1] "Drafts of this letter remained intact among Ivinskaya's papers: there are the initial autograph of Pasternak,* Polikarpov's typed copy with notes on the margins and separate paragraphs written by Pasternak's hand, stuck on the typed sheets and torn away."** "Boris Leonidovich wrote — at first it was not a letter of repentance. Then it was worked on hard, so that the final result was a lie and admission of guilt. And pointedly voluntary at that."***

* See part 1 of the present document.

** Elena and Evgeny Pasternak. *Continent*, 95, No. 1, 1998, p. 244.

*** O. V. Ivinskaya. *Ogonek* No. 37, 1988, p. 31.

[2] Most likely, Pasternak meant the rumor of his nomination for the 1954 prize (see note 1 to doc. 9.1–1).

[3] See doc. 9.11–I.
[4] In conformity with the resolution of the CPSU Central Committee of October 23,1958 (doc. 9.5, p. 2), the text of the critical response of the *Novy Mir* editorial board to the manuscript of the novel was published in the *Literaturnaya Gazeta* on October 25, 1958.
[5] This phrase is the only one in the published variant that was taken from the text of the letter prepared by Pasternak himself (part 1 of this document). Symptomatically, however, in the process of publication the word "legally" before the word "persecuted" was removed in the mentioned phrase. Polikarpov did not want to agree to Pasternak the deliberate narrowing of the wide enough notion of persecution to mere absence of legal persecution or a threat of it, which at that moment corresponded to the true state of affairs.
[6] The utterly false enthusiasm of the last phrase is the apotheosis of the hybrid of Pasternak's wishes, made up in the atmosphere of threats and promises, and a propaganda leaflet prepared in the Central Committee's Cultural Department, which, after long bargaining, Polikarpov made him sign.
[7] The idea to organize this letter, in addition to the one published on November 2, 1958 (doc. 9.11–1), came into being in the Central Committee, most probably, already on the next day. As we can see from doc. 9.13, the date November 3, 1958 is on Polikarpov's cover letter to the selection of protest telegrams from a number of European and US intellectuals, which was sent to the Secretariat of the Central Committee. The text that Pasternak was made to sign looks as if giving soothing answers to fears and appeals of the authors of the telegrams and at the same time admitting his own errors which "explain" the causes of the ongoing scandal.

Document 9.13

Telegrams from abroad received by the Soviet Writers Union on October 30 and 31, 1958 and sent from there to the Central Committee, for familiarization[1]

1

From the Head of the Cultural Department of the CPSU Central Committee D. A. Polikarpov, to the Central Committee, of November 3,1958

Please see telegrams from abroad coming to the Writers Union in connection with the case of Pasternak.

Note in the top left corner: "To be given to the commission members for familiarization."[2] E. Furtseva. Signatures of N. A. Mukhitdinov and M. A. Suslov.

Note in the bottom right corner: "To be kept in the archive. V. Gorbunov.5/XI."

RGANI, fond 5, opis' 36, ed. khr. 61, l. 46

[1] As all telegrams in this document were sent to the Central Committee translated into Russian, they are given below in translation from this language.
[2] The Ideological Commission of the Central Committee headed by M. A. Suslov, secretary of the Central Committee.

II

From Paris. To the Soviet Writers Union (received on October 30, 1958).

The International P. E. N. Club, being greatly worried about the rumors concerning Pasternak, asks you to protect the poet by providing a free atmosphere for his creative work. Writers of the whole world think about him as their colleague. Andre Chamson, President of the International P. E. N. Club, David Carver, Secretary.[1]

ibid, 1.47

[1] Pasternak's French correspondent Jacqueline de Proyart enclosed with her letter of December 12, 1958, a clipping from a French newspaper of October 31,1958 which mentioned, among other things, the above telegram of the P. E. N. Club leaders.*

* *Novy Mir* No.1, 1992, p. 154.

III

From London. To the Chairman of the Soviet Writers Union (received on October 30, 1958).

We are extremely anxious about the fate of one of the world's greatest poets and writers, Boris Pasternak. We regard the novel *Doctor Zhivago* as an exciting personal evidence and not as a political document. In the name of the great literary tradition that is behind you we appeal to you not to dishonor this tradition by persecuting the writer esteemed by the whole civilized world. Maurice Bowra, Kennedy Clerk, Thomas Eliot, E. Forster, Graham Greene, Aldos Huxley, Julian Huxley, Rose MacColey, Somerset Maugham, John Priestley, Alan Price, Jones Herbert Read, Bertrand Russel, C. P. Snow, Stephen Spender, Rebecca West.

ibid, 1.48.

IV

From London. The Writers Union, Moscow (received on October 30, 1958).

The writers community is highly indignant at Boris Pasternak being expelled from the members of the Writers Union and strongly insists on his reinstatement.[1]

ibid, 1.49.

[1] No signatures. This telegram must have been sent to Moscow shortly after the previous one, after those who had signed it learned from reports of information agencies about the expulsion of Pasternak from the Soviet Writers Union at the joint sitting of the Presidium of the Board of the Soviet Writers Union, the organizing committee of the Russian Federation Writers Union and its Moscow branch, on October 27, 1958.

V

From Stockholm. The Soviet Writers Union, Moscow
(received on October 30,1958)

The Society of Students of Humanities Departments of Stockholm University categorically protests against the persecution of Boris Pasternak after him being awarded the Nobel Prize for Literature. This proof of the complete lack of personal rights of Soviet writers filled us with the feeling of deepest sorrow and disappointment.[1]

ibid, 1.50

[1] No signatures.

VI

F*rom Washington. Moscow, the Soviet Writers Union*
(received on October 31, 1958)

I would like to make a protest against the infamous and ignorant attacks on Boris Pasternak. This recalls Hitler's refusal before the war to allow von Ossietzky to accept the Nobel Prize.[1] I. F. Stone.

ibid, 1.51.

[1] See doc. 3.16

Document 9.14

Chairman of the State Committee for Cultural Relations with Foreign Countries G. A. Zhukov, to the CPSU Central Committee, of November 4,1958.

I am sending you the record of the talk with the ambassador of India. K. P. S. Menon, which took place on November 3, this year.

Two notes under the cover text: "Show around the commission.[1] M. Suslov"; "comrade Polikarpov has seen it. November 10,1958. G. Dyakonov."

The record of the talk with the ambassador of India, K. P. S. Menon.

I met Menon in the evening of November 3, at a reception organized by the ambassador of Ethiopia. Menon started a talk about "the case of Pasternak" saying that, in his opinion, this "case" was, in fact, settled.[2] Menon said: "The other day Nehru received a copy of the letter addressed to Khrushchev by a group of writers.[3] In this letter English writers ask Khrushchev "to save" Pasternak and not to subject him to any persecution. Nehru enquired of me by telegram what I think about the case of Pasternak and how he should act.[4] I answered that, in my opinion, Pasternak is a lone intellectual who does not accept the October Revolution and in his novel promotes ideas alien to Soviet people.

"I have written that before Pasternak was awarded the Nobel Prize no one in the Soviet Union had spoken against him, although it had been known what his novel was. But when the Nobel Prize was awarded to Pasternak and a campaign of political character started abroad around this event, Soviet people could no longer keep silence — they saw in it a political operation aimed at doing moral harm to the Soviet Union. I told Nehru that I was deeply convinced that, in spite of everything, repressive measures would not be applied to Pasternak and his personal fate could not be a cause for concern. Attempts to make of him 'a victim of the Soviet regime' will not succeed."

RGANI, fond 5, opis' 36, ed. khr. 61, l.73, 74

[1] The Ideological Commission headed by the secretary of the Central Committee M. A. Suslov.

[2] He means Pasternak's rejection of the Nobel Prize.

[3] This is probably about the group of writers who also signed the telegram to the Soviet Writers Union (doc. 9.13-III).

[4] There is a different interpretation of the initiative of the prime minister of India, J. Nehru in connection with the campaign against Pasternak. As wrote E. B. Pasternak, with reference to what was said by Ilya Ehrenburg, "who came those days from Sweden, where, in the electric atmosphere of an international scandal, he presented the Swedish writer Lundkvist with the Lenin Peace Prize, a sudden change in the course of events happened after Javaharlal Nehru's telephone call about the persecution of Pasternak. TASS immediately guaranteed the immunity of Pasternak's personality and property and his unimpeded trip to Sweden."* Perhaps, there is more truth in Menon's information on his telephone talk with Nehru; that is, Nehru did not call Khrushchev.

* E. B. Pasternak. *Boris Pasternak .Biography,* Moscow, Tsitadel, 1997, p. 707.

Document 9.15

From B. L. Pasternak's letter to J. de Proyart, of December 22, 1959.

One Indian writer, a certain Prafulla,[1] asks me to write a foreword to Quasimodo, whom he wants to publish,[2] like last year he wanted to publish me.

I did not answer him anything, but I could support my refusal by the simple fact that I know nothing about Quasimodo, probably much less than he knows about me. Because when there was a storm raging around me, Quasimodo, who was at that time at the Kremlin hospital,[3] certainly must have had some reasons for writing half a page against me in the manner of *Pravda* or the *Literaturnaya Gazeta* and so on. I was shown this abusive judgement at that very time, at the CPSU Central Committee,[4] as a proof of the so-called opinion of the West, which was supposed to disarm me. It is not a secret, it can be told, if necessary.[5] Now a joke: it's not only me who undeservedly got the Nobel Prize; so did Quasimodo.[6]

Novy Mir No. 1, 1992, p. 180–181.

[1] See note 2 to doc. 9.12–1.
[2] In the *oria* language, in connection with Quasimodo being awarded the Nobel Prize, in 1959; for the same reason as for publishing *Doctor Zhivago* in this language. The publisher probably decided to mark the verdicts of the Swedish Academy by publishing works of the laureates...
[3] Quasimodo was treated at the ordinary Botkin Hospital, but in one of its special departments where only patients sent by the Party and government organs were admitted.
[4] November 1, 1958.
[5] Pasternak means that this information does not fall under the "non-divulgence" undertaking which he signed during the interrogation in the USSR Procurator's Office, on March 14,1959 (doc. 9.24).
[6] In the given context of the document this phrase is difficult to understand, and it looks as if it is separated from the idea which generated it and which Pasternak chose not to develop. As one of supposed reasons why this phrase appeared in the letter we can suggest the following. Pasternak suspected that the Swedish Academy was, to a certain extent, compelled to choose Quasimodo the next year after awarding the Nobel Prize to him. The world media repeatedly expressed the opinion that political reasons in the framework of the principle of "deterrents and counterbalances" often influence the decisions of the Swedish Academy. Such an approach to its verdicts was meant to emphasize that in the violent confrontation of the Cold War period their positions were neutral and balanced. From this point of view, indeed, it would have been difficult to select a more suitable candidate for the 1959 prize than Quasimodo, who publicly criticized Pasternak in 1958. Hence the common international reaction to awarding the Nobel Prize to the poet. whose works were little known even in his homeland. The bewilderment at this fact, expressed by the majority of critics, who thought that Quasimodo did not deserve such a high award, was probably what Pasternak meant in the mentioned phrase.

Document 9.16

From B. L. Pasternak's letter to Jacqueline de Proyart of November 28,1958

I hope that you have received my postcard about December 10 in Stockholm.[1] Without thinking of you being unbelievably busy and having no time, I suggested that you use your former power of attorney written before the storm, before the scandal.[2] I mentally painted a picture of you going, maybe, to Stockholm, not on my instruction but in accordance with your own rights, freely taking part, instead of me, in the giving ceremony and making, in your own way, a response speech addressed to the king.[3] But as far as I can see, the opportunity for that has passed. Nevertheless, I am enclosing for you the letter from Baroness von Golstein.[4] She is probably in the know of the Nobel affairs. Judging by her words, my refusal did not annul the sum which I declined, and it can still be disposed at our discretion.[5] Do you think I should take advantage of her advice? Could you undertake to do it and make the gift to the Red Cross in your name? For my part I will not touch upon the Nobel theme and will behave as if I did not exist.[6]

Novy Mir No. 1, 1992, pp. 152–153.

[1] "In the process of controlling Pasternak's correspondence", the postcard in question was confiscated by the KGB, along with his other mail (doc. 9.21).
[2] The power of attorney made up by Pasternak in the name of Jacqueline de Proyart enabled her to be his proxy in all affairs connected with publishing the novel *Doctor Zhivago* in foreign publishing houses.
[3] These were absolutely fantastical dreams, as the fact of declining the Nobel Prize, confirmed by the winner's signature, legally denied anybody's right to officially participate in the award-giving ceremony and certainly to make a "response" speech.
[4] Enclosed in the envelope with the letter being commented on were several sheets of the letter to Pasternak from the baroness Alice Schtal von Golstein, mailed from Stockholm on November 18, 1958. As says E. B. Pasternak, in this letter the baroness suggested what seemed to her a worthy and appropriate use of the sum of the Nobel Prize — giving it as a charity donation to the USSR Red Cross and Red Crescent Society. That was supposed to help eliminate misunderstanding and accusations, and she, with Pasternak's consent, was ready to act as intermediary.*
* *Novy Mir* No. 1, 1992, p. 153.
[5] Baroness von Golstein bore no relation to the Nobel Foundation institutions. Such a suggestion on her part was as fantastical as the suggestion of Pasternak (see note 3 to this document). Her supposition that the rejection of the Nobel Prize did not annul the monetary component of the prize showed that she was not familiar with the Nobel Foundation's charter regulations. The charter provides for the cases of declining the prize that in such cases the money allocated for the prize is given for increasing the Foundation's reserve capital.
[6] According to E. B. Pasternak,* his father liked very much the article *Pasternak's Victory* by the writer Roman Gul' published in the emigrant journal *Novy Zhurnal*. The article notes very subtly that in *Doctor Zhivago* the characters are not submitted to the logic of the ongoing events, everything is guided by unmotivated actions, and this is "extremely characteristic for Pasternak in general and inherent in his creative

nature."** Pasternak's abstraction from the logic of what happened manifested itself in the text of his commented letter, particularly in the last phrase about the possibility of receiving a big sum of money by an extraneous person for an extraneous purpose, as if its potential owner "did not exist".

* *Novy Mir* No. 1, 1992, p. 181.

** *Novy Zhurnal* (New York), 55, 1958, pp. 114–115.

Document 9.17

From B. L. Pasternak's letter to the Head of the Cultural Department of the CPSU Central Committee D. A. Polikarpov, of January 16, 1959

... I remember my painting a picture of not being pressurized or persecuted in any way and having voluntary turned down the luxurious trip (without leaving hostages)[1] kindly granted to me[2] — I shamelessly lied to your dictation only to find that I finally have got nothing in return. I understand, for I am grown-up, that I cannot demand anything, that I do not have any rights, that against the eyebrow movement of the highest authorities I am a midge which can be crushed without anybody saying a word — but it will not happen so simply, before that somewhere someone will feel sorry for me.[3]

Then again, I understand that if I am not in jail and not turned out from my *dacha,* it is infinitely much, but why is it that along with these facts, which are quite true, two institutions, the Ministry of Culture and the Ministry for Foreign Affairs, give assurances that I have received and am receiving orders for paid work...[4]

You have emphasized the affair of reconciliation with the Writers Union as the condition of existence... Why should I be a paragon of nobleness and keeping my word of honor amidst the overall double-dealing and pretense... The last words of my 2nd letter are sufficient to restore me in the Union. Actually, I was silly enough to expect signs of broadmindedness and magnanimity in response to these letters. Indeed, the terrible and cruel Stalin did not consider it beneath his dignity to comply with my requests about prisoners and to summon me to the phone, on his own initiative, about this question.[5] The tsar and his brothers, great princes, sent letters to my father expressing their gratitude in different matters not concerning the state. But of course, they cannot compete against the present loftiness and splendor...[6] I will not write to anybody higher than you and will not undertake anything else.

Continent 95, No. 1, 1998, pp. 246–247.

[1] To Stockholm, to the Nobel award-giving ceremony.

[2] See doc. 9.11-II.

[3] A part of this paragraph and one phrase from the third paragraph are quoted in the letter of the chairman of the KGB Shelepin to the CPSU Central Committee (doc. 9.21). It means that Polikarpov immediately informed the institution which had the right to investigate, arrest, etc. about Pasternak's letter addressed to him.

[4] Here Pasternak speaks about false assertions of the mentioned governmental institutions, in response to queries from abroad, that no one denies to Pasternak the opportunity to receive orders for translations and to be paid for his work in time.
[5] That is about Pasternak's letters to Stalin in 1934–1935, among them the letter in October 1935 with the gratitude "for the miraculously instantaneous release of Akhmatova's relatives,"* and also about Stalin's telephone conversation with him in June1934, in connection with his standing up for the poet Osip Mandel'shtam, arrested in May.**

* E. B. Pasternak, p. 439.

***ibid*, p. 494.

[6] In doc. 9.21 these three phrases are described as "vicious attacks".

Document 9.18

The commentary of the correspondent of the British newspaper Daily Mail Anthony Brown to Pasternak's poem The Nobel Prize, which he published

...In the interview[1] Mr. Pasternak told me that he realized fully that publication in the West of *The Nobel Prize* might further gravely endanger his personal safety. But in permitting publication of the moral stand he is taking,[2] it is unlikely that serious legal action will be taken against him without the personal assent of Mr. Khrushchev.

The best-informed sources in Moscow told me that the Soviet leader is emerging not as a defender of Mr. Pasternak but as a mediator in the fight[3] which is, at least in part, the outcome of personal jealousy of Pasternak's literary stardom.

I understand that Mr. Khrushchev has no wish to make a martyr out of Mr. Pasternak by allowing the Soviet Writers Union to force the aged ailing man to trial and sanction.

Realizing that by permitting the great controversy to continue Russia was giving the West invaluable anti-Soviet propaganda and was creating in Pasternak a symbol of hope and freedom in the minds of the young Soviet intellectuals, Mr. Khrushchev ordered restraint in the campaign.

The Russian premier is understood in Moscow to have called a conference of members of the Soviet Writers Union who were loudest in demanding that Pasternak should be deprived of his Soviet citizenship.[4]

The story goes that at this conference Mr. Khrushchev asked member by member: "Have you read *Doctor Zhivago*?" Each writer confessed that he had not, for the book had not yet been accepted for publication in the USSR.

"Then", Mr. Khrushchev is reported to have asked, "what is all the fuss about?" The meeting dissolved with Pasternak's critics nervously moderating their campaign.

Mr. Pasternak, still afraid that he would be banished, was reported by TASS to have appealed to Mr. Khrushchev not to banish him...[5]

So far as is known, Mr. Khrushchev gave Mr. Pasternak no formal assurance that he would be permitted to retain his citizenship.

But the Kremlin, extremely sensitive on the *Zhivago* issue, took the unprecedented step of publishing the formal letter of rejection sent to the author when he submitted the novel for publication in Russia.[6]

The letter, nearly 20 *Daily Mail* columns long, was unique because — for the first time in history, I am told — the Soviet government[7] had felt compelled to justify one of its actions in reasonable and human terms....

The book, which was published when Mr. Pasternak sent the script to an Italian Communist publisher during the post-Geneva thaw,[8] has become in the West a symbol of one man's struggle against the dictates of a rigid, unforgiving totalitarian order.

The explosion — for that was the word he used himself in his interview with me to describe his view of the impact of his book — embarrassed the Soviet Union not because of its critical view of Marxist and Communist ideology but because he was said to be attempting to infect the Soviet Union with the principles of individual Christian conscience.

Daily Mail, February 2, 1959, pp. 1,7.

[1] The interview and passing of the autograph of the poem *The Nobel Prize* to Brown took place on January 30, 1959.*

* E. B. Pasternak. *Boris Pasternak.Biography*, Moscow, Citadel, 1997, p. 713, also doc. 9.19

[2] As says E. B. Pasternak, referring to his father's words, the latter asked Brown to forward the poems he gave him, including *The Nobel Prize*, to Paris, to Jacqueline de Proyart, but he did not give him permission for publication.* That was confirmed by the worker of the Ministry for Foreign Affairs V. A. Kuzmishchev, who also happened to be at Pasternak's place at that moment (doc. 9.19). That was also reported to their newspapers by a British and an American correspondent in Moscow (doc. 9.20), as well as by O. V. Ivinskaya, who, as she said, saw Pasternak that very day, after his guests left. However, her version of the circumstances of passing the poem to Brown substantially differs from the one reported by Kuzmishchev and from the agent information given in the memo from the KGB Chairman Shelepin to the CPSU Central Committee.** Here is what she remembered three decades after those events: "...the telephone rang [in Ivinskaya's Moscow flat]. It was a call from the Central Committee: 'The trick that Boris Leonidovich has just played.' said Polikarpov with indignation, 'is even worse than the story of the novel... Now all waves are broadcasting his poem, which he passed to one foreigner. Everything that had grown quiet is now making noise again. Go [...] and do your best to keep him from new follies...' I started to get dressed when [...] Borya called... 'Yesterday, when you got angry with me, quite justly, and left, I did not want to believe it. I went to "The Big Dacha"*** and wrote a poem about the Nobel Prize. [...] I wrote this poem and went to your place',**** Borya was telling me, 'I could not believe that you left. And here I encountered a foreign correspondent. He was following me and asking if I wanted to tell him something. I [...] showed him the poem that I was bringing to you' [...]"***** What sounds the least plausible here is tying up the date

of passing the poem to Brown with the reports of Western radio companies. Such a time coincidence is absolutely impossible, since the *Daily Mail* published the poem, with copyright, only a week and a half later after it was received by Brown.

* E. B. Pasternak, p. 713.

** In the memo sent to the CPSU Central Committee on February 16, 1959, Shelepin reported: "This February Pasternak was visited by the American citizen Tailor and the permanent correspondent of the English newspaper the *Daily Mail* in Paris, A. Brown, who were in Moscow as tourists. According to our information, during his talk with Brown, Pasternak passed to him his poem *The Nobel Prize* written in the anti-Soviet spirit, which was published, on February 11, in the English press."

This memo was published by *Trud* newspaper on December 22, 1995, p. 8.

*** The *dacha* in Peredelkino, where Pasternak lived with his family.

**** That year Ivinskaya was renting a room in one of the houses in Peredelkino.

***** O. V. Ivinskaya. *Gody s Borisom Pasternakom. V Plenu u Vremeni,* Moscow, Libris, 1992, pp. 346–347.

3 This information quite corresponds to Khrushchev's regrets about "the case of Pasternak" expressed by him in his memoirs (doc. 9.27).

4 There is no authentic information about such a meeting of Khrushchev with the leaders of the Writers Union; most probably, it is merely a rumor.

5 Doc. 9.11-II.

6 In September, 1956 the *Novy Mir* journal refused to publish the novel *Doctor Zhivago* and gave the manuscript back to the author with a verbose negative review signed by five members of the editorial board.* After the fact of Pasternak having been awarded the Nobel Prize became known, the *Literaturnaya Gazeta* published the complete text of the review.

* E. B. Pasternak, p. 680.

7 It is an error: officially, the government did not take part in the story of the *Novy Mir* refusal to publish *Doctor Zhivago.* The secret contacts with the *Novy Mir*'s editorial board were carried out by the Cultural Department of the CPSU Central Committee.

8 That is, after the Geneva summit of the USSR, the USA, Great Britain and France, which took place from 18 to 23 July 1955. The summit took a number of important decisions which contributed to the considerable relaxation of tension between the USSR and Western powers.

Document 9.19

Information from the CPSU Central Committee about the Uruguayan journalist Albert Etcheperi visiting B. Pasternak's dacha, *of February 12,1959*[1] *(from the diary of V. A. Kuzmishchev, deputy head of the department of the countries of America of the USSR Ministry for Foreign Affairs)*

On January 30 I was accompanying the Uruguayan journalist A. Etcheperi visiting the *dacha* of B. Pasternak. B. Pasternak met us himself and led us into one of his rooms, where the correspondent of the English newspaper the *Daily Mail* Anthony Brown (staying at the National hotel, room 219) was at that time. When meeting us Pasternak told me that he was sick and tired of receiving all sorts of foreigners. A. Etcheperi did not have any talks with Pasternak. He only told him that he came to inform him that the Uruguayan poetess Susanna Sokka, with whom Pasternak had corresponded and who had made a report in Uruguay about him as a poet, had died two weeks before in a plane crash. [...]

The correspondent A. Brown asked Pasternak to give him his photograph with an autograph. B. Pasternak said that he would also give his photograph to A. Etcheperi, which surprised the latter, as he did not ask for one. It should be noted that B. Pasternak repeatedly asked me "if he committed a crime receiving foreigners".

Pasternak's talk with A. Brown was in the English language, which I do not know. However, from what little I understood I had an impression that A. Brown asked Pasternak if he wanted to go to England, and Pasternak answered that he "was quite all right here". However, judging by the fact that Brown asked Pasternak the same question several times, it seems to me that he had difficulty understanding him (Pasternak's spoken English is rather primitive). Pasternak handed over to Brown some poem, "from the old ones",[2] as he put it, for some friend of his in Paris.[3] The whole visit to B. Pasternak's dacha lasted about 40 or 50 minutes...

RGANI, fond 5, opis'36, ed. khr. 61, l.137,138.

[1] The information for the CPSU Central Committee is dated the day after the publication in the *Daily Mail.* Thus, the report about the publication came to Moscow on the day of the newspaper issue coming out. The surveillance organs must have known in advance about Kuzmishchev's accidental presence at Pasternak's *dacha* during Brown's visit. Since they had not asked him in advance to prepare the corresponding information, they were not worried about the poet's meeting with the English journalist. That is one more indirect piece of evidence that during the meeting there were neither Pasternak's offers, nor Brown's requests for publication of poems abroad.

[2] According to E. B. Pasternak, on January 30, 1959, along with *The Nobel Prize,* Brown was given for passing to Proyart three more poems from the so-called *January supplements* to the planned collection of verses *When it Becomes Clear.** The fact that the informer mentions these poems as "from the old ones" is, perhaps, connected with his insufficient knowledge of the English language, or maybe with the poet's unwillingness to say that the poems were new.

* E. B. Pasternak. Boris Pasternak. Biography, pp. 712–712.

[3] For Jacqueline de Proyart.

Document 9.20

*From report No. 44, for the CPSU Central Committee, of the telegrams sent abroad by foreign correspondents on February 13,1959, before 8 p.m.**

* As the telegrams for the report were translated into Russian and the English originals were not preserved, here they are given in translation from Russian.

I

Dobson (*Daily Express*). Today Pasternak turned 69. I wished him many happy returns at his snowed-in *dacha* near Moscow. However, I regret doing it because I upset Pasternak by telling him that published in London was one of his poems[1] which he had never intended to publish, as it could do almost as much harm to his relations with his colleagues here as has been done by the novel *Doctor Zhivago.* Pasternak told me with bitterness: "I gave the poem to a young man who came to visit me, so that he passed it as a personal message to a friend of mine in Paris. I did not give permission for its publication. Indeed, some present you've brought me for my birthday." There is the danger that Pasternak might again be accused of deliberate smuggling his work out of the Soviet Union, like he was accused of forwarding the book *Doctor Zhivago.* The publication of this book abroad aroused such anger here that the Writers Union made the decision to expel Pasternak from Russia, which would have meant death to him. Only personal interference of some high-ranking members of the Soviet government[2] curbed the passions. Pasternak is an unsophisticated person. He cannot understand why his novel *Doctor Zhivago* caused such a stir abroad. He wants to be left alone and given the opportunity to write....

RGANI, fond 5, opis' 36, ed. khr. 93, l.8

[1] From the text of Dobsom's telegram it should follow that it is from him that Pasternak first heard about the *Daily Mail* publication of the poem *The Nobel Prize.*
[2] These probably were K. E. Voroshilov,* A. I. Mikoyan** and N. S. Khrushchev.***

* See note 2, doc. 9.12-I.

** *ibid.*

*** See doc. 9.18 and note 3 to it.

II

Shapiro (*United Press International*). Pasternak bitterly complained of the British journalist, who betrayed his trust and without his permission published one of his poems. Pasternak said that this innocuous-looking young correspondent offered to pass the poem to Pasternak's friend Jacqueline de Proyart, curator of the Tolstoy Museum in Paris. Using expressions which cannot be used for publication, Pasternak condemned this correspondent and stated that from then on he would not receive any correspondents, "who only hamper my work and do me harm." Pasternak gave your correspondent a statement signed by him: "The poem was handed on to Brown as a gift to be given to my friend Madam Proyart, but not for publication." Below there is a signature: B. Pasternak. The story of publishing the poem without permission darkened and, in fact, upset the celebration of Pasternak's 69th birth anniversary, which he celebrated with his wife and several close friends...[1] Speaking of what he considered to be a method contradicting the journalist ethics, Pasternak said: "This poem is only one of the several poems intended for my friend,[2] and the fact that the gloomiest of them was chosen for publication shows what were the motives of this young man."...

ibid, l. 11, 12.

[1] Pasternak's birthday is on February 10, but it was the family tradition to celebrate it on the following day.* So the poet's 69th birth anniversary was celebrated on February 11. On the same day fell the publication in the *Daily Mail* newspaper and, accordingly, the visit of both foreign correspondents to the Peredelkino *dacha*. The fact that the telegrams sent by them were dated February 13 must have been the result of some circumstances — as, for instance, a delay in handing on to the Central Committee of reports on telegrams from abroad.

* According to the Julian calendar, which was in use in Russia, Boris Pasternak was born on January 29, 1890, which in the 19th century corresponded to February 10 by the Gregorian calendar. The day of January 29 had been celebrated in the family through 1918. After the governmental decree of January 24, 1918 introduced the transition to the Gregorian calendar from 1 (14) February, 29 January proved corresponding to 11 February. That is why the poet's family continued to annually celebrate this date as his birthday, although encyclopedias and other reference books always give the true date of his birth in Gregorian chronology, that is, as February 10.

[2] He means three more poems from *January Supplements*, which were handed on to Brown along with the poem "The Nobel Prize".*

* See also note 2 to doc. 9.19.

Document 9.21

From the Chairman of the State Security Committee (KGB) A. N. Shelepin to the CPSU Central Committee, of February 20, 1959

... In the summer of 1958 an intensive campaign started abroad for awarding the Nobel Prize to Pasternak, which was joined, on the American initiative, by many reactionary and anti-Soviet organizations and associations; specifically, the leaders of the PLU[1] tried to make of Pasternak the leader of the literary opposition allegedly existing in the Soviet Union and to present him as the embodiment of "the new democratic Russia." A certain P. P. Suvchinsky,[2] who pursues the studies and popularization of Pasternak's work in France, wrote to the latter, in July 1958, that in a very short time it had become possible to raise Pasternak's name in France to the highest level of recognition that he deserved, thus correcting one of the greatest injustices of the Russian history. In reply to this letter Pasternak wrote: "You should not think that I was something before the novel. It is only this book that is my beginning, all that was before is rubbish."

Pasternak did not take any active measures in order to cut short using the novel *Doctor Zhivago* in anti-Soviet propaganda, which preceded awarding the Nobel Prize to this work. He continued to make a point of his personal opinion that *Doctor Zhivago* is a work of a true artist but it cannot be recognized in the Soviet Union because of this country's social and political conditions. On October 23, 1958, when it became known that Pasternak had been awarded the Nobel Prize, he told one of the foreign correspondents that he was happy to receive this news and he thought that in the Soviet Union this award should be welcomed since it was a member of the Soviet society who was given such honor. He hoped for the positive reaction of the Soviet authorities and public opinion, but he did not rule out the possibility of trouble for him.

Not taking into account the indignation of Soviet public opinion, Pasternak did not want to reject the Nobel Prize, and his statements to the press were of a double-dealing character. In fact, as was ascertained in the process of controlling Pasternak's correspondence,[3] he attempted to mail abroad a number of letters, in which he confirmed his satisfaction with being awarded the Nobel Prize and gave authorization to receive it to his acquaintance, Countess de Proyart living in France.[4] In his letters to his numerous foreign correspondents Pasternak is always stressing that "relatively mild measures" taken against him allegedly result from the campaign launched in the West; he is also speaking about the difficult conditions of his being isolated from society, which, according to him, are specially created, and about the alleged wish to finally ruin him by different ways, up to financial limitations. For example, in his letter to a certain Mac Gregor, of January 3, 1959, Pasternak wrote: "I waited in vain for some signs of magnanimity and leniency in response to my two letters that had been published.[5] Magnanimity and tolerance are not in the nature of my addressees, abuse and humiliation will continue. The loop of obscurity, which is tightening more and more around my neck, has the purpose to forcefully bring me to my knees, in a financial respect, but it will never happen. I stepped across the threshold of the new year with suicidal mood and rage."

Of late Pasternak's bitterness has become stronger. This can be seen, among other things, from his letter to the CPSU Central Committee, the draft of which was read out by Pasternak's mistress Ivinskaya and became known to us: "...I understand, for I am grown-up, that I cannot demand anything, that I do not have any rights, that against the eyebrow movement of the highest authorities I am a midge which can be crushed without anybody saying a word — but it will not happen so simply, before that somewhere someone will feel sorry for me... Actually, I was silly enough to expect signs of broadmindedness and magnanimity in response to these letters." (He means the letter published in newspapers).[6] Further on the letter contains spiteful attacks...

Published in the journal *Istochnik* No. 4, 1993, pp. 108–109, from the original from APRF (fond 3, opis' 34, ed. khr. 269, l. 96–101.

[1] PLU (People's Labor Union) is an emigrant organization based in Paris.
[2] Suvchinsky was the first to draw Pasternak's attention, after coming out of foreign language editions of *Doctor Zhivago*, to procedural difficulties which might have arisen when his candidature was considered in the Nobel Committee, in connection with the absence of the published Russian original of the novel.* Its publication in Russian by the Hague publishing house Mutona removed this obstacle.**

* *Novy Mir* No. 1, 1992, p. 151.

** See also note 5 to this chapter.

[3] Information about Pasternak's contacts, whether really taking place or merely supposed, had been received by surveillance long before the story of the novel *Doctor Zhivago.* Specifically, a worker of the foreign department of the USSR Academy of Sciences reported, on January 4, 1955, in the "Information on the visit of a delegation of British scientists to the USSR Academy of Sciences," that a member of the delegation Christopher Hill, a historian from Oxford, the author of the fundamental work *Lenin and the Russian Revolution,** expressed his wish to meet with Pasternak and pass him a parcel from his sisters. As was further noted by the author of the memo, the scientist was informed that Pasternak was ill and, besides that, refused to meet him.**

* Christopher Hill. *Lenin and the Russian Revolution*, London, 1947.

** RGANI, fond 5, opis' 17, ed. khr. 509, l.12. According to E. B. Pasternak, the meticulous British scholar, who was able to estimate professionally the degree of trustworthiness of information received from security institutions in the homeland of Lenin and the Russian revolution, managed, in spite of everything, to find the opportunity to pass the parcel to Pasternak.

[4] See doc. 9.16.
[5] The letters to N. S. Khrushchev, with Pasternak's signature (doc. 9.11-I) and to the *Pravda* newspaper (doc. 9.12-II).
[6] See doc. 9.17. The fact that the phrases quoted here repeat word for word the autograph of Pasternak's letter shows that, apart from the agent information, the KGB also had its copy passed from the Central Committee's Cultural Department.

Document 9.22

From Procurator-General of the USSR R. A. Rudenko to the CPSU Central Committee, of February 20, 1959

I

Having familiarized myself with the materials of TASS and Glavlit[1] on B. L. Pasternak's passing his anti-Soviet poem *The Nobel Prize* to the correspondent of the

Daily Mail newspaper A. Brown,[2] I would consider it necessary to take the following measures:

1. to summon Pasternak to the USSR Procurator's Office for official interrogation;

2. before the beginning of interrogation to inform Pasternak that his actions which manifested themselves in composing and propagating abroad anti-Soviet literary works — the novel *Doctor Zhivago* and the poem *The Nobel Prize*, which contained slanderous inventions discrediting the Soviet social and state system and became the instrument of the international reaction in its hostile activities against the USSR — are incompatible with the norms of conduct of a Soviet citizen, make *corpus delicti* of a specially dangerous crime against the state, and by force of law entail criminal responsibility;

3. simultaneously to state that materials that the USSR Procurator's Office has at its disposal testify that Pasternak has misused the humane attitude towards him on the part of the Soviet government, and despite his public assurances about his patriotism and condemnation of his "errors and delusions", he (Pasternak), having taken the road of deception and double-dealing, secretly continued his antinational activities, consciously and intentionally detrimental to the Soviet society.

4. in the course of the interrogation to produce an issue of the *Daily Mail* newspaper in which Pasternak's poem is published with reproduction of his facsimile;[3]

5. to provide a complete reflection of Pasternak's testimony in the interrogation record and give it to him to sign;

6. on completion of the interrogation to announce to Pasternak that a proper investigation of his actions will be carried out by the USSR Procurator's Office;

7. I suppose that there should be no criminal proceedings against Pasternak and no coming of his case to trial. Conducting such a trial appears inexpedient in every respect.

Soviet public opinion, having condemned Pasternak's actions as traitorous, demanded that he should be deprived of citizenship and expelled from the boundaries of the USSR. Under the legislation in force the court is able to do neither of these;

8. Since Pasternak has committed an act of treachery with regard to the Soviet people and put himself, as a result of his political and moral degradation, beyond Soviet society, it would be more expedient to take a decision to deprive him of Soviet citizenship and expel him from the USSR. Such a decision can be taken by the Presidium of the USSR Supreme Soviet in conformity with p."b", art. 7 of The USSR Citizenship Law of August 19, 1938;

9. the grounds for consideration of this question by the Presidium of the Supreme Soviet of the USSR could be provided by the representation of the USSR Procurator-General. Presented below is a draft of the corresponding decree. I ask for it to be considered.

II

Decree of the Presidium of the Supreme Soviet of the USSR "On Deprivation of B. L. Pasternak of the Soviet Citizenship and Expulsion of Him beyond the Borders of the USSR" (draft)

It was ascertained by the investigation conducted by the Procurator- General of the USSR that B. L. Pasternak composed and propagated abroad anti-Soviet works containing slanderous inventions which discredit the Soviet state and social system and are used by the international reaction in their hostile activities against the USSR. Despite his public assurances about his patriotism and condemnation of his "errors and delusions", Pasternak, having abused the humane attitude towards him on the part of the Soviet government and taken the road of deception and double-dealing, secretly continued, until the very latest time, his anti-national activities. Such actions of Pasternak, incompatible with the norms of conduct of a Soviet citizen, place him beyond Soviet society.

Proceeding from the above, and also taking into account the demands of Soviet public opinion, the Presidium of the Supreme Soviet resolves: to deprive Pasternak Boris Leonidovich of the Soviet citizenship and to expel him from the boundaries of the USSR.

Published in the journal *Istochnik* No. 4, 1993, pp. 109–111,
from the original from AP RF (fond 3, opis' 34, ed. khr. 269, l. 104–108).

[1] The governmental organ of censorship.
[2] On January 30, 1959 Pasternak gave the poem *The Nobel Prize* to Anthony Brown, the correspondent of the British newspaper *Daily Mail* in Paris, who came to Moscow as a tourist, and asked him to pass it to Jacqueline de Proyart. Pasternak did not give permission for its publication.*

* E. B. Pasternak. *Boris Pasternak. Biography*, Moscow, Citadel, 1997, p. 713, see also doc. 9.19–9.20.

[3] *Daily Mail*, February 11, 1959, pp. 1,7 (doc. 9.18).

Document 9.23

Resolution of the Presidium of the CPSU Central Committee (record 207, point XII), of February 27, 1959 "On Pasternak"

To instruct the Procurator-General, comrade Rudenko to take measures in conformity with the exchange of opinions at the sitting of the Presidium of the Central Committee.

Published in the journal *Istochnik* No. 4, 1993, p. 111, from the original from APRF (fond 3, opis' 34, ed. khr. 269, l.102).

Later publications of documents of the Presidium of the CPSU Central Committee offer a different text given in record 207 of the sitting of the Presidium of February 27, 1959; "To give Pasternak a warning from the Procurator and to say that if he continues his hostile work, proceedings will be instituted against him."[1]

[1] The Archive of the Kremlin. Presidium of the CPSU Central Committee. 1954–1964. Vol. 1, Moscow, ROSSPEN, 2003, p. 347.

Document 9.24

Report of the Procurator-General of the USSR R. A. Rudenko to the CPSU Central Committee, of March 14, 1959

I am transmitting herewith a copy of the interrogation record of B. L. Pasternak. At the interrogation he behaved in a cowardly way. It seems to me that he will make the necessary conclusions from the warning of criminal responsibility.

Interrogation record

On March 14, 1959 the Procurator-General of the USSR, State Counselor for Justice R. A. Rudenko interrogated the citizen mentioned below, who gave the following information about himself:

Pasternak Boris Leonidovich, born in 1890, a Jew, having a family, non-Party man.

It was explained to Pasternak that he was summoned[1] in connection with his actions violating the laws of the Soviet state.

Question: It is common knowledge that your writing *Doctor Zhivago*, which contains slanderous inventions discrediting the social and state system of the USSR, has been used by the international reaction in its hostile activities against the USSR. Your actions make *corpus delicti* of a specially dangerous crime against the state and by force of law (article 1 of the law on crimes against the state) entail criminal responsibility. It is also well known that the Soviet public opinion unanimously condemns your actions, incompatible with the norms of conduct of a Soviet citizen. I am asking, first of all, if your statement about your patriotism and condemnation of your errors and delusions, which was made in November 1958, was sincere.[2]

Answer: Yes, this statement on my part was quite sincere.

Question: The USSR Procurator's Office has materials which show that you, having abused the humane attitude to you on the part of the Soviet government and, despite your public statement about your patriotism and condemnation of your errors and delusions, which, as you now claim, was sincere, took the road of deception and double-dealing, secretly continuing the activities consciously and intentionally detrimental to Soviet society. Please answer: after all these public assurances did you pass your anti-Soviet writings to any of the foreign correspondents?

Answer: Indeed, at the beginning of this February[3] I had an experience of an incautious passing of several poems, including the poem *The Nobel Prize*, to a correspondent of the English newspaper the *Daily Mail*, who visited me at my dacha. Handing on these poems to Brown, I said that I did not mean to publish them, but was giving them as an autograph I was asked for. My passing of the poem *The Nobel Prize*, whose content can be easily interpreted as slander of the really humane attitude to me on the part of the Soviet power, is a fatal carelessness, which can justly be regarded as double-dealing.

Question: Now being shown to you is the journal *Daily Mail* of February 11, this year, with your poem *The Nobel Prize* and the reproduction of your facsimile, where the correspondent Brown maintains that this poem was given by you to him for publication. What can you say about this claim?

Answer: If my memory does not fail me, I did not give him the poem for publication. Having familiarized myself with the newspaper shown to me, I got a convincing proof that my words and writings are used with the purpose of slanderous attacks and inventions against the Soviet society. This new experience is especially deplorable for me because it questions the sincerity of my decisions to serve my homeland. I condemn these actions of mine, and I am perfectly aware of the fact that they entail my legal responsibility as a Soviet citizen.[4]

In conclusion of the interrogation the procurator-general explained:

At this stage of the investigation I have to warn you that if the actions which, as was said above, make *corpus delicti* are not stopped, in conformity with the law criminal proceedings will be instituted against you. I also have to warn you, in conformity with article 96 of the Penal Code of the Russian Federation, of the responsibility for divulging the data of the investigation. Is it clear to you?

Answer: Absolutely clear and will be performed by me unreservedly.

Published in the journal *Istochnik* No. 4, 1993, pp. 111–112, from the original from the AP RF (fond 3, opis' 34, ed. khr. 269, l.110–113).

[1] It was not by a prior summons that Pasternak found himself in the Procurator's Office. On March 14, 1959, when he was having his usual walk in Peredelkino, a car drew up to him, he was put into it and taken to the meeting with the procurator.
[2] See doc. 9.12-II.
[3] According to E. B. Pasternak, the passing of the poems took place on January 30,1959.*

* E. B. Pasternak, p. 713.

[4] Even a quick glance at the artificial language of "Pasternak's answer" shows that their stylistic figures did not belong to him and at best could but approximately correspond to the meaning of his words.

Document 9.25

From the Head of the Cultural Department of the CPSU Central Committee D. A. Polikarpov and the worker of the Department E. Trushchenko to the Central Committee, of April 25, 1959

The Soviet embassy informs that the director of the Russian Institute in Stockholm Nils Oke Nilsson leaving, on April 17, to the Soviet Union as a member of a group of Swedish journalists, intends to meet with Pasternak in order to use the talks with him for new press publications hostile to us. The embassy requests to take measures for preventing Nilsson's meeting with Pasternak.[1] We consider it expedient that the Procurator's Office warn Pasternak that meeting with Nilsson should not be allowed. The warning can be passed through the worker of the Procurator's Office Preobrazhensky, who attends to Pasternak's case.[2]

The note on the lower part of the sheet: "The instruction has been passed to the secretary of the Soviet Writers Union (comrade Voronkov). April 21,1962." The signature is illegible.
RGANI, fond 5, opis' 36, ed. khr. 93, l.19

[1] The wish was fulfilled. Nilsson was detained at the railway platform in Peredelkino and it was suggested to him, without any explanations, that he should come back to Moscow. This attempt to see Pasternak cost Nilsson dearly. As this author was told, in January 2000 by the Slavonic scholar from Goteborg Magnus Unggren, who knew Nilsson well, for 17 years after the mentioned incident he was refused permission to visit the Soviet Union.
[2] The case was brought in connection with publication abroad of the poem *The Nobel Prize* (see doc. 9.24).

Document 9.26

From the deputy head of the Cultural Department of the CPSU Central Committee A. Petrov and the head of division of the Department I. S. Chernoutsan to the Central Committee, of June 4, 1960

The early hours of May 31 this year, at the dacha in the town of Peredelkino (near Moscow) saw the death of the writer B. L. Pasternak, who died in his 71st year after a serious disease (heart infarction, malignant tumor). [...] The foreign bourgeois press tried to use the disease and death of Pasternak for arousing public opinion in the anti-Soviet spirit. [...] The Swedish press was publishing information about Pasternak's state of health. Foreign correspondents for several days were in constant attendance at the dacha waiting for his death.[1]

On June 2 Pasternak's funeral took place, in accordance to his wish, at the cemetery in Peredelkino, where about 500 people gathered, among them 150–200 elderly people, apparently from old intelligentsia, and about the same number of young people. Out of prominent writers and arts personalities present at the funeral were K. Paustovsky,[2] B. Livanov, S. Birman. Wreaths were sent from several writers, cultural personalities, the Literary Fund and also from private persons. The correspondent of the Associated Press agency laid a wreath from "American writers". Orations from K. Paustovsky and the renowned actor B. Livanov were expected.

However, at the last moment both of them refused to speak, pleading illness.[3]

Professor Asmus spoke at the grave. He described Pasternak as a translator and writer of genius, saying in conclusion that while the Russian language and Russian poetry exist on earth, Pasternak's name will live. Asmus announced the funeral ceremony closed. However, while the casket was being lowered, one of the young men came to the hill and started to make a confused speech, calling Pasternak "a genius", "great" and so on, and finished by saying that "Pasternak's teaching about the love for man should scatter around the earth like beads and get into the soul of every inhabitant..."[4]

Suddenly there was a cry from the crowd: "I'm speaking in the name of workmen [...]", and a young man of "*stilyaga*"[5] type started to shout in a heart-rending voice something like this: "Pasternak, a great writer, was not allowed to publish his book in our country [...] No other Soviet writer was able to rise to such creative heights as our dear Pasternak." About 15 or 20 people standing next to him started to applaud, but the majority of those present disapproved of his cries. [...] After the casket was buried, the greater part of the public left the cemetery. A small group of young people remained at the grave. They recited poems dedicated to Pasternak, but without any political attacks. Among those who read their poems was the graduate of the Literary Institute Kharabarov, recently expelled from the Komsomol.[6] The foreign correspondents who gathered at the funeral were disappointed that there was no scandal and no sensational facts, which they had expected, even no workers of the militia to be photographed for their newspapers.[7]

Nevertheless, it would be worth drawing the attention of the Writers Union and the Ministry of Culture to the necessity of strengthening educational work among the creative youth and students, part of whom (a paltry number) is infected with unhealthy pseudo-rebellious moods and tries to present Pasternak as a great artist misunderstood by his epoque.[8]

RGANI, fond 5, opis'36, ed. khr. 119, l.63–64

[1] So it is in the text. The authors of the report did not take into account that the foreign correspondents might have been waiting for the news of Pasternak's state of health.

[2] The list of writers who were present at the funeral was published, 40 years later, by the poet Andrei Voznesensky (doc. 9.28).

[3] Paustovsky's words about his illness sound quite plausible, as he suffered from asthma, which made it difficult for him to speak loudly. The reason why Livanov with his well-trained voice of an actor, an old friend of the deceased, refused to speak is not so evident. Doing it might have been "not recommended" to him; as can be seen from the text of doc. 9.26, a number of officials from the Central Committee's Cultural Department were present at the funeral.

[4] Here it will be useful to cite the words of the playwright and literary critic A. K. Gladkov. In his memories about Pasternak published in France he gave a somewhat different picture of what was happening at the cemetery: "Professor Asmus is the first to speak. His task is rather difficult, but he copes with it perfectly. I remember his speech badly, but nothing in it seemed tactless, unnecessary, superfluous [...]

The actor Golubentsev reads Pasternak's poem: 'Oh, if I knew that it can happen [...]' And some other voice, unknown to me, modern and sincere, reads the poem *Hamlet,* not published until now, but widely known [...]* In response to the last lines of *Hamlet*** a noise is heard in the crowd. The atmosphere instantly becomes tense, but the same voice which announced the opening of the funeral ceremony, [...]*** hastily closes it. The noise and the voices of protest are growing louder. And immediately, still with the general noise and cries, some small sweet voice says something about the dew in which the poet will soon turn and some other sugary mystical nonsense of such kind. He has not finished yet, when a coarse and probably not quite sober voice cries that in the name of Peredelkino workmen he must declare [...] that "they" do not understand why 'Pasternak was not published and that he loves workmen'..."****

* The young theorist physicist M. K. Polivanov.

** I'm alone; all round me drowns in falsehood:
Life is not a walk across the field.
Translated by L. Slater. *Poems by Boris Pasternak.* Fairwarp, Sussex, Peter Russel Press, 1958, p. 30.

*** The organizing secretary of the Soviet Writers Union's Board K. V. Voronkov.

**** A. K. Gladkov. *Vstrechi s Pasternakom* (Meetings with Pasternak). Ymca Press, Paris, 1973, p. 158.

[5] In the Soviet Union of the1950s, *stilyaga* was an abusive name for young people whose dress and manners were regarded as uncritically fashionable and extravagant.

[6] The Siberian poet I. M. Kharabarov. In February 1960 he and his fellow student Y. P. Pankratov were sent on a one-day trip to the *kolkhoz* "Mayak" of Kasimov district, Ryazan *oblast.* The young poets did not find a common language with the chairman of the *kolkhoz* about the *kolkhoz* workers' achievements, which was reported to Moscow by the Ryazan *oblast* Party Committee. As a result the Central Committee's Cultural Department pointed out to the Writers Union that "[i]t is necessary that the work of ideological education among students of the Literary Institute should be strengthened",* and both students were expelled from the Komsomol.

* RGANI, fond 5, opis' 36, ed. khr. 119, l.40.

[7] Militia workers were effectively replaced by "people in civilian clothes". They were vividly described by the playwright A. K. Gladkov: "Some people, who did not look like idle observers, became noticeable in the crowd. They were listening to people's talks and clicked their cameras. I spotted one of them and watched him for a long time. He pretended that he was going with the crowd into the house, but he was staying in the same place all the time, now and then casting glances around; an unbuttoned checked shirt, low forehead and facial expression that cannot be concealed."*

These observations were confirmed, 47 years later, by the former journalist of the *Izvestiya* newspaper M. G. Sturua. He visited Pasternak in Peredelkino many times and, writing about the day of the poet's funeral, told of his talk with the newspaper editor-in chief A. I. Adzhubei: "Adzhubei summoned me and said: '[...] I know that you were close with Pasternak and are probably going to his funeral in

Peredelkino. So, my advice to you is: don't do it. I've just had a talk with Semichastny [then the KGB chairman], and he told me that there will be people shooting a film of all the participants of the funeral.'"**

* A. K. Gladkov, p. 155.

** *Izvestiya*, March 13, 2007, p. 4.

[8] The hopes of Pasternak's persecutors for "educational work among the creative youth" did not come true. Every year on May 30, the day of the anniversary of the death of the poet, who was understood and accepted by "his epoque", his grave become the place of pilgrimage of admirers of his talent, with poems read "from morning till late at night," as was noted by the poet' s son E. B. Pasternak. As he also said, on one of such days: "Jacques Chirac with a huge basket of roses came to the grave, [...] another time Nancy Reagan came." By the visit of the French guest the local authorities had made a gravel path around the cemetery fence, and by the day when the wife of the American President came the path had been asphalted.*

* *Izvestiya*, May 29, 2001, p. 12.

Document 9.27

From the memories of N. S. Khrushchev

"After Stalin's death he wrote *Doctor Zhivago* [...] I was informed and had an opportunity to influence the decision of whether or not to publish it — which boiled down to a question of whether or not to accept the advice of someone who was reporting to us — but I failed to act [...] and now I regret it. When dealing with creative minds, administrative measures are always most destructive and nonprogressive.

Pasternak worked hard on *Doctor Zhivago.* The manuscript found its way abroad, where it was published and caused a stir. It obtained recognition and was awarded the Nobel Prize,[1] though I can't say to what extent this work deserved it. Anyway, Pasternak was chosen to be a Nobel Prize laureate, while here (in the Soviet Union) there were administrative measures. His book was put into cold storage; it was banned. The decision to use police methods[2] put a whole different coloration on the affair and left a bad aftertaste for a long time to come.

People raised a storm of protest against the Soviet Union for not allowing Pasternak to go abroad and receive the prize.

I said, 'Let's go ahead and publish the book so that Pasternak will be able to go abroad and pick up his award. We'll give him a passport and some hard currency to make the trip.'[3]

Then, quite unexpectedly, Pasternak let it be known through a statement in the newspapers that he had no intention of going abroad and that he wasn't even going to raise the question.

To this day I haven't read the book and therefore can't judge it.[4] People who've spoken to me about it say they don't have any special admiration for the artistic aspect of this work, but that's beside the point. To judge an author and to judge his work are two different matters. [...] The main point is: readers should be given a chance to make their own judgments; and administrative measures, police measures shouldn't be used. A sentence should not be pronounced over our creative intellectuals as though they were on trial.

In connection with *Doctor Zhivago*, some might say it's too late for me to express regret that the book wasn't published. Yes, maybe it is too late. But better late than never."[5]

Khrushchev Remembers. The Last Testament. Translated and edited by Strobe Talbot.t. Boston, Toronto: Little, Brown and Company, 1974, pp. 76–77.

[1] That is not quite true. The Swedish Academy awarded the prize to Pasternak "for his important achievement both in contemporary lyrical poetry and in the field of the great Russian epic tradition".*

* See also note 8 to this chapter.

[2] After the death of Pasternak the authorities also had the need for "police methods". At that time it was the poet's widow, Zinaida Nikolaevna, who became their victim, with their dirty tricks being meanly emphasized. N. Kozhevnikova, who knew her, painted in her memoirs the following picture of the last years of her life: "her financial situation in the last years was extremely difficult. All her savings had been spent on doctors when Pasternak had been ill and dying, whereas the accounts in foreign banks proved to be frozen. Zinaida Nikolaevna made efforts to obtain a pension. But, as her dacha neighbor Fedin told her, her pension was 'a delicate matter'. 'And is it delicate if the widow of such a writer', she told her closest friend Nina Tabidze.* 'sells her last coat at the market for fifteen rubles. Who has given the sanction to kill me in broad daylight...'"**

* The widow of Pasternak's friend, classic of Georgian poetry Titsian Tabidze, who died in 1937 during the Great Terror.

** N. Kozhevnikova. *Sosed po Lavrukhe. Vospominaniya* (My Neighbour in Lavrukha. Memories), Moscow, AGRAF, 2003, p. 126.

[3] It is probably on reading these passages — defenselessly confused and at the same time painfully sincere — that the poet D. S. Samoilov felt the need to share his sensations with his old friend and correspondent L. K. Chukovskaya: "I am reading Khrushchev with interest. He himself is an extraordinary personality, a person fresh and ignorant, but with brains and some (though greatly confused) ideas of decency."*

* *Znamya* No. 7, 2003, p. 172.

[4] On October 13, 2004, the day of the 40th anniversary of Khrushchev's forced resignation, his daughter R. N. Adzhubei said in a radio program of the Rossia broadcasting company that her father had read the novel — probably, already after he had tape-recorded his memories. According to her, answering the question about his impressions from the novel, Khrushchev said: it is boring to read, and it was silly to make so much fuss about it...

[5] These words of the pensioner Khrushchev are to be trusted. In order to confirm it I shall give the following salient fact. The 1958 yearbook of the Nobel Foundation *Les Prix Nobel* was not sent to the special departments of the Soviet libraries, closed to the general public, as was done with its issues in the times of Stalin and Brezhnev;* it was kept in the open and was accessible to every reader.

* See note 26 to Chapter 12.

Document 9.28

From the memoirs of the poet A. A. Voznesensky

Thorns show themselves through the name of Pasternak. His funeral became the first demonstration of people's protest [...] The reader of today cannot even imagine how it can be possible to forbid people to come to the funeral of a poet, even if he is "a renegade" and recently expelled from the Writers Union.

Whereas the authorities considered it the main political crime of the year... So, June 1, 1960.[1] Alexandr Petrovich Mezhirov, a poet with a God-given talent, offered to give me a lift to Peredelkino [...] Suddenly, at the Goluboi Dunai, a joint where Pasternak used to have a drink with workmen, the car sharply braked to a halt. Mezhirov got out and said: "They are writing down the numbers. Go further by yourself. We'll be waiting for you here." There was fear in his sky-blue eyes: "I'm afraid. Because I'm a member of the Party [...]" Was it him, an officer who had been at war, taken part in battles, who had not been afraid of the Sinyavino marshes[2] — was it him who was afraid?! Here is what the System does to a man! His lips were trembling. He was muttering something like: "I can't... When Vas'ka Smirnov[3] and this... what's his name... the one who had syphilis, reeking of alcohol, from the tribune ask you questions in their hoarse voices..."

... Let's remember those who came to pay homage to the poet.

Their names are:[4] V. Asmus, V. Bokov, A. Gladkov, B. Gribanov, Y. Daniel, Vyach. Vs. Ivanov, T. Ivanova,[5] V. Kaverin, V. Kornilov, N. Korzhavin, I. Noneshvili, B. Okudzhava, K. Paustovsky, A. Sinyavsky, Ir. Ehrenburg[6]...

Obshchaya Gazeta No.28, 2000, p. 8.

[1] The funeral took place on June 2 (see doc. 9.26).

[2] Near the town of Sinyavino, Leningrad *oblast,* to the south of Lake Ladozhskoye, in September, 1942 the troops of the Volkhov and Leningrad Fronts were fighting grueling battles in their vain efforts to break through the blockade of Leningrad. However, these active efforts made the German command "fling their reserves here, which mitigated the situation of our troops in the South,"* that is, in the environs of Stalingrad and in the Northern Caucasus. The battle of Sinyavino remained in the history of the Great Patriotic War as one of the bloodiest diversion operations.

* D. I. Ortenberg. *God 1942* (The Year 1942), Moscow, the Publishing house of political literature, 1988, pp. 350–351.

[3] V. A. Smirnov, in 1954–1959 a secretary of the board of the Soviet Writers Union.

[4] See also note 64 to this chapter.

[5] The wife of the writer Vs. Ivanov, who had fallen ill.
[6] I. G. Ehrenburg, the daughter of the writer; she was present at the funeral with her mother. According to the memories of V. A. Kaverin: "Lyubov Mikhailovna Ehrenburg, who grew pale and even grey... (Ilya Grigorievich was abroad at that time)."*

* V. A. Kaverin. *Epilogue. Memories*, Moscow, Moscowskii rabochii, 1989, p. 378.

Document 9.29

From the Head of the USSR Council of Ministers' Chief Department of Military and State Secrets Protection in Press P. K. Romanov to the CPSU Central Committee, of May 5, 1962[1]

...I. Ehrenburg is especially zealous in assessing B. Pasternak's literary work and "social activities". He constantly argues that B. Pasternak "was one of the best lyrical poets of our time" (book 2), who received world recognition.

In the fourth book I. Ehrenburg writes that the Paris Congress of Anti-fascist Writers, in 1935,[2] gave an ovation to Pasternak, as one of "the most respected and honored" poets. "When Tikhonov moved on to the assessment of Pasternak's poetry, the audience stood up, by long applause greeting the poet who had proved that high mastery and high conscience are not at all enemies."

I. Erenburg is defending B. Pasternak's anti-patriotic action in connection with the publication abroad of his novel *Doctor Zhivago*, thus, in fact, expressing his disagreement with the appraisal given to this slanderous work by our press. I. Ehrenburg maintains that B. Pasternak "by a cruel play of fate, became a victim of the cold war..."[3]

RGANI, fond 5, opis' 36, ed. khr. 141, l.66–67.

[1] The letter was written in connection with the plans of the *Novy Mir* journal to continue publication of the further part of Ilya Ehrenburg's memories *Lyudi, Gody, Zhizn'* (People, Years, Life) (books 4–6)*.

* Book 4 mentioned here was first published, with numerous cuts, in the *Novy Mir* journal in 1962, No. 4–6. The first edition of the memories without censorship withdrawals came out in the Sovetskii Pisatel' publishing house in 1990, in three volumes.

[2] The name is inexact, as the 1935 Paris Congress was named the Congress in Defense of Culture.
[3] In connection with the cited letter of Romanov the head of the Central Committee's Cultural Department D. A. Polikarpov, the section head of the department and the Department worker A. Mikhailov, on June 14, 1962, among other things, reported to the Central Committee: "While apologetically assessing such writers as Mandelshtam, Tsvetaeva and Pasternak, the author of the memoirs tries as consistently to diminish the significance of writers and cultural personalities whose work established the most important principles of the socialist culture."*

* RGANI, fond 5, opis' 36, ed. khr. 141, l. 66–67.

Document 9.30

From the memo of the USSR Minister of Internal Affairs N. A. Shchelokov "Regarding the Question of Solzhenitsyn" directed to the General Secretary of the CPSU Central Committee L. I. Brezhnev; the document has no date, but it was prepared at the latest at the beginning of October, 1971

...5. Dealing with Solzhenitsyn we are repeating the same gross errors that we committed with Boris Pasternak. Pasternak is certainly a big Russian writer. He is even bigger than Solzhenitsyn, and the fact that his novel *Doctor Zhivago*[1] was awarded the Nobel Prize, against out wishes, was certainly our gross error, greatly aggravated by our wrong standing after awarding him the prize. *Doctor Zhivago* should have been "edited" here, in the country, and published. There is no doubt that in this case the West would not have shown any interest in the novel, especially as its ideological content could have been brought to the required level.[2] Anyway, *Doctor Zhivago* attracted rapt attention of readers in this country because of the great stir created around it. *Doctor Zhivago* was spread in manuscript form. It came to us in foreign editions, it received wide publicity through the radio.[3] Thus, given extensive connections with foreign countries and extensive possibilities to listen to broadcasts from the West,[4] in our time it is already impossible to hush up a work of literature. This reality should be taken into account. Unfortunately, no one takes it into account.

It is time our publishing houses understood that nowadays works of literature, instead of being rejected, should be transformed.

In the present time our literature has again turned its attention to Pasternak. A number of articles have been printed which appraise him as a great Russian writer. So, why should we have taken such a strange position on Pasternak, if now we are extolling him? The same can happen with Solzhenitsyn.

Kremlevskii Samosud (The Kremlin Lynch Law). A collection of documents. Compiled by A. V. Korotkov and others. Moscow, ROSSPEN, 1994, pp. 170–171. The original of the document is in AP_RF, fond 80.

[1] In the original, here and below, the surname of the protagonist of the novel is written as "Zhevago".

[2] The efforts to make Pasternak agree to the "editing" of the novel and "bringing it to the required level" could have eventually succeeded. By persuasion or by coercion, at last, the authorities might have achieved what they wanted. This cannot be said about Solzhenitsyn, about whom the author of this revealing document wrote that "under all circumstances, even against the author's will, such writings can be edited and published, in order to hush them up in future." He concluded his free-thinking and cynical passage about Solzhenitsyn by the phrase which Brezhnev underlined with satisfaction: "In this case, rather than executing enemies in public,

we should smother them with kindness. That is an elementary truth which should be known to those comrades who manage literature."*

* Kremlevskii Samosud, p. 170.

[3] The author meant Western radio companies giving broadcasts in the Soviet Union in Russian.

[4] In 1971 there was no jamming of Western broadcasts in Russian.

Chapter 10

The Nobel Prize and Mikhail Sholokhov
His Rivals: Paustovsky and Akhmatova

Boris Pasternak was first nominated for the Nobel Prize for Literature in 1946. A year later the Nobel Committee received the first nomination for Mikhail Sholokhov; his candidacy being offered by the professor of Stockholm University, afterwards a member of the Swedish Academy Henry Olsson (see Table 4.2).

The tandem of the two Soviet writers was not accidental. Both of them became known to the general Swedish reader as far back as in pre-war times, and Sholokhov's popularity was specially noted, in 1934, by the Soviet ambassador in Stockholm A. M. Kollontai; she noted in the diplomatic correspondence with her superiors: "He is read and his talent is highly regarded" (see doc. 3.9). That is why, when in informal discussions the question was raised about which of the actively working Soviet writers could be worthy of the prestigious award, both these names were the first to come to mind.

At the same time, it would have been difficult to choose an example with more different personalities than Pasternak and Sholokhov.

Indeed, they made a tandem of antipodes — by the differences in their biographies, their human qualities, their manner of creative thinking. The ways of a refined introspective intellectual on the one hand, and pointedly emphasized closeness to peasantry on the other hand; a natural gracefulness of style — and the style deliberately made rough; life in the atmosphere of internal underground and, therefore, complete absence of any outward signs of official favor — and being in an aura of innumerable awards, prizes, titles, articles in newspapers and magazines, everything up to the membership in the CPSU Central Committee, which brought the highest level of tangible benefits and immense possibilities. The only thing that these two outstanding figures had in common was the evident signs of a highest writing talent always attracting readers both at home and abroad.

Today we have no trustworthy evidence of personal contacts of Pasternak and Sholokhov, but between them there certainly was a mutual attraction, although it sometimes could assume not quite worthy forms.

There is an archive document connected with an interview given by Sholokhov to the newspaper *France-Soir* during his stay in Paris. It concerned Pasternak's *Doctor Zhivago* and aroused the unconcealed anger of the Communist Party officials (doc. 10.1). In it Sholokhov, while devastatingly criticizing not only the novel but "Pasternak's work on the whole", which, in his opinion, was "devoid of any importance whatsoever", at the same time said that his poetic translations from German and English were '"brilliant". That is, whether he wanted it or not, he did justice to his "brilliant" mastery as a poet. For Sholokhov's human nature such recognition of a colleague who was morally alien to him meant a lot.

On the part of Pasternak a rather peculiar episode was described in the memories of Olga Ivinskaya. Having come back, after Stalin's death, from imprisonment after being arrested in October 1949 for her close relationship with Pasternak,[1] she told him about the anti-Semite tricks of the KGB investigators who questioned her ("how could you, a Russian woman, have fallen in love with this old Jew"). Pasternak listened to her sullenly and then said "with a sigh how lucky Sholokhov was to have bio-data without anything 'vexed'". Another time he said facetiously (and later repeated this joke many times) that "it would be good to swap nationalities with Sholokhov."[2]

That is, if swapping with someone, it should be not Fedin, Leonov or Simonov, who were in the avant-garde of the Soviet literature of the mid-1950s, but Sholokhov. Was it not also an involuntary, subconscious recognition of the other writer's talent?

And here is independent evidence, which once again confirmed that in the national mentality there was something that naturally put together these two giants of Soviet literature — even with those who did not have any liking for one or the other participant of this tandem. The poet and playwright A. V. Sofronov, editor-in-chief of the *Ogonek* magazine for many years, one of the loyal drummers of the regime, when speaking at the meeting on "Pasternak's case" brought down a barrage of criticism on the First Secretary of the Soviet Writers Union A. A. Surkov. In his attacks he reproached him, quite undeservedly, for his alleged liberalism with regard to Pasternak and, as a sort of antithesis, claimed that at the same time "comrade Surkov humiliated, in every possible way, the world's number one writer, comrade Sholokhov."[3] This passage of Sofronov is directly linked to the circumstances of Sholokhov's speech at the Second Congress of Soviet Writers, on December 21, 1954, which aroused violent protests not only on the part of Soviet participants of the congress, but also of guests from abroad (doc. 10.2).

And one more example of the apparent link between Pasternak and Sholokhov. In the editorial *On the Occasion of the Jubilee* written for the January 1965 issue of the *Novy Mir* journal on the occasion of its forthcoming 40th anniversary, the editor-in-chief A. T. Tvardovsky noted that the journal published works of "writers with most different themes and styles, from Sholokhov to Pasternak."[4]

The appropriateness of opposing these two brightest representatives of Soviet literature to each other appeared so evident that it was often used in most common pieces of reasoning as one of the doubtless antitheses. For example, the well-known literary critic L. A. Anninsky, writing in his essay about the imagery of substantially different literary preferences existing in the common "Russian field", noted that "a Cossack [...] will prefer Sholokhov to Pasternak, and in Tolstoy will put *The Cossacks* before *Resurrection*."[5]

True, it must be noted that there is some essential similarity in the biographies of both antagonists. Both of them repeatedly found themselves in the situations when they were on the verge of being arrested. On both of them investigation organs had bulky files, and in the years of the Great Terror both had a lot of sleepless nights expecting a knock at the closed door (see doc. 10.3). However, the existence under unpredictable waves of the cruel ruler's baton became everyday reality for any member of the Writers Union who was talented enough to stand out against the general grey background. The unquestionable evidence of that are hundreds of innocent victims by which the country's writers' organization paid for the authorities' intolerance of any sign of free thinking.

The ambivalent attitude of the authorities towards Sholokhov also showed in the questions related to the possibility of him being awarded the Nobel Prize; in the West, including the Scandinavian countries, his high reputation as a classic of modern literature was unquestionable, and the novel *Quiet Flows the Don* became, already in the early 1930s, one of the internationally recognized bestsellers.[6] Nevertheless, in the 1940s and 1950s,[7] the Soviet press almost never complained of Sholokhov not being awarded the world's highest literary prize; even in periods of pseudo-patriotic rhetoric the sins of the Swedish academicians were remembered, if ever, only with regard to Leo Tolstoy and Maxim Gorky.

It was only a small article published, in 1946, in the *Literaturnaya Gazeta* (doc. 10.4) that can be mentioned as an exception. Although its core is, rather, based on conjectures, it quotes the words of the Swedish literary critic Eric Blomberg that "Sholokhov, like nobody else, is worthy of the Nobel Prize", which, of course, were not invented and could have drawn the attention of the Central Committee to the possibility of the first winning of the Nobel award by a Soviet writer. However, that did not happen. The Party officials in charge of propaganda were already thinking over the idea of setting up alternative international awards to counterbalance the Nobel Prizes (see Chapter 5), whereas the contacts with the Nobel institutions, first of all, dealing with science, which the country had had before the war, were being forcefully ruined. The propaganda was again presenting the country as a fortress under siege, and in such conditions the regime did not need a Soviet citizen to be a laureate of the ideologically alien award.

That is why Sholokhov's attempt to go to Sweden, made in 1947, was doomed to failure. The writer must have heard about his candidacy being nominated in the Nobel Committee, and it was not accidental that the date on his letter to Stalin with the request to give permission for this trip was July 29 (see doc. 10.5). August is the closing stage of preparing the Nobel Committee's recommendations for the Swedish Academy and the start of final discussions in the Academy itself. By his visit and his meetings with Swedish intellectuals Sholokhov hoped to create a favorable impression of himself and thus to make an indirect influence on the mood of the voting academicians. Whether this hope was reasonable is another question, but the idea appears evident enough.[8]

The reality wrecked his plans. In spite of the heartfelt content of his letter Stalin did not even answer it. There is no doubt that the letter reached him. On its original there is an office mark that the document was directed to the archive on October 8, 1947, nearly two and a half months after it had been received in the Kremlin; and any document can be sent to the archive only on the addressee's instruction.[9]

Stalin's death, on March 5, 1953 raised hopes in the West that the political climate in the Soviet Union would get warmer. The first steps of the country's new leaders gave grounds for expectations that these hopes would not prove false. One sign of the attempts to restore the contacts lost in the 1940s was a letter received in the fall of 1953 by the oldest Soviet writer Sergeev-Tsensky. It was a letter from the Nobel Committee for Literature with an invitation to nominate a candidate for the 1954 prize.

According to the archive documents, Sergeev-Tsensky wished to nominate Sholokhov and addressed the Writers Union in order to get permission for sending a letter with his proposal to Stockholm. The document with the corresponding request was only sent from the Union to the Central Committee on January 21, 1954, ten days before the deadline for accepting that year's nominations. Its revealing text is worth quoting: "There no need to remind you that the Nobel Committee is a reactionary organization", the document says to the Secretary of the Central Committee M. A. Suslov. Offered further are two mutually exclusive approaches — a true masterpiece of cynicism, when equally acceptable are both a plus and a minus, depending on which of them the country's chief ideologist will consider more desirable at the given political moment. As is said further: "It can be used for a corresponding political action: either for a public substantiated rejection with the denouncement of this organization [...] or a substantiated nomination of one of the writers as an active fighter for peace."

On January 27 in the Central Committee's Department of Science and Culture a decision is signed with the following conclusion: "We consider it expedient to recommend, through comrade Sergeev-Tsensky, taking advantage of the proposal..." Then there is a big interval — probably, used for coordinating the positions of the members of the Central Committee's Secretariat. Finally, four weeks later, on February 23, the following decision of the Secretariat appears in record No.54: "To accept the proposal of the Soviet Writers Union on the nomination of the writer Sholokhov M. A. for the Nobel Prize. To submit it for approval of the Presidium of the CPSU Central Committee." It was approved by the Presidium on February 25.[10]

Two days after this decision, on February 27, Sergeev-Tsensky sent a telegram to Stockholm from Simferopol, where he lived. He informed the Nobel Committee of his nomination saying that "the written justification is being sent by mail" (doc. 10.6-II). The text of the telegram, which now is in the Swedish Academy's archive among the documents of 1955, was stamped in Stockholm with the same date, February 27.

Of course, this nomination was hopelessly late for being considered for the 1954 prize, of which the nominator was promptly informed by the secretary of the Nobel Committee for Literature Hugo Villars — who did it in good Russian (doc. 10.6-III), writing it by hand, as the Nobel Committee did not have a typewriter with the Cyrillic alphabet.

The letter from Stockholm was stamped at the post on March 6. That is, Villars sent it already after receiving from Simferopol the written justification of nominating Sholokhov.

In 1955 the well-known Slavonic scholar from Sweden, professor of Stockholm University Oke Nilsson, became the nomination's reviewer. It was the same Nilsson who, four years later, in 1959, during his next visit to Moscow was forcefully prevented by the authorities from meeting with the dishonored Pasternak in Peredelkino (see Chapter 9 and doc. 9.25).

That year Nilsson did not support Sergeev-Tsensky's nomination — to a considerable degree owing to the fact that the Nobel Committee did not have trustworthy information of the nominee's literary activities for the last years. The 1955 prize was awarded to the Committee's nominee of many years, the Icelandic writer Halldor Laxness.

Nevertheless, Sergeev-Tsensky's initiative bore fruit. Nilsson's review was dated by May 1 1955,[11] and already on June 26 the Writers Union received, through diplomatic channels, the recomendation to prepare, "as soon as possible", and to forward to Stockholm, through the Scandinavian department of the Ministry of Foreign Affairs, some supplementary material on Sholokhov's work and publications of the last years. "As soon as possible" took almost two months and the documents to be passed to the Nobel Committee were not sent to the Soviet embassy in Stockholm until August 20 (doc. 10.8).

It is difficult to say whether it was because of the delays in sending the materials expected in Stockholm that the vote on the candidacy of the Soviet writer was negative. But it is quite possible that the Moscow delays were not accidental. In this case, though, they might have been caused not by political motives but by the increasing symptoms of the nominee's alcoholism. In the summer of 1955 his drinking reached a critical stage, which made the authorities of the medical institution taking care of the health of the members of the CPSU Central Committee insist that the Central Committee permitted to apply to Sholokhov measures of compulsory treatment.[12] Naturally, with such physical condition of the Nobel nominee nobody in Moscow was interested in his being awarded the prize and going to Sweden to receive it.

However, irrespective of these circumstances, which might have been unknown at the Swedish Academy,[13] it turned out that there were some people who opposed the candidacy of Sholokhov on principle. Kjell Espmark in his book mentioned that one of the Academy members, Dag Hammarskjold, at that time holding the position of the UN Secretary-General, had a strongly negative attitude to his candidacy. Afterwards, when touching upon the situation with the 1955 election, Hammardskjold said that he would have voted against Sholokhov — not only because his determination to do so was based on aesthetic purposes and on natural resistance to the attempts of putting pressure on him, but also because awarding a prize to a Soviet writer would have been a rather bad idea in itself, and that would have made others listen to him.[14]

The question of awarding the Nobel Prize to Sholokhov arose in the Central Committee again in 1958. It was connected with the rumor that the Swedish Academy might divide the prize of that year between him and Pasternak. True, there were no special reasons for such expectations. On August 30, 1958, nearly two months before the decision of the Swedish Academy to award the prize to Pasternak, the head of the department of Scandinavian countries of the Soviet Ministry of Foreign Affairs informed the Writers Union that the Soviet embassy in Stockholm had received serious information hampering the positive consideration of the candidature of Sholokhov. What was hampering was, first of all, the fact that the writer's second work, the novel *Virgin Soil Upturned* remained unfinished (doc. 10.9).

Nevertheless, for the case of the double awarding a special scenario was worked out in the Central Committee. Here is the main idea of this document prepared for the Party highest leadership by two departments of the Central Committee, cultural and of propaganda and agitation: "If Sholokhov is awarded the Nobel Prize of this year along with Pasternak, it would be expedient that in protest Sholokhov should pointedly reject it and state in press that he does not want to be a laureate of the prize which is used for anti-Soviet purposes. Such a statement from Sholokhov appears all the more necessary if the prize is divided between him and Pasternak." (see doc. 9.5, point 3).

The next year when Sholokhov's candidacy was considered in the Nobel Committee was 1962. Now a signal favorable for the candidate came from the committee's official, its secretary Hugo Villars (doc. 10.10–10.11). Here again a situation arose which showed that officials in power structures were not very happy about this nomination. Although Villars asked, in a confidential talk with a diplomat from the Stockholm embassy, to send the required documents before March 1, the secretary of the Writers Union Board Voronkov did not get round to send them to the Central Committee, for approval, until February 27, two days before the deadline. Explaining this fact by trivial slovenliness would be hardly appropriate. The person in question was not an ordinary writer but a member of the CPSU Central Committee, and such carelessness could have cost dear the culprit. Even the approval promptly prepared by the Central Committee's Cultural Department did not rouse Voronkov to sending the documents as promptly — they were not sent from Moscow until March 12.

The situation described reminds us of the unconcealed dawdling in preparation of replies to the requests received, in 1946 and 1947, from the Nobel Committee for Physiology and Medicine to send them the names of Soviet professors whom the Committee could use as nominators (see Chapter 4). Both then and in 1962 the sending of the necessary information was ill-intentionally delayed until it was no longer necessary in Stockholm.

The secret sabotage in 1962, however, had the same specific reason having nothing to do with ideological fears as such. It was again connected with the writer's ongoing inclination to alcoholism, so that the authorities were justly worried about possible incidents if he went abroad.[15] N. S. Khrushchev told about such circumstances in his memoirs related to that time (doc. 10,12).

Sholokhov's wish to go to Sweden at the beginning of 1962 might have been connected not only with the necessity to meet with his publisher G. Gidlund and his care about promoting books by young Soviet writers at the Swedish market (doc. 10.13). Like in 1947 (see doc. 10.5), he must have known that the Nobel Committee was going to consider his candidacy (doc. 10.10) and decided that it would be useful if he was in the Swedish capital after January 31, when the Nobel institutions started the campaign of discussing the 1962 nominations.

As can be seen from doc. 10.12, a foreign trip was refused to him. As for the supplementary documents asked for by the Nobel Committee, this request was openly sabotaged, as was noted above. The ban on Sholokhov's trips abroad was lifted only in the summer of 1963, and he went on a tour of six countries with his wife, daughter and daughter-in-law (doc. 10–14). Here it is appropriate to remember that a year later, in 1964, the Nobel laureate for physics A. M. Prokhorov was refused permission to take his son, a student, to the Nobel award-giving ceremony (see Chapter 8).

In 1965 Sholokhov was awarded the Nobel Prize. A significant part in it was probably played by the refusal of the French existentialist writer Jean-Paul Sartre to accept the 1964 prize. The writer, whose leftist views and inclination to shock the public were widely known, named as one of the indirect motives for taking such an extraordinary decision, the fact that the Nobel Prize had been awarded to Pasternak and not to Sholokhov (doc. 10.15).[16]

As the Swedish Academy's Nobel archives of 1965 are so far closed, we cannot familiarize ourselves with the list of the writer's nominators. However, there are quite a few reasons to suppose that the most substantiated of the proposals received were those from foreign literary critics, first of all, from Scandinavian countries. Proposals from Sholokhov's compatriots, if there were any, could only have had an auxiliary importance for the Nobel Committee members and the voting academicians.

Because of the sharply critical attitude to Sholokhov, predominant among the intellectual stratum of the Soviet society in the 1950s and 60s, it would have been difficult to expect that the country's most reputable literary critics would sincerely wish to help him get the century's highest literary award. The blame for that should be put, to a large extent, on the writer's human qualities — intolerance and bias in appraisals, deliberate rudeness in personal contact, little concealed disposition to xenophobia; arrogant speeches, deliberately meant to cause a scandal, at different forums — from party congresses to writers meetings and seminars.

But there was something else as well. The post-war years, first of all. The period after the second part of *Virgin Soil Upturned* was published, in 1960, was, in fact, a period of the writer's intellectual idling. A. T. Tvardovsky — a rather important figure in Soviet literature, and also editor-in chief of the country's leading literary and social journal *Novy Mir* and secretary of the Writers Union Board — wrote sternly about that in his working notebooks.[17]

One graphic example of unspoken refusal to nominate his candidacy was the reaction of the leaders of the Literature and Language Department of the USSR Academy of Sciences to an invitation from Stockholm, passed from the academy's Foreign Department, to nominate their candidate for the 1965 prize. The acting head of the Literature Department M. V. Khrapchenko was certainly informed that the academician Sholokhov, that is, a member of the structure he headed, had been repeatedly nominated for the prize and stood a very good chance for being awarded it. Nevertheless, he preferred to wash his hands and reported that the department was not going to nominate anybody that year (doc. 10.17 and note 2 to it).

True, there is information about a person from the Soviet Union who spoke in favor of Sholokhov. It was the country's leader N. S. Khrushchev, who was in Sweden, in June 1964, on an official visit. At the reception in his honor a member of the Swedish government Ulla Lindstrom addressed him with a spontaneously facetious question, which of the Soviet writers he would like to see among the Nobel Prize laureates. Having rejected the names of two 1964 prize nominees she named, Khrushchev pointed at Sholokhov (doc. 10.16–1).[18]

Like in the previous years, he knew in advance of the discussions of his candidacy which were going on at the Nobel Committee. The letter he sent to the CPSU Central Committee's General Secretaty L. I. Brezhnev is very revealing. It expresses unconcealed concern whether he would be made to reject the awarded prize, as it had been planned at the Central Committee (doc. 9.4). That time everything went smoothly for him, and having calmed down, he left on a hunting trip to Kazakhstan. It was there that he got the news about the decision of the Swedish Academy.[19]

The preparation for the writer's departure to the Nobel ceremony was amazingly quick — solving all problems, including financial, took just one working day! One and the same date, November 29, is on all the documents — the letter from the Writers Union, the letter from the Central Committee's Cultural Department to the Central Committee's Secretariat, the assent of the five Secretariat members, a new request from Sholokhov and a new letter from the Cultural Department, on which the repeated signatures of the highest officials appeared as promptly (doc. 10.19–10.22).Where the promptness and unquestioning obedience are concerned, that was perhaps an unprecedented case for the bureaucratic structures of the Central Committee.

No less unprecedented is the Central Committee's consent to sending a whole delegation to Stockholm together with the laureate. Apart from his wife, he was accompanied by his two daughters, a literary critic chosen by Sholokhov and intended for the role of speechwriter, and an interpreter. That day, perhaps, nobody in the Central Committee remembered the previous departures for Stockholm of Soviet prize-winners for sciences, who had been allowed to be accompanied only by their wives (see Chapters 7 and 8).

Before the departure for Stockholm the Swedish embassy, by tradition, arranged a reception in honor of the laureate, having invited, among other guests, representatives of Soviet creative intelligentsia. Their response clearly showed the attitude of a number of the Writers Union leaders to Sholokhov as a person. According to the memories of Gunnar Jarring, the ambassador of Sweden in the USSR, the invitation to dinner was declined, under various pretexts, by the Union's First Secretary K. A. Fedin and two other secretaries. Nonetheless, on the morning of the day of the reception the Union informed that four of its representatives would come to the dinner. "It seemed that some invisible hand directed them to us" in order "to prevent a writers demonstration against Sholokhov".[19a]

Sholokhov's stay in Sweden was quite adequate. At the ceremony on December 10 Sholokhov was introduced by the eminent member of the Swedish Academy Anders Osterling, the Academy's permanent secretary of 23 years. Here is what he said of Sholokhov's Nobel Prize: "This award is a tribute of justice and gratitude for your considerable contribution to Soviet literature, the contribution which in our country is as well known as in the whole world."[20]

After the award-giving ceremony Sholokhov read a traditional Nobel Lecture in the City Hall. After he finished, A. Osterling asked Yarring, who, by protocol, was accompanying the laureate in his trip to Stockholm, about his impressions of the lecture. Before hearing the answer Osterling said that he "was dissatisfied with his speech because he gave the lecture a political coloring". Jarring agreed with him, noting, in a more imaginative way, that the lecture made him think of a "prefabricated wooden house ready for assembling. Its drafts are made in advance by definite standards, and it is impossible to change them."[20a]

The day of February 13 is for the Swedes a day of merry-making, festive manifestations and beauty contests arranged in honor of the great martyr St. Lucia. Sholokhov spent this day in Uppsala University. Apart from meetings with students, who bombarded him with questions, he took part, together with Professor Gunnar Gunnarsson, in a seminar on Slavonic studies, participated in the ritual of coronation of the most beautiful university student democratically elected for the part of St. Lucia, and so on.

True, the writer's answers to the questions of students about women's literature (doc. 10.23),[21] despite being undoubtedly sincere, were not quite appropriate for a Swedish audience. However, according to the press, the listeners took them quite calmly. The assertions of the French writer Louis Aragon about the scandalous end of the meeting remained unsupported — for example, by M. Sohlman, who that day accompanied M. Sholokhov as interpreter (note 2 to doc. 10.23). It is most likely that these utterances reflected the long-standing mutual dislike of the two writers.[22]

On coming back to Moscow Sholokhov's Nobel Prize was hailed as his personal triumph. A reception in his honor was held in Moscow's most prestigious venue of the 1950s and 60s — in the House of Receptions on the Lenin Hills, where official ceremonies of receiving foreign guests of honor and eminent representatives of the public opinion were organized (doc. 10.24).

The writer's death was preceded by several years of complete intellectual incapability. He died on February 21, 1984. The news of the Nobel laureate's death revived discussions in the world press about the circumstances of awarding him the prize. One of such discussions was in the Swedish newspaper *Dagens Nyheter* between its Moscow correspondent Disa Hostad and the permanent secretary of the Swedish Academy Lars Gyllensten (doc. 10 26). Hostad in her article remembered the writer's participation, in Stalin times, in an ambiguous dispute about the writer's right to a pen name (see note 46 to Chapter 6) , which had a strong touch of anti-Semitism.

Sholokhov returned to the topic of the plot "of the world's Zionism, both foreign and internal" against the Russian people in 1978, more than a quarter of a century after the discussion of the issue of pen names and twelve years after receiving the Nobel prize. He touched upon this topic in his letter to the general secretary of the CPSU Central Committee L. I. Brezhnev (doc. 10.25). The letter was first published only in 1993, after the collapse of Communism in the country, but persistent rumors about its content had circulated among creative intelligentsia as far back as in the late 1970s.

The majority of the Soviet intellectuals, while paying tribute to Sholokhov's great contribution to the national literature and not being particularly bothered about the problems of the origin of *Quiet Flows the Don*,[23] were rather negative about the writer's human qualities. One of the consequences of such moods was the evident restraint of the country's leading writers where Sholokhov's nomination for the Nobel Prize was concerned. The above mentioned refusal of the head of the Language and Literature Department, in 1965, to enjoy the right of nominator accorded to him (see doc. 10.17) was only one episode of the latent protest.

Whether this silent opposition took active forms, whether any national nominator offered any other representative of Russian literature to counterbalance Sholokhov will be possible to say for certain only after the expiry of the 50-year term while the Nobel Committee archive documents cannot be available to the public. But indirectly, from publications in the world press or from personal memories, two most probable figures come to mind, whose names were certainly present in the lists of nominees of the early 1960s. Of prose writers it is, first of all, K. G. Paustovsky and of poets, A. A. Akhmatova.

The first mentions of Paustovsky as one of the Nobel Committee's nominees appeared in the Swedish press at the very beginning of 1962, even before the deadline of the Committee's receiving proposals for that year's prize. The Moscow correspondent of the newspaper *Stockholms Tidningen*, writer and translator Hans Biorkegren, responded to these publications with an article about the life and work of the aspirant to the prestigious award, who was unknown to the majority of Swedish readers. He gave this article a quite definite heading: "Russian Candidate for Nobel Prize". The beginning of the article was as follows: "Konstantin Paustovsky, 69, is called one the major contemporary Russian prose writers. The Nobel Committee now regards him as the most prominint figure in Soviet literature, on a par with Mikhail Sholokhov and Vera Inber."[24]

In the summer of the same 1962 the Swedish writer Artur Lundkvist came to Moscow. He was a figure of authority not only in literary circles but also among voting academicians; suffice to say that six years later, in 1968, he became member of the Swedish Academy, that is, one of the eighteen "immortals", whose votes determine the name of the new laureate. The apparent purpose of Lundkvist's visit was participation in the session of the World Peace Council, of which he was vice-president. However, the first question he asked A. M. Borshchagovsky, member of the Board of the USSR–Sweden Friendship Society, who came to meet him at the airport, was not about the forthcoming session. His main concern was whether he would be able to meet and talk with Puastovsky during this stay (doc. 10.27). This question and the events that followed make us suppose with a high degree of probability that it was not only problems of defense of peace that brought him to Moscow that time. He also had the task to meet with the aspirant to the 1962 Nobel Prize and form his own opinion of him. That is, he came to Moscow also as an unofficial emissary of the Nobel Committee, as one of its experts, whose judgement is further used for working out the final decision; such working practices of the Nobel institutions were several times mentioned in the previous chapters.

Lundkvist's reference must have been favorable, and Paustovsky was perhaps seen in the Nobel Committee as one of the most probable 1962 award-winners. Otherwise there would have been no grounds for persistent rumors about that, which filled the Western press. In his letter to L. N. Delektorskaya, an old friend of the Paustovsky family and translator of his works into French, the writer told her, on November 15,1963, that he had received from Paris a bundle of cuttings from French newspapers describing him as a possible Nobel laureate.[25]

In 1962 it was the American novelist John Steinbeck who became the winner of the Nobel Prize for Literature. But three years later Paustovsky's candidacy again appeared in the list of aspirants to the prize. And again the name of the Russian writer was often mentioned. Paustovsky was widely discussed in the press, not only Swedish, the more that he was present at the congress of the European Writers Association. The congress was held in Rome in October 1965, when the verdict of the Swedish Academy was about to be announced.[26]

According to the Russian emigrant writer Mark Slonim, who met with Paustovsky in Rome and told about his talk with him nine years later in the New York newspaper *Novoe Russkoe Slovo,* those days his interlocutor was "strongly disturbed by newspapermen". Slonim himself was at that time more interested in Paustovsky's attitude to the contemporary Soviet literature and its representatives, but he also did not pass over the Nobel theme, since he was commissioned by the *New York Times* to interview the potential Nobel laureate (doc. 10.28-I), just in case.

Italian newspapers were not behind American ones. Two weeks before the expected verdict Paustovsky was regarded in Italy as the most probable candidate for winning the Nobel award. The newspaper *La Fiera Letteraria* announced that the Russian writer was the certain favorite of the forthcoming vote.[27] The rumors from Stockholm drew a peculiar response from the Milanese publishing house, which published, in 1965, a collection of Paustovsky's works, with the words on the title page, under the title, and printed in the same type: "Candidato Nobel 1965."[28]

That time was also remembered, in 1991, by L. N. Delektorskaya mentioned above (doc. 10.28-II). She specially came to Rome from Paris, where she lived.

In all Paustovsky's nominations known to us he was invariably accompanied by the opposing shadow of Sholokhov. It did not concern only the episode with Lundkvist described by A. M. Borshchagovsky (doc. 10.27). According to the director of Paustovsky's Moscow museum in Kuzminki Park I. I. Komarov, one of the first nominators of Paustovsky was Louis Aragon, longstanding admirer of his works and Sholokhov's ill-wisher of many years (see doc. 10.2 and note 2 to doc. 10.23). He was supported by the German scholar of Soviet literature Wolfgang Kasack.[29]

In 2003 appeared one more solid piece of evidence of the likelihood of Paustovsky's nomination — this time for the 1963 prize. According to the above mentioned I. I. Komarov, in answer to their inquiry sent to Paris the museum received a letter from the president of the French P. E. N.-Club Jean Bloch, who visited Moscow in October of that year. On the day when the name of the next laureate of the Nobel Prize for Literature was to be announced in Stockholm, Bloch found himself in Paustovsky's flat on Kotelnicheskaya Naberezhnaya. Both of them were waiting with interest for the news from the Swedish capital. The Greek poet Gregor Seferis became the laureate. Undoubtedly, Sholokhov at the same time was having the same hopes. Having visited, in the summer of 1963, six European countries, including Sweden, he certainly also had information about discussing his candidacy in the Nobel Committee.

The same connection of Paustovsky and Sholokhov was mentioned again in 1992 by the poet B. A. Chichibabin. "Paustovsky", he wrote, "turned out to be aspirant to [...] the Nobel Prize for Literature, while prose was at that time represented by Sholokhov, who had been officially nominated for this prize for many years running [...] I do not want to reread *Quiet Flows the Don* [...] Sholokhov's epic style — tragically picturesque, thoroughly detailed, marked by pagan spontaneity — seems to me somewhat outdated, ethnographically archaic, belonging and interesting to the past rather than to the present and the future. Paustovsky's poetic prose, with its characters appearing in quick graphical sketches but lit by the light of eternity, which makes the reader think, imagine and dream, seems to me more modern and promising."[30]

The latest year when Paustovsky was mentioned as a suitable candidate for the next award was 1964. According to the ambassador in Moscow G. Jarring, who, in accordance with the traditional ritual, came to Stockholm to accompany the Soviet physicists N. G. Basov and A. M. Prokhorov, due to receive the Nobel Prize awarded to them, he got into conversation with the permanent secretary of the Swedish Academy A. Osterling. The latter unexpectedly turned the conversation to Paustovsky and asked his interlocutor if he considered him worthy of the prize for the memoirs he had written. Jarring answered that he had also read them, and not only once. "But nonetheless I don't think that awarding the prize to Paustovsky and ignoring Sholokhov will gain understanding in the Soviet Union. It will only give cause for new bitterness [...] Osterling promised to think about this problem and acquaint himself once again with Sholokhov's work."[30a] The result is well known — it was Sholokhov who became the winner of the 1965 prize for literature.

Related to the same period of time as described above, that is, the early 1960s, is the available information of the Nobel nomination of Anna Akhmatova.[31] One of the main characters in this plot was the eminent Swedish literary critic and expert of the Nobel Committee, a scholar of poetry in Slavonic languages, Erik Mesterton. He came to the Soviet Union many times and visited Anna Akhmatova, who, in the Russian manner, called him Erik Rikardovich.

She first heard of herself as the Nobel Committee nominee in the summer of 1962, that is, at the same time when in Moscow some people feared that the Nobel Prize might be awarded to Paustovsky (doc. 10.27-I). As Akhmatova noted in her notebook, Mesterton saw her at her *dacha* in Komarovo, near Leningrad, on July 14, and when saying goodbye mentioned that she was going to hear something interesting from the son of the writer Vera Panova, B. B. Vakhtin. It was the latter who informed her that she was that year's Nobel nominee.[32]

On July 24 she sent a letter to Mesterton, who was still in Leningrad. In it she touched upon decoding certain images in *The Poem without a Hero,* which she forgot to tell him when they met,[33] but, of course, said nothing about what Vakhtin told her. However, overflowing with this unexpected news, on August 13 she materialized it in a *vers libre*:

The Northern message found me in the North
Amid the heather now in full bloom,
The August jasmin and even this scarlet
Unfading evening glow.[34]

One more piece of evidence of Akhmatova being excited by the news passed in Mesterton's name are two notes in the list of the most memorable events of 1962 selected by her:

XIV. (August. Scandinavia announced the beginning of the struggle for the N.) and I was listening to the port.radio: "Premio Nobel..."

XV. About the Nob. Prize and the Ital. newspaper (September, 1962).[35]

What is behind the poetic image of the first line of the *vers libre* is evident and does not require any comment. A year later she is still excited by the thoughts of the North pointedly opposed to the three other cardinal points. In the sonnet dated July 30, 1963 and dedicated to "E. M.", that is, Erik Mesterton,[36] she writes about her "old friend", her "true North". However, she concludes the sonnet on a pessimistic note. As she does not really believe that the news from Stockholm would be positive, she makes the "iron Finland" say: "You will know everything, except // Joy. Even so, live!"[37]

One more piece of indirect evidence that she did have some faint hopes for receiving the Swedish Academy's attention is her brief note made in her notebook some time between September and November of 1964. Mentioned together in the note are Mesterton and the permanent secretary of the Swedish Academy Anders Osterling (in the notes called Andreas Osterling).[38]

Many years later the Swedish Slavonic scholar Johan Oberg, who was at that time the editor of the literary journal *Ord Och Bild,* in a conversation with Mesterton tried to make out what he had meant by the information intended for Akhmatova in 1962. The answer was disappointing. Mesterton told Oberg that when discussing literary topics with Vakhtin he imprudently asked him how awarding the Nobel Prize to Akhmatova would have been taken in the Soviet Union. Vakhtin answered: "that would be very good..." Then, as Mesterton put it, from one small feather a big hen was made, since the only grain of truth in this story was the above mentioned question to Vakhtin, and all the rest was invention.[39]

However, in this case Mesterton should not be entirely trusted. The question that he asked Vakhtin in 1962 was imprudent indeed, and further on he must have many times regretted his being so unrestrained. The requirement of complete confidentiality in the questions of contacts with the Nobel Committees is a fundamental condition of cooperation of the Nobel institutions with their nominators and experts.[40] Mesterton's contacts with the Nobel Committee for Literature are well known; among other things, he was one of the experts who considered Pasternak's candidacy in 1958, and he certainly could not have been unaware of what was going on in the committee in 1962. His repeated visits to Akhmatova give reason to suppose that they were connected not only with the interest of a translator to the poet he translated, but also with the assignment he received from the committee as one of its experts on Russian poetry.[41]

The next block of information relating mostly to the notes of Akhmatova herself is connected with 1965. It is worth noting that at that time nominators from Oxford University were ready to join the Swedes (doc. 10.31). Judging by the poetess's notes, this readiness was put into life, and when, at the beginning of June 1965, in Oxford they robed her in a mantle and a cap of the university's honorary doctor, her candidacy was already being considered by the Nobel Committee experts. Anyway, that is what might have been written in the newspaper given to her by the Englishman Williams, who came to see her in Komarovo, on June 11, 1965.

Her exclamations "Gibberish!!", "Nonsense?" are merely her defensive reaction to the unknown. There is no doubt that Akhmatova waited for the news from Stockholm and hoped that it would be favorable — both in 1962 and in 1965. The evidence to that is provided not only by her notes cited in doc. 10.31, but also by he drafts she prepared in 1966 for the supposed interview to M. V. Ardov (doc. 10.30).

There is no direct evidence of Ahmatova's nominations for the 1963 and 1964 prizes, but there are some indirect indications, especially for 1963. Among them there is the sonnet "The North" dated July 30 and dedicated to Erik Mesterton.[36]

In 1962 she had already responded to the news passed by B. B. Vakhtin in the name of the Swedish scholar, by the poem "The Northern Message Found Me in the North...". Why a year after should she have again turned to the image of "the true friend North", which for her had a strong association with the Nobel Prize theme, if it had not been for some new leakage of information, now already about the 1963 prize?

More evidence that rumors of her 1963 nomination reached her is given by the entry in L. K. Chukovskaya's diary dated 17 October 1963. Once her talk with the poetess had turned to Isaiah Berlin, British philologist and philosopher, whom Ahmatova met in Leningrad in the summer of 1945, Chukovskaya remembered that Akhmatova said in Moscow: "Why does this gentleman bother about me so much?" — thinking that her Nobel nomination and in general her fame in the West was Berlin's doing.[42]

Thus, the probability of Akhmatova being repeatedly nominated for the Nobel Prize can be regarded as a quite established fact. Among the Nobel Committee experts and members her candidacy was not taken as accidental and was regarded highly enough. Otherwise it would hardly have been mentioned with such respect in the book by Kjell Espmark, the present-day chairman of the Committee.[43] Let us dwell on some examples given in this book.

In 1966, after a half-century interval, the Nobel Prize was divided between two aspirants, the Israeli Schmuel Agnon and the German poetess Nelly Sachs, who lived in Germany.[44] The double award aroused a discussion in Swedish literary circles. They argued over who rather than Agnon would have made a suitable pair with Sachs, who, in 1965, occupied the second place on the Nobel Committee's reserve list, after the Guatemalan writer Miguel Asturias.[45] One of those who expressed their expert opinion was the member of the Swedish Academy Henry Olsson, who, incidentally, was the first nominator of Mikhail Sholokhov in 1947. In his letter to Erik Lindegren, translator of Nelly Sachs, to whom she owed her popularity in Sweden, he named three of the 1966 nominees, Ahmatova among them.

Still more categorical was the statement of Robert Vlakh, next year. At the 1967 Nobel symposium, speaking of the story of Sholokhov's Nobel Prize, he said that among all candidates of that period it was Akhmatova who best of all complied with the requirements for a Nobel laureate.[46] Kjell Espmark, the author of the mentioned book, placed Akhmatova's candidacy on the list of the Swedish Academy's evident mistakes over the post-war period. He mentioned the Russian poetess among 15 persons who deserved the Nobel Prize but did not get it.[47]

It is natural to suppose that for being the subject of such discussions one should have been at least included in the lists of the Nobel nominees, and not once at that.

Anatolii Naiman, one of Akhmatova's closest assistants in the last years of her life, gave in his memories a parable he had heard from her. "When after the war the allies were exchanging receptions", she said, "and Zhukov rode on horseback into the Western zone of Berlin, Montgomery and Eisenhower, both walking, took his horse by the bridle and brought it down the road. That was the end of him, because Stalin had imagined that he would ride in on a white horse, and the other two would be walking by his side." After that Naiman concluded: "For her the Nobel Prize was that very white horse of the true winner in the exhaustive half-century struggle."[48]

Akhmatova did not win the Nobel Prize. Of course, it was her ill luck. The Swedish Academy, when deciding on its awards, tries to somehow regulate the choice taking into account geography, in order to obtain the widest possible representation of writers from different states and continents of the world among the Nobel laureates. From 1962, supposedly Akhmatova's first nomination, to 1965, it was Sholokhov who was regarded in Stockholm as an unquestionable favorite among Soviet writers; *Quiet Flows the Don*, the 20th century great epic novel, could not have been left without an award, and the Swedish academicians considered Sholokhov to be its author.[49]

In the next year, 1966, Akhmatova died. As was noted in the memories of L. D. Bolshintseva, the poetess's close friend, in the casket she was wrapped up in a lace shawl, the one she was wearing when receiving the prize in Italy,[50] in 1964 (see doc. 10.31-II and note 1 to it).

How the official circles of the Writers Union took the death of the great poetess, who during her lifetime received from her homeland much less than she deserved, was scathingly and graphically described by Victor Nekrasov. He painted a portrait of "a tall, smiling, beaming representative of the Moscow writers organization" at Anna Akhmatova's funeral, who cheerfully said: "We're being late, we are indeed. That's not good."[51] And Vyacheslav Ivanov, already touching upon the Nobel motifs in her life, stated sadly: "The Nobel Committee was too slow with Akhmatova ...] and did not manage it in time, which is a shame."[52]

In conclusion let us return to the months of Mikhail Sholokhov's stardom, the time of awarding and receiving the long-awaited award. By coincidence, in September of 1965, when the Nobel Committee for Literature prepared and passed for consideration of the Swedish Academy their recommendation to elect the Soviet writer that year's prize winner, in Moscow the KGB arrested two writers, Andrei Sinyavsky and Yuly Daniel. The reason for their detainment were their publications in the West, under the pen names of Abram Terz and Nikolai Arzhak respectively.

The public opinion did not miss this event. The Soviet Union's highest authorities started to receive letters and telegrams of protest signed by prominent representatives of Soviet and foreign literature (doc. 10.33). Many letters with the request to interfere in person in the KGB's unlawful actions and help get the colleagues out of trouble were also addressed to Mikhail Sholokhov, who was just announced the Nobel laureate (doc. 10.34).

The Nobel laureate, however, kept completely silent. His voice was not heard until a month and a half after the trial of both writers was completed, both of them sentenced to long terms of imprisonment in corrective labor camps.[53] He spoke from the tribune of the XXIII CPSU Congress held in March and April 1966. In his demagogically eloquent speech he told the world that "humanism is not dithering" and "these werewolves" should have been tried "according to the revolutionary sense of justice" (doc. 10.35-I).

Comment is superfluous here. The writer himself brilliantly managed to comment upon his personality. As Lidia Chukovskaya said in her "Open letter to Mikhail Sholokhov, "Your disgraceful speech will not be forgotten by history" (doc. 10.35-II).

Many years later, already after the collapse of the Communist regime, the writer Fazil Iskander, giving his next book to his friend, participant of the Great Patriotic War Yuly Daniel, accompanied it with a dedicatory poetic inscription:

I'm so very glad for you,
As you did in your lifetime
All that in Russia is to do —
Fought at war, wrote books, did time.[54]

As can be seen from the above, the contacts of emissaries from Stockholm with all three Soviet writers, potential candidates for the prize, were most active in 1962. Such interest in all of them together could hardly have been a chance coincidence. Perhaps the assumption that the Nobel Committee had a preliminary idea to make that year "a Russian year" may be close to the truth. However, the badly concealed reluctance of the Soviet side to fulfill in time the request for additional information about Sholokhov made the Committee members change their plans. It was impossible to seriously choose from the Soviet candidates the most worthy one without considering the candidacy of the author of *Quiet Flows the Don.* As a result the initially planned Russian year transformed into the American year: the Nobel Prize was awarded to John Steinbeck. Taking into consideration the persistent anti-Americanism, which in the post-war Soviet Union turned into the pivot of endless ideological campaigns, a sort of religion of the Party officials, it would have been difficult to think of a brighter antithesis.[55]

On the other hand, if the Nobel Committee really had had any plans to make 1962 "a Russian year", we can, given the principle of dialectic unity of opposites, take the liberty of viewing the decision at which the Swedish Academy finally arrived from a somewhat paradoxical perspective.

The highlight of Steinbeck's work, the novel *The Grapes of Wrath,* was published in the Russian translation in 1940, merely a year after its coming out in the writer's homeland. In the Soviet Union the novel immediately won the readers' acclaim and was probably even more popular than in the US. And the title of the book, which, nearly a quarter of a century later, brought its author the prestigious award, soon became almost a buzzword in this country. Especially when the need arouse to use euphemisms for describing some new absurdities of everyday Soviet life.[56]

Notes

[1] According to the judgement passed in her absence by three members of the tribunal in and out of court examination, as it was done in Stalin time, Ivinskaya was sentenced to five years inprisonment in a camp of general regime "for closeness to persons suspected of espionage"*

* O. V. Ivinskaya. *Gody s Borison Pasternakom. V Plenu Vremeni* (The Years with Boris Pasternak. Captured by the Time). Moscow: Libris, 1992, p. 132.

[2] *ibid,* p. 162.

[3] When A. A. Surkov was the First Secretary of the Soviet Writers Union (1953–1959), his attitude to the personality of Sholokhov was restrained. It most clearly showed in the fact that he did not mention Sholokhov when naming the most prominent wtiters in the Union's official reports prepared for the 2nd Congress of Soviet Writers held in December 1954.* At the congress itself, after Sholokhov's scandalous speech with its heavily biased attacks on some Soviet writers, Surkov did nothing to stop the delegates who severely condemned the tonality of his words. ** But a year later, in January 1956, in connection with Sholokhov's speech at the national conference of young writers, Surkov spoke his mind more definitely, describing this speech as "clowning".*** Such stern reproach addressed to the eminent writer was probably caused by the following passage. Having reminded the audience, in passing, of Gorky's words concerning the way that to the writer's mastery, learning to bark lime-trees is one thing but making bast shoes of them is another, he decided to joke in the rough manner characteristic of him: "Talking of bast, it is better to bark lime-trees in one familiar place, doing it thoroughly and not by way of brief raids."****

* V. O. Osipov. *Tainaya Zhizn' Mikhaila Sholokhova* (The Secret Life of Mikhail Sholokhov) — Moscow: Libereya, Raritet, 1995, pp. 299–304.

** See doc. 10.2.

*** L. K. Chukovskaya. *Zapiski ob Anne Akhmatovoi* (Memories of Anna Akhmatova), vol. 2. — Moscow: Soglasie, 1997, p. 178.

**** *Literaturnaya Gazeta,* January 12, 1956, p. 1; the style of the utterance corresponds to the newspaper text.

[4] *Novy Mir* No.1, 1965, p. 4. The head of the CPSU Central Committee's Cultural Department D. A. Polokarpov, having read this article in page proofs, made a number of remarks — among other things, concerning the cited phrase, and A. T. Twardovsky wrote down their dialogue in his working notebooks in the following way: "Here [...] I can also say some words of my own... 'From Sholokhov to Pasternak — standing next to each other!' But 'from to' is not next, it's the other way round. But still..."*

* A. T. Tvardovsky. *Working Notebooks of the 60s. Znamya* No. 12, 2001, p. 134, entry of January 28,1965.

[5] L. A. Anninsky *Russkie plyus* (The Russians Plus) — Moscow: Algoritm, 2001, p. 172.

[6] The first three books of *Quiet Flows the Don,* which appeared in the Soviet Union between 1928 and 1932, were promptly translated into foreign languages. In all, by 1940, when the last, the fourth book of the novel came out, the novel had been published abroad in 12 languages, including Chinese, in 13 countries.* The first foreign language edition of the first book appeared in the fall of 1929 in Germany, and the next year the second book came out also there. In January 1930 the French translation of the novel appeared; in August,

Swedish; in October, Spanish; in November, Czeck; in December, Dutch, and so on. In Denmark all three books were published between 1932 and 1934;** in Great Britain, in 1934, and in January 1935 the writer visited both these countries at the invitation of their literary circles.***

* Bibliography reference book *Knigi Sholokhova na Yazykakh Narodov Mira* (Books by Sholokhov in the Languages of Peoples of the World) — Moscow, 1975, pp. 71–86.

** Interestingly, the first foreign language translation of the fourth book of the novel appeared in Denmark as early as in 1940, almost simultaneously with the publication of the Russian original; the next was the Swedish translation, in 1943.

*** See note 3 to doc. 10.5.

[7] Before 1940, the year when the final fourth book of *Quiet Flows the Don* was published, the novel could not be presented in the Nobel Prize nomination, because according to their regulations the Nobel Committee considers only completed works.

[8] It must be said that there were some cases when writers successfully visited Stockholm just before the beginning or during the period of discussing the new Nobel nominations. One of such lucky men was the Italian writer and playwright Luigi Pirandello, the 1934 Nobel laureate. He made a triumphant trip to the Swedish capital in December 1933. The Soviet ambassador A. M. Kollontai wrote in her diary, on December 14, 1933: "Pirandello has come, and there are different celebrations in his honor [...] In honor of Pirandello the Italian embassy [...] arranged a ball [...] really luxurious..."* In this case can we categorically deny at least a possibility that this much talked about visit might have had an indirect impact on the thoughts of the members of the Swedish Academy when they had the voting paper in their hands? So Sholokhov's desire to get to Stockholm in the summer of 1947 is quite understandable.**

* A. M. Kollontai. *Diplomaticheskie Dnevniki* (Diplomatic Diaries) 1922–1940 — Moscow: Academia, 2001, p. 2000.

** See also note 4 to doc. 10.5.

[9] There also is some evidence that Stalin's silent refusal of consent to Sholokhov's trip abroad was not accidental. It appears that such a request caused his steady irritation with Sholokhov, which burst out in the presence of the Yugoslav political personality M. Djilas. During his official visit to Moscow, the latter was invited to Stalin's *dacha* in Kuntsevo, near Moscow, for a late night supper. The conversation at table turned to contemporary Soviet literature. "'I, like more or less all foreigners, pointed at Sholokhov'", wrote Djilas in his memories. "Stalin said: 'Now we have better writers', and he named two names unknown to me, one of them a woman's."* Nonetheless, Sholokhov was not angry with Stalin and after his death used every opportunity to express his admiration of the deceased tyrant. A. T. Tvardovsky, when having a rest in Sukhumi in October 1968, wrote down in his diary what the Abkhazian poet B. V. Shinkuba had told him about Sholokhov's recent tour of Georgia. "On Lake Ritsa, at Stalin's house there was a banquet where Shinkuba and his spouse were invited. Sholokhov twice proposed a toast to Stalin ('we are now in the house where He...', and so on) using expressions which were possible only during Stalin's lifetime: a man of genius, wise, the savior of our homeland, and so forth."**

* M. Djilas. *Razgovory so Stalinym.*(Talks with Stalin). Frankfurt am Main, Possev-Ferlag, 1970, p. 149.

** *Znamya* No. 10, 2003, p. 153.

[10] Osipov, pp. 292–294. The documents mentioned in the last two paragraphs are stored in the RSAMH, fond 4*, opis' 9, ed. khr. 999, l.75–80. From 1992 through 1995 *fond* (fund) 4 was open to researchers, and in this period Sholokhov's biographer managed to acquaint himself with the documents he was interested in. In 1995 the fund was closed again for an indefinite time. That is why here we had to confine ourselves to repeating what was published by V. O. Osipov.

* Fond 4 stores materials on the works of the Secretariat of the CPSU Central Committee after the XIX Party Congress, that is, after October 1952.

[11] The Archive of the Swedish Academy, Stockholm. The Fund of the Nobel Committee for Literature for 1955.

[12] When the fund of the CPSU Central Committee's Secretariat in the RSAMH was available for researchers,* Sholokhov's biographer V. O. Osipov found there a document concerning one of the sad episodes in the writer's life. In October 1962, during the work of the XIX Party Congress, whose delegate Sholokhov was, he had his next period of hard drinking. The report from the Party office addressed to the secretary of the Central Committee G. M. Malenkov stated that, according to the letter from the secretary of the Party Committee of Rostov *oblast'* Kiselev, "with whom we had a talk, it is necessary to apply measures of compulsory treatment to comrade Sholokhov, as the measures taken by the oblast' committee do not have any impact on the writer."** Three years later, on September 3, 1955, the head of the 4th Department of the USSR Health Ministry informed the Cultural Department of the CPSU Central Committee that because of the chronic alcoholic intoxication "comrade Sholokhov has progression of liver cirrhosis, cardiosclerosis, general atherosclerosis and hypertension diseases. The usual admonition and attempts to make comrade Sholokhov carry out the necessary treatment were not successful. Comrade Sholokhov needs compulsory treatment in a special medical institution. Further alcohol abuse will lead to irreversible consequences and the patient's death."*** Sholokhov's painful addiction to alcohol had been widely known long before the years mentioned here. A. E. Bovin in his memories told a story from the war times, which he heard from L. I. Brezhnev: "Having a mandate from Malenkov, Sholokhov came to their formation [the unit where the prospective General Secretary of the CPSU Central Committee served as political instructor] in order to write a book about the war [*They Fought for Their Country*]. For two weeks everything was OK. And then a drinking bout started, with an accordion, broads and so on."****

* See note 10 to this chapter.

** Osipov, p, 282.

*** RSAMH, fond 5, opis' 17, ed. khr. 535, l.113.

**** A. E. Bovin *Dvadtsatyi Vek kak Zhizn'.* (The Twentieth Century as Life), Moscow: Zakharov, 2003. p. 157.

[13] Most probably, the fact of Sholokhov's addiction to drinking was known in Sweden. According to N. M. Lunkov, who in the mid-1960s was the acting ambassador in Stockholm, in May and June 1957 Sholokhov, accompanied by his wife, visited the Nobel capital. Moscow had warned the embassy in advance of the writer's bad inclinations. That is why at the festive dinner arranged on his birthday, 25 May, there were no alcoholic drinks on the table. The writer, "not bearing a grudge against us, snapped that those were dried-up sticks and not people who worked in the embassy, if even on his birthday they did not offer him at least one glass of wine. But after the Sholkhovs' departure quite a few empty bottles were taken out

from under his bed." * It is difficult to imagine that the consequences of consumption of the contents of these bottles could pass unnoticed by the eyes and the noses of the Swedes who met him.

By the way, it was probably because he appreciated the conditions of staying in the embassy, which were very comfortable for him, that in the year of receiving the Nobel Prize he refused luxurious suites of the Grand-Hotel, where the Nobel laureates traditionally stay, and again took advantage of the Soviet ambassador's hospitality.**

* N. M. Lunkov *Russkii Diplomat v Evrope* (A Russian Diplomat in Europe) — Mocow: OLMA-PRESS, 1999, p. 89.

** G. Jarring. *Do Glasnosti i Perestroiki* (Before Glasnost' and Perestroika) — Moscow: Progress-Polotika, 1992, p. 181.

[14] K. Espmark. *The Nobel Prize in Literature. A Study of the criteria behind the Choices.* Boston, Mass., G. K. Hall and Co., 1991, pp. 106–107.

[15] Although Khrushchev mentions only Norway, Sholokhov must have supposed to visit Sweden as well. Anyway, when in 1963 he was at last given permission to go abroad, apart from Norway, he visited five more countries, including Sweden (note 1 to doc. 10.12).

[16] Sartre's numerous utterances, where he praised everything that was done in "the first country of the victorious socialism", showed not so much the ideologically blinkered outlook of a leftist intellectual successfully taking advantage of all benefits of Western democracy at once, as the fully conscious cynicism of a prosperous bourgeois. He did not have any delusions about the true situation of the creative intelligentsia in the Soviet Union and the realities of its existence in the conditions of the totalitarian regime. Sartre understood all that perfectly well and interpreted it more than adequately. It can be seen, among other things, from the episode told afterwards, according to A. T. Tvardovsky, by his close friend V. Y. Lakshin. It took place in the summer of 1963, when the next session of the European Writers Association was held in Leningrad. After the end of the session its foreign participants and several major Soviet writers were invited to the resort Pitsunda on the Black Sea, where N. S. Khrushchev was having a rest at that time. "Brought by a special plane, they found themselves in a luxurious residence, where dinner was on the table for them [...] After the first toasts Khrushchev suggested, unexpectedly for those present: 'It seems that Alexander Trifonovich has prepared something to read to us...' Tvardovsky, without putting on airs, started to read a long poem [the anti-Stalin poem *Terkin na Tom Svete* (Terkin in Another World)], and Khrushchev spontaneously reacted to the reading — now frowning, now laughing at the top of his voice, in a village manner. On the way back Tvardovsky, willing to be polite, asked Sartre if he was bored listening to the reading without translation. Sartre answered: 'Not in the least, it was a wonderful show! I watched the changes of Khrushchev's facial expressions and how the faces of your writers simultaneously changed'; there were several leaders of the Writers Union."*

* It is worth noting that Sartre, who shockingly emphasized his leftism, was, in his mother's line, a great nephew of the great humanist, invariably respectable Albert Schweitzer, the 1952 Nobel Peace Prize laureate, and in his childhood he lived with his mother in his grandfather's aristocratic home.

V. Y. Lakshin, *Otkrytaya Dver'* (the Open Door) — Moscow: Moskovsky Rabochi, 1989, p. 218.

[17] "Yesterday there was Sholokhov's celebration.* Strangely, his 50th birthday was more exciting and significant than the present jubilee. And he probably does not have anything against such festivities after, in fact, an idle decade. Over these years he said so many silly things and banalities and so many times kept silent when one should not have kept silent. Who and what he is now is difficult to say. Official glory beyond measure during his lifetime — and an evident ebbing away of genuine popularity, glory by inertia. When he dies he will be a great writer, but while alive he is some incomprehensible buffoon."**

It was not the first time that Tvardovsky mentioned Sholokhov's "idle decade". In connection with the ceremonial trip of Khrushchev and his retinue to Sholokhov's place in the Cossack village of Veshenskaya, in the summer of 1959, *Pravda*'s board suggested to him writing an article about his colleague, who was given such an honor. Tvardovsky refused and explained his refusal, judging by the entry in his diary, in the following way: "I would write it, but what I would write you won't publish [...] It's better not to do it, it won't do any good to our literature, nor to Sholokhov himself, who is writing worse and worse, nor to *Pravda*'s prestige. If we can't tell the truth, at least keeping silent will be nobler."*** See also note 1 to doc. 10.35.

B. L. Pasternak, on meeting the writer V. T. Shalamov, frankly expressed his opinion of Sholokhov. Deliberately not dwelling on the problem of the authorship of the great epic novel, he said that "the first part of *Quiet Flows the Don* is magnificent — strong and fresh. He has not written a single line more. He is very far from humanism, from love of man. Human cruelty — that is his undercurrent theme."****

* On May 24, 1965 a grand ceremonial meeting on the occasion of Sholokhov's 60th birthday took place in the Column Hall of the House of Unions in Moscow.

** A. T. Tvardovsky. Working notebooks of the 1960s. *Znamya* No. 12, 2001, p. 143, entry of May 25, 1965. That day the author of the entry was in hospital and learnt of the jubilee meeting from newspapers.

*** A. T. Tvardovsly. Working notebooks (1953–1960). *Znamya* No. 9, 1989, p. 151, entry of September 1, 1959.

**** *LG Dossier* February 1990, p. 14.

[18] The fact that Ulla Lindstrom did not mention Sholokhov among the Soviet nominees probably means that nobody nominated him for the 1964 prize; in any case, there must have been no mention of Sholokhov in the Swedish press, where Lindstrom could only pick up this information. The Soviet nominees mentioned by Lindstrom might have been K. G. Paustovsky and A. A. Akhmatova (see doc. 10.27–10.31).

[19] V. O. Osipov, p. 338.

[19a] G. Jarring. *Do Glasnosti i Perestroiki* — Moscow: Progress-Politika, 1992, p. 179.

[20] *Pravda* December 23, 1965, p. 4.

[20a] G. Jarring, pp. 181–182.

[21] See note 1 to doc. 10.23.

[22] In the same interview to the *Express* journal, answering the journalist who reminded him that "Sholokhov wrote a lot against you", Aragon said: "There, in his country, he was even told to stop it."* It was not only Sholokhov with whom Aragon had strained a relationship in the Soviet Union. For example, A. A. Akhmatova, in connection with the allegation of

the American literary critic Sam Driver that N. S. Gumilev did not value her as a poet, on February 30, 1965 wrote down the following: "The remark that Gumilev did not like my poems anymore is somewhat perplexing. By the rules of conduct of that time it just did not befit him to praise his wife's poems in the press. It was not like now, when Aragon has corrupted everybody; ** and besides, I would not have tolerated it."***

* RSAMH, fond 5, opis' 63, ed. khr. 146, l.54.

** It is about Aragon's articles in the French press giving laudatory references of the poetic work of his wife, Elsa Triolet.

*** Anna Akhmatova's notebooks (1958–1966). Torino, Einaudi editore, 1996, pp. 617–618.

[23] For many years rumor had it that when creating the first three books of the novel Sholokhov used someone else's manuscript. Here we deliberately did not touch upon this question. There is a considerable bulk of special studies which examine the involved story of the priority.* For example, a noteworthy discussion on the subject "*Quiet Flows the Don*: enigmas imaginary and real" took place, in the last years of Gorbachev's *perestroika*, in the journal *Voprosy Literatury;* taking part in the discussion were the Soviet authors R. A. Medvedev and S. H. Semanov, who represented opposing stands, and the American literary critic of Russian origin, professor of Princeton University G. C. Ermolaev.** There is no doubt that all arguments for and against were thoroughly analyzed by the Nobel Committee. The fact that the Swedish Academy's decision was positive definitely proves that the view of Sholokhov as the true author of the 20th century bestseller finally gained the upper hand, and there is no point in discussing these arguments again, at least not here. However, I cannot help mentioning an original, appealingly paradoxical thought expressed by the writer Andrei Bitov.

Writing about the importance of inner energy of a creator of large-scale literary works, Bitov said that "maybe now a boy is sitting somewhere and writing a novel of genius [...] By the by, the only thing that speaks in favor of the idea that Sholokhov could have written all that is his age. Because only a young man was able to do it — to give an authentic picture of the world not visible to anyone but the author. Doing that requires tremendous energy..."*** True, when setting forth this idea that occurred to him Bitov cunningly did not mention the main doubts about *Quiet Flows the Don* — that for creating such a novel, apart from being a young genius, one also must have had some specific information, which Sholokhov definitely could not have.

Nor will Bitov's logic find support in this case in a textbook example about A. Einstein, who, as early as at the age of ten, thought about the question of how the vision of what one sees around would change if one ran up to the speed of the wind. Actually, it was this question that put into his mind the foundations of the future theory of relativity. But in the context of the story of the origin of *Quiet Flows the Don* this example is certainly irrelevant. Putting together the fantasies of an intelligent child with the potential of a future genius, on the one hand, and the genius of description of everyday life, on the other, would be incorrect.

These and similar doubts have not so far received any worthy explanations even after the scientists from the Institute of the World Literature of the Russian Academy of Sciences were unexpectedly lucky to find, in 1999, the handwritten originals of the first and the second books of *Quiet Flows the Don*, which are most questionable in the mentioned meaning. Nor were there any convincing answers to the "anti-Sholokhov" doubts in the fundamental work by F. F. Kuznetsov, which was published in 2005 and contained the most detailed literary analysis of the discovered manuscripts.

Concluding the brief review of opinions about the painful problem of *Quiet Flows the Don*'s publication, we shall mention in passing the strictly uncompromising stand of two more people who studied the genesis of the 20th century great epic novel, the Makarov spouses. It should be noted that their article came out already after the publication of the book**** by Kuznetsov, who has no doubt about Sholokhov's authorship. "We must", say the Makarovs in conclusion, "restore the name of the great Don writer Fedor Kryukov in the historical and literary landscape [...] Sholokhov's right can only be recognized as the right of a co-author, compiler of *Quiet Flows the Don*. Our task is to convince the common reader of the falseness of the picture of the world which the scholars of Sholokhov support in order to protect their positions ..."*****

* See, for instance, Osipov, pp. 394–411, and also: J. Barron. *KGB. The Work of Soviet Secret Agents*. Tel Aviv, Effect Publications, 1988, pp. 141–144.

** *Voprosy Literatury* No. 8, 1989, pp. 148–222.

*** *Izvestiya*, June 7, 2000, p. 11.

**** F. F. Kuznetsov. *"Tikhii Don": Sud'ba i Pravda Velikogo Romana* (*Quiet Flows the Don*: the Fate and the Truth of the Great Novel) — Moscow: IMLI RAN, 2005, p. 864.

***** A. G. Makarov, C. E. Makarova. *"Tikhii Don" — Nereshennaya Zagadka Ruskoi Literatury* (*Quiet Flows the Don*: Unsolved Mystery of Russian Literature). *Svobodnaya Mysl'* No. 6, 2005, p. 187.

[24] *Stockholms Tidningen*, January 1, 1962, p. 4. The mention of Vera Inber's name in this context may be an indicator of her nomination for the Nobel Prize, but there is no more or less reliable information of that. Most probably, the journalist was mistaken, and it could be none other than Akhmatova. Incidentally, Akhmatova herself felt hostile towards Inber. According to the emigrant writer G. V. Adamovich, with whom the poetess met in Paris on her way back from Oxford, where, on June 5, 1965, she was given the mantle of the university's honorary doctor (see note 3 to doc. 10.30), Akhmatova spoke of Inber "not with hatred but, rather, with contempt"*.

In 2004, in his 103rd year, the legendary caricature artist Boris Efimov disclosed a sensational piece of information about Inber, who turned out to be the niece of Leo Trotsky, who was Lenin's associate during the 1917 October *coup d'état* and after the latter's death became Stalin's sworn enemy. Discussing the question of how Inber managed to remain alive in the period of Stalin's Great Terror, Efimov noted philosophically: "Most probably [...] he just decided that the poetess may still be of benefit." And further: "Having ruined my brother [Mikhail Koltsov, well-known journalist of the 1920s and 30s], he spared me. Maybe he thought: Efimov is rather good at drawing, so let him bring benefits..."**

* *Sovremennye Otechstvennye Zapiski* No. 2, 1999, p. 56.

** *Literaturnaya Gazeta* No. 16, 2004, p. 11.

[25] "Do you see the Kodryanskys?* They seem to be in Paris. He has sent me a letter with cuttings from French newspapers, *Le Monde* among them, about the Nobel Prize. If I receive it (which. of course, is pure fantasy), all right, I'll give you something exceptional..."**

Paustovsky's addressee Lidia Delektorskaya was herself a rather notable figure in the history of the 20th century culture. In the period from 1932 through 1954 she was at first the secretary, and then the civil wife of the outstanding French Impressionist painter Henry Matisse. Forced to emigrate by the troubles of the civil war, she always was a patriot of her country. Starting from 1965 she repeatedly gave to the Hermitage museum the works of Matisse that she owned. It is only thanks to her that the Hermitage now has 26 completed easel paintings by this artist. Among them there are two portraits of Delektorskaya, painted in 1939 and 1947.

* The husband and wife I. V. and N. V. Kodryansky were Paustovsky's Paris acquaintances. N. V. Kodryanskaya was a specialist in literature, she knew in person the Russian emigrant writers I. A. Bunin and A. M. Remizov, who lived in Paris. She first met Paustovsky in Moscow, where she came in July 1963 in hope of republishing in the Soviet Union her book on Remizov published in Paris in Russian, in 1959. Afterwards their relashionship was supported by their mutual interest in Bunin's work. In 1947 the latter very highly regarded Paustovsky's short story *The Tavern on Braginka*; the facsimile of the postcard sent by Bunin to Moscow on September 15, 1947 can be seen in *Literaturnoe Nasledstvo* (Literary Inheritance), vol. 84, book 2, 1973, p. 405, and in the same book, pp. 341–362 there are N. V. Kodryanskaya's memories *Meetings with Bunin*. More details about Bunin's postcard sent to Paustovsky to Moscow, to the address of the Writers Union can be found in note 3 to doc. 10.27.

** K. G. Paustovsky, Collected Works, vol. 9 — Moscow: Khudozhestvennaya Literatura, 1986, p. 437.

[26] The stir around Paustovsky in Rome was vividly described in the diaries of other members of the Soviet delegation at the congress, K. I. Chukovsky and A. T. Tvardovsky. Chukovsky: "In Italy Paustovsky was intensively photographed for newspapers in connection with the rumors of his nomination for the Nobel Prize." * Tvardovsky: "The trip to Rome [...] The blissfully dissolved Paustovsky** expecting the Nobel Committee decision***, which seemed already known but not yet officially announced."****

* K. I. Chukovvsky. *Dnevnik 1930–1969* (Diary) — Moscow: Sovremennyi Pisatel', 1997, p. 381.

** In the epithet used here one can definitely feel the underlying irritation with the fuss around the colleague and even emphasized disdain for him. The fact that such an attack, which showed the author's ambivalent attitude to Paustovsky, was not accidental is confirmed by his diary entry made a day after the writer's death, on July 14,1968: "Today Paustovsky was favored with an obituary — in *Pravda*, with a portrait, in *Izvestiya*, without any. He is certainly a bad writer, but still he is a man of old literary faith, who had a reverential attitude towards it..." (*Znamya* No. 9, 2003, p. 147). The spark that started a conflict between the two key figures of the Soviet literature of that time was the incident when Paustovsky had to take away from the *Novy Mir's* office the manuscript of his tale *Vremya Bol'shikh Ozhidanii* (The Time of Great Expectations) describing life in Odessa in 1920–21. Tvardovsky, in his letter of November 26, 1958 insisted that Paustovsky brought some changes into the manuscript. These were mostly demands that would have been approved at the CPSU Central Committee, especially on including in the text "some kind, informal words about working people", "resolutely making shorter the apologetic story about Babel" and "removing 'the dispute about homeland' in the Writers Creative Work Home in Peredelkino." (*Pis'ma o Literature* (Letters about Literature) 1930–1970, Moscow: Sovetckii Pisatel', 1985, pp. 145–146). Paustovsky vehemently rejected these suggestions and demanded, on December 7, 1958, "to send immediately two copies of my manuscript to my Moscow flat" (Collected works, vol. 9, Moscow, Khudozhestvennaya Literatura, 1986, pp. 351–353). Shortly after the tale was published in the *Oktyabr'* journal. No. 3–5, 1959.

*** It means the decision of the Swedish Academy, the institution endowing the prizes for literature.

**** A. T. Tvardovsky. *Rabochie Tetradi 60-kh Godov* (Working notebooks of the 60s). *Znamya* No. 2, 2002, pp. 153–154.

[27] *La Fiera Letteraria* October 3, 1965, p. 8.

[28] Paustovski, *E'ancora giorno.* Milano. Nuova Accademia, 1965.

[29] That is a fact from the letter of the director of Paustovsky's museum in Moscow's Kuz'minki Park I. I. Komarov written in reply to the German specialist on literature and Slavonic scholar W. Kasack, who asked to inform him of the museum's materials connected with the writer's nominations for the Nobel Prize.* There is no date on the copy of the letter, but it certainly had been written before September 1997.** Convincing evidence of the writer's nomination for the 1965 Nobel Prize is also given by the authoritative encyclopedic edition of Russian literature prepared and published by Yale University in New Haven, Connecticut. ***

* From the fund of Paustovsky's museum in Kuz'minki. Park (Moscow).

** In the September issue of *Zeitschrift fur Literaturwissenschaft und Lingwistik,* H.107, 1997, published by the university in Siegen (Germany), in the article "Russian Laureates of the Nobel Prize" Kasack refers to the already mentioned letter from Komarov (pp. 54–55).

*** *Handbook of Russian Literature,* Edit. V.Terras, Yale University Press, New Haven and London, 1985. p. 333.

[30] *Literaturnoe Obozrenie* No. 11–12, 1992, pp. 43–44.

[30a] G. Jarring. *Do Glasnosti i Perestroiki* — Moscow: Progress-Politika, 1992, p. 176.

[31] See also *Literaturnaya Gazeta* No. 49, 1999, p. 10.

[32] From Akhmatova's note of July 16, 1962 in Komarovo. "The Swede came.* He amazed me by his unfading interest in the poem** and the fact that he found *The Queen of Spades* there [...] When saying goodbye, he said: 'Panova's son will tell you [...]' Two days later Panova's son (28, sinologist) turned up at the Gitovichs. I came there to have supper. Gitovich and his guest were drinking vodka. Alexander Yl'yich was already completely drunk and wandering in some countries of his. The guest was absolutely sober. They drank for me, and the guest said: 'Erik Mesterton asked me to tell you that this year you are nominated for the Nobel Prize.' I only wonder: why did not Erik tell me the news himself?"***

* Mesterton.

** *The Poem without a Hero* by A. Akhmatova.

*** *Zapisnye Knizhki Anny Akhmatovoy* (Anna Akhmatova's Notebooks).(1958–1966), Torino, Einaudi editore, 1996, p. 235.

[33] *Zapisnye Knizhki* ...(Anna Akhmatova's Notebooks), p. 112.

[34] The *vers libre* was first published in 1996: *Anna Akhmatova. Sochineniya v Dvukh Tomach* (Works in Two Volumes), vol. 2 — Moscow: Tsitadel, 1996, p. 76.

[35] *Zapisnye Knizhki...* p. 241. N. V. Koroleva, compiler and commentator of Akhmatova's collected works, interprets the abbreviation 'N' as "Nobel Prize" and "port.radio", as a portable radio. *

* A. A. Akhmatova. *Sobranie Sochnenii v Shesti Tomakh* (Collected Works in Six Volumes), vol. 2, book 2 — Moscow: Ellis Lak, 1999, p. 413 (comments).

[36] The West slandered and believed itself,
And the East luxuriously betrayed,
The South doled air for me stingily,
Grinning from behind clever lines.
But the clover stood as if on its knees,
The damp wind blew into a horn of pearl,
Thus my old friend, my true North
Comforted me as well as it could.
I languished in stifling lassitude,
I suffocated in stench and blood,
I couldn't bear this house anymore...
That's when iron Finland *
Declared: "You will know everything, except
Joy. Even so, live!" **

In the original, this sonnet, written on June 30, 1963 in Komarovo, near Leningrad, has the title with dedication to Erik Mesterton: "And at last you said the word. E. M."

In November 1965, shortly before her death, Akhmatova in Botkin Hospital returned again to the text of the sonnet. She rewrote it once more into her notebook and, in comparison with the text written on June 30, 1963, brought in several changes that certainly strengthened the sense of the poem. ***

* This line implies that the *dacha* village of Komarovo, where the Literary Fund gave Akhmatova a *dacha,* in which she lived not only in summer, is situated on part of the Karelian Isthmus, which until 1940 was owned by Finland. In her poems and notes she repeatedly remembered that her abode used to belong to the territory of the neighboring state: for instance, in the entry of September 11,1965: "It is raining — a good Finnish rain." (*Zapisnye knizhki...* p. 669.)

** *The Complete Poems of Anna Akhmatova.* Translated by Judith Hemschemeyer. Somerville (Mass.), Zephyr press, 1990, p. 713.

*** *Zapisnye Knizhki....* p. 716.

[37] The first poetic translation of the sonnet into Swedish, under the title *Finsk Sonnet,* was done by the poet Hans Bjorkegren in 1989 and published in the collection of poems of Russian poets *From Derzhavin to Brodsky.** A line by line, thoroughly accurate rendering of the text of the sonnet in Swedish, from its title to the mention of the changes in several lines brought in by the poetess in November 1965 was provided in 2001 by the expert on the work of Akhmatova, literary critic Johan Oberg.**

Here it is appropriate to say a few words about Oberg himself — the cultural councilor of the Swedish embassy in Moscow in 1994–1999, a courageous man, ready to take extraordinary steps if necessary. In 1989 in Chardjou the popular poet Shurali Nurmuradov was sent to prison for epigrams on the first secretary of the Central Committee of the Communist party of Turmenistan, C. Niyazov. According to the director-general of the Russian P. E. N.-Club A. P. Tkachenko, at that time a worker of its Moscow branch, the poet's friends informed them that the murder of the prisoner was being prepared and asked to interfere. With the help of the central authorities it became possible to transfer the poet into a prison in the Russian city of Penza and to release him half a year later. In the middle of the 1990s he was arrested again, now already in Moscow, where he had settled after coming back from Penza. He was arrested for the alleged illegal storage of drugs, which had been planted on him, and imprisoned in a town near Moscow. "It was a provocation of Turkmen authorities," Tkachenko said "... Johan Oberg and I went there and pulled Shurali out 'on bail' [...] Then he was sent to Sweden, where he has been living since."*** Indeed, it was both an unusual and worthy action on the part of a diplomat in the country of his stay!

* *Rysk dikter. Fran Derzjavin till Brodsky*. Stockholm, 1989, pp. 246–247. It must have been unknown to the translator to whom and on what occasion Akhmatova dedicated her sonnet. Otherwise he would not have changed the revealing author's title to his bland word combination "Finnish Sonnet'".

** *Skonhet men Jamval fornuft. Festskrift till Svezker Goransson*. Stockholm, 2001, p. 23.

*** *Voprosy Literatury* No.1, 2001, pp. 157–158.

[38] Andreas Osterling and Erik Mesterton, both form Sweden

That is not the only mention of Osterling in her notebooks. One more entry, made earlier, in January 1963, is made by someone else's hand, certainly foreign, which fills in a letter missing in the word "professor" and in the clumsy attempt to give the sounding of Osterling's name with the help of the Russian transcription.* It can be added that under the note about Osterling there is a phrase from J. Joyce's *Ulysses* made by the same hand.

* *Zapisnye Knizhki...* p. 557.

[39] *Ord och bild* No. 4, 1990.

[40] Having met, on January 21, 2000, with the permanent secretary of the Swedish Academy, Horace Engdal, this author told him about what Mesterton had told Vakhtin. "But he had no right to speak about that!" Engdal exclaimed. However, it is worth noting that such leakages of information still take place, especially among those who have something to do with the prizes for literature and for peace. With the prizes for natural sciences such breaches of confidentiality happened much more seldom, although, especially in the first years of the existence of the Nobel institutions, there were such cases, some of them looking rather funny. For instance, after the committees sent the next invitations to nominate their candidates for the 1905 prizes, a piece of information appeared in the international press, which was published, on 28 September 1904 by the Petersburg newspaper *Novoe Vremya*. It pompously announced that one of the professors of Prague University had received an invitation to nominate his candidate for the prize in physiology and medicine.*

* "The Committee of the Nobel Prize addressed the professor of the medical faculty of Prague University doctor U. Tomaer with the request to point out a candidate for receiving this prize for achievements in the area of medical knowledge."

[41] The practice of preliminary reconnaissance, from the point of view of the aspirant's personal interests — whether the prize awarded will not do him any harm, what is his physical condition and so on — is usual for the Swedish Academy. In his conversation with this author in Goteborg, on January 17, 2000, the 96-year-old Mesterton remembered that in 1958, after his return from Moscow, where he met with Pasternak in summer and tape-recorded his voice, Osterling asked him if, taking into consideration the political atmosphere in the Soviet Union, it was worth awarding the Nobel Prize to the poet. Mesterton answered in the affirmative, and his opinion certainly played a positive, and probably even decisive role in taking the final decision.

[42] L. K. Chukovskaya. *Zapiski ob Anne Akhmatotvoi* (The Memories of Anna Akhmatova), vol. 3 — Moscow: Soglasie, 1997, p. 83. At the same time the poet A. G. Naiman, touching

upon the same question, noted that both during his meeting with Akhmatova in England in June 1965 and in his later memories Isaiah Berlin "strongly denied" his taking part in her nomination for the Nobel Prize.* True, if we take into consideration the requirements of confidentiality existing in the Nobel institutions, his assurance should not be too strongly relied upon. We will not be able to learn the truth until the beginning of 2016. Nonetheless, Akhmatova had some grounds for such guesses. Among other things, the title of the honorary doctor of Oxford University was awarded to her in 1965 at the initiative of Isaiah Berlin.** And he himself, in one of his talks with Naiman, referred Akhmatova, without much thought, to the Silver Age's poets of genius: "...Blok was a genius. A real genius [...] Pasternak, *par excellence.* Mandelshtam was a genius. Akhmatova."***

* A. G. Naiman. *Rasskazy ob Anne Akhmatovoi* (Stories about Anna Akhmatova) — Moscow: Khudozhestvennaya Literatura, 1989, p. 199.

** *Oktyabr'* No. 11, 2000, p. 53.

*** *Oktyabr'* No. 12, 2000, p. 85.

[43] K. Espmark. *The Nobel Prize in Literature. A Study of Criteria behind the Choice.* G. K. Hall & Co, Boston, Mass., 1991, pp. 125, 159,161, 192.

[44] The last time before 1966 that the Nobel Prize was divided into two parts was in 1917, between the Danish writers Karl Giellerup and Henrik Pontopiddan.

[45] That Asturias had second place on the reserve list made by the Nobel Committee was also reported by the Italian newspaper *La Fiera Letteraria,* but in her version it was Paustovsky who was the "grand favorite" and not Sachs, as in Espmark's rendering. However, this difference enables us to make a conclusion that awarding the prize to Sholokhov perhaps was not planned in the committee that year.

[46] R. Vlach visited Akhmatova in the Soviet Union some time between September 27 and October 14, 1964, which is proved by his American address put down in her notebook by his own hand.* After that he continued to communicate with her by correspondence; in the small hours of January 5, 1965 Akhmatova writes down, under the heading "To bibliography" the title of his article in the Polish language.** We can suppose that he sent her the off-print of his article together with New Year wishes.

* Robert Vlach. *Books Abroad.* University Norman, Oklachoma, U.S.A.; Zapisnye Knizhki..., p. 492.

** *Vlach Robert: Z Nove Tvorly Achmatova. K Sesim 75 Narozenian* (From New Works of Akhmatova. To her 75th Birth Anniversary). ibid, p. 504.

[47] Among those passed over by the Nobel Prize, Espmark also named the German writer Gunther Grass; in 1999 he became the Nobel laureate. Of those writing in Russian. apart from Akhmatova, Espmark mentioned V. V. Nabokov. Concrete information about the nomination of the latter could be found in the book by Stacy Schiff, Nabokovs' long-standing friend. As nominator she confidently names the professor of the University of California, expert on Russian literature Simon Carlinsky, who later was the scientific editor of Nabokov's epistolary heritage. According to her, Carlinsky first nominated Nabokov for the award in January 1969, then repeated the attempt once again. In her memories she gave Nabokov's words said at table to one of his guests. He said that even if he shared the prize with Borges,* he would get 500 thousand dollars.** Against the background of such ingenuous calculation of the possible income, like counting one's chickens before they are hatched, serious doubts

should be raised by the information of the writer G. N. Vladimov, dated by 1980, without giving any source of this information. Allegedly Nabokov offered, in 1970, to withdraw his candidacy in favor of Solzhenitsyn, persecuted in his homeland. But Nabokov was too interested in calculating his share in case the award had to be shared with Borges.***

* The Argentine writer Jorge Luis Borges was also mentioned by Espmark in his list of the 15 most worthy candidates for the Nobel award of the 1950s–70s.

** S. Schiff. *Mrs. Vladimir Nabokov. Vera.* Moscow., Nezavisimaya Gazeta, 2002, p. 438.

*** *On Mezhdu Nami Zhil... Vospominaniya o Sakharove* (He Lived Among us. Memories of Sakharov) — Moscow: Praktika, 1996, p. 418.

[48] A. G. Naiman. *Rasskazy ob Anne Akhmatovoi* (Stories about Anna Akhmatova) — Moscow: Khudozhestvennaya Literatura, 1989, p. 105.

[49] There were strong suspicions among Soviet and foreign intellectuals about Sholokhov's authorship in creating the first three books of *Quiet Flows the Don* (see doc. 10.33-I).

[50] *Izvestiya* June 22, 2002. p. 8.

[51] *Druzhba Narodov* No. 8, 1968, p. 227. The given quotation describes the secretary of the board of the Soviet Writers Union and at the same time the first secretary of the Union's Moscow branch S. V. Mikhalkov, delegated as the Union's official representative at Akhmatova's funeral, which took place on March 10, 1966 in Komarovo (doc. 10.32).

[52] *Russkaya Mysl'* (Paris), May 25,1990, p. XI.

[53] Quite a few colleagues of the accused gave evidence at the trial confirming the charges against them. The trial's well-directed scenario went astray only once. The professor of the philological department of Moscow University V. D. Duvakin, once a teacher of A. D. Sinayvsky, when summoned to the court session as a witness said about him "what he would have said at home, at the chair or in a classroom [...] The judge had to stop the witness." As a result the members of the department's academic council "condemned him and demanded that he be dismissed as not corresponding to the position he held [...] — and that after twenty-seven years of faultless work."* The dismissal was done by the university authorities.

* V. V. Radzishevsky. *Beskonechnyi Viktor Dmitrievich.* (The endless Viktor Dmitrievich). In the book: *M. M. Bakhtin. Besedy s Duvakinym* .(The talks with Duvakin) — Moscow: Soglasie, 2002, pp. 9–10.

[54] See: N. Y. Rapoport. *To li Byl', to li Nebyl'* (Whether True or Invented), St. Petersburg: publishing house of Pushkin Fund ,1998, p. 116.

[55] It must be said that an antithesis as sharp can be provided by juxtaposition of the personalities of Steinbeck, who won the award in 1962, and Sholokhov, who was the most probable aspirant to it. Isaiah Berlin, who knew Steinbeck well, said in his talk with Anatoly Naiman that on hearing that he had been awarded the Nobel Prize Steinbeck exclaimed: "I do not deserve it."* It is truly impossible to imagine Sholokhov saying such words.

* *Oktyabr'* No. 11, 2000, p. 11.

[56] Here is, for example, a vivid picture of everyday life in a "communal" flat (a flat where several families lived, which was common in the Soviet Union) given by Viktor Nekrasov in his tale *Saperlipopet*: "[...] with half a dozen primus stoves in the kitchen, with separate light bulbs over kitchen tables and in the lavatory (looking at a grape of bulbs hanging in the ante-room, my friend said: "The Grapes of Wrath"), a pile of wash-tubs and other trash in the corridors [...]"*

* V. P. Nekrasov. In the book *V Samykh Adskikh Kotlakh Pobyval* (I have been in most hellish cauldrons) — Moscow: Molodaya Gvardiya, 1991, p. 290.

Document 10.1

From the head of the Cultural department of the CPSU Central Committee D. A. Polikarpov to the Central Committee's Secretariat, of April 23, 1959

According to TASS, the French bourgeois newspaper *France Soir* published an interview of its correspondent with Sholokhov staying in Paris. The interview says that answering the question about "the case of Pasternak" Sholokhov stated: "The Soviet Writers Union's collective leadership lost presence of mind. They should have published Pasternak's book *Doctor Zhivago* instead of banning it.[1] Pasternak should have been defeated by his readers instead of putting him to discussion. If they had acted in such a way, our readers, who are very demanding, would have long forgotten him. As for me, I think that Pasternak's work does not have any importance except for his translations, which are brilliant (Pasternak has translated Gothe, Shakespeare and works of major English poets)."[2] As for the book *Doctor Zhivago*, which he read in Moscow in manuscript, "it is a shapeless work, amorphous mass which does not deserve to be called a novel."

In connection with that, I would consider it necessary to charge the Soviet ambassador in France with checking the authenticity of the information of the *France Soir*, and if such an interview really took place, to draw the attention of M. Sholokhov to the fact that such statements contradicting our interests are inadmissible.[3] If the newspaper's information proves to be false, it should be recommended to comrade Sholokhov that he disprove it publicly.

RGANI, fond 5, opis' 36, ed. khr. 93, l.25.

[1] Even before Pasternak was awarded the Nobel Prize, the Writers Union suggested to the CPSU Central Committee publishing *Doctor Zhivago* urgently in a small number of copies "so that the Western press was devoid of opportunity to make a fuss about publication of a work banned in the USSR, but the workers of the Cultural Department [...] found this suggestion inexpedient."*

* *Continent*, 95, No.1, p. 238.

[2] Two days after Polikarpov wrote this memo, on April 25, 1959, a small article by George Adan about his talk with Sholokhov appeared in the Paris weekly *Figaro Littéraire*, in which the writer once again expressed his attitude to Pasternak's poetry, doing it even more sharply than before: "My attitude to him has long been negative. Even to his poetic work. But he is a magnificent translator; Georgian writers, among others, owe their popularity in Russia to him."*

* *"A za Mnoyu Shum Pogoni." Boris Pasternak i Vlast'*, 1956–1972 ("And the Hunters Are Gaining Ground." Boris Pasternak and the Power) — Moscow: ROSSPEN, 2001. p. 260.

[3] This statement of the head of the Cultural Department is revealing: for his superiors he cynically emphasizes “our interests”, not saying whether Sholokhov was right or wrong in his answers to the questions of the French journalist.

Document 10.2.

From the memo of the deputy editor-in chief of the Pravda newspaper, journalist Y. A. Zhukov to the Central Committee: “The utterances of foreign guests about the speech of M. A. Sholokhov at the Writers Congress on December 21, 1954.

After M. Sholokhov’s speech, on the evening of December 21 and the morning of December 22, the writers Anna Seghers (GDR), Luis Aragon (France) and Jorge Amado (Brazil) in their personal talks with me expressed their indignation at this speech.

Anna Seghers said, with the sharp straightforwardness inherent in her: “Your Sholokhov is a big talent but also a big boor. His speech is a cheap stuff. Seeking scandalous popularity and only thinking how to get applause by his scoffing, he had no wish to realize what response it could cause all over the world... When Sholokhov rose to the rostrum, everybody thought that he was going to say something important, but he spoke as a buffoon. I spoke with many foreign writers who were present at the congress, and Sholokhov’s speech left all of them with a distressing impression.”

Louis Aragon said: “I was ashamed for the talent of this man when listening to his demagogical speech. The man who has not written a single book in 20 years[1] has no right to lecture writers whose creative work is very intense, and to do that in such inadmissible tone at that. He sought applause with the help of easy tricks and obtained it....”[2]

Jorge Amado said: “Sholokhov spoke about his colleagues with the bitterness of a failure, as if he himself was a goner in literature. Hatred for writers showed through his speech. One could think that he feels envious of those who wrote so many books in the years when he kept silent. One cannot help thinking of that because he hurled without any hesitation grave accusations not only against separate writers but against the whole of literature. Especially incomprehensible are his dark and not completely revealed attacks against women writers.”[3]

Amado emphasized that he deeply respected Sholokhov as the author of world-famous books, which are popular in Brazil as well. “The more was I upset by this demagogical speech at the writers’ congress”, he said.

RGANI, fond 5, opis’ 30, ed. khr. 84, l.95–98.[4]

[1] That is not quite true: the fourth book of the novel *Quiet Flows the Don* was finished and published in 1940.

[2] Nearly two years later, in an article published on August 14,1956 in the Paris newspaper *Lettres Françaises*, Elsa Triolet, Aragon’s wife, added some more arguments to what her husband had said. The cause was Sholokhov’s speech at the XX CPSU Congress,

in which he once again expressed contempt of his colleagues at the Writers Union.* "This part of his speech", Triolet wrote, "upsets me, as it hampers our efforts here to gain understanding both with our friends and our adversaries [...] Denying the role of talent, he raises doubts whether the Soviet literature has talents."**

* The cause was given by the following words of Sholokhov: "The Soviet Writers Union has 3773 members... As you can see, it looks a great force. But .... it only 'looks', while in fact the list of members consists of 'dead souls'... It must be said directly that Soviet literature became leading not because it attained the tops of artistic perfection which had earlier been unattainable for writers, but because all of us promote the victorious ideas of Communism. That is what makes us successful!" *Pravda, February 21, 1956, p. 8.*

** See: *Lilya Brik — Elsa Triolet. Neizdannaya Perepiska* (Unpublished Correspondence) (1921–1970) — Moscow: Ellis Lak, 2000, p. 244.

[3] See also doc. 10.23.

[4] Perhaps it should be noted that in a substantial part of Soviet society Sholokhov's utterances met with a diametrically opposite reaction. According to the information of the Department of Science and Culture, prepared for the CPSU Central Committee three weeks later, on January 16, 1955, the writer F. V. Gladkov, who publicly condemned Sholokhov's speech, received after that quite a few letters threatening him, which made "a depressing impression on him". After the congress, at the meeting with some of its participants organized by the Lenin Military and Political Academy, the students actively supported those who defended Sholokhov and did not sympathize with those who condemned him.*

* RGANI, fond 5, opis' 17, ed. khr. 534, l.12.

Document 10.3

From the General Secretary of the Soviet Writers Union V. P. Stavsky to I. V. Stalin, of September 16, 1937

In connection with alarming reports about the conduct of Mikhail Sholohov I visited him in the Cossack village of Veshenskaya. [...] So what is the situation surrounding Sholohkov in Veshenskaya? Arrested three months ago was the former secretary of the Veshenskaya district party organization Lugovoi, the closest political and personal friend of Sholokhov. Arrested before that and after that were a number of district workers (Krasyukov, Logachev and others), all of them being charged with belonging to a counter-revolutionary Trotskyist organization. M. Sholokhov told me directly: "I don't believe that Lugovoi is guilty. If he is convicted, it means that I am also guilty and will also be convicted. Because we did everything together in the district."... M. Sholokhov said with bitterness that the district NKVD[1] worker folllows him and collects all sort of gossip about him and his relatives... The authorities of the *krai* (Rostov-Don) have an extremely suspicious attitude towards Sholokhov..."[2]

On the back of the last sheet of the document there is Stalin's resolution: "Comrade Stavsky! Try to summon comrade Sholokhov to Moscow for about two days. You can refer to me. I'm not averse to talking with him."

Sholokhov was received by Stalin on September 25, 1937; V. M. Molotov and N. I. Ezhov were present at the talk.

AP RF, fond 45, opis'1, ed. khr. 827, l.32–36, published in the *Istochnik* journal No.4, 1993, pp. 5–7.

[1] NKVD was the previous name of the government security body later to be called KGB.

[2] The talk with Stalin brought results. As Sholokhov informed him on October 5, 1937. "comrade Ezhov [...] is today bringing up the question of releasing Lugovoi and Krasyukov at the session of the Central Committee."*

* *Istochnik*, p. 7.

Document 10.4

Information in the Literaturnaya Gazeta in 1946.

M. Sholokhov is Nominated for the Nobel Prize.

On December 10 the next awarding of the Nobel Prize is taking place.[1] In the liberal circles of the Swedish intelligentsia, writers included, the question has been brought up many times that the Nobel Prize has never been awarded to representatives of Soviet science and culture. In the field of literature repeatedly nominated over the recent years has been the candidacy of M. Sholokhov, the writer who is well known and much liked in Sweden.[2] This year, expressing the opinion of Sweden's radical circles, the renowned Swedish poet and journalist Erik Blomberg nominates Sholokhov again[3] and publishes in the *Ny Dag*[4] a series of articles on Solokhov's work. M. Sholokhov, "like nobody else, is worthy of the Nobel Prize, which should be awarded not only for artistic merits but also for the ideological content",[5] writes E. Blomberg.[6]

Literaturnaya Gazeta, October 19,1946, p. 3.

[1] On December 10 the Nobel Prizes are annually given. In 1946, their awarding took place in October and November. For instance, the prize for literature was awarded to the Swiss writer German Gesse on November 14, 1946.

[2] This assertion corresponded to the facts; see doc. 3.9.

[3] The information that Blomberg "nominates Sholokhov again" did not correspond to the facts. The first nomination of Sholokhov was registered in 1947, the nominator being the member of the Swedish Academy H. Ollsson.

[4] The organ of the Swedish Communist Party.

[5] Such interpretation of requirements set down for the works aspiring for the Nobel award does not correspond to the text of Alfred Nobel's will, nor to the regulations of the Nobel Foundation's charter. Neither of these documents mentions "the ideological content" as a prerequisite for receiving the award.

[6] It appears that in the cited newspaper information the only true fact is that of the *Ny Dag* publication of a series of articles about Sholokhov's work, written by the historian of art and lyrical poet E. Blomberg, and his personal opinion that the writer is worthy of the Nobel Prize. The rest is the fruit of imagination of the anonymous author of the publication.

Document 10.5

From M. A. Sholokhov to I. V. Stalin, of July 29, 1947

Dear Iosif Vissarionovich! For republishing my books in England, America and Sweden foreign publishing houses have transferred some amount of foreign currency to my account at the Vneshtorgbank. In order to use the money I addressed comrade Zhdanov this winter[1] for permission to go to these countries. Then Zhdanov advised me not to go to these countries, and the subsequent events confirmed that his advice was right.[2]

Last time, with your permission, I was abroad in 1935 [...][3] Other writers, my comrades — Fadeev, Simonov, Ehrenburg, Gorbatov, Surkov, Kozhevnikov and others — have visited many countries after the war. In the last years I did not think about any trips as I worked hard on the novel *They Fought for Their Country,* and the work was difficult. Now, when I am already completing it, I would like, if it is possible, to go with my wife to Sweden, for a short while, using for the trip the money due to me for the translations.[4]

I ask you to permit me this trip. I have not seen you for five years, but I did not dare to ask you to receive me about such a petty question, so I am addressing you with this letter. Now I am in Moscow expecting your decision. I ask you to inform me of it through comrade Poskrebyshev.[5], who knows my address. Always yours. M. Sholokhov.[6]

The note in the bottom right corner of the sheet: "Archive.8. X. 47"
Published in the book by Y. Murin *Pisatel' i Vozd' Perepiska*
M. A. Sholokhova s I. V. Stalinym (The Writer and the Leader)
Correspondence of M. A. Sholokhov with I. V. Stalin) — Moscow: Raritet,
1997, p. 138; by the original from APRF (fond 45, opis' 1, ed. khr. 827).

[1] A Secretary of the Central Committee of the Communist party, the head of its Secretariat's office.

[2] The winter of 1947 brought to the country the fabricated "Klyueva–Roskin case" and the hysteria of an anti-espionage campaign coming together with it (see Chapter 6), which was accompanied by an almost complete rupture of scientific and cultural relations with the Western world, first of all, with the US and Great Britain.

[3] In January 1935 Sholokhov visited Denmark and Great Britain. The grounds for going abroad was given by the decision of the Political Bureau of the Communist Party's Central Committee, which was taken, on June 15, 1934 "by questioning the members of the Political Bureau" and signed by the Central Committee's general secretary Stalin.*

* The Russian State Archive of Social and Political History (RSASPH), fond 17, opis' 3, ed. khr. 947, l.11 (record 9, point 47).

[4] Sholokhov must have had the information of his being nominated for the 1947 Nobel Prize. He considered it expedient to support this nomination by personal impressions of voting academicians and members of the Nobel Committee from his stay in Sweden — from his meetings with Swedish readers, newspaper comments which are always published during such visits and which he expected to be positive, etc.* The date of the letter commented on here is also not accidental: the conclusive stage of considering the submitted nominations in the Nobel Committees is in August to September of the corresponding year.

* See note 11 for this chapter.

[5] Stalin's personal assistant.

[6] Stalin did not answer the letter, although he read it; it is shown by the note of the document being directed to the archive.*

* See also V. O. Osipov *Tainaya Zhizn' Mikhaila Sholokhova (*The Secret Life of Mikhail Sholokhov) — Moscow: Libereya, Raritet, 1995, p. 262.

Document 10.6

Material about the nomination of Sholokhov for the 1954 Nobel Prize by S. H. Sergeev-Tsensky.

I

S. N. Sergeev-Tsensky's telegram to the Nobel Committee for literature, from Simferopol, of February 27, 1954.

In reply to your kind offer to nominate a candidate for the 1953 Nobel Prize for Literature, I have the honor to inform you that I am nominating the outstanding writer of our time, Mikhail Alexandrovich Sholokhov. The written motivation is being sent by post. Full member of the Academy of Sciences, Sergeev -Tsensky.

The Archive of the Swedish Academy, Stockholm. The Foundation of the Nobel Committee for 1955.

II

From the secretary of the Nobel Committee for Literature Hugo Villars to S. N. Sergeev-Tsensky, of March 6, 1954.

Dear Mr. Sergeev-Tsensky,
The Nobel Committee of the Swedish Academy took interest in your proposal to award the Nobel Prize to Mikhail Alexandrovich Sholokhov. As proposals should be received by us before February 1 (see the regulations attached) your proposal reached us too late to be discussed this year. However, Sholokhov will be nominated as candidate for the Nobel Prize of 1955.

Expressing the Committee's gratitude for our proposal let me assure you of our great respect.

ibid

III

From the Secretary of the Soviet Writers Union Board A. A. Surkov to the Secretary of the CPSU Central Committee M. A. Suslov, of March 20, 1954.

Today the relatives of the academician Sergeev-Tsensky passed to us by telephone the answer of the Nobel Committee to the proposal about awarding the Nobel Prize to M. A. Sholokhov, which was addressed to Sergeev-Tsensky.

RGANI, fond 5, opis' 17, ed. khr. 486, l.67.

Document 10.7

From the head of the 3rd European department of the USSR Ministry of Foreign Affairs S. Lapin to the chairman of the Foreign Commission of the Soviet Writers Union B. N. Polevoy, of June 23, 1955.

According to the information from the USSR embassy in Sweden, the Swedish Committee for awarding Nobel Prizes for Literature intends to nominate the candidacy of the Soviet writer M. Sholokhov for the 1955 Nobel Prize. For this purpose the Committee needs works by M. Sholokhov in one of the European languages — English, French or German, and also materials about the writer's work, preferably also in foreign languages. If there are no objections to nominating Sholokhov for the Nobel Prize,[1] the USSR embassy in Sweden asks you to send the books and materials as soon as possible.

The Archive of Foreign Policy of the Russian Federation (AVP RF), fond 0140, opis' 44, papka 60, ed. khr. 23, l.35

[1] This cautiously expressed doubt is rather symptomatic. It is certainly connected with Sholokhov's regular periods of heavy drinking, which were no great secret.*

* See note 12 to this chapter.

Document 10.8

From the deputy head of the 3rd European department of the Ministry of Foreign Affairs M. Gribanov to the USSR Ambassador Extraordinary and Plenipotentiary in Sweden K. K. Rodionov, of August 20, 1955.

At the request of the Secretariat of the Board of the Soviet Writers Union, I am sending you the books by M. Sholokhov *Quiet Flows the Don* in the English and German languages and *Virgin Soil Upturned* in German,[1] and also bibliographical information about the publications of works by M. Sholokhov abroad in translations into foreign languages. The translations are mentioned in the publications from 1929 to 1955. The books not included in this information are in the attached book *Works by Soviet writers in translations into Foreign Languages. Separate Foreign Publicatins 1945–1953*, pp. 258–262 and 319–320. The books and the bibliographical information are meant for passing to the Swedish Committee on Nobel Prizes.[2]

AVP RF, fond 0140, op[is'44, papka 60, ed. khr. 23, l.49

[1] By 1955 only the first book of *Virgin Soil Upturned* had been published. The second book completing the novel did not come out until 1960. In the opinion of the director of the Russian Institute of Stockholm University N. O. Nillson, only the incompletion of *Virgin Soil Upturned* was the main obstacle for the Swedish Academy passing the verdict in favor of Sholokhov (doc. 10.9).

[2] The author means the Nobel Committee for Literature.

Document 10.9

From the head of the department of Scandinavian countries of the Ministry of Foreign Affairs K. K. Rodionov to the chairman of the Foreign Commission of the Soviet Writers Union B. N. Polevoi, of August 30, 1958.

.....The Director of the Russian Institute in Stockholm Oke Nilsson said in his talk with the third secretary of the USSR embassy in Sweden comrade Rymko, on July 22, that the question of awarding the Nobel Prize is complicated by the fact that the second part of his book *Virgin Soil Upturned*, which Sholokhov promised to write by the end of the previous year, has not come out yet. According to Nilsson, the delay with the writing of the second part of *Virgin Soil Upturned* is allegedly the only obstacle for awarding the Nobel Prize to Sholokhov.[1] That is for your information.

AVP RF, fond 0140, opis' 47, papka 70, ed. khr. 40, l.3

[1] The documents of the Nobel Committee for 1955 and 1956, now declassified, definitely confirm this version.

Document 10.10

From the Secretary of the Board of the Soviet Writers Union K. V. Voronkov to the CPSU Central Committee, of February 27, 1962.

According to the information from the Soviet embassy in Stockholm, the Swedish Royal Library[1] shows interest in the work of M. A. Sholokhov over the last years in connection with his probable nomination, in 1962, for the Nobel Prize.[2] Having prepared, at the request of our embassy, the information of publications and screen adaptations of M. A. Sholokhov's works from 1958 to 1962, the Board of the Soviet Writers Union asks for instructions about the stand which should be taken in the question of the possible nomination of the candidacy of M. A. Sholokhov to the Nobel Prize.[3]

The attached information will be sent by us to Stockholm immediately after receipt of the instrustions, as the embassy emphasizes the necessity of receiving it no later than on March 1 this year.[4]

RGANI, fond 5, opis' 36, ed. khr. 141, l.37.

[1] It might be a slip of the pen. Such interest could rather be shown by the library of the Swedish Academy, which accumulates the book production of potential Nobel aspirants.

[2] Erroneous formulation. Here it was not possible to speak about the nomination of Sholokhov's candidacy, since the receipt of 1962 nominations had been finished on January 31. The information from the embassy could refer only to the probability of awarding the prize to the Soviet writer that year, for which the Nobel Committee needed the latest information about publications of his works.

[3] A rather important evidence of the Central Committee's suspicious attitude towards the possible awarding the 1962 Nobel Prize to Sholokhov.

[4] Thus, the author of the letter directed his enquiry to the Central Committee two days before the deadline when the required document was to be received in Stockholm. That could hardly have been done at the personal initiative of Sholokhov's ill-wishers in the Writers Union (see note 3 to this document).

Document 10.11

From the deputy head of the Cultural Department of the CPSU Central Committee A. Petrov and the head of section of the department I. S. Chernoutsan to the Central Committee, of March 1,1962.

The Secretariat of the Board of the Soviet Writers Union asks for permission to direct to the Swedish Royal Library[1] information about publications and screen adaptations of works by M. A. Sholokhov in 1958–1961 in connection with the possible nomination[2] of his candidacy for the Nobel Prize, which has became known from a confidential talk of the Nobel Committee's secretary H. Villars with a councilor of the USSR embassy in Sweden.[3] Informed of Villars's request to send the list of Sholokhov's works published and filmed, the Soviet embassy supports this request.

The Cultural Department of the CPSU Central Committee considers it expedient to direct to Sweden the mentioned information about the works by M. A. Sholokhov. The answer to the Soviet Writers Union (comrade Voronkov K. V.) has been given.

At the bottom is the note by the head of the section of the Department I. S. Chernoutsan of March 12, 1962: "According to additional information from comrade Voronkov, the lists of works by M. Sholokhov published in foreign countries has been sent to the Soviet embassy in Sweden."

RGANI, fond 5, opis' 36, ed. khr. 141, l.44.

[1] See note 1 to doc. 10.10.

[2] See note 2 to doc. 10.10.

[3] The press-secretary of the embassy in Stockholm E. P. Rymko.

Document 10.12

From the memoirs of N. S. Khrushchev, Chairman of the USSR Council of Ministers and the First Secretary of the CPSU Central Committee.

...We knew he had a weakness for drink. I'd talked to him about it in the past. Once he came to me and said, "I've been invited to Norway, but the authorities won't let me go." I explained it wasn't because he was politically untrustworthy; rather, we were afraid he'd lose his wits and stumble about, perhaps inflicting physical injury on himself and moral injury on our country. He promised to restrain himself, so we let him go. He visited a number of countries — England, Sweden, Norway, and Finland[1] — and we didn't receive a single report against him from our ambassadors there. In such a case they would certainly have reported his making a fool of himself.

Khrushchev Remembers. The Last Testament. Translated and edited by Strobe Talbot. Boston, Toronto: Little, Brown and Company, 1974, p. 371.

[1] In the period when he was the First Secretary of the CPSU Central Committee, Khrushchev could mean three of Sholokhov's trips to Scandinavia: in 1957, 1960 and 1962.*

* M. A. Sholokhov. *Pis'ma* (Letters), Moscow, IMLI RAN, 2003, pp. 319, 353.

Document 10.13

M. A. Sholokhov's letter to N. S. Khrushchev, of December 9, 1961.

When we met in the days of the XXII CPSU[1] Congress I told you in brief about my wish to go to Scandinavia. It can be explained by the following motives. Last year my publisher in Denmark O. Lindhardt and his wife stayed ay my place in Veshenskaya,[2] and this year the Norwegian publisher K. Fjeld, also with wife, were my guests. From both these publishers my wife and I received an invitation to visit their countries. Through K. Fjeld I also received such an invitation from my Swedish publisher Gidlund, who could not come to Veshenskaya this year because of a heart disease. This time I would like to limit my trip to going only to Sweden and Finland. The thing is that in Sweden three books of mine have come out this year: *Virgin Soil Upturned* (2nd book), *Tales from the Don* and *The Fate of a Man.*[3] In Finland, published as a separate book, are chapters from the first book of *They Fought for Their Country.* In Denmark and Norway the second book of *Virgin Soil Upturned* has also come out, but the main thing for me is that in Sweden and Finland next spring the first book of *They Fought for Their Country* will be published completely, and it is necessary for me to stipulate with the publishing houses in advance both the dates of the books coming out and a translation of good quality, whereas in Norway and Denmark this book is not coming out until the next fall, and, in fact, I have nothing to do there now. Besides, Fjeld asks to transfer my trip to the next summer....Apart from that, I would like, with consent from the leadership of the Soviet Writers Union, to take with me several good books by our young writers and recommend them for publication in Sweden.[4] Scandinavian publishers listen to my opinion, and if I manage to persuade Gidlung and the books are published in Sweden, I will be happy simply as a comrade. Publication in Sweden also ensures publication in Norway and Denmark, as all three Scandinavian publishing houses have business relations with one another. It must be admitted that *Tales from the Don* is a weak, immature book, and I do not see any reasonable ground on the part of Gidlund to refuse publishing more mature books by the present day young Soviet writers, telling, the more that they tell not about the far past but about our today's reality...

I ask the CPSU Central Committee to permit to me and my wife a trip to Sweden and Finland at the beginning of the next year for about ten or twelve days.[5]

With all my heart I wish you good health, dear Nikita Sergeevich, Yours, Sholokhov.

Published in the book by B. O. Osipov *Tainaya Zhizn' Mikhaila Sholokhova* (The Secret Life of Mikhail Sholokhov) — Moscow: Libereya and Raritet, 1995, pp. 328–329.

[1] The XXII CPSU Congress took place from October 17 through 31, 1961.

[2] The Cossack village of Veshenskaya in Rostov *oblast'*, where Sholokhov permanently lived.

[3] Such wide publication of Sholokhov's works in the early 1960s in Sweden, including the "weak, immature" *Tales of the Don*, was certainly arranged in support of the campaign for awarding him the Nobel Prize.

[4] These years Sholokhov was responsible in the Writers Union for working with young writers.

[5] Sholokhov got the permission and visited Sweden and Finland at the end of January and the first days of February 1962.*

* M. A. Sholokhov. *Pis'ma* (Letters) — Moscow: IMLI RAN, 2003, p. 353.

Document 10.14

From the head of the Department of Scandinavian countries of the Ministry of Foreign Affairs A. G. Kovalev to the USSR ambassador in Sweden N. D. Belokhvostikov, of August 12, 1968.

Decisions taken:
....2. To permit to the writer Sholokhov M. A. a trip to France, England, Denmark, Norway, Sweden and Finland with his wife Sholokhova M. P., daughter Turkova C. M. and daughter-in law Sholokhova V. A. for a period of up to one month for negotiations with foreign publishers.

For your information.

AVP RF, fond 0140, opis' 54, papka 209, ed. khr. 27, l.59.

Document 10.15

From the statement of Jean-Paul Sartre to Swedish journalists, made in Paris, on October 23,1964.

I am very sorry that the matter took the form of a scandal: the prize is awarded, but it is declined.... When I read in *Figaro Litteraire* of October 15 the report of its correspondent in Stockholm saying that the Swedish Academy is inclined to choose my candidacy but the final choice has not been made yet, it seemed to me that by writing a letter to the Academy — I sent it the next day — I would be able to rectify the situation so as not to return to that anymore. Then I did not know yet that the Nobel Prize is awarded irrespective of the opinion of the prospective laureate and thought that it could still be prevented.[1] But I perfectly understand that after the Swedish Academy has made its choice it already cannot refuse it. As I explained in the letter addressed to the Swedish Academy, the reasons for my declining the award do not concern the Swedish Academy nor the Nobel Prize as such...

At present the only possible form of struggle at the cultural front is the struggle for peaceful coexistence of two cultures, Western and Eastern. I do not mean that fraternization of the cultures is necessary. I fully understand that the mere juxtaposition of these two cultures inevitably must take the form of a conflict. But this juxtaposition should take place between people and cultures without interference of any institutions...[2] That is why I do not want to accept any awards either from Eastern or from Western institutions, although I perfectly understand that they exist. In spite of the fact that all my sympathy lies with socialism,[3] I would be equally unable to accept, for example, the Lenin Prize,[4] should it be offered to me.[5]

I clearly understand that the Nobel Prize itself is not a literary prize of the Western block, but it was made such, thus making possible events beyond the control of the Swedish Academy. That is why in the present situation the Nobel Prize is in fact an award intended for writers from the West or for "rebels" from the East.[6] For example, among those who were not awarded the prize was Neruda, one of the greatest poets of South America.[7] Nor has the candidacy of Aragon ever been seriously discussed, although he quite deserves this prize. It is regrettable that the Nobel Prize has been awarded to Pasternak and not to Sholokhov[8] and that the only Soviet work which has won the prize was the book published abroad and banned in its home country....[9]

Translated from the Russian translation published in the *Za Rubezhom* weekly No. 45, 1964, p. 23.

[1] There was a similar situation in 1958 in connection with B. Pasternak. Then the German writer Renate Schweitzer forwarded to the Nobel Committee the poet's letter addressed to her, where he expressed concern about the possibility of being awarded the Nobel Prize. She asked in her own name to postpone the awarding for a year. Then the Swedish Academy also did not think it possible to yield to the pressure from the outside.*

* See also note 2 to doc. 9.1-III.

[2] In the part of Sartre's statement omitted here it was explained that "if Jean-Paul Sartre, Nobel Prize winner", speaks in defense of the Venezuelan Resistance, it will mean "personifying the very institution of the Nobel Prize", and so a writer "should not allow himself to be turned into an institution, even if it [...] takes honorary forms."

[3] The first socialist country for a long time did not respond to the enthusiastic signs of the French existentialists' love of socialist ideals. Here is what was said by the General Secretary of the Soviet Writers Union A. A. Fadeev on August 25, 1948, fifteen years before Sartre's shocking rejection of the Nobel Prize. Speaking from the rostrum of the World Congress of Personalities of Culture in Defense of Peace in Vroclav, Poland, he declared: "Spiritual debauchery is inspired by the 'newly

brought to light philosophy' of Sartre's kind, which tries to make man go down on all fours."* In the following years, in the atmosphere of the "anti-cosmopolitanism" campaign organized at the Central Committee and definitely directed against the West, what was said about Sartre sounded even stricter: "Sartre's philosophy of renegades, pimps, spies, provocateurs, which spits upon everything human in man and mankind, should be denounced as an undisguised ideological weapon of Anglo-American imperialism preparing a world war."**

* *Literaturnaya Gazeta* August 29, 1948, p. 2.

** *Bolshevik* No.12, 1950, p. 71.

[4] All that was merely demagogical rhetoric with the purpose of adopting apublicity pose. As the Nobel Foundation's executive director Michael Solhman told this author, in 2000, some time after Sartre's shocking refusal of the award people close to the writer started to sound out in Stockholm if it was possible to repudiate the refusal and receive the prize. One can suppose that even if they did it on their own initiative, the Nobel laureate must have been in the know.

[5] Three years later, in 1967, the future Nobel Prize winner A. I. Solzhenitsyn, remembering the prize awarded to Sholokhov, hurled vicious words against Sartre: "...it was that fall when the Nobel Prize was thrust into [...] Sholokhov's hands... Half a year later the man who obtained this prize for Sholokhov, thus giving most painful offence to Russian literature, Jean-Paul Sartre, was in Moscow and expressed through his interpreter his wish to see me... Did Satrtre read in my refusal the true depth of our non-acceptance of him?"* In the circles of Soviet intelligentsia Solzhenitsyn was not the only one who had such non-acceptance of Sartre. The cardinal change in perception of the French existentialist's political face and philosophy, which took place in post-Stalin times in the party and ideological spheres, received rather caustic comments in private conversations. D. D. Shostakovich and his son Maxim looked in at the authors rights protection department. "Having come up to the cash office we saw Jean-Paul Sartre standing there, carefully counting a rather thick wad of banknotes. It must be noted that at that time in the Soviet Union foreigners were not paid fees. It was only done in exceptional cases, for rewarding the personalities who proved particularly useful to the bolshevist regime. Sartre was probably one of them. Father cast a glance at the Frenchman and whispered into my ear: 'We do not deny that there can be financial interest in coming from the reactionary camp to the camp of progress.'"**

* A. I. Solzhenitsyn. *Bodalsya Telenok s Dubom* (The Oak and the Calf) — Moscow: Soglasie, 1996, p. 132.

** *Hovy Mir* No. 6, 2002, p. 130.

[6] In her *Zapisnye Knizhki* (Notebooks) Akhmatova responded to the passage about "rebels from the East" by the following entry dated October 25, 1964: "There is a strange statement by Sartre in newspapers. Who does he mean, besides Pasternak?"* This note shows once again that Akhmatova knew about her nomination for the

Nobel Prize in those years and in this case suspected that Sartre implied that one of the "rebels" was her.

* Zapisnye Knizhki, p. 498.

[7] Pablo Neruda was awarded the Nobel Prize in 1971.

8. On October 26, 1964 Akhmatova returns again to Sartre's statement: "*Znamya* No.10 has been brought. There are two poems of mine.* For some reason it is pleasant to me that they are not about love. And *The Death of a Poet* is greatly needed right now, after Sartre's statement.** Some ruler of men's minds! It's shameful to attack the young grandchildren and such a grave."*** About the poem *The Death of a Poet* see Chapter 9, with notes 65–67 to the corresponding text.

* *Znamya* No.10, 1964, p. 91; poems *Ballade* (1943) and *The Death of a Poet* (1960).

** The poem is dedicated to the death of B. Pasternak, on May 30, 1960, and written on June 1, 1960 in Botkin Hospital, where the poetess was staying at that time.

*** *Zapisnye Knizhki,* p. 499. Akhmatova means Pasternak's three grandchildren, who were one, three and seven in 1964.

[9] Actually, Sartre's grudge against the Swedish Academy for not having awarded the prize to Sholokhov was merely mentioned in passing. However, the Soviet propaganda presented it as a dominating idea, which made the Academy take their decision.*

* Here is an example of official wording in this connection: Sartre refused the prize, "referring to the fact that the committee which awards it disregards the achievements of the 20th century revolutionary writers"; *Great Soviet Encyclopedia,* 3rd edition, vol. 22, 1975, column 1794.

Document 10.16

I

From the memoirs of N. S. Khrushchev, Chairman of the USSR Council of Minister and First Secretary of the CPSU Central Committee, about his official visit to Sweden in June 1964.

... I should mention a woman who was the Minister of Culture[1]... She came up to me at a reception and was very cheerful: I think she was a drunk.

"Mr. Khrushchev," she said, "I'd like to ask your advice about something. Pretty soon we're going to be awarding the Nobel Prize for literature. There are two candidates in the Soviet Union." She named them.[2] "What do you think? I'm inclined to support this one" — and she named her favorite.

"Why are you asking me? I can't influence your decision."

"I'd still like to hear your opinion."

"Well, in that case," I said, "both writers certainly deserve the prize, but neither of them enjoys very wide support in my country. We have more acceptable candidates. I think our public would be much more satisfied if the prize were awarded to a third writer."

"And who would that be?"

"Mikhail Alexandrovich Sholokhov. If you took a poll in our country, I think the majority would vote for Sholokhov. I know our intelligentsia and our literary community would certainly be pleased if he won the prize."

She let the matter drop there, and I decided not to say anything more about it myself. After all, the Nobel Prize is an internal question for the Swedish government. Besides, for us to beg for the prize would be beneath our dignity — it would be degrading. We have our own prizes. The Nobel Prize, for instance, cannot be compared to the Lenin Prize.[3]

A year or so later, after I'd already retired,[4] I learned that the Swedes[5] had awarded the prize to Sholokhov. I think that the Swedish government must have taken into consideration my remarks to the Minister of Culture.[6]

Khrushchev Remembers. The Last Testament. Translated and edited by Strobe Talbott. Boston, Toronto: Little, Brown and Company, 1974, pp. 526–527.

[1] See note 1 to M. Solhman's comment (doc. 10.16-II).

[2] Most probably named were A. A. Akhmatova and K. G. Paustovsky (see doc. 10.27–10–31 and notes 24–52 to this chapter).

[3] Khrushchev's way of thinking is meaningfully similar to Stalin's earlier utterances; the only difference is that the latter spoke about significance and prestige of prizes of his own name (see Chapter 5 and notes 21 and 22 to it).

[4] As a result of a plot of several members of the CPSU Central Committee's Presidium, Khrushchev was dismissed from the posts of Chairman of the Council of Ministers and First Secretary of the CPSU Central Committee and forced into retirement in October 1964.

[5] Not being well enough informed, Khrushchev regarded the Nobel Prize as a state award, like the Stalin or the Lenin Prize.

[6] Additional proof of what was said in note 5.

II

Comment to Khrushchev's memoirs from the Nobel Foundation's executive director Michael Solhman given by him to this author.

Discussing whether Mr. Khrushchev's version is convincing it should be remembered that in 1964 Sweden had no female Minister of Culture and no Minister of Culture at all, as cultural affairs were taken care of by the Minister of Education.[1]

Mr. Khrushchev also mentioned that the lady with whom he spoke drunk. In this case she must have been in good company. But the main thing — and, of course, it was very difficult for Mr. Khrushchev to understand — was that the Swedish Academy had never received instructions from anybody and the Swedish government was the least possible source from which they could be received.[2]

[1] Khrushchev's interlocutor was Ulla Lindstrom, minister without portfolio (see doc. 10.26).

[2] The post-Soviet 21st century inherited the conviction asserted by the Soviet propaganda for several decades that such a prestigious event, attracting attention all over the world, as awarding the Nobel Prizes is under the hopelessly biased patronage from the Swedish authorities. An example is a writing by the intellectual journalist Maxim Sokolov. In one of his Saturday columns in the *Izvestiya* newspaper, willing to support by a bright comparison his idea of the deficiency of political thinking of some Russian public personalities involved in building their parties, Sokolov sarcastically remarked that "the Swedish king, having learnt about that, will give all of them the Nobel Peace Prize."* In comparison to such a fundamentally incorrect concept the journalist's ignorance of the fact that giving the Nobel Peace Prize is a prerogative not of the Swedish but of the Norwegian king will seem but a nice, and therefore pardonable, slip of a glib pen.

* *Izvestiya*, April 24, 2003, p. 3.

Document 10.17

From the Acting Academician-Secretary of the Language and Literature Department of the USSR Academy of Sciences, Associate Member of the Academy M. V. Khrapchenko and the Academic Secretary E. F. Trushchenko to the Academy of Sciences' Foreign Department, of January 8, 1965.

To your No. 35–2005 of 22. XII. 64.[1]

The Language and Literature Department of the USSR Academy of Sciences is not planning to nominate a candidate for the 1965 Nobel Prize for Literature.[2]

The Archive of the USSR Academy of Sciences, fond 579,
opis' 1, ed. khr. 640, l.135.

[1] The letter from the Academy's Foreign Department of December 22, 1964 is not published.

[2] The anti-Sholokhov implication of the given document is obvious, especially if we take into consideration that Sholokhov, as Member of the USSR Academy of Sciences, was a member of the department headed by the author of the memo to the Foreign Department. The fact that the CPSU Central Committee's ideological structures were interested in promotion and support of Sholokhov's nomination for the Nobel Prize was no secret to the department's academician-secretary. In this case Khrapchenko

showed himself as a person of principle, since he knew well that in wide circles of Soviet intellectuals the attitude towards the writer was negative.* Thus he did not confirm the opinion of him formulated by a party official from the Central Committee's personnel department as far back as in 1946: "He is an indecisive and timid man, inclined to defense and not to attacking actions [...] at the level of the tasks set before him."**

* L. K. Chukovskaya, who was in the Writers' Creative Work Home in Komarovo, near Leningrad, on the day when the Swedish Academy's decision was announced, graphically described in her diary how awarding the Nobel Prize to Sholokhov was taken by the literary community: "... the weather was so nasty in Komarovo on the October day when we read about the Nobel Prize. The meanest wind blowing from the sea, whether snowing or raining from the sky, and people are insulted, humiliated, embarrassed, not looking each other in the eye. Shame — both in nature and in souls. I heard the news when running from the sleeping building to the canteen. I felt as though hit in the face with a dirty towel." (L. K. Chukovskaya. *Zapisky ob Anne Akhmatovoi* (Memories of Anna Akhmatova) Vol. 3, Moscow, Soglasie, 1997, p. 307). Among Western creative intelligentsia the intensity of emotions was also high. Peter Norman, well-known British poet and translator of Pasternak, said: "If it were not Sweden but my country that gave the prize to Sholokhov, I would emigrate." (*Zhamya* No.10, 2001, p. 137).

** RGASPI, fond 17, opis' 117, ed. khr. 617, l.92,93. See also: N. I. Balashov, Y. N. Karaulov. *Vestnik RAN* (Journal of the Russian Academy of Sciences), 75, No. 12, 2005, pp. 1123–1131.

Document 10.18

From M. A. Sholokhov to the General Secretary of the CPSU Central Committee L. I. Brezhnev, of July 30, 1965.

Dear Leonid Ilyich! A Vice-President of the Nobel Committee was recently in Moscow.[1] In his talk at the Writers' Union he gave to understand that this year the Nobel Committee will probably be discussing my candidacy. After Jean-Paul Sartre refused (last year) to receive the Nobel Prize saying that the Nobel Committee was not objective in their appraisal and that this committee, among other things, should have long awarded the Nobel Prize to Sholokhov,[2] the Vice-President's visit cannot be regarded as other than reconnaissance.[3] Just in case, I would like to know what attitude will be taken by the Presidium of the CPSU Central Committee if this prize is (contrary to the class convictions of the Swedish committee)[4] awarded to me and what will my Central Committee[5] advise me?

The prizes are usually awarded in October, but I would like to be in the know of your attitude to this question in advance. At the end of August I am going to Kazakhstan[6] for 2 or 3 months, and I would be glad to hear from you before my departure. With the most friendly feelings, M. Sholokhov.

In the bottom right corner of the sheet with the typed copy of

Sholokhov's letter is a typed text with the signature of the deputy head of the CPSU Central Committee's Cultural Department G. Kunitsyn, addressed to the Central Committee:

"In the opinion of the CPSU Central Committee's Cultural Department, awarding the Nobel Prize for literature to comrade Sholokhov M. A. would be a just recognition on the part of the Nobel Committee of the fact that the work of this outstanding Soviet writer has worldwide significance. In this connection the Department does not see any reason to refuse the prize if it is awarded. I ask your permission to inform comrade M. A. Sholokhov of that."

In the top left corner of the sheet is a handwritten note: "To agree with the Department" — and signatures.

In the bottom left corner of the sheet is a handwritten note: "Information. Comrade Sholokhov M. A. is informed. G. Kunitsyn. 16. VIII.65."

RSAMH, fond 5, opis' 36, ed. khr. 148, l.59.

[1] The secretary of the Nobel Committee for literature Hugo Villars. Probably, not having received in time, in 1962, the required additional information from the Writers Union (doc. 10.10 and 10.11), in 1965 he preferred to come to Moscow in person.

[2] The given interpretation of the motives of Sartre's rejection of the Nobel Prize in Sholokhov's rendering is rather far from the true content of his statement (doc. 10.15), although Sholokhov certainly read its full text.

[3] Sholokhov's guess was quite right. Here is what was told 36 years later by the then press-secretary of the Soviet embassy in Stockholm E. P. Rymko: "The secretary of the Swedish Academy's Nobel Committee Hugo Villars tried to elicit from me, carefully but insistently, whether Mikhail Alexandrovich would refuse the prize, as Sartre did in 1964, and soon asked Sholokhov's address."*

* *Obshchaya Gazeta* No. 42, 2001, p. 5.

[4] Sholokhov went on playing up the topic of "class convictions" of the Nobel institutions eight years later in his scandalous speech at the XXIII CPSU Congress. Touching upon the role of the writer in social life, he remembered that "it held a prominent place in my speech in Stockholm's City Hall during the Nobel celebrations last year." He concluded his ideas with the following phrases, which were twice interrupted, as can be seen from the shorthand report, by prolonged applause from the audience responsive to passages like that: "The audience there was quite different from today's. And accordingly, the form of rendering my ideas was different. The form! Not the content."*

* *Pravda* April 2, 1966, p. 5.

[5] The word combination "my Central Committee" firmly established in Sholokhov's subconscious in the decade of his contacts with the highest party organ. According to his biographer V. O. Osipov, the writer's daughter C. M. Turkova "told me that

when he was dying, her father tossed in bed and repeatedly half pronounced, half moaned: 'Where is my Central Committee? Where is my Central Committee?'"*

* V. O. Osipov. *Tainaya Zhizn' Mikhaila Sholokhova* (Mikhail Sholokhov's Secret Life), Moscow: Libereya, Raritet, 1995, p. 374.

[6] It was in Kazakhstan that he learnt about the Swedish Academy's decision.*

* V. O. Osipov, p. 338.

Document 10.19.

From the secretary of the Board of the Soviet Writers Union K. V. Voronkov to the CPSU Central Committee, of November 29, 1965.

The laureate of the 1965 Nobel Prize for Literature M. A. Sholokhov is to leave for Sweden to receive the prize. According to the USSR embassy in Stockholm, the Swedish Academy of Sciences[1] is going to give the Nobel Prize to comrade Sholokhov M. A. on December 10 this year.[2] M. A. Sholokhov is ready to leave for Stockholm on December 4 this year and asks to direct members of his family together with him. Besides that, he expressed his wish for sending to Sweden a literary critic who could give him necessary consultations when preparing speeches on Soviet literature, and an interpreter. At the suggestion of a number of organizations, Sholokhov gave his consent, along with making a speech at the award-giving ceremony,[3] to speak in universities before students, at a literary evening organized by a publishing house, and also to give a press conference. The USSR embassy is arranging, on December 7, a reception in honor of the Nobel laureate, and asks to allocate 500 rubles in foreign currency for this purpose. After the end of the visit to Sweden comrade Sholokhov M. A. would like to go to Finland for creative purposes.

The secretariat of the Board of Soviet Writers Union makes a proposal: to permit the departure for Sweden and Finland for a month to comrade Sholokhov Mikhail Alexandrovich with his wife Sholokhova Maria Petrovna, daughters Svetlana Mikhailovna and Maria Mikhailovna, sons Alexander Alexandrovich and Mikhail Alexandrovich,[4] the writer Lukin Yuri Borisovich and an interpreter.

RGANI, fond 5, opis' 36, ed. khr. 140, l.183–184.

[1] This is an error: he means the Swedish Academy.

[2] The Swedish Academy, being the institution which endows the Nobel Prize for Literature, awards the prize, which is given by the king of Sweden. The award-giving ceremony takes place every year on one and the same day fixed for that purpose, December 10. So the words "is going to" are not quite appropriate in this context.

[3] An error: according to the regulations, the laureates of the year, one from each category of the prize, make their speeches not at the award-giving ceremony, but at the end of the Nobel banquet, in the evening of the same December 10.

[4] These are not sons but grandsons of the Nobel laureate.

Document 10.20

From the deputy head of the Cultural Department of the CPSU Central Committee G. Kunitsyn to the CPSU Central Committee, of November 29, 1965. Rather urgent.

In connection with the forthcoming giving the Nobel Prize to comrade Sholokhov M. A., which is taking place on December 8 this year[1] in Stockholm (Sweden), comrade Sholokhov asks to permit to his family members (wife and 4 children), the literary critic Lukin Y. B. and an interpreter going together with him to Sweden and Finland for the period of up to 1 month. In connection with the fact that, according to the established ritual, at the award-giving ceremony comrade Sholokhov should be wearing a special tailcoat, and in our conditions making such a tailcoat in the time left appears impossible, comrade Sholokhov asks to give him[2] three thousand American dollars for purchasing the tailcoat and the equipment for his companions in Helsinki, with the subsequent returning of this money from the Nobel Prize. The Finance Ministry [...] has no objections. Comrade Sholokhov's departure is planned for December 2 this year.

The Cultural Department of the CPSU Central Committee supports comrade Sholokhov's requests. Please give consent.

The note below: "The business trip and the giving of the money are registered. Chief accountant of the Managing Department of the CPSU[3] Central Committee (signature). XII. 3,. 65."

The note in the top left corner: "To agree" and signatures.

RGANI, fond 5, opis' 36, ed. khr. 148, 1.185.

[1] An error: the Nobel Prize award-giving ceremony was to take place and did take place on December 10, 1965.

[2] A tailcoat and other equipment necessary for taking part in the ceremony could have been found in Stockholm just as well, but there during the Nobel festival it would have cost much more than in Helsinki.

[3] Communist Party of Soviet Union Central Committee Administration.

Document 10.21

From M. A. Sholokhov to the CPSU Central Committee, of November 29, 1965.

In connection with my forthcoming trip to Stockholm for receiving the Nobel Prize I ask to permit to accompany me on this trip, apart from my family members, comrade Melentiev Y. S[1] and some interpreter.

RGANI, fond 5, opis' 36, ed. khr. 148, 1.186, written by hand

[1] At the last moment Sholokhov decided to substitute his consultant and speechwriter for the period of his staying abroad, the expert on his work Y. B. Lukin, for the critic Y. S. Melentiev, who mostly studied the literary work of the 19th century

Russian revolutionary democrats, but at that moment was an official of the CPSU Central Committee's Cultural Department.

Document 10.22

From the deputy head of the Cultural Department G. Kunitsyn to the CPSU Central Committee, of November 29,1965. Rather urgent.

Comrade Sholokhov M. A. asks the CPSU Central Committee, in connection with giving him the Nobel Prize due for December 8[1] this year, to permit the deputy head of the Cultural Department of the CPSU Central Committee comrade Melentiev Y. S. to accompany him on this trip. The Cultural Department of the CPSU Central Committee considers it possible to support the request of comrade Sholokhov M. A. for sending comrade Melentiev Y. S. to Sweden for the period of 10 days. The departure of comrade Sholokhov M. A. is planned for December 2 this year.

The notes below: on the left "Executed. 3.12.65"; on the right "The business trip is registered. The Chief Accountant of the Managing Department of the CPSU Central Committee (signature). 12.3, 65."

RGANI, fond 5, opis' 36, ed. khr. 148, l.187

[1] See note 1 to doc. 10.20.

Document 10.23

M. A. Sholokhov's answers to the questions of students of Uppsala University during the meeting with him at the St. Lucia festival on December 13, 1965.[1]

"What do you think of women's literature?"

"I do not like women writers as such. The writing trade is men's business. In literature it's the same as on a shop floor in a foundry — for women working there is dangerous. In addition I would describe what women do in literature, with rare exceptions, as 'ladies' needlework'", the Nobel laureate Mikhail Sholokhov told Uppsala students.

Someone asked:

"So do you think that Selma Lagerlof undeservedly received the Nobel Prize for her needlework?"

"But I said that there are some exceptions", Sholokhov answered.

Among the motley mosaic of questions there was the following: does the Nobel laureate hold the opinion of the inequality of the sexes?

"Yes, I think that the wife should be firmly kept in check. In this respect I'm conservative, although I'm a communist. But among our communists there are different points of view."[2]

From the newspaper *Uppsala Nye Tidning* of December 14, 1965.

[1] Saint Lucia's Day, on December 13, is a traditional Swedish festival in memory of a Christian girl from Syracuse, executed in 304 and announced a saint and great martyr in the 15th century. On that day all over Sweden processions are arranged, headed by a crowned girl playing the part of St. Lucia, beauty contests are organized, etc. In Uppsala in 1965 the right to crown the most beautiful girl of the university was given to the guest of honor at the festival, Sholokhov.

[2] In his interview to the *Express* journal the French writer Louis Aragon touched upon the episode of Sholokhov's meeting with Uppsala students:*

... Sholokhov was speaking before the students. Among them there were young men of leftist views, who shouted: "Bravo, Sholokhov!" They asked him several questions about women's literature (don't forget that in Sweden there are many women writers). And this man replied that women's literature did not mean anything, had never meant anything and had never been of any use, and women had better stay at home and darn their men's socks. The students got furious. It was a bolt out of the blue. Just imagine: the man who in their eyes is a representative of the first socialist state in the world, October Revolution, Lenin and all that. This man is among them, and here is what he says. It's incredible! Sholokhov has always used his authority for humiliating others**, doing all that in the name of Marxism and Leninism, which would greatly surprise Marx and Engels... ***

* The translation of the interview from French into Russian was done for the CPSU Central Committee's Cultural Department, on October 13, 1971, and is stored among the Department's archive documents: RSAMH, fond 5, opis' 63, ed. khr. 146, l. 53–54.

** This trait of Sholokhov's character was also noted by L. K. Chukovskaya in her "Open Letter to Mikhail Sholokhov", written on May 25,1966, in response to the writer's shameful speech at the XXIII CPSU Congress after the trial of the Soviet writers A. D. Sinyavsky and Y. M. Daniel (doc. 10.35). "It is already a long time", she wrote, "that you, Mikhail Alexandrovich, have been in the habit of speaking about writers with disdain and rude jibes." (in the collection *Tsena Metafory ili Prestuplenie n Nakazanie Sinyavskogo i Danielya* (The Price of a Metaphore or the Crime and Punishment of Sinyavsky and Daniel) — Moscow: Kniga, 1989, p. 503).

*** From the letter of the literary critic and translator Noemi Wallenius, who, at the request of this author, reviewed the newspapers of those days: "Aragon certainly embellished this episode. Sholokhov was bombarded with questions, but the 'women's issue' was not at all dominating. He was asked about the living conditions of Soviet writers, about how foreign literature is read in the Soviet Union, there was even a question about building socialism in Sweden, to which he answered: 'That is your business.' In Uppsala he was accompanied by the Soviet ambassador N. D. Belokhvostikov, and the interpreter was Mikhael Solhman. I talked with him, and he does not remember any scandal. Naturally, Sholokhov's answers about women's literature did not gain wholehearted approval from the audience, but if there had been some peculiar response, newspapers would not have missed the chance to report it. I even had the impression that his answers were taken as something amusing..."

Document 10.24

From the secretary of the Board of the Soviet Writers Union K. V. Voronkov to the CPSU Central Committee, of December 12, 1965.

One of these days M. A. Sholokhov is coming back home from Sweden, where the 1965 Nobel Prize for Literature has been given to him. The Secretariat of the Board of the Soviet Writers Union is planning to arrange in Moscow a reception in honor of the laureate on his coming back from abroad. We would consider it expedient to invite to this reception writers, personalities of science and culture, representatives of social organizations and cultural institutions, as well as of Soviet and foreign press, 400 people in all. The expense of organizing the meeting will be covered by the Soviet Writers Union (the Literary Fund).

We ask the Central Committee to permit conducting the meeting with M. A. Sholokhov at the Receptions House in the Lenin Hills.[1]

We ask you to give instructions.

RGANI, fond 5, opis' 36, ed. khr. 148, l.241

[1] The meeting of representatives of the Moscow public with Sholokhov in Receptions House took place on December 22, 1965.

Document 10.25

From M. A. Sholokhov's letter to the General Secretary of the CPSU Central Committee L. I. Brezhnev, March 1978.

... At the present time one of the main objectives of the ideological offensive of the enemies of socialism is Russian culture, which makes the historical foundation, the main wealth of our country's socialist culture. Belittling the role of Russian culture in the historical spiritual process, distorting its high humanistic principles, denying its progressive character and creative originality, the enemies of socialism are thus trying to discredit the Russian people as the main international force of the Soviet multinational state, to show that it is spiritually feeble, incapable of intellectual creative work. Not only do they propagandize the idea of the nation's spiritual degeneration, but they also intensify their efforts to create favorable conditions for that.

And all that is done to prove, firstly, that socialism in our country allegedly is socialism with an "inhuman face",[1] created by barbarians and for barbarians, and, secondly, that this socialism has no future, as its death is predetermined by the national inferiority of the Russian people, the leading force of the Soviet state.

Russian culture is especially violently, actively attacked by world Zionism, both foreign and internal. It is common practice to promote anti-Russian ideas through cinema, television and press, discrediting our history and culture, as well as to oppose the Russian to the socialist. In this respect the appearance on the Soviet screen of the film by A. Mitta *How tsar Peter married off the Arab*, overtly humiliating the dignity of the Russian nation, spitting upon progressive initiatives of Peter I, ridiculing the Russian history and our people, is symptomatic.[2] Up till now many topics dealing with our national past remain forbidden. It is extremely difficult, and often just impossible, to arrange an exhibition of a Russian artist of patriotic trend, working in the tradition of the Russian realistic school. At the same time organized one after another are mass exhibitions of the so-called "avant-garde", which does not have anything to do with the traditions of Russian culture, with its patriotic spirit. In spite of the government's resolutions, destruction of Russian architectural monuments is going on, restoration of monuments of Russian architecture is done most slowly and often with the conscious distortion of their initial look...

The note in the left top corner: "To the Secretariat of the Central Committee. I ask you to consider it, with the subsequent consideration at the PB.[3] 3.14.78. Brezhnev."[4] Under the note are signatures of the Central Committee's secretaries who familiarized themselves with the document — Suslov, Zimyanin, Kapitonov; and the heads of ideological departments — Demichev, Markov, Tyazhelnikov.

Published in the supplement to the newspaper *Vechernyaya Moskva,* of December 15,1993: *Gradskie Vesti* No.10, 1993, p, 2; the original of the document is stored in RGANI in the "Special File" of the CPSU Central Committee. See also: M. A. Sholokhov *Pis'ma* (Letters) — Moscow: IMLI RAN, 2003, pp. 429, 430.

[1] This is the antithesis to the slogan of the late 1960s movement in Czechoslovakia for renovation of the social system, whose motto was "building socialism with a human face". Soviet propaganda presented this movement as a consequence of subversive activities of "world Zionism", and Sholokhov below is using this propaganda clich– for rendering his own ideas.

[2] *The Tale of How Tsar Peter Married off the Arab*; Mosfilm studio, 1976. This is how the director Alexander Mitta characterized the genre of the film he made: "Talking of the genre, our picture is closest to poetic tragic farce; closely intertwined here are tragedy and comedy, drama and melodrama, the methods of psychological cinema and of adventure film."* Concrete episodes that might have given the author of the letter the grounds for the ideas he sets forth in it are most probably those that feature launching from the stocks of a holey frigate, uncompleted as a result of the boyar Rtishcnev's sloppiness, or the scenes where a faulty cannon received in the army is being tested with the help of a fourfold charge — the cannon smashed to smithereens, everything topsy-turvy, however, in accordance with the rules of the genre, everybody is alive, although with some bruises. There certainly can be seen some allusion which the authors of the film do not try to conceal, but in our

everyday life the viewer could watch lots of similar situations; and for Sholokhov himself absurdities of our existence were not at all a secret.

* From the fund of newspaper clippings of the library of the Cinema Centre, Moscow.

[3] At a sitting of the Political Bureau of the CPSU Central Committee.

[4] The resolution adopted by the Political Bureau of the Central Committee said: "To explain to comrade Sholokhov the true state of affairs with regard to the development of culture in the country and in the Russian Federation and the necessity of a deeper and more exact approach to the questions he raised in the highest interests of the Russian and Soviet people. Not to start any open discussions on the issue of the Russian culture arisen by him."*

* See V. O. Osipov. *Tainaya Zhizn' Mikhaila Sholokhova* — Moscow:, Libereya, Raritet, 1995, p. 371.

Document 10.26

Discussion in the Swedish newspaper Dagens Nyheter in connection with M. A. Sholokhov's death.

I

From the newspaper's Moscow correspondent Disa Hostad, of February 22, 1984.

The Soviet writer, Nobel Prize winner Mikhail Sholohov died at the age of 79...

Did M. Sholokhov write *Quiet Flows the Don*? As far back as in 1928 a number of writers accused him of plagiarism, considering it impossible that such a young man (Sholokhov was 22 in 1927) could have written such a mature work and known so much about the Don Cossacks, although not being one of them. In 1974 Solzhenitsyn joined this opinion and spread the theory in the West, according to which Sholokhov simply used a work by a writer of an older generation, and that explained the fact that Sholokhov was never again able to create a new masterpiece... Sholokhov joined the party, became a dedicated party member and because of that had to do some dirty things. One of the most disgusting of them was his initiative to disclose the pen names of Jewish writers during the 1949 Stalin anti-Semite campaign.[1] In Khrushchev's time Sholokhov became a progressive critic flagellating Soviet writers who loved convenience and comfort, traveled only between Moscow and their *dacha*, and were too old of age at that.

But perhaps Sholokhov can be understood as a writer who worked in hard times, some Swedish critics said after he was awarded the Nobel Prize — which was the Swedish Academy's special move in order to balance awarding the Nobel Prize to Pasternak. What Khrushchev said in his memoirs — that it was him who (through Ulla Lindstrom) advised the Swedish Academy to give the prize to Sholokhov if the Academy wanted to reward the Soviet writer loved by the people[2] — did not improve the situation, but confirmed once again, apart from the rumors of plagiarism, that awarding the prize to Sholokhov was one of the most joyless episodes in the history of the Nobel Prizes.

[1] The article about pen names was written by Sholokhov later, in 1951; see note 46 ** to Chapter 6. About his anti-Semitic views see also doc. 10.25.

[2] See doc. 10. 16–1.

II

The response of the Swedish Academy's permanent secretary Lars Gyllensten in Dagens Nyheter, of February 27,1984.

In the *Dagens Nyheter* newspaper of February 22 Disa Hostad comments on the demise of Mikhail Sholokhov. At the end of the article she writes that Khrushchev says in his memoirs "that it was him who through Ulla Lindstrom advised to give the prize to Sholokhov". Further on she writes that it confirmed "that awarding the prize to Sholokhov was one of the most joyless episodes in the history of the Nobel Prizes". Among the correspondence, records or other documents of the Swedish Academy there is no trace of any pressure on the part of Khrushchev.[1] Generally, people often address us with good and inappropriate advice and remarks about awarding the Nobel Prizes. These are most different people — from the heads of states to schoolchildren, who send us letters written in block letters with drawings, and other participants of discussions on cultural issues. If the Academy yielded to pressure of political, economic. opportunistic, ideological, racist or any other character, the very basis of the activities of awarding the Nobel Prizes would be undermined, and the prize itself would soon lose respect, and the interest in it would disappear. However, the Nobel Prize has an extraordinary and ever increasing authority, which, of course, does not rule out occasional blatant attempts of gross interference. Often it is the way of making themselves known used by representatives of various political trends, when such activities serve the purposes of their own propaganda.

In the communist countries academies are controlled by the government, and their activities express the political will of the state authorities. Our Swedish academies, including the Swedish Academy, are private associations, and the freedom from being politically governed is for them a matter of primary importance. The government, in its turn, does not wish to interfere in the matters connected with awarding the Nobel Prize by one or another organization, as it would only lead to political and diplomatic complications. That is difficult to understand in communist and totalitarian countries. Whereas for the Swedish Academy, as well as for other institutions endowing the Nobel Prize, it goes without saying that the influence of representatives of political power must be avoided. Whatever the attitude to awarding the Nobel Prize to Sholokhov, Nikita Khrushchev cannot be praised nor shamed for it.

[1] There were not supposed to be any traces. The talk between Khrushchev and Lindstrom was private, it was started at the latter's initiative, and Khrushchev in his memoirs did not mention any attempt to address the Nobel Committee.

Document 10.27

Materials related to K. G. Paustovsky's nomination for the 1962 Nobel Prize.

I

From the memories of the writer A. M. Borshchagovsky,[1] *in 1962 a member of the Board of the USSR-Sweden Friendship Society, shared with this author on October 22, 1999.*

In the summer of 1962 I was having a rest on the Oka river, in the Rysan *oblast'*. It was there that the telephone call found me. Calling was Korolev, secretary of the Board of the Society where I was the chairman of the literature and culture section. He said that the Swedish writer Artur Lundkvist, vice president of the World Peace Council, was coming to Moscow for the session of the Council, to be held in the Kremlin Palace of Congresses, and I was asked to meet him and accompany him till the end of the session.

I was coming back from the airport in the car together with Lundkvist and his wife, poetess Maria Wine. His first question to me was "Is Paustovsky in Moscow?" Having received an affirmative answer, he went on: "Will I be able to make his acquaintance at last?" It turned out that during his previous visits to Moscow his requests concerning this question invariably received evasive answers from people who certainly did not want this meeting to take place. Now they told him that the writer was away, now that he was ill, and so on. "Do you know Mr. Azhaev and Mr. Surkov?" he inquired emotionally." So, when I was here last time and came to the Writers Union, they started to ask me what I found in Paustovsky and added that here already nobody read him..."[2]

Right at that time Paustovsky was making his final recovery after the infarction in a governmental sanatorium in Barvikha, near Moscow. On coming back from the airport I immediately called the writer's wife Tatyana Alexeevna, as I knew well both her and her husband. I asked her to ask Konstantin Georgievich if we could see him. The answer was in the affirmative, and I addressed Korolev with the request to promptly solve the question of auto-transport and admission to the sanatorium, which was in a f special regime zone. And here I got the opportunity to see for myself the evasiveness suspected by Lundkvist. Time was going on, but things didn't move. Eventually Korolev had to speak frankly with me. "You should know," he said conspiratorially, "that the Swedes are going to give Paustovsky the Nobel Prize this year, but he has a competitor there, Sholokhov. Lundkvist has a great authority in the Nobel Committee, and he himself wants Paustovsky to be the laureate. Neither the Writers Union, nor the instance are interested in such an outcome.[3] Hence all the obstacles — because they are like that: if they have voted for someone, they won't accept any protests.

Two days before the end of the Peace Council session Lundkvist, a tall and powerful man, told me, almost stammering from embarassment, that if today the question of going to Paustovsky is not satisfactorily solved, tomorrow he cuts short his visit and goes back home. The ultimatum was quite a success. The talk took place at noon, and already at 2 or 3 p.m., Lundkvist with his spouse, I and a brilliant translator from Swedish, Tamara Tumarkina, made for Barvikha.[4]

[1] It is worth noting, incidentally, that in 1949 Borshchagovsky turned out to be one of the first and foremost victims of the ideological campaign of the "struggle against cosmopolitanism and servility before the West". Apart from being banished from the positions he held in literary journals and in the Soviet Writers Union, he was stripped of his PhD degree, which he had been awarded on June 30, 1947, after defending successfully his thesis "Tobilevich's Drama". As was written in the resolution of the plenum of the Highest Attestation Commission of November 12, 1949, it was "due to the fact that the work of the aspirant is a superficial popular essay on the life and work of the named playwright, whereas Borshchagovsky himself sides with the antipatriotic cosmopolitan group of theatre critics denounced by Soviet public opinion."*

* RSASPH, fond 17, opis' 132, ed. khr. 352, l.29

[2] Mr. Azhaev and Mr. Surkov were blatantly lying. This is not difficult to establish referring to their superiors from the Central Committee. Four years before that episode, on August 12, 1958, in the memo on drawbacks in books publishing policy, which was prepared for the Central Committee's Secretariat, the heads of propaganda and agitation departments in the union republics and the heads of cultural departments stated with irritation: "According to the plan, the works of K. Paustovsky [*Collected works in six volumes*] were to be puiblished in an edition of 75 thousand copies, but the subscription to them was 225 thousand copies. However, 4 months after the subscription had already been finished and 4 volumes had been printed, the department of books trade demanded to print an additional 75 thousand copies. Thus the edition of Paustovsky's works reached 300 thousand copies." Several paragraphs below the Central Committee's secretaries were informed, with no less disappointment, that "[f]or his *Collected Works* K. G. Paustovsky is to receive the royalty of 1 million 200 thousand rubles."*

* *Kultura i Vlast' ot Stalina do Gorbacheva. Ideologicheskie Komissii TsK KPSS. 1958–1962.Dokumenty* (Culture and Power from Stalin to Gorbachev. Ideological Commissions of the CPSU Central Committee. 1958–1962. Documents). — Moscow: ROSSPEN, 1998, pp. 82,83.

[3] As was noted in this chapter, neither the Writers Union, nor "the instance", that is, the Central Committee, were interested in receiving the prize by Sholokhov either. The interest appeared the next year, when he underwent a course of treatment and was permitted to make a trip around several European countries (see doc. 10.14).

The obstacles posed in 1962 for Lundkvist, who wanted to meet with Paustovsky, were connected, first of all, with the personality of the writer, who had recently found himself in the center of a political scandal. The scandal was caused by the coming out in Kaluga, in 1961, of the literary almanac *Tarusskie Stranitsy* (The Pages from Tarusa), which the authorities regarded as ideologically criminal.*

The schemes of "the instances" continued, however, in the next year as well, when N. S. Khrushchev, at the sitting of the Presidium of the CPSU Central Committee on April 25, 1963, suddenly came down on Paustovsky with rude attacks because of his article in the *Pravda* newspaper, published seven (!) years ago, on June 26, 1956. The rage of the first person of the state was caused by the writer's appeal to protect the picturesque surroundings of the town of Tarusa from the barbarous construction of stone quarries. "Somewhere near to the writer's *dacha*", Khrushchev demagogically exclaimed, "gravel is being worked out... And, he says, they are spoiling the landscape, the place is beautiful there. And when he said it, it was explained to him that here a cubic meter is 2 kopecks cheaper than in other places. Look, 2 kopecks, he says... But just think how many fewer apartments the people will get and how longer will be the terms of meeting the requirements of the people..."** and so on. There can hardly be any doubt that the stale material for this passage was offered to the First Secretary of the Central Committee by the writer's "well-wishers" from the Central Committee's Cultural Department. The only question is — what for? Was it for the case if Paustovsky is awarded the Nobel Prize in 1963? So that the party cultural officials would not be reproached by their senior comrades that they did not warn, did not keep an eye on...

* See *Mir Paustovskogo* (Paustovsky's World) No. 11–12, 1998, pp. 15–19 and doc. 10.29.

** *Arkhiv Kremlya. Presidium TsK KPSS. 1954–1964. Dokumenty* (The Archive of the Kremlin. Presidium of the CPSU Central Committee. 1954–1964. Documents) Vol. 1. — Moscow: ROSSPEN, 2003, p. 709.

[4] The comment on the first minutes of the meeting in Barvikha was given, two years later, by Lundkvist himself. In the letter he sent to the Secretariat of the Soviet Writers Union on the occasion of the 70th birthday of the world-known writer I. E. Babel, who died in Stalin's torture chambers, he wrote, among other things: "When several years ago I had the chance to meet with a big and subtle writer, K. Paustovsky, we were taken up with one another at once, having found that we both admired Babel."* When at the end of the same year of 1962 Paustovsky was in Paris, there he also expressed his admiration for Babel's work. Speaking before the students and teachers of Sorbonne, he mentioned Babel, along with B. L. Pasternak and A. A. Akhmatova, among the best Soviet writers. The USSR ambassador in France S. A. Vinogradov did not fail to inform the CPSU Central Committee, by cable, of that definition, which was strongly at variance with what was declared a number of "the best Soviet writers" by the official ideology.**

* See *Voprosy Literatury* No. 6, 2002, p. 329.

** *Archiv Kremlya.Presidium* TsK KPSS. 1954–1964. Vol. 1 — Moscow: ROSSPEN, 2003, p. 1122.

II

From the preface to the Swedish translation of K. G. Paustovsky's book The Distant Years, by the Swedish writer Artur Lundkvist[1]

Today Konstantin Paustovsky occupies a prominent place among the Soviet writers of the older generation. He is one of those who have a voice unlike others; his stylistic mastery is full of feeling, that is why he is so close to his readers. His name is often mentioned alongside Chekhov and Gorky...

But Paustovsky has not always been so popular. When in 1952, having read the writer's selected stories in the English translation, I first became interested in his literary work, no one wanted to speak of him in detail, and he himself refused personal meetings.[2] His life at that time was in many aspects like the fate of Pasternak: they had to recognize him but tried to edge him out into the background.[3] His only luck was not being awarded the Nobel Prize, which would have meant world fame so excruciating in his home country.[4]

But when I came to Moscow again in 1962 and again asked about Paustovsky, the response was quite different.[5] People at once started to admire him and take pride in him. One of his enthusiastic colleagues[6] was ready to arrange a meeting with him at once, although there were some difficulties in it. Shortly before that Paustovsky had had a heart attack and at that time he was under medical supervision in a rest home several dozen kilometers from Moscow.

Very well dressed and courteous as a man of the world, he met us at the rest home's veranda. His sincerity was amazing, he was open and at that time uncommonly tactful, so unlike a typical Soviet man. Paustovsky had the strongly chiseled features of a man who had experienced much but was not harsh. He had the eyes of a dreamer and a poet, but their look remained clear. In spite of his heart condition, he looked young and full of strength for his years, it was easier to imagine him on a walk or in the thick of things than at the writer's table... Konstantin Paustovsky has every reason to be in the number of the Russian writers who have deserved the love and admiration of Swedish readers.

[1] K. Paustovsky. *Ungdomsar.* Tolereds folkbibliotek . Tidens forlag, Stockholm, 1963, Inledning, s.7–9.

[2] What is told by the author of the preface cannot be regarded as a sign of Paustovsky's work being unpopular in the early 1950s. What Lundkvist saw in Moscow in 1952 was one of the numerous consequences of the persecution of creative intelligentsia carried out in the last years of Stalin's rule, especially with regard to those of its representatives who had contacts with foreigners.*

* See Chapter 6.

[3] The authorities, including "the generals" from the Soviet Writers Union, did not generally favor Paustovsky, who understood perfectly and tried to avoid crossing their paths whenever possible. As said A. I. Vaksberg, according to the emigrant writer G. V. Adamovich, whom he met in Paris in May 1968, Kostantin Georgievich felt guilty about Bunin for not having answered his enthusiastic judgement of the first part of *The Tale about Life* sent through the Writers Union.* Allegedly, Fadeev — as Paustovsky explained to Adamovich ** — did not permit him to answer. And later when he found himself in Paris, Paustovsky immediately asked to take him to Saint Genevieve de Bois cemetery, where he laid a bunch of roses on Bunin's grave.***

* According to the information of I. I. Komarov, an expert on Paustovsky's life and work, he might have meant Bunin's response to one of the chapters of the first part of *The Tale about Life* entitled "Korchma na Braginke" (A Tavern in Braginka).

** The meeting of Paustovsky and Adamovich took place on December 24, 1962.

*** *Sovremennye Otechestvennye Zapiski* No. 2, 1999, p. 57.

[4] Lundkvist's notion of the possible reaction in the Soviet Union to receiving the Nobel Prize by Paustovsky rendered here is incorrect. The consequences of its being awarded to Pasternak, in 1958, were the result of circumstances which did not threaten Paustovsky — that is, Pasternak's passing the manuscript of the novel *Doctor Zhivago*, which they refused to publish in the writer's home country, to a foreign publisher.* Paustovsky's situation was altogether different; all his works were published by Soviet publishing houses, and if he had been awarded the Nobel Prize, the reaction of the Soviet authorities to the Swedish Academy's decision would certainly have not been so painful and harsh.

* See Chapter 9.

[5] Lundkvist, of course, did not know what was happening in 1962 about his wish to meet with Paustovsky and what was told by Borshchagovsky (see part 1 of this document).

[6] A. M. Borshchagovsky.

Document 10.28

Some materials related to K. G. Paustovsky's nomination for the 1965 Nobel Prize.

I

From the 1974 memories of the Russian emigrant writer M. L. Slonim A Talk with Paustovsky.

In the fall of 1965, in Rome, I met Paustovsky, and we had a talk, which I immediately wrote down. It took place in a spacious sitting-room of the Plaza hotel, on the noisy Korso street described by Gogol. For some reason Soviet writers like to stay in this old-fashioned hotel, and Paustovsky stayed in it during his visit to the Eternal City. We sat opposite each other, sunk in massive Victorian mastodon armchairs, with velvet elbow-rests and small lace covers on the tops of the backs, and Konstantin Georgievich seemed to me small and frail. He was not tall. The wrinkled face showed his age — he was in his 74th year — and only his eyes, somewhat sad, were shining vividly and in a young way. I knew that of late he had been ill, he had a heart disease, angina... In those days he was greatly disturbed by newspapermen. They knew only too well that his name was mentioned in connection with the award of the Nobel Prizes, which was due in several weeks. But his rival was Sholokhov, most actively supported by all Soviet institutions and communist parties of Europe. The victory of the author of *Quiet Flows the Don* appeared inevitable to me, and when I received a cable from America with the request to interview Paustovsky, "just in case"' for the *New York Times,* I decided not to even mention it to him. But at the very beginning of our talk he asked me directly how I estimated his chances. Not willing to upset him, I said, as mildly as it was possible, and in cautious expressions, that prominent American journalists emphasized the political advantages of Sholokhov. By choosing him the Swedish Academy could sort of counterbalance, or smooth over, the effects of their brave action in 1958, awarding the prize to Pasternak. I also told him about the campaign, overt and behind the scenes, of the Kremlin and its European partisans and about Sholokhov's recent trip to Stockholm. Paustovsky shrugged his shoulders and said: "I am a shy man, I cannot elbow" and after being silent for some time, he asked in a somewhat muffled voice: "So, it's not even worth dreaming of it?" I answered: "You can dream, but you shouldn't hope." Afterwards, as I learnt from our common friends, Paustovsky believed until the last moment in what he called "the triumph of justice"; he considered Sholokhov's success to be "the result of a political intrigue", and it was difficult for him to put up with it.

The New York emigrant newspaper *Novoe Russkoe Slovo* of March 30, 1974.

II

From the letter of L. N. Delektorskaya[1] *to Irina Butylskaya from Paris, of September 26, 1991.*

... But about the Nobel Prize, perhaps, no one can give you an explanation more trustworthy than what is known to me. At the moment of the election[2] K. G..was in Rome at the international writers congress. I also came there (and spent a month with him and Tatyana Alexeevna — at first in Rome and then on Capri). In Rome, on the eve of the election for the Nobel Prize for Literature, all press and television got ready to interview next morning the elected laureate Paustovsky. In the corresponding community in Rome nobody doubted that it would be him who would be elected.[3] Only he himself was categorical, with a smile convincing the excited me that it was simply impossible. "Don't be upset, Lidia Nikolaevna. You should understand that it is in fact impossible. The old men" (as he generalized the electors[4] for the Nobel Prize) cannot afford to choose me. Last year they gave the prize to Pasternak, which caused a scandal. The second year running they won't dare, with respect to the Union, to give the prize to a Soviet writer, again unorthodox.[5] It is out of the question!"

The prize was awarded to Sholokhov... Although the choice of Sholokhov was commented on in a negative way...

From the funds of Paustovsky's museum in Kuz'minki Park, Moscow

[1] According to the film director E. A. Ryazanov, "Lidia Delektorskaya, Matisse's girlfriend, [...] was his sitter, maid, clerk, took care of his sick wife, became his model and beloved. Matisse painted her a lot in the last years of his life, and she bought his paintings on her salary. In 1945 she bought six paintings as a gift to the USSR, Matisse giving the seventh..."*

* *Izvestiya* March 30, 2001, p. 7.

[2] She means the election of the next Nobel laureate by the Swedish Academy.

[3] I. I. Komarov in his already mentioned letter to W. Kasack* wrote, among other things, that he happened to hear that the general secretary of the International Writers Association G. Vigorelli as the chairman of the opening of the writers congress, suggested "greeting Konstantin Paustovskly, who is present in the audience, as a new laureate. of the Nobel Prize for Literature.**" Whether these rumors corresponded to the reality or not, they were probably giving the atmosphere of expectation of the possible news from Stockholm authentically enough.

* See note 29 to this chapter.

** This is certainly a slip of the pen. Vigorelli could have spoken only about the presence in the audience of one of the most probable candidates for the 1965 Nobel Prize.

[4] She meant the 18 members of the Swedish Academy electing the next Nobel laureate.

[5] An obvious example of a memory aberration. Pasternak was awarded the Nobel Prize seven years before the described events, and Paustovsky could have spoken only about Pasternak's prize in general. As for the "unorthodox", Paustovsky had many reasons to suppose that the authorities continued to regard him as one of those. An employee of the *Novy Mir* journal, L. A. Levitsky, wrote in his diary, on July 15, 1963: "The day before yesterday I managed to visit Konstantin Georgievich.* He is depressed... In Moscow Semen Gekht died, whom K. G. loved and knew well. He told me about the obituary that he had written. They phoned him five times from the *Literaturka* **, cutting out words and phrases. ... K. G.'s text said that in the

years of Stalin regime Gekht was arrested." "Stalin regime" was substituted for "the epoch of the personality cult".*** K. G. got furious and asked the employee who was working with his material: "You do not want to mention this name because you are getting ready to bring it back?"****

* To the village of Tarusa, where the writer usually spent summer months.

** *Literaturnaya Gazeta.*

*** The phrase from Paustovsky's obituary "Our Youth", which was changed by the *Literaturnaya Gazeta* (June 13, 1963, p. 4) finally sounded as follows: "In the years of the personality cult fate did not spare him..."

**** *Znamya* No. 7, 2001, p. 121.

Document 10.29

From the letter of the head of the Chief Department of Protection of Military and State Secrets in Press of the USSR Council of Ministers P. K. Romanov to the Russian Federation Bureau of the CPSU Central Committee, of December 23, 1961.

This October the Kaluga *oblast'* book publishing house issued the illustrated literary collection *Tarusskie Stranitsy* (Pages from Tarusa) Familiarization with the content of the collection shows that it includes a number of works deficient in terms of their ideological and artistic qualities, distorting the life of our village and of Soviet people...[1]

The *Nedelya* No. 47 of 1961, publishing a report on the writers meeting in this weekly's office, quotes the speech of K. Paustovsky, in which he explains that the reason for publication of the collection was the wish of a group of old writers to give young writers an opportunity to publish their works, as, according to him, "our literary youth has difficulty publishing."

In his memo to our department the head of the Department of Protection of State Secrets in Press of the Kaluga *oblast'* comrade Ovchinnikov reported that the Kaluga department, where the collection *Tarusskie Stranitsy* had been submitted for control, considered its publication in such a shape undesirable,[2] of which they informed the CPSU *oblast'* committee in written form. The *oblast'* committee conducted a meeting for discussing the collection. In spite of the fact that at this meeting a number of works supposed to be published were seriously criticized by the literary community of the city of Kaluga, owing to the pressure of the authors and members of the editorial board headed by the writer K. Paustovsky, the collection was published.[3] At the present time, out of the 75 thousand copies planned for the edition only 30 thousand have been produced,[4] the production of the remaining part of the edition being suspended because of a lack of paper.[5]

Taking into account the above, we think that having published the collection *Tarusskie Stranitsy*, the Kaluga publishing house made an error. Considering it expedient not to carry out the further printing of the collection, we ask to commission the corresponding departments of the CPSU Central Committee to examine this question.[6]

RGANI, fond 18, opis' 2, ed. khr. 383, l.74–77, published in the journal *Mir Paustovskogo* No. 11–12, 1998, pp. 18–19.

[1] The resolution of the CPSU committee of the Kaluga *oblast'* gave a still stricter formulation — the almanac was called an assemblage "of ideologically pernicious works".*

* See *Mir Paustovskogo*, p. 18.

[2] The circumstances connected with the ban on publication of the almanac *Tarusskie Stranitsy* are described in detail in the article by Ilya Milshtein: the *Ogonek* No. 14, 1989, pp. 23–25.

[3] The leaders of the Kaluga publishing house had serious grounds for fearing that printing of the almanac's edition could be suspended any moment. That is why they "took the decision to print quickly a thousand copies. And then to print quietly the rest of the edition... So, they printed quickly a thousand, and then 30 thousand more. Unfortunately, they did not have time to do any more." *

* *Ogonek*, p. 24.

[4] As is seen from note 3, 31 copies were printed.

[5] The greater part of the printed 31 thousand copies did not reach the reader, having been destroyed. Symptomatically, in the Russian State Library, which received two compulsory copies of every book published in the USSR, the catalogue mentions only one copy of the almanac.*

* As wrote Andrew Field in the preface to the collection of translations from *Pages from Tarusa* published in the USA in 1963, according to reliable sources, only several thousand copies fell in the hands of readers. See *Mir Paustovskogo*, p. 28.

[6] The authorities started to harbor a grudge against Paustovsky long before 1961. As early as at the beginning of 1957 the writer was mentioned in the CPSU Central Committee's closed letter entitled "On strengthening the political work of party organizations in the masses and stopping sorties of anti-Soviet hostile characters". The letter was prepared in connection with the protests of students and creative intelligentsia after the suppression of the Hungarian revolution by the Soviet Army, in October and November 1956. When the letter was being discussed at the Party organization of the Matematical Institute of the USSR Academy of Sciences, a revealing dialogue arose. The associated member of the Academy A. A. Markov asked the man who was making the report: "How did Paustovsky criticize the Soviet reality?" The answer was rather unintelligible but energetic: "The *Literaturnaya Gazeta* wrote, although briefly but clearly that his speech was anti-Soviet."*

* CASPHM, fund 7334, opis' 1, ed. khr. 9, l.1.

Document 10.30

A. A. Akhmatova's notes, in connection with preparing for an interview with M. V. Ardov, and Ardov's comment received by the author on February 18, 2000.

January 28, 1965 (Hospital).[1]
For Misha

1. About Modigliani's first prose. Appeared in America and in England.[2]
2. Enekava translated as early as before the war. Nabory in 1927.
3. From America, Frost and R. Yakobson, and Malia. Dissertation also there.
4. From symposium — N. Sarraute and Czechs.
5. Delegation of Poles. (To Budka).[3]
6. Bagryana and a Serb.
7. Dissertations (starting with Landolfi).
8. In 1962, a Swede, doctor Eric Mesterton.[4]
9. Einaudi found and visited in Moscow for taping the voice (in Sweden there are translations of poems).[5]

9.[6] Having nothing to do with poetry, being in USSR on business visit David Carver (general secretary of the PEN-Club) asked to take him to Budka, and a simply renowned traveler Miguel Otero Silva (Venezuela) visited Budka this summer, and Boll with wife and two sons. It was to him that Akhmatova said: "A poet is a person from whom nothing can be taken away and to whom nothing can be given."[7]

However, are all these examples necessary if we remember that Harvard University included the collection *Three Poets* (Gumilev, Akhmatova, Mandelshtam) in the series *The Craftsmen of Word.* We could also look into Prince Dmitry Obolensky's Oxford Russian anthology, where he calls Akhmatova the first of the living Russian poets, and into the *Monde* about the Russian poets' visit to Paris (1965, November-December). (L'auteur du Requiem).[8]

Zapisnye Knizhki Anny Akhmatovoi (Anna Akhmatova's Notebooks) (1958–1966), Torino, Einaudi edit., 1996, p. 691.

[1] The notes are a synopsis for the forthcoming interview to M. V. Ardov (see doc. 10.30-II). It was made in Botkin Hospital in Moscow, where Akhmatova stayed for more than three months after her fourth infarction (from November 9, 1965 to the second half of February 1966). In the original there is a slip of the pen in the date — instead of January, it should be December 28, 1965. The words which were abbreviated in the original are given here in their complete form, in accordance with the published text of Anna Akhmatova's notebooks.

[2] Akhmatova means her memories of Modigliani and the countries where their translations into English were published.

[3] Budka (the Booth) is what Akhmatova called the *dacha* in Komarovo given to her by the Literary Fund.

[4] She means the news of her nomination for the Nobel Prize, which reached her in 1962 through Mesterton (see this Chapter), which also follows from her mention below, also in this notes, of her "formulation " (see note 7 below and doc. 10.30-II).

[5] Mesterton was the first translator of Akhmatova's poems into the Swedish language.

[6] The author's slip of the pen in the numeration of points.

[7] Akmatova invariably pronounced this phrase when a talk touched upon the topic of the Nobel Prize in connection with her name (see doc. 10.30-II). Probably, in this case H. Boll, who visited her on August 17,1965 in Komarovo, said something on the same subject. Here referring to herself in the third person, Akhmatova appears to be offering Ardov the already ready phrase for publication.

[8] The information from the *Monde* would be again mentioned by Akhmatova a couple of weeks later, between January 6 and 19, 1966 (see note 5 to doc. 10.31-I).

II

For these notes marked "for Misha", that is, for me, I can give the following explanations.

As I remember, at the beginning of or in spring of 1965, the Novosty Press Agency, or to be more exact, an old friend of mine, Alexander Alexandrovich Avdeenko, who was the editor of the Agency's Cultural Bulletin, asked me to interview Akhmatova. I fulfilled this request. The work was laborious. Anna Andreevna said that she never allowed journalists to give her direct speech in their publications. However, for me, taking into account our special relationship,[1] she was making an exception. Along with direct quotation, I included into the interview some short additions of different kind, which could be interesting for the reader, because at that time Akhmatova appeared to many people a half-banned figure. Here is one of such additions. In the 1960s, until her death, rumors permanently circulated in the society that she might be awarded the Nobel Prize. That is why I considered it necessary to mention one of her well-known phrases, with which she always responded to such gossip. Here is this phrase (she called such utterances "formulations"): "A poet is a person from whom nothing can be taken away and to whom nothing can be given."[2]

The interview entitled "Attitude to My Country" was published, with reference to the Novosty Press Agency, by 15 or 20, or may be even more, provincial newspapers (unfortunately, not a single copy of this publication remained in my archive). Akhmatova was satisfied. There was talk of the possibility of another, more extensive interview. I suppose that the notes discussed here are connected with it. In them she enumerates the modern writers with whom she met those years — Robert Frost,[3] Roman Yakobson, Nathalie Sarraute, Elisaveta Bagryana, whom she translated. In these records Akhmatova wanted to show the contemporary reader that she was still alive and still continuing to take part in the literary process. And that she is far from being some figure infected with plague, removed from literature by the Soviet power.[4] That is why she reminds indirectly of information of her nomination for the Nobel Prize, coming from different sources, or, to be more exact, of her invariable reaction to rumors like that. That is why she adds that her name is in the Oxford Russian anthology and the Harvard series *The Craftsmen of Word.*

In my opinion, that is the sense of Akhmatova's record given here. In it we see a draft of what was supposed to become the content of the next interview. Which, alas, never took place. M. Ardov.

[1] Coming from Leningrad to Moscow, Akhmatova repeatedly received accommodation and stayed for a long time, in the 1930–1960s, at the apartment of M. Ardov's parents on Bolshaya Ordynka Street * and had known M. Ardov since his early childhood.

* "On the legendary Ordynka", as Akhmatova herself called the Ardovs' place. "Thanks to Anna Andreevna, we started to use this expression in our family", M. V. Ardov wrote later. "I was brought to this apartment when I was one year old and I lived there until I was thirty, so for me the word combination "the legendary Ordynka" means, among other things, my childhood, adolescence and youth." (*Novy Mir* No. 4, 1994, p. 3). M. V. Ardov's half brother, well-known actor Alexei Batalov, in the interview on the occasion of his 75th birthday told the story of the apartment with which Akhmatova associated the idea of "the legendary Ordynka": "It [the apartment] was in the first Moscow house of writers, now pulled down... Mama and Vitya (Mikhail Ardov's father and Batalov's stepfather) raffled this small apartment in a card game with Shostakovich: they were given the coupons, and who would live where was decided by the game of sixty-six." (*Izvestiya,* November 20, 2003, p. 11).

[2] Here is an example of how the "formulation" was used in communication with Akhmatova's interlocutors. According to the entry of February 22, 1963 in L. K. Chukovskaya's diary, when an argument arose whether the awarding of Lenin Prizes was always fair, Akhmatova said, not only about these prizes: "The Lenin Prizes, like all prizes, are sometimes given rightly, but more often, wrongly", and added: "Let us agree once and for all: a poet is a person from whom nothing can be taken away and to whom nothing can be given."*

* L. K. Chukovskaya. *Zapiski ob Anne Akhmatovoi* (Memories of Anna Akhmatova) Vol. 3, Moscow: Soglasie, 1997, p. 35.

[3] The meeting with the American poet Robert Frost took place on September 4, 1962 in Komarovo, at the *dacha* of Akhmatova's neighbour, literary academician M. P. Alexeev. According to L. K. Chukovskaya, this is how Akhmatova remebered this event arranged at the official level: "It was, of course, impossible to show Budka to foreigners; the banquet was arranged at Alexeev's... We were sitting in comfortable chairs opposite each other — two old people.* I was thinking: when he was admitted somewhere, I was expelled from somewhere; when he was awarded, I was vilified, but the result is the same: we are both candidates for the Nobel Prize.**

Here is material for philosophical thought." Further on, in response to Akhmatova's words about the Nobel Prize, Chukovskaya continued: "I confessed that I was both glad for her and worried about her nomination for the Nobel Prize. I was afraid that the Pasternak history might repeat."***

* Frost was 14 years older than Akhmatova and died soon after coming back to his homeland, on January 29, 1963.

** As is clear from the letter of the Swedish Academy member Dag Hammarskjold of March 11, 1955, Frost was among the nominees for the 1954 Nobel Prize. (K. Espmark. *The Nobel Prize in Literature. A Study of the Criteria behind the Choices* — Boston, Mass., 1991, p. 106).

*** L. K. Chukovskaya. *Zapiski ob Anne Akhmatovoi* Vol. 2 — Moscow: Soglasie, 1997, pp. 509, 511.

[4] The poetess was not wrong thinking like that. Here is an example showing that she still retained her popularity in the society, especially among the young people. In November 1956, soon after the defeat of the Hungarian uprising, in a number of Moscow higher educational establishments, komsomol meetings took place, where students spoke emotionally in favor of the speedy liberalization of social life in the country, against the party's paternalism over the komsomol and so on. Answering these speeches and calling for "responding by fight to the attacks on socialist realism", the literary critic D. D. Blagoi, future academician, noted that he was "surprised by the loss of patriotism", when "instead of our native, national melodies our youth learns different Western melodies," and "university newspapers are full of the names of Tsvetaeva, Akhmatova, Pasternak."*

* CAODM, fund 2472, opis' 1, ed. khr. 18, l.74–75.

Document 10.31

Evidence of rumors of A. A. Akhmatova's nomination for the 1965 Nobel Prize.

I

Zapisnye Knizhki Anny Ahmatovoi (*1956–1966*), Torino, Einaudi editore, 1996
P.177. July 31, 1964.

Totya brought Peter.[1] He passed to me Oxford University's offer about Prix Nobel.

P. 483. August 26, 1964.

Yesterday Faina came and Peter Norman... Peter told Faina and Nina[2] that I am invited to Oxford[3] because they want to nominate me for the Nobel Prize. Gibberish!!

P. 636. July 10, 1965.

The Czech tried his voice... On June 11 the Englishman Williams came.[4] He left the newspaper about the Nobel Prize. Nonsense?

P. 695. Between January 6 and 19, 1966.

... the *Monde*. 1965. November. (About the Soviet writers' visit to Paris) and about Prix Nobel.[5]

[1] Totya — A. N. Izergina, art critic. Piter — P. Norman, British art critic, translator of Akhmatova.

[2] Faina — F. G. Ranevskaya, renowned Soviet actress. Nina — N. A. Olshevskaya, M. A. Ardov's mother.

[3] Akhmatova was in Oxford at the beginning of June 1965. She was given the mantle and diploma of Oxford University's honorary doctor on June 5, 1965.

[4] An unidentified person.

[5] Having visited Akhmatova in Botkin Hospital on December 19, 1965, L. K. Chukovskaya that very day made the following entry in her diary: "'The *Monde* newspaper', Anna Andreevna said in a low voice, 'has published my biography... To my surprise, everything is correct, to the smallest detail. And it ends by saying that *this man** — you understand who I mean? — should guess that this year's prize should have been given to the author of *Requiem.* ** Whereas he has been given the prize simply for fear: if it is given to Akhmatova, will the story of Zhivago repeat itself?'"***

* Mikhail Sholohov, who became the 1965 Nobel Prize winner.

** A graphic example of the naivety of the author of the article in the Paris newspaper.

*** L. K. Chukovskaya. *Zapiski ob Anne Akhmatovoi.* Vol. 3 — Moscow: Soglasie, 1997, p. 307.

II

A. A. Akhmatova. Collected Works in six volumes. Vol. 2. Book 2. — Moscow: Ellis Lak, 1999.

P. 222. Akhmatova's two-line poem:

But who could think that nineteen sixty four
Deep down had such things in store for me.

P. 490. Comment from N. V. Koroleva.

"That is about awarding the Etna-Taormina Prize and the visit to Italy in December,[1] about the nomination for the Nobel Prize and the intention of Oxford University to give Akhmatova the honorary doctor degree."[2]

[1] The prize was given in the city of Catania (Sicily) on December 12, 1964. On the trip the poetess was accompanied by A. T. Tvardovsky. His speech at the ceremony was published in the *Literaturnaya Gazeta* of December 17,1964.

[2] See note 3 to doc. 10.31-I.

III

L. K. Chukovskaya. Zapiski ob Anne Akhmatovoi. Vol. 3 — Moscow: Soglasie, 1997, p. 243.

Page 243.

"The Oxfordians took it into their heads to nominate me for the Nobel Prize, they asked me if I allowed that. But at that moment I was having a fit of negativism from which I have been suffering since childhood. I answered: 'A poet is a person to whom nothing can be given and from whom nothing can be taken away.'"

IV

A. G. Haiman. *Stories of Anna Akhmatova.* — Moscow: Khudozhestvennaya Literatura, 1989.

P. 121.

"The Swedes demand the Nobel Prize for me", she said to Ranevskaya and took out a newspaper cutting from her bag. "Here, printed in Stockholm." "Stockholm", said Ranevskaya. "How provincial!" Akhmatova laughed: "I can show you the same from Paris if you like it better."

Document 10.32

From the memories of V. Kardin[1] *about A. A. Akhmatova's funeral, on March 10, 1966.*

... At breakfast we took the decision: Viktor Nekrasov goes to Leningrad to the burial service. I stay in Komarovo so that his mother, Zinaida Nikolaevna, would not feel lonely when waiting for the funeral. According to the will, the funeral is to be at the cemetery here. Its time is not fixed, the place for the grave is not determined. Trying to make inquiries, I looked in on the director of the Writers' Creative Work Home. He sat burying his bold head in his hands in despair... Only by lunch time did they [the Leningrad authorities] deign to grant permission...

Viktor Platonovich[2] came back after I had already been to the cemetery twice. Soon a packed minibus with residents of the Writers' Home, groaning and creaking, made for the graveyard. Crowds of people were coming from commuter trains, arriving from Leningrad and Vyborg. Black greatcoats of navy officers arrested attention. Muscovites also appeared. The official envoy Sergei Mikhalkov,[3] the unofficial Arseny Tarkovsky, Lev Kopelev and someone else... Today Tarkovsky looked completely exhausted. He had not been able to sleep on the train at night. Now, in the late afternoon, his short coat did not protect him from the cold. He had difficulty moving on a prosthesis and was shivering all over.

Exercising my right of an aborigine,[4] I squeezed Tarkovsky into the minibus. Mikhalkov, also lightly clad, in a cloth cap, stamped his feet to get warm, clapped his hands on his arms, saying repeatedly: "When will they bring the little casket?" He flung open the door of the minibus: "Girls, who will warm me?" One of the "girls", a writer's widow, beckoned him. Nekrasov stood up: "I'll warm him. I'll punch his mug in." I objected: "You won't do it." Contrary to his nature, Nekrasov submitted... Later, in K. Chukovsky's diary I read that Mikhalkov approved of the operation of Akhmatova's secret burial:[5] "Thank God, we have GPU."[6]

The funeral cavalcade arrived when it was still light. I kept close to Tarkovsky. "Won't they let me speak?" — he repeated feverishly. They won't. They won't let Kopelev either. And someone else who was not included on the list approved in advance. The ceremony was conducted by a children's writer whose name nobody knew. Mikhalkov, stammering, uttered a dozen standard phrases.

However — such things happened as well — the bureaucratically formal ritual was unexpectedly broken. The Leningrad professor Makogonenko,[7] an expert on Karamzin, Radishchev and Fonvizin, found the right words about Akhmatova's greatness and the disgrace in store for her persecutors. The elementary truth sounded like a challenge. People at the village cemetery had been waiting for it. It was saving them from one more humiliation. The authorities had tried to obtain a wholesome plot. Akhmatova's ever poisoned years should have been completed by vacuous formal speeches. Professor Makogonenko broke the plot...

People still continued to arrive. Someone climbed a tree. Someone sitting on the branches was filming the funeral. I nodded at Nekrasov: and yet the film will remain. It will be seen some day. He did not share my hopes — most probably, those were foreign correspondents. However, separate shots remained intact and much later flashed on the TV screen.

Several decades later details came to light of the filming of Anna Akhmatova's funeral in Leningrad, in Nikola Marine Cathedral.[8] At one time a fuss was made around them The director and two cameramen were persecuted as if they were to blame for something. It is difficult to say how all that would have ended if the secretary of the Cinema Workers Union A. Karaganov had not firmly taken the side of the "plotters". Why don't we appreciate the lone persons, who, in spite of their position, their status, sometimes went against the stream...[9]

Znamya No. 2, 2003, pp. 148–150.

[1] The literary critic E. V. Kardin signed his publications as V. Kardin.

[2] V. P. Nekrasov.

[3] At that time S. V. Mikhalkov was the secretary of the Board of the Soviet Writers Union and the First Secretary of the Board of the Moscow organization of the Russian Federation's Writers Union.

[4] When Akhmatova died, Kardin and Nekrasov were in Komarovo, near Leningrad, where they were having a rest in the Writers Creative Work Home belonging to the Soviet Writers Union.

[5] Chukovsky's daughter Lidia Korneevna, in her turn, immortalized in her diary the utterance on the same occasion of another official of the Soviet Writers Union'

Board, K. V. Voronkov. Before the civil funeral ceremony in Moscow, after which the casket was sent to Leningrad, he gave the following instructions to his subordinates, Mikhalkov among them: "The main thing is to do everything quickly."*

This disgusting petty bustle around the casket of the great poetess confirms once again the deliberately shocking truth of the words said by the master of creating hyperbolical type, poet and translator of the 1920s and 30s V. G. Shershenevich, an old friend of V. V. Mayakovsky. After the latter's funeral he wrote in his notebooks with cynical frankness: "The death of a big man can never be without surrounding triviality." He confirmed this idea by the description of the following episode which he involuntarily witnessed: "Mayakovsky lay in the casket, and some official attached to literature was bustling around him and, dusting off his shoes with a handkerchief, demanded that the guard close the windows, saying a phrase which was truly a stroke of genius in its triviality: 'Don't you see that 'he' can become dusty?'"** Was there a great difference between the trivial cares of the "official attached to literature" and Mikhalkov's similar triviality ("When will they bring the little coffin?")?

* *Znamya* No. 1, 2003, p. 156.

** See L. Fleishman. *Boris Pasternak v Dvadtsatye Gody* (Boris Pasternak in the Twenties) — Saint-Petersburg: Academicheskii Proekt, 2003, p. 345.

[6] GPU — the Chief Political Department, the state security organ of the 1920s.

[7] G. P. Makogonenko, professor of Leningrad University.

[8] A slip of the pen. In that cathedral Akhmatova's funeral service took place.

[9] About A. Karaganov see note 55 to Chapter 4.

Document 10.33

From Danish writers to the Minister of Culture E. A. Furtseva, about awarding M. A. Sholokhov the Nobel Prize and about the "case of Sinyavsky and Daniel".

We got great satisfaction from the news of awarding the Nobel Prize to Mikhail Sholokhov, which we received a month ago. We in Denmark were pleased to note that owing to this event Sholokhov, and together with him Soviet literature occupied their natural and deserved place in Europe. However, we are watching with concern some current developments in the Soviet Union which undermine the beginning cooperation and, among other things, cloud the joy of awarding the Nobel Prize. The arrest of Andrei Sinyavsky (Abram Terz) and Yuli Daniel (Nikolai Arzhak) means using the methods unworthy of the Soviet Union and incompatible with establishing positive international cultural contacts, including those which found their expression in awarding the Nobel Prize... Under the telegram there are 21 signatures.

Tsena Metafory ili Prestuplenie i Nakazanie Sinyavskogo i Danielya (the Price of a Metaphore, or Crime and Punishment of Sinyavsky and Daniel) — Moscow: Kniga, 1989, p. 16.

Document 10.34

Appeals of writers and cultural figures of foreign countries to the Nobel laureate M. A. Sholokhov about the "case of Sinyavsky and Daniel".

I

A telegram from Italy

We were puzzled and shocked on learning of the arrest of two prominent Soviet writers, Sinyavsky and Daniel, accused, as far as is known, of publishing their works abroad, which is not a crime according to Soviet laws. We believe that this question will soon be solved in conformity with liberalization, whose symptoms we were pleased to watch in the last years in Soviet cultural life. We are sure that freedom and tolerance uniting cultural workers of the whole world will make you interfere in order to give back your colleagues Sinyavsky and Daniel the freedom to work and communicate with people from other countries. Fifteen signatures under the telegram.

II

A telegram from Mexico

Mexican writers ask the Nobel laureate Sholokhov to make his kind effort for favorable resolution of the case of the Soviet writers Andrei Sinyavsky and Yuli Daniel.

Thirty-five signatures under the telegram.

III

A telegram from Chile

Chilean writers and literary critics ask for the interference of the Nobel laureate, their colleague Mikhail Sholokhov with the Soviet authorities in order to obtain immediate release of the writers Andrei Sinyavsky and Yuli Daniel, whose imprisonment seriously compromises the prestige and significance of Soviet literature.

Seven signatures under the telegram.

Tsena Metafory..., pp. 17–18.

Document 10.35

From Mikhail Sholokhov's speech at the XXIII CPSU Congress, on April 1, 1966 and Lidia Chukovskaya's Open Letter addressed to him, of May 25, 1966.

I

...We call our Soviet homeland Mother... I think that any person will understand: there nothing more blasphemous and loathsome than to slander your mother, to vilely insult her, to lift up your hand against her! (Tumultuous applause.) I'm ashamed not for those who slandered our Motherland and flung dirt at everything that is most dear to us. They are amoral. I'm ashamed for those who tried and are trying to protect them, whatever motives for this protection. (Prolonged applause)... Some people, using words about humanism as a cover, are moaning that the sentence is too strict. Here I see delegates from the party organizations of our dear Soviet Army. What would they do if in some unit of theirs traitors appeared? It is well known to them, our warriors, that humanism is far from being dithering. (Prolonged applause.) And there is one more thing I'm thinking about. Should these characters with black conscience have been caught in the memorable twenties, when trials were not based on strictly differentiated articles of the Penal Code but were "guided by the revolutionary sense of justice" (applause), the punishment for these werewolves sure would have been different! (Applause.) But still, some people say that the sentence was too "severe"...[1]

Pravda, April 2, 1966, p. 5.

II

At the XXIII party Congress you, Mikhail Alexandrovich, rose to the rostrum not as a private person but as a "representative of Soviet literature". Thus you gave each writer the right [...] to pronounce their judgement about the ideas you conveyed allegedly on behalf of all of us.

Your speech at the congress can indeed be called historic. In all the centuries-old existence of Russian culture I cannot remember another writer who, like you, would have publicly expressed regret not about the sentence imposed by the judges being too severe but about it being too mild... The business of writers is not to persecute people but to stand up for them. That is what we are taught by the great Russian literature in the person of its best representatives. That is the tradition you broke when loudly expressing regret that the court sentence was not severe enough!...

The trial of the writers Sinyavsky and Daniel was outwardly conducted with observation of all formalities required by the law. From your point of view, it was its drawback; from mine, a virtue. However, I object to the sentence... Because the criminal prosecution of Sinyavsky and Daniel was unlawful in itself... A writer [...] can and must be tried in a criminal trial for any wrongdoing — but not for his books. Literature is not under criminal jurisdiction. Ideas should be opposed by ideas, and not by prisons and camps... That is what you should have told the audience if you had really taken the floor as a representative of Soviet literature. But you delivered your speech as its apostate. Your disgraceful speech will not be forgotten by history... And no honors, money, national and international prizes will avert this sentence from your head.[2]

Tsena Metafory..., pp. 502–506

[1] Three years later the prominant literary critic, associate member of the USSR Academy of Sciences V. R. Shcherbina made an attempt to strengthen the sense of Sholokhov's shameful utterances; true, it was not in public but once at table:

"They say we have disorderliness, vacillation — it would be worth shooting some millions of people, and everything will be quiet, without any disorder." And touching upon "The case of Sinyavsky and Daniel", he said: "In general our KGB is more like an almshouse. They come to the IMLI* and say, it would be good if your representative went to Sinyavsky, learnt what he is doing, what he is going to do after he has been released. And that instead of adding a new term of imprisonment to him."**

* IMLI — the Institute of World Literature of the USSR Academy of Sciences.

** From diary notes of the literary critic V. Y. Lakshin, of August 29, 1969; *Druzhba Narodov* No.5, 2003, pp. 171,172.

[2] The reaction of A. T. Tvardovsky, chief editor of the *Novy Mir* journal, proved to be still more emotional. On April 3, 1966, a day after Sholokhov's speech, he wrote down in his diary: "Corrupt old man, but he could have done so much good for literature, were he — this is no slip of my pen — a bit more independent, free and humane. Bitterness and shame."* Nonetheless, a year and a half later Sholokhov again attracted the attention of the writers community to himself by his unseemly utterances. This time they are about A. I. Solzhenitsyn, in connection with expelling him from the Soviet Writers Union, which was considered at the sitting of the Union's Secretariat at the end of 1967. In his letter to the Secretariat Sholokhov stated: "Solzhenitsyn is either a mentally ill person dangerous to society, a malicious graphomaniac or, if he is healthy, a malicious anti-Soviet person, a direct enemy."** Demanding to expel Solzhenitsyn from the Union, he suggested "not allowing him to write", which can be practically carried out only in conditions of the strictest imprisonment. Dwelling upon Sholokhov's suggestions, A. T. Tvardovsky in his letter to K. A. Fedin, in January 1968, claimed that Sholokhov "discredited himself in the eyes of his readers and admirers,.. instead of showing an example of a worthy [...] artistic attitude to the artist."***

* A. T. Tvardovsky. *Rabochie Tetradi 60-kh Godov* (Working Notebooks of the 60s). *Znamya* No. 4, 2002, p. 167.

** *ibid*, No.10, 2002, p. 160.

*** *ibid*, p. 186

Chapter 11

Solzhenitsyn, Sakharov, Brodsky, Gorbachev: Unaccepted by Society

The personalities of these four Soviet Nobel Prize winners, their fates connected with the two decades which finished the existence of the Soviet Union have little correlation with each other. A yesterday's prisoner of the Stalin empire of terror, deprived of rights, and the General Secretary of the CPSU Central Committee, whose opportunities for limitless absolute power objectively exceeded all thinkable prerogatives of His Majesty the Emperor; a poet, of Brezhnev's *dolce vita* period, also without rights and without shelter, and a brilliant scientist and thinker, to whom eternally indebted are, among other people, the homespun national patriots, who hated him in his lifetime and got the enviable possibility to frighten the rest of the world by the image of the deadly weapon created by his genius.

There is only one thing that unites them. In their social, creative or political activities that brought each of them the century's most prestigious award, all four were rejected by the greater part of the society to which they belonged. More exactly, in those activities they remained, for a consistently large number of their compatriots, beyond positive recognition of their merits and were perceived as an alien element unacceptable to the majority. As it was figuratively formulated in a prose book by the poet Igor Guberman, the famous author of caustic satirical quatrains, "the people is great and ruthless Procrustes in relation to everything and everybody who is out of the common."[1]

The greater part of the documents in this chapter which concern Alexander Solzhenitsyn (doc. 11.1–11.18) are written by him, in his characteristic, rigid and passionate manner. There is almost no need to precede them with my own comments, like in the previous chapters. It is impossible to find a better commentator on Solzhenitsyn than himself, especially in the epic context of the Nobel Prize. Whatever question might arise, in his books you can find all the necessary explanations.

Anna Akhmatova, noting her first impressions from her personal meeting with him in January 1963, almost immediately after the sensational appearance of the story *Odin Den' Ivana Denisovicha* (A Day in the Life of Ivan Denisovich)[2] in *Novy Mir*, noted the clarity and simplicity of communication with the future Nobel laureate, easiness "from the first minute."[3] Ten yeas later Lidia Chukovskaya in her records colorfully generalized the reader's emotional perception of Solzhenitsyn's work, of his writing. She asks herself a rhetoric question: "What does *one* Solzhenitsyn mean for our country?" And immediately answers it herself: "Only lips parched with thirst will drink fresh water with such frenzied greed as that with which readers devoured Solzhenitsyn's published and unpublished short stories, novels and tales, where they found the long-awaited words, expected by millions of people and finally uttered."[4]

The authorities perceived him quite differently. His unyieldingness, which was often an aim in itself not to tolerate any compromises, his persistence in moving towards the chosen goal, even when it seemed better to stop for a while; his complete non-acceptance of any concession if it required waiving personal dignity — all these qualities of a fighter helped him to survive in a dramatic and unequal combat with the powerful machine of totalitarian power and emerge as the victor.

But the authorities, with their pathological intolerance of heterodoxy, did not leave his activities unanswered. They insistently formed a rigid zone of 'no man's land' around the disfavored writer, repeatedly using against him the method of vilification long worked out by the communist propaganda, where collective protest letters were made up and respected people made, by force or by threats, to put their signatures to them.

In case of Solzhenitsyn, the organizers of such acts once took the liberty of using a completely exotic method of coercion. Here is a phantasmagorical story told by the literary critic and translator L. P. Sinyanskaya, who heard it from the secretary of the Board of the Soviet Writers Union K. M. Simonov: "He was going to the South, by train. At some small station the train stopped, and Simonov was strolling along the platform. No sooner had he wondered why the train had stopped so long, and not according to schedule, that he saw the chief of the train running along the carriages as if looking for something. It turned out that he was looking for Simonov. He told him that he had just received a radiogram with the order to get Simonov's signature under the letter against Solzhenitsyn. 'We won't be able to move further until you sign', the chief of the train told me. 'And I signed'"...[5]

In their confrontation with the disobedient writer the authorities did not stop even at blasphemous trampling on elementary requirements of human ethics. The journalist from the *Literaturnaya Gazeta* A. B. Borin told in his memoirs about one such shameful example from the late 1960s. A brilliant oncologist, and participant in the Great Patriotic War, was unexpectedly fired from a Leningrad medical institution. As it became clear afterwards, it happened on the instruction of the Leningrad *oblast'* committee of the CPSU headed by the party official V. S. Tolstikov; earlier he had managed to become famous through his participation in the scandalous trial of the poet I. A. Brodsky (see this chapter and note 51 to it). It turned out that Solzhenitsyn asked the doctor as an expert in his area of medicine to consult him privately about his suspicions that he had symptoms of cancer. Their meeting took place. When it became known at the *oblast'* party committee they "phoned the director of the institute and ordered him to immediately turn this worker out. The director obeyed without any objections."...[6]

Having formulated his philosophy of confrontation in November 1967, Solzhenitsyn stuck to it invariably and consecutively. "There is nothing to compare with this state of relief from having spoken out," he wrote. "Because for almost half a century you have to bend, bend, bend, to be silent, silent, silent — and now you straighten yourself up and bellow — not from the roof, not to the square but to the whole world — in order to feel how the already pacified and harmonious universe comes back into your chest. And already there are no more doubts, nor rushing about, nor repentance — only the pure light of joy!.. And all your perception of the world has so brightened that you are flooded with placidity, although nothing has been achieved."[7]

Here Solzhenitsyn meant, first of all, his uncompromising struggle with Brezhnev and Suslov's camarilla. But he was also like that at the time of the Khrushchev thaw. As is known, it was only owing to Khrushchev's help that the *Novy Mir* managed to publish, in 1962, his first tale *A Day in the Life of Ivan Denisovich*,[8] and a year later, several short stories, the splendid *Matryona's Farm*, whose plot was used by a Swedish artist in the allegorical picture on the left side of Solzhenitsyn's Nobel diploma. But in 1964, even before Khrushchev's dismissal, the writer's situation changed dramatically, which showed most clearly in the episode of his nomination for the Lenin Prize for Literature.

After the list of aspirants for the award was published, a laudatory article, *Truthful Tale* by S. Y. Marshak, appeared in the *Pravda* of January 30, 1964. But two and a half months later, on April 11, just before the decisive vote in the committee, the same paper published something directly opposite. The editorial announced that the writer was not worthy of the award, which, of course, could not have been unknown to Khrushchev.[9]

One of the consequences of the growing pressure on the writer was the delay by the censorship, in 1967, of coming out of the completely ready fifth volume of A. T. Tvardovsky's collected works. The poet's daughter, V. A. Tvardovskaya, explains the action by her father's refusal to withdraw the articles mentioning Solzhenitsyn from the literary criticism materials selected for the volume.[10] The fifth volume appeared in 1971, four years after having been handed in for composition on September 21, 1970, that is, shortly before the Swedish Academy's verdict was made public, when it became clear that this verdict could not be avoided. Being aware of the propagandist Sabbath awaiting the country after that, Tvardovsky preferred to give in. Anyway, in the published volume there is not a single mention of Solzhenitsyn.[10a]

Winning the Nobel Prize for Literature, in 1970, and his struggle for the right to receive the award not secretly but in an auditorium where he could, in accordance with his credo, feel "relief from having spoken out", is described in detail in one of the chapters of his book, *The Oak and the Calf*.[11] In the same chapter he confessed that he had dreamed of the Nobel Prize as far back as when he was a GULAG prisoner, with the tag "É-854" on his camp padded jacket, and when in 1958 Boris Pasternak was forced to decline the award, he, a school teacher from Ryasan', whom nobody knew, "writhed with shame for him as if for myself" (doc. 11.1).[12] True, it must be added that at that time Solzhenitsyn could not know of the measure of pressure put on Pasternak not only from his adversaries but, which was still more intolerable, on the part of his near and dear ones, first of all, of Olga Ivinskaya: "they won't do anything to you, but I will be completely ruined"...[13]

According to Solzhenitsyn's information, his first nominator was François Mauriac.[14] The French writer and Nobel Prize laureate probably felt an urge to do that because of the campaign of vilification of Solzhenitsyn launched in the Soviet Union. Its peak was the expulsion of Solzhenitsyn from the Russian Federation Writers Union, on November 4, 1969, which was followed by the writer's open letter addressed to the Union's Secretariat, with a demand for complete openness as the first condition of "the health of any society, including ours,"[15] and the threats to strip him of his citizenship and deport from the country.

Mauriac's initiative evoked a wide response in the Western society. Intellectuals of the world started an unprecedented campaign of collecting signatures in support of Solzhenitsyn's candidacy (doc. 11.2, 11.3). Rumor had it that Heinrich Boll, who gave Solzhenitsyn shelter for some time in Germany in 1973, after the latter's forceful deportation from the USSR, addressed the Nobel Committee for Literature, expressing his wish to withdraw his candidacy for the 1970 prize in favor of his Russian colleague.[15a]

Retaliatory measures planned at the Central Committee could not change anything — as warned the Ministry of Foreign Affaires and also the writer Boris Polevoi, who already had his own negative experience of the attempt to block awarding the Nobel Prize to Pasternak in 1958 (see doc. 9.3).

Twelve years later it turned out that the heads of the Soviet literature from the Central Committee's ideological departments had not learnt anything from the story of Boris Pasternak's prize. Once again they made the same errors, as if guided by their sworn enemies. The juxtaposition of the texts of doc. 11.4–11.5 with epistolary and journalistic assertions of 1958 clearly shows almost word-for word coincidences in the given arguments and suggested methods of struggling, which were undoubtedly doomed to failure.[16]

Nevertheless, there were some significant differences between the situations of 1958 and 1970. First of all, the personalities of the victims of the Central Committee's persecutors were totally unlike. Instead of an introspective intellectual tormented by his natural need to agree, for the peace of his near and dear ones, with the rules of "expediency" game being imposed on him by sophisticated party officials and his fellow writers, ruthless in their subservience to the authorities, now was an altogether different personality. The Central Committee suddenly found in their new adversary a man of another temperament altogether — the man who was hardened by the lawlessness of Stalin's empire of suppression, who had the experience of confinement cells, of *sharashkas* (secret research laboratories where prisoners worked) and "cancer wards", able not only to take a blow but also to attack corageously, driving into a corner the complacent party authorities not used to frontal attacks.

Solzhenitsyn got into a straight fight on a full scale, his refusal to compromise sometimes making him go to extremes on the way to achieving the set objective. He often perceived his strategic allies as enemies, demanding from them only the maximum of everything he wanted,[17] not accepting the tolerant philosophy of "small businesses".[18] But as a result he emerged from this uncompromising combat as a victor over the decrepit regime.

All that could have been foreseen, long before the Nobel Prize. As early as September 15, 1967, in connection with the Central Committee's ban on publishing *The Cancer Ward* in the *Novy Mir*, the editor-in-chief of the journal, A. T. Tvardovsky, noted prophetically in his diary: "I think that here we have a big, a great writer, whom we are treating abominably unjustly. But it may end tragically and this end will cost us dear — dearer than *Zhivago* and Sinyavsky and Daniel — well, actually we are already suffering losses."[19]

All foregoing, from the dreams of the Stalin camp prisoner without rights about his own Nobel Prize to his moral grudges against Pasternak for his refusal to accept it, absolutely does not fit in with the information made public by V. V. Nabokov's biographer Brian Boyd. Allegedly, on being awarded the Nobel Prize in 1970, Solzhenitsyn wrote in his letter to Nabokov, the copy of which he sent to the Swedish Academy, that the latter deserved it more than him. Here one can only shrug one's shoulders. Such melodramatic emotions are incompatible with Alexandr Isayevich's true nature. Maybe these were wild fantasies of Vladimir Vladimirovich, for whom this prize had long become *idée fixe*.

Solzhenitsyn pointedly refused going to Stockholm to the award ceremony, as he was not given any guarantee that he would be able to come back (doc. 11.6). Instead of that, he sent, through Joran Lundstrom at the Swedish embassy in Moscow,[20] a text of gratitude to be read at the Nobel banquet (doc. 11.7).[21] Whereas he himself concentrated his effort on organizing the giving of the Nobel distinctions (a diploma and a gold medal) to him officially in Moscow, setting out as an indisputable condition the presence at the ceremony of the guests he invited and representatives of the press and his right to read the Nobel lecture before them. (doc. 11.8–11.10).

The authorities, from the KGB to the CPSU Central Committee, did not permit the event by refusing a visa to the Swedish Academy's permanent secretary Karl Gierow (doc. 11.12, 11.13). The impetus for such decisive actions was added by the writer's extensive interview given on March 31, 1971 to correspondents of two leading American newspapers (doc. 11.11 and note 1 to doc. 11.11-II).

Of course, Solzhenitsyn's uncompromising struggle against the authorities found support from people, especially in creative workers communities. However, this support did not as a rule take any public form. With rare exceptions, the press, subject to censorship, managed to succeed in deceiving vigilant censors with the help of a luckily found hint or a resourcefully found euphemism.

One such case when a poet and the journal in which published his poem certainly outwitted the censorship occurred in the fall of 1971, the time of the very peak of the laureate's efforts to organize in Moscow a public ceremony of his acceptance of the Nobel regalia. On October 7, two weeks before his letter to the authorities of the Swedish Academy and the Nobel Foundation, in which he discussed the possibility "to receive the Nobel signs in Moscow but... not confidentially" (doc. 11.8), the October issue of *Novy Mir* was sent to printing.

Placed modestly somewhere in the middle of the journal was a selection of unpretentious poems by the little known poet from Ryasan', A. F. Markin. But one of the poems published there, entitled *The White Buoy,* contained explosives in its unsophisticated stanzas. Here the poet addresses the girl who is in love with a luckless buoy-keeper at a river crossing, about whom "ill fame" goes round:

You're looking at him. Perhaps it's
Known only to you
That his lot is so absurd
And his eyes are so pure.

That he moves through murky waters,
Though it earns him no acclaim,
Putting buoys for ships to warn them
Of the dangers on their way.

But from passing boats, as sign of
The respect to which he's due
Strangers wave: "Hello, Isaich!"
Their hands raised, like in salute.

In order to support his secret scheme, Alexei Markin put at the end of the poem the year when it was written, "1970". That is, the year when Alexandr Isayevich Solzhenitsyn received the Nobel Prize.[22]

Actually, these rhymes can hardly be regarded as true poetry. But owing to the character's name "Isaich" (a most informal but quite respectful way of addressing a man), along with the idea of saluting him and the mention that his activity "earns him no acclaim", the poem instantaneously became an outstanding occurrence, its implied meaning being clear to every reader. This author does not have any information how much this blunder cost the censor attached to the *Novy Mir* and whether it was made involuntarily or deliberately. But there is no doubt that with this issue of the journal coming out the Central Committee's ideological army proved to be a complete failure.

However, at the highest levels of power there also were people who did not agree with the persecution of Solzhenitsyn. For instance, the Minister of Internal Affairs, N. A. Shchelokov, stood up for Solzhenitsyn. It must be said at once that in protection of Shcelokov involved some personal interest, having for many years friendly relations with L. I. Brezhnev — the first person of the state (see note 68 to Chapter 9). His interest lay in his longstanding hidden rivalry with his competitor from the parallel power structure, the Chairman of the KGB, Y. V. Andropov.[23] Convincing evidence which confirms this supposition is given by another memo of the Minister of Internal Affairs to the CPSU Central Committee. It is dated February 8, 1980, which means it was prepared more than nine years after his 1971 letter to the Central Committee, and puts forward completely opposite proposals (doc. 11.16). Nevertheless, the argument of the 1971 memo, pointedly liberal in itself, is rather typical for adequate perception of hidden relationships in the circle of the people closest to Brezhnev, and for that reason it is worthy of attention (doc. 11.14).

As one might have expected, this notable message did not get any constructive response from the Central Committee's Secretariat. And no wonder — the persecution of the writer had gone too far, and he himself was too uncompromising. The discussion of the document, judging by the working record of the sitting of the Secretariat of October 7, 1971, came to an unproductive and hasty exchange of opinions on questions of secondary importance touched upon by Shchelokov (doc. 11.15).

In February 1974, under the decree of the Presidium of the USSR Supreme Soviet, Solzhenitsyn was stripped of Soviet citizenship and forcefully exiled from the country. The immediate cause of the exile was the publication in Paris, on December 28, 1973,[24] of his book *The GULAG Archipelago* (doc. 11.17). After he heard the news on the foreign radio, being aware that some actions with regard to him were inevitable, he made a ten-point forecast of what could be expected from the authorities as revenge for what he had done. In point four he supposed "exile from the country", assessing such a measure as "possible."[25]

Having found himself abroad against his will he, as he wrote afterwards, "from the first hours, from the first minutes understood his situation in Europe as activities no longer constrained... Only now can I promptly and carefully reap my harvest."[26] One of the landmarks of "reaping the harvest" was receiving the Nobel Prize, four years after it was awarded. On December 10, 1974 the King of Sweden gave him the diploma and the laureate's gold medal,[27] and several hours later, standing on the steps of the Stockholm City Hall, where the traditional Nobel banquet took place, Solzhenitsyn made his speech of thanks (doc. 11.18). The deep civil content of this speech made it quite different from laureates' traditional speeches annually pronounced here. For several years a videotape of Solzhenitsyn's speech was shown at every meeting of the Nobel laureates before December 10, arranged by the Nobel Foundation in one of the halls of the Grand Hotel[28] for their first acquaintance with one another.[29]

What was the reaction to Solzhenitsyn's Nobel award in Soviet Union? As in the case of Pasternak, it was ambiguous. There was some quiet joy that one more representative of Russian literature had entered the Nobel laureates' community. However, the greater part of his compatriots, stupefied by the official propaganda, quite sympathized with the utterances overflowing newspaper pages.

The same situation arose five years later, when the Nobel Peace Prize was given to academician A. D. Sakharov. Everything repeated itself almost up to the finer details. Both in the clumsy attempts of the official propaganda to ground their position with regard to "renegades" and in the public perception of boisterous newspaper campaigns. While in 1958 and 1970 the border of confrontation divided the writers' community into two obviously unequal parts, the same distinct division took place in the academic community in connection with the human rights activities of the great scientist, which brought him the Peace Prize.

It started to take shape in the late 1960s, when Sakharov passed to *samizdat* (clandestine circulation) his article (or "memorandum", as he called it) *Thoughts on Progress, Peaceful Coexistence and Intellectual Freedom* (note 2 to doc. 11.24-IV), in which, among other things, he developed the idea of convergence of the socialist and capitalist systems. At that time permission for public statements on Sakharov's human rights initiatives was not given yet, and the matter was confined to private exchange of opinions between members of the Presidium of the USSR Academy of Sciences (doc. 11.24-IV) and the scientist's transfer from the classified establishment in Sarov, Gorky *oblast'*, to the Physical Institute of the USSR Academy of Sciences (FIAN), where he had worked before going to Sarov.[30]

The public campaign against Sakharov started after he convened, on August 21, 1973, in his apartment on ulitsa (street) Chkalova (to which its historical name "Zemlyanoi Val" has now been returned), a press conference for a group of Western correspondents accredited in Moscow. The cause for it was given by the warning from the USSR Procurator's Office about the inadmissibility for him as a bearer of state secrets (see doc. 11.19 and note 3 to it) to have any unsanctioned contact with foreign citizens. By that time he had already had two meetings with foreign journalists: the first took place in the fall of 1972, with the correspondent of the American journal *Newsweek* Jay Axelbank, and the second, which gave the immediate impetus for summoning him to the Procurator's Office, on August 16, 1973, with Ulle Stenholm, who represented Swedish radio and television, in June or July of the same year.

The subject of the latter talk was the key questions of the functioning of the Soviet society; the nature of the ruling regime; the possibility of its transformation under the influence of the dissident movement. The problem, rather painful for the powers that be, of secret economic privileges carefully hidden from the people, was also touched upon. Sakharov prophetically predicted the dangers for the country posed by the strengthening of nationalistic tendencies on its outskirts, and by the humbled situation of the intelligentsia, which formed in society an "anti-intellectual atmosphere when an intellectual profession, the profession of a teacher or a doctor does not command due respect."[31]

The first sign of the anti-Sakharov campaign launched in the press was the article under the screaming headline "Supplier of Slander" published in the *Literaturnaya Gazeta* on July 18, 1973 in response to Stenholm's interview. But the main signal sounded after Sakharov's visit to the Procurator's Office was on August 22, when the *Letter of Members of the USSR Academy of Sciences*, signed by 40 academicians appeared in the *Pravda* newspaper.[32] As soon as the next day the newspaper pages were full of carbon copy-like letters of scientists, writers, composers, cinema workers, turners, peasants and so on and so forth. They had the same arguments and the same conclusions, although hardly any of those who signed these letters had read or had the opportunity to read either the text of the interview or the materials of the press conference.[33]

Of course, when signing the document, not all forty members of the Academy of Sciences felt that they had fulfilled their civic duty. Some of them were simply afraid to refuse to sign the document, others regarded it as a habitual ritual. However, there were also quite a few who put their signature quite consciously, sincerely condemning Sakharov's human rights activities.[34]

The boisterous campaign of collecting signatures under group condemnations, which started after the "Letter of Forty", involved practically all scientific institutions in the country. Associate Member of the USSR Academy of Sciences E. L. Feinberg vividly described how it was at the Physical Institute, where the academician Sakharov worked (doc. 11.19).

The surveillance structures secretly made their contribution into the "expression of the Soviet people's will", sometimes even resorting to bold lies. For instance, in order to discredit the academician still more among a certain, rather large part of the Soviet society, the KGB launched disinformation that the last name of Sakharov, who was in fact the grandson of an orthodox priest, was not real, and his real name was Tsukerman. This invention spread widely not only in the Soviet Union but also in the countries of Eastern Europe; for instance, in Czechoslovakia (doc. 11.20).

The following question arises: taking into consideration the scientist's unique intellectual abilities, were his human rights activities an inadmissible waste of his intellectual potential? The answer was given by the outstanding philosopher of the second half of the 20th century, M. K. Mamardashvili. The structure of the society in which Sakharov lived, he wrote, could not but bring him, because of his inner requirement for freedom of thought, to social and political movement. For he himself was a "phenomenon of self-worthiness of thought, which did not need leaning against anything — whether against some outward authority, or power, or a collective."[35]

A. I. Solzhenitsyn was the first to whom the vilification campaign gave the impetus to nominate Sakharov for the Nobel Peace Prize. He directed his nomination to Oslo, in the form of the article *Peace and Violence*, which was published, on September 10, 1973, by the Norwegian newspaper *Aftenposten*. Naturally, the Nobel Committee could not accept a publication as a legal nomination. But the writer's initiative quickly found wide response. On September 11 Sakharov was nominated by British members of parliament, the next day they were followed by the liberal faction in the Danish parliament, "then a Munich group of physicists, then others, ... to give Sakharov the prize in 1974!"[36]

Andrei Dmitrievich Sakharov received the prize in 1975.[37] The verdict of the Nobel Committee, announced on October 9, caused a new paroxysm of newspaper hysteria, its incandescence no less than it had been in response to the Nobel awards to Boris Pasternak and Alexandr Solzhenitsyn, and much stronger than in the 1973 anti-Sakharov campaign. Again, like two years earlier, in the avant-garde of organized letters was a collective letter of members of the USSR Academy of Sciences, this time already with 72 signatures (see doc. 11.21 and note 6 to it). And again the prize-winner was denied permission to go to the Nobel ceremony.[38] The only difference was that Pasternak and Solzhenitsyn had been formally granted such an opportunity, but with a real danger that return entry would be prohibited. Sakharov was directly informed of the refusal, on the grounds that he was involved in state secrets (see doc. 11.22 and note 5 to it).[39]

The Nobel Prize was accepted in Oslo, on December 10, 1975, by the laureate's wife, E. G. Bonner. Sakharov himself was tin Vilnius, where he went in connection with the beginning of the trial of his human rights activities associate, S. A. Kovalev. In the evening, with his Lithuanian friends, he listened to the live broadcast of the Nobel ceremony given by a foreign radio company, like Solzhenitsyn had done five years before (doc. 11.22).

A year later a foreign correspondent asked the academician whether anything changed in his life after receiving the Nobel Prize; he could not say anything gratifying in response (doc. 11.23). His cheerlessness quite fit into the moods predominating in society. The authorities enjoyed strong support in their activities with regard to Sakharov. The conviction that the Nobel Prize all but tarnished the reputation of the laureate was widely spread among the people — and not only among philistines who took on trust everything that was written in the newspapers.[40]

On January 22, 1989, less than a month after the Soviet invasion of Afganistan, which Sakharov actively condemned, he was forced to go into exile to the city of Gorky, closed for foreigners. Soon, distributed in manuscript, appeared *The Open Letter* to academician P. L. Kapitsa, as Nobel laureate. The letter was written by the human rights activist Anatoly Marchenko. The letter asked why the famous scientist was keeping silence, not responding to the arbitrary punishment (doc. 11.24-I).

Kapitsa, as well as a number of other scientists, did not keep silence (doc. 11.24-II–IV). However, it did not become possible to bring Sakharov back from exile until December 23, 1986, after the beginning of *perestroika*.

The initiator of his coming back to Moscow was M. S. Gorbachev, whose decision was secretly resisted by power structures, first of all, by the KGB. About two years later, in the fall of 1988, Andrei Dmitrievich was elected, from the Physics and Mathematics Department of the USSR Academy of Sciences, as a member of the Academy's Presidium. The initiator of such choice was academician R. Z. Sagdeev, whose candidacy for this position had already been agreed upon at the very top of the power. Sagdeev refused to run for the post and suggested Sakharov instead, which was actively supported by the academicians in the ballot. The chairman of the KGB, V. M. Chebrikov, did not fail to comment on this arbitrariness. Speaking at the sitting of the Political Bureau of the CPSU Central Committee, he said, without any trace of doubt: "How immature is our Academy".

Gorbachev's assistant, A. S. Chernyaev, considered it his civic duty to put these words down in his diary entry and afterwards to make them public, so that future researchers would be able to analyze objectively this tumultuous epoch.[40a]

In 1987 the Nobel Prize for Literature was received by Joseph (Iosif) Brodsky. He received it as a citizen of the United States. However, the work of the Nobel laureate cannot be severed from his native roots, from his formation as a poet and essayist, first of all, under the Russian sky. The most serious Western researchers of his creative strength unreservedly link the main features of Brodsky's unique poetical style with the atmosphere of the Soviet poetical heritage.[41] Brodsky was and remained to the last a great Russian poet, whom the ignorant power made in 1972, fifteen years before his Nobel triumph, leave his home country forever.

Before that there had been a trial, the content of which was unprecedented. The poet was tried for being a poet. The starting point was a satirical article *A Drone Attached to Literature*, published in the newspaper *Vechernii Leningrad* on November 29, 1963. Here are several phrases from this article: "He imitated poets whose work is filled with pessimism and lack of faith in man, his poems are a mixture of decadent modernism and ordinary gibberish. Brodsky's wretched imitated attempts looked miserable... He lacked the knowledge of culture. But can we speak of any knowledge if he has not even graduated from high school?" At the end of this writing there was a direct, uncovered threat: "Literary idlers like Iosif Brodsky should meet with the sharpest rebuff. They should be taught not to stir up trouble."

At a meeting of Leningrad party activists the city's chief host, first secretary of the CPSU Committee of Leningrad *oblast'*, V. S. Tolstikov, joined in the vilification of the poet as a person "not engaged in socially useful labor". When many years later the scriptwriter of a film about Brodsky, N. Y. Yakimchuk, found the pensioner Tolstikov and asked him what was his current attitude to the condemnation of the poet who became a Nobel laureate, the retired party official snapped: "But you tell me — who are these prizes given to? Well, Solzhenitsyn received it — but he is a traitor, he surrendered to the Germans. Here is to whom these prizes are given!"[42] In this way he repeated the old lie about the writer, launched by the Central Committee or from the KGB five years before he was awarded the Nobel Prize (see note 91 to Chapter 6).[43]

Brodsky was brought to trial, with a preliminary period in custody, on the grounds of the decree of the Presidium of the Russian Federation Supreme Soviet adopted on May 4, 1961 "about strengthening the struggle against persons evading socially useful labor and having a parasitic way of life." Conviction, especially after Tolstikov's rousing speech, was inevitable. Brodsky was convicted on March 13, 1964. Nothing helped — neither references, nor evidence of his work at a plant and in geological expeditions,[44] nor his contracts with state publishing houses for poetical translations from the Polish and Serbian languages and selections of these translations already published.[45] Having crushed the poet, the punitive machine continued to move in the set direction. As a result he was sentenced to five years of exile in a remote district of Arkhangelsk *oblast'*.

Documentary evidence of the level of this mock trial is provided by the dialogue between the judge E. A. Savelieva and the poet, which took place at the hearing of February 16, 1964, after which Brodsky was directed to a coercive psychiatric examination. The text is given by the record made by the writer F. A. Vigdorova, who attended the trial:

Judge: Who recognized you as a poet? Who listed you among poets?

Brodsky: No one. Who enlisted me among members of the human race?

Judge: Did you study this?

Brodsky: This?

Judge: How to become a poet. You did not even try to finish higher school, where they prepare... where they teach...

Brodsky: I didn't think... I didn't think you could get it from school.

Judge: How then?

Brodsky: I think that it comes from God.[46]

As was sternly but deservedly formulated by the venerable poet Pavel Antokolsky in his diary, "the most horrible thing is that these proceedings of the Soviet People's Court reflect, like a mirror, the whole epoch of the so-called reality: the defenselessness of truth, conscience, intellect and gift in the face of lie, baseness, idiocy..."[47]

The future laureate's last words in the court were prophetic: "Far from being a social parasite, I'm a poet who will bring glory to his native country." These words caused scoffing laughter from the judge, the assessors and members of people's patrol and seasonal workers specially brought to make the audience.[42]

Tolstikov was not the only one whom Yakimchuk managed to meet; he also found the easily amused judge E. A. Savelieva, already retired as well. Her answer to the same question was milder than that of the hardened party official, but, perhaps, even more cynical. "I think", she said, "that this court was of benefit to him, it was beneficial for his poetic development, and now he has even received the Nobel Prize — well, that does him credit!"[48]

Quite a few writers came to the defense of the young poet's insulted honor. The letter with the demand to release him was signed by about 20 poets and writers including A. A. Akhmatova, K. G. Paustovsky, K. I. Chukovsky, and S. Y. Marshak. A sharp address to the meeting of Leningrad writers was delivered by B. B. Vakhtin (doc. 11.25). From abroad came the letter from Jean-Paul Sartre addressed to the Chairman of the Presidium of the USSR Supreme Soviet A. I. Mikoyan (doc. 11.26). Mikoyan, or to be more exact, the reception of the Presidium of the Supreme Soviet, was also addressed by telephone by A. A. Ahkmatova.[49]

The core of the court sentence was so absurd that after it had been passed the foreign media repeatedly commented, according to circulating rumors, on its abrogation.[50] There were sure grounds for these rumors indeed.

The letters from some outstanding representatives of Soviet literature and the reaction of foreign intellectuals did find response of those involved in this case, up to workers of the Procurator's Office. As soon as 1964 the deputy head of the USSR Procurator General submitted a protestation to the presidium of the Leningrad city court. The protestation was declined, first of all, because of the fear of the omnipotent Tolstikov.[51] Only in the middle of 1965 did the Supreme Soviet reduce the term to the period in fact served, that is, to one year and five months, and one day in July

Brodsky returned to his native city.[52] However, the poet still had a conviction.[53] It was lifted by the resolution of the Russian Federation Supreme Court only on June 26, 1989, at the peak of the Gorbachev *perestroika*, when Iosif Alexandrovich was already a citizen of the United States of America, world-famous poet, and Nobel Prize laureate.[54]

Brodsky left the Soviet Union on June 4, 1972 and never again returned to his country. His famous lines "Neither country nor graveyard // Do I want to choose, // To Vasil'evsky island // I will come to die..." remained merely a poetical image.

On December 28, 1987 David Samoilov, a poet with a God-given talent, wrote to Lidia Chukovskaya: "I had a phone call from *Litgazeta* asking me to name the most important event in Russian literature for their New Year questionnaire. I answered: awarding the Nobel Prize to Iosif Brodsky."[55] That was no spontaneous impulse. According to the *Izvestiya* of April 18, 2003 (p. 12), in Brodsky's memorial museum, created by Petersburg enthusiasts, there is a collection of Samoilov's verses presented by him in 1963. Written on the front page is the prophetic note: "Hold on, Brodsky, you are a wonderful poet, and you will have enough trouble with that matter for your entire life".

The Nobel Prize could not pass him by. As early as in 1986, Joe Ellis, the dean of women's Mount Holyoke College in Western Massachusetts, where Brodsky spent in total 15 years of his time in the United States, trying to obtain for Brodsky the position of professor, in spite of the fact that the latter had not even graduated from high school and had no certificates of any education, used a good trump card in his negotiations with the administration: "He will soon receive the Nobel Prize". Next fall Brodsky himself phoned Ellis from London and stated quite definitely: "In a week it will be announced that the Nobel Prize has been awarded to me."[56]

As a result, so it happened. On December 10, 1987, on the stage of the Concert Hall, the King of Sweden handed him the deserved Nobel regalia.

After that, every year until his death in 1996, Brodsky considered it his duty to visit for a couple of weeks the northern country that he took a liking to. He preferred to rent small rooms in town or modest *dachas* in the country. His permanent companion in Sweden was the translator of his Russian language prose, writer Bengt Jangfeldt, who is until now cherishing memories of his great friend.

True, opinions still differ about what was the pivot of awarding him the most prestigious prize. Brodsky called himself a Russian poet and an English language essayist. According to K. V. Pleshakov, who interviewed many of Brodsky's acquaintances in Mount Holyoke, it was his essays that brought him the Nobel Prize.[57] We will be able to gain the authentic knowledge only when the Nobel Committee's documents are declassified, that is, not until 2038.

V. L. Serduchenko, a literary critic from Lvov, Ukraine, being greatly interested in the problem of "extremities in unity", gave the following description of the motives for giving the Nobel awards to Brodsky and to in many aspects his antipode, Solzhenitsyn. "In my opinion, much more worthy is Iosif Brodsky's being an outcast free from any political bustle. He cut his Russian umbilical cord forever: he preferred the existential cold of world culture to the Russian oven warmth." Solzhenitsyn, "having settled at the new place,... now came down upon Western democracies. His speeches and letters... became bristling with reproaches for selfishness, criminal light-mindedness, lack of understanding..." and so on. The critic completed his juxtaposition analysis with the following conclusion: "Both Brodsky and Solzhenitsyn became Nobel laureates, but the reason of the awards given to them was quite different. One was awarded the prize as a poet, citizen of the world, and the other, as a star of a transient political intrigue."[58]

However, discussions on whether Brodsky would have become a Nobel laureate if he had remained in his home country have not petered out. It appears that the most reasonable explanation of this fact was given by the St. Petersburg expert on his work, Y. A. Gordin: "If Brodsky had remained in Russia, not having died in a camp or a mental hospital, he would have still remained an important poet. He was already so by the time he departed. But, of course, he would have been a different poet. And as for the Nobel Prize... most probably, he would not have received it. There would have been no such fame, no English language prose..."[58a]

In 1990, announced the laureate of the Nobel Peace Prize was the President of the USSR and, simultaneously, the General Secretary of the CPSU Central Committee, M. S. Gorbachev. That was the Norwegian Nobel Committee's appraisal of his contribution to changing the political climate in Europe and creating the basis for fundamental changes in the relationship between countries, members of the Warsaw Treaty with the Western world (doc. 11.27). However, the reaction to giving the award to "the Soviet Union's chief communist"[59] in the laureate's home country and in the rest of the world were polar opposites. While in the West the decision of the Nobel Committee met with almost unanimous approval, the greater part of Soviet society took it without any enthusiasm. As Gorbachev's drama was aphoristically described by the British newspaper, *The Times*, he was extolled in the world and damned in his country. The President's assistant on international affairs, A. S. Chernyaev, in his diary responded to this collision with a melancholy phrase: "One more surprise for the civilized world."[60]

The secret discontent was headed by officials of the CPSU Central Committee, the most reactionary part of the Soviet bureaucracy. Something like that had already happened three and a half decades before, after Khrushchev's sensational debunking of Stalin's cult in his report at the XX CPSU Congress. As was afterwards said by a delegate of the congress, a future secretary of the Central Committee and the creator of the openness of the Gorbachev period, then a young worker of one of the Central Committee's departments, A. N. Yakovlev, workers of the Central Committee met Khrushchev's exposures at the congress "negatively, but did not speak about that openly", although in future contrived to successfully "sabotage the decisions of the congress" (doc. 11.28).

However, in 1990, contrary to 1956, the hostile attitude of the Central Committee's officials to the international recognition of the merits of the head of the communist party found resonance in the perception of the same event by the greater part of Soviet society. And when the KGB agents informed Moscow that the population had a negative view of awarding the prize to Gorbachev, this information reflected the true picture of the people's frame of mind (doc. 11.29).

Of course, that was to a certain extent due to the mass propaganda, which had been for decades knocking into the Soviet people's conscience the infallible idea of the Nobel institutions being anti-Soviet. The boisterous campaigns connected with the Nobel Prizes to Pasternak, Solzhenitsyn and Sakharov had also left a trace. So when the head of the USSR was crowned with such an award, the response of the society was easy to foresee. It formed on the feedback principle, that is, the reverse effect of the information which the propaganda had been fixing in the subconscience of millions of citizens by way of systematic "brain drain". The system crisis of *perestroika,* which had arisen by the fall of 1990 and brought empty store shelves and soaring inflation, was firmly associated in the eyes of the people with Gorbachev's name and thus only added wood to the fire of people's discontent (doc. 11.29-III).

The laureate himself could not but take into account the moods predominant in society. Hence his indecisiveness, his painful doubts when the time came for going to Oslo for the award-giving ceremony. Hence his attempt to evade the Nobel lecture, which he was to read according to the charter regulations of the Nobel Foundation (see doc. 11.29-III).

The name of Mikhail Gorbachev will forever remain in the history of the twentieth century. His lifetime biography *President Lear,* as the tragic figure of the first and the last USSR President was surprisingly exactly described in one Western publication, is typical for great reformers. From a short distance it is always difficult to realize the fruits of their activities adequately and assess them justly.[61]

In conclusion I would like to quote Alexandr Bovin, one of Russia's most perspicacious journalists of the last three Soviet decades: "Gorbachev... is certainly a great figure, one of the great political figures of the twentieth century. He destroyed the prison, the barracks in which we had lived for decades... Like Peter I, he made Russia rear, but in contrast to Peter, he did not manage to control the reins. This was not so much his fault and his choice, it was his trouble, his fate. History needed him in order to destroy the old, to tear off the fetters from Russia... From the ranks of his retinue the most controversial voices were heard, some admiring him, some cursing. Intuitively, without proof, I am on the side of the former...".[62]

Our descendants will certainly find kind words for the reformer Gorbachev and will do justice to the wise verdict of the Norwegian Nobel Committee, which judged his selfless effort so highly. Really, he did deserve the Peace Prize like no one else, in spite of all the inconsistencies and deviations that accompanied his decisions.[63]

Notes

[1] I. M. Guberman. *Kniga Stranstvii* (The Book of Journeying) — Saint Petersburg: Retro, 2002, p. 116. For Solzhenitsyn his loneliness in society had never been a secret, though. In 1972, in an interview to the Moscow correspondent of the *Washington Post* newspaper, he claimed that "a writer who thinks differently from the majority of society represents an asset to that society and not a disgrace" (see doc. 11.11-II).

[2] On the day of the 40th anniversary of sending to press the eleventh issue of the *Novy Mir* with *One Day in the Life of Ivan Denisovich* the *Izvestiya*'s observer A. H. Arkhangelsky, noting the volcanic impact of the tale on Soviet society, remembered the parable, popular with the intelligentsia of that time, about the biblical Moses, who "wandered with his people around the desert for 40 years, so that the first free population of the future Israel could be born and grow up", and completed this passage with the solemn assertion that we would like to believe: "... we embarked on this path on November 18, 1962. Exactly 40 years ago... The preparatory class is finished. The desert is still around us, but we are already ready to say: Hello, Promised Land. Hello, freedom."*

* *Izvestiya*, November 18, 2002, p. 1.

[3] "Yesterday [the 28th]* I was visited [at Marusia's in Moscow]** by Ryasansky [Solzhenitsyn].*** An impression of clarity, simplicity, great human dignity. With him you feel at ease from the first minutes. He said: "I only was afraid to go mad in prison."****

* January 28, 1963.

** The Moscow poetess M. S. Petrovykh, at whose home Akhmatova was staying at that time, having come to Moscow from Leningrad.

*** "Ryasansky" was the name Moscow intellectuals called Solzhenitsyn at first, as he was then a resident of Ryasan'. The first to call him so was the chief editor of the *Novy Mir* A. T. Tvardovsky. It was under this pen name that he directed the tale for a closed review to K. I. Chukovsky, who was delighted with what he read and prophetically foretold: "... with this story a very strong, original and mature writer has entered literature"; see the *Izvestiya*, November 18, 2002, p. 2.

**** *Zapisnye Knizhki Anny Akhmatovoy* (Anna Akhmatova's Notebooks), Torino, Einaudi Editore, 1996, p. 253.

[4] From the entry of October 3–4, 1973; *Druzhba Narodov*, No. 9, 2001, p. 199.

[5] L. P. Sinyanskaya. *Vo Sne i Nayavu sredi Glyb* (Among Boulders, in Dreams and in Reality). *Znamya* No. 3, 2003, p. 145. This talk might have taken place in 1976, when the author of the memoirs found herself on a business trip to Spain together with Simonov. The occasion for it might have been given by Simonov's kind offer to his companion before coming back to the USSR, the acceptance of which as a result also proved to be connected with Solzhenitsyn: "Konstantin Mikhailovich asked me if I wanted to take with me something that was dangerous to bring; he was ready to put it in his suitcase — as a deputy of the Supreme Soviet he was not subject to customs examination. Of course, he meant books. I gave him *The Oak and the Calf* by A. Solzhenitsyn, and he brought it for me."* It would be just to add here, however, that Simonov was not one of those who obediently signed collective letters. An example of his overt "disobedience" can be provided by the story of the preparation of the *Open Letter to Writers of Czechoslovakia* published by the *Literaturnaya Gazeta* on October 30, 1968, after the invasion of Czechslovakia by the armies of the Warsaw Treaty countries.

Under the letter, organized by the CPSU Central Committee's ideologists, were signatures of all working members of the Secretariat of the Soviet Writers Union except three — Simonov, Tvardovsky and Leonov. As Tvardovsky wrote down in his diary, hot on the heels, the Writers Union's Board received a telegram from Simonov: "I do not have the moral right to teach Czechoslovakian writers, as I myself do not quite understand what is what." Something like that was directed to the same address by Tvardovsky himself: "... I certainly cannot sign the letter to the writers of Czechoslovakia, as its content appears to me rather unfavorable for the honor and conscience of a Soviet writer. I very much regret it."**

* *Znamya,* No. 3, 2003, p. 165.

** A. T. Tvardovsky. *Rabochie Dnevniki 60kh Godov* (Working Diaries of the 60s). *Znamya,* No. 10, 2003, p. 148.

[6] A. B. Borin. *Vospominaniya Literatora* (Memories of a Writer). *Zvezda,* No. 8, 2002, pp. 178–179.

[7] A. I. Solzhenitsyn. *The Oak and the Calf* — Moscow: Soglasie, 1996, p. 166. In Tvardovsky's mind the events around Solzhenitsyn gave rise to more pessimistic thoughts. In the same 1967, on March 11, he wrote in his diary: "About Solzhenitsyn we'll be able to say clearly comparing his fate with those of Platonov, Zoshchenko, Pasternak, Akhmatova... In their lifetime we persecute them, slam them, drive them to the grave, and after their death pick up their sheets and lines."* But, as the future has shown, it was Solzhenitsyn who proved to be the winner in the combat with the cruel authorities, his fate being luckier than the fates of his colleagues named by Tvardovsky.

* A. T. Tvardovsky. *Rabochie Tetrady 60kh* (Working Notebooks of the 60s. *Znamya,* No. 9, 2002, p. 177.

[8] The unexpected joy of the support from the first person of the state and the difficulties Khrushchev faced in overcoming fierce resistance in his immediate surroundings was colorfully described by A. T. Tvardovsky in his records of that time.* No less picturesquely did he feature his own highly emotional reaction when he heard over the phone, the day before the record mentioned above, the information from Khrushchev's assistant on literary questions V. S. Lebedev of Khrushchev's positive attitude to Solzhenitsyn's tale: "I'm happy that I am starting this new ... notebook by recording the fact significant not only for my everyday life and not only having in it, as it seems to me, the importance of a turning point, but also promising serious consequences in the general course of the literary (therefore, not only literary) affairs: Solzhenitsyn (*One Day*) is approved by Nikita Sergeevich... I even... cried with joy,.. but in this case I took pleasure in my ability to burst into tears in a sober state"**. The odyssey of the publication of Sozhenitsyn's tale in the *Novy Mir* was told in detail in the memoirs of V. Y. Lakshin, Tvardovsky' s comrade-in-arms at the *Novy Mir.****

* A. T. Tvardovsky. *Rabochie Tetradi 60kh Godov* (Working Notebooks of the 60s). *Znamya,* No. 7, 200, pp. 131–133, 135–136; the entries of September 20, 21 and 23 and October 21, 1962.

** *ibid,* p. 129, the entry of September 16, 1962.

*** V. Y. Lakshin. *Otkrytaya Dver'* (The Open Door) — Moscow: Moskovsky rabochii, 1989, pp. 192–203.

[9] Such a dramatic change of landmarks in the position of the organ of the CPSU Central Committee, which, in fact, meant the Central Committee's direct instruction to refuse the award, was analyzed in the book by Wolfram Eggeling, a professor at the Lotman Institute of Russian and Soviet Culture at Ruhr University in Bohum.* The story of the discussions around

the candidacy of Solzhenitsyn, which took place at the sittings of the Lenin Prizes Committee, was told by V. Y. Lakshin. He heard it from Tvardovsky, who was present at these sittings as a member of the committee.** Touching upon the article in the *Pravda* of April 11, 1964, Lakshin stressed that Khrushchev "was at that time on one of his journeys, and running the whole show in Moscow was his deputy, Leonid Ilyich Brezhnev." It was his initiative to impose on *Pravda* the duty of publishing a review of readers' letters with the emphasis on those whose authors did not accept Solzhenitsyn's tale."*** Tvardovsky himself, in his diary entry of June 4, 1969, a day before the appearance of the next entry with a draft of his statement about his readiness to quit the *Novy Mir*, considered it useful to note in passing: "In fact, I was being dismissed for three or four years, dismissed in parts. Also when there was a fight over Solzhenitsyn at the Lenin Committee, from which I was crossed out after that..."**** That is, he was expelled from the committee. One of the main persons to provoke such a radical anti-Solzhenitsyn move, as Tvardovsky noted, was the first secretary of the Komsomol Central Committee S. P. Pavlov, who had succeeded V. E. Semichastny, the 'hero' of the anti-Pasternak hysteria (see doc. 9.10). Pavlov directed a letter, in his name, to the Lenin Prizes Committee, maintaining that Solzhenitsyn was convicted, in 1945, not under political clauses but as a criminal, which was a downright lie. The ill-intentioned insinuation launched by the new Komsomol leader became a starting point for spreading rumors about Solzhenitsyn allegedly abetting the enemy, participating in battles on the side of the Vlasov army and so on.*****

* V. E. Eggeling. *Politika i Kultura pri Khrushcheve i Brezhneve* (Politics and Culture under Khrushchev and Brezhnev) — Moscow: AIRO-XX, 1999, pp. 160–162.

** V. Y. Lakshin. *Otkrytaya Dver'*, p.p. 209–212.

*** *ibid*, pp. 210–211.

**** A. T. Tvardovsky. *Rabochie Tetradi 60kh Godov. Znamya,* No. 5, 2004, pp. 151–152.

***** *ibid,* No. 9, 2002, pp. 189–198.

[10] *Znamya,* No. 10, 2002, p. 190.

[10a] Here it is appropriate to give a record from the diary of A. A. Kapitsa dated February 11, 1976: "We should learn to be tolerant... If our people are intolerant of Solzhenitsyn, he is as intolerant of them. That is our colossal difference from the West, where the liberal view of other ideas is predominant among intelligentsia.*

* *Dvadtsatyi Vek Anny Kapitsy Dnevniki. Zametki.* (Anna Kapitsa's Twentieth Century. Diaries. Notes) — Moscow: Agraf, 2005, Vol. 2, p. 121.

[11] *ibid*, pp. 268–306.

[12] Olga Ivinskaya, who once chanced to be Mstislav Rostropovich's fellow traveller in a first-class sleeping car compartment, remembered afterwards what Rostropovich, at that time giving shelter to the homeless Solzhenitsyn at his *dacha* in Zhukovka, near Moscow, was telling her of the writer's reaction to Pasternak's declining the prize: "Solzhenitsyn was simply indignant at his disgraceful, cowardly behavior, at this foolish letter of rejection, which Pasternak permitted himself to write!.. One should not be afraid of 'them' — show that you are afraid — and you are done for!.. No! Alexandr Isaevich is not a wretched intellectual. He has really declared war for the truth and managed to stand up for it! He was really ready to die for it!"*

* Olga Ivinskaya. *Gody s Borisom Pasternakom. V Plenu u Vremeni* (The Years with Boris Pasternak. Captured by the Time). — Moscow: Libris, 1992, p. 288.

[13] E. B. Pasternak. *Boris Pasternak. Biography.* — Moscow: Tsitadel, 1997, p. 704.

[14] "And here,* Mauriac, may he rest in peace, started his campaign of obtaining the Nobel Prize for me." (see also note 47 to Chapter 10). "And again our people had their game upset: to deport me now — as if in response to Mauriac — would be silly. And if I am given the prize, turning me out because of the prize would again be silly."** V. Y. Lakshin noted in his diary, in November 1969, that at that time rumors circulated in the writers' community that Solzhenitsyn nearly got the Nobel Prize in 1969. "It is said", he wrote, "that the Nobel literary committee voted for awarding the prize to him. But the Swedish Academy at the final vote changed the candidate, firstly because there are too many Russian laureates and secondly, things might turn bad for him in Russia because of that. Conversely, with us they waited and, as soon as it became clear that there would be no prize, decided to flog him to death"*** — that is, to banish him from the Writers Union, which was done on November 4, 1969.

* After Solzhenitsyn was expelled from the Russian Federation Writers Union, on November 4, 1969.

** Solzhenitsyn, p. 276.

*** *Druzhba Narodov*, No. 5, 2003, p. 186.

[15] Solzhenitsyn, p. 629.

[15a] From A. T. Tvardovsky's diary entry of April 3,1970, about the visit of the German writer Heinrich Boll to Moscow: "With Boll everything turned out all right and was easy — of course, if you do not think that Room 8 in the old TsDL* club, where we dined, was destined for some special purposes. He is nice and tactful... At table Melnikov** told me that Boll, to whom it was planned that year to give the Nobel Prize, withdrew his candidacy in favor of Solzhenitsyn. I was moved almost to tears, but I did not tell him about that (with the approval of Melnikov, who might have told a lie). But there has been something like that..."***

* Tsentral'nyi Dom Literatorov (the Central House of Writers, that is, the club of the Writers Union).

** D. E. Melnikov, an expert on the history of Germany and the history of the second third of the 20th century; during the meeting acted as interpreter.

*** *Znamya,* No. 10, 2005, p. 155.

[16] True, something new appeared in the propaganda, which had not been seen in 1958 — it was a more reserved attitude to the anti-Nobel rhetoric. The Swedish Academy was stigmatized for the "organization of provocation", but the value of the idea of the Nobel Prize laureate as such was not questioned, even if the discussion took place behind closed doors and was not meant to be made public. When the possibility of deporting Solzhenitsyn was being discussed at the sitting of the Political Bureau, on March 30, 1972, the Chairman of the Presidium of the USSR Supreme Soviet, N. V. Podgorny, who considered this action inexpedient at that moment, said that "he should not be turned out. He is the Nobel Prize laureate, and the bourgeois propaganda will certainly use this fact against us to the full."*

* From publications of declassified working records of sittings of the Political Bureau, where "the case of Solzhenitsyn" was discussed; quoted in the article *Sakharov and Solzhenitsyn* by the historian R. A. Medvedev in the journal *Svobodnaya Mysl'*, No. 8, 2001, p. 69.

[17] Here is an example of Solzhenitsyn's maximization. With the idea to prepare a collection of articles, *From under the Boulders,* whose objective would be to demand political freedom in the Soviet Union, he offered to academician D. S. Likhachev to become jointly with

himself the founder and one of the authors of this collection. Likhachev evaded this offer, and Solzhenitsyn broke off the relationship with him.* This trait of Solzhenitsyn's character, of not tolerating objections even from a person of the same views, was perspicaciously noticed by the poet David Samoilov. After their transient meeting, which took place once, he gave the following description of him in his "day-to-day records": "Admiring him, I made sure once again that there might be circumstances when we'll find ourselves in different camps." Concluding his idea he stated: "The worse the circumstances, the better Solzhenitsyn, the more significant his personality. He is a man of will and honor. The combination of these qualities, irrespective of his views, makes him a paramount figure of our time."**

L. K. Chukovskaya, who had known both the author of *Day-to-Day Records* and Solzhenitsyn for many years, after a midnight talk in Samoilov's home at a New Year table, half a month later, considered it necessary to return, contrary to the assertions of the latter, to her own arguments concerning the personal qualities of the disfavored Nobel laureate as shown in his everyday life. In an extensive letter dated January 15, 1975, she resolutely rejected assertions of Solzhenitsyn's intolerance in ordinary life, portraying him as an attentive, well-wishing and modest neighbor, always ready to understand the needs of other people and if possible, to help him. Lidia Korneevna knew that from her own experience — Alexandr Isaevich had spent a lot of time in the Chukovskys' hospitable home, both at *Peredelkino dacha* and in their Moscow apartment.***

* From the answers of D. S. Likhachev's daughter Lyudmila to the questions of the *Radio Russia* host, on November 6, 1999.

** *Novy Mir*, No. 1, 2004, pp. 203–204.

*** *ibid*, No. 6, 2006, pp. 163–167.

[18] Symptomatic are the differences between Solzhenitsyn and Sakharov, which arose in 1973. The former reproached the latter that the form adopted by him "of defending not all 'mankind' or 'people' at once but each oppressed person separately was rightly perceived by our society ... as a wonderful, curative approach for our country quest for the truth and love of man. But this form ... was exhausting, it took a disproportionate amount of Sakharov's strength and health, in comparison to the results (almost zero). And this approach also, owing to innumerable appeals bearing his signature, was already starting to flicker, breaking into pieces, across reports by the world press..."* Sakharov in turn noted that when reading Solzhenitsyn's articles he "could not but mentally express my solidarity with very many things of what he writes and says. However, I am often dissatisfied with his judgement being too categorical, with the absence of nuances and lack of tolerance of the opinion of others."** Afterwards, when living in exile and on coming back to Russia, Solzhenitsyn repeatedly made statements which were critically perceived by readers and listeners, including those who, like Sakharov, certainly held the same views where the writer's basic activities were concerned. Having touched upon this painful point in his interview for the *Izvestiya*, Ignat Sozhenitsyn made an attempt to interpret such peculiarities of his great father's statements. As he put it, "people try to apply my father's utterances and appraisals to a certain episode, to an instant. But his view, his judgements are broader than the features of the current moment."*** It is probably because of such a peculiar frame of mind that Solzhnenitsyn is predominantly perceived, especially beyond the boundaries of his home country, as a fighting politician rather than as a writer, an "engineer of human souls", using the common expression of the Soviet times. This can be seen from the results of a survey conducted in 1976 among senior pupils of American schools. 19.5% of respondents qualified him as a public figure drawing the greatest

admiration, which made thrice as many votes as the number received by the then President of the United States.****

* Solzhenitsyn, p. 339.

** A. D. Sakharov. *Vospominaniya v Dvukh Tomakh* (Memories in Two Volumes). Vol. 1 — Moscow: Prava Cheloveka, 1996, p. 582.

*** *Izvestiya*, May 22, 2001, p. 8.

**** *Voprosy Literatury*, No. 2, 2003, p. 52.

[19] A. T. Tvardovsky. *Rabochie Tetradi 60kh Godov. Znamya*, No. 10, 2002, p. 159.

[19a] B. Boyd. *Vladimir Nabokov. The American Years* — Moscow: Nezavisimaya Gazeta, and Symposium, 2004, p. 742.

[20] "As it turned out, with his own money, he went privately for a weekend to Finland and sent it from there."*

* Solzhenitsyn, p. 284.

[21] "This is how several of my close friends and I celebrated this unusual — Nobel — evening of mine: we were sitting in Rostropovich's* attic "tavern", at an unpainted ancient table with glasses and several candelabra of candles burning, and from time to time listened to reports of the Nobel ceremony from different radio stations. At last it came to broadcasting the banquet speeches. One program was slurred by jamming**, but I had the impression that my last phrase was not there.*** We waited for the repetition of the speech in the latest news — no, it wasn't!"****

* At Rostropovich's *dacha* in Zhukovka, near Moscow, where Solzhenitsyn was staying at that time.

** In 1970, after the 1968 invasion of Czechoslovakia by Soviet troops, the jamming of broadcasts of foreign radio stations in Russian, stopped earlier, was renewed.

*** See note 1 to doc. 11.7.

**** Solzhenitsyn, p. 284.

[22] *Novy Mir*, No. 10,1971, pp. 96–98.

[23] "He [Shchelokov] was known as a person close to Brezhnev. Under his protection he allowed himself to interfere with the questions which were totally under the authority of the KGB. That can be judged by Shchelokov's letter addressed to Brezhnev,... where the Minister of Internal Affairs gave grounds for his opinion that Solzhenitsyn's exile from Russia was unnecessary... Let us remember that Andropov was a staunch advocate of Solzhenitsyn's exile."*

* R. G. Pikhoya. *Sovetskii Soyuz: Istoriya Vlasti.* (The Soviet Union: The History of Power) 1945–1991. — Novosibirsk: Sibirsky Khronograph, 2000, pp. 380–381.

[24] "On December 28, at the Chukovskys' *dacha* in Peredelkino, which since the fall had been my deserted winter shelter, during my usual daytime listening to the BBC, I unexpectedly heard that in Paris the first volume of *Archipelago* had come out in Russian."*

* Solzhenitsyn, p. 345.

[25] Solzhenitsyn, p. 348. In Brezhnev times forceful deportation became a rather often used means of fighting against heterodoxy. This measure was not always applied on the initiative of power authorities. According to the memories of the artist and sculptor Mikhail Shemyakin,

whose work was beyond the framework of the official art, he was arrested in 1971 on a tip-off from the Soviet Artists Union, "and the state security general offered me a choice of three proposals: a nuthouse, places far from my favorite city [Leningrad], or I quit my country — noiselessly and forever." At the end of his talk with an observer of the *Izvestiya* newspaper, Shemyakin noted that "the general engaged in preparations for my exile... said that the Artists Union would not let me live in Russia."*

* *Izvestiya*, July 18, 2001, p. 7.

[26] Solzhenitsyn, p. 399.

[27] Demonstrating his idiosyncratic personality, Solzhenitsyn carried out one more stage of "reaping the harvest" during the Nobel ceremony. Walking in the rank of the laureates along the stage of the Stockholm Concert Hall to their chairs, he was pointedly keeping his hands behind his back — in a prisoner's manner. As the observer of the *Radio Liberty*, B. M. Paramonov* said in one of his programs that he was even going to appear at the ceremony dressed in his camp quilted jacket. His wife nearly fell at his feet begging him to abandon the wild idea. He put on a tailcoat — but still without a bow tie.

* In the program broadcast on September 1, 2002.

[28] As is known, the Grand Hotel is the hotel for Nobel laureates and honorary guests of the Nobel Foundation coming for the award-giving ceremony.

[29] P. E. Rubinin, academician P. L. Kapitsa's assistant, taken by him to Stockholm to the 1978 award ceremony, told this author about the episode, characteristic of that time, of playing the record of Solzhenitsyn's speech in 1974.* The executive director of the Nobel Foundation, Baron Stig Ramel, who knew well the situation in the Soviet Union, asked Kapitsa in advance if he would be embarrassed by demonstration of a video film with Solzhenitsyn's speech, if because of that he would have problems at home on returning. Petr Leonidovich, who on principle had never signed any collective statements about Solzhenitsyn, answered that even if such problems arose, they would be of little concern to him.

* See doc. 11.18.

[30] "In May 1969 I was summoned by Slavsky.* He asked me whether I would object if they transferred me to permanent work at the FIAN** (where I had started my scientific work in 1945–1948). I said I would be happy... I became senior research worker of the Theoretical Department... It was clear that nobody expected from me any concrete scientific results — it was important to get rid of me at the establishment in a decent way." ***

* The minister of the USSR Medium Engineering E. P. Slavsky, under whose command was the establishment in Sarov; the ministry's name camouflaged its true activities — guiding the industry of extracting and refining uranium raw material for atomic production.

** The Russian abbreviation for the Physical Institute of the Academy of Sciences.

*** A. D. Sakharov. V*ospominaniya d Dvukh Tomakh* (Memories in Two Volumes). Vol. 1 — Moscow: Prava cheloveka, 1996, p. 413.

[31] *The Interview to Ulle Stenholm*: Sakharov. Vol. 2, pp. 449–456.

[32] The round number of signatures was, most probably, not accidental, being the minimum set for the Academy's Presidium by the Science Department of the CPSU Central Committee. Academician A. M. Obukhov, who also found himself among those who signed the letter and

afterwards tormented himself for "getting into a mess", had not been included in the initially prepared list of the candidacies for signing. But someone could not be found, and at the last moment before sending the document to the newspaper the final signature was taken from him.* A quite exotic was of obtaining a signature for a collective letter, this time against Solzhenitsyn, was used by the Soviet Writers Union with regard to the secretary of the Union's Board, K. M. Simonov (see this chapter and note 5 to it).

* *On mezhdu Nami Zhil... Vospominaniya o Sakharove* (He Lived Among Us. Memories of Sakharov). — Moscow: Praktika, 1996, p. 791.

[33] Corresponding Member of the USSR Academy of Sciences V. I. Goldansky, when asked to write a letter to the press with the criticism of Sakharov, asked to give him a possibility to get acquainted with Sakharov's foreign publications, "to make the grounds for criticism more exact." The request was not fulfilled, but the demand was dropped.*

**ibid*, p. 507.

[34] These thoughts were shared by many in the academic community. At the turn of the 1970s, I knew well a worker of the FIAN, who was my neighbor. At that time a young scientist with a doctor's degree, now an academician, he spoke about Sakharov's human rights activities without concealing his irritation.

[35] *Priroda* , No. 8, 1990, p. 87.

[36] Solzhenitsyn, p. 324.

[37] Using the leakage of information, the press reported that Sakharov's main competitor for the 1975 Peace Prize was the President of Finland, Urho Kekkonen, who in the final ballot received two votes from the members of the Nobel Committee, with three votes given for the Soviet aspirant.*

* *Izvestiya*, September 12, 2000, p. 4.

[38] On receiving the news of Saharov having been refused an exit visa the Norwegian Minister of Foreign Affairs invited the Soviet ambassador in Oslo and, having expressed a deep regret over this occasion, asked to have this decision reconsidered. In reply the ambassador stated that he regarded this request as "an interference in the internal affairs of the Soviet Union."*

* the *Russkaya Mysl'* (Paris), November 27, 1975, p. 2.

[39] It was not until June 1989, half a year before his unexpected death, that Sakharov managed to go to Oslo and to be received at the Norwegian Nobel Peace Committee, where a dinner in his honor was held.*

* On mezhdu Nami Zhil, p. 309.

[40] The mathematician N. A. Dmitriev, Sakharov's colleague in Sarov, remembering their repeated arguments about political questions wrote that "to be supported by Western media, to receive the Nobel Peace Prize means to agree to a certain humiliation..."* Fifteen years later, M. S. Gorbachev would have to face similar moods (see doc. 11.29).

**ibid*, p. 279.

[40a] A. S. Chernyaev. *Na Staroi Ploshchadi. Iz Dnevnikovykh Zapisei 1988 Goda* (On Staraya Square. From Diary Records of 1988) *Novaya i Noveishaya Istoriya,* No. 3, 2006, p. 94.

[41] For example: D. MacFadyen. *Joseph Brodsky and the Soviet Muse.* Montreal, 2000.

[42] N. A. Yakimchuk. *Kak Sudili Poeta. Delo I. Brodskogo.* (How the Poet Was Tried. I. Brodsky's Case). Leningrad, 1990, p. 31.

[43] The persecution of Brodsky is completely on the conscience of local party officials; first of all, the arch-reactionary first secretary of the Leningrad *oblast'* CPSU party committee, V. S. Tolstikov. Four and a half years after the poet had been convicted, Tolstikov provoked one more notorious ideological scandal. This time his victims were Leningrad literary critics. On October 30, 1968 L. K. Chukovskaya wrote her father on this occasion: "The chairman was Tolstikov himself... For the meeting a special resolution had been prepared, where Orlov and his associates were to be only reprimanded. But Tolstikov was not satisfied with that, and he suggested turning all of them out. Orlov's collection and Etkind's foreword are announced to be ideological sabotage. Gumilev is regarded as an English spy (!!!). In a word, the Middle Ages night has come — for long? Who knows! And, as usual, in Leningrad it is the most terrible."* Commenting on that letter, E. Ts. Chukovskaya and Zh. O. Khavkina, who prepared the journal publication of the correspondence between Kornei Ivanovich and Lidiya Korneevna, noted that E. G. Etkind in *Zapiski Nezagovorshchika* (Notes of a Non-Plotter) published in 1977 in London and republished in Russia in 2001, dwelt in detail on the mentioned story. In the chapter "The Case of the Phrase" he wrote that in the preface to the two-volume collection *Mastera Stikhotvornogo Perevoda* (Masters of Verse Translation) he wrote, among others things: "In Soviet times translation has reached the unprecedented level. The social reasons for this process are understandable. *In a certain period, especially between the XVII and XX party congresses, Russian poets, deprived of the possibility to express themselves to the full in their original creative work, spoke with the reader in the language of Goethe, Orbeliani, Shakespeare and Hugo.*" "It was the italicized phrase", wrote the publication compilers, "that was described as ideological sabotage."**

* *Druzhba Narodov,* No. 11, 2001, p. 185–186.

** *ibid,* p. 186.

[44] In the opinion of Petr Vail' and Alexandr Genis, perspicacious explorers of the Soviet society of the Khrushchev thaw times, Brodsky's getting a job of seasonal worker in a field geological party was, to a certain extent, a result of euphoria of romance experienced by the young people of that time. That is, we can say that he started to work with geologists "from force of inertia — like everyone".* In this respect revealing is Brodsky's dialogue with a woman worker of the All-Union Geological Institute in Leningrad: "This year we have three expeditions: to Kol'skiy Poluostrov, to the region behind the Urals and to Magadan. Where would you like to go?" "Absolutely doesn't matter", snorted Iosif grabbing at his chin. "Really? And what do you like better — mapping or search and prospecting of mi..." "I don't give a damn", interrupted her Brodsky, "I only want to get out of here!"**

* P. L. Vail', A. A. Genis. *The 60s. Mir Sovetskogo Cheloveka* (The World of the Soviet man). — Moscow; *Novoe Literaturnoe Obozrenie,* 2001. 0.338.

** L. Shtern. *Po Mestu Zhitelstva* (The place of Residence). New York, 1980. pp. 55–56.

[45] According to the memories of M. V. Ardov, who was present at the trial, the court rejected some information on royalties from translations; as a result the amount of earnings confirmed by documents proved to be tiny. That enabled the judge to sum up that an adult person cannot live on such money and therefore Brodsky is a social parasite. In answer Brodsky said: "I was held in militia custody for two days and given a paper to sign that I ate food which cost forty kopecks a day. It means that an adult person can exist on a sum far less than is made by my royalties."*

* *Novy Mir,* No. 5, 1994, p. 139.

[46] *Neva*, No. 1, 2001, p. 179.

[47] P. G. Antokolsky. *Diary. 1964–1968.* St. Petersburg, the Pushkinsky Fond publishing house, 2002, p. 15. This record by the prominent poet can be added by official information submitted to the court, in response to their inquiry, by the Leningrad branch of the Soviet Writers Union: "Iosif Brodsky is not a poet."*

* *Sovetskaya Kultura*, May 3,1988, p. 2.

[48] Yakimchuk, p. 23.

[49] Akhmatova's record of September 16, 1965: "If only Brodsky had come [to Komarovo] ... I am glad that I remained loyal to him to the end (phone to Mikoyan), although it is certainly not worth being proud of."*

* *Zapisnye Knizhki Anny Akhmatovoi* 1958–1966. Torino, Einaudi editore, 1996, p. 669.

[50] "11.7.64, Komarovo. Gitovich came and said that the London BBC announced five minutes ago that Brodsky's sentence had been quashed."*

* *ibid*, p. 472.

[51] "Here is the reason why the sentence was not quashed: otherwise the Leningrad party authorities headed by Tolstikov would have looked silly. Because at the meeting of active party members Tolstikov allowed himself to say the most awful things about Brodsky..."*

* Yakimchuk, p. 27.

[52] The poet P. G. Antokolsky mentioned in his diary entry of September 19, 1965, the rumors that Brodsky was coming back to Leningrad and the chief editor of the Goslitizdat A. I. Puzikov ordered to renew preparations for publishing a number of his translations which had been removed from the schedule of publications on the authorities' demand. In conclusion the poet wrote: "That, as rumor has it, is connected with a certain infuriated letter from Alexei Surkov addressed to Brezhnev, where he refuses to ever go abroad and represent our power, in which such outrageous things as in the case of Brodsky are done."*

* P. G. Antokolsky. *Diary. 1964–1968.* St. Petersburg, Pushkinsky Fond, 2002, p. 40.

[53] K. I. Chukovsky wrote, on September 20, 1965, from Peredelkino to his daughter in Leningrad: "As Brodsky was released on the account of his 'psychical inferiority', his enemies will always have the possibility to put him into a lunatic asylum. That is why his friends here believe that it is very dangerous for him to appear in Leningrad; he should go to Tarusa* or some other place near Moscow."** Happily, the friends' fears proved unfounded; Brodsky did not quit Leningrad, except for being absent for a short while.

* A picturesque town on the banks of the Oka river, nor far from Moscow. Paustovsky lived in Tarusa for a long time (he is also buried there), around him there was a circle of writers, which Chukovsky had in mind when naming this town.

** *Druzhba Narodov* No. 11, 2001, p. 179.

[54] Yakimchuk, pp. 33–34.

[55] *Znamya*, No. 7, 2003, p. 157.

[56] *Druzhba Narodov*, No. 3, 2001, p. 174.

[57] *ibid*, p. 186.

[58] *Svobodnaya Mysl'*, No. 1,2001, pp. 121–122.

[58a] *Zvezda*, No. 1, 2006, p. 160.

[59] The very fact of the relatively young Gorbachev being elected General Secretary of the CPSU Central Committee was sensational against the background of the steady predominance of gerontocracy at the heights of power, which had developed in the years of Brezhnev's rule. This crucial choice became possible, first of all, owing to the mutual understanding between the 54-year-old secretary of the Central Committee M. C. Gorbachev and the 75-year-old Minister of Foreign Affairs of the USSR, member of the Political Bureau A. A. Gromyko, which developed in 1984, after Y. V. Andropov died and the senile and mortally ill K. U. Chernenko was elected General Secretary. The fundamental role in their coming closer was played by the directors of three institutes of the Academy of Sciences: Gromyko's son Anatoly Andreevich, director of the Institute of Africa, E. M. Primakov from the Institute of Oriental Studies and A. N. Yakovlev from the Institute of World Economy and International Relations, and also by the deputy head of the KGB V. A. Kruchkov. Summing it up, the well-known archivist and explorer R. G. Pikhoya wrote: "Perhaps, it was the only case in the history of the Soviet Academy of Sciences when three directors of institutes happened to directly influence the country's political development."*

* R. G. Pikhoya. *Sovetskii Soyuz: Istoriya Vlasti. 1947–1991.* (The Soviet Union: the History of Power. 1947–1991). — Novosibirsk: Sibirsky Khronograph, 200, p. 402.

[60] A. S. Chernyaev. *Shest' Let s Gorbachevym. Iz Dnevnikovykh Zapisei* (Six Years with Gorbachev. From Diary Records). — Moscow: Progress-Kultura, 1993, p. 384.

[61] In this case the idea of "close distance" is, of course, rather vague. To confirm it we can give a sad example taken from everyday life and involving a certainly respected man. Eleven years after awarding the prize to Gorbachev and shortly before the festivities organized in Oslo and Stockholm to mark the centenary of the first Nobel award-giving ceremony, the head of the TV-6 channel, Evgeni Kiselev, made an attempt to organize, on October 6, 2001, a live broadcast of a meeting with two laureates, Zhores Alferov and Mikhail Gorbachev, and the Nobel Foundation director Michael Sohlman. The subject of the talk was to be appraising the place of the Nobel movement in the 20th century history. Sohlman, despite being busy, gave his consent for coming to Moscow, but eventually had to cancel his visit. As he told this author, he did it on learning that Alferov had refused to sit at the same table with the Peace Prize laureate. The refusal of the prize winner for physics to meet Gorbachev was quite logical, though. At the moment when Kiselev was discussing the planned project with Sohlman, at the office of the *Literaturnaya Gazeta* newspaper lay a big article by Alferov written for the column "Ten Years That Shook..." * The article published several days before the planned television program repeated the widely spread idea — which the author passed for his own — that if the plotters "had held the power, Mikhail Sergeevich would have come back from Foros and headed the country." Here the Nobel laureate did not notice his own blunder repudiating his utterance: about thirty lines above, in connection with another topic, he made the reader remember the trivial maxim that "history does not know conditional mood."**

To that we can add that such utterances of Zhores Ivanovich were to a great extent based on his pain understandably caused by today's difficulties of Russian fundamental science, whose budget financing leaves much to be desired, to put it mildly. But he directly connected the existing situation with the *perestroika* romance of the Gorbachev times. Although living in the Soviet times he perfectly understood that the existing command and administrative system was organically unable to survive and perfect itself. Responding in 1994 to P. L. Kapitsa's

100th jubilee and remembering his "brilliant administrative abilities", which he showed when heading the Glavkislorod industrial association, in the middle of the1940s, Alferov honestly stated that Petr Leonidovich managed to demonstrate these exclusively individual qualities in the conditions of "the economic system which practically does not work."***

* In this column the newspaper published materials connected with the tenth anniversary of the 1991 events — that is, the attempt of a putch of Moscow in August and the subsequent collapse of the USSR in December.

** *Literaturnaya Gazeta,* No. 39, 2001, p. 3.

*** *Vestnik RAN,* 64, No. 6, 1994, p. 513.

[62] A. E. Bovin. *XX Vek Kak Zhizn'* (XX Century as Life) — Moscow: Zakharov, 2003, p. 513.

[63] In this respect a short-sighted judgment was demonstrated by the Norwegian physicist Christoffer Jotterud, who compared the prize awarded to Gorbachev to Sakharov's prize by saying: "Regrettably, the Nobel Peace Prize Committee does not always keep to this high standard. The Committee, sad as it is, did not demand respect for human rights of the last laureate. That is why awarding the 1990 Nobel Peace Prize to Gorbachev caused so many disputes."* Unfortunately, the physicist was not able to see the radically different starting positions of the two laureates when making decisions — the position of a human rights activist guided by the voice of his conscience, and the situation of the leader of the country which had existed for three quarters of a century not under the aegis of the law but in the conditions of the superiority of "political expediency".

* *On mezhdu Nami Zhil* (He Lived Among Us), p. 308; from the speech at the International Sakharov Physics Conference in Moscow, on May 22,1991.

Document 11.1

A. I. Solzhenitsyn's dreams of the Nobel Prize

...As our scientists also did not receive these overseas prizes very often, they were almost never mentioned here; before the Pasternak tempest few people knew of their existence. I don't remember from whom I lerned about the exsitance of the prise, it must be from someone in the camps. And determined at once, in the spirit of our country, quite politically: that is exactly what I need for my future breakthrough...

In 1958, being a Ryasan' teacher, how envious I was of Pasternak: here is a man who was lucky to win the lot that I thought of! He is the one who will fulfill it! — now he will go there, and make a speech, and print all the rest of his works, what was secret, what was impossible to risk if you lived here! It is clear that his trip will not be only for three days. It is clear that they won't let him in back here, but meanwhile he will change the whole world and change us — and will come back, but with triumph![1]

After my camp schooling I, sincerely, was not able to expect that Pasternak would choose a different line of action, that he had a different aim. I measured him by my aims, my measures and writhed with shame for him like for myself: how could he have got frightened of some newspaper swearing, how could he have weakened when facing the threat of exile and humbly ask the government, and mutter about his "errors and delusions", his "own fault" put into the novel — to renounce his own ideas, his own spirit — only in order not to be exiled? And "the glorious present", and "pride in the time in which I live" and, of course, "radiant belief in our common future",[2] — and it is not a professor being flogged in a provincial university, it is our Nobel laureate speaking for the whole world to hear! Well, we are hopeless, indeed!.. Well, if you are called for the battle, especially under such superb circumstances, go and serve Russia! I condemned him cruelly and reproachfully, finding no excuse. That affections can outweigh duty I could not forgive or understand even when I was young, the more so when I was a brutalized prisoner...

All the more clearly I understood, making a wish for the future: I need this prize! As a step in a position, in a battle! And the earlier I get it and become firmer, the harder will I strike! And here I will do everything the opposite way round to Pasternak: I will firmly accept, firmly go, and make the firmest speech. So, the way back will be closed. But: I'll have everything printed! I'll speak everything out! The whole charge accumulated from the my detention in Lubyanka cells throughout going to work in the camp in winter, for all those strangled, shot, killed by hunger and cold! To hold out till the Nobel rostrum — and burst out! For all that the lot of an exile is not an exorbitant price...

However, *Ivan Denisovich*[3] ... did not bring me much closer to the Nobel Prize. It was simply because I thought of it, that I believed, mixing the idea with presentiment, and expected it as something inevitable. Although Pasternak, by his renouncement and then by his death shortly after, closed the way for the next laureate to come from Russia: how can Russians be given the prize if it kills them?..

A. I. Solzhenitsyn. *Bodalsya Telenok s Dubom* (The Oak and the Calf) — Moscow: Soglasie, 1996; from the part "The Third Supplement. December, 1973", pp. 269–271.

[1] Surprising foresight! That is exactly how it happened. Not with Pasternak, but with Solzhenitsyn himself, though. Two decades after it had been written.

[2] Not quite exact quotations from the published version of Pasternak's letter to the *Pravda* (doc. 12-II).

[3] The first tale by Solzhenitsyn, *One Day in the Life of Ivan Denisovich*, published in the *Novy Mir* journal No. 11, 1962.

Document 11.2

From the Chairman of the State Security Committee (KGB) of the USSR Council of Ministers Y. V. Andropov to the CPSU Central Committee, of July 8, 1970.

According to the information received by the State Security Committee, a campaign for awarding Solzhenitsyn the Nobel Prize is being conducted in the West. In particular, in June 1970 the French Arts and Progres society directed letters with the request to support this proposal to the addresses of over thirty Soviet writers. It is known that representatives of the Arts and Progress society have been lately making attempts to establish connection with some anti-social characters in Moscow, P. Y. Yakir among them. Measures have been taken by the State Security

Committee to prevent mass sending of the mentioned document to the territory of our country.

Reported for your information. A copy of the letter is attached..

RGANI, fond 5, opis' 62, ed. khr. 84, l. 116.

Attachment: the letter of the General Secretary of the Arts and Progres society Therese Baskin, Paris, June 3, 1970; directed to the addresses of thirty Soviet writers.[1]

The name of Alexandr Isaevich Solzhenitsyn is certainly known to you. The majority of writers, specialists and literary critics both in Western countries and in the USSR believe that already today he is one of the most prominent Russian writers. Solzhenitsyn is an author hugely popular among readers of the whole world. In 1969 in France he was awarded the Prize for the Best Foreign Novel. Solzhenitsyn continues the best traditions of Russian literature. In his novels *One Day in the Life of Ivan Denisovich, The Cancer Ward, The First Circle,* in *Matryona's Farm,*[2] *Tiny Stories* and other short stories and plays he concentrates on the human side of man, on the problems of depth, humanity, justice. His works reflect the cast of mind of the Russian people. His style, original and vivid, is worthy of the most outstanding prose writers of our time. Solzhenitsyn's work deserves worldwide recognition, which is the Nobel Prize for Literature.

The Swedish Academy responded favorably to our proposal and informed us that candidates can be nominated by members of academies of all countries, with support from writers and personalities of these countries' science, art and culture. In this connection, in response to our appeal, we have already received several dozens of signatures of prominent French writers, academicians and representatives of progressive intelligentsia, who have unreservedly supported Solzhenitsyn's candidacy... Appealing to Soviet writers, many of whom have already spoken in support of A. Solzhenitsyn, we ask them to support our initiative and add their voice to the voices of outstanding representatives of the French intelligentsia promoting the idea of giving the Nobel Prize to Solzhenitsyn in literary and intellectual communities of the whole world, expressing it at meetings and in private conversations, submitting individual and group petitions to corresponding institutions...[3]

ibid, l. 117, with turn over.

[1] The typed text of the original of the letter is in Russian. It is typed on the form with the letterhead: "Arts and Progres, 37, av. Eugene Adam (78). Maisons-Laffitte. Association regie par la loi Juillet. 1901-".

[2] The short story *Matryona's Farm* is Sozhenitsyn's second work published in the USSR, after *One Day in the Life of Ivan Denisovich..* In connection with its coming out Chukovskaya noted in her diary that there were rumors that Sholokhov, in Febrauary 1963, was allegedly going "to fulminate" against the story. Here is Akhmatova's response to these rumors: "Excellent. In this way he will definitively do himself in!"*

* L. K. Chukovskaya. *Zapiski ob Anne Ahkmatovoi.* Vol. 3. — Moscow: Soglasie, 1997, p. 36.

[3] A copy of the letter received by the writer S. V. Mikhalkov is attached to the text of the resolution of the Secretariat of the CPSU Central Committee of October 8, 1970 (record 11, point 10c), the content of which is as follows: "To commission the Cultural Department of the CPSU Central Committee to prepare the necessary material on this question in accordance with the exchange of opinions at the sitting of the Secretariat of the Central Committee."* Here also is Mikhalkov's answer of July 7, 1970: "Madam, you ask me... to express my opinion on your initiative of nominating A. Solzhenitsyn for the Nobel Prize. Personally I regard this initiative as none other than another provocation directed against Soviet literature and having nothing in common with true concern for the development of literature."**

* RGANI, fond 4, opis' 20, ed. khr. 770, l. 71; published in the journal *Istoricheskii Arkhiv*, No. 1, 1992, p. 147.

** *ibid*, p. 158.

Document 11.3

From the head of the Cultural Department of the CPSU Central Committee V. F. Shauro to the Central Committee, of October 1, 1970.

In July this year a number of French personalities from the Art and Progress association (the writer Armand Lanou, film director Rene Claire, sociologist Raymond Aron and others, about 50 people in all) nominated A. Solzhenitsyn for the 1970 Nobel Prize.[1]

The organizers of this nomination sent letters to many personalities of literature and art of different countries, including the USSR, with the request to support the candidacy of A. Solzhenitsyn... The secretaries of the Soviet Writers Union, comrades Markov, Voronkov, the chairman of the Board of the Novosti Press Agency[2] Udaltsov, representatives of some other organizations are bringing forward a number of proposals, the implementation of which could prevent awarding the Nobel Prize to Solzhenitsyn. The following measures are supposed to be implemented...

b) to use the trip to France of the director of the Gorky Institute of World Literature, comrade Suchkov B. L. for personal contacts with prominent French writers and explaining to them the position of the Soviet Writers Union on the question of A. Solzhenitsyn's work and social behavior;

c) maybe, the Soviet ambassador in Sweden should be instructed to urgently clarify the situation in the Nobel Committee, to give detailed information in connection with A. Solzhenitsyn's nomination and to make his proposals;

d) in case it is necessary, the Soviet ambassador in Sweden could also be instructed to make an oral unofficial statement to the Swedish authorities noting that the Soviet public opinion regards the Committee's intention to award the prize to Solzhenitsyn as an unfriendly act, which could hamper the development of cultural relations between writers of our countries;

e) through the Soviet ambassador in France to address our friends with the request to take possible measures for neutralization of the propagandist campaign in support of A. Solzhenitsyn among a part of the French intelligentsia.

At the same time there are other opinions. For example, the deputy Minister of Foreign Affairs Kozyrev, the secretary of the Soviet Writers Union Polevoy, think that in this situation articles in the Soviet press and governmental action will hardly make any influence on the Nobel Committee. In their opinion, articles in the Soviet press will be appropriate only in case the Nobel Committee makes the decision of awarding the prize to Solzhenitsyn. The KGB representative in the USSR Council of Ministers (comrade Bobkov) is of the same opinion, but he thinks that in the Soviet press a satirical article could appear deriding certain circles in the West which are interested in organizing this provocation...[3]

The records on the first sheet of the document: "To be given for familiarization to secretaries of the CPSU Central Committee. P. Demichev." and "To exchange opinions at the CC Secretariat. M. Suslov".

RGANI, fond 4, opis'20, ed. khr. 770, l. 72–87; published in the journal *Istoricheskii Arkhiv*, No.1, 1992, pp. 147–149.

[1] See doc. 11.2.

[2] Novosti Press Agency.

[3] Having considered the memo of the head of the Central Committee's Cultural Department V. F. Shauro, the Secretariat of the Central Committee adopted a resolution, on October 8, 1970* (record 11, point 10), of the following content: "To instruct the Cultural Department of the CPSU Central Committee to prepare necessary materials on this question in accordance with the exchange of opinions at the CC Secretariat."**

* On that day the decision of the Swedish Academy on awarding the Nobel Prize to Solzhenitsyn was published, so the quoted resolution lost its sense.

** RGANI, fond 4, opis'20, ed. khr. 770, l.71: published in the journal *Istoricheskii Arkhiv*, No. 1, 1992, p. 147.

Document 11.4

"On measures in connection with the provocative act of awarding the 1970 Nobel Prize for Literature to A. Solzhenitsyn"; memo from the Cultural Department and the Propaganda Department of the CPSU Central Committee to the Secretariat of the Central Committee, of October 9, 1970.

On October 8 this year the Nobel Committee (Stockholm, Sweden) awarded the 1970 prize for literature to A. Solzhenitsyn with the formulation: "For the ethical force with which he develops invaluable traditions of Russian literature." As is known, Solzhenitsyn's candidacy had also been nominated for this prize in the previous years.[1]

Solzhenitsyn's works and behavior have long been used by the bourgeois propaganda for anti-Soviet purposes. Awarding him the Nobel Prize is aimed at further intensifying this campaign. We would consider expedient the following:

1. To publish in the Soviet press (the newspapers *Izvestiya, Trud, Komsomolskaya Pravda, Literaturnaya Gazeta*) a brief report from the Secretariat of the Board of the

Soviet Writers Union with the explanation of the fact that this action of the Nobel Committee has a political and not a literary character. The report of the Secretariat can be made in the form of their answer to the question of the *Izvestiya* correspondent (attached).[2]

2. In the *Literaturnaya Gazeta* to publish a lampoon revealing the essence of political speculations around Solzhenitsyn's name and work in the West.[3]

3. To give instructions to the State Committee of Television and Broadcasting of the USSR Council of Ministers (comrade Lapin) and the APN[4] (comrade Udaltsov), in connection with the decision of the Nobel Committee, to prepare and spread in foreign countries, through corresponding channels, the necessary propagandist materials.

4. To inform the heads of local party organizations (in oral form) about the provocative nature of awarding the Nobel Prize to Solzhenitsyn.

RSAMH, fond 4, opis'20, ed. khr. 781, l. 2–3; published in the
journal *Istoricheskii Arkhiv*, No. 1, 1992, p. 150.

[1] The assertion was not confirmed by any documental evidence. Solzhenitsyn wrote in the book *The Oak and the Calf* that in the fall of 1969 "Mauriac ...started his campaign of obtaining the Nobel Prize for me"*, that is, the 1970 prize, which he eventually received. From these words it is evident that before that there had been no leakage of information about nominating the writer for the prizes of the previous years. That means that the phrase was merely a far-fetched supposition of the leaders of the Central Committee's ideological departments in order to emphasize that they were well-informed.

* See note 14 to this chapter.

[2] Published in the *Izvestiya* newspaper of October 9, 1970 (the Moscow evening issue) and October 10, 1970 (the morning issue for the rest of the country).

[3] *Literaturnaya Gazeta*, October 14, 1970, a short article entitled *K Voprosu o Prioritete* (On the Question of Priority).

[4] Novosti Press Agency.

Document 11.5

From the Chairman of the State Security Committee of the USSR Council of Ministers Y. V. Andropov to the Central Committee, of October 10, 1970.

According to the information received by the State Security Committee, the news of awarding the Nobel Prize to Solzhenitsyn, on October 8, 1970, has noticeably stimulated the activity of foreign correspondents in Moscow and caused a number of congratulatory letters and telegrams addressed to him. The correspondent of the Norwegian newspaper *Aftenposten* Hegge contacted Solzhenitsyn and received from him the following statement: "I am grateful for being awarded, and I accept this award. I am going to go and accept it in person as tradition requires, if only it depends on me. I am well, and the journey will not do any harm to my health."

In the community of the Soviet intelligentsia the Nobel Committee's decision has mostly met with disapproval. Many writers, cinema workers, artists, theatrical personalities, composers consider awarding the Nobel Prize to Solzhenitsyn another anti-Soviet action and are discussing the question of how to react to it... Some representatives of intelligentsia have expressed a positive attitude to the fact of awarding the Nobel Prize to Solzhenitsyn. Academician A. Kolmogorov: "Solzhenitsyn has been awarded the Nobel Prize of 1970. It is good that he has been given it; he deserves it. I wonder: will Solzhenitsyn be allowed to go abroad to accept this prize?"[1] The literary critic L. Levitskii, of the *Novy Mir*: "For us it's a great occasion for celebration, as it was the *Novy Mir* that first discovered and printed Solzhenitsyn.[2] A deserved award."[3]

About receiving the prize Solzhenitsyn himself says that he would leave for Sweden only if he is guaranteed the re-entry visa. For your information.

RSAMH, fond 5, opis'62, ed. khr. 84, l. 163–165; published in the journal *Istoricheskii Arkhiv*, No. 1, 1992, pp. 150–151.

[1] These words are, of course, not from a public speech. They were said somewhere in a private conversation with colleagues and reported to the KGB by one of them. The agents' interest in the world-famous academician was also stimulated by the fact that in 1970 he still was the tutor of N. D. Svetlova, Solzhenitsyn's second wife, in her postgraduate studies.

[2] The tale *A Day in the Life of Ivan Denisovich*; Novy Mir No. 11,1962.

[3] Here is one more example of the attitude of "some representatives of the intelligentsia" to Sozhenitsyn being awarded the Nobel Prize; it was given in the memoirs of M. V. Ardov , to whom it had been told by T. M. Livinova, daughter of the famous Soviet diplomat. On the day she came, on literary business, to K. I. Chukovsky to Peredelkino, a group of writers was going around the village collecting signatures under a collective letter condemning Solzhenitsyn. Chukovsky knew about the "action" and "at odd moments followed the movement of this group... We were continuing our studies but at some moment Chukovsky told me: "Tanya, whatever happens now, whatever you hear from me, don't be surprised." Just three minutes after a doorbell was heard downstairs, and the housemaid opened the door. At this moment Chukovsky jumped out into the staircase and yelled in a frightening voice: "What bastard has woken me up?! I didn't sleep all night! I've just dozed off! Throw them out! Out on their ear! Everybody out!" The entrance door banged shut. The hapless collectors of signatures left in embarrassment. And Kornei Ivanovich very quietly sat down in his armchair at the table and said: "So, where were we?"*

* *Novy Mir*, No. 6, 1999, pp. 134–135.

Document 11.6

From A. I. Solzhenitsyn to the Swedish Academy, of November 27, 1970.

In the telegram addressed to the secretary of the Academy I already expressed[1] and am now expressing again my gratitude for the honor given to me by awarding me the Nobel Prize... In the same telegram I expressed my intention to come to Stockholm, although I could imagine that I would have to go through the humiliating procedure which is usual in our country for all those going on a foreign trip — filling in special forms, getting character references from party organizations, even for a non-party man, and being instructed on how to behave. However, over the last weeks the hostile attitude to my prize shown by the national press and the continuing persecuted state of my books (for reading them people are fired from work and expelled from universities) make me suppose that my trip to Stockholm will be used for severing me from my native land, simply for barring my coming back home...[2]

Having weighed all said above and taking advantage of your kind explanation that arrival at the ceremony in person is not an indispensable condition for accepting the prize, I would prefer at the present time not to submit an application for the trip to Stockholm. I could receive the Nobel diploma and medal in Moscow, if you will find such a form acceptable, from your representatives at a time mutually convenient for you and for me. As it is established by the Nobel Foundation statute, within half a year I am ready to deliver the Nobel lecture or to submit it in written form.

A. I. Solzhenitsyn. *The Oak and the Calf*, p. 632.

[1] "...giving the telegram in answer to the Swedish Academy: 'I regard the Nobel Prize as a tribute to Russian (certainly not Soviet) literature and our difficult history'"...*

* *ibid*, p. 378

[2] "*Our people* anxiously waited for my departure, were on the watch for it!.. Later I learnt for certain: the resolution was prepared that I was deprived of the Soviet citizenship. It only remained to roll me over the border."*

* *ibid*, p. 280.

Document 11.7

The text of A. I. Solzhenitsyn's brief greeting to be read in his absence at the Nobel banquet in the Blue Hall of the Stockholm City Hall on the evening of December 10, 1970.

Your Majesty! Ladies and gentlemen! I hope that my involuntary absence will not darken the completeness of today's ceremony. In the sequence of brief greetings my words are also expected. I would still less like my words to darken the celebration. However, I cannot disregard the significant coincidence that the day of giving of the Nobel Prizes coincides with the Day of Human Rights. The Nobel Prize winners cannot but feel responsibility for this coincidence. All those who have gathered in the Stockholm City Hall cannot but see a symbol here. So, let us not forget at this festive table that today political prisoners are on hunger strike asserting their disparaged or altogether trampled rights.[1]

ibid, p. 633.

[1] When reading the text at the Nobel banquet the organizers of the ceremony preferred to omit the last phrase. As Solzhenitsyn wrote afterwards, "whether they were shy of the ceremony or, as they say, feared for me (for all of them feel sorry for me. As said the Swedish academician Lundkvist, the Lenin Prize winner:* 'The Nobel Prize will do harm to Solzhenitsyn. Such writers as him are used to poverty and should live in poverty')."**

* In 1952 Lundkvist became laureate of the international Stalin Prize "For strengthening peace between people". In 1956 the prize was renamed the Lenin Prize.

** A. I. Solzhenitsyn. *The Oak and the Calf,* p. 284.

Document 11.8

From A. I. Solzhenitsyn to the permanent secretary of the Swedish Academy K. R. Gierow and the executive director of the Nobel Foundation N. S. Stahle, of October 22, 1971.

....Last year, indeed, the ambassador Mr Jarring, choosing from several options preferred to give me the Nobel diploma and medal in the Swedish embassy in Moscow. Having already understood by the moment of our talk that I would not be able to go to Stockholm, I wanted to accept the option he suggested, thinking that the ceremony would take place openly, with a certain number of people present, and I would be able to read my Nobel Lecture before them. However, Ambassador Yarring categorically objected to it saying that the award giving could only be confidential, "like now, in the study".[1] It seemed to me that agreeing to such a proposal would be humiliating for the Nobel Prize itself; as if it were something vicious that should be concealed from people. As I understand, the Nobel award ceremony is public exactly because this ceremony has public meaning.

It is this natural interpretation that I meant when I was writing to you, on November 27, 1970,[2] that I was ready to accept the Nobel signs in Moscow. Since then neither my situation nor my point of view has changed. This year, like last year, I am ready to accept the Nobel signs in Moscow, but certainly not confidentially. But if, like last year, such award giving is considered undesirable or inconvenient, I will again ask you to leave my Nobel signs for further storage at the Nobel Foundation, especially as that does not contradict your rules... In this case I will together with you keep a patient hope that in some year the circumstances would at last become favorable for my participation in the traditional Nobel ceremony in Stockholm.[3] Personally I ask both of you to accept my profound apologies for having involuntarily become the cause of unnecessary trouble and cares, which you did not have with the majority of my predecessors.

ibid, p. 637.

[1] " ... And what about the diploma and the medal? Is it possible to arrange a reception at your embassy?"

" It's impossible. It's never been like that."

" But there's also never been a case like mine. Don't be so sure, Mr Yarring, let the Academy think."*

Yarring answers confidently: either by mail or we'll give it to you in the study, like now, without anybody present."**

* Solzhenitsyn in this case does not understand the main point. In the place of Gunnar Yarring any ambassador of any state would take a similar decision. Even in the most liberal state conducting a public anti-governmental event in the building of a foreign embassy would inevitably result in the demand of immediate recall of the ambassador from this country and, most probably, the wreck of his diplomatic career in his homeland. In this connection Yarring quoted in his memoirs the words of Bjorn Alender, who had lived in Moscow for a long time. These words were published in a Goteborg newspaper on October 14, 1971, just at the peak of the crisis of Solzhenitsyn's relationship with the ambassador: "A diplomat who offers resistance to the rules of the game of a dictatorial state may be subjected to isolation so effective that he will be no longer able to be useful to his government." G. Yarring. *Do Glasnosti i Perestroiki* (Before *Glasnost* and *Perestroika*).- Moscow: Progress-Politika, 1992, p. 199.

** Solzhenitsyn. *The Oak and the Calf*, p. 282.

[2] See doc. 11.6.

[3] The Nobel Prize was presented to Solzhenitsyn in Stockholm on December 10, 1974, after he had been forced into exile from the Soviet Union, in February that year.

Document 11.9

The permanent secretary of the Swedish Academy K. R. Gierow to A. I. Solzhenitsyn, of November 22, 1971.

Nils Stahle and I have just met with Gunnar Jarring. Our talk has not brought anything new, but we could hardly expect positive results. We have to say that the embassy does not have any suitable hall for a public lecture and the Academy at present does not have the ability to arrange such a hall somewhere else in Moscow. But we will have to summon up patience, as you write, in the hope that circumstances will later allow us to fulfill our wishes which we now have to refuse. The signs of honor still remain here. But of course, I am always ready to go to Moscow so as to give you, in a worthy form, the Nobel diploma and medal at the embassy or at any other place convenient for you, as far as possible. In this case I, perhaps, could take with me a copy of your lecture in order to have it published in *Les Prix Nobel*[1] pending an opportunity for you to read it. That is merely a proposal which I would like to mention.[2]

ibid, pp. 637–638.

[1] Solzhenitsyn's Nobel lecture was published in the Nobel Foundation's yearbook *Les Prix Nobel* in 1974, when it was delivered by him in Stockholm.

[2] Solzhenitsyn's attempts to obtain a public ceremony of giving him the diploma and gold medal of the Nobel prize-winner in Moscow were no secret to security services. On November 16, 1971 a memo with information on that question, with the signatures of the KGB Chairman Y. V. Andropov and the Minister of Foreign Affairs A. A. Gromyko, was directed to the CPSU Central Committee. Three days later, on November 19, 1971, the Central Committee's Political Bureau adopted a resolution (record 26, point 44) approving the draft of instructions to the Soviet ambassador in Sweden, which had been prepared by the Ministry of Foreign Affairs. The ambassador was to immediately meet with the Swedish Minister of Foreign Affairs and state to him that organizing a presentation of the insignia of the Nobel laureate in the building of the Moscow embassy will be regarded here "as an unfriendly act with regard to the Soviet Union, detrimental to the relationship between our countries." At the same time it was recalled that on December 1, 1970 the Ministry of Foreign Affairs of Sweden assured the Soviet party that "the interference of the Swedish government in the matters connected with awarding the Nobel Prize is ruled out."* Probably, when sending his letter (doc. 11.9) three days after the Political Bureau's approval of the instructions to the ambassador in Stockholm, Gierow was already informed of the official warning received from Moscow.

* *Kremlevskii Samocud. Sbornik Dokumentov* (The Kremlin Lynch Law. Collection of Documents). Compiled by A. V. Korotkov and others. — Moscow: ROSSPEN, 1994, pp. 175–177.

Document 11.10

From A. I. Solzhenitsyn to the permanent secretary of the Swedish Academy K. R. Gierow, of December 4, 1971.

The last four letters of yours (of October 7 and 14, November 9 and 22)[1] clarify more and more if it is possible to give me the Nobel signs in Moscow, in a worthy, as you write, atmosphere... I highly and even cordially appreciate the unbending intention expressed by you to come to Moscow in person, at any time and under any circumstances so that this ceremony can take place. I am sincerely grateful to you for this decision and, frankly speaking, I think that it comes like a ray of light through this situation full of obstacles... But here I... remember your words that the Swedish Academy and the Nobel Foundation are independent and inviolable in their activities and their decisions, and an official ceremony organized "as if" by the Swedish state[2] could do damage to this fact. Fully understanding and sharing this feeling of yours, but, on the other hand, not knowing in Moscow any social or cooperative organization which would agree to give us a building for our purpose, I take the liberty to suggest a different option: conducting the whole ceremony in Moscow at a private apartment, notably, at the address to which you send letters to me.[3] True, this apartment is certainly not more spacious than the Swedish embassy, but it can accommodate forty or fifty people quite freely, according to the Russian standards. The ceremony may lose a bit of its official character but it will gain some home warmth... I do not know the Nobel annals but I can suppose that in the past there already might have been a case when a Nobel prize-winner was tied to one place — for instance, by an illness — and a representative of the Foundation or of the Academy went to him and gave him the prize directly at his home? But what if we find all those options impeded? Well, then we shall submit to fate; let my Nobel signs continue to be stored at the Nobel Foundation, and some day, even after my death, your understanding successors will present these signs to my son?..

ibid, pp. 638–639.

[1] Gierow's letters of October 7 and 14 and November 9, 1971 were not published by Solzhenitsyn; for the letter of November 22, 1971 see doc. 11.9.

[2] That is, conducting the public ceremony in the building of the embassy of Sweden in the USSR, on which Solzhenitsyn had insisted earlier. The given argument was suggested by the Swedish Academy in order to help both parties find the way out of the delicate situation of using the embassy building for organizing a public political action inadmissible by the rules of diplomatic etiquette.

[3] In the apartment of the laureate's wife, N. D. Svetlova.

Document 11.11

From the Chairman of the State Security Committee of the USSR Council of Ministers Y. V. Andropov to the CPSU Central Committee, of April 4, 1972.

I

Directed hereby is the translation of the report by the Moscow correspondent of the *Washington Post* newspaper Robert Kaizer about his meeting with Solzhenitsyn.

The record in the top left corner: "to comrade Demichev." In the bottom right corner are signatures of A. N. Yakovlev, G. L. Smirnov and V. A. Medvedev and the date: 25.4.1972.

RGANI, fond 5, opis 64, ed. khr. 135, l. 29[1]

[1] Andropov's letter is preceded by the letterhead sheet of the Central Committee's Cultural Department (l.28) with the handwritten record: "to comrades Shauro V. F., Yakovlev A. N. 4.IV"; no signature, but the record was most probably made by the secretary of the Central Committee P. N. Demuchev, to whom the document was directed.

II

Robert G. Kaiser. *Washington Post.*[1]

Moscow. Alexandr I. Solzhenitsyn, the author and winner of the Nobel Prize, accused Soviet authorities of conceiving an elaborate plan "to suffocate me". Making his first personal statement to the press in nearly ten years, Solzhenitsyn said that since 1965 he has been slandered systematically by an official propaganda campaign intended to "drive me out of society or the country, throw me in a ditch, or send me to Siberia, or have me dissolve in an alien fog" ...

Solzhenitsyn made these observations in a Moscow apartment during four hours last Thursday afternoon, just ten days before he hopes to receive his Nobel Prize medal in a small ceremony here...[2]

He refused to answer questions which he thought were political... His 32-year-old wife Natalia contributed vigorously to family deliberations on whether or how to answer various questions... He asked that his words not be printed on April 1 for fear they might be taken for a joke of some kind...

Solzhenitsyn's lonely position in this society, emphasized by his decision to speak out in this way, was also a subject of his own reflections. "It really never occurs to them", he said of "those who direct the campaign against me", that "a writer who thinks differently from the majority of society is an asset to that society and not a disgrace or defect."

Yet such thoughts apparently occur to some Russians. Last week, when Solzhenitsyn appeared at the concert given by his friend, Mstislav Rostropovich, the cellist,[3] smiling people crowded around him and asked for an autograph...

He announced that the royalties that he has earned in the West — substantially more than $1 million — will be spent for "the general welfare of my country" under the terms of a will he had written.[4] He said he did not know how it could be arranged, but he hoped it would become possible in the future.

Solzhenitsyn said that he and his family will live on the $78,000 of the Nobel Prize, although the Soviet Ministry of Foreign Trade had made it "degrading, difficult and uncertain for me" to get that money.

He is currently living near Moscow at Rostropovich's country home, where he works "all the time when he is not sleeping"... He needs "fresh air and quiet" to work. His wife, a mathematician, spends most of her time in Moscow, at the her family's flat, though they see each other often. Solzhenitsyn said she was fired from her job in a Moscow institute when the director found out she was his wife...[5]

In his discussion Solzhenitsyn devoted the most attention to the "plan to suffocate me", as he called it. He claimed that party propagandists had told a long series of lies against him in lectures and meetings... This began in 1966, he said. Solzhenitsyn mentioned these as some of the false charges that had been made against him:

His books were criminal, he was "improperly rehabilitated" during Nikita Khrushchev's destalinization, he smuggled his works abroad, he helped the Nazis and worked for the Gestapo in World War II, and — most serious of all, Solzhenitsyn said — he was a "traitor to his country".[6]

Moreover, he claimed, lecturers slandered his family life (his second marriage is still not legally consummated, as he has not been finally divorced from his first wife) and tried to suggest he was Jewish because of his Russian middle name or patronymic, Isayevich, which means "son of Isiah".[7] All these lectures were closed events, actually or in effect, and there was no opportunity to answer the accusations, Solzhenitsyn said....

Solzhenitsyn refused to say that harassment and persecution of talented writers was a permanent condition of Russian life, under the czars as well as under communism. This is not a subject "for a brief, superficial conversation", he said — it would require detailed research to comment on it...

Speaking of the ceremony at which he hopes to receive the Nobel medal and deliver his Nobel lecture next Sunday,[8] Solzhenitsyn said he would invite the Soviet minister of culture, his close friends, representatives of the Soviet art world and journalists.

Dr. Karl Gierow of the Nobel Academy[9] announced in Stockholm that he would come to Moscow next weekend and bestow the prize on Solzhenitsyn in a "private ceremony" on Sunday. Solzhenitsyn said that the Soviet authorities could still put some obstacle in the way of this ceremony, but that it would be a "shameful absurdity" if they did...

His first two main novels, *The Cancer Ward* and *The First Circle*, were not published here [in the USSR]. By 1969 both works had appeared abroad, and the *Literaturnaya Gazeta*, paper of the Soviet Writers Union, called him "slanderer against the Soviet Union". Solzhenitsyn's opponents became considerably stronger than his friends, and he was expelled from the Writers Union. In 1970 Solzhenitsyn was awarded the Nobel Prize. Official Soviet commentators stated that the Swedish Academy committed an anti-Soviet act. Solzhenitsyn said that he was afraid to go to Stockholm for receiving the prize, as he was afraid that he would not be allowed to come back to Russia.

During the war Solzhenitsyn served in the army in the position of an artillery officer at the front. In 1945 he was arrested for criticizing Stalin in a letter to his friend and sentenced to eight years...[10] In 1953 he came out of prison and was exiled. The exile ended in 1956, and in 1957 he was completely rehabilitated, as, according to the decision of the military tribunal, "there was no *corpus delicti* in his actions"... Since he was expelled from the Writers Union, he has lived most of the time at the cellist Mstislav Rostropovich's, who has a house near Moscow.[11]

True: the head of the State Security Department of the USSR
Council of Ministers F. D. Bobkov

RGANI, fond 5, opis' 64, ed. khr. 135, l.30–38.

The abridged text of the interview is quoted from the English original published in the *Washington Post*, April 3, 1972, pp. 1–17.

[1] Sozhenitsyn's meeting with R. Kaiser and the correspondent of the *New York Times* H. Smith took place on March 31, 1972.* The articles about the meeting** and the text of the writer's interview prepared in advance in the written form were published in the named newspapers on April 2–4, 1972***.

* For his own memories of the meetings with both American correspondents see: Solzhenitsyn, pp. 303–304.

** Solzenitsyn was dissatisfied with the publications: "... as I am not used to it, the loosened incoherent form in which those correspondents strewed my thoughts proved unbearable for me", *ibid*, p. 304.

*** The complete text of the prepared interview: *ibid*, pp. 640–652.

[2] The ceremony of giving the Nobel signs was to take place at N. D. Svetlova's apartment on April 9, 1972 (see doc. 11.12).

[3] The concert of the Moscow State Philharmonic Society Symphony Orchestra under the guidance of K. P. Kondrashin, with M. L. Rostropovich's participation, took place on March 27, 1972, at the Grand Hall of Conservatory.

[4] The will, made in 1971, was forwarded to the West with the help of Heinrich Boll, who came to Moscow in February 1972. He certified each page of its text with his signature. Official notarizing of the will through state notary offices was out of the question as its content would certainly be regarded in Moscow as anti-Soviet.

[5] Solzhenitsyn's second wife N. D. Svetlova was by the time of her marriage a postgraduate of the Moscow State University under the guidance of academician A. N. Kolmogorov.*

* See note 1 to do.11.5.

[6] See note 91 to Chapter 6.

[7] Also at briefings, as if pulled by a string: "He is, by the way, not Solzhenitsyn, but Solzhenitser..."* The same was written in the diary of the member of the *Novy Mir*'s editorial board A. I. Kondratovich: "January 9, 1970. A rumor is persistently being spread that Solzhenitsyn's real name is Solzhenitser. They want to carry out a 'Black Hundred' ** attack on this line. Monakhov, from the MVD journal*** phoned me and said: we have an argument — Solzhenitsyn or Solzhenitser. I told him as it is and advised him to have a look in the Historical Library into the sixth velvet book, where the Solzhenitsyns' name is mentioned as that of noblemen."****

* Solzhenitsyn, p. 285.

** The name of armed anti-revolutionary groups in Russia, active in 1905–07.

*** The organ of the Ministry of Internal Affairs.

**** *Novy Mir*, No. 2, 1990, p. 224.

[8] The ceremony arranged for April 9, 1972, the first day of the orthodox Easter, did not take place because of the fact that the Soviet embassy in Stockholm refused to give a visa to enter the Soviet Union to the permanent secretary of the Swedish Academy, Karl Gierow* (see doc. 11.12).

* "...the Soviet embassy in Stockholm made a reservation, though, that 'it does not rule out giving Gierow a visa at some other, more convenient time', in order to assuage irritation... The Swedish Ministry of Foreign Affairs followed suit... The ban on Gierow's arrival closed the whole ceremony, making it senseless." Solzhenitsyn, pp. 304–305.

[9] He means the Swedish Academy.

[10] The veiled attacks on Stalin that the letter contained were deciphered by the military censorship, which in the period of the Great Patriotic War 1941–1945 was given legal grounds to open and look through all mail correspondence sent within the Soviet Union.

[11] Solzhenitsyn lived at Rostropovich's *dacha* in Zhukovka, near Moscow, from the fall of 1969* through May 1973.** V. Y. Lakshin, a member of the *Novy Mir*'s editorial board, in his diary entry of January 29, 1970, picturesquely described a scene in the study of the secretary of the CPSU Central Committee P. N. Demichev, where Rostropovich was urgently summoned for the talk about the rightfulness of Solzhenitsyn's living at his *dacha*: "He declared, crossing the threshold: if it's about Solzhenitsyn, I warn you that he will live at my place as long as he wishes."***

* Solzhenitsyn, p. 253: "I do not remember who has ever given me a bigger present than Rostropovich by this shelter. He had invited me already in the last year, 1968, but I was somehow afraid to inconvenience them. But this year moving and settling there was most appropriate and timely."

** *ibid*, p. 31: "I did not dare to spoil Rostropovich and Vishnevskaya's lives any longer. And in May 1973 I left their place..."

*** *Druzhba Narodov*, No. 6, 2003, p. 159. See also doc. 11.15, which shows that two years later Rostropovich was already tired of the troublesome guest's long stay under his roof.

Document 11.12

From the Chairman of the State Security Committee of the USSR Council of Ministers Y. V. Andropov to the Central Committee of April 4, 1972, incoming No. 10780 of April 4, 1972.

Reactionary circles in the West continue the anti-Soviet stir around Solzhenitsyn, using preparations for handing him the signs of the Nobel Prize. In order to disrupt this action planned for April 9, the entry to the USSR of the secretary of the Nobel Foundation[1] Gierow, who is to award the signs, is closed.[2] Also annulled is a tourist visa to the lawyer Heeb representing Solzhenitsyn's interests abroad.[3] Under these conditions, however, it cannot be ruled out that the Nobel Foundation might use diplomatic or other channels for delivering the signs to the USSR and charge some unknown representative of Sweden or another foreign state to award them, which will seriously hinder the efforts to prevent this scheme being carried out.

The document is directed, on April 4, 1972, by the secretary of the Central Committee, P. N. Demichev to the head of the Cultural Department V. F. Shauro, who passed it to his deputy, A. N. Yakovlev; *ibid*, l.28. The next day, April 5, 1972, A. M. Alexandrov, the assistant to the General Secretary of the Central Committee, L. I. Brezhnev, directed the document to the head of a department of the Central Committee, K. U. Chernenko, with the record on a separate sheet: "To No. 10780. Konstantin Ustinovich! Will you find it possible to report this material to comrade M. A. Suslov?"; *ibid*, l.26.

RGANI, fond 5, opis' 64, ed. khr. 135, l. 27; published in the journal *Istoricheskii Arkhiv*, No. 1, 1992, p. 157.

[1] An error: K. R. Gierow was a permanent secretary of the Swedish Academy.

[2] "The date of the Nobel ceremony, April 9, the first day of orthodox Easter, was announced by Gierow when submitting an application for a visa — it seems that it was on March 24."*

* Solzhenitsyn, p. 302; see also note 8 to doc. 11.11.

[3] Solzhenitsyn established contact with the lawyer from Zurich, Heeb, through intermediaries in 1970–71. By April 1972 he had already managed to forward to Heeb his "main will".*

* "My main will (impossible to be submitted to a Soviet notary office) was sent to the lawyer, Heeb"; Solzhenitsyn, p. 293.

Document 11.13

Solzhenitsyn's statement in connection with the cancellation of the Nobel ceremony fixed for April 9, 1972, of April 8, 1972.

Gierow and I have made concessions wherever possible: his trip was planned as a private one, to a private apartment, for conducting an almost private ceremony. The ban on the ceremony even in such a form is an irrevocable and final ban on any form of handing me the Nobel Prize on the territory of my country.[1] That is why the belated concession of the Swedish Ministry of Foreign Affairs is already unnecessary. But it is also insulting: the Swedish Ministry of Foreign Affairs continues to regard giving me the Nobel Prize not as a cultural phenomenon but as a political event and so it imposes a condition that would again lead either to a "closed" option of the ceremony or to a special selection of guests and somehow prohibiting them to express their attitude to what is going on, as all that could be interpreted by someone as "political demonstration".[2]

Besides, after Gierow was refused a visa I would consider accepting the Nobel signs from some other hands than those of the Permanent Secretary of the Swedish Academy humiliating both for him and for me. Finally, with our modest forces we already made all the difficult preparations: invitations were sent, not only to Moscow addresses, but to about 20 writers whom I see as the flower and creative force of our literature of today and to about the same number of actors, musicians, academicians; because of that, many of them fixed or canceled their trips, or rehearsals, or other responsibilities. Now all of these 40 guests have been offered the insult of refusal, and cancelations of the invitation have been sent. Both they and I are people busy enough to undertake such a procedure a second time.

According to the rules of the Swedish Academy, which were explained to me, the Nobel signs can be stored by it indefinitely. If my lifetime is not enough, I bequeath their acceptance to my son.

A. I. Solzhenitsyn. *The Oak and the Calf,* p. 651.

[1] "With our usual cunning skill of saying no, the Soviet embassy in Stockholm made a reservation that 'it does not rule out giving Gierow a visa at some other, more convenient time', in order to assuage irritation, create an illusion and smooth transition of chances to get visa to zero. The Swedish Ministry of Foreign Affairs followed suit. But we here understand such a game only too well! — and I swiftly stoped it by a special statement."*

* Solzhenitsyn, p. 304.

[2] See footnote * to note 1 in doc. 11.8, and note 2 to doc. 11.9.

Document 11.14

Memo from N. A. Shchelokov, Minister of Internal Affairs of the USSR "On the question of Solzhenitsyn".[1] No date, but no later than the beginning of October 1971.

When solving the question of Solzhenitsyn it is necessary to take into account the following situation.

1. Solzhenitsyn has become a major figure in the ideological struggle. It is a reality which cannot be disregarded. This fact cannot be hushed up or passed over. Anti-Soviet forces have given him their reactionary banner.

2. Objectively, Solzhenitsyn is talented. It is a phenomenon in literature. From this point of view he is certainly of interest to the Soviet power. If his pen served the interests of the people, it would be highly advantageous. A right solution to "the problem of Solzhenitsyn" would make quite real the task of turning his creative interests to ideologically faultless subject matter.[2]

3. When solving the question of Solzhenitsyn it is necessary to analyze the errors made in the past with regard to creative workers... It is difficult to understand why Solzhenitsyn had to be expelled from the Writers Union for the book, *The Cancer Ward,* written from the same ideological point of view as *One Day in the Life of Ivan Denisovich.* In the first case he was admitted to the Writers Union. In the second case he was expelled from it. Such inconsistency weakens our positions in the struggle for the ideological purity of our literature. It makes the attitude to Solzhenitsyn himself incomprehensible and inconsistent. "The problem of Solzhenitsyn" was created by foolish administrators in literature.[3]

...8. So, how should "the problem of Solzhenitsyn" be solved at the present time? Firstly, there should be no obstacles in the way of his going abroad for accepting the Nobel Prize. Secondly, by no means should the question of stripping him of his citizenship be raised. Work should be done to ensure Solzhenitsyn's worthy behavior abroad.[4] Solzhenitsyn should be made to expressing the following ideas in public: " I do not have any disagreements with the Soviet power. I do not have any divergences with the party. I am a Soviet writer. I am proud of what is going on in this country. I have some divergences with my literary colleagues. It is professional divergence. It was, is and will be, as long as literature exists."[5] It is the professional character of divergence that should be stressed.[6] Solzhenitsyn should be given an apartment urgently. He should get official registration and be given attention. Someone of our prominent executives should talk to him and remove all that bitter aftertaste that the persecution of him has certainly left. In short, we should struggle for Solzhenitsyn rather than throw him out. It should be the struggle for Solzhenitsyn and not against Solzhenitsyn.

Kremlevskii Samosud. Sbornik Documentov (The Kremlin Lynch Law. Collection of Documents.). Compiled by A. V. Korotkov and others. — Moscow:ROSSPEN, 1994, pp. 169–172.

[1] The part of the text of this document concerning B. L. Pasternak is given in doc. 9.30.

[2] As it was easy to understand even in 1971, these idle plans were built on sand. (See also note 2 to doc. 9.30).

[3] In the original of the document the first and the last phrase in the given part of point 3 are underlined by Brezhnev.

[4] In connection with this phrase a curious parallel with an earlier event springs to mind. In 1905 the Nobel Prize for Literature was awarded to the Polish writer and Russian citizen H. Sienkiewicz, who did not have any reverence for the Russian empire and its rulers. In spite of that, before Sienkiewicz's trip to Stockholm the authorities did not even think of posing before him something like the preliminary conditions that Shchelokov formulated, six and a half decades later, and from the highest motives at that... In Stockholm in 1905 everything turned out happily (see doc. 1.20). But even if Sienkiewicz, in his Nobel report or in any other speeches before the Swedish audience, had made any attacks on the Polish policy of the St. Petersburg authorities, these utterances would certainly not have provoked any shock in governmental circles, whereas the Kremlin rulers' level of tolerance in similar cases has always been measured on a completely different scale.

[5] Isn't it a real masterpiece of sweet and groundless reverie?

[6] After this phrase the concluding text of point 8, as well as the first phrase of this point, are in the original underlined by Brezhnev.

Document 11.15

From the working record of the sitting of the CPSU Central Committee's Secretariat, of October 7, 1971.

SUSLOV. There is comrade Shchelokov's memo about Solzhenitsyn. You have familiarized yourselves with it. The gist of the matter is solving the question of Solzhenitsyn's place of residence.[1] The musician Rostropovich, at whose *dacha* in Zhukovka he is now living, raises the question of evicting Solzhenitsyn. But in order to be evicted, he should be permitted to live somewhere. Now he does not have registration in Moscow. His apartment is in the city of Ryasan', but in that apartment lives his wife, who he has divorced. Now Solzhenitsyn is married to another woman, who lives in Moscow and has a two-room apartment, and in fact Solzhenitsyn lives all the time at her place...

KATUSHEV. Shall we permit him to build a *dacha* on a garden plot near Maly Yaroslavets,[2] where he has a small summer house. That will hardly be expedient, as many workers from Moscow and other cities live there.

SUSLOV. So far we could do it like this: not to solve this question now, ask the KGB for advice what would be better: whether to exile Solzhenitsyn outside Moscow or to permit him to live in the Moscow apartment of his new wife, which will allow easier surveillance.

DECISION. To confine themselves to the exchange of opinions.

ibid, pp. 172–173.

[1] Everything concerning political aspects of the attitude to Solzhenitsyn that was brought up in Shchelokov's memo, was resolutely severed from the debate by M. A. Suslov, who was chairing the sitting.

[2] Maloyaroslavets is a town to the south of Moscow, in the north of the Kaluga *oblast'*.

Document 11.16

The memo from the Minister of Internal Affairs of the USSR N. A. Shchelokov to the CPSU Central Committee, "On considering A. Solzhenitsyn's telegram",[1] of February 6, 1980.

In connection with the assignment[2] we report that A. Solzhenitsyn's telegram has been considered by us jointly with the KGB. We have decided that it will be expedient to refuse the Soviet citizen Shcherbak N. I., a distant relative of the author,[3] exit to the USA. It is taken into account that at the time of Solzhenitsyn's deportation from the USSR all his wishes that members of his family would go abroad were taken into consideration[4] and that the decision was taken not to have official relations with this person in the future. It has also been decided that it will be expedient not to give Solzhenitsyn any answer in connection with his appeal.

Kremlevskii Samosud. — Moscow:ROSSPEN, 1994, pp. 569–570.

[1] The telegram addressed to the secretary of the CPSU Central Committee K. U. Chernenko and sent on December 20, 1979, said: "The Soviet embassy in Washington informed me that my only relative, Aunt Irina Ivanovna Shcherbak was categorically refused a visa for visiting me in the United States. Is the disgrace already made public not enough that it has to be added to by unjust and inhuman treatment of a ninety-year old, blind, deaf, hunched, illiterate old woman? Give instruction, please, let the old woman go, don't force me to make it public. Alexandr Solzhenitsyn."*

**ibid*, p. 569.

[2] Chernenko passed the telegram for taking measures to Shchelokov, who readdressed it, on December 28, 1979, to his subordinate B. K. Elisov with the assignment "urgently... to decide together with the KGB, having received the agreement at the Central Committee."*

**ibid*, p. 569.

[3] I. I. Shcherbak was the wife of Solzhenitsyn's uncle on his mother's side. In his early childhood Solzhenitsyn lived and was brought up in her house for several years.

[4] This is definitely a false assertion. The head of the Ministry of Internal Affairs could not have been ignorant of the fact that, after being detained, on February 13, 1974 and familiarizing himself with the decree of the Presidium of the Supreme Soviet of the USSR, "On Depriving of Soviet Citizenship and Deportation from the USSR" of February 12, 1974 (published on February 14, 1974), Solzhenitsyn immediately wrote a petition addressed to the Chairman of the Presidium N. V. Podgornyi about his departure together with members of his family — his wife, their four children (including N. D. Svetlova's child from her first marriage) and his mother-in law, E. F. Svetlova. These wishes were indeed completely fulfilled. But in the same petition Solzhenitsyn simultaneously asked to "give permission and a practical opportunity to leave for the place of my new stay to my aunt, Shcherbak Irina Ivanovna (resident of the town of Georgievsk, Stavropolskii krai, 93 ulitsa Lineinaya)."* This wish was silently refused both in 1974 and in the subsequent years, of course, with the knowledge of Minister Shchelokov.

**ibid*, p. 498.

Document 11.17

From Solzhenitsyn's memories about his deportation from the USSR on February 14, 1974.

In the reception compartment[1] they gave me back my watch and cross and asked me to put my signature. The lieutenant-colonel was worried — why was my pocket bulging. I showed him bread. After some hesitation — well, he let it be... We come out. Again a circle in the yard, again to the back seat between two guards... The damned gate swung open, we are going. Two cars, four people in each one, so eight of them again... Again a half-circle — now around the boarding staircase... The staircase leads to the front cabin. Inside it there are seven people in civilian clothes, and the eighth is the doctor accompanying me. Except for the doctor, all the rest have been changed again. The seat where I must sit is pointed out to me — by the aisle in the middle row... I am encircled as if by a belt... The plane is taxiing along the gloomy airfield covered with dirty snow... the last thing that I see in Russia... When the plane shudders, taking off, I cross myself and bow to the land going away...

We are coming down. Lower. A big city can be discerned. On a river. The river is not very wide but not small either. The Danube? Who knows... Now here is the airfield. We are running along landing strips. Amid several buildings one is high, and on it is the inscription *Frankfurt-am-Main...*[2] We have stopped. Outside: the staircase is being brought. And suddenly from the pilots' cabin a command is heard — loud and sharp: "Dress him! Take him out!"... However, I put on my coat with my own hands. Then the cap... Suddenly at the threshold of the cabin one of the eight runs into me face to face, chest to chest — and from stomach to stomach passes me five banknotes — five hundred German marks... And he vanished, I had not even distinguished his face, had not noticed it. And in general — the road is free for me. KGB people are standing by the sides, the pilot has also come out here. A voice: "Go!"... About 200 people are standing in a broad circle — apparently, behind the line that is not to be crossed — applauding, taking pictures or filming... At the foot of the staircase stands a very likeable man, smiling, who speaks rather good Russian — Peter Dingens, representative of the Ministry of Foreign Affairs of the Federal Republic. A woman comes up and gives me flowers... In a police car I am taken out from the airport through a side exit, my companion suggests going to Boll, and we are tearing along the freeway, already talking about *this* life: it has already begun. We are driving 120 kilometers per hour, but another police car overtakes us and orders us to turn to the side. A red-haired young man jumps out from it and presents me with a huge bunch of flowers, explaining: "That is from the Minister of Foreign Affairs of the Rhine-Pfaltz land. The minister expresses the guess that it is the first bouquet you have ever received from a minister of internal affairs"...

In the evening in Boll's small village we were forcing our way through two rows of correspondents' cars already standing along the narrow streets. Under photo flashes we jumped into the house, but until late at night and then in the morning we heard the hubbub of correspondents near the house... In the morning, as it was explained to me, it was inevitable to go out, become the prey of photographers and say something. What to say?... And it came out of me by itself: "I spoke enough when I was in the Soviet Union. And now — I'll keep silence"...

Keep silence — I meant keeping silence before microphones, but my state in Europe I realized, from the first hours, the first minutes as activities not constrained at last...

Solzhenitsyn, pp. 391–399

[1] In the Lefortovo investigatory prison, after the announcement of the decree of the Presidium of the Supreme Soviet of the USSR about the stripping of Soviet citizenship and deportation from the Soviet Union.

[2] According to the employee of the *Novy Mir* journal Evgeniya Katseva, Solzhenitsyn must have not been surprised by the landing of the plane in Germany. In her memoirs about that time, touching upon the silent ban, in 1974, on the publication of works by Heinrich Boll, the 1972 Nobel Prize winner, which was all the more difficult to understand because he was very popular with Soviet readers, she wrote: "Boll vanished from the pages of periodicals and from publishers' plans. The reason is known: when being thrown out from the country in 1974, Solzhenitsyn, in answer to the 'magnanimous' question where to send him, named Boll's address." The first attempts to break through the wall of silence around the key figure of the post-war European literature were made, in January 1983, by the chief editor of the *Literaturnaya Gazeta*, A. B. Chakovsky. But only on July 3, 1985 did the newspaper manage to publish an abridged version of *Letters to My Sons* — "a confessional story written by the 40th anniversary of the end of the war." Boll learnt about the publication several days before his death, from Lev Kopelev; on July 16 he passed away.*

* Evgenia Katseva. *Moi Lichnyi Voennyi Trofei* (My Personal Spoils of War). *Znamya* No. 3, 2002, p. 157.

Document 11.18

A. I. Solzhenitsyn's speech at the Nobel Banquet at the City Hall in Stockholm, December 10, 1974.

Your Majesty, Your Royal Highnesses, Ladies and Gentlemen,

Many Nobel Prize laureates have appeared before you in this hall, but the Swedish Academy and the Nobel Foundation have probably never had as much bother with anyone as they have had with me. I have already been here on at least one occasion, although not in the flesh;[1] once the honorable Karl Ragnar Gierow was already on his way to meet me; and now, at last, I have arrived out of turn to occupy an extra seat. Four years had to pass to give me the floor for three minutes, and the secretary of the Academy is being forced now to address the third speech to the same writer.[2]

I must ask your forgiveness, therefore, for having caused all of you so much trouble, and thank you especially for the ceremony in 1970, when your king and all of you welcomed here an empty chair.[3]

But you will agree that it has not been so simple for the prize winner either: carrying his three-minute speech around with him for four years. When I was preparing to come to you in 1970 no room in my heart, no amount of paper was sufficient to let me speak my mind on the first free tribune of my life. For a writer from a land without liberty his first tribune and his first speech is a speech about everything in the world, about all the torments of his country, and it is pardonable if he forgets the object of the ceremony, the persons assembled there, and fills the goblets of joy with his bitterness. But since that year when I was unable to come here, I have learned to express openly practically all my thoughts in my own country as well. So finding myself expatriated to the West, I have acquired all the better this unhindered possibility of saying as much as I want and where I want, which is something not always appreciated here. I needn't, therefore, further burden this short address.

However, I find a special advantage in not responding to the award of the Nobel Prize until four years later. For example, in four years it is possible to experience the role this prize has already played in your life. In my life it has been a very large one. It has prevented me from being crushed in the severe persecutions to which I have been subjected. It has helped my voice to be heard in places where my predecessors have not been heard for decades. It has helped me to express things that would have otherwise been impossible.

In my case, the Swedish Academy have made an exception, and a rather rare one, awarding me the prize when I am middle-aged and my open literary activity is a mere child of some eight years.[4] For the Academy there was a great hidden risk in doing so: after all, only a small part of the books I had written had been published.

But perhaps the finest task of any literary or scientific prize lies precisely in helping to clear the road ahead.

And I would like to express my heartfelt gratitude to the members of the Swedish Academy for the enormous support their choice in 1970 has given my works as a writer. I venture to thank them on behalf of that vast unofficial Russia which is prohibited from expressing itself aloud, which is persecuted both for writing books and even for reading them. The Academy has heard for this decision of theirs many reproaches implying that such a prize has served political interests. But these are the shouts of raucous loudmouths who know of no other interests. We all know that an artist's work cannot be contained within the wretched dimension of politics. For this dimension cannot hold the whole of our lives and we must not restrain our social consciousness within its bounds.[5]

Les Prix Nobel 1974, Stockholm, 1975.

[1] See doc. 11.7.

[2] The first speech was read, from the text sent, on December 10, 1970, also in the Stockholm City Hall.* The second was prepared to be delivered in Moscow at the ceremony of giving Solzhenitsyn the Nobel diploma and the laureate's gold medal, planned for April 9, 1972.** At the 1970 Nobel award-giving ceremony in the Concert Hall building Gierow said: "Solzhenitsyn has expressed in his writings the things that nowadays are more important to hear than at any time before. He told us about the unexterminable dignity of an individual. No matter where, why and by what methods this dignity is trampled on, Solzhenitsyn's word will always sound as a word of denunciation of the tramplers and a word of consolation to the trampled, giving them a firm assurance that justice will invariably triumph and the only person who will remain humiliated will be the insulter himself."***

* *Les Prix Nobel 1970.* Stockholm, 1971, pp. 62–64.
** See doc. 11.12 and 11.13.
*** Les Prix Nobel 1970, p. 63.

[3] It was not for the first time that one of the chairs at the award-giving ceremony in the Concert Hall remained empty. Also empty was the next chair to the left of the wife of the 1956 Nobel laureate for Chemistry, N. N. Semenov, sitting in the same row with family members of other laureates of that year. That chair was meant for their daughter, Lyudmila, who had also not been permitted to go to Stockhlom from the Soviet Union.*

* See doc. 7.36.

[4] That is, from 1962, the year of publication of the tale, *One Day in the Life of Ivan Denisovich.*

[5] Solzhenitsyn' speech met with wide response in Sweden and later its videotape was repeatedly shown during meetings of Nobel laureates of subsequent years, organized by the Nobel Foundation.*

* See note 29 to this chapter.

Document 11.19

From the memoirs of Associate Member of the USSR Academy of Sciences R. L. Feinberg about academician A. D. Sakharov at the beginning of the 1970s.

The atmosphere around A.D. at that time was already very tense. Meeting him in the foyer during the General Meeting of the Academy at the House of Scientists, most of his colleagues, academicians who knew him personally, quickly greeted him without stopping. Those who passed by him did not turn round.

At the sittings of the General Meeting, entering the crowded hall, I saw more and more often (especially after 1973)[1] that the chairs next to him remained empty. He began more and more often (I think deliberately) to sit in the very last row, which was usually half-empty. The sight of him being isolated was painful, and I would take a seat next to him. Our department[2] received the close malevolent attention of our superiors, was sometimes stung and, of course, was "under surveillance"... I can say that I had a right to speak of the pride that the department's workers felt. All years, including the years of the Gorky exile, all of them, from senior researchers to typists, deeply felt that they had some, even if flimsy, connection with Andrei Dmitrievich's fate and took pride in it...

The tension around A.D. was mounting. In September 1973, after his first interview to a foreign correspondent,[3] an explosion occurred. It started from a newspaper campaign, from the memorable first letter of 40 academicians.[4] At the FIAN, of course, signing a collective protest against A.D.'s political position and activities also started. Despite the great respect that he commanded among scientific workers, there were quite a few people who signed the "protest" easily and even with pleasure.

Like everywhere, there were people simply brainwashed by the propaganda, which had influenced them all their life from early childhood, and so sincerely ignorant[5] But the greatest impact was made by fear, which was already in people's nature.

To many who refused to sign, the institute's party committee had to apply the usual "arm-twisting" technique to make them give up. In one big research laboratory the chief together with the party organizer and the trade union organizer decided: OK, let the three of us disgrace ourselves signing this letter, but let them leave the other workers alone. A worker of another laboratory was "processed" for a long time in the party committee. When he said: in 1937 also many people were condemned who are now reabilitaed, he was answered: think about what you are saying, think about your children. The next day he came to the party committee and signed the "protest". If I am not mistaken, about two hundred signatures were collected in such a way. But with the Theoretical Department... they could not do anything. However, the atmosphere around the department was becoming heavier. The document — the text of the "protest" remained in the party committee and whenever appropriate was used for blackmail... A senior scientific worker of the department, G. F. Zharkov, received an invitation to a conference in Paris. But for refusing to sign the "protest" he was refused a character reference.[6] Corresponding Member (now Member) of the USSR Academy of Sciences E. S. Fradkin, a major scientist internationally known, who fought at the front the whole war and was wounded near Stalingrad, ... received a prestigious invitation to deliver a report at the Nobel symposium. He was summoned to the sitting of the party committee and told directly: sign, and you will go. He refused. A talented young theoretician (now Corresponding Member) brilliantly defended his thesis. Before sending it for approval to the VAK[7] they wrote a "bad" character reference to accompany it... Although we were not close, he asked my advice what to do. I said: "decide it for yourself, but I understand your situation. Even if you yield to this shameful blackmail, the attitude to you in the department will not change." He yielded. He was the only one of about 50 scientific workers who signed.

On mezhdu Nami Zhil. Vospominaniya o Sakharove. (He Lived among Us. Memories about Sakharov) — Moscow: Praktika, 1996, pp. 673, 675–677.

[1] After the persecution in newspapers, which started from publication in *Pravda* on August 29, 1973, of the letter signed by 40 members of the Academy of Sciences.

[2] The Theoretical Department of the Physical Institute of the Academy of Sciences (FIAN), where after 1969 Sakharov again worked in the position of senior scientific worker.

[3] The interview was given at a press conference arranged, on August 21, 1973, in Sakharov's apartment and attended by about 30 correspondents of foreign agencies and major newspapers. The press conference was organized after he was summoned to the USSR Procurator's Office, on August 16, 1973, and given a warning, among other things, in connection with his contacts with foreigners.

[4] See note 1.

[5] See also note 34 to this chapter.

[6] To go abroad, approval of a doctorate and so forth, a person needed a positive character reference signed by the leaders of a scientific institution.

[7] The Higher Attestation Commission.

Document 11.20

An anonymous letter from a Prague resident to the Czech theoretician physicist Frantisek Janouch, of September 26, 1973.

I am writing to you as a stranger because I cannot keep silent about what you are doing. I heard on the *Voice of America* that you are supporting as a nuclear physicist the candidacy of the Soviet Zionist dissident Sakharov for the Nobel Prize... In such a way the Zionist dissident Sakharov receives in your person an ally in the CSSR, acting as a Jewish saboteur. I know that the Jewish international (Zionism) has given you the order to start the cause of sabotage in the CSSR. Here, as the Jews usually do, they veil their intentions with pompous words about freedom and democracy. Why do you and Sakharov demand human rights and the freedom of information? It is clear that the freedom of information and human rights proclaimed by the French revolution brought civil liberty to the Jews as well. But how the Jews abused it! After leaving the ghetto, they completely concentrated on seizing leading positions in the economies of all countries and started the unreserved exploitation of non-Arians and all half-Jews... The Jew Sakharov and the Jew Janouch offer their hands to each other to combat, on the orders of the international Zionism, for disintegration of the workers state, where the people are at power. You will not convince me that you are inspired by other motives than those that I have pointed out. I know the Jewish baseness and cunning in preempting many-sided arguments...[1]

He Lived Among Us, p. 800

[1] There is no doubt that this anonymous letter was a response to the total campaign of vilification of Sakharov launched in the Soviet Union, which included hints at his allegedly Jewish origin. Materials from the Soviet press were widely reprinted by the press of Czechoslovakia and other socialist countries. Concluding the publication of the anonymous letter Janouch writes: "Under no circumstances was I going to argue with the author of the letter. But still — did he know that Andrei Dmitrievich was Russian, and Frantisek Janouch, Czech?"*

* This anti-Semite passage by the anonymous Czech author surprisingly echoes the inventions of his Russian counterpart, who 30 years later put forward similar assertions but concerning different personalities. His name is Alexandr Andryushkin. He published, in 2003, in the St. Petersburg publishing house Svetoch, in a print-run of 200 copies, a noteworthy opus under the eye-catching title: *Jews in the Russian Literature of the XX Century. A Book without Concealed Meaning,* (*Iudei v*

Russkoi Literature XX Veka. Kniga bez Podteksta). By the force of his fevered imagination, indeed, without any concealed meaning but directly, he added on to Jews Anna Akhmatova, Mikhail Zoshchenko, Alexandr Tvardovsky, whom he "caught" because he defined Vasily Terkin born by his talent as a man "with an obscure Russian-Jewish last name" (p. 178) and thus substituted "a Russian soldier and toiler for a joker Jew" (p. 4). His sharp-sighted eye also did not pass over Alexandr Solzhenitsyn, with his "Judaic patronymic"; his literary activity, according to his namesake, "enables us to rather confidently put him into this well-known category: a Jew posing as an anti-Semite" (p. 215). The rest is of the same sort. True, contrary to Frantisek Janouch's anonymous correspondent, he pointedly did not relate Sakharov to Jews.

In conclusion it is worth noting that the writer of the quoted delirious nonsense is not some graphomaniac freak, but a quite self-sufficient person who even knows foreign languages. Alexandr Pavlovich Andryushkin is the author of the two editions of the textbook on business English, which came out in St. Petersburg in 1992 and 2000, and of the novel, *The Politician*, published in 1994 also in this city, and he is also the compiler and editor of the translation from German into Russian of *Selected Works* by Martin Luther, which appeared simultaneously with the novel. That is, he should be completely aware of what is pouring out from his pen and how much it corresponds to the reality. This would be, of course, if only he had not lost his mind in the last years, having got stuck on his blind hatred and internal malice, of which, by the way, he surprisingly managed to suspect even the great poet Iosif Brodsky — who, as Andryushkin sees it, in his work "never found any image and died from malice" (p. 10)...

Document 11.21

F. Janouch's memories of awarding A. D. Sakharov the Nobel Peace Prize.

The news that academician Sakharov had been awarded the Nobel Peace Prize reached me in Copenhagen, on my way from Kyoto.[1]

I must have been one of the first to phone him that day, as Andrei Dmitrievich had a number of questions about the Nobel Prize and also about the place where the award ceremony would be held. Soon after that I received a letter from Oslo, where the Nobel award ceremony annually takes place on December 10. I was invited to take part in it, as one of the few personal guests of Andrei Dmitrievich. I was leaving for Oslo with a tight feeling of expectation as to whether Sakharov would at last be given permission to come and participate personally in the ceremony. Unfortunately, that did not happen. In the history of the existence of the Nobel Peace Prize it was only the second case where the prize winner could not receive it in person from the hands of the Norwegian king.[2] In 1935, refused permission to come to Oslo by Hitler was the well-known German pacifist leader Carl von Ossietzky, who was at a concentration camp at that time. Having refused to give a foreign passport to Sakharov, Brezhnev thus placed himself in this unflattering company...

The reaction of the Soviet press to awarding the Nobel Peace Prize to Sakharov was almost hysterical — moreover, using the method that even Goebbels could have used... It was difficult to expect anything else from the Soviet press of the mid-1970s. I was most of all upset by the collective letter of 72 members of the USSR Academy of Sciences[3] condemning Andrei Sakharov. I responded to it with the short article "The Absent Academicians", which I published in the *Russkaya Mysl'*[4] newspaper in December 1975, and in *Nature*... The best-known academic physicists, who certainly know about Sakharov's moral and scientific qualities better that the rest, do not feature on the list. Among the physicists who did not sign this statement are P. L. Kapitsa, N. N. Bogolyubov, A. B. Migdal, E. M. Livshits, Y. B. Zeldovic, B. M. Pontekorvo, M. A. Leontovich, V. L. Ginzburg, G. I. Budker, S. T. Belyaev, I. M. Frank, P. A. Cherenkov, B. B. Kadomtsev, Y. B. Khariton, I. K. Kikoin, V. P. Linnik, R. Z. Sagdeev, C. N. Vernov. Similarly,.. absent are the major Soviet mathematicians, such as A. N. Kolmogorov, P. S. Alexandrov, A. D. Alexandrov, I. G. Petrovsky, L. S. Pontryagin, S. L. Sobolev, I. M. Vinogradov, L. V. Kantorovich,[5] I. M. Gelfand, I. R. Shafarevich.[6] It should also be noted that the statement has not been signed by T. D. Lysenko and M. A. Sholokhov. It can probably be explained by the fact that many of those who agreed to sign the anti-Sakharov statement would not have done it if Lysenko and Sholokhov had.

So it can be said that the anti-Sakharov statement is evidence of the real situation and the moods at the Soviet academy, to a greater extent than the condemnation of Andrei Sakharov.

On mezdu Nami Zhil, pp. 804, 807–808.

[1] In August 1975, Kyoto was the city where the 25th Paguosh symposium, in connection with the 30th anniversary of the atomic bombardment of Hiroshima and Nagasaki, was held. At the symposium, on Sakharov's instructions, Janouch read his message to the participants.

[2] That is inexact: it happened for the third time. The 1961 Nobel Peace Prize was awarded to the UN Secretary-General Dag Hammarskjold, who perished in a plane crash, on September 18, 1961, and the king of Norway handed the award to his widow.

[3] The concluding paragraph of the *Statement of Soviet Scientists*: "The Nobel committee's decision to award Sakharov the Peace Prize, which flatly contradicts the letter and the spirit of the main regulations of this prize, is unacceptable for true partisans of peace. The Soviet scientists think that awarding the Nobel Prize to Academician Sakharov has an unworthy and provocative character, is a blasphemy against the noble ideas of humanism and peace, justice and friendship between people of all countries, which are dear to all of us."*

* *Izvestiya*, October 26, 1975, p. 3.

[4] *Russkaya Mysl'* (Paris), December 18, 1975.

[5] In the same 1975 Kantorovich received the prize in memory of Nobel for economics, and he had to withstand serious confrontation with the leadership of the USSR

Academy of Sciences so as not to put his signature under the letter condemning the awarding of the Nobel Peace Prize to Sakharov (see doc. 8.48).

[6] The lists of those who signed the collective letter or, *vice versa,* were absent from these lists, should not be analyzed straightforwardly. Each person in each group could have their own special circumstances and reasons for that. The academician C. V. Vonsovsky signed both letters, in 1973 and in 1975, but in April 1989, at the meeting for electing the USSR people's deputies from the Academy of Sciences, he apologized to Sakharov in public*, objectively admitting in such a way that both times he had done it under brutal coercion. Besides, in the given list there are the names of the academicians Bogolyubov, Sobolev, Frank, Cherenkov, Khariton, who managed to evade signing the 1975 collective letter, but had not managed to do so two years earlier. On the other hand, Y. B. Zeldovich did not sign any collective letters, but in a telephone conversation with Sakharov and a personal letter which followed it "demanded that I should not accept this 'provocative' prize — having called it, quite in the spirit of the *Trud* newspaper, the 'thirty silver coins'..."**

* *On mezhdu Nami Zhil,* p. 237.

** Sakharov. Vol.1, p. 194.

Document 11.22

From A. D. Sakharov's memories of his being awarded the Nobel Peace Prize.

On October 9, I in Moscow and Lyusia in Italy[1] simultaneously learnt that I had been awarded the Nobel Peace Prize... Soon she saw me on television and accepted congratulations from the whole world — like me in Moscow. These events interrupted her, in fact, merely medical stay in Italy; she suddenly found herself snowed under with a lot of things to do and finally, after many events, she had to participate in my name in the Nobel ceremony... In the morning, representatives of the Norwegian embassy, headed by the first secretary, came to congratulate me — they brought the ambassador's congratulatory letter and marvelous roses in a beautiful vase...

The official reaction in the USSR to the award was of irritation, nervousness... Again, like in 1973,[2] many articles appeared where my activities were "dethroned", my articles caricaturized and ridiculed, and the Nobel Committee's decision was characterized as a hostile and provocative act. The *Izvestiya* published a new collective letter with signatures of academicians, associate members of the Academy and directors of academic research institutes,[3] similar to the letter of 40 academicians two years before...

I decided to apply for permission to go to Norway for the Nobel ceremony. Of course, refusal was the most probable result, but the refusal after I had made an application gave the Nobel committee the opportunity to conduct the ceremony, and if I had done nothing at all, they would have found themselves in a very difficult situation. As for not being allowed to come back, this possibility I considered ruled out. I sent the application to the OVIR[4] on October 20 and started to wait calmly for the results.... On November 14 I was summoned to the Moscow OVIR. In the study on the first floor, where the deputy head usually sits, this time was the head, Fadeev himself. It seemed to me that he was very nervous. He announced that I was refused permission to go to Norway, being "a person having the knowledge of a state secret".[5] I said I would dispute this decision.

Interestingly, a week before that the English newspaper *Evening News* published an article by Victor Louis saying, with reference to some anonymous executives, that I would be refused with this very reasoning. That must have been a check of the force of the public opinion reaction. Victor Louis — a Soviet citizen and a correspondent of an English newspaper (an unprecedented combination) — is an active KGB agent of many years carrying out the most delicate and provocative assignments... Before that, on November 7, it had been reported on Italian television, the origin of this news unknown, that I was permitted to go to the ceremony. Lyusia ordered a tailcoat for me. But this order had to be cancelled...

It was natural for me that now it was Lyusia, the person closest to me, who was to represent me at the Nobel ceremony. On the day of my receiving the refusal (or the next day) Lyusia phoned me from Rome, and I told her about this assignment... Throughout November I gave a lot of interviews. Among them was an interview for a Japanese newspaper... In this interview I used the formula: *peace, progress, human rights*, which soon became the principal idea of the Nobel lecture. It seems to me that these words, indeed, formulate well my position. I wrote the Nobel Lecture easily, with elation. It reflects not only my social views on the issues of safeguarding peace, the necessity of rapprochement of the socialist and capitalist systems, disarmament and strategic balance, progress, the openness of society and human rights, but to a certain extent also my inner emotional world... The text of the speech at the Nobel ceremony proved to be worse — I just felt I was not up to it, having already put everything into the lecture. Having received the text several hours before her flight from Italy to Norway, Lyusia edited it. It was almost the only case when she changed something in my documents without agreeing it with me[6] — there was no other way out. The speech became much brighter, more logical and emotional...

On December 10, 1975, the day of presenting the Nobel Peace Prize... we were invited by the Lithuanians to their place...[7] The radios are on. We hear the sound of fanfares. Mrs. Bonner-Sakharova, representing her husband at the ceremony, is asked to come to the place for participation in the Nobel Peace Prize award-giving ceremony. Aase Lionaes, Chairman of the Nobel Committee, takes the floor. She reads out the Nobel Committee's decision to award the 1975 Nobel Peace Prize to Andrei Sakharov. I hear the sound of Lyusia's steps — she is coming up the steps. And here she starts speaking. I understand the meaning of the words afterwards, several minutes later. At first I perceive only the timbre of her voice, so close and dear and at the same time sounding as if raised into some other world, solemn and radiant. A low and deep voice, at some moment ringing with emotion!...

On December 11 Lyusia conducted the Nobel press conference. These were three hours of questions and off-the-cuff answers, without any preliminary preparation. The press conference was broadcast live and transmitted by radio and television to many countries. Lyusia's extraordinary ability to be ready for action in a difficult situation helped her cope with this trial not worse than many predecessors, who, unlike her, represented themselves. One of the questions was about the positions of Sakharov and Solzhenitsyn, what was the difference between them. Lyusia answered by the parallel with the Westernists and Slavophiles in 19th century Russia, that is, quite accurately and correctly with regard to both, as it seems to me.[8] On that day Lyusia also read out the Nobel Lecture. It seems to me that it was broadcast as well...

A very strong emotional experience for Lyusia was a nocturnal torch-lit procession, which, as was explained to her, is spontaneous and, far from happening frequently, takes place only when the people approve of awarding the prize to a certain person. Lyusia cried when it was translated to her that on many posters was written: "Sakharov is a good person." Much later, on learning about that I, too, was deeply moved...

A. D. Sakharov. *Vospominaniya v Dvukh Tomakh* (Memories in Two Volumes), Vol.1 — Moscow: Prava Cheloveka, 1996, pp. 620–637.

[1] Elena Georgievna Bonner, Sakharov's wife, left for Italy on August 16, 1975 to have an operation on her eyes.

[2] The campaign of vilification of Sakharov in the press started from publication in the *Pravda*, on August 29, 1973, of the "Letter of Members of the USSR Academy of Sciences" with forty signatures. The letter condemned his statement to foreign correspondents made on August 21, 1973, in response to the warning about his human rights activities, which he received at the USSR Procurator's Office.

[3] The so-called "Letter of 72" published on October 26, 1975 in the *Izvestiya* (see also note 3 to doc. 11.21).

[4] The Department of Visas and Registrations, where the documents for going abroad were given to Soviet people.

[5] The state secret here means Sakharov's participation in creating the hydrogen bomb. According to the legal regulations existing in the USSR, the ban on going abroad for citizens having access to state secrets remained for three years after the access stopped. With Sakharov, by the moment of his submitting an application to the OVIR, this period was already over by ten years.

[6] This phrase contains a veiled answer to the assertions that Sakharov was under his wife's thumb and all his statements were written to her dictation.

[7] Sakharov came to Vilnius, on December 9, 1975, where the trial of his friend and associate in human rights movement, S. A. Kovalev, began. The court sentenced the latter to seven years' imprisonment and a subsequent three years of exile. In Sakharov's opinion, "it was not a coincidence — the authorities pursued some goals

But for us this coincidence of the trial and the solemn international ceremony had an exciting symbolic character."*

* Sakharov, Vol.1, p. 629.

[8] In this context the idea of Solzhenitsyn being a Slavophile implies first of all his strongly negative attitude to the problems of free emigration from the USSR. Sakharov's active support of the right to go to other countries for permanent residence was seen by him perhaps as a decisive "dislocation... which broke our battle, deprived us of the main success..."*

* Solzhenitsyn, p. 341; see also Sakharov, Vol.1, pp. 572–578.

Document 11.23

From A. D. Sakharov's answers to questions from foreign correspondents, on October 30, 1976 (I) and December 6, 1976 (II).

I

Has you situation improved after you received the Nobel Peace Prize?

The last year's decision by the Nobel Committee ... has a great significance of a long-term character. It brought great joy, honor and support to me and my friends, including those who are in prison and those who only secretly sympathize with us, not being able to bring themselves to show it openly... Unfortunately, I cannot say anything gratifying about the change of the authorities' position with regard to me personally, my friends and relatives, my social activities and the conditions under which they are carried out. I, as well as my friends, am still not let in the halls of political trials. Nor do the authorities show great respect for my appeals and letters, which they simply do not answer. My telephone and mail connections with the West are still completely blocked... In all that I see evidence of the disrespect of the Soviet authorities for authoritative international organizations, the Nobel Committee and the International Human Rights League, which recently elected me its honorary vice-president...

A. L. Sakharov. *Trevoga i Nadezhdy* (Concern and Hopes) — New York: Khronika, 1978, p. 31.

II

How has your life changed after you received the Nobel Prize?

My life, if you mean my personal life... To begin with, for ten and a half months of this year I was devoid of registration. True, now this registration has been given to me at last, but I think that it was some forced action, I do not know yet what will come next... There is no progress in providing me with accommodation. Now I do not own any accommodation and live in the apartment of my wife's mother.[1]

The Academy of Sciences alleges that granting any accommodation whatsoever is impossible, because the Moscow City Council objects. That is what the Academy of Sciences say, although they usually solve such problems very quickly and easily...[2]

Do you think that the pressure has increased after you received the Nobel Prize?

There were also some other peaks of pressure, for instance, in 1973.[3] Just before awarding the prize there seemed to be some calm. But maybe they did not want to attract attention to me before awarding the prize. Maybe it was purely tactical relaxation. In general, I have been under high enough pressure already for many years...

ibid, pp. 45–46.

[1] Form E. G. Bonner's statement of January 22, 1976: "Sakharov had not lived in the apartment where he had been registered previously for four years, and where I live my husband is refused registration, so now he does not have an apartment or registration in Moscow."*

* *ibid*, p. 169

[2] The managing department of the Academy of Sciences had formal grounds for such refusal. After the death of his wife in the spring of 1969, Sakharov started a new family, in August 1971, having left his big apartment in the Oktybrskoye Pole district to his children from his first marriage. In such case an official, using contradictory resolutions of the Moscow City Council could make any decision, whether on his own initiative or on recommendation of the authorities. Not only could he decide whether to give a new apartment, but he could also stop the exchange of an apartment already owned for a more suitable one.*

* From Sakharov's television interview for the American CBS channel, on February 10, 1977: "The recent refusal of the exchange of apartments is very sensitive for our family. Without affecting the housing fund of the state, this exchange was supposed to create desirable housing conditions for five more families besides ours. The district executive committee, no doubt on the instruction of the KGB and blatantly violating the law, did not approve this exchange. Exchanges of apartments are a special part of life in our country, very graphi cally showing the lack of rights of Soviet citizens." (*ibid*, pp. 82–83).

[3] See note 2 to doc. 11.22.

Document 11.24

Materials connected with the "Open Letter" by the human rights activist A. T. Marchenko, addressed to the academician P. L. Kapitsa, in connection with A. D. Sakharov's administrative exile from Moscow to Gorky.

I

From A. T. Marchenko's[1] "Open Letter" to P. L. Kapitsa, of March 1, 1980.

... I think that Andrei Dmitrievish Sakharov is a great phenomenon, who has exceeded the national limits. He has outgrown the predestination prepared for every person on Earth. It seems to me that he can be neither panned nor praised... But why are you, a worthy and respected scientist with a world-known name, keeping silence?.. Isn't it insulting for you, Nobel laureate, to see the Nobel Prize interpreted as a handout for anti-Soviet activities?[2]

Sakharovskii Sbornik (The Sakharov Collection) — New York: Khronika, 1981, p. 58.

[1] A. T. Marchenko was one of the founders, in May 1976, of the Moscow Helsinki Group on monitoring human rights in the Soviet Union. Of the 49 years of his life he spent 20 behind bars. The last, sixth time, he was arrested in March 1981 and condemned to ten years in a special regimen camp under the penal article "Anti-Soviet agitation and propaganda". He died in a prison hospital on December 7 the same year.

[2] The author of the letter means propagandist clichés of the Soviet press; in particular, the newspaper *Trud* likened the Nobel Peace Prize received by Sakharov to the "thirty silver coins".

II

From the letter by the academician P. L. Kapitsa to the Chairman of the State Security Committee of the USSR Council of Ministers, Y. V. Andropov, on December 11, 1980.

...the dialectic of the development of human culture is in the grip of contradictions between conservatism and heterodoxy. It happens in all times and in all areas of human culture. If we look into the behavior of a man like Sakharov, we shall see that the basis for his creative activities is also provided by his discontent with the things that exist. When it concerns physics, where he has a great talent, his activities are exceptionally useful. But when he spreads his activities to social issues, that does not lead to as useful results and causes a strong negative reaction from people with bureaucratic casts of mind, who usually lack creative imagination. As a result, instead of simply... not paying attention to signs of heterodoxy in this area, they try to suppress it by administrative measures, not paying attention to the fact that they simultaneously ruin the scientist's useful creative activities as well...

It is known that administrative pressure on heterodox scientists has existed since ancient times and even recently happened in the West. For example, the well-known philosopher and mathematician Bertrand Russell was put in prison twice — true, only for short periods.[1] But on seeing that it only caused indignation among intellectuals and did not have any influence on Russell's behavior, the Englishmen stopped using this method of pressure. I cannot imagine what other ways of influencing our dissident scientists we suppose to use. If we are going to increase methods of coercion even more, it will not bring any gratifying results. Isn't it better to simply go in reverse?

Communist journal No. 7, 1991. pp. 53–55.

[1] Russell was first put in prison for six months in 1918, for publishing articles maintaining that the British government might use American soldiers staying in Britain as strike-breakers, which had been repeatedly done in the USA. The second time he found himself in prison was in 1961, when the 89-year-old philosopher was convicted for organizing anti-war marches for nuclear disarmament and survival of mankind, appeals for civil non-obedience to the law and circulation of leaflets.

III

From Y. V. Andropov's letter in response to P. L. Kapitsa, of December 19, 1980.

... It is known that in our country people are not tried for "non-conformism", and Soviet law does not prescribe all citizens to think in the framework of some synonymous stereotypes. Read what is written about this subject by Leonid Ilyich Brezhnev. He repeatedly emphasized that with us it is not prohibited "to think differently" from the majority, to give critical appraisal of these or other sides of social life... But when the so-called "thinking differently" takes the road of subversion of Soviet power, subversion of the Soviet social system, takes the road of treachery, does damage to the security of our country, that is another matter.... Since 1968, the academician Sakharov has systematically been undertaking subversive activities against the Soviet state. He has prepared and spread in the West more than 200 different materials containing falsification and gross slander against the foreign and internal policy of the Soviet state. His materials are used by imperialists for stirring up anti-Sovietism, for carrying out a policy hostile to our system and state... In breach of the existing rules, he, being a bearer of especially important state secrets, dozens of times visited the American embassy. You, as a person of experience, understand that imperialist services are eager to know our secrets. When Sakharov was sent for residence to the city of Gorky this factor was also taken into account.[1]

Those are some ideas I would like to share with you in connection with your letter... Actually, the questions you posed are not within my competence, nor within the competence of the organization I head. Responding to your letter, I was guided, Petr Leonidovich, by a feeling of respect for you...

ibid, pp. 56–57.

[1] Gorky was a city closed to foreigners.

IV

From the commentary of P. E. Rubinin, Kapitza's personal assistant, "To the Story of one Letter by Kapitza"

...This message [from A. T. Marchenko]... was immediately passed, by some person close to Marchenko, to Petr Leonidovich personally... After Kapitza's death, when I stared to sort out his archive and put it in order, I found this letter. It was stored in the file where materials about Sakharov's social "dissident" activities were collected...

Giving me a draft of his letter [to Andropov] Petr Leonidovich had said: "I would not like anybody in the institute to know about this letter"... That is why I put a copy of the letter in my safe. And then I also put Y. V. Andropov's letter there. After Petr Leonidovich's death ... The idea of the possibility of sending Kapitza's letter to *samizdat* (clandestine circulation) did not occur to me... However, Andrei Dmitrievich was still languishing in Gorky, and April 1985 has already come. Who knows, maybe then, at the very beginning of our *perestroika*, this letter could have "worked"? And Sakharov would have come to Moscow a year or even a year and a half earlier?..

Sakharov said in some interview that Kapitza had never helped him...[1] I remember feeling awfully hurt... also because I knew what the attitude to Andrei Nikolaevich was in P. L. Kapitza's home, with what admiration.... they spoke of Sakharov's outstanding bravery...

And Kapitza always tried to support... often very resourcefully. Here are two examples of his "secret" support of Sakharov.

The fall of 1968. A sitting of the Presidium of the USSR Academy of Sciences. Its president, M. V. Keldysh, suggests condemning Sakharov for publication abroad of his *Reflections on Progress, Peaceful Coexistence and Intellectual Freedom*.[2] Kapitsa stands up and says: "I, like probably many other members of the Presidium, have not read this work, and so we cannot seriously discuss it, let alone condemn it. Sakharov's article should be circulated among the members of the Presidium, so that they can get acquainted with it, then Andrei Dmitrievich should be invited to a sitting of the Presidium for us to listen to his report. And if we decide that he is deluded, we shall try to make him change his mind..." Deathly hush. Embarrassment... M. V. Keldysh has to change the subject and take the delicate question off the agenda...

August 1973. M. V. Keldysh in his study has been persuading Kapitsa for an hour an a half to sign the collective letter stigmatizing Sakharov.[3] At the end of the conversation, feeling that he will not manage to obtain Petr Leonidovich's signature, Keldysh asks: "Do you remember any case of an academician being expelled from the Academy? I have not heard anything of such cases..." "But of course," said Kapitsa, "there is such a precedent. In 1933 Hitler expelled Einstein from the Prussian Academy of Sciences..."[4] That was the end of the conversation in the study of the President of the Academy of Sciences...[5]

So, when I learned about Sakharov's interview, I felt hurt for Petr Leonidovich... I phoned Sakharov, and he soon came. It was on February 19, 1988. And here he is sitting at Kapitsa's writing table and reading his letters — a long one of November 11, 1980 and a quite brief one of December 4, 1981, when Andrei Dmitrievich was having a hunger strike in Gorky and we feared so much for his life... Andrei Dmitrievich read both these letters. He was sitting silently at Kapitsa's table and looking at me. "Yes," he said, "I did Petr Leonidovich an injustice..."[6]

ibid, pp. 59, 67–68.

[1] This is probably about a phrase from Sakharov's memoirs written as far back as during his exile in Gorky: "I appealed to Academitian Kapitza to interfere in

Marchenko's fate: it was to him that Tolya's letter was addressed, but I did not receive any answer!"*

* In the collection *Vegetarianskaya Epokha* (Vegetarian Epoch) — Moscow: Pik, 2003, p. 281.

[2] *Reflections* was written from February through mid-April 1968.The article defended the idea of convergence of the socialist and capitalist systems, it contained the rejection of intolerance of revolutionary movements, an appeal to compromise, to combine progress with reasonable conservatism, etc. The manuscript was given by Sakharov to Roi Medvedev, who spread it in *samizdat.* As a result several copies made it abroad. The first publication of *Reflections* appeared in a Dutch evening paper, on July 6, 1968.*

* Sakharov, Vol.1, pp. 390–396.

[3] The author means the letter of members of the Academy of Sciences published in *Pravda* on August 29, 1973 and signed by 40 academicians. Beside Kapitza, signing was consciously evaded by A. P. Alexandrov, I. M. Lifshits, M. A. Leontovich.

[4] Some inaccuracy, also noted by Rubinin, who saw in it Kapitza's conscious step in order to influence Keldysh and those who were putting pressure on him from above."* In 1933 Einstein himself withdrew from membership in the Prussian Academy of Sciences, to show his non-acceptance of Hitler's regime. However, taking into account that the scientist was soon outlawed by the Nazi authorities, his being deprived of the membership in the Academy was pre-determined even without Einstein's initiative.

* *Communist,* p. 68.

[5] The situation of making an attempt to expel Sakharov from the Academy of Sciences arose again after he had been deprived of state awards and exiled to Gorky. As remembered the theoretician physicist M. E. Perelman, at the opening of the Academy's general meeting, President A. P. Alexandrov announced that "we have to consider the question of Sakharov's expulsion from the Academy. True, there has not been such a precedent at the Academy yet."

Here P. L. Kapitsa again cited the story of Einstein and the Prussian Academy of Sciences. Alexandrov, after a slight hesitation, concluded: "So, let us move on to the next question of the agenda."*

* *On mezdu Nami Zhil. Vospominaniya o Sahkarove* (He Lived Among Us. Memories of Sakharov). — Moscow: Praktika, 1996, pp. 512–513.

[6] "In 1987 to 1988 it became known to me that Kapitsa twice took secret actions in my defence."*

* Sakharov, Vol.1, p. 778.

Document 11.25

From the writer B. B. Vakhtin to the presidium of the meeting of Leningrad writers, on January 14, 196.

I have always thought that there is no vocation higher than that of a writer, that there is no business more significant, more vitally important for society than the business done by writers, and there is no occupation more difficult, responsible, injurious to health and even, I would say, mortally dangerous than the occupation of a writer. The work of a writer is difficult and arduous.

That is why it was a shock for me when the People's Court legally condemned the young poet and translator I. Brodsky, who was just beginning his heavy literary work, as an idle drone and a loafer, a social parasite. Maybe, for the first time in the history of our literature the writer's work was denied the name of "work"; for the first time the business was called idleness. I have no doubt that this unprecedented court error will be corrected and Brodsky will be brought back to literary work, but it is not about Brodsky that I want to speak.

I want to speak about poets and prose writers who were elected by writers to guide our writers union and who, far from helping the court to avoid this error, on the contrary, did their best to help them commit it. Why did the head of our union A. A. Prokofiev, why did the secretariat on the whole not help the people's judge and the people's assessors understand that literary work is really work, that the business of a writer, the business to which we have devoted out lives is not "idleness"? Why did you conceal from the court that Brodsky is a brilliant translator, that he does not drink hard, does not loaf but works his fingers to the bone? Why did you brush off what was said and written by such experts on him as S. Y. Marshak, K. I. Chukovsky, A. A. Akhmatova? More than that — in this case you have done almost all you could do to drag through the mire the high title of a writer. In this occupation you collaborated with such figures as Lerner,[1] and it was you who helped, in this case again, to represent in a false light, before the whole world, the situation of a writer in out homeland...

Collection *Iosif Brodsky: Tvorchestvo, Lichnost', Sud'ba* (Iosif Brodsky: Creative Work, Personality, Fate) — St. Petersburg; the publishing house of the *Zvezda* (Star) journal, 1998, p. 285.

[1] Y. M. Lerner, a shady character, was one of the leaders of the so-called "people's patrols", created in the years of Khrushchev to help the militia in maintaining order. He was the principal organizer of the persecution of Brodsky; later he was convicted in a criminal trial. *

* See: N. A. Yakimchuk. *Kak Sudili Poeta. Delo I. Brodskogo.* (How the Poet Was Tried. The Case of I. Brodsky). — Leningrad, 1990.

Document 11.26

J.-P. Sartre's letter to the Chairman of the Presidium of the Supreme Soviet of the USSR A. I. Mikoyan, of August 17, 1965.[1]

I venture to address this letter to you because I am a friend of your great country. I often visit you country, meet many writers and know perfectly well that what the Western opponents of peaceful coexistence are already calling "the case of Brodsky" is merely an incomprehensible and regretful exception. But I would like to inform you that the anti-Soviet press has taken advantage of that in order to start a massive campaign and is presenting this exception as an example typical of Soviet justice; it has come to reproaching the authorities for dislike of intellectuals and anti-Semitism. Until the first months of 1965 it was easy for us, partisans of broad comparison of different cultures, to respond to this unscrupulous propaganda — for our Soviet friends assured us that the attention of the highest judicial organs would be attracted to Brodsky's case and the court ruling should be reversed. Unfortunately, time went by, and we learned that nothing had been done. Attacks of the enemies of the USSR, who are also our enemies, are becoming more and more vigorous. For instance, I would like to note that it has been repeatedly suggested to me that I express my position in public. Until now I have been refusing to do so, but keeping silence is becoming as difficult as answering.[2]

I would like to inform you, Mr President, of the concern that all of us feel. We cannot be ignorant of the difficulties, under any social system, entailed by the revision of decisions already taken. But knowing your deep humaneness and your interest in strengthening cultural relations between East and West within the framework of ideological struggle, I have ventured to send this exclusively personal letter in order to ask you, in the name of my sincere friendly feelings for socialist countries, on which we pin all our hopes, to come to the defense of a certain young man, who already is or may become a good poet.

RGANI, fond 5, opis' 36, ed. khr. 148, l. 152–153.

[1] Sending this document to Moscow, on August 17, 1965, Sartre probably did not know that after the corresponding ruling of the Supreme Court of the Russian Federation, Brodsky had already been brought back to Leningrad from his exile in the village of Norinskoe of the Konoshsky district of Arkhangel'sk *oblast'*.

[2] In this case, when speaking about his position with regard to what was going on in the Soviet Union, including the time of the wild burst of Stalinism, Sartre was quite sincere. Among other things, he refused for many years to acknowledge the existence of "the GULAG Archipelago", considering such discussions inadmissible. In the West, scandalous passages of such kind were so well known that in 1954 they made one of the plots in Simone de Beauvoir's bestselling novel *Les Mandarins,* which won the Goncourt Prize. In the novel Sartre is portrayed under the name of Robert Dubreuilh, and his real-life antipode, Albert Camus, under the name of Henri Perron. The Russian translation of the novel was first published in 2005, by the publishing houses Ladomir and Nauka.

Document 11.27

The extraordinary and plenipotentiary ambassador of Norway, Dagfinn Stenset, on M. S. Gorbachev's Nobel Prize, in his talk with this author on March 25, 1992.

... The ambassador specially dwelt on Mikhail Gorbachev's personality and his fundamental role in the development of historical processes at the end of our century. In his opinion, awarding Gorbachev the Nobel Prize was one of the most just decisions made by the Nobel Committee in Oslo. He drew a parallel with Willy Brandt receiving the Nobel Peace Prize in 1971, who in 1970 signed a treaty between the Federal Republic of Germany and the Soviet Union. This historic act radically changed the political atmosphere in Europe, created a basis for entirely new relations between the West and the East-European countries. The ambassador was in Bonn at that time and he remembers well that the chancellor's courageous decision caused fierce resistance in conservative circles of the FRG (Westen Germany). Mr. Stenset expressed the noteworthy idea that the greatest changes in the Soviet Union that took place a decade and a half later, might have failed to occur but for Willy Brandt's courageous step, bordering on a political heroic deed.

Published in the newspaper *Delovoi Mir* on April 10, 1992, p. 11.

Document 11.28

From the memoirs of the academician A. N. Yakovlev, in 1956 a delegate of the XX Congress of the CPSU, about the reaction in the CPSU Central Committee to the exposure of Stalin's crimes by N. S. Khrushchev.

The scale of Khrushchev's "secret report", its radicalism and the danger for the system coming from it made it material of explosive force, a turning point in history... I heard it at that famous session, sitting on the balcony. At that time I was merely an ordinary worker of the Central Committee, who came from the provinces... There was deathly silence in the auditorium. Neither the sound of chairs creaking nor cough, nor whisper. Nobody looked at one another — whether because what happened was so unexpected, or because of confusion and fear, which seemed to have forever settled in the Soviet man. I came across the assertions that the report was accompanied by applause. There was none. The applause were marked in due places by Khrushchev's assistants in the shorthand report, to show that the report was supported by the congress ... Within the Central Committee Khrushchev's report met with negative reaction, but it was not expressed openly. People whispered in the corners. Nikita does not understand it, they said. The party may not survive after such a blow! Saying "the party", they meant themselves. In practical matters the mechanism of the Central Committee started right off to sabotage the decisions of the congress. It did exactly the same in the years of *perestroika.*[1]

... The atmosphere "above" was like that before a storm. Already soon after the XX congress the frightened party leadership directed three letters around the party,[2] with severe demands to strengthen the struggle against anti-party and anti-Soviet moods. These letters are a graphic example of the party machine beginning a struggle against the decisions of the XX Congress, which also meant the struggle against Khrushchev.

The third letter was the most panicky one. It was typically Stalinist — rude, raging, full of threats. It ended with the words: "The Central Committee makes a point of emphasizing that with regard to enemy rabble we cannot be in two minds about how to fight against it. Dictatorship of the proletariat with regard to anti-Soviet characters must be merciless. The communists working in the organs of the procurator's office, court and state security must be vigilant in safeguarding the interests of our socialist state, must be alert to the intrigues of hostile characters and, in accordance with the laws of the Soviet power, stop criminal actions in good time"...

A wave of arrests and severe sentences for "slander against Soviet reality" and "revisionism". Only in the first months of 1957 criminal proceedings were instituted against several hundred people. Thousands of people, who somehow showed themselves as partisans of renovating society were thrown into the GULAG's camps. The Central Committee made more severe its control over the activities of ideological institutions, unions of creative workers, scientific centers and mass media. The Central Committee's special resolutions sharply condemned the position of newspapers and journals which "understood too straightforwardly" the ideas in Khrushchev's report.

Obshchaya Gazeta No. 24, 2000, p. 8.

[1] That is, in the period of Gorbachev's reforms. In his very frank book *Gor'kaya Chasha* (Poisoned Chalicc), which came out six years before the quoted newspaper publication, Yakovlev noted that only in 1987 was he able to understand that "society built on violence and fear cannot be reformed." And his conclusion: "Unfortunately, too much time and effort has been spent on the policy of 'modernizing'; as a result, *perestroika* has "eaten itself up", having not spent its potential efficiently enough.*

* A. N. Yakoblen .*Gor'kaya Chasha*- Yaroslavl': Verkhne- Volzhskly publishing house, 1994, p. 417.

[2] That is about the Central Committee's secret appeal letters on issues of politics, economics and so on, which were prepared from time to time at the Central Committee' s ideological divisions for being read at closed meetings of CPSU members and not intended to be published in the media.

Document 11.29

A. S. Chernyaev, assistant of the President of the USSR, about M. S. Gorbachev's Nobel Prize.

I

From A. S. Chernyaev's memoirs

... The train of Soviet history had already passed the main switch and went, although in the same direction, not along the rails that had been carefully and painfully laid by Gorbachev. Symbolically, these very days he was awarded the Nobel Prize. It was a signal that "history itself" was beginning to appreciate his heroic deed which, in fact, had already taken place.[1]

With us this act of world recognition was taken at best ironically, if not with scoff...

A. S. Chernyaev. *Shest' Let s Gorbachevym. Po Dnevnikovym Zapisyam.* (Six Years with Gorbachev. From Diary Records). — Moscow: Progress-Kultura, 1993, p. 384.

[1] Taking Gorbachev's Nobel Prize as a starting point, the author of the memoirs tried, 13 years later, to consider the potential possibility of awarding it to Brezhnev, of whom mass Soviet propaganda persistently created the image of a "convinced and steadfast champion of peace". Being aware of the fact that "imagining Brezhnev as a Nobel laureate is for that time [the 1970s] pure fantasy", he nevertheless, in his concluding remarks to his 1976 diary records, came to a most extravagant conclusion: "But for 'Czechoslovakia-68'*, to which he reluctantly agreed (now it is known), most probably because he did not feel quite confident in the highest leadership**, but for 'Afghanistan-79' ***, to which three members of the Political Bureau**** pushed him, taking advantage of his physical and mental helplessness (he had almost lost the awareness of what was going on around him), he would have quite deserved, as I now think, the Nobel Peace Prize."*****

* The invasion of Czechoslovakia by the armies of the Warsaw Treaty countries on August 21, 1968.

** Brezhnev had been the country's leader for over three years, after Khrushchev's dismissal.

*** The invasion of Afghanistan by the "limited contingent" of Soviet troops on December 27, 1979.

**** In his 1980 records the author of the memoirs names A. A. Gromyko as the main initiator of the Afghanistan, being actively supported by D. F. Ustinov: the *Svobodnaya Mysl'* No. 12, 2002, p. 40. The third member of the Political Bureau, as the author most probably means, must have been Y. V. Andropov.

***** *Svobodnaya Mysl'* No. 5, 2003, p. 99.

II

From A. S. Chernyayev's diary of 1990

October 22... KGB agents from different parts of the Soviet Union inform that the news of the awarding of the Nobel Prize to Gorbachev has been received negatively by the greater part of the population.

October 23... Gorbachev had long been turning the pages of letters and telegrams on his table on the occasion of awarding him the Nobel Prize and reading some of them out to me with his comments. The letters were, for instance, like this: "Mr. (!) General Secretary of the CPSU Central Committee, I congratulate you on the imperialists' prize for your making a mess of the USSR, selling Eastern Europe, destroying the Red Army, giving all resources to the United States and the mass media to Zionists." Or: "Mr. Nobel laureate, we congratulate you for your utterly ruining your country, for obtaining a prize from world imperialism and Zionism, for your betrayal of Lenin and the October revolution, for ruining Marxism–Leninism." There were dozens of such letters and telegrams. I asked Gorbachev a question: why does Kryuchkov collect all that and put it on your table?[1] Why do they regularly count and bring you the results of polls in *oblasts* and working collectives with a 90 per cent negative response to the Nobel Prize? He answered: "You suppose I did not think about that?"[2]

And he went on turning the pages. I tell him: "Mikhail Sergeevich, what makes you spend your nerves and time on this rubbish? It's about time you 'soared' in your president's position over this preposterous ignorance." He remained silent.[3]

A. S. Chernyayev. 1991. *Dnevnik Pomoshchnika Presidenta SSSR* (The Diary of the Assistant of the President of the USSR) — Moscow: Terra, 1997, pp. 52–54.

[1] Just a rhetorical question. All that is a time-tested system of collecting and transforming information advantageous to power organs for the moment. As a vivid illustration to that we can offer an archive document from the time of the Great Patriotic War. It contains agents' information collected in November 1942 at the session of the USSR Academy of Sciences in Sverdlovsk, where academic institutions had been evacuated. The information "On the political moods of academic intelligentsia", meant for the head of the Propaganda and Agitation Department of the communist party's Central Committee was definitely having denunciative character in one part of it, and the other part was openly complimentary for the addressee (doc. 11.30).

[2] Kryuchkov, an incurable conservative of the Soviet pattern, did not fail to show himself as such when he was already long past retirement age. In 2004, after his book *Personality and Power* had appeared, he told a correspondent of the *Izvestiya* newspaper about his torments when he was working on the chapter about Stalin. As the complexity of the leader's figure was evident, "I thought I would write 60% positively and 40% negatively, but when the work... was completed 100% of the things there were positive."* That is, millions of his compatriots guiltlessly ruined in prisons and in the GULAG; the thoughtless disorganization of agriculture, which provoked terrible famine in the early 1930s; a number of unforgivable errors on the eve of Hitler's preparations for aggression; the unprecedented betrayal of an infinite number of Soviet soldiers and officers who were taken prisoner by the Germans, first of all, through his personal fault as Supreme Commander-in-Chief**; the felonious mass deportations of people of certain nationalities, and so on — all that, to the mind of the former Chairman of the KGB, proved not enough even for one per cent of negative points.

A year later, on November 7, 2005, the anniversary of the 1917 October revolution, Kryuchkov's revelations were logically echoed by one of the slogans shouted from the columns of participants of a demonstration walking under red banners along the streets of St. Petersburg: "On reforms we'll pull the plug: Stalin, Beria, GULAG! "*** Thus offering once again, together with the KGB's former chief, one more example of the pathological amnesia of historical memory — the memory of the leader's unprecedented cruelty to his own people, who from time to time arranged, as a warning, blood baths even for his most loyal satraps and admirers.

* *Izvestiya*, September 15, 2004, p. 5.

** That can be confirmed by the bitter memories of a participant of the war, V. Ivanov, from the village of Nikolskoe of Voronezh *oblast'*, who was in one of the two armies encircled near the town of Uman', but after four days in captivity managed to escape, cross the front line and reach the Soviet troops; the letter was published by the same *Izvestiya*, on March 10 (p. 8), when the country was preparing to celebrate the 60th anniversary of the victory. Having reminded the readers of the humiliating interrogations on the part of Kryuchkov's colleagues, and also the "cynical and misanthropic" question posed by officers from special service, "Why didn't you shoot yourself?", the veteran strictly concludes: "The fault for millions of people having been captured ... lies with those who beheaded the army before the war had begun, who did not provide it with tanks, planes, arms and ammunition, who overlooked the beginning of the war, who missed the German offensive on the South-Western front." And something else, which is directly addressed to the shadow of Kryuchkov's idol: "The English premier, Winston Churchill, who in his younger years... was captured by Boers and escaped; he received an order for that. But we were spat upon for decades by the Soviet power, which brought the postwar generations in the same spirit."

*** *Izvestiya*, November 8, 2005, p. 2.

[3] After his resignation, at the end of 1991, Gorbachev had to go through a number of demonstrative humiliations vindictively organized by a host of petty officials and their lackeys — from the same category as those who mockingly sent the already mentioned "congratulations" to the Nobel laureate. As Gorbachev himself remembered a decade and a half later, especially painful for him and his wife, Raisa Maksimovna, was "the caddishness with which we were thrown simultaneously from the president's apartment and from his *dacha*. It was announced to us: 'Immediately. In one day'... That day some characters were following the Gorbachevs, making an insulting inspection — suspecting that the Gorbachevs would steal the state property. They were asking rudely: 'Where is the furniture?'... My God, what furniture? What sort of people they are! Such petty vindictiveness".... *

* *Izvestiya* January 12, 2007, p. 8.

III

From A. S. Chernyaev's talk with this author, on November 17, 1999

The report on awarding Gorbachev the Nobel Peace Prize aroused mixed feelings around him.

On the one hand, there was sensation that it was right and he deserved it, on the other hand, suspiciousness. We knew well of the negative attitude towards the

Nobel institutions, which had hardened in our society with time. They were seen as organizations of anti-Soviet character, biased against our people, especially where the humanitarian prizes were concerned — for literature, for social activities.[1] In the years of *perestroika* this prejudice did not disappear. It was thought that we were undeservedly passed over — for political, ideological reasons. We foresaw that the reaction in society might be negative and did not conceal our apprehensions from Gorbachev. Despite that, we were convinced that the recognition of his achievements by the Nobel Committee is a positive fact confirming his outstanding role in changing the whole international situation for the better. He usually brushed off the discussions, he did not want to go deep into the matter. It was clear that he was flattered to receive the prize.

However, feeling himself the leader of this country, he, nevertheless, projected on himself both the opinion of party members around him and the public mood. It was already the fall of 1990. The crisis of *perestroika* had already reached a very high level, dissatisfaction with Gorbachev was growing and becoming more and more malicious, more and more open. People did not mince their words at the Congress of People's Deputies, nor in the Supreme Soviet, nor in the press, not to mention the Central Committee and its officials. He understood that the Nobel Prize would give a spur to irritation, to grudges against him: you receive awards from the West, but what have you done for the country? Queues everywhere, deficit of many things, sometimes even lack of bread in shops. The economic reform did not move any further. Just at that time the famous economic program "Five Hundred Days" was put forward by Shatalin and Yavlinsky. At first Gorbachev actively supported it, but later he abandoned it. That also gave rise to dissatisfaction. Intelligentsia, too, was losing faith.

In a word, from the political point of view, being awarded the Nobel Prize was an inconvenient event for him. But as long as it took place, we thought that he should go to Norway and accept the award in person. However, Gorbachev categorically refused to take part in the ceremony. He chose for a long time whom to send as his representative. At last someone advised Anatoly Gavrilovich Kovalev (at that time he was the first deputy of the Minister of Foreign Affairs). He accepted all that was due there. The money from the prize was at once passed to charity; here everything is all right, there are documents that confirm it.

We started to receive responses [see doc. 11.29-II]. Kryuchkov[2] collected abusive, insulting information and laid it out before Gorbachev, not without some intent. In the press ambiguous reasoning predominated as well,[3] although there were also positive responses in which Gorbachev's Nobel Prize was regarded as an appropriate, worthy award.

Then the stir subsided, people no longer thought about the prize. But the Nobel Committee did not forget about their new laureate. They insistently and repeatedly reminded Gorbachev of the statute requirements: within half a year from the day of giving the award each laureate has to deliver the Nobel Lecture. Gorbachev delayed for a long time, not reacting to the signals. I told him many times that, after all, he should decide whether to go or not. .. At last, at the end of May 1991 he let it drop and said that he would go, because otherwise it would simply be indecent. But, he said, if I have to go, the Nobel rostrum should be used for the conceptual rendering of my ideas of the country, the world and my own place in this world.[4]

The lecture — recently we remembered it together — took place on June 5 in the building of the Oslo mayoralty, where the peace prizes are given annually. It turned out to be at the level of his speech at the U.N. General Assembly, in December 1988[5] — by the depth of his appraisal of the situation in the world, his hopes, intentions and vision of the prospects. By the warning given by him to the West, which underestimated the complexity of the situation, the acuteness of the turning point, when the whole future of international development depended on whether the Soviet Union would get help in good time in order to overcome the crisis.[6]

In Norway Gorbachev received a splendid welcome, both at the top level and from ordinary people. At the hotel, where the Soviet delegation was staying, an orchestra of volunteer musicians was permanently playing. Russian music also sounded, and some Norwegians were dressed in Russian costumes. When Gorbachev came out into the street, he was greeted enthusiastically. Not to mention the official part, which was adequately conducted and became a true reflection of the attitude of the Norwegian and international public towards the President of the USSR.[7]

[1] "...Bunin, Landau, Solzhenitsyn, Sakharov — all of them were from the 'emigrant or dissident province', and the prizes given to them were regarded as provocative acts of the 'cold war' (Sholokhov was not included in this number, being considered an exception)."* Mentioning Landau in this context is symptomatic. Awarding him the Nobel Prize, soon after the terrible car crash, the consequences of which he never was able to get over, did not meet with any irritated comments in the Soviet press. Nor did the authorities pose any obstacles to conducting the solemn ceremony of giving the Nobel signs in the hospital where the laureate was staying [see doc. 8.30–8.32]. In this case the appearance of Landau's name next to the 'White Guard' Bunin and the 'traitors' Solzhenitsyn and Sakharov might have been involuntary evidence of the fact that the ideological structures of the CPSU Central Committee retained political distrust of Landau even after the death of the academician, whose independent behavior had repeatedly caused the irritation of the party officials responsible for science. **

* Chernyaev, 1993, p. 471.

** Here is what was reported, on May 30,1958, to the Secretariat of the CPSU Central Committee by an employee of the Department of Science, Higher educational institutions and Schools, A. S. Monin: "In the Department of Physical and Mathematical Science of the USSR Academy of Sciences an improper situation has arisen, when a number of scientists, who are not members of the party... ignore the opinion of the party organizations and try to oppose their high scientific authority to the party influence... The most active is the group headed by Academician Landau, who is an open nationalist (his nationality is Jewish) and, according to the information from the KGB, shows anti-Soviet moods." (RSAMH, fond 5, opis' 35, ed.khr.68, l.80).

[2] V. A. Kryuchkov was the chairman of the USSR KGB. After the death of the General Secretary of the CPSU Central Committee K. U. Chernenko, in 1985, being the deputy chairman of the KGB, he was one of those who made for the election of M. S. Gorbachev to the highest position in the Soviet Union (see note 59 to this chapter). In August 1991 he took an active part in the putch aimed at dismissing Gorbachev from power. His conservatism and natural inability to perceive adequately the dictates of the time* showed even in very small things. When in the late 1980s the Political Bureau permitted live broadcasts on the television, during the first such broadcasting of the popular television game "What? Where? When?", according to the memories of one of its participants, "behind the back of such player stood agents of the special services and waited — what if something happens."**

* See also note 2 to doc. 11.29-II.

** *Izvestiya* June 9, 2003, p. 9.

[3] It is curious that some foreign observers, who were used to the fact that previously the whole Soviet press had expressed only unanimous opinion, overlooked such ambiguity on the newspaper pages of 1990. The theoretician physicist Frantisek Janouch, a Czech dissident promoting the same ideas as Sakharov, remembering that "the Soviet press reacted to awarding Sakharov the Nobel Prize almost hysterically, moreover, by the method which could have been used even by Goebbels", added: "I am writing these lines on the very day that the same prize was received by Gorbachev. With a feeling of sad satisfaction I am comparing the reaction of the Soviet press then and now."*

* *On mezdu Nami Zhyl... Vospominaniya o Sakharove* (He Lived Among Us. Memories of Sakharov) — Moscow: Praktika, 1996, p. 807.

[4] From Gorbachev's Nobel Lecture: "*Perestroika,* which once again is returning our people to common sense, has enabled us to open up to the world... But all this takes a lot of hard work. To a people which believed that its government's policies had always been true to the cause of peace, we proposed what was in many ways a *different policy,* which would genuinely serve the cause of peace, while differing from the prevailing view of what it meant and particularly from the established stereotypes as to how one should protect it...

Perestroika has now entered its most dramatic phase. Following the transformation of the philosophy of *perestroika* into real policy, which began literally to explode the old way of life, difficulties began to mount. Many took fright and wanted to return to the past...It was not only those who used to hold the levers of power... but also many people... who, during the preceding decades, had forgotten how to take initiative and to be independent, enterprising and self-reliant."*

* *Les Prix Nobel 1990.* Stockholm — published by World Scientific Publishing.

[5] From Gorbachev's Nobel Lecture: "We are now approaching what might be the crucial point when the world community and, above all, the States, with the greatest potential to influence world developments, will have to decide on their stance with regard to the Soviet Union, and to act on that basis.

The more I reflect on current world developments, the more I become convinced that the world needs *perestroika* no less than the Soviet Union needs it. Fortunately, the present generation of policy-makers, for the most part, are becoming increasingly aware of this interrelationship, and also of the fact that now that *perestroika* has entered its critical phase, the Soviet Union is entitled to expect large-scale support to assure its success."*

From Chernyaev's diary entry of June 5, 1991; "We had a lot of bother with the preparation [of the content of the Nobel Lecture]. The main idea was thrown out by him: *perestroika* has given to the world more than to the country, and if you are civilized, you, the West, must pay us back when the country is in such a crisis which can turn into trouble for everyone! The idea remained implied in the text, and I finished this nine-page insertion with the words: 'I want this warning to remain in the annals of the Nobel Committee!!'"**

* *ibid.*

** Chernyaev, 1997, p. 147.

[6] "The lecture evoked general and often enthusiastic response in the international press. The Soviet press kept silent about it, just as could have been expected."*

* Chernyaev, 1993, p. 472.

[7] During the "Direct line" organized by the *Izvestiya* newspaper in December 1999, a reader asked Gorbachev on the telephone why the country had not supported him. Gorbachev's answer was philosophical: "You know, it is not our fault, it is our trouble. Because we are like that, we have such history, such mentality. It is very difficult to get out of that."*

* *Izvestiya,* December 4, 1999, p. 1.

Document 11.30

From the head of the NKVD[1] Department of Sverdlovsk oblast' Borshchev to the head of the Propaganda and Agitation Department of the Central Committee of the communist party G. F. Alexandrov, on November 17, 1942, No.24839

On the political moods of the academic intelligentsia

I

Military events in Libya and Africa...[2]

The doctor of geological and mineralogical sciences A. A. Saukov, non-party man: "The events in Africa certainly mean the beginning of a second front. We will soon witness Italy's withdrawal from the war followed by the inevitable defeat of Hitler..."

The senior research worker at an institute of the USSR Academy of Sciences N. I. Ginzburg, non-party woman: "At last we are coming to the time when Hitler is really facing the threat of a crushing defeat. The Americans are real allies making war with an efficiency characteristic of them."

Professor I. K. Kikoin, non-party man: "Africa is the second front. But still we should be ready for a long drawn-out war, which may last until 1945.[3] We mustn't set our minds at rest."

Academician M. V. Shevyakov, non-party man: "The Germans have reacted to the events in Africa very quickly and successfully, having occupied France[4] and made troop landings in Africa. The Germans are still rather strong, and that has to be reckoned with."

Academician M. V. Kirpichev, non-party member: "The situation of the Soviet Union is difficult, but the most dangerous things are behind. The crushing defeat of Hitler is inevitable. It is England and America that will get the results of the victory, as our Union will have been very exhausted by the end of the war. There won't be any radical breakdown of our system, which has been talked about. Only minor changes will take place."

On the session of the Academy of Sciences[5]

The fact that the session was convened is regarded by all participants as a big political event. About the reports of Professor G. F. Alexandrov and Academician A. F. Ioffe. All those who participated in the session on November 16 assessed the content, form and rendering of the report by Professor G. F. Alexandrov as very good. The speaker showed himself to be quite well prepared for an academic audience. Professor Alexandrov's modesty inspires sincere respect.[6] The Vice-President of the Academy Ioffe's attitude towards the report was quite different.

Academician I. M. Vinogradov, non-party man: "The report, ... although it was better that that by Academician Baikov,[7] can be compared with a blanket made of multicolored chintz shreds. Ioffe has never been and cannot be objective. And adding to this that he knows only physics, it is easy to imagine what he could say about exact sciences in the last 25 years."[8]

RSASPH, fond 17, opis'125, ed. khr. 84, l.5,7,8

[1] NKVD is a Russian abbreviation for the People's Commissariat of Internal Affairs, the predecessor of the KGB.

[2] Discussed is the successful offensive of the British Eighth Army, which banished the German troops from Egypt in October–November 1942, and entered the territory of the Italian colony Libya, and also a big American landing in the French Northern Africa (Morocco, Algeria, Tunisia), on November 7,1942, guided by the general Dwight Eisenhower. There were some reasons why personal opinions of participants of the academic forum attracted specific interest from surveillance services. In May and June that year, as a result of the visit of the People's Commissar of Foreign Affairs, V. M. Molotov, to London and Washington the allies reached an agreement on opening a second front in 1942. The landing of the allied troops on the northern coast of Africa received an enthusiastic response from many people

in Soviet society as the beginning of opening a second front. However, Stalin rigidly connected the idea of a second front only with an Anglo-American invasion of France and the contiguous Benelux countries. That is why the information about the talks in the lobby of the academic session, received by the NKVD from their secret agents, was reported to the Central Committee of the communist party as compromising the mentioned persons.

[3] Amazing accuracy of foresight!

[4] In retaliation for the landing in Northern Africa, German and Italian troops entered the territory of the non-occupied zone of France, on November 11, 1942.

[5] The jubilee session of the USSR Academy of Sciences on the occasion of the 25th anniversary of the October revolution took place in Sverdlovsk on November 15–18, 1942.

[6] At the session of the Academy G. F. Alexandrov delivered the report "The 25th Anniversary of the Great October Socialist revolution and the Patriotic War of the Soviet people". The compliments paid by the provincial chief of surveillance services with regard to the high-ranking party official give his memo a farcical touch, especially when directing the agents' information to the head of department of the Central Committee G. F. Alexandrov he writes about "Professor Alexandrov's modesty". In fact, the addressee was famous for quite different traits of his nature, which led to his dismissal, in 1948, from a key position in the Central Committee. He went on visiting his haunts of vice and for that was sent, in 1955, to the provinces to do an ordinary job (see also Chapter 5 and note 4 to it). As a antithesis to the compliments to the academician and the head of the Propaganda and Agitation department from the NKVD worker, it is worth citing an example of a radically different impression from a fleeting acquaintance with Alexandrov of a rather perspicacious Yugoslav political figure, M. Jilas, which took place during the latter's official visit to Moscow in March 1944: "Alexandrov did not make any impression on me — vagueness, almost impersonality was his main distinctive feature."*

* M. Jilas. *Razgovory so Stalinym* (Talks with Stalin). — Frankfurt am Main: Possev-Ferlag, 1970, p. 27.

[7] On the first day of the session Academician A. A. Baikov delivered the introductory report, "The Academy of Sciences over 25 Years".

[8] The subject of the report made by Academician A. F. Ioffe was "The Development of Exact Sciences in the USSR and their Role in the Patriotic War". The assessment of the report by Academician I. M. Vinogradov, a mathematician, and his characteristics of his colleague have an imprint of his exclusively personal attitude; Vinogradov's dislike of Ioffe, which also spread to quite a few of his other colleagues, was no secret and was widely known in the academic community (see doc. 8.23 and note 3 to it).

Chapter 12 • 1901–2000

Gone with the Wind. Instead of an Epilogue

Some statistics to start...

In all Nobel Prize categories (including the Prize in Economic Sciences in Memory of Alfred Nobel) from 1901 to 2000, 695 people (Table 12.1) became laureates, four of whom were awarded twice. Moreover, the Nobel Peace Prize was also given to sixteen international bodies, the Red Cross being awarded three times and the UN refugees committee twice. Of the 695 people seventeen are Russian citizens, which constitutes less than 2.5%.

The criteria for awarding the Prizes in science categories are more objective and weighed, reflecting the true value of what the human mind could achieve. So, correlating the number of Prizes in these classes with the laureates' nationalities, we have the following data (Table 12.2).

According to the number of Prizes counted in this way, the 8–9 places in the table are shared by Russia, with its vast territory and huge human resources, and small Austria. Further, it was only in the last year of the twentieth century that Russia could break away from tiny Denmark, due to the Nobel Prize for physics being awarded to the academician Zhores Alferov[1] on October 10, 2000. Also ahead of Russia were Holland and Switzerland, whose territories and population are far from being outstanding. And if one compares the nine Russian prizes with the awards given in the twentieth century to the researchers of the United States, Great Britain and Germany, the gap will appear incredible. In the United States the amount of prizes is seventeen times as big, in Great Britain 6.5, and Germany 6.3 times...[2]

Such a great lag cannot be taken as just accidental, even granted that there were some evident failures on the part of Nobel institutions. The reasons are rooted not in somebody's intrigues, but mainly in the peculiarities of the historic development of Russia in the last one and a half centuries. That is, not only 75 years of Soviet-like totalitarian regime are to blame, but also birthmarks of the Russia "we had lost" in the cataclysms of the time of the revolution.

There is a scope of reasons why Russia occupied its low position in the sphere of fundamental discoveries of the twentieth century in the field of natural sciences. And of these reasons we can extract one which deals closely with the obvious subject of our consideration, that is the problem of "brain-drain", of mass emigration, to be more accurate.

Table 12.1. The number of Nobel Prize laureates in different categories, 1901–2000 inclusive

Nobel prize categories	Years	Number of prize laureates
Prize for Physics	1901–2000	161
Prize for Chemistry	1901–2000	133
Prize for Physiology or Medicine	1901–2000	172
Prize for Literature	1901–2000	97
Peace Prize	1901–2000	86
Prize in Economics	1969–2000	46

Table 12.2. Total number of Nobel Prizes in science categories, received by the researchers from different states by the year 2000 inclusive

States citizens of which are nobel laureates in sciences	Total number of prizes receive	States citizens of which are nobel laureates in sciences	Total number of prizes receive
USA	152	Russia	9
Great Britain	58	Denmark	8
Germany	56	Italy	7
France	21	Canada	6
Sweden	17	Japan	6
Switzerland	14	Australia	4
Holland	12	Belgium	4
Austria	9	Norway	3

Notes to the table.

1. The citizens of Argentina, India and Hungary received two awards each; the citizens of Ireland, Spain, China, Pakistan, Portugal, Finland, Czechoslovakia and South Africa — one.
2. In the table account was taken of the number of Prizes awarded to citizens of corresponding countries. If one Prize was divided between the scientists of three countries, it was entered in the table three times; two countries — two times accordingly. But if the award was jointly given to two or three researchers from the same country, it was considered only once.

It is sometimes well-forgotten that hundreds of thousands, even millions of people tried to escape from Russia not only because of the horrors of the Civil war, the problems of displaced persons during the Second World War or absurdities of the Brezhnev–Suslov administration. The mass overseas departure of hundreds of thousands of émigrés in the reign of the emperors Alexander III and Nicholas II was the disgrace of the autocratic power as well. The people fled from the *pogroms*, from despair, from discrimination, from not being able to provide their children with a decent education in their mother country. Those who fled were the poor and the families with modest income. But the Jewish intelligentsia incorporated into the Russian civil society preferred to believe in the better future of their motherland, and abstained from emigration...[3]

The future takes severe revenge on the society unable to cherish its own intellect. If Nobel Prizes in sciences be considered as a relative standard, and their amount could reflect the society's ability to provide its citizens with the best conditions to express themselves, you would be astonished by the following quantitative correlations. While Russian researchers had nine Prizes, Russian émigrés and their descendants received almost two and a half times as many. How many great discoveries could have enriched mankind, if the intellect that brought them into being was born and fed under the skies of Russia! If only the power, either monarchal or bolshevist, had not been contaminated so much with xenophobic and ideological prejudices...

Let us then get acquainted with the life of some of the celebrities who, though of Russian origin, earned fame for other countries with their outstanding discoveries.

Of those "gone with the wind", the first to receive the Nobel Award was Selman Waksman, born in 1888 in Priluka in Poltava province. His elder married sisters having emigrated before, he found himself in the USA at the age of 22. He hadn't had an opportunity to get a serious secondary education because of the percentage rate established in the reign of Alexander III. According to it there had to be a correlation of not less than ten Christians, studying in educational institutions to only one Jewish student. That's why the future Nobel laureate covered the gymnasium course with the help of private tutors. In 1910, after he had passed his final exams[4] he left his native country forever.[5]

What could have been the destiny of the young man if he hadn't had the certificate of secondary education? Would mankind have acquired streptomycin in 1944, the first antibiotic effective against tuberculosis owing to which millions of people could escape early deaths? In 1952 would a little Swedish girl have been alive, who Waksman mentioned in his speech at the foundation banquette? He said, "This morning I received from her an award even more important than the Nobel Prize itself — five flowers, one for each year of her life to which she owes streptomycin."

This discovery made by "the Russian doctor" as the Americans called him, was a real scientific feat. Based on Pasteur's concept of possible presence of natural antibiotics in soil and on the works of S. N. Winogradsky, the Pasteur Institute researcher and one of the founders of microbiology in soil (see Chapter 2 and notes 15, 16 to it), Waksman started a broad programme for their search. Over 10 thousand samples of soil microorganisms had been tested, until in 1944 fungi was found having a radiant thread-like composition called Streptomyces (streptos — thread, myces — fungi). It had the properties searched for. It was only in 1946 after several years of careful studies that streptomycin could be used in everyday medicine.[6] A new term, "antibiotics", was coined by our compatriot in 1942, long before this discovery. Now this word is known to everyone from childhood.

In the summer of 1946, being at the peak of his fame as a discoverer, Waksman came to his motherland for the first time after emigrating at the invitation of the USSR Academy of Sciences to give a series of lectures on "Antibiotics, their nature, isolation and practical application". The following year the State Publishing House of foreign literature released his monograph *The antagonism of microbes and antibiotic substances* in Russian translation. In August 1947, when a new wave of "witch-hunting" started in the country, B. L. Suchkov, the director of the Publishing House, later arrested, was fired and charged with "confusing the business of subscribing to foreign literature, thus opening channels for sending counter-revolutionary books to the USSR". The books by "non-returners and traitors", Selman Waksman's monograph among them, were attributed to the latter.[7]

There were, however, additional reasons for entering his name on the proscription list. Less than a year after Waksman's visit to Moscow there was unleashed in the Soviet Union an anti-intellectual campaign provoked by higher authorities. It was called the "Klueva–Roskin case" (see Chapter 6). The campaign was designed against the country's scientific intelligentsia, and Agitprop executives found nothing better than to involve the American bacteriologist, naming him in the so-called "CK VKP(b) closed letter on the Klueva and Roskin case" (see note 12 to Chapter 6). Developing further these absurdly concocted claims, Waksman was also accused of having espionage motives (document 12.1).

We can only add here that the spy story featuring Waksman and fabricated Agitprop was soon developed into a film released in January 1949 under the title *A Trial of Honour.* The prototype of American Professor Carter was considered to be the streptomycine discoverer as he was depicted in "The Closed Letter of CK VKP(b)". Besides other coincidences eliminating all doubts about who this Carter really was, there was an episode, without any sign of veiling, indicating clearly the personality of the future Nobel laureate. Appealing to the vigilance of one of his colleagues, the hero of the film announces, "Are you aware, by the way, what price they [the Americans] are selling streptomycin to us for? Only one dose for treating a single patient is enough to erect a castle [...][8] And you are playing up to them, thrusting the doors wide open [...]" (see doc. 12.2).

Did Waksman himself know about the campaign secretly[9] spread around his name? His correspondence with Winogradsky reveals that he did.

As early as spring 1947, when the winds of the Cold War began to blow heavily around the whole world, when the American press was permeated with invectives towards the Soviet Union politics, the scientist never changed his affectionate attitude to his native land. In his letter to Paris, after having touched upon the hard life of people in the Europe destroyed by war, he further pointed out that "Here [in the USA] nobody suffered, they even indulged themselves. But they are still persecuting Russia. Nobody knows what may come out of it all, but sometimes I become a bit worried about the destiny of mankind. The Germans managed to spread so much hatred, but hatred is much easier to be grafted than love." Later, though, when his addressee complained about the difficulties of corresponding with the Soviet colleagues, he himself notes down, "The correspondence with them is still very difficult. You write to them receiving no answer at all, or receive one eight months later." But this is still nothing more than complaints about the difficulties of everyday life.

The letter quoted was dated March, 2, 1947. Waksman was totally unaware that half a month before that date, W. W. Parin, the AMN USSR academician-secretary, whom he knew personally, had been arrested on the accusation of being an American spy. He happened to just return from a business trip to the United States (see note 1 to doc. 4.34) when the fabrication of "the Klueva–Roskin case" was in full swing, and he was attributed to it. But in his letters that followed, one could trace notes of bitterness, the origin of which were quite easy to reveal. On 12.10.1948 he is asking his senior colleague, "Do you have any news from Russia? I have none. They all have already fallen silent there. Who knows what's going on there." Half a year later, on 7.04.1951 in his reply about the preparation of Winogradsky's monograph for publication in Russian (see doc. 4.40), he stated, "I personally get no information from the Soviet Union. Even closest friends don't write. And no scientific works are received from there. I don't send anything either. God knows what's going on and what will come out of it all."[10]

It was in 1973 when our compatriot died, without having a chance to visit the Soviet Union once again (see doc. 12.1 and note 2 to it). The years when Selman Waksman actually did return to his native country were the 1990s, the time of the destruction of the totalitarian regime. In the Ukrainian museums of regional studies there started to appear stands devoted to the memory of the outstanding scientist. And in 2003 the son of the streptomycin discoverer Byron Waksman,[11] who followed in his father's footsteps and became a professor in microbiological studies, also visited Vinnitsa to pay tribute to the land of the ancestors.

As to Sheldon Lee Glashow, who was born in 1932 in New York, he had no problems whatsoever in getting a proper education. His future father and mother still being children, their families left their native Bobruisk at the beginning of the 20th century, trying to flee from the Jewish *pogroms* who had seized the Mogilev province. In the United States Leiba Glukhovsky, using all his abilities, managed to found a prosperous company for repairing water pipes, and became a well-to-do person. In the motherland that he had deserted he was deprived of a decent secular education. But he could afford it now, in his new country, for his three children. When having noticed that his youngest son in his school years had a yearning for natural sciences, he personally arranged for the fifteen year old Sheldon a private chemical laboratory in the cellar of their house. And later, in all possible ways, the father encouraged his son's interest in the scientific field chosen. As was common among Russian émigré Jews, the father of the family anglicised their original surname to make it "Glashow".

Sheldon Glashow received the Nobel Prize for physics in 1979 together with Abdus Salam and Weinberg "for their contribution to the theory of the unified weak and electromagnetic interaction between elementary particles". As he metaphorically noted in his Nobel lecture, earlier "the theory of elementary particles looked like a patchwork quilt." Electrodynamics, strong and interaction, were absolutely separate disciplines, taught and studied in alienation from each other.

So being born in the United States and thus its citizen, Sheldon Glashow in other circumstances might have stayed our compatriot, if not for the inherent signs of the Russian Empire. At the border of the two centuries its birthmarks mentioned above were still medieval forms of discrimination and violence provoked by autocratic power. Together with Glashow, almost fifteen Nobel laureates fall into this category of our could-be compatriots, whose parents had in due time emigrated from the Russian Empire. They are physicists Douglas Osheroff and Arthur Schawlow, chemists Paul Berg, Herbert Brown (Brovarnik),[12] Melvin Calvin, medics and biologists John Vane, Bernard Katz, Stanley Cohen, André Lwoff,[13] Daniel Nathans, Ernst Boris Chain (Hain), Gertrude Elion... They were all united in their natural interest in the historic motherland, and many attempted to visit it, failing to be successful at times.

At the end of the 50s, during the Khrushchev thaw, Sheldon Glashow also made such an attempt, expecting to combine business and interest. Being a postgraduate student at Harvard, he received a grant from the Copenhagen University to work under the supervision of Niels Bohr. Supported by his postgraduate instructor Professor Julian Schwinger and some other Harvard professors, he wrote to academician I. E. Tamm expressing a wish to work at his laboratory.[14] He received a positive reply from Tamm without any delay. But Glashow never got a reply, either positive or negative, to his application for an entrance visa to the Soviet Union.[15]

It could be for the traditional hatred of the authorities towards the emigrants from Russia and their descendants. Or the fear of State Security bodies that the Soviet scientist wouldn't be able to fight the temptation of passing atomic secrets known to him over to the young American...

The Soviet period of Russian history was full of excesses, which also contributed to filling the list of Nobel laureates with scientists born in Russia but bestowed the Award being in the position of immigrants.

Ilya Prigogine, the 1977 Nobel Prize winner for chemistry was born in Moscow on January 25, 1917, five weeks before the overthrow of the tsarist government. His father, a qualified chemistry engineer, had founded a lacquer and paint factory in the former suburbs of Moscow, which provided the family with sufficient means for a decent life. In 1918 after the October Revolution, the factory was nationalised and converted into a soap production plant, the owner being appointed its director.[16] In the spring of 1921 there started the arrests of the proprietors, and Roman Prigogine considered it best for his family to leave his motherland. They spent their first year of emigration in neighbouring Lithuania,[17] moving then to Berlin where his younger brother was living, and finally settling down in Brussels by the early thirties.

Despite all this migration the scientist-to-be was brought up in a Russian language environment with never ending reminiscences of the lost motherland. At the beginning of the Khrushchev Thaw, when the borders of the Soviet Union started to open up, as early as 1955 his father came to Moscow to visit his numerous kin, relations with which had been broken since the 1930s. In 1957, the future Nobel Prize winner also arrived there, having become by that time a regular professor and administrator of the Free University of Brussels, a corresponding member of the Royal Academy of Belgium, the author of several fundamental works, in which his early ideas of the creative role of reversible equilibrium in nature could be traced. These ideas were later awarded the Nobel Prize and made him famous as one of the most outstanding thinkers of the second half of the twentieth century.

During this first visit[18] he could still find the house in Bolshaya Pereyaslavskaya St., abandoned by his family 36 years before (when he was only four). The house was later destroyed due to the reconstruction of the district. Like his father he could only visit the families of his two aunts. The rest of the Prigogines preferred to stay away from meeting the newly found foreign relative.[19]

The reason for such strange behaviour was quite understandable. When applying for a job it was safer to conceal the existence of relatives abroad. Hundreds of thousands, not to say millions, of their fellow citizens did the same thing. A positive answer to the corresponding question in the application form immediately made a person "politically unreliable", created obstacles for promotion, for access to jobs requiring specific reliability, and so on.[20]

Even the gradual liberalisation of the regime couldn't extinguish the fear for the future. In 1971 Roman Prigogine wrote a letter to his sister Lyubov Abramovna intending to inform her of her nephew's future visit to Moscow, saying that as he was coming on an official invitation there was no danger for his cousins to see him. Having lived in a totally different society for half a century he could hardly realise that the very next day after the supposed meeting, the Moscow cousins would have to explain to their staff management why their application forms contained unreliable information.[21]

The Moscow Prigogines only had the opportunity to first meet their famous relative after the downfall of the communist regime, when he came to the city of his birth for the seventh time. It happened a week before 1993. The VIP hall of Sheremetyevo airport was full. When, having crossed an imaginary border, Ilya Romanovich slowly started for a crowd of people seeing him in, there stepped forward an elderly man almost his replica. It was Vladimir Borisovich Prigogine, his cousin, eight years older, having always protected and defended him in their childhood games, and never forgotten over all these seventy years. All the rest of those present at the heart-warming reunion tried to master their overflowing feelings...

At the Nobel Prize awarding ceremony Ilya Prigogine, a symbolic figure of the twentieth century, was introduced by the academician Stig Clæsson, who called him "the poet of thermodynamics". A crucial question may have resulted from his triumph of whether he could have achieved such an enormous success, should the family not have left the Soviet Union. The answer is supposed to be more than just pessimistic. The outcome would have been poor...

Poor would have it become as early as the beginning of the 1930s when in his last year at school he had to start thinking of his ongoing education. His social origin was that of the son of a former factory-owner. He either had had to falsify his personal information or be considered as one of the "deprived", for whom entering a university had been practically impossible.

Even if he could have found a way out of such complexities of Soviet life he would have had to face another problem of the post-war reality, which could have been dramatic for him as an abstract-thinking scientist. Ilya Prigogine would never have been left out of the witch-hunting propagandist campaigns, so numerous in the last decade of the Stalin regime. His scientific philosophy seemed so vulnerable to ignorant charges by the party ideologists with adherence to "idealism" and other imaginary sins, that he would have been a prime target for severe attacks in the times of a hysterical search for "bourgeois cosmopolitans", "Weismanists", "Morganists", etc. The dramatic consequences for the thinker's inner world would have been inevitable.

Science is a specific sphere for the manifestation of the human spirit. A conservative way of thinking is implicit in scientific circles. According to the outstanding scientific and philosophic essay "Order out of Chaos" by Ilya Prigogine and Isabel Stengers, researchers try steadily to uncover the secrets of nature using in their studies the same techniques, the same notions, ignoring the stubborn resistance of matter. "When nature finally rejects the use of the universally accepted language, there comes a crisis... And only then do all intellectual forces start concentrating on the search for a new language."[22]

"Prigogine language" as a phrase is now an inherent part of science. But the job of creating a new scientific basis for mutual understanding needs a completely free spirit. When entangled in ideological tenets that are not supposed to be subject to any possible doubt, a thought is then either lost or seized by everyday petty problems.

That is why Ilya Prigogine could hardly have been accomplished in his mother country. This is why the 1977 Nobel Prize for chemistry would have certainly gone to some other researcher, and the world would have been deprived of a mighty thinker.

...When this great son of Russia died he was not forgotten in his reformed motherland. In Brussels on 6 June 2003, the day of Ilya Prigogine's state funeral, an enormous wreath of roses and lilies sent by the Moscow University in memory of its honoris causa had been attached to the back of the catafalque, paying homage to the native land of the deceased.

Two other Soviet émigrés Simon Kuznets and Wassily Leontief, Nobel Memorial Prize winners in economic sciences, had a scientific fate similar to Prigogine's.

Kuznets' father, a Kharkov fur dealer, had left for the United States soon after the 1905 revolution followed by the Jewish *pogroms* in many parts of the Russian empire. He meant to find a niche for himself in business in a couple of years, intending then to send for his family left in Kharkov. The outbreak of World War I, followed by the cataclysms in Russia, interrupted his original plans. Meanwhile, his son finished school and entered the economics department of the university in Kharkov.

It was not until 1922 that, at the age of 21, together with his younger brother, Kuznets could join his father overseas. During the summer of that same year, he taught himself English and entered Columbia University to continue his studies in economics.

From his very first years in science he manifested an unusual way of thinking, which in the 1920s and 30s helped him discover natural regularities in the periods of macroeconomic recession and long-term economic development. The analysis of such interrelations, which hadn't been tackled before, made a great contribution to the economic thought of that time, establishing him as a prolific young scientist.[23] After it had been established in 1969, in 1971 he became one of the first to be awarded the Prize in economic sciences.

Just two years later in 1973, Wasily Leontief was awarded with the same Prize. He was born in 1906 in Munich,[24] at which university his mother had studied. Among his most vivid childhood memories there were three events; when he was four, the death of Leo Tolstoy who, according to him, everyone mourned deeply; bullets whistling in the streets in the days of the February revolution; and Lenin on a rostrum in front of the Winter Palace whose speech he was listening to at the mass rally in Dvortsovaya (Palace) Square.[25]

Leontief entered the Petrograd (Petersburg) University in 1921 and graduated from the economics department in 1925. While still a student he participated in working out the Soviet Union national economy development plan for 1924. In 1921, being in his first year, he was held in custody for a short time on the charges of "anti-Soviet activity". There was a censorship prohibition on the publication of his first scientific article, though it contained no political allusions at all. This made him start considering emigration. In 1925, after undergoing serious jaw surgery with a rib bone graft, he was allowed to move to Germany. "It was primarily because of that", he remembered, "that I was given permission to leave. They supposed me to die in any case."[26]

Having received a Ph.D. in economics at the University of Kiel, he was offered a position of economic adviser to the Chinese government in 1929, settling down in the USA two years later. The world's richest industrialised state was then gripped in the Great Depression, and the young scientist was immediately recruited to help overcome the situation. By the mid-thirties, his input-output method of analysis based on the system of transactions tables had been completed and in 1936 he provided a detailed description of 41 sectors of the American economy worked out under his supervision. Almost 40 years later he was awarded the 1973 Alfred Nobel Memorial Prize[27] in economic sciences, the fifth for this fundamental work. By that time its practicability in the extreme conditions of World War II had been absolutely clear. Sticking to the concept devised by Wassily Leontief, American industry could provide with the accuracy of clockwork both the inner needs of the huge state and the demands of the ally armies at the fronts from the Atlantic Ocean to the Pacific.

Returning again to the question of Prigogine, whether Leontief could have realised his ambitions had he stayed in his motherland with the ever penetrating power of its ideological supervisors, the Nobel laureate himself gives the answer. After the breakdown of the Soviet regime had begun, Wassily Wassilievich offered his assistance in converting the Soviet planned economy into the market one, having established an iternational centre for social economic studies in St. Petersburg. In this very same year of 1991 he was asked to become Yeltsin's personal economic adviser. The answer was unambiguous. "Good Lord", he said, "I would never become an economic consultant for the party authorities."[26] It would not be impossible to imagine his sad destiny as a Soviet scientist if this evident hostility towards the party's bureaucratic guidance of the economy could be taken together with the information he would have had to always mention in the numerous questionnaires, which all USSR citizens had had to fill in, such as of his grandfather — a textile manufacturer in pre-revolution St. Petersburg, his own detention on anti-Soviet charges and so on.

And here comes the finale.

The ideological authorities of the Soviet Union were suspicious not only of the activity of the Nobel Prize bodies but also the Nobel Fund publications, especially the yearbook *Les Prix Nobel.* Having been received in the Soviet Union, every one of its current issues was carefully checked. Any ideologically hostile facts found would cause the whole issue to become inaccessible in the libraries, being sent for "safe keeping" to the so-called "Spetskhran". Only a very limited number of readers could be admitted to this literature, having provided an official demand from the research institute they were employed at, detailing reasons for such interest.

The biggest book depository of the Soviet Union, the Lenin Library as it was then called (the Russian State Library now) had in its "Spetskhran" section 17 issues of *Les Prix Nobel,* 1937 to 1984,[28] the 1985 issue being sent there in the Gorbachev era. Only a very careful study of these issues could provide an explanation for this "imprisonment", though some reasons are quite obvious. Let us first look at the 1977 edition containing the information on Ilya Prigogine's Nobel award. The life of Prigogine's works in the Soviet Union happened to be quite lucky. He was one of the rare foreign natural scientists whose works were translated into Russian[29] almost entirely. One of his earliest books *Introduction to Thermodynamics of Irreversible Processes,* published in 1954, was translated into Russian as early as 1960. His Nobel speech[30] was printed in *The Progress of Physical Sciences* academic journal. But *Les Prix Nobel,* containing the same speech as in the original, had been destined for the shelves of "Spetskhran". In the same volume of the journal there was published his short autobiography in which Prigogine added mildly that "for his family it was difficult to adjust to the new regime, so in 1921 they had to leave Russia."[31] This statement was considered by the Soviet censor to be "an anti-Soviet attack", and the book was dispatched for safe keeping.

To finalise his story we can only admit that, though Ilya Prigogine himself had a filial affection for his native country, he had never been misled about the intellectual level of its Soviet leaders. So what can be considered as preventive suspiciousness in the censor of the Brezhnev epoch could really be far from being politically groundless.

In June–July 1975, more than two years before he was awarded the Nobel Prize, Prigogine was a member of the state delegation, with the king of Belgium Baudouin at the head, visiting the Soviet Union. During the talks concerning the draft version of the Soviet–Belgian declaration, collaboration in protecting the environment, development of mutual tourist industry, etc., Prigogine was introduced to the USSR leaders and personally to L. I. Brezhnev. Talking to the author of this book about two decades later he gave his earlier impressions straight. "I couldn't perceive them as politicians. All suggestions beyond the habitual and foreseen were met with a stony silence. For me, they bore a resemblance to the farmer who knew exactly when to fodder the cattle and feed the hens but kept totally away from anything besides his everyday chores..."

The 1937 yearbook got to "Spetskhran" much later, after the outstanding Hungarian biologist Szent-Györgyi, the 1937 Nobel laureate in Physiology or Medicine, illegally crossed the border of Communist Hungary in 1947, emigrating to the United States (see note 1 to doc. 6.24). The 1946 issue appeared to be hidden behind the steel doors of "Spetskhran" because of the Nobel laureate Hermann Muller, who had worked at the USSR Academy of Sciences Institute of Genetics from 1933 to 1937, together with the academician N. I. Vavilov, who was later arrested. Being a corresponding member of the USSR Academy of Sciences, in 1948 he openly manifested his protest against the destruction of USSR genetics,[32] organized by Lysenko under Stalin's protection, and rejected the membership.

In 1947 the party ideologists could not conciliate themselves with the Nobel award of André Gide, the anti-Soviet and reactionary, who had published his impressions of the 1936 tour of the USSR, which were far from flattering. In the following year of 1948 the Department of Propaganda and Agitation of the Central Committee of (15) the Communist Party (bolsheviks) was enraged by the prize being given to Thomas Eliot, the poet, who, as the newspapers argued, was awarded for his links with the external world (see attachment 4 to the book). In 1949 the Nobel speech "Science and the World" by Lord Boyd Orr, the Peace Prize laureate, was found unacceptable for the Soviet reader. Besides Lord Orr, the same year the Nobel Prize for Physics was given to the Japanese scientist Hideki Yukawa, which also outraged the Soviet propaganda of those times, according to which "the science for him was but a means for homicide" (see attachment 3). Yukawa, though, was awarded the Prize for "his prediction of the existence of mesons on the basis of theoretical work on nuclear forces".

The two last issues of *Les Prix Nobel* before the end of the Stalin era contained information on the 1950 and 1951 Prize winners. The 1950 Nobel Prize winner in literature was Bertrand Russell. The war in the Korean Peninsula having started the same year, in his Nobel speech the laureate uncovered everything he thought about the actual perpetrators and reasons for its origin.[33] At the same time the award for Peace was given to Ralph Bunche, "the Negro renegade, playing a shameful part in concealing the racial discrimination policy in the United States." (See attachment 4.)[34] In 1951 the Nobel Peace Prize laureate was Léon Jouhaux, an outstanding leader of the French trade-union movement. In 1947 he refused to collaborate with the Universal Labor Confederation, it being totally under Communist pressure, and created a new social democratic organization "Force Ouvrière". This was enough for *Les Prix Nobel 1951* to be dispatched to "Spetskhran".

After a 20-year interval that started at the time of the Khrushchev Thaw and having mechanically gone on for several years after the forced retirement of N. S. Khrushchev, the first yearbook of the Nobel Fund to find its way to the shelves of the safe keeping body was *Les Prix Nobel 1970.* The reason for that was absolutely obvious. It was the Nobel Award in literature given to Aleksandr Solzhenitsyn. The following ban on free access to the 1974 issue of the yearbook was due to the same reason. Having been thrown out of the Soviet Union, Solzhenitsyn arrived in Stockholm and, after presenting his Nobel speech at the Swedish Academy as prescribed by the Statutes, received his award together with the 1974 Nobel Prize laureates. Then came 1975, which was the year of Andrei Sakharov's award.

The underlying reason for the 1980 and 1983 issues to be sent to "Spetskhran" is also quite obvious. For the former it is that the Prize winner for literature was the Polish poet and essayist Czesław Miłosz, who refused to return to Poland in 1951. He explained his decision two years later in a brilliant pamphlet *The captive mind,* the subject of which was a thorough analysis of the totalitarian regime's devastating impact on a creator's personality.[35] For the latter the reason was that the 1983 Nobel Prize for Peace was given to Lech Wałesa, a "Solidarity" legendary leader.

Whereas in 1980 Miłosz's award caused dissatisfaction with the ideologists in the Soviet Union but not in the laureate's native country, the reaction to the same award given in 1984 to the Prague poet Jaroslav Seifert was the same with the both the Soviet and the Czechoslovakian authorities. Seifert never concealed his rejection of the totalitarian regime, which responded by banning the publication of his works. In the 1970s his new poems circulated predominantly in Samizdat. The news that this most prestigious award was given to their compatriot was silenced in Czechoslovakia, and it was only after a collaborative campaign in the Western media that the information about this event appeared in the press. So it could have been predicted with utmost certainty that access to the 1984 Nobel Foundation yearbook would also be denied.[36]

The keeping of the 1978 volume could also have been predicted, despite the fact that it contained the materials concerning the Soviet academician P. L. Kapitza, that year's laureate of the award for physics, including his Nobel Lecture "Plasma and the governed thermonuclear reaction", the speech at the banquet, his autobiography, etc.

Simultaneously with Kapitza receiving the award from the king of Sweden, in Oslo there was a ceremony of awarding the president of Egypt Anwar Sadat and prime minister of Israel Menachem Begin with the Nobel Prize for Peace. They were bestowed the Prize for a very courageous act of having given rise to negotiating the

Middle East problem, which led to the two-frame agreement on peace in the Middle East and peace between the two countries. In the Soviet Union this grand event was considered as nothing more than "the shot in the back of the Palestinian people's liberation movement", and hence there was a ban placed on *Les Prix Nobel 1978*.[37]

Besides the materials on awarding the Prize for Peace, in that same yearbook there existed another irritant for a scrupulous censor, the Nobel lecture of the literature laureate Isaac Bashevis Singer, about Yiddish, the language of "banishment, deprived of state and borders, [...] which no existing government is supporting...".[38]

In the years when the main CPSU ideologist Michail Suslov[39] was still in power, to send the book to "Spetskhran"[40] for having found in the lecture "Zionist propaganda" alone would face no problem at all. The 1982 issue of *Les Prix Nobel*, which also found its way to "Spetskhran", justifies these suspicions.

Among the nine 1982 Nobel Prize laureates in all its six categories there was no one to arouse suspicion with the Soviet censor either on political or ideological grounds. The Prize for literature for that year was given to one of the most well-known contemporary foreign writers in the Soviet Union Gabriel García Márquez, a man of left views. Both Peace Prize laureates, Swedish and Mexican diplomats Alva Myrdal and Alfonso García Robles, awarded for disarmament activity, were also out of suspicion. It would also be improbable to doubt the ideological positions of six other laureates for sciences. A careful review of their speeches and biography features revealed nothing reprehensible, but for one catch...

The family of Aaron Klug, the Nobel Prize winner for chemistry, born in Lithuania in 1926, emigrated to South Africa when he was still very young. When writing in his autobiography about his parents, he mentioned his father saying that he was qualified as a harness-maker after finishing a secular technical school, and also had a traditional Judaic education.[41] Could this fact have frightened the censor scanning the text of the publication for samples of ideological impurity?

At the end of the twentieth century Russia found itself in a torturous state of transition, trying to overcome the consequences of the 75-year Communist reign. The choice of targets in this process was often far from being correct, which led to numerous failures both in economy and in science and culture.

Hope is always around somewhere, though. It is living in the soul of every Russian, this hope for the resurrection of a strong and wise state, free from economic and moral perplexities and an equal member among other civilized nations. To achieve this it is necessary to get rid of deeply rooted masochist complexes, the never ending supply of permanent suspicion and far-fetched resentment, which in the end can turn into a complex of exclusiveness, a disease most dangerous for the well-being of a society.

All this correlates naturally with the maxim stated by A. N. Yakovlev, in which he prophesied the danger of criticising the seventy years of Soviet power alone in a one-sided way, cut off from reality, "as if there hadn't existed thousands of previous years having predetermined the Bolsheviks' victory in 1917." At the end of his maxim Yakovlev argues that with another approach "the history of the state could be artificially torn into pieces".[42]

One of the indirect but symbolic signs of the resurrection of democratic Russia could become the notable representation of Russians among 21st century Nobel Award laureates, the elite of the world of science and culture, the representation symbolising the increased potentiality of the great state than in the past twentieth century.

Notes

[1] One of Alferov's nominators was his teacher, the 1964 Nobel Award laureate academician A. M. Prokhorov. As he had told the *Izvestia* correspondent S. L. Leskov, on the whole he submitted Alferov's name seven times.*

* "Izvestia — Nauka", 11.01.2002, p. 1.

[2] By 2003 the number of awards given to scientists of the United States had risen to 163, of Great Britain to 63, of Germany to 57, of Switzerland to 15, and of Japan to 9. In the table of ratings Japan moved from 12–13th to 8–10th place in the first two years of the 21st century, as the Japanese scientists' representation among the Nobel laureates for natural sciences became 1.5 times greater, which was extraordinary in itself. But when Russian academicians A. A. Abrikosov and V. L. Ginzburg received the 2003 Nobel Prize for physics "for pioneering contributions to the theory of superconductors and superfluids"* Russia could safely occupy the eigth place, breaking away from its immediate neighbours in the table. If we compare its position in the last year of the 20th century with the three leading countries, at the beginning of the 21st it slightly improved. With the United States, the lag decreased from 17 to 16.3, with Great Britain from 6.5 to 6.3, and with Germany from 6.3 to 5.7.

Let us review simultaneously what the 2003 Nobel laureate Ginzburg in his 1979 memoirs wrote *a propos* about the Nobel Award itself, when contemplating the troubles concerning the Prizes for scientific discoveries: "I'm in no hope of receiving it, there's such a rush there... But I am writing now because if it's simply a 'no', then it's OK. But if others get (such things happen) what I must have received, it'd be a pity." He also provides an example of such possible injustice: "...if it were given to Abrikosov, having solved the Ginzburg–Landau equations..."** As we now know, 24 years later the Royal Academy of Sciences did not let this injustice happen.

* For both 2003 laureates as personalities and scientists, see Chapter 8 and docs. 8.16–18 for Abrikosov, and 3, 8.1, 12.4 and 49 for Ginzburg.

** W. L. Ginzburg. About Science, myself, and others. Moscow, Physmatlit, 2001, p. 377. See also note 1 to document 8.18.

[3] Quantitative analysis of Jewish mass emigration from Russia at the end of the 14th and the beginning of the 20th centuries was given by A. I. Solzhenitsyn in his well-known but dubiously received research, which he based on a great number of already published historic documents.* The writer emphasised though that "the intelligentsia was not inclined to emigrate, blaming it as a deviation from the fate and mission of Russia." According to the data given, the number of "Jews in educated levels was in 1899 to 1907 [among the emigrants] a little over one percent".** That is, many thousands of their descendants who made their name in

science and achieved outstanding results were by origin mostly of badly educated lower and middle class layers of Russian immigrants.

* A. I. Solzhenitsyn. Two hundred years together. (1795–1995). — M.: Russkiy Put', 2001. Part 1, pp. 308–312.

** Same, pp. 311–312.

[4] For entering high schools and universities of the Russian Empire, in July 1887 there was established a percentage rate for people professing Judaism. In the circular letter sent by the minister of public education I. D. Delyanov to those in charge of educational districts, it was stipulated that the rate of admission of Jews to state educational institutions could not be more than 10% in the Jewish ghettoes with the highest density of Jewish population, 5% outside, and only 3% in both capitals — Moscow and St. Petersburg. After the 1905 revolution, during the period of liberalisation, the government tried to improve the negative impression concerning this discrimination by "widely practising final examinations for high school students without their attending classes".* Young Waksman, a future Nobel laureate, was lucky to get into this group. But as early as February 1911 this conception had been eliminated by an Act of the government approved by the tzar on the 11th of March 1911, with the percentage rate only applying since then.** The number of Christians using this concession was very small as there was no need in it for them. So this new Act actually blocked Jews from the possibility of using this loophole. Though such limitations were quite legal for high schools and universities of the Empire, they could be unexpectedly and voluntarily changed, especially when having in mind education in prestigious educational institutions. Thus, in 1902, just before the period of entrance applications, it was published in newspapers that "in the academic year-to-come, those professing Judaism will not be admitted to the Moscow University."***

* Solzhenitsyn, p. 505.

** Complete collection of laws of the Russian Empire. Collection three, vol. 31, sec. 1, 1911, No. 34900 (11.03.1911) St. Petersburg, 1914.

*** "Novoe Vremya", 21.05.1902, p. 4.

[5] It was not only because of religious discrimination that young people in autocratic Russia were forced to leave their motherland in order to get a proper education. The fate of young women having high school certificates was even more hopeless. It was exceptionally rare that they could receive a university education.* Maria Curie-Skłodovska, twice a Nobel laureate, who was born in 1867 in Warsaw (then part of the Russian Empire) to a family of Polish intellectuals, later entered the natural sciences department of Paris University, and never returned to her country. If she had, she would have repeated the destiny of the outstanding Russian mathematician Sofya Kowalewskaya. While being a Doctor of Sciences at the Göttingen University in Germany, back home she could neither get a position at St. Petersburg University nor receive a masters degree at Moscow University. So having been invited by the famous Swedish mathematician M. G. Mittag-Leffler she moved to Sweden in November 1883. As early as June 1884 she was already the first female professor in the history of Swedish universities, later receiving two prestigious scientific awards, those of the French and Swedish Academies of Sciences.

Kowalewskaya's close relative A. I. Kosich, the Saratov governor-general, standing out strikingly among Russian high-ranking officials by his anxious attitude towards public education and science and decent female representation, appealed to Great Prince Konstantin, President of the Academy of Sciences of the Empire, asking him to return Kowalewskaya to

"Russia and Russian Science".** Academician P. L. Chebishev had to explain to the Great Prince that according to Russian laws the highest position for a woman was that of a professor at the high school for girls. Therefore, "having a very important and honorable position of a Stockholm University professor, Sofya Wassilyevna would hardly agree to return."*** The only thing the President could do was to have her elected a corresponding member at the Department of Physics and Mathematics, which was fulfilled on December, 2 (14), 1889.

* Here are the figures illustrating how eagerly high school female graduates used their only chance to get university education. As was published in the newspaper *Russkie Vedomosti* of 19.05.1900, of 543 female students studying in Swiss universities in 1899 there were 321 Russian subjects, 78 Swiss, 60 German etc.

** PFA RAN, fund 6, unit 1/1, pp. 56–57.

*** Same, pp. 59–60.

[6] In the same year of 1946, Waksman was first nominated for the Nobel Prize by Anderson, the professor of pharmacology from San Francisco. In 1947 the scientist was proposed by five American researchers. In 1948 he had six nominations, two from Europe; in 1949 — seven, one of which was from a New York professor of medicine P. Reznikoff, whose surname obviously shows the Russian origin of him himself or his parents and grandparents. In 1950 the number of American "Russian doctor" nominations rose to seventeen. Besides 16 proposals sent to the committee on Medicine and Physiology, one more was received in Stockholm from the Harvard professor H. Shapley, but for the Prize in chemistry. He repeated his nomination the following year, 1951. It is worth observing that in 1950 the number of Waksman's European nominators for the first time exceeded their American colleagues ten to seven. And in the next two years the nominations of the streptomycin discoverer were exclusively European. In 1951 there were only three proposals that were obtained by the Nobel Committee: from Germany, Italy and Spain. But in 1952, which was successful at last for our compatriot, there were four, the authors of the three being professors of medicine from Salamanka in Spain, Naples and Belgrade. The fourth was submitted by undoubtedly the key figure in the story, the Lund University professor Sune Bergström, a respected member of the Nobel Committee, who received his Nobel Award only three decades later, in 1982. It was his proposal that could be considered a sign to regard Waksman's nomination as favourable, that is, given the most priority.*

* By the way, in 1951 a Stockholm microbiology professor H. Bergstrand was the only nominator of the American microbiologist Max Theiler from the laboratory of the department of medicine and health services of the Rockefeller Foundation in New York. And like Waksman a year later, Theiler did receive the 1951 Nobel Prize. Suffice it to say, it's rather easy to follow the similarly busy activity of respected Swedish nominators in the year when their candidates became the recipients of the award by checking the nomination lists of the Royal Academy of Sciences' Nobel Committees. (See Chapter 6 and Table 6.3).

[7] RGASPI, fund 17, unit 117/917, pp. 36–41; see also note 24 to Chapter 4.

[8] Here we can see the emphasised hyperbolization of the subject only too appropriate for political satire, in other words, caricature. The following is the same fabrication but presented as reliable information: "Being in Moscow the American scientist Waksman valued the secret of streptomycin at $150 thousand. You can easily understand what limited number of people can use such medication abroad."* This announcement where the cost of selling the right to produce streptomycin was ill-naturedly confused with the retail price for an

individual consumer was made in August 1947 at a party meeting of the editorial board of the political magazine *Novoye Vremya* when discussing the closed letter of the Central Committee of the Communist Party concerning "the Kluyeva–Roskin case".

* CAODM, fund 69, unit 21/88, p. 72.

[9] All information of "the Kluyeva and Roskin case" as regulated by CK CP(b), was not liable for open press; those breaking the instruction were severely punished. (see doc. 6.12.)

[10] Archives of the Russian Academy of Sciences, fund 1601, unit 1/127, pp. 56, 60, 66.

[11] As is testified by the authors of the biographic encyclopedia of Nobel Prize laureates in Medicine and Physiology released in Uzhgorod, the place of Waksman's birth was not the town of Priluka, but the village Nowaya Priluka, situated now in the Lipovetski district of the Vinnitsa region, not far from Berdichev and Vinnitsa. The information was provided by the Nobel laureate's son, Professor Byron Waksman,* who attended the scientific conference in Vinnitsa in the autumn of 2003.**

* W. I. Kimakovitch, I. D. Gherich, O. O. Kutsch. Nobel Prize laureates in Medicine and Physiology. (173) Uzhgorod, "Zakarpattya", 2003, p. 57.

** The author received the information about B. Waksman's visit to Ukraine from W. I. Tomchani, the expert on Nobel Awards from (173) Uzhgorod.

[12] When changing the language environment the surnames transformation also took place in the movement from the West to the East of Europe. One of the striking examples is how the family name of one more Nobel laureate, Ilya Prigogine, was transformed. His ancestors, together with the tribes migrating in the 17th to the beginning of the 18th century from the territories of future Germany, settled down in the western parts of Russia. Their original surname was Schein, "schön" in German, which meant beautiful, wonderful, comely ("prigogiy" in the Russian transcription). One branch of the Scheins gradually made the family name sound Russian by adding the Russian ending and changing in into Scheinin, which is not rare in Russia. The others who settled down in the Moghilev region, where the scientist's grandfather comes from, replaced their name with "Prigogine".

Prigogine's Moscow kin happened to be acquainted with one of the relatives from the Scheinin branch. It was he who told this story at the end of December 1992 sitting at dinner during the scientist's visit to his native town.

[13] André Lvoff's mother is known only too well to every educated Russian due to the famous painting *A Girl Lit by the Sun*, which is one of the treasures of the Tretyakov Gallery in Moscow. W. A. Serov, an outstanding Russian painter portrayed in it Maria Yakovlevna Simanovitch, his beloved cousin and eventual mother of the Nobel laureate, the painter's second nephew.*

* *Izvestia*, 3.1.1991, p. 3.

[14] On the envelope of the letter sent by Glashow to Tamm of 8.5.1958 the address of the USSR Academy of Sciences Physics Institute was printed in English with the handwritten address in Russian to the right of it, which shows that Glashow could have known basic Russian, presumably spoken in the family.* All the above data about Sheldon's parents' leaving Russia, their settling down in the new country, his own young years, are taken from the 1979 yearbook of the Nobel Foundation.**

* *Priroda* (Nature),No. 5, 1993, p. 126.

** Les Prix Nobel. 1979. Stockholm, 1980, p. 57.

[15] The reply from the department for North America of the Foreign Office of Russia to the request of the Russian Academy of Natural Sciences about whether there existed any evidence of Glashow's application for an entry visa to the Soviet Union was negative. It was also mentioned that "the success [of further research] is unlikely because [...] the reason for refusing entrance to the country was not to be explained at the time."*

* *Priroda* (Nature), p. 128.

[16] In the first years of the Soviet power such behaviour in respect of the former owners of nationalized property was widespread, but sometimes it could become dramatically humorous. M. M. Bakhtin, an outstanding literary critic, provides an example. At the time of revolution his close relative was mayor of the town of Oryol. He "was arrested, but nothing bad was done to him. Moreover there were such enterprises which he alone could be the manager of [...] and he was, so to say, released at times to run them properly..."*

* M. M. Bakhtin. "Besedy s (Talks to) V. D. Duvakinym". M.: Soglasie, 2002. pp. 24–25.

[17] That Lithuania was chosen as a transition place at the beginning of the 1920s was due to certain circumstances which made this state the most popular legal zone for those wishing to emigrate. This was greatly assisted by well-known poet J. K. Baltrušaitis, who was a habitué of the Moscow literary elite and a friend of A. W. Lunacharsky, an eminent member of the Soviet government of those times. Having such close connections and being appointed head of the Moscow consulate of the independent Lithuanian Republic, Baltrušaitis managed to obtain the most emigration quota, and thousands of people succeeded in using it. Besides the Prigogines, the "Lithuanian route" was also used for leaving Russia by the symbolist poet K. D. Balmont, a 1923 Nobel Committee nominee in literature. As B. K. Zaytsev, a writer-émigré, remembered, "in 1920 [...] Baltrušaitis, his true friend [...] arranged his legal departure, thus actually saving his life. One autumn morning in Nikolopeskovsky [a side street in Moscow, near Arbat St.] we, some men of letters and ladies, waved goodbye to Balmont and his kin, leaving for the railway terminal in a pick-up truck of the Lithuanian embassy."*

* B. K. Zaitsev. "*Golubaya Zvezda* (A Blue Star)". — M.: Moskovsky Rabochiy (Moscow worker), 1989. p. 477.

[18] His first trip to the native town in the summer of 1957 was at the invitation of academician A. I. Oparin to take part in the international symposium on the problems of the origin of life on Earth.

[19] Here are the reminiscences of A. A. Kamionskaya, Ilya Prigogine's cousin, which she shared with the author, "My uncle [R. A. Prigogine, father of I. R. Prigogine] came to Moscow as soon as it became possible. It happened in 1955. But he could only visit my younger sister, now dead, who then lived with her family in his flat in Bolshaya Pereyaslavskaya St., and his own sister, my aunt Lyubov. My mother had already died, my husband, a military man, and I living in Moldavia. The other Prigogines, his brother among them, were worried about the possible troubles at work for themselves and their children, and decided not to see him. During the second visit I came to Moscow deliberately, but as for the others, he couldn't see them again."*

* This conversation took place in December 1997.

[20] The following are but several of tens of thousands of recorded examples of how the newly come information about relatives abroad could affect the future life of a person. In 1950 a foreign policy commission reviewer was fired by the ordinance of the Central Committee of

the Communist Party when it became known that he had a brother in New York who had emigrated 30 years before.* In the same year of 1950, after a special check of the work of one of the central aviation research institutes had been conducted by the Party Controlling Committee, it was stated that "the leading positions there were occupied by the persons who either had relatives abroad or political convictions by appropriate Soviet bodies." Business qualities of these "persons" didn't matter much for the conptrollers and their bosses. This is how they wrote about it themselves: "To our mind, the idea of irreplaceability is a far-fetched thing."** That there are no irreplaceable individuals and that the idea is "far-fetched", can be illustrated by the destiny of the outstanding Polar researcher V. U. Wise, corresponding member of the USSR Academy of Sciences, notwithstanding the fact that his name had been given to an island in the Kara Sea, to the glacier, the cape and bay in Novaya Zemlya (the New Land) in the Arctic Ocean. In October 1949 a similar check took place of the staff of the Arctic Institute in Leningrad. In the act it was stated in particular that "socially alien and politically unreliable Professor Wise, whose relatives live abroad and who he keeps in touch with", was not entitled to keep the position of a senior researcher. As a result, he had lost his post and could only stay at the institute "as an adviser on a limited range of problems concerning Arctic research".***

Astonishing as it might seem, having relatives abroad can still be considered an obstacle to being admitted to a secret job. The following is evidence of how viable these disgusting remnants of the Soviet discrimination system are. According to the existing legal system, the Russian Agency for State Resources is a closed institution. The *Izvestia* newspaper made public the announcement of one of its leaders: "We employ people as in any other closed enterprise. Besides professional skills, we need a résumé free of undesirable facts and no relatives abroad..."****

* RGASPI, fund 17, unit 118/836, p. 24.

** Same, unit 908, p. 41, 43.

*** Same, unit 964, p. 125–129, 156.

**** *Izvestia,* 8.9.2000, p. 9.

[21] Soviet citizens were afraid of keeping in touch with the relatives and close friends living abroad, and human tragedies connected with this were many and multiplying from day to day, from hour to hour. Here comes one more example, typical as it is. In 1964 from Israel to Moscow to see her brother there came the aunt of the famous poet Boris Slutsky and his brother, a colonel of the Soviet army. But the nephews preferred to abstain from the meeting.*

It is necessary to point out that everyday situations apt to become dangerous for humans, especially in Stalin times were innumerable, and absolutely anyone could find themselves inside their vague borders. For instance, one of the staff of the USSR Academy's Institute of World Literature who studied at school together with Rikov's daughter in 1933–34, several times visited her at her home. A. I. Rikov, the head of the Soviet government at the time, in 1938 was announced the enemy of the Soviet people. So about fifteen years later she was reprimanded at the party board meeting of the Moscow Regional Committee of the Communist Party for not mentioning this "offence" in her autobiography at entering the party rows. And only in 1954, a year and a half after Stalin's death, the institute's party board dared to claim on remitting the absurd punishment.**

* *Izvestia,* 2.12.1999, p. 1.

** CAODM, fund 2472, unit 1/14, p. 155.

[22] I. Prigogine, I. Stengers. Order out of chaos. — M.: Progress, 1986, p. 381.

[23] Further extending his ideas, Kuznets was later first to formulate the notion of "periods of modern economic growth". Now it is substituted by the term "global transitional period". Being the director of the institute of the transitional period E. I. Gaidar explains that it should be taken as contemporary "periods of radical changes in the standard of living, technological organization, system of employment, demographic behavior and social security".*

* "Izvestia", 20.03.2006, p. 2.

[24] Not long after the baby was born the family left Minsk and settled down with the young father's parents in St. Petersburg. The baby was baptised there, which fact could be the source of confusion, and so St. Petersburg was mentioned in the Nobel Foundation Directory as the future Nobel laureate's place of birth.

[25] Les Prix Nobel 1973. Stockholm, 1974, p. 235.

[26] Literary Gazette, No. 12, 1991, p. 7.

[27] Along the same lines as Leontief, there worked the Soviet economist L. V. Kantorovich. That is why the Nobel Committee vigorously debated the problem of whether to share the 1973 award between Leontief and his Moscow counterpart. But for political reasons the idea had then been rejected, Kantorovich sharing the Nobel Prize two years later, in 1975, with the Dutch scientist T. Koopmans.*

* See document 8.46 and note 1 to it.

[28] In the Lenin Library "Spetskhran" department there were kept *Les Prix Nobel* volumes for the years of 1937, 1946–1952, 1970, 1974, 1975, 1977, 1978, 1980, 1982–1984. Thus eight volumes, up to 1952 inclusive, were published during the Stalin reign, the other nine being shut from free access in the USSR's following historic periods. It's quite obvious that in Khrushchev's time, from 1953 to 1964, no issue of *Les Prix Nobel* was dispatched for "Spetskhran", including the 1958 volume with the materials concerning Boris Pasternak's Nobel Prize.

[29] Of the ten most significant scientific works written by Prigogine with his co-authors, which were mentioned in the 1990 *Who's Who* directory eight were translated into Russian:

1. Introduction to Thermodynamics of Irreversible Processes. — M.: FL, 1960.
2. Non-Equilibrium Statistical Mechanics. — M.: Mir, 1964.
3. Chemical Thermodynamics, with R. Dafay, Novosibirsk, "Nauka", 1966.
4. Thermodynamic Theory of Structure Stability and Fluctuations, with P. Glansdorff, — M.: Mir, 1973.
5. Self-Organization in Non- Equilibrium Systems. From dissipative structures to order through fluctuations, with G. Nicolis. — M.: Mir, 1979.
6. From Being to Becoming: Time and Complexity in the Physical Sciences. — M.: Nauka, 1985.
7. Order out of chaos. A New Dialogue with Nature, with I. Stengers. — M.: Progress, 1986.
8. Exploring Complexity. Introduction. with G. Nikolis. — M.: Mir, 1990.

[30] I. Prigogine. Time, Structure and Fluctuations. "Success of Physical Sciences", vol. 131, issue 2, 1980, pp. 185–207.

[31] *Les Prix Nobel* 1977, p. 121.

[32] He had to leave the Soviet Union for Spain not only because of the persecution by Lysenko's followers like I. I. Present and B. M. Zavadovsky, testifies academician V. I. Vernadsky, who was closely acquainted with him. The other reason was that on totally absurd grounds he was denied membership in the USSR Academy of Sciences.*

* V. I. Vernadsky. 1935–1941 diaries in two volumes. Volume I, 1935–1938. Moscow, "Nauka", 2006, p. 133. Entry of 04.04.1937.

[33] For the Soviet propaganda the very name of B. Russell was for a long time identified with some kind of an evil spirit, an enemy of the whole of mankind. Even after Stalin's death, when the political situation was getting milder and milder in Soviet society, stereotyped accusations concerning him were duplicated now and then. Here is what we encountered in the archived minutes of a regular party meeting of the USSR Academy's Institute of Philosophy, one of the most ideologically oriented academic bodies. At the time when the state had been recovering after Stalin's reign for already over six months, when the media had begun publishing very cautious but clear deliberations that the cult is unbearable "for the advanced socialist society", the speaker allowed the following utterance: "...We [the Soviet Union] do not participate in international congresses. It may happen because it doesn't become us to share a round table with Russell, with other aliens to the people and science..."*

* CAODM, f. 2501, unit 1/27, p. 165.

[34] The Nobel Prize for Peace was given to R. Bunche for his great contribution to stopping the 1948–1949 Arab–Israeli war and the signing of the armistice agreements between Egypt, Jordan, Lebanon and Syria as one party, and Israel as the other, in 1949. The citation is likely to be connected with his activity as director of the UN department for custody and information in dependent territories. He then vigorously followed the course for decolonization and against the separation of custodial rights over the former Italian colonies between the Soviet Union, the United States and Great Britain, which the USSR was striving for.

[35] The Soviet censorship rejected free access to *Les Prix Nobel 1980* according to the instructions of the CK CPSU department of culture. They certainly overdid it, because when on his way to Stockholm Miłosz visited his native country after almost 30 years of emigration, he was welcomed as a national hero at all levels, including the governmental one. In 1982 a stamp with his portrait was issued by the Post Office of Poland. There was given an official party to honour the Nobel laureate in Warsaw, but the fact was never mentioned in the Soviet press. Moreover, there was a ban on the sale of the stamp, even to the Poland philately subscribers.

[36] Quite opposite to Poland, the stamp with Seifert's portrait was issued in his mother country only in 1991, after the destruction of the Communist system.

[37] As Kapitsa's reviewer P. E. Rubinin remembers Pyotr Leonidovich had not received the 1978 volume of *Les Prix Nobel* sent to him from Stockholm. It was only after his death in 1984 that his widow could get this yearbook for the library of the scientist's memorial museum, with the assistance of the Nobel Foundation.

[38] *Les Prix Nobel* 1978, p. 248.

[39] Suslov was universally known as "a grey cardinal"*, which definition, borrowed from the French, manifests on the one hand his seemingly secluded position in the country's political

life. On the other was his extraordinary ability to keep his top positions through all the cataclysms from the Stalin tyranny to Khrushchev's experiments to the Brezhnev stagnation, the end of which he only just did not live to see by some ten and a half months. His latent aggression and pathological misanthropy, inherent xenophobia and unbearable domestic despotism hidden behind four walls, were not only inborn features but were also aggravated by the sickly state which he had suffered from for almost all his life.** All this could not but drastically affect his moral health. So here is one more definition given to Suslov by S. N. Lakshina, a literary critic and the wife of V. Y. Lakshin, who himself was A. T. Twardovsky's fellow-worker in *Novy* Mir magazine. She touched upon the Russian national idea that originated and started to develop in the Soviet society of the 1960s, mentioning that the uppermost governing layers were also concerned, the names of a number of the CPSU Central Committee Politbureau members given, who were either in line with the movement or opposed to it. In conclusion she spoke about "the grey cardinal": "Suslov's role in this movement is not quite clear to me. He either was a pure official and nothing more or, as others say, a Freemason. I'm not sure…"***

* R. A. Medvedev, D. A. Ermakov. "Grey Cardinal". M. A. Suslov — A political portrayal — M. Respublica, 1992.

** From the age of 35 Suslov had diabetes accompanied with the early-stage failure of the whole vascular system and of the brain vessels.

*** "Druzhba Narodov", No. 5, 2003, p. 171.

[40] Here are two examples showing though indirectly that the supposition is likely to be true. In the 1970s, live on TV academician D. S. Likhachev was talking about the valuable contribution to Russian culture made by different nationalities, Jewish among them. Suslov happened to be watching the programme. He called the studio, ordering them to immediately stop the broadcast. For a long time after that Likhachev had been banned from appearing on TV.* Simultaneously there was a similar event concerning the Moscow Literary Institute. V. F. Pimenov, its director, when preparing by the order from the KGB the dismissal of Konstantin Kedrov, the professor of the department of Russian literature and allegedly a dissident, sent there a memo accusing Kedrov of mentioning in his lectures on Russian folklore, ancient Russian literature and in specialist courses on L. N. Tolstoy and F. M. Dostoyevsky sources provided by Jewish researchers.** An episode described by the famous singer Iosif Kobzon correlates quite well with the above. In 1983 he was offered participation in the concert "dedicated to a jubilee of friendship with foreign countries". The general secretary of the "Israel–USSR" society was among the spectators. "To show how hospitable we are to him in our country" Kobzon decided to sing some Jewish songs. "Arab delegates", also present at the jubilee, "got up and left immediately". After the analysis of the incident at the CPSU Central Committee session the singer was for a year prohibited from performing in radio and television programmes.***

* "Izvestia", 12.05.1990, p. 5.

** "Literaturnaya gazeta", No. 33, 1991, p. 9.

*** "Izvestia", 17.12.2001, p. 9.

[41] *Les Prix Nobel* 1982, p. 89.

[42] A. N. Yakovlev. Gorkaya Chasha (Hard Lot). Bolshevism and Reformation in Russia. Yaroslavl, Verhne-Volzhskoe izdatelstvo (publishing house), 1994, p. 131.

Document 12.1

From the report of the party of committee member V. I. Zorin at the party meeting on 15.01.1948 at the Presidium of the USSR Academy of Sciences.

...In the letter from VKP(b) there was mentioned the name of Waksman. We actually overdid it when welcoming him here. He is an ordinary dealer who came here by the order of his master, the dollar.

In the premises of the Presidium a report was made, a book was written[1] and other things were done in an American way: much pronounced, nothing said. Hospitality is a characteristic feature of Russian, Soviet people, but together with being hospitable one should be watchful, as it is not only with good will that they visit us but with certain spying tasks.[2] Waksman also appeared to be interested in the Klueva–Roskin case.[3]

CAODM, fund 992, unit 2/18, p. 9.

[1] The speaker must have meant Waksman's monograph *The antagonism of microbes and antibiotic substances* published in Russian in 1947.

[2] As to "hospitality to those coming to the USSR with certain spying tasks", everything was prepared in due time and without the usual red tape. On his next visit to Europe Waksman was going to go to his native country* on the way from Copenhagen, the place of the International Congress on Microbiology in July, to France where his colleague S. N. Winogradsky was waiting for him in Paris. But the Soviet consulate in the US refused his wife and him visas, of which Waksman informed Winogradsky in a letter of 21.06.1947.**

* RAN archives, fund 1601, unit 1/127, p. 57.

** Just at the end of June 1947 party organisations everywhere in the Soviet Union started discussing "The secret letter of the VKP(b) Central Committee" concerning "the Klueva–Roskin case" fabricated by Stalin and Zhdanov. Waksman was mentioned in "The letter" as one of those concerned, which was the reason why he was refused the entrance visas. "The Klueva–Roskin case" was initiated at the USSR government session of 17.02.1947, presided over by Zhdanov with Stalin present. (Attachments I-II, item 1)

[3] On 26.06.1947 in the USSR Academy of Sciences at the party meeting hearings of "The secret letter of the VKP(b) Central Committee" concerning "the Klueva–Roskin case" the Presidium business manager I. W. Zubov extended the idea that "we shouldn't try to court foreign scientists, but the foreign scientist should learn from us", stating in conclusion that: "It was the head of the US embassy Smith and academician Waksman, who were trying to persuade Klueva and Roskin to collaborate when a really significant scientific discovery was at stake."* Science historian W. D. Esakov, who carefully studied all existing documents connected with "the Klueva–Roskin case", stated that Waksman never visited Klueva's laboratory when in

Moscow. He only had a nodding acquaintance with her as with some other members of the USSR Academy of Medical Sciences when he was on a short visit to the premises of the AMN Presidium. That's why the statements made by the two employees of the Presidium could not have been based on any factual material but were totally true to the spirit and contents of "The closed letter..." heard.*

* CAODM, fund 992, unit 2/14, p. 247.

Document 12.2

An excerpt from the review of the film *A Trial of Honour*, based on a story by Vadim Kozhevnikov.

An American professor Carter[1] who had signed a life-long agreement on all his works with the medicinal substances firm "Radley and Company", having made him an actual slave, visits an Institute. Together with him there come an American intelligence agent, Wood,[2] and a manager of its department, Wilby, both disguised as scientific workers. Their target is some missing information about an anaesthetic[3] attracting American imperialists for military purposes.

Blindly loyal to the idea of the so-called "universal" science having pure aims and being shared by all, Dobrotworsky (the name can be translated as "creating good"), encouraged by Lossev[4] lets the spies into the secret laboratory. To keep the scientific secret from the foreign agents, an employee of the Institute destroys the device where the preparation of the substance is taking place. The American spies are still striving to get the missing data, and Dobrotworsky is just about to uncover it as "a form of scientific exchange". But when the Institute's scientific researcher asks Carter for the formula of his substance, important for the Soviet scientists, Carter is forced to confess that all the rights for his discoveries belong to the capitalist company,[5] not himself. The scene vividly depicts the slavish dependence of the scientists in capitalist countries on their masters.[6] The film is convincing in disclosing cosmopolitism as a tool for hiding predatory theft by the imperialists of scientific discoveries with the aim of using them against democratic states.[7]

"Pravda", 25.01.1949, p. 3.

[1] Based on Prof. Waksman.

[2] The representation here was of Smith, the head of the US embassy in the USSR who did visit N. G. Klueva's laboratory (see Chapter 6).

[3] The medicinal substance "KR" is meant here, which N. G. Klueva and G. I. Roskin were working on.

[4] "Professor Dobrotworsky is developing an anaesthetic about to become a discovery of universal significance. Professor Lossev, his co-author, is far from having standards of morality characteristic of a Soviet scientist. While being in America Lossev was flattered by the attention of capitalist companies and their agents and betrayed

his own land and science by publishing the results of scientific discovery in the capitalist press."*

* "Pravda", 25.01.1949, p. 3. Dobrotworsky was based on N. G. Klueva and G. I. Roskin, the main figures of the fabricated propaganda campaign, while Lossev was based on the USSR AMN's secretary-academemician W. W. Parin. Having returned from his business trip to the USA, Parin was arrested and sentenced to 25 years in prison for "treason against his motherland". He was rehabilitated in 1953 (see note 1 to doc. 4.34).

[5] This statement was included in the screenplay, having been taken from "The Closed letter of the VKP(b) Central Committee on the case of Professor Klueva and Professor Roskin", and was widely used in party meetings all over the state as an example of perfidy on the part of "the servants of American imperialism". There were also doubts expressed as to the verity of this information. At the party meeting held on 28.07.1947, W. P. Peshkov, a researcher at the USSR Academy's Institute of Physical Problems, complained about the lack of officially released documents on the case, which made it difficult to clear up the case for the scientists — non-Party members. The words of A. I. Shalnikov, a non-Party corresponding member of the USSR Academy of Sciences, served an example: "Being invited to the Academy's session Waksman disclosed the whole technological process of streptomycin manufacture. In the Trial of honour materials concerning the Klueva–Roskin case it is argued that Waksman concealed the whole matter because it was the company's trade secret."*

* CAODM, fund 995, unit 1/4, p. 115.

[6] The episode describing "Wood" and "Carter's" visit to the laboratory working at the anti-cancer anaesthetic was purely an invention.* After finishing his work in the USSR as the head of the US embassy, Bedell Smith published the memoirs about his three-year stay in Moscow, touching in detail upon "the Klueva–Roskin case". He happened to be one of those concerned and to have a private meeting with them in June 1946, which had been approved by the highest authorities.** Waksman was never mentioned by the author.

* See note 3 to document 12.1.

** W. B. Smith. Moscow mission. 1946–1949. Melbourne, 1950, pp. 283–286.

[7] The review in question was published in *Pravda* three days before an editorial in the same newspaper, which gave a start to the ideological campaign "against cosmopolitism and the servility to the West".

Document 12.3

From the letter of R. A. Prigogine from Brussels to his sister L. A. Ostrovskaya in Moscow, 20.11.1971.

...Here comes the latest news: Ilya and Marina[1] are coming to Moscow for 10 days. Their flight is on December, 12. He was invited for a series of lectures on his speciality (atomic physics).[2] He's got an invitation from your government.[3] He's going to make reports in Moscow, Leningrad and Tiflis.[4] Please, on receiving the letter, reply as soon as you can, I can't find your telephone number. Send it to me, he'll call you when in Moscow. Anyway, could you stay at home as much as possible on 12, 13 and 14 of December, as he'll be very busy. Let his cousins know about it. I hope they won't fear to see him as they did me — he has been invited by the Soviet government. It must be very exciting for them to see their cousin, whose name in science is widely known in the world, as there are Prigogine's theorems and even laws by Ilya Prigogine. And he is a member of eleven Academies of Sciences. He was nine times appointed the honoris doctor of the biggest European and American universities,[5] and there is a great possibility of his becoming a Nobel Prize winner.[6] It's a real miracle: one of our kin appears to be a scientist of genius, being at the same time a very modest and nice person in everyday life.[7] As to Marina, she's also a scientist in the sphere of physics. And her character! She is more of a piece of gold, not just a human being.[8] God bless them...

From the family archive of A. A. Kamionskaya, Ilya Prigogine's cousin.

[1] Ilya Prigogine and his wife Marina.

[2] An error. Prigogine never dealt with atomic physics.

[3] It was from the USSR Academy of Sciences that Prigogine received his invitation in 1971. His visit on governmental invitation took place from 23.06 to 02.07.1975 when among others he was accompanying the king of Belgium, Baudouin.

[4] The former name of the city of Tbilisi.

[5] As was published in the directory of the Free University of Brussels, which was released by the 80th anniversary of the scientist's birth and the 20th anniversary of his Nobel Award, by 1997 Prigogine had been a member of 70 academies and scientific societies of 21 countries on all continents, held honorary doctoral degrees from 38 universities of 19 countries and had been bestowed 22 scientific awards and the same number of medals from scientific institutions.*

* Hommage au professeur Ilya Prigogine. 1977–1997: Vingt ans de Prix Nobel de Chimie. ULB, 1997.

[6] Roman Abramovich didn't live to see his son receive the Nobel Prize.

[7] See also: "Priroda (Nature)", No. 2, 2004, p. 69–75.

[8] In the epilogue to his autobiography for the Nobel Prize yearbook Prigogine emphasised that for the theorist "the awareness of the lucidity of reasoning helps select the most correct route to the truth among the diversity of others. I owe this clarity of mind to my wife, Marina."*

* *Les Prix Nobel 1977*, Stockholm, 1978, p. 129.

Document 12.4

Academician V. L. Ginzburg and the 2003 Nobel Prize for physics.

I

Colleagues on Ginzburg
himself and the award.

When the news of the Nobel Prize being awarded to academician Vitaly Lazarevich Ginzburg came to our editorial office, all the staff's faces were lightened and smiling as if it were their award as well... Much had been done for 63 years of non-stop work at the theoretical department of the P. N. Lebedev Physical Institute. Though retired of his own free will from the head of the department he still keeps on working here in "the position of Ginzburg" (physicists' jokes!). So his post is the highest of all possible. He is — Ginzburg. Nothing to be added.

"Vestnik RAN", No. 12, 2003, p. 1060.

II

Ginzburg on the possibility of the award for himself
a quarter of a century before receiving it.

... By the way, I sometimes come to think about the Nobel Prize.[1] I'm in no hope of receiving it, there's such a rush there. I've got no megalomania of any kind and don't think I must be awarded, only that I might be. But I am writing now because if it's simply a "no", it's OK. But if others get (such things happen) what I must have received, it'd be a pity, e.g. if the award were given to Abrikosov, trying to solve the Ginzburg–Landau equations, or Shklovsky for radioastronomy and the origin of space rays, though my contribution is in no way less.

V. L. Ginzburg. About Science, Myself, and others.
M.: Physmatlit, 2001, p. 377.

[1] These words were part of his autobiographical notes, while in hospital after a serious surgery.*

* *ibid.*, p. 341.

Epilogue by the Nobel Prize laureate Ilya Prigogine

It is a privilege to write a short epilogue for this remarkable book. I would like to be the first to congratulate the author, A. M. Blokh. It is a work of love. It certainly has needed years of research.

In fact, there are three aspects to this book. First, the main text, which is written with great clarity and elegance. Then hundreds of documents, many of them published for the first time, and finally interviews with the families of Nobel Prize winners.

The outcome is like a window on the relations between Russia and the West; a very special window, which is written from the point of the relations between Russia and the Nobel Foundation. However, this book leads to a greater perspective. In Russian history in the twentieth century, political regime and science cannot be separated. We all know the anti-Western tendencies of the Soviet Regime.

The book contains the burlesque and the tragic. We hear about Stalin's questions to his co-workers: What is higher, a Nobel Prize or a Stalin Prize? On the other hand we have a detailed account of the ordeal of Pasternak. We hear about the action of courageous people such as P. Kapitsa. In retrospect some decisions of the Nobel Foundation appear strange. How did it happen that neither Tolstoy nor Mendeleev received the Nobel Prize? The intrigues behind this are interesting to read.

The Nobel Prize is the most prestigious award, especially in sciences. But how to make the choice? Whom to give this exceptional award? The Nobel Prize leads perhaps to a too sharp distinction between scientists who have it and scientists who do not. It is sad that great Russian scientists like Gamov, Bogolyoubov, Zeldovitch or Kolmogorov,[1] who are certainly domineering figures of the twentieth century, have not received the Nobel Prize.

A particularly revealing aspect of this book is the place devoted to scientists who received the Nobel Prize and who were born in Russia or came from Russia with their parents. They are classified as citizens of the country where they were living at the time of the Nobel Prize. It is curious to read how many Nobel Prize winners belong to this category, including myself.

What is also remarkable is how much Russians contributed to the culture of the twentieth century. Science is certainly an aspect of the cultural impact of Russia but we may also quote Russian poets, musicians or painters.

Let me go back to the book. It clearly belongs to the archives of the past century. It will be read with great interest by all people who want to follow the intriguing history of the European culture of the twentieth century. The past is necessary to understand the present and perhaps also to prepare for the future.

[1] According to what Ilya Prigogine had told the author, of those Russian scientists mentioned above who were worth receiving the award, Y. B. Zeldovitch* and A. N. Kolmogorov were the ones nominated by Prigogine himself. He also submitted the names of physical chemists M. W. Wolkenstein and A. M. Zhabotinsky.**

*Y. B. Zeldovitch received three Gold Star medals of the Hero of Socialist Labour for his most widely known, outstanding developments in creating atomic and nuclear armaments. But these were not the object of the nomination.

Prigogine got acquainted with the scientist in the summer of 1983 at the symposium on self-organisation in the science town Pushchino, in the suburbs of Moscow. It was there that he first learned about Zeldovitch's striking achievements in the sphere of physical knowledge so dear to himself.

** Refer also to the newspaper *Poisk*, No. 10, 1998, p. 14 for the nomination of Zhabotinsky, one of the discoverers of the famous reaction BZ (Belousov–Zhabotinsky), or "chemical clock". Prigogine considered this achievement an empirical confirmation of the theory of non-equilibrium thermodynamics, which brought him the 1977 Nobel Prize for chemistry. The details of the discovery can be found in the book *Heroes and villains of the Russian science* by S. E. Shnol. Moscow, Publishing House "Cron-press", 1997, p. 144–166.

List of Abbreviations

AVP RI	Archives of Foreign Politics of the Russian Empire
AVP RF	Archives of Foreign Politics of the Russian Federation
Agitprop	Administration (from July 1948 — Department) of Propaganda and Agitation under the CK VKP(b) — CPSU (Central Committee of All-Russian Communist Party (bolsheviks) — Communist Party of the Soviet Union)
AMN USSR	Academy of Medical Sciences of the USSR
AN USSR-RAN	Academy of Sciences of the USSR; from 1992 — Russian Academy of Sciences
AP RF	Archives of the President of the Russian Federation
BCS-theory	Theory of superconductivity by Bardeen-Cooper-Schrieffer
BSE	Big Soviet encyclopedia
CAODM	Central Archives of Public Movements of Moscow
CGIA RF	Central State History Archives of the Russian Federation
CK VKP(b) — CPSU	Central Committee of All-Russian Communist Party (bolsheviks) — CPCU
No. SSR	Czecho-Slovak Socialist Republic
ed. hr.	unit (for archives documents)
FIAN	USSR Academy of Sciences Physics Institute
FSB RF	Federal Security Service of the Russian Federation
GARF	State Archives of the Russian Federation
GIHL	State Publishing House of Fiction
AUCP(b)	All-Union Communist Party (bolsheviks)
Glavlit	Main Administration for Censorship in the USSR
Goshimtehizdat	State Publishing House of Chemical and Technological Literature
GULAG	Main Administration of labour correction camps and labour settlements under the USSR Home Affairs Ministry (MVD USSR)
INO AN USSR	foreign department under the Presidium of the USSR Academy of Sciences
Inoizdat	State Publishing House of foreign literature

IEM	Institute for Experimental Medicine in St. Petersburg
KAN	Royal Academy of Sciences in Stockholm
KGB USSR	Committee for State Security of the USSR
Litfond	Literary Fund under the USSR Writers' Union
NEP	New Economic Policy
NKID	MID USSR — People's Commissariat of Foreign Affairs of the USSR; from March 1946 — the USSR Ministry of Foreign Affairs
NKVD	MVD USSR — People's Commissariat for Home Affairs of the USSR; from March 1946 — the USSR Ministry for Home Affairs
OFMN	Department of Physical and Mathematical Sciences of the USSR Academy of Sciences
OSO	Special consultation body under NKVD — MGB (Ministry for State Security) USSR
OVIR	Office of visas and registrations, where documents for the Soviet people going abroad were being prepared
PB	Political Bureau of CC AUCPP(b)
PFA RAN	St. Petersburg branch of the Archives of the Russian Academy of Sciences
Polpred USSR	Plenipotentiary (head of embassy) of the USSR in foreign countries
RAEN	Russian Academy of Natural Sciences
RGALI	Russian State Archives of Literature and Arts
RGANI	Russian State Archives of Latest History
RGASPI	Russian State Archives of Social and Political Investigations
RSFSR	Russian Soviet Federative Socialist Republic
SNK	SM USSR — Council of People's Commissars; from March 1946 — the USSR Council of Ministers
SP USSR (RSFSR)	USSR (RSFSR) Writers' Union
SSRU	Soviet Socialist Republic of Ukraine
UK RSFSR	Criminal code of the RSFSR
USSR	Union of Soviet Socialist Republics
VASHNIL	All-Union Academy of Agricultural Sciences named after Lenin
VCIK-CIK USSR	All-Union Central Executive Committee; from December 1922 — Central Executive Committee of the USSR
VLKSM	All-Union Lenin Young Communist League
VOKS	All-Union Society for Cultural Relations with Foreign Countries
VTO	All-Union Theatrical Society

Name Index